WORLD *of* MATHEMATICS

WORLD *of*

MATHEMATICS

Brigham Narins, *Editor*

Volume 1

A-L

GALE GROUP

Detroit
New York
San Francisco
London
Boston
Woodbridge, CT

STAFF

Brigham Narins, *Editor*

Zoran Minderovic, *Associate Editor*

Ellen S. Thackery, *Associate Editor*

Mark Springer, *Editorial Technical Consultant*

Margaret A. Chamberlain, *Permissions Specialist*
Shalice Shah-Caldwell, *Permissions Associate*

Mary Beth Trimper, *Composition Manager*
Evi Seoud, *Assistant Production Manager*
Stacy Melson, *Buyer*

Kenn Zorn, *Product Design Manager*
Michelle DiMercurio, *Senior Art Director*

Barbara Yarrow, *Manager, Imaging and Multimedia Content*
Randy Bassett, *Imaging Supervisor*
Robert Duncan, *Senior Imaging Specialist*

Indexing provided by Julie Shawvan.
Illustrations created by Electronic Illustrators Group, Morgan Hill, California.

ISBN: 0-7876-3652-5 (set)
ISBN: 0-7876-5064-1 (Volume 1)
ISBN: 0-7876-5065-X (Volume 2)

Printed in the United States of America
10 9 8 7 6 5 4 3 2 1

Library of Congress Cataloging-in-Publication Data
World of mathematics / Brigham Narins, editor.
    p. cm.
    Includes bibliographic references and index.
    ISBN 0-7876-3652-5 (set : alk. paper)—ISBN 0-7876-5064-1 (v. 1)—ISBN 0-7876-5065-X (v. 2)
    1. Mathematics-Encyclopedias. I. Narins, Brigham, 1962-
QA5 .W67 2001
510'.3—dc21
                            00-051408

Contents

INTRODUCTION

The works of mathematics are all around us. Born into the age of technology, we don't think twice about relying on machines whose interiors are mysteries to us, and that depend on mathematics in order to function. From eyeglasses and contact lenses, whose shapes depend on the principles of geometric optics, to CD players, which owe their existence to the ideas of quantum mechanics, the fruits of our mathematical understanding have improved the quality and length of our lives dramatically. One of the oldest and deepest of enigmas is, Why is our physical world governed by the laws of mathematics? The nature of the connection between mathematics and physics is enormously complex and well beyond the scope of this introduction; but that connection is an essential ingredient in our understanding of the world. Nobel physicist Richard Feynman said, "To those who do not know mathematics it is difficult to get across a real feeling as to the beauty, the deepest beauty of nature. If you want to learn about nature, to appreciate nature, it is necessary to understand the language she speaks in."

Most mathematicians, however, did not choose to study mathematics because they wanted to build a better VCR. Instead, many were drawn to mathematics by its aesthetic qualities—its purity, its spare elegance. Beauty in mathematics is hard to define, but mathematicians instantly understand what their colleagues mean when they call a proof "beautiful," and they think of their calling almost as an art. The great mathematician Henri Poincaré said, "A scientist worthy of his name, above all a mathematician, experiences in his work the same impression as an artist; his pleasure is as great and of the same nature." So strong is this feeling among mathematicians that they have an instinctive distrust for awkward or complicated mathematical work. G. H. Hardy wrote in his *Mathematician's Apology,* "The mathematician's patterns, like the painter's or the poet's must be beautiful; the ideas, like the colors or the words must fit together in a harmonious way. Beauty is the first test: there is no permanent place in this world for ugly mathematics."

The great Hungarian mathematician Paul Erdös tried to explain his sense of the aesthetic quality of mathematics by the allegorical idea that God has a Book that contains all the most beautiful, most perfect proofs of mathematical truths. For Erdös, the highest achievement for a mathematician is to find a proof worthy of being in the Book. "You don't have to believe in God, but you should believe in the Book," he said. In *World of Mathematics,* you will read about proofs—such as the proof of the four-color theorem—so large and unwieldy that they are almost impossible to check, and you will see other proofs—such as the proof of the Pythagorean theorem or the irrationality of the square root of 2—so simple and elegant that they perhaps deserve a place in the Book. You will also find an entry in *World of Mathematics* called "Beauty in Mathematics" in which this issue is discussed at greater length.

Erdös's Book metaphor emerges from a question that has stimulated some of the most energetic debates of mathematical philosophy: Is mathematics invented or discovered? Do mathematicians build an edifice of theorems from a foundation of their own design, or do mathematical truths exist in some realm that human beings can only brush against, occasionally being fortunate enough to knock some small piece of the truth within human reach?

This question dates back to almost the earliest days of mathematical thought. In the 6th century B.C., Pythagoras, awed by the unexpected patterns he found in numbers, concluded that mathematical objects have a natural harmony, one that is only partially understood by mathematicians. Later, Plato put forth the idea that mathematical truths have their own existence in an ideal world of essences, one that human beings can only try to discover; since then, this belief has been referred to as Platonism. Equally influential on the other hand was the work of the great geometer Euclid. His *Elements,* written in the 4th century B.C., stripped down the ideas of geometry to a few basic building blocks called axioms, and then constructed a huge structure of theorems on this foundation, following rigorous logical arguments. Euclid's work, which spanned 13 books, was an intellectual triumph that has set the standard for logical thought for more than two millenia. The idea that human beings invent mathematics using sheer logical power has since been termed reductionism.

Towards the end of the 19th century, the advantage in the debate seemed to be with the reductionists. With the development of the principles of computers and algorithms, the reductionists set themselves a new goal: to construct a system of axioms from which all of mathematics could be derived, following the strict laws of logic. Their efforts culminated in a monumental opus by Bertrand Russell and Alfred North Whitehead published between 1910 and 1913, the *Principia Mathematica*; in it, Russell and Whitehead tried to construct a system of axioms and rules of deduction from which every mathematical truth could be proved. They could not, of course, include proofs of every theorem, but the proofs that they did present followed the laws of logic to the smallest detail; they were so careful that it took several hundreds of pages before they got to the definition of the number 1. While this might seem like absurd overkill, to the reductionists it was worth the effort to be certain that their work rested on a solid foundation. This foundation was swept out from under their feet, however, when in 1930 the young mathematician Kurt Gödel proved a theorem that shook the mathematical community to its core. Gödel's theorem says that in any axiomatic system powerful enough to prove significant mathematical theorems, there will be statements that are true but which cannot be proved using the axioms and deductive laws of the system.

Gödel's "Incompleteness" theorem reopened the question, What is mathematical truth? The response of the Platonists to Gödel's theorem is to say that truth is not the same as provability; that mathematical truth exists independent of any particular system of axioms. The reductionists, on the other hand, say that mathematical truth simply does not exist—that the only thing mathematicians can say is whether a statement is provable in a given axiomatic system, not whether it is true or false. The mathematical community continues to be divided on this question.

In the nearly 800 entries in this book, you will find not just ammunition for such philosophical arguments, but also the everyday, bread-and-butter ideas of mathematics—the patterns and puzzles, theorems and proofs, shapes and symmetries that have fascinated both professional and amateur mathematicians over the centuries. Read about prime numbers, or fractals, or the nature of infinity, and form your own opinion about whether mathematics is invented or discovered.

You will also find biographies of the great mathematicians whose lives span the whole of human history, and whose stories range from comical to tragic, from humble to heroic. You will read of duels, suicides, ritual sacrifices, murders, and also quiet lives devoted to reflection and study. Through the centuries, mathematical scholars have come from the aristocracy and from the slums; they have dazzled the world as youthful prodigies or have worked long years in patient obscurity. The scope of their experiences shows above all that mathematics is for everyone—not just the Fields medalist at

Harvard or Princeton, but also the child who looks at the night sky with wonder, or who happens upon an unexpected pattern in numbers, and is struck by its beauty.

Erica Klarreich
Assistant Professor of Mathematics
University of Michigan
May 2000

How to Use the Book

This first edition of *World of Mathematics* has been designed with ready reference in mind.

- **Entries are arranged alphabetically**, rather than by chronology or scientific field.
- **Boldfaced terms** direct reader to related entries.
- **Cross-references** at the end of entries alert the reader to related entries that may not have been specifically mentioned in the body of the text.
- A **Sources Consulted** section lists many worthwhile print and electronic materials encountered in the compilation of this volume. It is there for the inspired reader who wants more information on the people and concepts covered in this work.
- The **Historical Chronology** includes over 500 important events in the history of mathematics and related fields spanning the period from around 35,000 B.C. through 1995.
- A **comprehensive general index** guides the reader to all topics and persons mentioned in the book. Boldface page references refer the reader to the term's full entry. Page numbers in italics refer to illustrations. Page numbers followed by *f* indicate figures.

Advisory Board

In compiling this edition, we have been fortunate in being able to call upon the following people—our panel of advisors—who contributed to the accuracy of the information in this premier edition of *World of Mathematics,* and to them we would like to express sincere appreciation:

Kristin Chatas
Department Chair—Mathematics
Pioneer High School
Ann Arbor, Michigan

Erica Klarreich, Ph.D.
Assistant Professor of Mathematics
University of Michigan

Dana Mackenzie, Ph.D.
Freelance Mathematics and Science Writer
Santa Cruz, California

The editor would like to thank the advisors for all of their good work reviewing and compiling the entry list, for their patience and good will answering my many questions, and for their outstanding writing. You're the best math teachers I never had. Thanks are also due to all the writers who contributed to this book; in particular, William Arthur Atkins, Lewis Bowen, Fran Hodgkins, Elisabeth Morlino, and Stephen R. Robinson. Finally, thanks go to Zoran Markovic of the Institute of Mathematics in Belgrade for his assistance early in the project; and to Julie Shawvan for her indexing and critical review at the end.

Cover

The image on the cover represents the symmetry operation called inversion.

ACKNOWLEDGMENTS

Abacus, drawing by Hans and Cassidy. Gale Group.-Abel, Niels Henrik, screen print. Corbis-Bettmann. Reproduced by permission.-Agnesi, Maria Gaetana, engraving. Corbis-Bettmann. Reproduced by permission.-Aikcn, Howard, photograph. Corbis-Bettmann. Reproduced by permission.-Anaxagoras with Pericles, in his palace, engraving. Corbis-Bettmann. Reproduced by permission.-Archimedes, sculpture. Corbis-Bettmann. Reproduced by permission.-Archytas, engraving. Corbis-Bettmann. Reproduced by permission.-Aristotle, photograph. Corbis-Bettmann. Reproduced by permission.-Bernoulli, Daniel, woodcut. Corbis-Bettmann. Reproduced by permission.-Bernoulli, Jacques, engraving by P. Dupin. Corbis-Bettmann. Reproduced by permission.-Bernoulli, Jean I, engraving. Corbis-Bettmann. Reproduced by permission.-Boolean algebra, illustrations by Hans & Cassidy. Gale Group.-Calculation of the shortest distance between two points on a graph, using the Pythagorean theorem, diagram by Hans & Cassidy. Gale Group.-Calculator, Babylonian digits, diagram by Hans & Cassidy. Gale Group.- Calculus, bar graph by Hans & Cassidy. Gale Group.-Calculus, graph approximating line segment, illustration by Hans & Cassidy. Gale Group.-Calderon, Alberto P., Israel, 1989, photograph. Wolf Foundation. Reproduced by permission.-Carnap, Rudolph, 1963, photograph. AP/Wide World Photos. Reproduced by permission.-Carroll, Lewis, photograph. The Granger Collection. Reproduced by permission.-Cartan, Elie Joseph, c. 1936, photograph. AP/Wide World Photos. Reproduced by permission.-Cartesian coordinate plane, graphs by Hans & Cassidy. Gale Group.-Cartesian plane, illustrations by Hans & Cassidy. Gale Group.-Chandrasekhar, Subrahmanyan, with King Carl Gustaf of Sweden, 1983, photograph. AP/Wide World Photos. Reproduced by permission.-Chern, Shiing S., photograph. Wolf Foundation. Reproduced by permission.-Complex numbers figure, acute angle, photograph by Hans & Cassidy. Gale Group.-Conic section, cone, oval, circle, and arch, illustrations by Hans & Cassidy. Gale Group.-Continuity, graphs by Hans & Cassidy. Gale Group.- Coulomb, Charles August de, photograph. Corbis-Bettmann. Reproduced by permission.-Courant, Richard, 1961, photograph. AP/Wide World Photos. Reproduced by permission.-Cramer, Gabriel, 18th century, painting. The Granger Collection, New York. Reproduced by permission.-De Morgan, Augustus, photograph by Ernest Edwards. Corbis-Bettmann. Reproduced by permission.-Determinants, four small boxes, variables in corners, illustration by Hans & Cassidy. Gale Group.-Determinants, graph by Hans & Cassidy. Gale Group.-Determinants, six small boxes, photograph. Hans & Cassidy. Gale Group.-Diagram showing addition on an abacus, drawing by Hans and Cassidy. Gale Group.- Dipolar vortices, collision, photograph by G. Van Heijst and J. Flor. Photo Researchers, Inc. Reproduced by permission.-Dirichlet, Peter Gustav Lejeune, engraving. Archive Photos, Inc. Reproduced by permission.-Durer, Albrecht, self-portrait. AP/Wide World Photos. Reproduced by permission.-Electrons, classical and quantum, diagram by Hans & Cassidy. Gale Group.-"Euclid," painting by Justus van Ghent. Corbis-Bettmann. Reproduced by permission.- Fermat, Pierre, illustration. Corbis-Bettmann. Reproduced by permission.-Fibonacci, Leonardo, engraving. The Granger Collection, New York. Reproduced by permission.-Figurative numbers, illustrations by Hans & Cassidy. Gale Group.-Figurative numbers represented by pentagons, illustration by Hans & Cassidy. Gale Group.- Figurative numbers, three squares in a row, diagram by Hans & Cassidy. Gale Group.-Fractal, three lines, illustration by Hans & Cassidy. Gale Group.-Freedman, Dr. Michael, La Jolla, California, 1982, photograph. AP/Wide World Photos. Reproduced by permission.-Galois, Evariste, drawing. Corbis-Bettmann. Reproduced by permission.-Gauss, Carl Friedrich, drawing. Corbis-Bettmann. Reproduced by permission.-Germain, Sophie, engraving after a life mask. The Granger Collection, New York. Reproduced by permission.-Gödel, Kurt, 1951, photograph. AP/Wide World Photos. Reproduced by permission.-Hadamard, Jacques, 1920, photograph. Reuters/Corbis-Bettmann. Reproduced by permission.-Hermite, Charles, photograph. Corbis-Bettmann. Reproduced by permission.-Hilbert, David, photograph. Corbis-Bettmann. Reproduced by permission.-Hipparchus, engraving. Archive

Photos, Inc. Reproduced by permission.- Hypatia, conte crayon drawing. Corbis-Bettmann. Reproduced by permission.-Hyperbola, circle, graph by Hans & Cassidy. Gale Group.-Hyperbola, directrix, focus, illustration by Hans & Cassidy. Gale Group.-Hyperbola, graphs and planks, photographs by Hans & Cassidy. Gale Group.-Inequality, graph, left half shaded, photograph by Hans & Cassidy. Gale Group.-Inequality, line graph, photograph by Hans & Cassidy. Gale Group.-Jacobi, Karl Gustav, painting. The Granger Collection. Reproduced by permission.-"Karl Weierstrass," etching by Hans Thoma. Corbis-Bettmann. Reproduced by permission.-Keen, Linda, photograph. Reproduced by permission of Linda Keen.- Khayyam, Omar, drawing. Corbis-Bettmann. Reproduced by permission.-Kolmogorov, Andrei, Rome, 1963, photograph. UPI/ Corbis-Bettmann. Reproduced by permission.-Kovalevskaya, Sonya, engraving. Corbis-Bettmann. Reproduced by permission.-Locus figures and line graphs, illustrations by Hans & Cassidy. Gale Group.-Maclaurin, Colin, illustration. Corbis-Bettmann. Reproduced by permission.-Mandelbrot, Benoit, photograph. Copyright © Hank Morgan. Photo Researchers, Inc. Reproduced by permission.-Marquise du Chatelet, 1873, wood engraving. The Granger Collection, New York. Reproduced by permission.-Mersenne, Marin, 18th century, engraving by P. Dupin. The Granger Collection. Reproduced by permission.-Minkowski, Hermann, photograph. Corbis-Bettmann. Reproduced by permission.-Newton, Sir Isaac, engraving. Archive Photos, Inc. Reproduced by permission.-Parabola, cones and graphs, illustrations by Hans & Cassidy. Gale Group.-Peirce, Charles Sanders, photograph. Corbis-Bettmann. Reproduced by permission.-Poincare, M. Henri, photograph. Hulton-Deutsch Collection/Corbis-Bettmann. Reproduced by permission.-Polar coordinates, acute and obtuse angles, diagrams by Hans & Cassidy. Gale Group.-Polya, Dr. George, with his book "How To Solve It," 1978, photograph. AP/Wide World Photos. Reproduced by permission.- Poncelet, Jean Victor, marble sculpture. Corbis-Bettmann. Reproduced by permission.-Pythagoras, engraving. Corbis-Bettmann. Reproduced by permission.-Queneau,

Raymond, photograph by Jerry Bauer. © Jerry Bauer. Reproduced by permission.-Ramanujan, Srinivasa, photograph. The Granger Collection. Reproduced by permission.-Real numbers, line and triangle, illustration by Hans & Cassidy. Gale Group.-Regiomontanus, Johannes, illustration. Corbis-Bettmann. Reproduced by permission.-Riemann, G.F.B., 1868, engraving after a photograph. Corbis-Bettmann. Reproduced by permission.-Rotation (pie slice), graph by Hans & Cassidy. Gale Group.-Set of points on a coordinate plane (right angles), diagram by Hans & Cassidy. Gale Group. Set theory, difference, photograph by Hans & Cassidy. Gale Group.-Set theory figures, illustrations by Hans & Cassidy. Gale Group.-Snell, Willebrord, painting. Photo Researchers, Inc. Reproduced by permission.-Stevin, Simon, painting. Corbis-Bettmann. Reproduced by permission.-Taylor, Brook, engraving. Archive Photos, Inc. Reproduced by permission.-Texas map, illustration by Hans & Cassidy. Gale Group.-Thales of Miletus, engraving by Ambroise Tardieu. Corbis-Bettmann. Reproduced by permission.-Thales of Miletus, lithograph. Corbis-Bettmann. Reproduced by permission.-Third-order determinants (graph and equation), illustration by Hans & Cassidy. Gale Group.-Translation of a triangle, in one and two directions, diagrams by Hans & Cassidy. Gale Group.-Turing, Alan, finishing second in a 3 mile race at Dorking, England, photograph. The Granger Collection, Ltd. Reproduced by permission.-Two points plotted in a perpendicular coordinate system, graph by Hans & Cassidy. Gale Group.-Two triangles and an acute angle in a rotation, diagram by Hans & Cassidy. Gale Group.-Two-dimensional complex-number plane, graph by Hans & Cassidy. Gale Group.-Von Neumann, John, testifying before the Atomic Energy Commission, photograph. AP/Wide World Photos. Reproduced by permission.-Weyl, Hermann, 1954, photograph. AP/Wide World Photos. Reproduced by permission.-Whitehead, Alfred North, photograph. AP/Wide World Photos. Reproduced by permission.-Wiles, Andrew J. , photograph. Wolf Foundation. Reproduced by permission.-Young, Grace Chishol, 1923, photograph. Reproduced by permission of Sylvia Wiegand.

A

ABACUS

The abacus is an ancient calculating machine. This simple apparatus is about 5,000 years old and is thought to have originated in Babylon. As the concepts of **zero** and Arabic number notation became widespread, basic math **functions** became simpler, and the use of the abacus diminished. Most of the world employs adding machines, calculators, and computers for mathematical calculations, but today Japan, China, the Middle East, and Russia still use the abacus, and school children in these countries are often taught to use the abacus. In China, the abacus is called a suan pan, meaning counting tray. In Japan the abacus is called a soroban. The Japanese have yearly examinations and competitions in computations on the soroban.

Before the invention of counting machines, people used their fingers and toes, made marks in mud or sand, put notches in bones and wood, or used stones to count, calculate, and keep track of quantities. The first abaci were shallow trays filled with a layer of fine sand or dust. Number symbols were marked and erased easily with a finger. Some scientists think that the term *abacus* comes from the Semitic word for dust, *abq*.

A modern abacus is made of wood or plastic. It is rectangular, often about the size of a shoe-box lid. Within the rectangle, there are at least nine vertical rods strung with movable beads. The abacus is based on the decimal system. Each rod represents columns of written numbers. For example, starting from the right and moving left, the first rod represents ones, the second rod represents tens, the third rod represents hundreds, and so forth. A horizontal crossbar is perpendicular to the rods, separating the abacus into two unequal parts. The moveable beads are located either above or below the crossbar. Beads above the crossbar are called heaven beads, and beads below are called earth beads. Each heaven bead has a value of five units and each earth bead has a value of one unit. A Chinese suan pan has two heaven and five earth beads, and the

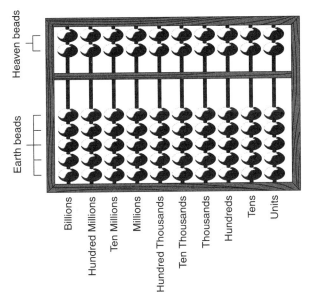

A Chinese abacus called a *suan pan* (reckoning board).

Japanese soroban has one heaven and four earth beads. These two abaci are slightly different from one another, but they are manipulated and used in the same manner. The Russian version of the abacus has many horizontal rods with moveable, undivided beads, nine to a column.

To operate, the soroban or suan pan is placed flat, and all the beads are pushed to the outer edges, away from the crossbar. Usually the heaven beads are moved with the forefinger and the earth beads are moved with the thumb. For the number one, one earth bead would be pushed up to the crossbar. Number two would require two earth beads. For number five, only one heaven bead would to be pushed to the crossbar. The number six would require one heaven (five units) plus one earth (one unit) bead. The number 24 would use four earth beads on the first rod and two earth beads on the second rod. The number 26 then, would use one heaven and one earth bead

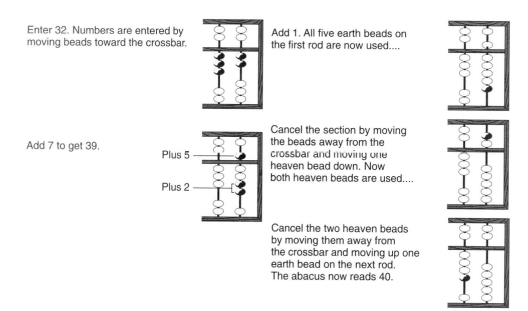

Enter 32. Numbers are entered by moving beads toward the crossbar.

Add 1. All five earth beads on the first rod are now used....

Add 7 to get 39.

Plus 5

Plus 2

Cancel the section by moving the beads away from the crossbar and moving one heaven bead down. Now both heaven beads are used....

Cancel the two heaven beads by moving them away from the crossbar and moving up one earth bead on the next rod. The abacus now reads 40.

An example of addition on a *suan pan.* The heaven beads have five times the value of the earth beads below them.

on the first rod, and two earth beads on the second rod. **Addition**, **subtraction**, **multiplication**, and **division** can be performed on an abacus. Advanced abacus users can do lengthy multiplication and division problems, and even find the **square** root or cube root of any number.

ABBE, ERNST (1840-1905)

German astronomer, mathematician, and physicist

Ernst Abbe made significant contributions in the fields of optics, mathematics, and microscope design. He attended both the University of Jena and the University of Göttingen, earning his doctorate from the latter in 1861. He was appointed to a teaching position at Jena in 1863, and eventually became professor of physics and mathematics and the director of both the astronomical and meteorological observatories there.

In 1866, Carl Zeiss appointed Abbe research director of his Zeiss Optical Works, a manufacturer of microscopes. Abbe remained in his teaching and directorship positions at the University, and used his combined resources to develop major advances in optical science.

Among Abbe's contributions to optical theory were the discovery of the optical formula known as the Abbe **sine** condition, which uses mathematics to define the optical characteristics a lens must have to form a clear image. Abbe also made major advances in the development of the microscope, including the apochromatic lens system and the condenser.

In 1876 Zeiss made Abbe a partner in his business, which also made Abbe a wealthy man. After Zeiss' death in 1888, Abbe reorganized the optical works and set up the Carl-Zeiss-Stiftung (or Carl Zeiss Foundation) for research in science and social improvement. The University of Jena received a portion of the profits, as did employees of the optical works.

Abbe also instituted a number of progressive human resource policies, including sick pay, profit sharing, and eight hour work days, that were almost unheard of at the time. In 1891, Abbe bequeathed his remaining shares in the company to the Stiftung. He passed away 14 years later shortly before his 65th birthday.

ABEL, NIELS HENRIK (1802-1829)

Norwegian algebraist

Niels Henrik Abel was a Norwegian mathematician who is most famous for having proved that fifth and higher order **equations** have no algebraic solution. Had he not died prematurely at the age of 26, it is speculated that he might have become one of the most prominent mathematicians of the 19th century. He provided the first general **proof** of the binomial theorem and made significant discoveries concerning elliptic **functions**.

Abel was born in Finnöy, on the southwestern coast of Norway, on August 5, 1802. He was the second son of Sören Georg Abel, a Lutheran minister, and Anne Marie nee Sorensen, the daughter of a wealthy merchant. In 1804, Abel's father was appointed to a new parish and the family moved to Gjerstad, in southern Norway. Abel received his early education from his father, and in 1815 he was sent to the Cathedral School in Oslo. There, he rapidly developed a passion for mathematics. In 1818, a new instructor, Berndt Holmboe, arrived at the school and fueled Abel's interest further, introducing him to the works of such European masters as **Isaac Newton, Joseph–Louis Lagrange,** and **Leonhard Euler.** Holmboe was to become a lifelong friend and advocate, eventually helping to raise money that allowed Abel to travel

Ernst Abbe

Niels Henrik Abel

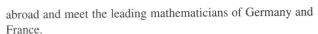

abroad and meet the leading mathematicians of Germany and France.

Abel graduated from the Cathedral School in 1821. His father had died a year earlier and his older brother had developed mental illness. The responsibility of providing for his mother and four younger siblings fell largely on Abel and to make ends meet he began tutoring. Meanwhile, he took the entrance examination for the university. His performance in **geometry** and **arithmetic** was distinguished and he was offered a free dormitory room. In an exceptional move, the mathematics faculty, who were already aware of Abel's promise, contributed personal funds to cover his other expenses. Abel enrolled at the University of Kristiania (Oslo) at the age of 19. Within a year he had completed his basic courses and was a degree candidate.

During his final year at the Cathedral School, Abel had become intrigued by a challenge that had occupied some of the best mathematical minds since the 16th century, that of finding a solution to the "quintic" problem. A **quintic equation** is one in which the unknown appears to the fifth power. Abel believed he had discovered a general solution and presented his results to his teacher Holmboe, who was wise enough to realize Abel's mathematical reasoning was already beyond his full comprehension. Holmboe sent the solution to the Danish mathematician Ferdinand Degen, who expressed skepticism

but was unable to determine whether Abel's argument was flawed. Degen asked Abel to provide examples of his general solution, who, in the end, discovered the error in his approach. Abel would remain obsessed with the quintic problem for the next few years. Finally, at Christmas time in 1823, he hit upon the realization and derived a proof that an algebraic solution was impossible. Abel sent a paper describing his proof to **Johann Karl Friedrich Gauss**, who reportedly ignored the treatise. Meanwhile, Abel began working on what would become the first proof of an integral equation, and went on to provide the first general proof of the **binomial theorem**, which until then had only been proved for special cases. He also investigated elliptic integrals and developed a novel way of examining them through the use of inverse functions.

In 1825, Abel left home and traveled to Berlin, where he met **August Leopold Crelle**, a civil engineer and the builder of the first German railroad. Crelle had a strong reverence for mathematics, and was about to publish the first edition of *Journal for Pure and Applied Mathematics*, the first periodical devoted entirely to mathematical research. Recognizing in Abel a man of genius, Crelle asked if the young man would contribute to the premiere edition. Abel obliged, providing Crelle with a manuscript that described his proof that an algebraic solution to the general equation of the fifth and higher degrees was impossible. The paper would insure both Abel's fame and the success of Crelle's fledgling journal. From Germany, Abel toured southern Europe, then traveled to France, where he made the acquaintance of **Adrien Marie Legendre, Augustin Louis Cauchy**, and others. In their company, he wrote the *Memoir on a General Property of a Very Extensive Class of Transcendental Functions*, which was submitted to the Paris Académie Royale des Sciences. The memoir expounded on Abel's earlier work on elliptical functions, and proposed what has come to be known as Abel's **theorem**. Unfortunately, it was received poorly, rejected by Legendre because it was "illegible," then temporarily lost by Cauchy. Two years after Abel's death, the manuscript finally resurfaced, but it was not published until 1841.

By 1827, Abel had run out of money and was forced to return to Norway. He had hoped to take up a university post, but could only find work as a tutor. Meanwhile, he had discovered he had contracted tuberculosis. Later in 1827 he wrote a lengthy paper on elliptic functions for Crelle's journal and began working for Crelle as an editor.

Abel died on April 6, 1829, while visiting his Danish fiancée, Christine Kemp, who was living then in Froland. A few days later, as yet unaware of Abel's death, Crelle wrote to say he had secured a position for him at the University of Berlin. Abel was honored posthumously, in 1830, when the French Académie awarded him the Grand Prix, a prize he shared with **Karl Jacobi**.

ABSTRACT LINEAR SPACES

Abstract linear spaces are also called **vector spaces**, and they occur in a wide variety of mathematical settings. The most familiar examples are the finite dimensional Euclidean spaces. Let N be a positive integer. Then $\mathfrak{R}^N$ can be represented as the set of all column vectors

$$X = \begin{pmatrix} x_1 \\ x_2 \\ \vdots \\ x_N \end{pmatrix}$$

in which each coordinate x_n, $1 \leq n \leq N$ is an element from the **field** $\mathfrak{R}$ of **real numbers**. Sometimes it is more convenient to represent the elements of $\mathfrak{R}^N$ as row vectors, but for the purposes of this article column vectors will be used. There are two basic algebraic operations that are defined in $\mathfrak{R}^N$. The first is **addition** of vectors. If x and y are two vectors in $\mathfrak{R}^N$ then x + y is the vector obtained by adding the corresponding coordinate of x and y. Thus if we write the vectors as columns we find that

$$\begin{pmatrix} x_1 \\ x_2 \\ \vdots \\ x_N \end{pmatrix} + \begin{pmatrix} y_1 \\ y_2 \\ \vdots \\ y_N \end{pmatrix} = \begin{pmatrix} x_1 + y_1 \\ x_2 + y_2 \\ \vdots \\ x_N + y_N \end{pmatrix}.$$

The second operation is scalar **multiplication**, that is multiplication of a vector by a scalar. A scalar simply means an element from the field $\mathfrak{R}$. If α is a real number and x is a vector then αx is the vector defined by

$$\alpha \begin{pmatrix} x_1 \\ x_2 \\ \vdots \\ x_N \end{pmatrix} = \begin{pmatrix} \alpha x_1 \\ \alpha x_2 \\ \vdots \\ \alpha x_N \end{pmatrix}.$$

Thus αx is obtained by multiplying each coordinate of the vector x by the real number α. It is easy to verify that the set $\mathfrak{R}^N$ together with the operation + on vectors is an Abelian group. The **zero** vector

$$\mathbf{0} = \begin{pmatrix} 0 \\ 0 \\ \vdots \\ 0 \end{pmatrix}$$

is the **identity element** of the group. The operation of scalar multiplication satisfies the identities

$0x = 0$, $1x = x$, $(\alpha\beta)x = \alpha(\beta x)$,

and also the two distributive laws

$(\alpha + \beta)x = \alpha x + \beta x$, and $\alpha(x + y) = \alpha x + \alpha y$.

the triple consisting of the set $\mathfrak{R}^N$ and the two operations of addition and scalar multiplication form a **linear space** or vector **space** over the field $\mathfrak{R}$ of real numbers. Note that we have determined a vector space for each positive integer N, but, for example, the spaces $\mathfrak{R}^3$ and $\mathfrak{R}^4$ are distinct.

Here is another example of a vector space, but in this case the field of scalars is the field **C** of **complex numbers**. Let $\Delta = \{z \in \mathbf{C}: |z| < 1\}$ be the open unit disk in **C**. Then let A(Δ) denote the set of all bounded analytic **functions** $f: \Delta \to \mathbf{C}$. Thus an analytic function $f: \Delta \to \mathbf{C}$ belongs to A(Δ) if there exists a nonnegative constant C_f such that

$|f(z)| \leq C_f$

for all complex numbers z in Δ. If f and g belong to A(Δ) we define their sum $f + g$ to be the function

$(f + g)(z) = f(z) + g(z)$.

Notice that the + sign on the left defines addition of elements from the set A(Δ), while the + sign on the right is addition of complex numbers. The sum $f + g$ defined in this way is again an analytic function on Δ. If C_f and C_g are nonnegative constants such that

$|f(z)| \leq C_f$ and $|g(z)| \leq C_g$.

for all complex numbers z in Δ, then, by the **triangle** inequality for complex numbers, we have

$|(f + g)(z)| = |f(z) + g(z)| \leq |f(z)| + |g(z)| \leq C_f + C_g$.

This shows that $f + g$ is a bounded analytic function and so is an element of A(Δ). In a similar manner we define scalar multiplication of the function f by the complex number α by

$(\alpha f)(z) = \alpha f(z)$.

Then it is obvious that αf is a bounded analytic function on Δ. The set A(Δ) together with the operations of addition and scalar multiplication satisfy algebraic identities analogous to those satisfied by the vector spaces $\mathfrak{R}^N$. In this case, however, $A(\Delta)$ is a vector space over the field **C** of complex numbers. In this example we could drop the requirement that the analytic functions f are bounded, and the resulting set would continue to form a vector space over **C** with the same definitions for addition and scalar multiplication. However, the most important examples of vector spaces often impose such additional restrictions and then, of course, one must verify that the restriction is preserved by the algebraic operations of addition and scalar multiplication. The space $A(\Delta)$ is typical of this sort of construction.

The concept of an abstract linear space is a generalization of the constructions just described for the spaces $\mathfrak{R}^N$ and $A(\Delta)$. Let **F** be a field and $(V, +)$ an Abelian group. It will be convenient to write $0\mathbf{F}$ for the additive identity element of **F**, $1\mathbf{F}$ for the multiplicative identity element of **F**, and **0** for the additive identity element in the group V. Now assume that there exists a function $(\alpha, \mathbf{v}) \to \alpha \mathbf{v}$ from $\mathbf{F} \times V \to V$, denoted by juxtaposition, which satisfies the following conditions:

$0_\mathbf{F}\mathbf{v} = \mathbf{0}$, $1_\mathbf{F}\mathbf{v} = \mathbf{v}$, $(\alpha\beta)\mathbf{v} = \alpha(\beta\mathbf{v})$,

and the two distributive laws

$(\alpha + \beta)\mathbf{v} = \alpha\mathbf{v} + \beta\mathbf{v}$, and $\alpha(\mathbf{u} + \mathbf{v}) = \alpha\mathbf{u} + \alpha\mathbf{v}$.

Here α and β are elements of the field **F**, and **u** and **v** are elements of the group V. Then V is a vector space or linear space over the field **F**. The elements of V are called vectors and the elements of **F** are called scalars.

Notice that the special spaces $\mathfrak{R}^N$ and $A(\Delta)$ are examples of vector spaces over the fields of real numbers and complex numbers, respectively. However, the concept of an abstract linear space is much more general and includes, for example, linear spaces for which the field **F** of scalars is a finite field. Here is an example of a vector space over a finite field. Let p be a prime number and write $\mathbf{F}_p$ for the field with p elements. We can use the field of **integers** modulo p for $\mathbf{F}_p$ because every finite field with p elements is isomorphic to the integers modulo p. Then $\mathbf{F}_p{}^N$ can be represented as the set of all column vectors

$$\mathbf{a} = \begin{pmatrix} a_1 \\ a_2 \\ \vdots \\ a_N \end{pmatrix}$$

in which each coordinate a_n, $1 \le n \le N$, is an element from the finite field $\mathbf{F}_p$. Addition and scalar multiplication in $\mathbf{F}_p{}^N$ are defined by exactly the same formulas that we used in $\mathfrak{R}^N$, but now the scalars and the entries in the vectors are elements of $\mathbf{F}_p{}^N$. It is easy to verify that $\mathbf{F}_p{}^N$ is a vector space over the finite field $\mathbf{F}_p$.

ADDITION

In the simplest sense, addition is a process of combining groups of things to form a new, larger group. Consider two baskets of apples: if apples are transferred from one basket into the other, a new bunch of apples is formed that is understood to be larger than either of the original two bunches. This process lead ancient peoples to create what is probably the earliest branch of mathematics, namely **arithmetic**. Eventually people abstracted from groups of objects (like bunches of apples) the concept of number that represents the quantity of objects in a particular group (e.g., five apples). In this process, the meaning of addition transformed from "combining groups of objects" to "combining the numbers associated with groups." Simply, the addition of numbers allowed for the counting of objects by increments greater than one. For instance, in order to add together two apples and three apples, all could be placed into a bunch and then counted individually from 1 to 5. However, arithmetic can more efficiently represent this procedure as "2 + 3 = 5." The concept of addition thus allowed people to determine the combined number of groups of things more readily, especially useful for larger quantities.

Addition is one of the four basic operations of arithmetic (the others being **subtraction**, **multiplication**, and **division**). Addition operates on the set of natural numbers $\{0, 1, 2, 3,...\}$ such that for any two numbers added together, a (unique) third number is determined. For the expression "2 + 3" there is a unique natural number assigned, namely "5" (called the sum). For "2 + 3," the operation of addition is denoted by "+" (or "plus" sign), while the pair of natural numbers "2" and "3" are called the summands. Historically, the word addition is derived from the Latin "additio," used by Italian mathematician **Leonardo Pisano Fibonacci** (1170-1250). The "+" symbol appeared in print in 1489 under Johann Widman. However, Widman used the symbol to indicate excess, and not as a mathematical operation. Dutch mathematician Vander Hoeche is known to have used "+" as an algebraic operator in 1514.

The properties of addition in relation to the natural numbers can be expressed in the form of the following four laws that hold for any natural numbers "a" and "b":

- "a + b" is a natural number, closure law.
- "a + b = b + a," commutative law.
- "a + (b + c) = (a + b) + c," associative law.
- "a + 0 = a," identity law.

It is important to realize that the ancient peoples empirically determined these addition laws, like the number system itself. As the whole number system was expanded to include **fractions**, **negative numbers**, **irrational numbers**, etc., culminating in the construction of the real number system, the addition laws were found to still hold true. Thus, the four addition laws are just as valid when adding **integers**, fractions, or **real numbers**. With the advent of negative numbers, the inverse law of addition was added to these laws; so that for every real number "a" there exists a corresponding unique real number "-a" such that "a + (-a) = 0."

Particular rules have also been developed to aid the addition of particular types of numbers. For example, the addition of fractions requires a common denominator. To add the **rational numbers** "a / b" and "c / d," the unit fraction "d / d" is multiplied into the first term, and "b / b" into the second term, resulting in "ad / bd" and "cb / bd." With a common denominator "bd", the two can be added together "(ad / bd) + (cb / bd)" to result in "(ad + cb) / bd." As an example, "(1 / 2) + (2 / 5) = (1/ 2) · (5 / 5) + (2 / 5) · (2 / 2) = (5 / 10) + (4 / 10) = 9 / 10."

Addition can also be defined as an operation upon mathematical objects such as matrices, vectors, **complex numbers**, and **functions**. Some of the rules governing the operation of addition on these quantities are now laid out.

The addition of matrices is valid only if the matrices to be added are of the same order (i.e., contain the same number of rows and columns). Adding together the corresponding elements of each **matrix** performs addition on the matrices. The resulting matrix is of the same order as the summed matrices.

The addition of vectors is valid when corresponding components of vectors are added to form a resulting vector. Specifically, adding the vectors "$\mathbf{a} = a_x\mathbf{i} + a_y\mathbf{j}$" and "$\mathbf{b} = b_x\mathbf{i} + b_y\mathbf{j}$" results in the vector "$\mathbf{r} = \mathbf{a} + \mathbf{b} = (a_x + b_x)\mathbf{i} + (a_y + b_y)\mathbf{j}$," where $\mathbf{i}$ and $\mathbf{j}$ are unit vectors in the x-direction and y-direction (respectively) of the x-y plane, and a_x, a_y, b_x, and b_y are scalar components of vectors $\mathbf{a}$ and $\mathbf{b}$ (respectively).

Complex numbers are expressed with a real part followed by an imaginary part. The operation of addition is defined as adding together the individual real parts and the individual imaginary parts. Adding the complex numbers "$x = a + i(b)$" and "$y = c + i(d)$" results in the sum "$x + y = (a + c) + i(b + d)$." For example, the sum of "$2 + i(3)$" and "$1 + i(6)$" is "$(2 + 1) + i(3 + 6) = 3 + i(9)$."

The addition of functions is accomplished by simply adding common terms. For example, the functions "$f(x) = x^2 + 3x - 4$" and "$g(x) = x^3 - 2x^2 + x + 3$" are summed to "$f(x) + g(x) = (x^2 + 3x - 4) + (x^3 - 2x^2 + x + 3) = (x^3 - x^2 + 4x - 1)$."

One of the triumphs of modern mathematics was the construction of a theory dealing with collections (or **sets**) of things, called **set theory**. Formulated in the late 1800s by German mathematician Georg Cantor (1845-1918), set theory has penetrated nearly every branch of mathematics. Early in the twentieth century it was shown that much of mathematics, including the natural numbers, were derivable from the concepts of set theory. Under set theory, the operation of addition on the natural numbers can be recast as the combining, or union, of sets, and the various addition laws can therefore be derived.

See also Additive notation; Arithmetic; Georg Ferdinand Ludwig Philipp Cantor; Complex numbers; Division; Leonardo Pisano Fibonacci; Fractions; Functions; Integers; Irrational numbers; Matrix; Multiplication; Negative numbers; Numbers and numerals; Rational numbers; Real numbers; Set theory; Subtraction; Sums and differences; Vector algebra; Whole numbers

ADDITIVE NOTATION

Addition is one of the four basic operations of **arithmetic** (the others being **subtraction**, **multiplication**, and **division**). In arithmetic, addition operates on the set of **real numbers** such

that for any real numbers added together, another real number is uniquely determined. Additive notation is defined as the system of symbols used in the operation of addition to represent numbers and actions. For an expression like "2 + 3" the operation of addition is denoted by "+" (plus), while the numbers "2" and "3" are called summands. The familiar "+" sign has only been used to denote addition during the last four to five hundred years. Prior to that time other symbols or techniques were utilized. The "+" symbol appeared in print in 1489 under German mathematician Johann Widman (1462-1498) in his book *Mercantile Arithmetic*. However, Widman used the symbol to indicate excesses in business dealings, not as a mathematical operation. The "+" symbol is known to have been used as an algebraic operator by Dutch mathematician Vander Hoecke in 1514, but it is believed the symbol was used that way even earlier.

As described previously, addition on the real numbers is an operation that takes a pair of numbers and yields a third unique number; stated symbolically "$a + b = c$." Thus, addition is defined as a binary procedure (the sum of two values represents a third). For example, one could sum "$a_1 + a_2 + a_3 + a_4 + a_5 +... + a_n$," where each subscripted "a" represents a different number and "n" terms are added. A compact notation for representing a long expression of added terms is the summation symbol, denoted by the Greek capital sigma (Σ). The sigma notation was introduced to facilitate the writing of sums. Swiss mathematician **Leonhard Euler** (1707-1783) first used the summation symbol (Σ) in 1755. It was later used by French mathematician **Joseph-Louis Lagrange** (1736-1813) but was largely ignored during the eighteenth century. A typical summation could be of the form

$$\sum_{i=1}^{i=n}$$

where "i" is the index of summation, "n" is the upper limit, and both symbols have only **integer** values. The summation can be written out explicitly as the sum of "n" terms. For example, to indicate the combined weights of 20 boxes one could write "total weight = $a + b + c +... + t$," where each letter stands for the weight of a different box. One could also write

$$\text{total weight} = \sum_{i=1}^{i=20} x_i = x_1 + x_2 + x_3 +... + x_{20},$$

where each "x_i" represents an individual weight.

Most often in mathematics "Σ" is used in the expression of a series. A series is a sequence of quantities, called terms, in which the relationship between consecutive terms is the same. Series can be used to find the value of constants like π, and to construct tables of **logarithms** and trigonometric **functions**. When the upper bound of a summation is **infinity**, the result is called an **infinite series**.

Another additive notation involves the integral sign "∫". A series

$$\sum_{k=1}^{k=n} f_k(x),$$

like the one above, can be extended to calculate an **area** under a curve by summing the areas of rectangles that together closely approximate the area under the curved region. An area is reached that is defined as its limit by reducing the size of the rectangles and increasing their numbers toward infinity (called the limit). This limit is denoted

$$\lim_{k \to +\infty} \sum_{k=1}^{k=n} f_k(x)\, \Delta x_k = \int_a^b f(x)\, dx;$$

where the curve is continuous on the closed **interval** [a, b], the area is contained under the curve, above lines of reference, and between the closed boundary, and the area is divided into n rectangles that approach infinity. This **definite integral** is thus the limiting value of a sum. It is used to evaluate the length of and the area under plane curves, the area of surfaces of revolution, the **volume** of solids of revolution, and for other problems.

The notation laid out in the preceding paragraphs use expressions whose values are numbers, particularly real numbers. The same notation can also be applied to the addition of **complex numbers**, vectors, matrices, etc. That is, addition is defined as an operation on these **sets** of mathematical objects, which means that a pair of elements of some set (e.g., the complex numbers) yields a third unique element of the set. The sigma and integral notation may be applied to these various mathematical objects; for example, within complex **analysis** and vector **calculus**.

In the nineteenth and twentieth centuries, an abstraction of **algebra** took place that possessed enormous implications for algebra and for mathematics in general. The use of additive notation within abstract algebra and **set theory** was, as a result, very important. Within abstract algebra, the **definition** of an operation is generalized to encompass sets possessing any type of element. So one finds the addition operator "+" used in regards to sets known as groups and fields, all of whose elements conform to certain rules under the operation of addition.

In the late nineteenth century, **Georg Cantor** formulated set theory to deal with collections (or sets) of things. Since then set theory has penetrated nearly every branch of mathematics. Early in the twentieth century it was shown that much of mathematics, including the natural numbers, were derivable from the concepts of set theory. Under set theory, the operation of addition on the natural numbers can be recast as the combining, or union, of sets. Additive notation within set theory relies on the symbol "∪" for the union of sets. For sets A and B, the union of A and B is represented with the symbol "∪", where "A ∪ B" is defined as "the set of all elements x such that x is an element of either set A or B, or both." For

instance, if A = {a, b, c} and B = {c, d, e}, then "A ∪ B = {a, b, c, d, e}."

See also Addition; Algebra; Arithmetic; Georg Ferdinand Ludwig Philipp Cantor; Complex numbers; Definite integral; Division; Leonhard Euler; Infinite series; Integral calculus; Joseph-Louis Lagrange; Limits; Logarithms; Matrix; Multiplication; Numbers and numerals; Real numbers; Sequences and series; Set theory; Subtraction; Sums and differences; Trigonometric functions; Vector analysis

ADELARD OF BATH (CA. 1075-CA. 1160)
English translator

Among the series of medieval English translators, mathematicians, and natural philosophers of England who traveled extensively in search of Arabic texts was Adelard of Bath. He is responsible for the conversion of Arabic-Greek learning into Latin.

Abelard was born in approximately 1075 in Bath, England. An extensive traveler, he went to study at Tours and later taught at Laon. Leaving Laon, for the next seven years he journeyed to various cities and countries, including Salerno, Syria, Cilicia, Spain and possibly Palestine before returning to Bath in 1130. During his travels, however, it was possible that he learned Arabic in Sicily and received Spanish-Arabic texts from other Arabists who had lived in or visited Spain.

Adelard made significant contributions to the field of philosophy with the writing of two treatises: *De eodem et diverso*, dedicated to William, bishop of Syracuse and written before 1116, and *Quaestiones naturales* written prior to 1137 and perhaps even earlier. Speaking as a quasi Platonist in *De eodem et diverso*, Adclard draws on the major themes of Platonism, but opposing the Platonic doctrine of realism in his theory of universals. His second treatise *Quaestiones naturales* consists of a dialogue with an unnamed nephew of 76 scientific discussions derived from Muslim science. In Adelard's first work, there is no trace of Arabic influence; however, there are descriptions of a pipette-like vessel mentioned in other Arabic works by the unnamed nephew in *Quaestiones naturales*.

While Adelard's contributions to the area of philosophy were significant, he is more renowned for his translations of Arabic scientific texts. With these germinal translations, Adelard was also able to culminate the use of algorithms and Arabic numerals, as opposed to the intractable Roman numerals.

Using *Ysagoga minor Iapharis matematici in astronomicam per Adhelardum bathoniensem ex arabico sumpta*, a translation of Abu Ma shar's *Shorter Introduction to Astronomy*, Adelard gave the first sampling to the Latin Schoolmen. This contained some astrological rules and axioms, and was abridged by Abu Ma shar for his longer *Introductorium maius*. Adelard's translation of this work proved pivotal, as the longer version was translated twice thereafter into Latin. Additionally, Adelard translated *Liber prestigiorum Thebidis (Elbidis) secundum Ptolomeum et*

Hermetem per Adelardum bathoniensem translatus, an astrological work on images and horoscopes by Thabit ibn Qurra.

Adelard also produced a variety of works involving **arithmetic**, the earliest being *Regule abaci*. Another work, *Liber ysagogurm Alchorismi in artem astronomicam a magistro A. compositus*, is composed of books concerning arithmetic, **geometry**, music, and astronomy.

Before Adelard translated Euclid's *Elements* into Latin, there were only a few incomplete versions, such as that of Boethius, in existence from the Greek. While Adelard's name is associated with three different versions, each somewhat incomplete, each codex has been pieced together to present a full version.

Among the most important works he translated were the Astronomical tables of Al-Majriti, and a treatise on Arabic arithmetic by the mathematician al-Khwarizmi, whose name later became synonymous to the mathematical system of algorithms.

Adelard's writings and translations remain an important bridge between Muslim science and Western learning.

Maria Gaëtana Agnesi

AGNESI, MARIA GAËTANA (1718-1799)
Italian algebraist, geometer, logician, physicist, and philosopher

One of the great figures of Italian science, Maria Gaëtana Agnesi was born and died in Milan, an Italian city under Habsburg rule. In early childhood, she demonstrated extraordinarily intellectual abilities, learning several languages, including Greek, Latin, and Hebrew.

Agnesi's father, who taught mathematics at the University of Bologna, hired a university professor to tutor her in mathematics.

While still a child, Agnesi took part in learned discussions with noted intellectuals who visited her parents' home. Her knowledge encompassed various fields of science, and to any foreign visitor who was not a Latinist (the discussions were held in Latin), she spoke fluently in his language. Her brilliance as a multilingual and erudite conversationalist was matched by her eloquence as a writer. When she was 17 years old, Agnesi wrote a memoir about the **Marquis Guillaume F. A. de l'Hôspital's** 1687 article on **conic sections**. Her *Propositiones philosophicae*, a book of essays published in 1738, examines a variety of scientific topics, including philosophy, logic, and physics. Among the subjects discussed is **Isaac Newton**'s theory of universal gravitation.

Following her mother's death, Agnesi wished to enter a convent, but her father decided that as the oldest child, she should supervise the education of her numerous younger siblings. As an educator, Agnesi recognized the educational needs of young people, and eloquently advocated the education of women.

Agnesi's principal work, *Instituzione analitiche ad uso della gioventu' italiana* (1748), known in English as her *Analytical Institutions*, is a veritable compendium of mathematics, written, as the Italian title indicates, for the edification of Italian youth. The work introduces the reader to **algebra** and **analysis**, providing elucidations of both and of integral and differential **calculus**. Praised for its lucid style, Agnesi's book was translated into English by John Colson, Lucasian Professor of Mathematics at Cambridge University. Colson, who learned Italian for the express purpose of translating Agnesi's book, had already translated Newton's *Principia mathematica* into English. Among the prominent features of Agnesi's work is her discussion of a curve, subsequently named the "**Witch of Agnesi**," due in part to an unfortunate confusion of terms. (The Italian word *versiera*, derived from the Latin *vertere*, meaning *to turn*, became associated with *avversiera*, which in Italian means *devil's wife*, or *witch*.) Studied previously by **Pierre de Fermat** and by Guido Grandi, the "Witch of Agnesi" is a cubic curve represented by the Cartesian equation $y(x^2 + a^2) = a^3$, where a represents a parameter, or constant. For $a = 2$, as an example, the maximum value of y will be 2. As y tends toward 0, x will tend, asymptotically, toward $\pm\infty$.

In 1750, Pope Benedict XIV named Agnesi professor of mathematics and natural philosophy at the University of Bologna. As David M. Burton explains, it is not quite clear whether she accepted the appointment. Considering the fact that her father was gravely ill by 1750, there is speculation that she would have found the appointment difficult to accept. At any rate, after her father's death in 1752, Agnesi apparently lost all interest in scientific work, devoting herself to a religious life. She directed charitable projects, taking charge of a home for the poor and infirm in 1771, a task to which she devoted the rest of her life.

AHLFORS, LARS V. (1907-1996)
Finnish-born American mathematician

Lars V. Ahlfors was a mathematician whose major area of research was **complex analysis**. In 1936, he was one of the first to receive a Fields Medal. Often considered the equivalent of the Nobel Prize, the Fields Medal is given every four years to a mathematician under the age of forty who has achieved important results in his or her work. Ahlfors received this award for his work on Riemann surfaces, which are schematic devices for **mapping** the relation between **complex numbers** according to an analytic function. Ahlfors's results led to new developments in the field of meromorphic **functions** (functions that are analytic everywhere in a region except for a finite number of poles); the methods he developed to obtain these results created an entirely new field of analysis.

Lars Valerian Ahlfors was born on April 18, 1907 in Helsingfors (now Helsinki), Finland. His mother, Sievä Helander Ahlfors, died giving birth to him. His father, Axel Ahlfors, was a mechanical engineering professor at the Polytechnic Institute. Even as a child, Ahlfors was interested in mathematics; his high school did not offer **calculus** courses, but Ahlfors taught himself by reading his father's engineering books.

He did not have access to mathematical books until he began his studies in 1924 at the University of Helsingfors, where he was taught by Ernst Lindelöf and Rolf Nevanlinna. Lindelöf worked in complex analysis and was known as the father of mathematics in Finland—mostly because, in the 1920s, all Finnish mathematicians were his students. Ahlfors received his degree in the spring of 1928, and he also began his graduate work that year. Although there were no official graduate courses in mathematics at the university, Lindelöf supervised students' advanced readings.

Ahlfors took his first official graduate course in mathematics in the fall of 1928, when he accompanied Nevanlinna to Zürich. The class Nevanlinna taught was on contemporary function theory. Topics included the major parts of Nevanlinna's theory of meromorphic functions and Denjoy's conjecture on the number of asymptotic values of an entire function, as well as Carleman's partial **proof** of it. During his study of this subject, Ahlfors proved the full Denjoy conjecture after he discovered a new approach based on conformal mapping. A conformal map is a function in which, if two curves intersect at an **angle**, then the images of the curves in the map will also intersect at the same angle.

When the course ended, Ahlfors travelled to Paris, where he continued his work for three months before returning to Finland. His research there led to a geometric interpretation of the Nevanlinna theory, which he would publish in 1935. Although this interpretation was also discovered inde-

pendently in Japan, it was the beginning of Ahlfors's concentration on meromorphic **functions**.

When he returned to Finland, Ahlfors was given the position of lecturer at Åbo Akademi, Finland's Swedish-language university. He also began work on his thesis, the subject of which was conformal mapping and entire functions. He had finished his thesis by the spring of 1930, and received his Ph.D. in 1932. Ahlfors was named a fellow of the Rockefeller Institute in 1932, which allowed him to live and do research in Paris for a year. In July of 1933, he married Erna Lehnert; they would have three daughters. He returned to the University of Helsingfors that same year as an adjunct professor and taught there until the fall of 1935, when he began a three-year assignment as assistant professor at Harvard University.

For his research in Riemann surfaces of inverse functions in terms of covering surfaces, Ahlfors was awarded the Fields Medal by the International Congress of Mathematicians in 1936. Ahlfors was attending the ceremony in Oslo, but he learned only hours before it began that he had been chosen as the recipient. In the talk about Ahlfors's work he gave to the congress, German mathematician Constantin Carathéodory specifically noted the contribution of Ahlfors's paper, "On the Theory of Covering Surfaces," which explained the methods Ahlfors had developed in his work on Riemann surfaces. Carathéodory pointed out that these methods were also the start of a new branch of analysis, which he termed "metrical topology."

In the spring of 1938, Ahlfors left the United States and returned to Finland to take a position as a professor at the University of Helsinki. World War II soon spread to Finland, however, and the university closed because there were not enough students. Although his family was evacuated to Sweden, Ahlfors stayed in Helsinki. He was not called for military duty because of a physical condition, but he participated in the military's communications setup.

In the summer of 1944, the University of Zürich offered Ahlfors a professorship, and he accepted the position. After an arduous journey from Sweden to Switzerland because of the war, he began teaching in the summer of 1945. He was not happy there, however, so when Harvard University asked him to return he gladly accepted. He began teaching there in the fall of 1946 and became a naturalized United States citizen in 1952. In 1953, Ahlfors's book *Complex Analysis* was published. It is still widely used as a basic text in graduate courses. In 1964, he was named William Caspar Graustein Professor of Mathematics. Ahlfors remained at Harvard until his retirement as Professor Emeritus in 1977, subsequently moving to Boston. He died in Pittsfield, Massachusetts.

AHMES (18TH CENTURY B.C.)
Egyptian scribe

Relatively few mathematical writings have survived from the time the rulers of Egypt built the pyramids. This is largely because writing at that time was done on laminated sheets of the papyrus plant, and such materials readily decompose under adverse environmental conditions.

One of the documents that has survived is known as the **Rhind papyrus**, named for the Scottish antique dealer, Henry Rhind, who purchased it in 1858. This document, about one foot high and 18 feet long, is now held by the British Museum (except for a few fragments held by the Brooklyn Museum). The document is also known as the Ahmes papyrus, after the individual who authored it. The papyrus is believed to date from about 1650 B.C., but according to information contained in it, the material contained in it is derived from an earlier version from the Middle Kingdom (about 2000 to 1800 B.C.

Although other mathematical writings from that time have been found, including the Moscow papyrus (now held in Moscow) and fragments of others, the Ahmes papyrus is one of the longest. These mathematical papyri are believed to have been written by ecclesiastical and government scribes.

The opening lines of the papyrus, written in the phonetic hieratic script (a simplification of pictorial hieroglyphics) that was used for everyday writing, reads "Directions for Obtaining the Knowledge of All Dark Things." The actual text appears to record the types of problems that business and administrative clerks frequently had to solve.

Because the document does not actually give any rules for solving arithmetical problems, there has been speculation that the Ahmes papyrus may have been an advanced arithmetical text for students, written in much the same tone as modern texts that leave proofs as an exercise for the reader. Others have speculated that it was the work of a student.

In the text, all **fractions** were decomposed into what are known as unit fractions. The author includes a table that expresses fractions with numerator 2 and odd denominators from 5 to 101 as sums of fractions with numerator 1. For example, Ahmes writes $2/5 = 1/3 + 1/15$.

To express a fraction like $7/29$ as a sum of unit fractions, Ahmes first notes that $7 = 2 + 2 + 2 + 1$, and then proceeds to convert each $2/29$ to a sum of fractions with numerator 1. In this way, he comes up with the following result: $7/29 = 1/6 + 1/24 + 1/58 + 1/87 + 1/232$.

In the papyrus, Ahmes' sixty-third problem reads "Directions for dividing 700 breads among four people, 2/3 for one, 1/2 for the second, 1/3 for the third, 1/4 for the fourth." (In its modern representation, this problem would read $2x/3 + x/2 + x/3 + x/4 = 700$, where one is to solve for x). Ahmes' solution was as follows: "Add $2/3 + 1/2 + 1/3 + 1/4$. This gives $1 + 1/2 + 1/4$. Divide 1 by $(1 + 1/2 + 1/4)$. This gives $1/2 + 1/14$. Now find $(1/2 + 1/14) * 700$. This is 400."

In Ahmes' papyrus, there is only minimal use of symbols. **Addition** and **subtraction** are represented by the legs of a man coming and going, respectively.

Some have speculated that one of the reasons that the Egyptians never developed advanced arithmetical or algebraic methods was that, although the method of fractions proved adequate for carrying out basic arithmetical operations, use of the method required extensive and complicated manipulations.

One of the exercises in Ahmes' papyrus, most certainly intended for students, simply reads: "seven houses, 49 cats, 343 mice, 2401 ears of spelt, 1607 hekats." Modern interpreters believe that the author was describing a problem in which each of seven houses contained seven cats, and that each cat would eat seven mice, and each mouse seven ears of grain, with each ear producing seven measures of grain. The statement and solution of the problem, which called for a calculation of the total number of houses, cats, mice, ears of spelt, and measures of grain, has not a little in common with our nursery rhyme that begins "As I was going to St. Ives, I met a man with seven wives," which some have interpreted as an indication that Ahmes had something of a sense of humor.

AIKEN, HOWARD (1900-1973)
American computer scientist and inventor

A noted physicist and Harvard professor, Howard Aiken designed and built the Mark I **calculator** in the late 1930s and early 1940s. The first large-scale digital calculator, the Mark I provided the impetus for larger and more advanced computing machines. Aiken's later conceptions, the Mark II, Mark III, and Mark IV, each surpassed its previous model in terms of speed and calculating capacity.

Howard Hathaway Aiken was born on March 8, 1900, in Hoboken, New Jersey, and was raised in Indianapolis, Indiana. Because of his family's limited resources, he had to go to work after completing the eighth grade. He worked twelve-hour shifts at night, seven days a week, as a switchboard operator for the Indianapolis Light and Heat Company. During the day he attended Arsenal Technical High School. When the school superintendent learned of his round-the-clock work and study schedule, he arranged a series of special tests that enabled Aiken to graduate early. In 1919 Aiken entered the University of Wisconsin at Madison and worked part-time for Madison's gas company while he attended classes. He received his bachelor of science degree in 1923 and upon graduation was immediately promoted to chief engineer of the gas company. Over the next twelve years he became a professor at the University of Miami and later went into business for himself. By 1935, however, he had decided that he wanted to return to school to work on his Ph.D. He began his graduate studies at the University of Chicago before going on to Harvard. He received a master's degree in physics in 1937 and was made an instructor. He wrote his dissertation while he was teaching and received his doctorate in 1939.

As a graduate student in physics, Aiken completed a great deal of work requiring many hours of long and tedious calculations; it was at that time that he began to think seriously about improving calculating machines to reduce the time needed for figuring large numerical sequences. In 1937, while at Harvard, Aiken wrote a 22-page memorandum proposing the initial design for his computer. His idea was to build a computer from existing hardware with electromagnetic components controlled by coded sequences of instructions, and one that would operate automatically after a particular process had been developed. Aiken proposed that the punched-card calculators then in use (which could carry out only one **arithmetic** operation at a time) could be modified to become fully automated and to carry out a wide range of arithmetic and mathematical **functions**. His original design was inspired by a description of a more powerful calculator in the work of English mathematician **Charles Babbage**, who devoted nearly forty years to developing a calculating machine.

Although Aiken was by then an instructor at Harvard (and was to become an associate professor of **applied mathematics** in 1941 and a full professor in 1946), the university offered little support for his initial idea. He therefore turned to private industry for assistance. Although his first attempt to muster corporate support was turned down by the Monroe Calculating Machine Company, its chief engineer, G. C. Chase, approved of Aiken's proposal and suggested he contact Theodore Brown, a professor at the Harvard Business School; Brown, in turn, put Aiken in touch with IBM. Aiken's idea impressed IBM enough that the company agreed to back the construction of his Mark I. In 1939 IBM President Thomas Watson, Sr., signed a contract that stated that IBM would build the computer under Aiken's supervision and with additional financial backing from the U.S. Navy. At the time IBM only manufactured office machines, but its management wanted to encourage research in new and promising areas and was eager to establish a connection with Harvard. During that same year, Aiken became a school officer of the Naval Warfare School at Yorktown, and when the Mark I contract was worked out he was made officer in charge of the U.S. Navy Computing Project. The Navy agreed to support Aiken's computer because the Mark I offered a great deal of potential for expediting the complex mathematical calculations involved in aiming long-range guns onboard ship. The Mark I provided a solution to the problem by calculating gun trajectories in a matter of minutes.

With a grant from IBM and a Navy contract, Aiken and a team headed by IBM's Clair D. Lake began work at IBM's laboratories in Endicott, New York. Aiken's machine was electromechanical—mechanical parts, electrically controlled—and used ordinary telephone relays that enabled electrical currents to be switched on or off. The computer consisted of thousands of relays and other components, all assembled in a 51-foot-long and 8-foot-high (1554 cm x 243 cm) stainless steel and glass frame that was completed in 1943 and installed at Harvard a year later. The heart of this huge machine was formed by 72 rotating registers, each of which could store a positive or negative 23-digit number. The telephone relays established communication between the registers. Instructions and data input were entered into the computer by means of continuous strips of IBM punch-card paper. Output was printed by two electrical typewriters hooked up to the machine. The Mark I did not resemble modern computers, either in appearance or in principles of operation. The machine had no keyboard, for instance, but was operated with approximately 1,400 rotary switches that had to be adjusted to set up a run. Seemingly clumsy by today's computer standards, the Mark I nevertheless was a powerful

Howard Aiken

improvement over its predecessors in terms of the speed at which it performed a host of complex mathematical calculations. Many scientists and engineers were eager for time on the machine, underscoring the project's success and giving added impetus for continued work on improved models. However, a dispute developed with IBM over credit for the computer, and subsequently the company withdrew support for all further efforts. A more powerful model was soon undertaken under pressure from competition from ENIAC, the much faster computer then being built at Columbia University.

Mark I was to have three successors, Mark II through IV. It was with the Mark III that Aiken began building electronic machines. Aiken had a conservative outlook with respect to electronic engineering and sacrificed the speed associated with electronic technology for the dependability of **mechanics**; only after World War II did he begin to feel comfortable using electronic hardware. In 1949 Aiken finished the Mark III with the incorporation of electronic components. Data and instructions were stored on magnetic drums with a capacity of 4,350 sixteen-bit words and roughly 4,000 instructions. With Aiken's continued concern for reliability over speed, he called his Mark III "the slowest all-electronic machine in the world," as quoted by David Ritchie in *The Computer Pioneers: The Making of the Modern Computer*. The Mark III's final version, however, was not completely electronic; it still contained about 2,000 mechanical relays in addition to its electronic components. The Mark IV followed on the heels of the Mark III and was considerably faster.

Aiken contributed to the early computing years by demonstrating that a large, calculating computer could not only be built but could also provide the scientific world with high-powered, speedy mathematical solutions to a plethora of problems. Aiken remained at Harvard until 1961, when he moved to Fort Lauderdale, Florida. He went on to help the University of Miami set up a computer science program and a computing center and became Distinguished Professor of Information there. At the same time he founded a New York-based consulting firm, Howard Aiken Industries Incorporated. Aiken disliked the idea of patents and was known for sharing his work with others. He died on March 14, 1973.

AJDUKIEWICZ, KAZIMIERZ (1890-1963)

Polish philosopher and logician

The Polish philosopher and logician Kazimierz Ajdukiewicz was born on December 12, 1890, in Tarnopol, Poland, then part of Austria-Hungary (now Ternopil, Ukraine). Ajdukiewicz's work was devoted to understanding how knowledge and the conception of knowledge depend on language.

Ajdukiewicz completed his secondary education in Lvov, Poland and subsequently studied philosophy, **mathematics and physics** at the university there. He earned his Ph.D. in philosophy from the University of Lvov in 1912 by writing a thesis on Kant's philosophy of **space**.

He then went to Göttingen University where the mathematician **David Hilbert** and the philosopher **Edmund Husserl** were lecturing. But one year later, he was drafted into the Austrian army and assigned to the Italian front. By the end of the First World War, however, he had joined the Polish army, from which he was demobilized in 1920.

Beginning in 1920 and until the outbreak of World War II, Ajdukiewicz was a first a university lecturer and then professor of philosophy at the Universities of Lvov and Warsaw. During this time, he was primarily concerned with his doctrine of radical conventionalism, which held that some conceptual apparatuses exist that are not intertranslatable, and that growth in scientific knowledge takes place through the replacement of one of these conceptual apparatuses by another. This idea was in opposition to the theories of the French mathematician **Henri Poincaré** (1854-1912), but it was important in establishing a precedent for ideas later developed by the philosopher and historian of science Thomas S. Kuhn.

In 1921, he completed a work entitled "From the Methodology of Deductive Science," that included structural definitions of such concepts as "proof", "theorem", "consequence", "logical theorem", "logical consequence." The same year, he married Maria Twardowska, the daughter of his philosophy teacher at the University of Lvov.

Under German occupation of Poland during World War II, Ajdukiewicz was forced to earn his living as a clerk, although he did manage find time to teach in underground Polish schools.

At the end of the war, he accepted the Chair of the Methodology and Theory of Science in the Faculty of Mathematical and Physical Sciences at the University of Poznan. Between 1948 and 1952, he served as Rector at the university. In 1955, he left Poznan to become Professor of Logic at the University of Warsaw, where he remained until his retirement in 1961. Toward the end of his life, Ajdukiewicz developed the outlines of a new research program in epistemology and the philosophy of language. He also served as head of the **Division** of Logic in the Institute of Philosophy of the Polish Academy of Science, a position he retained even after his retirement. Ajdukiewicz died unexpectedly in his sleep on April 12, 1963, in Warsaw, of heart failure.

ALBERTI, LEON BATTISTA (1404-1472)

Viennese architect, painter, musician, and mathematician

Leon Battista Alberti was born in Vienna on February 18, 1404. Alberti was something of a protypical Renaissance Man, having achieved success as an architect, painter, musician, and mathematician. In architecture, Alberti is remembered for designing the classical churches of San Francesco at Rimini and Sta Maria Novella at Florence.

As a mathematician, Alberti is most remembered for formulating the laws of perspective, which were later to have a major influence on later styles of painting. Because of his work on mathematical perspective, he has sometimes been called the originator of projective **geometry**, even though his contribution to that field was limited to establishing a starting point for further investigations.

As a painter, Alberti knew that in normal vision, the artist sees the same scene with each of his two eyes, but from two slightly different positions, and that the brain reconciles the two images to create the perception of depth. By using **light**, shading and color modification techniques, Alberti attempted to create an illusion of depth in his paintings.

The technique that Alberti finally came up with for creating perspective was to interpose a glass screen between himself and the scene of interest, and then imagine lines of light extending from the eye to each point in the scene. Where these lines intersected the screen, he imagined a set of points (constituting a *section*) mapped out. This section had the same effect on the eye as the scene itself because the same lines of light originated from the section as from the scene itself.

Alberti described many of his mathematical ideas in his book *Della pictura* in 1435 (printed in 1511). Although Alberti supplied some of the mathematical rules for creating the illusion of perspective in the book, he clearly intended that work to be a summary of his findings rather than a set of rigorous proofs. A later book of his, *Ludi mathematici* (1450) described the applications of mathematics to the fields of **mechanics**, surveying, time-reckoning, and artillery.

Alberti has sometimes been credited with the invention of the *camera obscura*, but that instrument more probably the invention of the Arabian scholar Alhazen (965-1038).

ALEKSANDROV, PAVEL S. (1896-1982)
Russian topologist

Pavel S. Aleksandrov laid the foundation for the field of mathematics known as **topology**. In addition to writing the first comprehensive textbook on the subject, Aleksandrov introduced several basic concepts of topology and its offshoots, homology and cohomology, which blend topology and **algebra**. His important work in defining and exploring bicompact (compact or locally compact) spaces laid the groundwork for research done by other mathematicians in these fields.

The youngest of the six children of Sergei Aleksandrovich Aleksandrov and Tsezariia Akimovna Zdanovskaia, Pavel Sergeevich Aleksandrov was born in Bogorodsk, Russia, on May 7, 1896. A year later the family moved to Smolensk, where Aleksandrov's father became head doctor in the state hospital. Although educated mainly in public schools, Aleksandrov learned German and French from his mother, who was skilled in languages.

In grammar school Aleksandrov developed an interest in mathematics under the guidance of Aleksandr Eiges, his **arithmetic** teacher. Aleksandrov entered the University of Moscow in 1913 as a mathematics student, and achieved early success when he proved the importance of Borel **sets** after hearing a lecture by **Nikolai Nikolaevich Luzin** in 1914. Aleksandrov graduated in 1917 and planned to continue his studies. However, after failing to reach similar results on his next project—**Georg Cantor**'s continuum hypothesis (since acknowledged unsolvable; that is, it can be neither proved nor disproved)—Aleksandrov dropped out of the mathematical community and formed a theater group in Chernigov, a city situated seventy-seven miles north of Kiev, in the Ukraine. Besides participating in the theater group, he lectured publicly on various topics in literature and mathematics. He also was involved in political support of the new Soviet government, for which he was jailed briefly in 1919 by counterrevolutionaries.

Later that same year, Aleksandrov suffered a lengthy illness, during which he decided to return to Moscow and mathematics. To help himself catch up, he enlisted the help of another young graduate student, Pavel Samuilovich Uryson. The two immediately became close friends and colleagues. After a brief, unsuccessful marriage in 1921 to his former teacher's sister, Ekaterina Romanovna Eiges, Aleksandrov joined some fellow graduate students in renting a summer cottage. There, he and Uryson began their study of the new field of topology, the branch of mathematics that deals with properties of figures related directly to their shape and **invariant** under continuous **transformation** (that is, without cutting or tearing). In topology, often called rubber-sheet **geometry**, a **cylinder** and a **sphere** are equivalent, because one can be shaped (or transformed) into the other. A doughnut, however, is not equivalent to a sphere, because it cannot be shaped or stretched into a sphere. No textbooks were available on the subject, only articles by **Maurice Fréchet**, **Felix Hausdorff**, and a few others. Nonetheless, from these articles, Uryson and Aleksandrov came up with their first major topological discovery: the **theorem** of metrization. Metrization is the process of deriving a specific measurement for the abstract concept of a topological **space**. In order to do this, Aleksandrov and Uryson first had to develop definitions of topological spaces. They initially defined a *bicompact* space (now known as compact and locally compact spaces), whose property is that for any collection of open sets (or groups of elements) that contains it (the interior of a sphere is an example of an open set), there is a subset of the collection with a finite number of elements that also contains it. Prior to their work, the concept of space was too abstract to be applicable to other mathematical fields; Aleksandrov and Uryson's research led to the acceptance of topology as a valid field of mathematical study.

With this result, the pair rose to fame within the mathematical community, gaining the approval of such notable scholars as **Emmy Noether**, **Richard Courant**, and **David Hilbert**. In 1924 Uryson and Aleksandrov went to Holland and visited with **Luitzen Egbertus Jan Brouwer**, who suggested that they publish their studies on topology. Aleksandrov and Uryson went on to the seaside in France for a spell of work and relaxation that ended tragically when Uryson drowned while swimming. In the aftermath of his friend's death, Aleksandrov lost himself in his work, conducting a seminar on topology that he and Uryson had begun organizing in 1924, and spending 1925 to 1926 working with Brouwer in an attempt to get his research into a form suitable for publication. During this time he further developed his theories of topology and compact space, with an eye to applying topology to the investigation of complex problems.

In 1927 Aleksandrov left Europe for a year to continue his work with a new friend and colleague, Heinz Hopf, at Princeton. Aleksandrov had met Hopf during the summer of 1926 in Göttingen, which along with Paris was considered to be the mathematical hub of Europe. It was in Göttingen in 1923 that Aleksandrov and Uryson first presented their results outside the U.S.S.R., and it was Aleksandrov's preferred summer residence until 1932. There he worked with others, including Noether, who gave the topological work of Aleksandrov and Hopf its algebraic bent. This may have led to Aleksandrov's growing interest in homology, the offshoot of topology incorporating algebra. Homology had first been developed by the French mathematician **Jules Henri Poincaré**, but only for certain types of topological spaces. In 1928 Aleksandrov made a major step in expanding the field when he was able to generalize homology to other topological spaces.

In 1934 Aleksandrov at last received his doctorate from the University of Moscow. The next year, he would issue his most famous work. After much difficult research, the first volume of Aleksandrov and Hopf's still-classic work *Topologie* was published (the remaining two volumes would not be published until after World War II, though they were completed sooner). In the tome they outlined, often for the first time, many basic concepts of this branch of mathematics. They also introduced the **definition** of cohomology, which is the "dual" theory, or mirror image, of homology. Cohomologists consider the same topics as homologists, but from a different vantage point, providing different results. The publication achieved, Aleksandrov settled in a small town outside of Moscow with his friend and colleague **Andrey Nikolayevich Kolmogorov**. They stayed together here, teaching at the University of Moscow, for the rest of their lives.

Always concerned with the younger generation of mathematicians, Aleksandrov in later years crafted groundbreaking textbooks in the fields of topology, homology, and **group theory**, which studies the properties of certain kinds of sets. He guided his students—noted mathematicians such as A. Kuros, L. Pontriagin, and A. Tikhonov—to great heights. He also led the mathematical community in Moscow, presiding over that city's mathematical society for more than thirty years. In 1979 Aleksandrov wrote his autobiography. He died three years later in Moscow on November 16, 1982.

ALGEBRA

Algebra is often referred to as a generalization of **arithmetic**. As such, it is a collection of rules: rules for translating words into the symbolic notation of mathematics, rules for formulating mathematical statements using symbolic notation, and rules for rewriting mathematical statements in a manner that leaves their **truth** unchanged.

The power of elementary algebra, which grew out of a desire to solve problems in arithmetic, stems from its use of variables to represent numbers. This allows the generalization of rules to whole **sets** of numbers. For example, the solution to a problem may be the **variable** x or a rule such as ab=ba can be stated for all numbers represented by the variables a and b.

Elementary algebra is concerned with expressing problems in terms of **mathematical symbols** and establishing general rules for the combination and manipulation of those symbols. There is another type of algebra, however, called abstract algebra, which is a further generalization of elementary algebra, and often bears little resemblance to arithmetic. Abstract algebra begins with a few basic assumptions about sets whose elements can be combined under one or more binary operations, and derives theorems that apply to all sets satisfying the initial assumptions.

Elementary algebra

Algebra was popularized in the early ninth century by Al-Khowarizmi, an Arab mathematician, and the author of the first algebra book, Al-jabr wa'l Muqabalah, from which the English word algebra is derived. An influential book in its day, it remained the standard text in algebra for a long time. The title translates roughly to "restoring and balancing," referring to the primary algebraic method of performing an operation on one side of an equation and restoring the balance, or equality, by doing the same thing to the other side. In his book, Al-Khowarizmi did not use variables as we recognize them today,

but concentrated on procedures and specific rules, presenting methods for solving numerous types of problems in arithmetic. Variables based on letters of the alphabet were first used in the late 16th century by the French mathematician **François Viète**. The idea is simply that a letter, usually from the English or Greek alphabet, stands for an element of a specific set. For example, x, y, and z are often used to represent a real number, z to represent a complex number, and n to stand for an integer. Variables are often used in mathematical statements to represent unknown quantities.

The rules of elementary algebra deal with the four familiar operations of **addition** (+), **multiplication** (×), **subtraction** (-), and **division** (÷) of **real numbers**. Each operation is a rule for combining the real numbers, two at a time, in a way that gives a third real number. A combination of variables and numbers that are multiplied together, such as $64x^2$, $7yt$, $s/2$, $32xyz$, is called a monomial. The sum or difference of two monomials is referred to as a binomial, examples include, $64x^2+7yt$, $13t+6x$, and $12y-3ab/4$. The combination of three monomials is a trinomial ($6xy+3z-2$), and the combination of more than three is a **polynomial**. All are referred to as algebraic expressions.

One primary objective in algebra is to determine what conditions make a statement true. Statements are usually made in the form of comparisons. One expression is greater than (>), less than (<), or equal to (=) another expression, such as $6x+3 > 5$, $7x^2-4 < 2$, or $5x^2+6x = 3y+4$. The application of algebraic methods then proceeds in the following way. A problem to be solved is stated in mathematical terms using symbolic notation. This results in an equation (or inequality). The equation contains a variable; the value of the variable that makes the equation true represents the solution to the equation, and hence the solution to the problem. Finding that solution requires manipulation of the equation in a way that leaves it essentially unchanged, that is, the two sides must remain equal at all times. The object is to select operations that will isolate the variable on one side of the equation, so that the other side will represent the solution. Thus, the most fundamental rule of algebra is the principle of Al-Khowarizmi: whenever an operation is performed on one side of an equation, an equivalent operation must be performed on the other side bringing it back into balance. In this way, both sides of an equation remain equal.

Applications

Applications of algebra are found everywhere. The principles of algebra are applied in all branches of mathematics, for instance, **calculus**, **geometry**, **topology**. They are applied every day by men and women working in all types of business. As a typical example of applying algebraic methods, consider the following problem. A painter is given the job of whitewashing three billboards along the highway. The owner of the billboards has told the painter that each is a rectangle, and all three are the same size, but he does not remember their exact dimensions. He does have two old drawings, one indicating the height of each billboard is two feet less than half its width, and the other indicating each has a **perimeter** of 68 feet.

The painter is interested in determining how many gallons of paint he will need to complete the job, if a gallon of paint covers 400 square feet. To solve this problem three basic steps must be completed. First, carefully list the available information, identifying any unknown quantities. Second, translate the information into symbolic notation by assigning variables to unknown quantities and writing **equations**. Third, solve the equation, or equations, for the unknown quantities.

Step one, list available information: (a) three billboards of equal size and shape, (b) shape is rectangular, (c) height is 2 feet less than 1/2 the width, (d) perimeter equals 2 times sum of height plus width equals 68 feet, (e) total **area**, equals height times width times 3, is unknown, (f) height and width are unknown, (g) paint covers 400 sq.ft. per gallon, (h) total area divided by 400 equals required gallons of paint.

Step two, translate. Assign variables and write equations.

Let: A = area; h = height; w = width; g = number of gallons of paint needed.

Then:(1) h = 1/2w - 2 (from [c] in step 1) (2) 2(h+w) = 68 (from [d] in step 1) (3) A = 3hw (from [e] in step 1) (4) g = A/400 (from [h] in step 1) Step three, solve the equations. The right hand side of equation (1) can be substituted into equation (2) for h giving 2(1/2w-2+w) = 68. By the **commutative property**, the quantity in parentheses is equal to (1/2w+w-2), which is equal to (3/2w-2). Thus, the equation 2(3/2w-2)=68 is exactly equivalent to the original. Applying the **distributive property** to the left hand side of this new equation results in another equivalent expression, 3w-4 = 68. To isolate w on one side of the equation, add 4 to both sides giving 3w-4+4 = 68+4 or 3w = 72. Finally, divide the expressions on each side of this last expression by 3 to isolate w. The result is w = 24 ft. Next, put the value 24 into equation (1) wherever w appears, h = (1/2(24)-2), and do the arithmetic to find h = (12-2) = 10ft. Then, put the values of h and w into equation (3) to find the area, A = 3x10x24 = 720 sq. ft. Finally, substitute the value of A into equation (4) to find g = 720/400 = 1.8 gallons of paint.

Graphing algebraic equations

The methods of algebra are extended to geometry, and vice versa, by graphing. The value of graphing is two-fold. It can be used to describe geometric figures using the language of algebra, and it can be used to depict geometrically the algebraic relationship between two variables. For example, suppose that Fred is twice the age of his sister Irma. Since Irma's age is unknown, Fred's age is also unknown. The relationship between their ages can be expressed algebraically, though, by letting y represent Fred's age and x represent Irma's age. The result is y = 2x. Then, a graph, or picture, of the relationship can be drawn by indicating the points (x,y) in the **Cartesian coordinate system** for which the relationship y = 2x is always true. This is a straight line, and every point on it represents a possible combination of ages for Fred and Irma (of course negative ages have no meaning so x and y can only take on positive values). If a second relationship between their ages is given, for instance, Fred is three years older than Irma, then a second equation can be written, y = x+3, and a second graph

can be drawn consisting of the ordered pairs (x,y) such that the relationship y = x+3 is always true. This second graph is also a straight line, and the point at which it intersects the line y = 2x is the point corresponding to the actual ages of Irma and Fred. For this example, the point is (3,6), meaning that Irma is three years old and Fred is six years old.

Linear algebra

Linear algebra involves the extension of techniques from elementary algebra to the solution of systems of **linear equations**. A linear equation is one in which no two variables are multiplied together, so that terms like xy, yz, x^2, y^2, and so on, do not appear. A system of equations is a set of two or more equations containing the same variables. **Systems of equations** arise in situations where two or more unknown quantities are present. In order for a unique solution to exist there must be as many independent conditions given as there unknowns, so that the number of equations that can be formulated equals the number of variables. Thus, we speak of two equations in two unknowns, three equations in three unknowns, and so forth. Consider the example of finding two numbers such that the first is six times larger than the second, and the second is 10 less that the first. This problem has two unknowns, and contains two independent conditions. In order to determine the two numbers, let x represent the first number and y represent the second number. Using the information provided, two equations can be formulated, x = 6y, from the first condition, and x-10 = y, from the second condition. To solve for y, replace x in the second equation with 6y from the first equation, giving 6y-10=y. Then, subtract y from both sides to obtain 5y-10=0, add 10 to both sides giving 5y=10, and divide both sides by 5 to find y=2. Finally, substitute y=2 into the first equation to obtain x=12. The first number, 12, is six times larger than the second, 2, and the second is 10 less than the first, as required. This simple example demonstrates the method of substitution. More general methods of solution involve the use of **matrix** algebra.

Matrix algebra

A matrix is a rectangular array of numbers, and matrix algebra involves the formulation of rules for manipulating matrices. The elements of a matrix are contained in square brackets and named by row and then column. For example the matrix has two rows and two columns, with the element (-6) located in row one column two. In general, a matrix can have i rows and j columns, so that an element of a matrix is denoted in double subscript notation by a_{ij}. The four elements in A are $a_{11} = 1$, $a_{12} = -6$, $a_{21} = 3$, $a_{22} = 2$. A matrix having m rows and n columns is called an "m by n" or (m × n) matrix. When the number of rows equals the number of columns the matrix is said to be square. In matrix algebra, the operations of addition and multiplication are extended to matrices and the fundamental principles for combining three or more matrices are developed. For example, two matrices are added by adding their corresponding elements. Thus, two matrices must each have the same number of rows and columns in order to be compatible for addition. When two matrices are compatible for addition, both the associative and commutative principles of elementary algebra continue to hold. One of the many applications of matrix algebra is the solution of **systems of linear equations**. The coefficients of a set of simultaneous equations are written in the form of a matrix, and a formula (known as **Cramer's rule**) is applied which provides the solution to n equations in n unknowns. The method is very powerful, especially when there are hundreds of unknowns, and a computer is available.

Abstract algebra

Abstract algebra represents a further generalization of elementary algebra. By defining such constructs as groups, based on a set of initial assumptions, called axioms, provides theorems that apply to all sets satisfying the abstract algebra axioms. A group is a set of elements together with a binary operation that satisfies three axioms. Because the binary operation in question may be any of a number of conceivable operations, including the familiar operations of addition, subtraction, multiplication, and division of real numbers, an asterisk or open **circle** is often used to indicate the operation. The three axioms that a set and its operation must satisfy in order to qualify as a group, are: (1) members of the set obey the associative principle $[a \times (b \times c) = (a \times b) \times c]$; (2) the set has an **identity element**, I, associated with the operation $\times$, such that $a \times I = a$; (3) the set contains the inverse of each of its elements, that is, for each a in the set there is an inverse, a', such that $a \times a' = I$. A well known group is the set of **integers**, together with the operation of addition. If it happens that the commutative principle also holds, then the group is called a commutative group. The group formed by the integers together with the operation of addition is a commutative group, but the set of integers together with the operation of subtraction is not a group, because subtraction of integers is not associative. The set of integers together with the operation of multiplication is a commutative group, but division is not strictly an operation on the integers because it does not always result in another integer, so the integers together with division do not form a group. The set of **rational numbers**, however, together with the operation of division is a group. The power of abstract algebra derives from its generality. The properties of groups, for instance, apply to any set and operation that satisfy the axioms of a group. It does not matter whether that set contains real numbers, **complex numbers**, vectors, matrices, **functions**, or probabilities, to name a few possibilities.

ALGEBRA OF SETS

Set **algebra** is fundamental to all of modern mathematics. Indeed, **sets** form the very foundation of mathematics. The algebra of sets consists of the operations union, intersection, and complement. Suppose S and T are sets. The union of S and T is written $S \cup T$ and equals the set that contains every element of S and every element of T and no other elements. The intersection of S and T is written $S \cap T$ and is the set that con-

tains only those elements that are contained in both S and T. S and T are said to be disjoint if their intersection is the empty set. The complement of S in T is the set of all elements in T that are not in S. It is sometimes denoted by T - S or T\S. If S is a subset of T then the complement of S in T can be denoted by S^C. The symmetric difference of S and T is the set of all elements that are either in S or T but not in both S and T.

For example, suppose $S = \{1, 2, 3\}$ and $T = \{2, 3, 4\}$. Then, $S \cup T = \{1, 2, 3, 4\}$. $S \cap T$ equals $\{2, 3\}$. The complement of S in T = $\{4\}$. The symmetric difference of S and T is $\{1, 4\}$.

The curly brackets { } stand for the words "the set containing" and sometimes "the set of all". For example "S union T = {x in S or T}", means T equals the set of all elements x that are in S or in T. Parentheses (), in this context, mean "the set equal to". For example "$(S \cup (T \cap U))$" means the set equal to S union the set equal to the intersection of T with U. The Greek letter epsilon, $\in$ means 'is in'. For example $x \in S$ means x is in S, or x is an element of S. The symbol $\subseteq$ means "is a subset of" or "is contained in", as in $S \subseteq T$. The symbol $\supseteq$ means "is a superset of" or "contains", as in $T \supseteq S$.

Here are the rules for how to combine unions, intersections, and complements.

- $S \cup (T \cap U) = (S \cup T) \cap (S \cup U)$.
- $S \cap (T \cup U) = (S \cap T) \cup (T \cap U)$.
- The complement of $(S \cup T)$ = (the complement of S) $\cap$ (the complement of T).
- The complement of $(S \cap T)$ = (the complement of S) $\cup$ (the complement of T).

A **Boolean algebra** of a set B is the set of all its subsets together with the three operations union, complement and intersection. The study of Boolean algebras is important to the design of computer chips and integrated circuits.

See also Boolean algebra; Set theory; Sets

ALGEBRAIC GEOMETRY

With the introduction of Cartesian coordinates by **René Descartes** and **Pierre de Fermat** in the seventeenth century, it was soon realized that **equations** in two variables generally define curves in the plane. Descartes had already made the distinction between curves that can be described by polynomial equations and those that cannot. He called the former geometrical and the latter mechanical curves. Nowadays, these are called algebraic and transcendental curves, respectively.

Algebraic **geometry** is the area of mathematics that studies the properties of **sets** (or loci) defined as the set of common zeros of a collection of polynomial equations on the coordinates of the points of some **Cartesian coordinate system**. Such sets are called algebraic sets or algebraic varieties. If they are one-dimensional, they are called algebraic curves, and if two-dimensional, algebraic surfaces. For example, a subset of the plane defined as the solution set of an equation f(x,y)=0, where f(x,y) is a polynomial in two variables, is an algebraic curve.

The total degree of the polynomial is called the degree of the curve. For example, lines are algebraic curves defined by **linear equations**, so they have a degree of one. The equation $x^2+y^2=r^2$ defines a **circle** of radius r, and thus the circle is an algebraic curve of degree two. More generally, conics are algebraic curves of degree two, cubics are algebraic curves of degree three, and so on.

Some of the earlier questions in algebraic geometry had to do with counting points on curves with certain properties; for instance, determining how many points two curves can have in common. **Bézout's theorem** states that two plane algebraic curves of degrees m and n intersect in at most mn points unless they have a common component; that is, unless there exists an algebraic curve which is a subset of both curves. Other questions had to do with how many inflexion points a curve can have or for how many points on a curve the tangent line goes through a specific point. Several of these questions are answered by the Plücker formulas, obtained by **Julius Plücker** in 1834.

Some of these counting questions led to realizations that there were some "missing points," since the problems would have a certain number of solutions in general, but in special cases some of these solutions would disappear. This motivated the introduction of the complex projective plane. The complex projective plane is defined in the following way. First one considers a Cartesian plane in which the (x,y) coordinates are allowed to take **complex numbers** as values. Second, one introduces "points at infinity." These are ideal points, one for each direction, and are introduced in such a way that parallel lines meet at these points at **infinity**. With the introduction of these extra points, it turns out that the "missing points" can be accounted for. The points in the complex projective plane can be described by projective coordinates (x:y:z), which are not all **zero**, by requiring that (x:y:z) and (ux:uy:uz) represent the same point for any nonzero number u. The points (x:y:1) then correspond to the point (x,y) in Cartesian coordinates, and the points of the form (x:y:0) give the points at infinity.

Parallel to the study of algebraic curves was the study of algebraic **functions**, due to **Niels Abel**, **Karl Jacobi**, and, most importantly, **Bernhard Riemann** with his theory of Riemann surfaces. This work provided an intrinsic theory of algebraic curves and later was used to study their geometric properties.

The intensive study of algebraic surfaces came much later, toward the end of the nineteenth century, first with M. Noether and then with the Italian school of algebraic geometry, some of whose most prominent members were C. Segre, F. Enriques, G. Castelnuovo, and F. Severi. They developed the theory of algebraic surfaces and higher dimensional varieties. One of their major achievements was Enriques's classification of algebraic surfaces. The members of the Italian school were criticized, however, for being imprecise, in that their intuition was not backed up by rigorous mathematical **proof**.

In the twentieth century there was a push for developing solid foundations for algebraic geometry and for fully justifying the results of the Italian school. This work was carried out initially by B. L. van der Waerden, **Andre Weil**, O. Zariski, and others, building upon the Abstract **Algebra** developed by E.

Artin and **Emmy Noether**. Meanwhile, there were also new results, such as H. Hironaka's resolution of singularities, a process for smoothing out algebraic varieties. A major event occurred in the fifties when **Alexander Grothendieck** unified **number theory** and algebraic geometry with his theory of schemes. He also showed how to use methods from **topology**, such as cohomology groups, in an algebraic setting. These novel techniques solved many problems and, at the same time, provided a general unified foundation for the field.

In the late twentieth century, some of the main developments were the work of S. Mori and others on the classification of three-dimensional algebraic varieties and the introduction of methods derived from string theory, an area of physics.

See also Bézout's theorem

ALGEBRAIC NUMBERS

The term **algebra** comes from the Arabic *al-jabr*, meaning to combine. *Al-Jabr* refers to the mathematician's process of combining like terms to solve an equation.

Historical references to algebra date back to Greek history, spanning the years 540 B.C.-250 B.C. **Euclid**, **Pythagoras**, and their followers used algebraic problems with geometric proofs. This method of mixing algebra and **geometry** led to complex constructions. The Greek system of mathematics used layers of numbers, letters, and punctuation marks written above each other to express these complexities.

The separation of Greek algebra from geometry did not occur until around A.D. 250, when **Diophantus of Alexandria** demonstrated algebra. Diophantus was the first to use symbolic notation for algebraic expressions. This was similar to Babylonian algebra, except Babylonian mathematicians were usually limited to approximate solutions. Diophantus worked out **equations** with exact solutions. He developed a system that offered symbols for frequently used numbers, operations, and variables.

The Greeks and Babylonians influenced the practice of mathematics in India, where **Brahmagupta**, a Hindu scholar, practiced circa A.D. 630. Brahmagupta understood **negative numbers**, **zero**, second-degree equations, and equations with **variables**.

In A.D. 825, **Al-Khwarizmi** (c. 780-c. 850) wrote *Hisab al-jabr w-al-musqabalah (The Science of the Transposition and Cancellation)* in Baghdad. This was the title that gave algebra its name. Al-Khwarizmi's work helped to bring about the acceptance of algebraic equations in Europe.

During the thirteenth and fourteenth centuries, Arabic, Persian, and Hindu advances in algebra were received in Europe as the Renaissance in art, the Reformation in religion, and the discovery of the New World helped to bring about a revolution in mathematics in the fifteenth and sixteenth centuries. Algebra had to be developed before advances could be made in **calculus** and analytical geometry in the seventeenth century. Italy was the center of this rebirth, with over 200 mathematical texts published between 1472 and 1500.

In 1545 the Italian mathematician **Girolamo Cardano** (1501-1576) wrote *Ars magna sive de regulis algebraicis*, which sparked interest in algebra by offering what, by today's standards, would be a tedious collection of similar problems intended to illustrate the same principle. *Ars magna* also demonstrated a solution to a quartic equation by manipulating it into a form of a cubic equation.

François Viète, the author of an algebra text published in 1591, is recognized as "the father of modern algebraic notation." Viète's text is similar to modern texts of elementary algebra. Viète worked out a standard procedure for solving **quadratic equations**, **cubic equations**, and **quartic equations**.

The system of **analytic geometry** of **Descartes** (1596-1650) used letters at the end of the alphabet to indicate unknown quantities (x, y, z) and letters at the beginning of the alphabet (a, b, c) to indicate known quantities. Descartes' system of analyzing geometric problems algebraically helped to advance both algebra and geometry.

In 1797, German mathematician **Carl Frederich Gauss** (1777-1855) published his **proof** of the **fundamental theorem of algebra**. A cascade of algebraic discoveries would continue through the nineteenth and early twentieth centuries. In his *A Treatise on Algebra*, a textbook published in 1830, the British mathematician George Peacock claimed that algebra had been thought of as **arithmetic** using symbols, with letters replacing numbers. Peacock proposed new ways of thinking about algebra and different kinds of algebra altogether. During the nineteenth century, algebra became very abstract.

In another significant advance, mathematicians Ernest E. Krummer (1810-1893), **J. W. Richard Dedekind** (1831-1916), and **Leopold Kronecker** (1823-1891) created a theory of algebraic numbers, which laid the groundwork for modern abstract algebra.

The work of **Nikolay Lobachevsky** (1793-1856) and **Janòs Bolyai** (1802-1860) in the early 1830s led to advances in non-Euclidean geometry and abstract algebra. These developments fundamentally affected the approach to deductive reasoning. Elementary algebra was formalized during the 1830s and there was a ripple effect in scientific discovery.

The English mathematician and logician **George Boole** (1815-1864) invented **Boolean algebra** in 1847, basing it on the concept that ideas have only two possible states when stable: on/off, closed/open, yes/no, true/false. Boolean algebra and its close relative, binary theory, opened the door for the development of computer science.

B. Pierce (1809-1880) said in 1870 that "all relations are either qualitative or quantitative" in algebra and these two relations may be considered independently of one another or in some cases combined. This was a step toward abstract mathematical theory. The trend was away from devising elaborate theories, with mathematicians looking more and more at the interrelationships among theories.

Toward the end of the nineteenth century, **linear algebra** seemed to move in promising new directions but without significant advances. A new era of algebraic discovery began around 1910 with a breakthrough in the development of general methods of linear algebra. The 1920s and 1930s marked the modern development of abstract algebra. During this period, there was a competitive relationship between scholars in algebra and **topology** (the specialized study of geometric configurations) for dominance in mathematics.

After World War II, much of the research in mathematics became so abstract that it was impossible for nonmathematicians to understand. Abraham Adrian Albert studied nonassociative algebras, and there were advances in **probability theory** and factoring. Discoveries in four-dimensional spaces and computer science were taking mathematics in entirely new directions. In 1955, **Henri-Paul Cartan** (1904-) and Samuel Eilenberg brought a certain **symmetry** to the study of mathematics by devising the system of homological algebra, a combination of abstract algebra and algebraic topology that reunited the two competitive fields.

ALGORITHM

An algorithm is a set of instructions that indicates a method for accomplishing a task. If followed correctly, an algorithm guarantees successful completion even without the use of any intelligence. The term algorithm is derived from the name Al-Khwarizmi, a ninth-century Arabian mathematician who is credited with discovering **algebra**. With the advent of computers, which are particularly adept at utilizing algorithms, the creation of new and faster algorithms has become an important consideration in the study of theoretical computer science.

Algorithms can be written to solve any conceivable problem. For example, an algorithm can be developed for tying a shoe, making cookies, or determining the **area** of a **circle**. In an algorithm for tieing a shoe, each step, from obtaining a shoe with a lace to releasing the string after it is tied, is spelled out. The individual steps are written in such a way that no judgement is ever required to successfully carry them out. The length of time required to complete an algorithm is directly dependent on the number of steps involved. The more steps, the longer it takes to complete. Consequently, algorithms are classified as fast or slow depending on the speed at which they allow a task to be completed. Typically, fast algorithms are usable while slow algorithms are unusable.

AL-KHWARIZMI, ABU JA'FAR MUHAMMAD IBN MUSA (CA. 780-CA.850)
Arabian astronomer

The Arab astronomer Abu Ja'far Muhammad ibn Musa Al-Khwarizmi was the author of about a half dozen astronomical works, including a book entitled *Al-jabr w'al muqabala* (written in 830 AD) that gave the name *al-jabr* to the branch of mathematics that is now known by its modern spelling as **algebra**.

The word *al-jabr* is usually translated as "restoring," with reference to restoring the balance in an equation by placing on one side of an equation a term that has been removed from the other. For example, given the equation $x^2-5=4$, a balance is restored by writing $x^2=9$. The second part of the title, *al muqabala*, probably meant "simplification," as in the case of combining $2x+5x$ to obtain $7x$, or by subtracting out equivalent terms from both sides of an equation.

Al-Khwarizmi's algebra was based on the earlier work of the Hindu mathematician **Brahmagupta** (b. 598), but also contained influences from **Babylonian** and Greek mathematics. Some of the operations that Al-Khwarizimi describes are the same as those described by the Greek mathematician **Diophantus** (c. 250 AD), although it is unlikely that Al-Khwarizimi was familiar with the latter's work. In the case of an equation is several unknowns, for example, both Diophantus and Al-Kwarzimi reduce the equation to one unknown and then solve it. Some of Al-Khwarizimi's terminology clearly reflects Diophantus' nomenclature; for example, both mathematicians use the term "power" to describe the **square** of an unknown. [Describing the powers of an unknown, Al-Khwarizimi designated the unknown as the root (as of a plant), which is the origin of our use of term.]

In the book, Al-Kwarizimi identifies the product of (x +/- a) and (y +/- b). Given an expression of the form $ax^2 + bx +c$, he describes how to add and subtract terms. When treating linear and quadratic **equations**, he followed Diophantus in retaining six separate forms in his solutions, i.e., $ax^2 = bx$, $ax^2 = c$, $ax^2 + c = bx$, $ax^2 + bx =c$, and $ax2 = bx + c$, where a, b and c were always positive. In this way, Al-Kwarizimi was able to avoid the problem of subtracting a larger number from a smaller one. Although Al-Khwarizimi recognized that **quadratic equations** could have two **roots**, he only listed the real positive ones (which could, of course, be irrational).

When Al-Khwarizimi wanted to describe the solution of a quadratic equation, he, like other Arab mathematicians, resorted to using geometrical constructions and **completing the square**.

It is of interest to linguists that the term *al-jabr* later came to mean "bonesetter," i.e., the person who restores broken bones. After the term reached Spain with the Moorish invasion, it became *algebrista*, and continued to be a term for one who sets bones. At one time, it was common for Spanish barbers to post a sign outside their establishments reading "Algebrista y Sangrador" (meaning bonesetter and bloodletter), it being the custom for barbers to practice those medical arts. And the tradition of barbers performing these treatments continued for some centuries. In sixteenth-century Italy, the term algebra continued to be used for the practice of bonesetting.

When Al-Khwarizimi's book was translated into Latin in the twelfth century, the title was translated *Ludus algebrae et almucgrabalaeque*. The name for the branch of mathematics, however, was eventually shortened to algebra.

Another of Al-Khwarizimi's books on **arithmetic** and algebra, *De numero indorum* (Concerning the Hindu Art of Reckoning) has survived only in its Latin translation. In it, the author gave such a complete account of the Hindu numeral system that that system of numbering is now mistakenly known as Arabic numerals. When Latin translations began spreading throughout Europe, readers began crediting the mathematical notation, which became known as *algorismi*, to Al-Khwarzimi. Later, the scheme that made use of the Hindu numerals became known as **algorithm** (a corruption of al-Khwarzimi). Today, of course, the word algorithm refers to a set of well-defined rules for solving a problem in a finite number of steps.

ALTERNATING SERIES

Alternating series—an **infinite series** whose terms alternate sign, i.e. whose terms are alternately positive and negative. Examples are the series $1 - 1 + 1 - 1 + 1...$ and the series $1 - 1/2 + 1/3 - 1/4 + 1/5 - 1/6 +...$, which is called the alternating harmonic series.

In general, it is more difficult to determine whether a series with terms of varying sign converges or diverges than it is with a series whose terms are all of the same sign. For a series to converge, its sequence of partial sums (that is, the sequence obtained by taking just the first term, then taking the sum of the first two terms, then the sum of the first three terms, and so on) must have a limit. If all the terms in a series are positive, the partial sums will all be positive and will get larger and larger as more terms of the series are added on. Therefore there are only two possibilities: either the partial sums will eventually level off, in which case the series converges, or they will grow without bound, in which case the series diverges. However, if the terms are not all positive, there are more possibilities. For instance, with the series of alternating 1's and -1's above, if we group the numbers in pairs, we get $(1 - 1) + (1 - 1) + (1 - 1) +...$, which would lead us to think the series converges to 0. On the other hand, if we group them in pairs starting with the second term, we get $1 + (-1 + 1) + (-1 + 1) + (-1 + 1) +...$, giving us a sum of 1. To resolve this inconsistency, we look at the sequence of partial sums:

$$1$$
$$1-1=0$$
$$1-1+1=1$$
$$1-1+1-1=0$$
$$1-1+1-1+1=1$$

and so on. Since the sequence of partial sums does not have a unique limit, the series diverges. Thus an alternating series can diverge even though its partial sums do not grow unboundedly large.

There is one test available to determine if an alternating series converges, called, unsurprisingly, the Alternating Series Test. An infinite series $((-1)^{k+1}a_k$ with $a_k > 0$ for all k (*i.e.* the alternating infinite sum $a_1 - a_2 + a_3 - a_4...$) converges if

- $a_1 > a_2 > a_3 >...$ and
- $\lim_{k \to 0} a_k = 0$.

Thus, for example, the alternating harmonic series $1 - 1/2 + 1/3 - 1/4 +...$ converges. Note that the **theorem** does not say that a series without the two properties will diverge. (It is true that any series, alternating or not, will diverge if it does not satisfy the second property, but a series can converge without satisfying the first property.)

The alternating harmonic series is an example of a conditionally convergent series, that is, a series that converges but which would not converge if all the negative terms were changed to positive. In other words, the alternating harmonic series is conditionally convergent because it converges but the series $1 + 1/2 + 1/3 + 1/4...$ diverges. A series that does converge if all the negative terms are changed to positive is said to be absolutely convergent. For example, the series $1 - 1/2 + 1/4 - 1/8 + 1/16 - 1/32 +...$ is absolutely convergent. Conditionally convergent series have a very interesting property, which has to do with how their terms can be rearranged. A rearrangement of a series has the same numbers being added, but in a different order, and therefore has a different sequence of partial sums. Consider the following rearrangement of the alternating harmonic series (which converges to the natural logarithm of 2, or about .69). First we take just enough positive terms so that the partial sum exceeds 1.2 (a number chosen at random). So our rearrangement starts $1 + 1/3 = 1.33333....$ Now we add on just enough negative terms so that the partial sum falls below 1.2:

$1 + 1/3 + -1/2 = .833333....$

Now we add on positive terms to get the partial sum above 1.2:

$1 + 1/3 + -1/2 + 1/5 + 1/7 + 1/9 = 1.28730....$

Now negative terms, to get the partial sum below 1.2:

$1 + 1/3 + -1/2 + 1/5 + 1/7 + 1/9 - 1/4 = 1.03730....$

Continuing in this fashion, we obtain the following partial sums (rounded off):

- 1.20513, 1.03847, 1.21659, 1.09159, 1.22269, 1.12269, 1.22646, 1.14313,....

It is not hard to see that these partial sums will approach 1.2—since the numbers we are adding on are shrinking in size, the amounts by which we "overshoot" 1.2 as we add on positive and negative terms shrinks as well.

Thus we have taken a series that converges to approximately .69 and rearranged it into a series that converges to 1.2. It is easy to see that we could just as easily have rearranged it to give any other number of our choosing. In fact, we can even rearrange it to get a series that diverges: first add enough positive terms to get above 10, say, and then enough negative terms to get below -10, then enough positive terms to get above 10 again, then enough negative terms to get below -10, and so on. Thus, against all our intuition, we have shown that **addition** can sometimes fail to be commutative once we shift from the finite addition we learned in grade school to the addition of infinite collections of numbers.

See also Convergence; Infinite series

ALTITUDE

The altitude of a **triangle** is the perpendicular line which joins one vertex of a triangle to a point on the opposite side. (A line joining the vertex to an arbitrary point is called a cevian.) Each acute triangle has exactly three altitudes. The three altitudes of such a triangle always meet at a central point known as the orthocenter. Obtuse and right triangles also have altitudes. However, for a right triangle, one of the legs is an altitude, and the orthocenter lies along the edge of the triangle. For obtuse triangles, the sides must be extended a certain **distance** to allow for a perpendicular to be formed. In this case, the orthocenter will lie outside the triangle itself.

The altitude is most commonly used in calculations of the **area** of a triangle: the area can be found by halving the product of a side with the altitude perpendicular to that side. The altitude can also be calculated indirectly if one already knows the area or if one knows the lengths of all the sides (from which Heron's formula is used to calculate the area; see **Hero's formula**). The altitude is among the geometric properties commonly used in proofs and theorems, as well as in construction of figures using compass and straightedge.

AMICABLE NUMBERS

Two numbers are said to be amicable (i.e., friendly) if each one of them is equal to the sum of the *proper* divisors of the others, i.e., **whole numbers** less than the given numbers that divide the given number with no remainder. For example, 220 has proper divisors 1, 2, 4, 5, 10, 11, 20, 22, 44, 55, and 110. The sum of these divisors is 284. The proper divisors of 284 are 1, 2, 4, 71, and 142. Their sum is 220; so 220 and 284 are amicable. This is the smallest pair of amicable numbers.

The discovery of amicable numbers is attributed to the neo-Pythagorean philosopher Iamblichus (c. 250- 330), who credited **Pythagoras** with the original knowledge of their nature. The Pythagoreans believed that amicable numbers, like all special numbers, had a profound cosmic significance. A biblical reference (a gift of 220 goats from Jacob to Esau, Genesis 23: 14) is thought by some to indicate an earlier knowledge of amicable numbers.

No pairs of amicable numbers other than 220 and 284 were discovered by European mathematicians until 1636, when the French mathematician **Pierre de Fermat** (1601-1665) found the pair 18,496 and 17,296. A century later, the Swiss mathematician **Leonhard Euler** (1707-1783) made an extensive search and found about 60 additional pairs. Surprisingly, however, he overlooked the smallest pair, 1184 and 1210, which was subsequently discovered in 1866 by a 16-year-old boy, Nicolo Paganini.

During the medieval period, Arabian mathematicians preserved and developed the mathematical knowledge of the ancient Greeks. For example, The polymath Thabit ibn Qurra (836-901) formulated an ingenious rule for generating amicable number pairs: Let $a = 3(2^n) - 1$, $b = 3(2^{n-1}) - 1$, and $c = 9(2^{2n-1}) - 1$; then, if a, b, and c are primes, $2^n ab$ and $2^n c$ are amicable. This rule produces 220 and 284 when n is 2. When n is 3, c is not a prime, and the resulting numbers are not amicable. For n = 4, it produces Fermat's pair, 17,296 and 18,416, skipping over Paganini's pair and others.

Amicable numbers serve no practical purpose, but professionals and amateurs alike have for centuries enjoyed seeking them and exploring their properties.

ANALYSIS

The word analysis comes from the Greek "analyein," which means "to break up." The basic meaning of the word is the separation of a whole into its component parts. While it is used generally in both the natural and the social sciences in several different ways, in mathematics the term analysis generally refers to the examination of **functions** of real or complex variables arising from differential and integral **calculus** including the individual elements of the functions and their relationships to each other. Technically speaking, this branch of mathematics deals with how functions converge, toward **limits** and with other limit processes. This means that analysis may be regarded as a branch of **topology**.

Analysis includes a number of techniques which act as tools to allow scientists to determine the significance of values, actions, and reactions. The various ways natural and social scientists analyze the results of experiments, observations, and so forth in the different disciplines usually include mathematical reduction of data gathered by measurement of physical phenomena, observation of processes, and statistical analysis of behavior patterns, changes, and other trends. Thus, mathematical analysis has practical applications in many fields. As phenomena from these fields are studied in detail, predicted results, reactions, or patterns are compared to real outcomes. From these comparisons, **equations** often result that can be used to explain the observations or predict future results when new conditions are introduced into a situation.

ANALYTIC GEOMETRY

Analytic **geometry** is a branch of mathematics which uses algebraic **equations** to describe the size and position of geometric figures on a coordinate system. Developed during the seventeenth century, it is also known as Cartesian geometry or coordinate geometry. The use of a coordinate system to relate geometric points to **real numbers** is the central idea of analytic geometry. By defining each point with a unique set of real numbers, geometric figures such as lines, circles, and conics can be described with algebraic equations. Analytic geometry has found important applications in science and industry alike.

During the seventeenth century, finding the solution to problems involving curves became important to industry and science. In astronomy, the slow acceptance of the heliocentric theory of planetary **motion** required mathematical formulas which would predict elliptical orbits. Other areas such as optics, navigation and the military required formulas for things such as determining the curvature of a lens, the shortest route to a destination, and the trajectory of a cannon ball. Although the Greeks had developed hundreds of theorems related to curves, these did not provide quantitative values so they were not useful for practical applications. Consequently, many seventeenth-century mathematicians devoted their attention to the quantitative evaluation of curves. Two French mathematicians, Rene Descartes (1596-1650) and **Pierre de Fermat** (1601-1665) independently developed the foundations for analytic geometry. Descartes was first to publish his methods in an appendix titled *La geometrie* of his book *Discours de la methode* (1637).

The link between **algebra** and geometry was made possible by the development of a coordinate system which allowed geometric ideas, such as point and line, to be described in algebraic terms like real numbers and equations. In the system developed by Descartes, called the rectangular **Cartesian coordinate system**, points on a geometric plane are associated with an ordered pair of real numbers known as coordinates. Each coordinate describes the location of a single point relative to a fixed point, the origin, which is created by the intersection of a horizontal and a vertical line known as the x-axis and y-axis respectively. The relationship between a point and its coordinates is called one-to-one since each point corresponds to only one set of coordinates.

The x and y axes divide the plane into four quadrants. The sign of the coordinates is either positive or negative depending in which quadrant the point is located. Starting in the upper right quadrant and working clockwise, a point in the first quadrant would have a positive value for the abscissa and the ordinate. A point in the fourth quadrant (lower right hand corner) would have negative values for each coordinate.

The notation P (x,y) describes a point P which has coordinates x and y. The x value, called the abscissa, represents the horizontal **distance** of a point away from the origin. The y value, known as the ordinate, represents the vertical distance of a point away from the origin.

Using the ideas of analytic geometry, it is possible to calculate the distance between the two points A and B, represented by the line segment AB which connects the points. If two points have the same ordinate but different abscissas, the distance between them AB = $x_2 - x_1$. Similarly, if both points have the same abscissa but different ordinates, the distance AB = $y_2 - y_1$. For points which have neither a common abscissa or ordinate, the **Pythagorean Theorem** is used to determine distance. By drawing horizontal and vertical lines through points A and B to form a right **triangle**, it can be shown using the distance formula that AB = $- (x_2 - x_1)^2 + (y_2 - y_1)^2$. The distance between the two points is equal to the length of the line segment AB.

In addition to length, it is often desirable to find the coordinates of the **midpoint** of a line segment. These coordinates can be determined by taking the average of the x and y coordinates of each point. For example, the coordinates for the midpoint M(x,y) between points P(2,5) and Q(4,3) are x = (2 + 4)/2 = 3 and y = (5 + 3)/2 = 4.

One of the most important aspects of analytic geometry is the idea that an algebraic equation can relate to a geometric figure. Consider the equation 2x + 3y = 44. The solution to this equation is an ordered pair (x,y) which represents a point. If the set of every point which makes the equation true (called the *locus*) were plotted, the resulting graph would be a straight line. For this reason, equations such as these are known as **linear equations**. The standard form of a linear equation is $Ax + By = C$, where A, B, and C are constants and A and B are not both 0. It is interesting to note that an equation such as $x = 4$ is a linear equation. The graph of this equation is made up of the set of all ordered pairs in which $x = 4$.

The steepness of a line relative to the x axis can be determined by using the concept of the **slope**. The slope of a line is defined by the equation $m = (Y_2 - Y_1) / (x_2 - x_1)$.

The value of the slope can be used to describe a line geometrically. If the slope is positive, the line is said to be rising. For a negative slope, the line is falling. A slope of **zero** is a horizontal line and an undefined slope is a vertical line. If two lines have the same slope, then these lines are parallel.

The slope gives us another common form for a linear equation. The slope-intercept form of a linear equation is written $y = mx + b$, where m is the slope of the line and b is the y intercept. The y intercept is defined as the value for y when x is zero and represents a point on the line which intersects the y axis. Similarly, the x intercept represents a point where the line crosses the x axis and is equal to the value of x when y is zero. Yet another form of a linear equation is the point-slope form, $y - y_1 = m(x - x_1)$. This form is useful because it allows us to determine the equation for a line if we know the slope and the coordinates of a point.

One of the most frequent activities in geometry is determining the **area** of a polygon such as a triangle or **square**. By using coordinates to represent the vertices, the areas of any polygon can be determined. The area of triangle OPQ, where O lies at (0,0), P at (a,b), and Q at (c,d), is found by first calculating the area of the entire rectangle and subtracting the areas of the three right triangles. Thus the area of the triangle formed by points OPQ is $= da - (dc/2) - (ab/2) - [(d-b)(a-c)]/2$. Through the use of a determinant, it can be shown that the area of this triangle is ½ |ab / cd|.

This specific case was made easier by the fact that one of the points used for a vertex was the origin. The general equation for the area of a triangle defined by coordinates is represented by the equation ½ $|x_1y_1 / x_2y_2|$ + ½ $|x_2y_2 / x_3y_3|$ + ½ $|x_3y_3 / x_1y_1|$.

In a similar manner, the area for any other polygon can be determined if the coordinates of its points are known.

In addition to lines and the figures that are made with them, algebraic equations exist for other types of geometric figures. One of the most common examples is the **circle**. A circle is defined as a figure created by the set of all points in a plane which are a constant distance from a central point. If the center of the circle is at the origin, the formula for the circle is $r^2 = x^2 + y^2$ where r is the distance of each point from the center and called the radius. For example, if a radius of 4 is chosen, a plot of all the x and y pairs which satisfy the equation $4^2 = x^2 + y^2$ would create a circle. Note, this equation, which is similar to the distance formula, is called the center-radius form of the equation. When the radius of the circle is at the point (a,b) the formula, known as the general form, becomes $r^2 = (x - a)^2 + (y-b)^2$.

The circle is one kind of a broader type of geometric figures known as **conic sections**. Conic sections are formed by the intersection of a geometric plane and a double-napped cone. After the circle the most common conics are parabolas, ellipses, and hyperbolas.

Curves known as parabolas are found all around us. For example, they are the shape formed by the sagging of telephone wires or the path a ball travels when it is thrown in the air. Mathematically, these figures are described as a curve created by the set of all points in a plane which are a constant distance from a fixed point, known as the focus, and a fixed line, called the directrix. This means that if we take any point on the **parabola**, the distance of the point from the focus is the same as the distance from the directrix. A line can be drawn through the focus which is perpendicular to the directrix. This line is called the axis of **symmetry** of the parabola. The midpoint between the focus and the directrix is the vertex.

The equation for a parabola is derived from the distance formula. Consider a parabola which has a vertex at point (h,k). The linear equation for the directrix can be represented by $y = k - p$, where p is the distance of the focus from the vertex. The standard form of the equation of the parabola is then $(x - h)^2 = 4p(y - k)$. In this case, the axis of symmetry is a vertical line. In the case of a horizontal axis of symmetry, the equation becomes $(y - k)^2 = 4p(x - h)$ where the equation for the directrix is $x = h - p$. This formula can be expanded to give the more common quadratic formula which is $y = Ax^2 + Bx + C$ such that A not equal 0.

Another widely used conic is an **ellipse**, which looks like a flattened circle. An ellipse is formed by the graph of the set of points the sum of whose distances from two fixed points (foci) is constant. To visualize this **definition**, imagine two tacks placed at the foci. A string, which is longer than the distance between them, is tied to each tack. The string is pulled taut with a pencil and an ellipse is formed by the path traced. Certain parts of the ellipse are given various names. The two points on an ellipse intersected by a line passing through both foci are called the vertices. The chord which connects both vertices is the major axis and the chord perpendicular to it is known as the minor axis. The point at which the chords meet is known as the center.

Again by using the distance formula, the equation for an ellipse can be derived. If the center of the ellipse is at point (h,k) and the major and minor axes have lengths of 2a and 2b respectively, the standard equation is

$[(x - h)^2 / a^2] + [(y - k)^2 / b^2] = 1$. Here the major axis is horizontal.

$[(x - h)^2 / b^2] + [(y - k)^2 / a^2] = 1$. Here the major axis is vertical.

If the center of the ellipse is at the origin, the equation simplifies to $(x^2/a^2) + (y^2/b^2) = 1$.

The "flatness" of an ellipse depends on a number called the **eccentricity**. This number is given by the ratio of the distance from the center to the focus divided by the distance from the center to the vertex. The greater the eccentricity value, the flatter the ellipse.

Another conic section is a **hyperbola**, which looks like two facing parabolas. Mathematically, it is similar in definition to an ellipse. It is formed by the graph of the set of points, the difference of whose distances from two fixed points (foci) is constant. Notice that in the case of a hyperbola, the difference between the two distances from fixed points is plotted and not the sum of this value as was done with the ellipse.

As with other conics, the hyperbola has various characteristics. It has vertices, the points at which a line passing through the foci intersects the graph, and a center. The line segment which connects the two vertices is called the transverse axis. The simplified equation for a hyperbola with its center at the origin is $(x^2/a^2) - (y^2/b^2) = 1$. In this case, a is the distance between the center and a vertex, b is the difference of the distance between the focus and the center and the vertex and the center.

Geometric figures such as points, lines, and conics are two-dimensional because they are confined to a single plane. The term two-dimensional is used because each point in this plane is represented by two real numbers. Other geometric shapes like spheres and cubes do not exist in a single plane. These shapes, called surfaces, require a third **dimension** to describe their location in **space**. To create this needed dimension, a third axis (traditionally called the z-axis) is added to the coordinate system. Consequently, the location of each point is defined by three real numbers instead of two. For example, a point defined by the coordinates (2,3,4) would be located 2 units away from the x axis, 3 units from the y axis, and 4 units from the z axis.

The algebraic equations for three-dimensional figures are determined in a way similar to their two-dimensional counterparts. For example, the equation for a **sphere** is $x^2 + y^2 + z^2 = r^2$. As can be seen, this is slightly more complicated than the equation for its two-dimensional cousin, the circle, because of the additional **variable** z^2.

It is interesting to note that just as the creation of a third dimension was possible, more dimensions can be added to our coordinate system. Mathematically, these dimension can exist, and valid equations have been developed to describe figures in these dimensions. However, it should be noted that this does not mean that these figures physically exist and in fact, at present they only exist in the minds of people who study this type of multidimensional analytic geometry.

ANAXAGORAS OF CLAZOMENAE (500 B.C.-428 B.C.)

Greek geometer and philosopher

Anaxagoras of Clazomenaewas a Greek philosopher who made contributions in astronomy and physics. He was the first philosopher to live in Athens and the first to propose several important theories about the cosmos. Among these are theories about the Earth and moon, including the reason for an eclipse and accurate descriptions of the lunar surface. In mathematics, Anaxagoras was the first to attempt **squaring the circle**.

Not much is known about Anaxagoras' life, though records of his theories are preserved. The Greek philosopher was born on the Ionian coast in the town of Clazomenae in Asia Minor (in what is now Turkey). His parents were wealthy but Anaxagoras chose to forsake his life of leisure to study philosophy. In 462 B.C., he moved to Athens, which was rapidly becoming an intellectual center. There he attracted the attention of the politician Pericles as well as the playwright Euripedes. Pericles welcomed Anaxagoras into his circle of friends.

Anaxagoras was the first to propose a molecular theory of matter. He believed that matterwas infinitely divisible. The universe, he said, began as a great whirling jumble of matter, which was controlled by the Mind. The Mind, however, was not a god, or a spiritual or mental essence. "It was the most delicate and purest of all things," Anaxagoras wrote. In Anaxagoras' theory of "nous" the Mind caused the dark and the **light** to form, creating air, water, and earth. In another stage, animal and plant seeds (which were part of the original mixture) came together to form flesh and vegetation. The growth of all living things occurred because they had portions of the Mind within them, which could attract nourishment.

Anaxagoras believed that objects of the natural world were elemental. That is, they could not have been derived from elements simpler than themselves, from things that were not made of the same material. Every single piece of matter came from something like itself. "How could hair come to be from what is not hair and flesh from what is not flesh," he asks in his writings. He believed that every element in the world—hair, skin, bone, and an infinite number of other things—pre-existed in our food

The Greek philosopher was also a great astronomer. He was the first to propose the reason for an eclipse and the first to theorize that the moon shone by reflected light. He also described the moon's surface accurately as a series of flat areas and depressions. The moon was not a **sphere**, Anaxagoras said. This was a contention that was confirmed 2,000 years later when **Galileo** trained his telescope on the moon. Direct **proof** of this theory also came through the American astronauts' first trip to the moon.

Anaxagoras, however, created real problems for himself and his friends when he proposed that the sun was a red hot stone. All the planets and stars were, in fact, made of stone, he said. His belief may have been suggested by the fall of a huge meteorite near his home when he was young.

Anaxagoras' belief about the sun, however, made him a prime target for his and Pericles' enemies. They resented the attempts of Pericles, philosophers like Anaxagoras and artists like Pheidias to bring a higher culture to Athens. As Pericles grew aged, his enemies began to attack his friends. They accused Pheidias of stealing some of the gold used on his artistic statues. They campaigned for a law that permitted prosecution of those who did not believe in religion and taught theories about celestial bodies. Under this law, they brought Anaxagoras to trial.

Anaxagoras of Clazomenae

It's not certain what was the result of the trial (records are not preserved), but we do know that while he was in jail, Anaxagoras made the first attempt to square the **circle**. In other words, he used a compass and a ruler to try to construct a square with the same **area** as a certain circle. This was the first time that such an effort had been made and preserved on record.

Pericles was able to get Anaxagoras released from prison. But Anaxagoras was forced to return to Ionia. There he started a school and was celebrated as a hero. He died in 428 B.C. and the anniversary of his death was celebrated for a century afterward in Ionia.

ANGLE

An angle is a geometric figure created by two line segments that extend from a single point or two planes which extend from a single line. The size of an angle, measured in units of degrees or radians, is related to the amount of **rotation** required to superimpose one of its sides on the other. First used by ancient civilizations, angles continue to be an important tool to science and industry today.

The study of angles has been known since the time of the ancient Babylonians (4,000-300 B.C.). These people used angles for measurement in many areas such as construction, commerce, and astronomy. The ancient Greeks developed the idea of an angle further and were even able to use them to calculate the **circumference** of the Earth and the **distance** to the moon.

A geometric angle is formed by two lines (rays) that intersect at a common endpoint called the vertex. The two rays are known as the sides of the angle. An angle can be specified in various ways. If the vertex of an angle is at point P, then the angle could be denoted by∠P. It can be further described by using a point from each ray. For example, the angle ∠OPQ would have the point O on one ray, a vertex at point P, and the point Q on the remaining ray. An angle can also be denoted by a single number or character which is placed on it. The most common character used is the Greek letter θ (theta).

Units of measurement of an angle

An angle is commonly given an **arithmetic** value which describes its size. To specify its this value, an angle is drawn in a standard position on a coordinate system, with its vertex at the center and one side, called the initial side, along the x axis. The value of the angle then represents the amount of rotation needed to get from the initial side to the other side, called the terminal side. The direction of rotation indicates the sign of the angle. Traditionally, a counterclockwise rotation gives a positive value and a clockwise rotation gives a negative value. The three terms which are typically used to express the value of an angle include revolutions, degrees, or radians.

The revolution is the most natural unit of measurement for an angle. It is defined as the amount of rotation required to go from the initial side of the angle all the way around back to the initial side. One way to visualize a revolution is to imagine spinning a wheel around one time. The distance traveled by any point on the wheel is equal to one revolution. An angle can then be given a value based on the fraction of the distance a point travels divided by the distance traveled in one rotation. For example, an angle represented by a quarter turn of the wheel is equal to.25 rotations.

A more common unit of measurement for an angle is the degree. This unit was used by the Babylonians as early as 1,000 B.C. At that time, they used a number system based on the number 60, so it was natural for mathematicians of the day to divide the angles of an equilateral **triangle** into 60 individual units. These units became known as degrees. Since six equilateral triangles can be evenly arranged in a **circle**, the number of degrees in one revolution became $6 \times 60 = 360$. The unit of degrees was subdivided into 60 smaller units called minutes and in turn, these minutes were subdivided into 60 smaller units called seconds. Consequently, the notation for an angle which has a value of 44 degrees, 15 minutes, and 25 seconds would be 44° 15' 25".

An angle may be measured with a protractor, which is a flat instrument in the shape of a semi-circle. There are marks on its outer edges which subdivide it into 180 evenly spaced units, or degrees. Measurements are taken by placing the **midpoint** of the flat edge over the vertex of the angle and lining the 0° mark up with the initial side. The number of degrees can be read off at the point where the terminal side intersects the curve of the protractor.

Another unit of angle measurement, used extensively in **trigonometry**, is the **radian**. This unit relates a unique angle to each real number. Consider a circle with its center at the origin of a graph and its radius along the x-axis. One radian is defined as the angle created by a counterclockwise rotation of the radius around the circle such that the length of the arc traveled is equal to the length of the radius. Using the formula for the circumference of a circle, it can be shown that the total number of radians in a complete revolution of 360° is 2π. Given this relationship, it is possible to convert between a degree and a radian measurement.

Geometric characteristics of angles

An angle is typically classified into four categories including acute, right, obtuse, and straight. An acute angle is one which has a degree measurement greater than 0° but less than 90°. A right angle has a 90° angle measurement. An obtuse angle has a measurement greater than 90° but less than 180°, and a straight angle, which looks like a straight line, has a 180° angle measurement.

Two angles are known as congruent angles if they have the same measurement. If their sum is 90°, then they are said to be complementary angles. If their sum is 180°, they are supplementary angles. Angles can be bisected (divided in half) or trisected (divided in thirds) by rays protruding from the vertex.

When two lines intersect, they form four angles. The angles directly across from each other are known as vertical angles and are congruent. The neighboring angles are called adjacent because they share a common side. If the lines intersect such that each angle measures 90°, the lines are then considered perpendicular or orthogonal.

In addition to size, angles also have trigonometric values associated with them such as **sine**, **cosine**, and tangent. These values relate the size of an angle to a given length of its sides. These values are particularly important in areas such as navigation, astronomy, and architecture.

APÉRY'S THEOREM

Apéry's theorem is the discovery that the number

$$\zeta(3) = \sum_{n=1}^{\infty} \frac{1}{n^3}$$

is an irrational **real number**. Here ζ is the Riemann **zeta-function**. This important **function** is defined in the right half plane $\{s = \sigma + it \in \mathbf{C}: 1 < \sigma\}$ by the convergent series

$$\zeta(s) = \sum_{n=1}^{\infty} \frac{1}{n^s}$$

and also by the convergent infinite product

$$\zeta(s) = \prod_{p} (1 - p^{-s})^{-1},$$

where the product is over the set of all **prime numbers** p. The function $s \rightarrow \zeta(s)$ is an analytic function of the complex **variable** $s = \sigma + it$ in the half plane $1 < \sigma$. Because the zeta-function is defined by both the **infinite series** and the infinite product, it provides a means for using techniques from **complex analysis** in order to investigate questions about prime numbers.

There are many elementary problems of interest about the zeta-function that have no immediate implications for the distribution of primes. One of these is to determine the values of $\zeta(s)$ at the **integers** $s = 2, 3, 4,\ldots$. Of course there is no *a priori* reason why the values of $\zeta(s)$ at $s = 2, 3, 4,\ldots$ should be related to other familiar numbers in a simple way. However, **Leonhard Euler** discovered that the numbers $\zeta(2)$, $\zeta(4)$, $\zeta(6),\ldots, \zeta(2n),\ldots$ can all be written as a rational number multiplied by π^{2n}. For example,

$$\zeta(2) = \sum_{n=1}^{\infty} \frac{1}{n^2} = \frac{\pi^2}{6} \quad \text{and} \quad \zeta(4) = \sum_{n=1}^{\infty} \frac{1}{n^4} = \frac{\pi^4}{90}.$$

More generally, define rational numbers $B_0, B_1, B_2,\ldots$ by means of the **power series** expansion

$$\frac{z}{1 - e^{-z}} = \sum_{n=0}^{\infty} \frac{B_n z^n}{n!},$$

which converges for $|z| < 2\pi$. Then it is not difficult to check that

$$B_0 = 1, \; B_1 = \tfrac{1}{2}, \; B_2 = \tfrac{1}{6}, \; B_3 = 0, \; B_4 = -\tfrac{1}{30}, \; B_5 = 0, \; B_6 = \tfrac{1}{42}, \; B_7 = 0,$$
$$B_8 = -\tfrac{1}{30}, \; B_9 = 0, \; B_{10} = \tfrac{5}{66}, \; B_{11} = 0, \; B_{12} = -\tfrac{691}{2730}, \; \ldots.$$

As is suggested by this list, the value of B_{2m+1} is 0 for $m = 1, 2, 3,\ldots$, while the value of B_{2m} is a nonzero rational number for each $m = 0, 1, 2, 3,\ldots$. The numbers B_n are called the Bernoulli numbers, (but some authors use a slightly different definition.) One of Euler's most famous discoveries is the identity

$$\zeta(2m) = \sum_{n=1}^{\infty} \frac{1}{n^{2m}} = (-1)^{m-1} \frac{2^{2m-1}\pi^{2m}}{(2m)!} B_{2m}, \quad \text{where } m = 1, 2, 3, \ldots,$$

which evaluates the Riemann zeta-function at the positive even integers. On the other hand, the value of the Riemann zeta-function at the odd integers 3, 5, 7,... is much more mysterious, and even today we know very little about these numbers.

In 1978 the French mathematician Roger Apéry obtained the first nontrivial result about the number $\zeta(3)$. Apéry proved that this number is **irrational**. His proof is rather elementary and, indeed, it uses mathematics that was available to Euler. But it must also be said that Apéry's proof uses some very surprising formulas that are not easy to discover. So far it has not been possible to generalize Apéry's argument to establish the irrationality of any of the numbers $\zeta(5)$, $\zeta(7)$, $\zeta(9),\ldots$. Thus we do not know if the number $\zeta(5)$ is **rational** or irrational. It is generally conjectured by experts in this field that the numbers $\zeta(3)$, $\zeta(5)$, $\zeta(7),\ldots$, are all **transcendental** real numbers—but this is only a conjecture.

APOLLONIUS OF PERGA (CA. 262 B.C.-CA. 190 B.C.)
Greek geometer

Apollonius was one of the founding fathers of mathematical astronomy in ancient Greece. He also originated the geometric shapes and terms that would become central to Newtonian astronomical physics nearly 20 centuries later. He may have even prefigured **Christiaan Huygens'** 1673 use of the "evolute," the **locus** of the centers of curvature in a given curve. Certainly the projective **geometry** of **Gérard Desargues** and **Blaise Pascal** owe their genesis to Apollonius. He also invented his own counting system for very large numbers. Still considered the greatest achievement of Greek geometry, the *Conics* earned Apollonius the moniker "The Great Geometer," according to a later mathematician, Eutocius. These books quickly supplanted the works of **Euclid** as authoritative texts.

Estimations of the time frame in which Apollonius lived have varied over the years, so much so that one reference will only place him in the second half of third century B.C. Others place him according to the reigns of Ptolemy Euergetes, who was king of Egypt beginning around 247 B.C., or of Ptolemy Philopator ending in 210–205 B.C. Apollonius was born in what was then the Greek town of Perga, south Asia Minor, now part of Turkey. He was apparently a second generation Euclidean scholar in Alexandria. Legend has it he was nicknamed "Epsilon" there, because that Greek letter looks like the half moon he studied. Apollonius is also said to have visited Pergamum, where there was a new library and museum like the ones in Alexandria, and traveled to Ephesus. At least one source speculates that he may have been employed as the treasurer–general to Ptolemy Philadelphus. While he lived around the same time as **Archimedes**, there is no direct proof that the two either influenced each other or had contact. Apollonius did, however, improved upon Archimedes' calculation of the value of π.

Apollonius set forth in his eight books on **conic sections**, along with roughly 400 theorems, a new idea on how to subdivide the cone to produce **circles**. He also catalogued new kinds of closed curves that he named **ellipse**s, **parabola**s, and **hyperbola**s. The Pythagorean distaste for infinities, **infinitesimal**s, and infinite **sets** was put aside in this new frame of mind, which paved the way for the eventual discovery of the infinitesimal **calculus**. Additionally, his epicircles, epicycles, and eccentrics replaced the concentric spheres of **Eudoxus** and influenced Ptolemaic cosmology. That framework would stand until **Johannes Kepler** finally reformed the geometry of astronomical **modeling** for the current day.

The first half of the *Conics* surveys and completes all inherited Greek geometry, including early efforts of Euclid. Apollonius boasted that a Euclidean problem such as finding the locus relative to three or four lines was completely solvable for the first time, thanks to his new **propositions**. It is perhaps this style of presentation that led **Pappus** to accuse him of envy, and for Archimedes' biographer Heracleides to accuse him of plagiarism. The material on conic sections took up the last four volumes, laying the foundation for modern–day astronomy, ballistics, rocketry, and space science. Conic sections, it was only discovered many hundreds of years afterwards, are the shapes formed by the paths or "loci" of projectiles and other objects in **orbit**.

The *Conics* covers both pure and applied geometry. In it, Apollonius considered the problem of finding "normals" on points along curves, which involves **trigonometry**, though he could not apparently figure the focus of a parabolic curve the way he could for an ellipse or hyperbola. He also presented a method of figuring at what points a planetary orbit takes on apparent retrograde **motion**. Finally, his still famous "problem of Apollonius" calls for the construction of a circle tangent to three given circles. His most important contribution in pure terms was how he generalized the means of production. From one cone he could derive all conic sections, whether perpendicular to it or not. Apollonius used this standard cone in a way that prefigured **analytic geometry** by splitting it along two fixed lines called the *latus transversum* and the *latus erectum*. These "conjugate diameters" became a coordinate system and frame of reference, making geometry do the work now done by **algebra**.

A number of writers have attempted to restore or recreate the lost eighth book of the *Conics* and Apollonius' other writings, including Alhazen, **Edmond Halley** and **Pierre de Fermat**. The *Conics* is all that survived, perhaps because most of Apollonius' writings were considered too obscure or outrageous to be worth preserving by his contemporaries. That one masterwork influenced the next generation of mathematicians such as **Hipparchus**, and later commentators, including **Hypatia of Alexandria** and Eutocius, reinforced its reputation. Some of Apollonius' ideas and writings are mentioned in other ancient writings, which document some of his other conclusions. In his work on "burning mirrors," for instance, Apollonius disproved the notion that parallel rays of **light** could be focused by a spherical mirror, and he also noted properties of the parabolic mirror. Titles of his lost works include *Quick Delivery*, *Vergings*, and *Plane Loci*, as well as *Cutting–off of a Ratio* and *Cutting–off of an Area*. The subjects of these and some of their formulae and comments were summed up by Pappus.

Although his works were undervalued by many commentators over the years, beginning with his Greek contemporaries, Apollonius has recently undergone a revisionist examination. One academic, Wolfgang Vogel of Massey University in New Zealand, believes that after two millennia the ideas of Apollonius can be applied to current, significant problems related to intersecting conics.

APPEL, KENNETH I. (1932-)

American mathematician

The American mathematician Kenneth I. Appel was born in Brooklyn, New York, on October 8, 1932. He earned a B.S. degree from Queens College in 1953, then served in the U.S.

Army for two years following his graduation. In 1955, he enrolled in graduate studies at the University of Michigan, subsequently earning his M.A. and Ph.D. degrees there in 1956 and 1959, respectively. In 1959, he married Carole Stein.

Between 1959 and 1961, Appel was on the technical staff of the Institute for Defense Analyses. In 1961, he joined the faculty of the Mathematics Department at the University of Illinois as Assistant Professor, later being advanced to Associate Professor in 1967, and to Professor in 1977. In 1993 he became Chairman of the Department of Mathematics at the University of New Hampshire.

In 1976, Appel and a colleague, **Wolfgang Haken**, succeeded in proving that any map in a plane or on a **sphere** can be colored with only four colors in such a way that no two neighboring countries are of the same color.

When Appel came up with his **proof**, mathematicians had been attempting to prove the **theorem** for over a hundred years. In 1852, Francis Guthrie had written to his brother asking him whether he knew of any proof that four colors are always sufficient. The brother relayed the question to the noted British mathematician **Augustus De Morgan** (1806-1871), but De Morgan did not know the answer. In 1878, the Cambridge mathematician **Arthur Cayley** (1821-1895) brought the problem before the London Mathematical Society. Very soon thereafter, Alfred Bray Kempe published a proof that the conjecture is true. But in 1890 Percy John Heawood discovered a flaw in Kempe's proof. The theorem thus remained unproven until Appel and Haken came up with their proof in 1976.

Appel and Hagen's proof, which required a very large amount of computer time, was described in the book-length article by Appel and Haken entitled "Every Planar Map is Four Colorable," *Contemporary Mathematics*, vol. 98, American Mathematical Society, 1989. A summary of the proof and a history of work on the problem was given in an article by Appel and Haken entitled: "The Solution of the Four-Color-Map Problem," *Scientific American*, vol. 237, No. 4, pp. 108-121 (1977).

In 1979, Appel was awarded the Fulkerson Prize in Discrete Mathematics by the American Mathematical Society and Mathematical Programming Society.

See also Four color map theorem

APPLIED MATHEMATICS

Applied mathematics is a collection of theories, techniques, and terminology that have practical application in various fields of science, including, but not limited to, astronomy, chemistry, dynamics, engineering, physics and even mathematics itself. These techniques allow scientists and mathematicians to use **equations** and models to help explain natural and theorized phenomena, create measurement parameters for physical objects and actions, and perform analysis in various disciplines. Through applied mathematics many new theories and models have been created that have dramatically advanced our understanding of the physical world.

A very simple example of the use of applied mathematics in physics is the equation **F=kma** where **F** is the **force** to be determined, **k** is a constant for a particular system of units. **K** is usually set to unity (in other words, **k**=1) when determining the force exerted by an object without consideration of the pull of the earth's gravity. The **variable m** is the mass of the object and **a** is the acceleration the object exhibits at a given moment in time. From this equation, once values for two of the variables have been determined, and assuming k=1, the remaining value can be determined. This equation, then, provides further desired information for the physical phenomenon being studied. The uses of applied mathematics through much more complex equations provide ways to quantify many observed properties of the universe as well as matters of theory that are as yet impossible to observe directly.

See also Theorem

ARBOGAST, LOUIS FRANÇOIS ANTOINE (1759-1803)

French number theorist and mathematical historian

Louis François Antoine Arbogast made a threefold contribution to mathematics. He is primarily known as a mathematics historian who organized **Marin Mersenne**'s papers, as well as letters and miscellany of other scientists. Arbogast was also a noteworthy mathematician in his own right; he did the earliest work in discontinuous **fractions** and predated some developments in **calculus**. He also participated in local politics (including the Commune of Strasbourg) and was influential in the development of certain schools and their curriculums in the 1790s.

Much of Arbogast's early life and education is unclear other than his date of birth, October 4, 1759 in Mutzig, Alsace. It is known that in 1780, he was listed as a lawyer in Alsace, and he also taught mathematics at the Collège de Colmar in 1787. Arbogast's life becomes more documented after moving to Strasbourg, France, in 1789. In that same year, he became a mathematics instructor at l'École d'Artillerie and a physics professor at Collège Royal. When the Collège became nationalized, he was director for seven months in 1791. In 1794, Arbogast was appointed to a professorship in Paris, at l'École Central de Paris, though he only taught at l'École Préparatoire, a temporary institution of higher learning. He began planning its replacement, l'École Centrale du Bas–Rhin, in 1795. (Arbogast was given this responsibility in part because he had experience in this area. He designed a program for public schools to a legislative assembly in Alsace circa 1792.) After the school's establishment in 1796, he served as mathematics chair until 1802.

As a mathematician, Arbogast's lasting contribution was in the collection and arrangement of manuscripts by other scientists. He hand copied the writings of such people as Marin Mersenne, **René Descartes, Jean Bernoulli, Guillaume de L'Hospital, Pierre Varignon**, and **Pierre de Fermat**. Arbogast's collection are now in the Bibliothèque Nationale in Paris and the Laurenziana Library in Florence.

Arbogast made his own mark on mathematics in 1787 with his work on discontinuous fractions and arbitrary **functions**. In 1789, he wrote a report that was never published outlining new principles of **differential calculus**. In his mathematical writings, he demonstrated what came to be known as operational calculus. In this area, he was years ahead of his time, implicitly demonstrating that operation and function are inherent differences in calculus. This is one of many mathematical areas Arbogast anticipated.

Arbogast's contributions to mathematics did not go unnoticed by his colleagues. In 1792, the Académie des Sciences made him a corresponding member. In 1796, the Institut National elected him an associate nonresident member. Arbogast died in Strasbourg seven years later on April 18, 1803.

ARCHIMEDES OF SYRACUSE (287 B.C.-212 B.C.)

Greek geometer

Archimedes of Syracuse is considered one of the greatest thinks of the ancient world. He established the principles of plane and solid **geometry**, discovered the concept of specific gravity, conducted experiments on buoyancy, demonstrated the power of mechanical advantage, and invented the Archimedes Screw, an auger–like device for raising water.

Archimedes was born in the Greek city of Syracuse, on the island of Sicily, in 287 B.C. He was the son of Phidias, the astronomer and mathematician. What we know of Archimedes' life comes from his extant writings, and from the histories authored by Plutarch, Cicero, and other historians several centuries after his death. Due to the length of time between Archimedes' death and his biographers' accounts, as well as inconsistencies among their writings, details of his life must remain subject to question.

Plutarch records that Archimedes was a relative of King Hieron, but Cicero claims he was of low birth. It is believed he obtained his early schooling in Syracuse, then traveled to Alexandria to study with Conon, the Egyptian mathematician and astronomer. Archimedes was close friends with both Conon and the custodian of the Alexandrian library, **Eratosthenes.** He corresponded with them about his mathematical and scientific discoveries long after he had completed his formal studies and returned to Syracuse. Although much of Archimedes' work was applied to practical ends, he himself was more interested in pure thought, and supposedly believed things connected with daily needs were ignoble and vulgar. Reports of his personal habits reflect his lack of concern with the mundane. He was known for becoming so engrossed in his thoughts that he would forget to eat and bathe. Many of the **equations** he developed were scratched out first in ashes or traced with after–bath oil on his skin.

Archimedes' contributions to mathematical knowledge were diverse. On the subject of **plane geometry** three of the treatises he wrote have survived, *Measurement of a Circle*, *Quadrature of the Parabola*, and *On Spirals*. In *Measurement*

Archimedes of Syracuse

of a Circle, he described his method for calculating π, the ratio between the **circumference** of a **circle** and its **diameter**. By a method that involved measuring the **perimeter** of inscribed and circumscribed **polygons**, Archimedes correctly determined that the value of π was somewhere between 3.1408 and 3.1428. In the same treatise he set forth the formula π *r²* for determining a circle's **area**. In *Quadrature of the Parabola* and *On Spirals*, Archimedes advanced his technique for determining the area under curves, a sophisticated version of the method of exhaustion, originally developed by the Egyptians. His use of this technique, elaborated upon in another volume, The *Method*, anticipated the development of integral **calculus** by two thousand years.

Archimedes dealt with the topic of **solid geometry** in his writings *On the Sphere and Cylinder* and *On Conoids and Spheroids. On the Sphere and Cylinder* contains several famous proofs, including his demonstration that the volume of a **sphere** is equal to *4/3π r³*. Archimedes also showed in this work that the volume of a sphere is two–thirds the volume of a **cylinder** surrounding it, as long as the cylinder's height and width are equal to the sphere's diameter. So proud was he of this latter discovery that he requested its illustration be engraved on his tombstone, a wish that was eventually fulfilled.

On the subject of **arithmetic** Archimedes wrote several essays, of which only *The Sand Reckoner* remains. Addressed

to the son of King Hieron, it proposed the problem of determining the number of grains of sand in the universe, and contained a special notation for estimating and expressing very large numbers. Archimedes was famous for another complicated arithmetic puzzle called the Cattle Problem, in which one had to determine the number of bulls and cows of various colors, given that each cattle color was represented in a particular ratio to the others. There are an infinite number of possible solutions to the problem, but deriving them was especially challenging to the ancient Greeks, who had no knowledge of **algebra**.

According to the ancient historians, Archimedes was frequently called upon by King Hieron to solve practical problems. Perhaps the most famous of these was the task of determining whether King Hieron's crown was made of pure gold. Hieron believed his jeweler had stolen a portion of the gold intended for the crown, substituting an equal weight of another metal. Legend has it that Archimedes stepped into a bath and, noticing the displacement of water, conceived of a vertical buoyancyforce. Realizing that this concept would allow him to measure the density of the king's crown in comparison to pure gold, he supposedly jumped from the tub and ran naked through the streets yelling "eureka, eureka!" Archimedes not only proved the crown was tainted, but went on to describe the idea of specific gravity and develop a generalized concept of buoyancy known today as Archimedes' Principle.

The Archimedes Screwwas reportedly invented in order to empty water from the hold of one of King Hieron's ships. This device consisted of a screw, encased in a cylinder, that was turned by a hand–crank. As the screw spiraled upward it carried water. A similar device is still used today to lift water in the Nile Delta of Egypt. Archimedes recognized the mechanical advantage that could be gained by using levers, and it is said he boasted that with a long enough lever he could move the Earth. He determined the inverse mathematical relationship between the effort required to raise a load with a lever and the **distance** of the load from the lever's pivot point or fulcrum. The story is told that King Hieron, skeptical of the power of mechanical advantage, challenged Archimedes to move a three–mast ship, laden with passengers and freight, that lay aground near Syracuse Harbor. To meet the challenge Archimedes is said to have designed a system of compound pulleys. With a relatively effortless pull of a rope he was able to guide the vessel into the water. Archimedes was also interested in astronomy and made several accomplishments in this field. He built a device to estimate the size of the sun and constructed a **model** planetarium to demonstrate the **motion** of the planets.

During Archimedes' lifetime the first two of the three Punic Wars between the Romans and the Carthaginians were fought. Syracuse allied itself with Carthage, and when the Roman general Marcellus began a siege on the city in 214 B.C., Archimedes was called upon by King Hieron to aid in its defense. The historical accounts of Archimedes' war–faring inventions are vivid and possibly exaggerated. It is claimed that he devised catapult launchers that threw heavy beams at the Roman ships, grappling cranes that hoisted ships out of water, and burning–glasses that reflected the sun's rays and set ships on fire. Marcellus had given orders that when Syracuse was finally conquered, Archimedes, whose reputation was widely known, should be taken alive. When the Romans finally sacked the city in 212 B.C., a soldier found Archimedes quietly etching equations in the sand, absorbed in a mathematical problem. Reportedly, Archimedes ordered the soldier not to disturb the figures in the sand. Enraged, the soldier drew his sword and impaled him.

ARCHIMEDES' SPIRAL

In the third century B.C., **Archimedes of Syracuse** created a special spiral-shaped curve by pulling the legs of a compass apart while turning it. By performing both actions at a steady rate, he found that the resulting **spiral** moved outward by the same amount with each turn of the compass. The groove in an old-style LP record is an example of such an Archimedean spiral.

The most significant mathematical use to which Archimedes tried to put his spiral was to create a better method of determining the **area** of a **circle**. Using a spiral to figure out the area of a circle seems a waste of energy today since anyone with a **calculator** can do so by pressing a few buttons. However, in ancient Greece either a physical measurement of the **circumference** of the circle had to be made or a critical factor in the still not widely known equation for determining the area had to be used. Measuring the length of a circle or any other curved shape is difficult and the area of the circle that is determined as a result can only be as accurate as the measurement. For these reasons, calculating the area of a circle presented a major problem for the mathematicians of Archimedes' time. Back then, they knew the area was related to the ratio of the circumference of the circle to its **diameter**, but this ratio, called **pi**, was (and even today still is) not known with complete accuracy. Today we can calculate its value much more closely, but in ancient times mathematicians used a value for pi that was inaccurate enough that their determination of the area of a circle was unsatisfactory for many critical applications.

The Greeks and others before them had tried a number of methods for determining pi and figuring out the area of a circle. One of these involved constructing a right **triangle** that had one side with a length equal to the circumference of the circle and another with the length of the radius of the circle. Such a triangle has approximately the same area as the circle. The straight sides of the triangle can be measured accurately and the area of the triangle determined, but this method only gives an approximation of the circle's area because, again, it is dependent upon measurement of the circumference. Archimedes tried to use his spiral to improve upon this method. He started the drawing of the spiral at the center of a circle and rotated and opened the compass in such a way that the spiral reached the **perimeter** after one turn. This meant that the point where the spiral intersected the circle provided the point for one corner of the triangle. Since this triangle method can be carried out with equal accuracy with or without Archimedes' spiral, his method was really only of mathematical interest. However, Archimedes went on to determine a

much more accurate value for pi, which advanced the determination of the area of circles in another way.

See also Circle; Circumference; Triangle

ARCHYTAS OF TARENTUM (CA. 428 B.C.- 350 B.C.)
Greek geometer, philosopher, and statesman

Archytas of Tarentum was a Greek mathematician of the Pythagorean school who formulated the harmonic **mean** and was the first to integrate mathematics and **mechanics**. He also developed an ingenious geometric solution for the ancient Greek problem of doubling the cube. A contemporary of the famous Greek philosopher **Plato**, Archytas was also famous in his own time as a philosopher, statesman, and military leader.

Only a few fragments identified as the work of Archytas have survived. As a result, most of what we know about his life and work comes from ancient Greek writers, such as **Aristotle** and Proclus. Archytas was probably born in Tarentum (now Taranto, Italy) around 428 B.C., possibly into an aristocratic family. Nothing else is known about his early life.

In the beginning of the fourth century B.C., Dionysius the Elder, a tyrant of Syracuse, had driven the Pythagoreans out of most southern Italy's cities. Tarentum was the last city in the region where the Pythagorean school of philosophy and mathematics maintained a strong presence in education and politics. A close friend of Plato, Archytas may have been his chief teacher of Pythagorean science and philosophy. More importantly, Archytas saved this great philosopher's life by obtaining a pardon for him through a letter he wrote to Dionysius the Younger, who wanted to execute Plato for subversive activities.

Archytas was a powerful statesman and an influential leader in Tarentum and throughout the Greek city–states. Immensely popular, he served as general of his city's citizen army for seven years, despite a law forbidding anyone to hold the position for more than one year. The confederation of Hellenic cities of Magna Grecia also appointed him commander, with full autocratic authority, over the confederation's armies. Troops under his command reportedly were never defeated in battle. According to some accounts, Archytas eventually gave up his command because of envious detractors, and his troops were immediately captured. Archytas was also greatly admired for his virtues and noble character, which included a love of children and the just and kind treatment of his slaves.

Archytas' creativity and ingeniousness were grounded in his recognition of how the sciences interconnect, especially the disciplines of mathematics, **geometry**, music, and astronomy. For example, he was the first to apply mathematics to the realm of mechanics and wrote a systematic treatise on the subject. He invented the simple pulley and screw and wrote on the mathematical basis of astronomy. Through his work in the theory of means and **proportions** he differentiated three basic means: the arithmetic **mean**, the geometric mean, and the harmonic mean. Seven other means were eventually added by Archytas and others. In his theory of music, Archytas developed numerical **ratios** representing intervals of the tetrachord on which he based his three musical scales: the enharmonic, the chromatic, and the diatonic. **Claudius Ptolemy** the astronomer credited Archytas as the most important Pythagorean to delve into the theory of music.

According to Proclus, Archytas also increased the number of theorems in geometry, developing them into a systematic body of knowledge, and influenced many other Greek mathematicians. Archytas is credited with most of the geometry contained in Book VIII of **Euclid**'s *Elements*, which served as the primary textbook of elementary geometry and logic for more than a thousand years.

Archytas' most famous mathematical achievement was to provide an elegant geometric solution to the Delian problem, known as duplicating or doubling the cube, or enlarging a cube according to a given ratio. Although Plato had complained that the Greeks knew little about three–dimensional geometry, Archytas exhibited a comprehensive knowledge of this area in his remarkable solution to the problem. By inventing a new type of three–dimensional curve through the intersection of a **cylinder**, a cone, and a **torus** (or doughnut shape), he was able to find the two mean proportionals (or geometric means) between two lines, a method first proposed by **Hippocrates** of Chios for doubling the cube.

Although he was a pioneer in mathematics and geometry, Archytas has been criticized for not following his contemporaries' lead in applying clear and logical explanations for his theories. His inability to cope well with the logical aspects of his work also affected other interests. For example, in the largest fragment of his extant works, Archytas proposed a complicated theory of sound. He correctly theorized that faster **motion** produces higher sounds or notes. However, based solely on empirical observations without the application of mathematical theories, Archytas wrongly concluded that higher sounds reach the listener faster than lower sounds.

Archytas' philosophy was probably based on his training as a Pythagorean, leading him to hold a strong belief in numbers as a mystical and basic part of nature. In a small fragment of one of his works, Archytas also reveals that he believed the universe was infinite in extent. Although Aristotle wrote three books on the philosophy of Archytas, none of them have survived the passing of time.

While Archytas is rightly remembered today for his seminal contributions in mathematics and geometry, he should also be admired as a man of action who applied his keen intellect to affairs of state. A successful army general, Archytas was an influential leader who helped forge alliances among Greek city–states to provide greater protection against foreign powers. Archytas also had his playful side; his interest in mechanics led him to create two mechanical devices: a mechanical wooden pigeon that could fly and, according to Aristotle, a

Archytas of Tarentum

type of rattle to amuse and occupy infants. In an ode, the famous Greek poet Horacerecounts Archytas' death by shipwreck in the Adriatic Sea.

AREA

Area is the measure of the size of a surface, or of a region on such a surface. In analytical **geometry**, the letter "A" commonly denotes formulas for the determination of the areas of simple two-dimensional plane figures, while the letter "S" denotes formulas for the determination of surface areas of three-dimensional solids. The concept of area is dependent upon the more fundamental quantity of length. In ancient civilizations, people measured various distances by determining the number of times a standard unit of length must be added together to yield the **distance** in question. For example, the ancient Romans used the "stadium", with about 8.7 stadia equal to one statute mile, as their "reference" length to measure large distances. About 1,045 stadia (or 120 statute miles) added together equals the straight-line distance between the two Italian cities of Rome and Naples. In the same way, a surface is measured by denoting the number of times a standard unit of area must be added together to equal the area in question. For instance, the playing area of a football field is equal to 48,000 **square** feet (300 feet of length times 160 feet of width), or about 5,333.33 square yards. As implied in this example, the standard geometric shape for measuring area is the square. Since all four sides of a square are the same length, the length of one side uniquely identifies the size of the square. To determine the area of a rectangle, the lengths of its width and height are measured and multiplied together.

The resulting quantity represents the number of identical squares which, when added together, equals the area of the rectangle. For instance, if a rectangle's width is nine centimeters (cm) and its height is 4 cm, then the rectangle's area is "9 cm x 4 cm" (36 cm²), or is equal to the combined area of 36 identical squares, each square having sides of 1 cm.

For other simple plane figures (like triangles) and the surface areas of simple solids (like spheres), formulas have been derived to calculate the area of each using their linear dimensions. For instance, the area of a right **triangle** can be calculated with the formula "1/2ab" where "a" represents the length of the triangle's base and "b" represents the length of the triangle's **altitude**. If a triangle's base equals nine centimeters and its height equals six centimeters, its area is "1/2 x 9 cm x 6 cm," or 27 cm².

As an especially important example of area determination, the area of a regular polygon (a figure of three or more straight sides of equal length joined at equal angles to one another) is defined as the sum of the areas of triangles into which it can be decomposed. Around the third century B.C., **Archimedes of Syracuse**, who established the determination of areas (then called the "method of exhaustion" by its inventor **Eudoxus of Cnidus** in the fourth century B.C.), utilized the method of inscribing equilateral **polygons** inside a **circle**. Then, upon increasing the number of their sides, the areas of the polygons (which Archimedes could calculate) approached the area of the circle as a limit. Using this result together with a similar idea involving circumscribed polygons, Archimedes was able to find the area of the circle as "A = πr²," in which "r" is the radius of the circle and "pi" is a constant that Archimedes calculated to possess a value between 3-1/7 and 3-10/71. Then, in 1635 **Bonaventura Cavalieri**, a professor of mathematics at the University of Bologna, formulated a systematic method for the determination of areas. His method involved complicated geometric considerations. Later, **Isaac Barrow**, the Lucasian professor of mathematics at Cambridge, published in 1670 a treatise that helped to unify formulations of areas within the mathematical branches of **analytic geometry** and integral **calculus**. Barrow used the traditional approach of geometry to calculate areas and was thus prevented from taking the final step to calculate areas with the use of **integral calculus**, as was later developed by Sir **Isaac Newton** of England and **Gottfried Wilhelm Leibniz** of Germany.

The areas of irregular figures, both of planar figures and the surfaces of solids, can be computed by the use of calculus. In two dimensions, this mathematical tool involves the computation of the area bounded between a curve having the equation "y = $f(x)$," the horizontal x-axis, and the distance between the vertical lines "x=a" and "x=b". This type of computation is commonly called the limit between the points "a" and "b", where a<b. The area of an irregularly shaped figure can be found by subdividing it into rectangles of equal width. If the number of rectangles is made larger and larger (therefore, their bases [or widths] become smaller and smaller), the sum of their areas (found by multiplying base by height) approaches the required area as a limit. By using this general method, integral calculus provides a systematic way for obtaining an exact calculation of many areas of irregular figures. The notation used in integral calculus to indicate total area of a region by summing the areas of these discrete rectangles is of the form:

$$\int_a^b f(x)\ dx.$$

This expression calculates the surface area of the region bounded by the continuous curve described by the function $f(x)$ and the intervals "x = a" to "x = b" (where a<b), and the x-axis. Where the integral of a function (of the form just described) cannot be "analytically" (i.e., explicitly) solved, numerical approximations can be found using recursive (or algorithmic) methods. The digital computer has, of course, found great application to such problems.

The system of measurement most commonly used by scientists and engineers throughout the world is the System International (SI), commonly referred to as the "metric system". Reference units of length, mass, charge, etc. are precisely defined within this system. The standard unit of length is the meter, from which the basic unit of area in the **metric system** is derived; namely, the square meter (denoted m²). The square meter is defined as the area of a square whose sides are one meter long. Larger or smaller units of area can be formulated from the square meter. Two examples are the square kilometer, 1 km², which equals (or is composed of) 10^6 m² and the square centimeter, 1 cm², which equals 10^{-4} m².

See also Analytic geometry; Archimedes of Syracuse; Isaac Barrow; Bonaventura Cavalieri; Eudoxus of Cnidus; Integral calculus; Limits; Metric system; Polygon

ARGAND, JEAN ROBERT (1768-1822)

Swiss number theorist and geometer

Jean Robert Argand invented a method of geometrically representing **complex numbers** and their operations. The Argand diagramis a graphic representation of complex numbers as points on a plane and their additions. He is also credited with giving **proof**, although with a few gaps, of the fundamental **theorem** of **algebra**. In 1814, Argand published a proof of the **fundamental theorem of algebra**, which may be the simplest of all proofs of this theorem.

Argand was born in Geneva, Switzerland, on July 18, 1768, to Jacques Argand and Eves Canac. Historians have limited knowledge of Argand's background and nothing is known of his education. Apparently, he was a self–taught mathematician, belonging to no mathematical societies or organizations. Mathematics appeared to be just a hobby to Argand. He has often been confused with Aime Argand, the physicist and chemist who had invented the Argand lamp, however, they are not related.

In 1806, Argand, his wife and children moved to Paris. He was working as an accountant when he published his method in a book entitled *Essai sur une maiere de representer les quantities imaginaires*. The book had been published in a small, privately printed edition but it did not include Argand's name. No one knew he wrote the book until sometime later.

How it came to light that Argand was the author is an unusual story. He may never have been credited with writing the book, had a set of curious circumstances not occurred. Around the time the book was published, two other mathematicians, **Casper Wessel**, a Norwegian and **Karl Gauss**, from Germany, were working on the same idea. However, neither had put their ideas into print, and Gauss is sometimes credited with writing the *Essai*.

Argand had spoken of his new method to **Adrien M. Legendre** before the book was published. Legendre spoke of the method in a letter to the brother of J.R. Français. Français was a lecturer at the Imperial College of Artillery at Paris. Français found the letter after his brother's death. In an essay published in the journal *Annales de Mathematiques*, Français discussed the idea of the new method and even developed it further. At the end of the piece, Français called for the author of the book to come forward and be recognized. Argand acknowledged his works by writing an article that was published in a later edition of the same journal.

Argand died in Paris on August 13, 1822, having contributed nothing more to the science of mathematics.

ARISTARCHUS OF SAMOS (CA. 320 B.C.-CA. 250 B.C.)

Greek geometer and astronomer

As with many of his contemporaries, the only extant facts about the life of Aristarchus involve remarks about him and his work written by others. Only one of Aristarchus' writings survived, *On the Magnitudes and Distances of the Sun and Moon,* but in it he articulated the reasoning behind what later became modern **trigonometry**, and how it might be employed in astronomy and navigation. Typical of Greek mathematics was Aristarchus' primarily geometric method of approximating this strategy of triangulation. From **Archimedes**, it is known that Aristarchus had proposed the sun be considered a fixed star, with the Earth circulating around it. This view was ridiculed at the time and remained dormant until **Nicolaus Copernicus** devised his heliocentric theory.

Birth and death dates for Aristarchus vary, but it is agreed that he was born on the island of Samos in Greece and studied under Strato (or Straton) of Lampsacus in Alexandria, Egypt. Strato went on to succeed Theophrastus as head of the lyceum founded by **Aristotle** at Athens, so it is probable that Aristarchus circulated among highly intellectual and influential men. Certain of his activities can be roughly dated, as he made observations of the summer solstice around 281–280 B.C., according to Ptolemy.

Aristarchus favored a mathematical approach to astronomy over the descriptive one, which tended to rely on intuition and rhetoric rather than observation. An example of his commitment to observation is his reported correction of Callippus' estimate of the length of the year, adding 1/1623 of a day to it. His own observations led to six astronomical hypotheses, from which Aristarchus drew eighteen **propositions**. These regarded measuring the sizes and distances of various celestial objects relative to the known **diameter** of the Earth. He correctly concluded that the **orbit** of the Earth was dwarfed by the overall size of the universe and **distance** of its furthermost visible stars. Archimedesproved this immensity was calculable with his famous "sand reckoning," counted at the time in myriad–myriads.

By studying the relative positions of the sun, moon, and Earth, Aristarchus concluded that during the half–moon each of them occupy respective points on a right **triangle**. He then reasoned that the Pythagorean theoremcould be applied to determine the ratio of the sun–Earth distance and the moon–Earth distance. In fact, his **proof** of this is best expressed today as a trigonometric formula.

Because Aristarchus did not have the tools to measure angular distances of heavenly bodies, he consequently underestimated these distances. Likewise, his estimate of the size of the moon relative to the Earth, and the size of the sun relative to the moon were inaccurate as well. Those figures were improved during the next century by **Hipparchus**, though it is only later that we learned Aristarchus underestimated the sun's size by nearly 400 times his original estimation.

Unfortunately, the idealistic Greek **model** called for circular orbits, which did not account for the unevenly distributed changing of seasons. These are now attributable to elliptical orbits. While Aristarchus could not completely free himself of the Greek intellectual loyalty to mathematical harmonies, he went a long way towards letting experiment rule theory rather than idealism.

It is from the writings of Archimedes and Plutarch that Aristarchus' heliocentric hypothesis of 260 B.C. became known. As articulated by Aristarchus, the hypothesis accounted for the apparent **motion** of the heavenly bodies and diurnal motion of the stars. He not only proposed that the sun is fixed and that the Earth revolves around it, but also that the Earth rotates on its own axis. Aristarchus was roundly criticized—his contemporaries marshaled **Aristotelian logic** to refute his premise as untenable—although he was apparently never persecuted.

Aristarchus died at the earliest estimate around 250 B.C. in Alexandria. Debunked in his own time, his contributions as a scientist and mathematician have since been reevaluated. He may have also been an inventor, making an improved design of sundial called a "skaphe" or "scaphion," which seems to have placed the shadow–throwing pointer within a hollow hemispherical base.

Aristarchus was not acknowledged by Copernicus himself, having struck out a passage referring to his distant precursor during the editing of his manuscript *De revoluntionibus orbium coelestium*. Aristarchus' most ambitious ideas could not be confirmed or denied until the time of **Isaac Newton**, when it became possible to test the effects of the **rotation** of the Earth and the phenomenon of stellar aberration.

ARISTOTELIAN LOGIC

Aristotle (384-322 B.C.), a student of **Plato** and tutor of Alexander the Great, invented a system of logic which remained essentially unchanged in Western European philosophy for more than two millennia following his death. Although his predecessors, Socrates and Plato, had placed great emphasis on correct reasoning, Aristotle was the first philosopher to set forth a carefully worked out system, which, he believed, if followed, could eliminate false reasoning. Some have suggested that Aristotle created the "science" of logic, but, to Aristotle, logic was not simply one of the sciences; rather it was the art and method of correct reasoning, a prerequisite to the study of any science. Followers of Aristotle gave the collection of his writings on logic the name *Organon*, which means "tool." Logic, for Aristotle, was the principal "tool" for reasoning correctly.

At the heart of Aristotle's logic is the syllogism. Perhaps the most famous syllogism is the following: "All men are mortal. Socrates is a man. Therefore, Socrates is mortal." Aristotle introduced variables into logic to emphasize the general form of the syllogism. Thus the above syllogism has the form "All A is B. Some C is A. Therefore, some C is B." In this form, "All A is B" is called the major premise; "Some C is A" is called the minor premise; and "Some C is B" is called the conclusion. Note the use of the word "some" in this syllogistic form distinguishes it from the form "All A is B. All C is A. Therefore, all C is B." This points out the difference between what Aristotle called "particulars" and "universals." So in the first syllogism above, "Socrates" is the name of a particular member of a class, say, Greeks. This syllogism "proves" only that one Greek, namely Socrates, is mortal; whereas, the syllogism "All men are mortal. All Greeks are men. Therefore, all Greeks are mortal," would be a syllogism purporting to "prove" that all Greeks, not just Socrates, are mortal. From a logical point of view, it does nothing of the sort, because the minor premise is false. Approximately half of all Greeks are not men. (Aristotle would not have made this objection because his use of "man" or "men" was meant to include women.) Using these and several other syllogistic forms, Aristotle thought that all deductive **inference** could ultimately be stated in terms of syllogisms. Thus he believed that the path to infallible reasoning consisted in stating arguments in one or more of the syllogistic forms and proceed from there according to the laws of syllogistic reasoning. Aristotle's development of the syllogistic method of argument formed the basis for the teaching of formal logic until the beginning of the twentieth century when the mathematical logicians **Giuseppi Peano** (1858-1932), **Gottlob Frege** (1848-1925), **Bertrand Russell** (1872-1970), and **Kurt Gödel** (1906-1978) discovered numerous shortcomings in the Aristotelian system. The work of these men and others led to the establishment of modern **symbolic logic** as the cornerstone of mathematical **deduction**.

Russell, in particular, criticized Aristotle's overemphasis on the syllogism as the fundamental form of deductive argument, pointing out that deductive proofs in mathematics are rarely given in syllogistic form. Russell also noted that Aristotle was often not careful about the distinction between particulars and universals in the statement of syllogisms, a problem symbolic logic addresses through the use of existential ("There exists an x") and universal ("For all x") quantifiers. Thus "All men are mortal" could be written as "For all x in the class 'men,' x is mortal," whereas "Socrates is mortal" could be expressed as "There exists an x in the class 'men' such that x is mortal and x is Socrates." Now it may seem that such strange statements complicate language rather than simplify it, but it turns out that introducing such innovations as quantifiers and classes was necessary to eliminate some troubling paradoxes in logical reasoning. Another modern criticism of Aristotle's logic is that it gave too much prominence to deduction and not enough to **induction**, although it should be noted that compared to his mentor, Plato, Aristotle did give strong credence to induction as a method of arriving at the "probable" **truth** of a premise such as "All men are mortal."

Although modern logicians may see Aristotle's largely syllogistic logic as merely a primitive beginning to their field, they cannot deny the authoritative position that Aristotelian logic held for more than two thousand years. It is difficult to overstate the enormous influence Aristotle's logic had on the generations of philosophers, mathematicians, scientists, and educators who came after him.

ARISTOTLE (384 B.C.-322 B.C.)
Greek philosopher

As a formidable student, researcher, teacher, and philosopher in virtually all scientific disciplines, Aristotle had a profound impact on the way science and mathematics is practiced and investigated today. His analytical method, now known as **Aristotelian logic**, is the backbone of not only mathematics, but of all the natural sciences.

Aristotle was born in 384 B.C. in Stagirus, Greece. His father Nicomachus, a doctor, was appointed as the King of Macedonia's personal physician while Aristotle was a child. Nicomachus passed away around 374 B.C., leaving Aristotle to be raised and educated by a guardian, Proxenus of Atarneus.

At age 17, Aristotle moved to Athens where he enrolled in Plato's Academy. Here he excelled in the study of rhetoric and dialectic, which he also taught when he became an instructor there at the end of his studies. Aristotle would spend twenty years at the Academy before leaving for the city of Assos. The exact reason for Aristotle's departure is unknown, but it was speculated that political tensions between Athens and neighboring Macedonia may have sparked the move. **Plato** had also just died, and Aristotle was known to have conflicting views with his successor at the Academy, Speusippus (Plato's nephew).

While in Assos, Aristotle continued his philosophical and scientific pursuits in the company of other philosophers, studying anatomy, zoology, and the biological sciences. Aristotle also met and married his wife, Pythias, the niece of Assos' ruler, Hermias. Hermias gave him a daughter, but passed away several years later at an early age. Aristotle stayed in Assos until Persia attacked the city, when he then left for the neighboring island of Lesbos. After a year-long stay on Lesbos, he moved to Macedonia in 343 B.C., where he was well-received by King Philip, the son of King Amyntas who had employed Aristotle's father as royal physician many years before. Three years later, after Aristotle was passed over to succeed Speusippus as director of the Academy, he returned to his birthplace, Stagirus, to continue his studies. While in Stagirus, Aristotle had a relationship with Herpyllis, who gave him his second child, a son named Nicomachus.

Aristotle returned to Athens to establish a new school, the Lyceum, in 335 B.C. The Lyceum would become a rival institution to the Academy, and instructed students in a wider range of disciplines than the Academy traditionally had. The natural sciences, logic, physics, astronomy, zoology, metaphysics, theology, politics, economics, ethics, rhetoric, poetics, mathematics, and other subjects, were all part of the curriculum of the Lyceum. It was from Aristotle's work at the Lyceum that we have much of the existing documentation of his work today, for his lecture notes were preserved for many years after his death, in some cases to teach from, and many of his notes were eventually published.

Although Aristotle's most significant contributions were in the areas of philosophy, physics, and biological sciences, as one of the leading thinkers and teachers of his day and a man who researched and taught in virtually all known scientific and humanistic disciplines, he also had a recognized impact on modern mathematics. Many of the mathematical works of Aristotle and his students have been lost, such as a biography of Pythagorus he is said to have written, and a historical account of **geometry** authored by one of his students. However, some works still exist, such as Aristotle's *Physics*, which contains a discussion of the infinite that Aristotle believed existed in theory only, and sparked much debate in later centuries. Aristotle is also thought to have authored a treatise entitled *On Indivisible Lines*, which disputed Xenocrates' claims of a doctrine of indivisibles. Aristotle also advanced the study of mathematics by discussing and recording many mathematical concepts and theorems of his contemporaries and their predecessors. However, his biggest contribution to the field of mathematics was his development of the study of logic, which Aristotle termed "analytics," as the basis for mathematical study.

Aristotle believed that analytical methods should be applied to every branch of science and learning, including mathematics. Analytical methods were necessary to develop the axioms, or unshakable rules, of these disciplines from which all further discovery would take place. Aristotle wrote extensively on this concept in his work *Prior Analytics*, which was published from Lyceum lecture notes several hundred years after his death. It was this work on **scientific method** that laid the foundation for the development of a mathematical discipline based on **syllogisms and proofs**.

Alexander, the son of Philip who had succeeded his father on the Macedonian throne and had assisted Aristotle in his establishment of the Lyceum, died in 323 B.C. Soon after, Aristotle retired amidst renewed anti-Macedonian sentiment in Athens. He returned to Stagirus and died a year later at age 62.

See also Aristotelian logic; Scientific method

ARITHMETIC

Arithmetic is a branch of mathematics concerned with the numerical manipulation of numbers using the operations of **addition**, **subtraction**, **multiplication**, **division**, and the extraction of **roots**. General arithmetic principles slowly developed over time from the principle of counting objects. Critical to the advancement of arithmetic was the development of a positional number system and a symbol to represent the quantity **zero**. All arithmetic knowledge is derived from the primary axioms of addition and multiplication. These axioms describe the rules which apply to all **real numbers**, including **whole numbers**, **integers**, and **rational** and **irrational numbers**.

Early development of arithmetic

Arithmetic developed slowly over the course of human history, primarily evolving from the operation of counting. Prior to 4000 B.C., few civilizations were even able to count up to ten. Over time however, people learned to associate objects with numbers. They also learned to think about numbers as abstract ideas. They recognized that four trees and four cows had a common quantity called four. The best evidence suggests that the ancient Sumerians of Mesopotamia were the first civilization to develop a respectable method of dealing with numbers. By far the most mathematically advanced of these ancient civilizations were the Egyptians, Babylonians, Indians, and Chinese. Each of these civilizations possessed whole numbers, **fractions**, and basic rules of arithmetic. They used arithmetic to solve specific problems in areas such as trade and commerce. As impressive as the knowledge that these civilizations developed was, they still did not develop a theoretical system of arithmetic.

The first significant advances in the subject of arithmetic were made by the ancient Greeks during the third century B.C. Most importantly, they realized that a sequence of numbers could be extended infinitely. They also learned to develop theorems which could be generally applied to all numbers. At this time, arithmetic was transformed from a tool of commerce to a general theory of numbers.

Numbering system

Our numbering system is of central importance in the subject of arithmetic. The system we use today called the Hindu-Arabic system, was developed by the Hindu civilization of India some 1500 years ago. It was brought to Europe during the middle ages by the Arabs and fully replaced the Roman numeral system during the 17th century.

The Hindu-Arabic system is called a decimal system because it is based on the number 10. This means that it uses 10 distinct symbols to represent numbers. The fact that 10 is used is not important because it could have just as easily been based on another number of symbols like 14. An important feature of our system is that it is a positional system. This means that the number 532 is different from the number 325 or 253. Critical to the invention of a positional system is perhaps the most significant feature of our system: a symbol for zero. Note that zero is a number just as any other and we can perform arithmetic operations with it.

Axioms of the operations of arithmetic

Arithmetic is the study of mathematics related to the manipulation of real numbers. The two fundamental properties of arithmetic are addition and multiplication. When two numbers are added together, the resulting number is called a sum. For example, 6 is the sum of 4 + 2. Similarly, when two numbers are multiplied, the resulting number is called the product. Both of these operations have a related inverse operation which reverses or "undoes" its action. The inverse operation of

addition is subtraction. The result obtained by subtracting two numbers is known as the difference. Division is the inverse operation of multiplication and results in a quotient when two numbers are divided. The operations of arithmetic on real numbers are subject to a number of basic rules, called axioms. These include axioms of addition, multiplication, distributivity, and order. For simplicity, note that the letters a, b, c, denote real numbers in all of the following axioms.

There are three axioms related to the operation of addition. The first, called the commutative law, is denoted by the equation a + b = b + a. This means that the order in which you add two numbers does not change the end result. For example, 2 + 4 and 4 + 2 both mean the same thing. The next is the associative law which is written a + (b + c) = (a + b) + c. This **axiom** suggests that grouping numbers also does not effect the sum. The third axiom of addition is the **closure property** which states that the equation a + b is a real number.

From the axioms of addition, two other properties can be derived. One is the additive **identity property** which says that for any real number a + 0 = a. The other is the additive inverse property which suggests that for every number a, there is a number -a such that -a + a = 0.

Like addition, the operation of multiplication has three axioms related to it. There is the commutative law of multiplication stated by the equation a × b = b × a. There is also an associative law of multiplication denoted by a × (b × c) = (a × b) × c. And finally, there is the closure property of multiplication which states that a × b is a real number. Another axiom related to both addition and multiplication is the axiom of distributivity represented by the equation (a + b) × c = a × c + b × c.

The axioms of multiplication also suggest two more properties. These include the multiplicative identity property which says for any real number a, 1 × a = a, and the multi-

plicative inverse property that states for every real number there exits a unique number (1/a) such that (1/a) × a = 1.

The axioms related to the operations of addition and multiplication indicate that real numbers form an algebraic **field**. Four additional axioms assert that within the set of real numbers there is an order. One states that for any two real numbers, one and only one of the following relations is true: either a < b, a > b or a = b. Another suggests that if a < b, and b < c, then a < c. The monotonic property of addition states that if a < b, the a + c < b + c. Finally, the monotonic property of multiplication states that if a < b and c > 0, then a × c < b × c.

Numbers and their properties

These axioms apply to all real numbers. It is important to note that a real numbers is the general class of all numbers which include whole numbers, integers, rational numbers and irrational numbers. For each of these number types only certain axioms apply.

Whole numbers, also called natural numbers, include only numbers that are positive integers and zero. These numbers are typically the first ones to which a person is introduced, and they are used extensively for counting objects. Addition of whole numbers involves combining them to get a sum. Whole number multiplication is just a method of repeated addition. For example, 2 × 4 is the same as 2 + 2 + 2 + 2. Since whole numbers do not involve **negative numbers** or fractions, the two inverse properties do not apply. The smallest whole number is zero but there is no limit to the size of the largest.

Integers are whole numbers which also include negative numbers. For these numbers the inverse property of addition does apply. For these numbers, zero is not the smallest number but it is the middle number with an infinite number of positive and negative integers existing before and after it. Integers are used to measure values which can increase or decrease such as the amount of money in a cash register. The standard rules for addition are followed when two positive or two negative numbers are added together and the sign stays the same. When a positive integer is added to a negative integer, the numbers are subtracted and the appropriate sign is applied. Using the axioms of multiplication it can be shown that when two negative integers are multiplied, the result is a positive number. Also, when a positive and negative are multiplied, a negative number is obtained.

Numbers to which both inverse properties apply are called rational numbers. Rational numbers are numbers which can be expressed as a ratio of two integers, for example, 1/2. In this example, the number 1 is called the numerator and the 2 is called the denominator. Though rational numbers represent more numbers than whole numbers or integers, they do not represent all numbers. Another type of number exists called an irrational number which cannot be represented as the ratio of two integers. Examples of these types of numbers include **square** roots of numbers which are not perfect squares and cube roots of numbers which are not perfect cubes. Also, numbers such as the universal constants π and e are irrational numbers.

The principles of arithmetic create the foundations for all other branches of mathematics. They also represent the most practical application of mathematics in everyday life. From determining the change received from a purchase to calculating the amount of sugar in a batch of cookies, learning arithmetic skills is extremely important.

ARITHMETIC SERIES

A sequence of numbers is said to be **arithmetic** if the difference between any two successive terms is the same. For example the sequence 1,3,5,7,9,... is arithmetic because the difference between any two consecutive terms is 2. This difference is usually called the common difference. Thus the common difference in the arithmetic sequence 2,6,10,14,... is 4. If the terms of an arithmetic sequence are added together, this sum is called an arithmetic series. So 1+3+5+7+... and 2+6+10+14+... are arithmetic series. Arithmetic series may be infinite or finite. The finite arithmetic series 1+3+5+7+9 has a sum of 25. There is a formula for computing the sum of a finite arithmetic series which is useful when the number of terms is large. The young **Carl Freidrich Gauss** (1777-1855), who would later become one of the greatest mathematicians of all time, is said to have discovered this formula when his elementary school teacher assigned the class the task of adding up all the **whole numbers** from 1 to 100. While the other students began the drudgery of adding the numbers one by one, Gauss suddenly realized that there was a simpler way. The sum 1+2+3+4+...+97+98+99+100 is an arithmetic series with common difference 1. Gauss noticed that the sum of 1+100=101, 2+99=101, 3+98=101, 4+97=101, and so on through 50+51=101. So, he reasoned, there are 50 pairs of numbers each of whose sum is 101; therefore, the sum of this series is 101x50=5050. Gauss, of course, finished before the other students and was the only one to get the correct answer. Notice that 101 is the sum of the first and last terms of the series and 50 is the number of pairs or 100/2. So if 101x50 is written as 101x100/2 it can be seen as a special case of the formula (first term + last term)x(the number of terms) divided by 2. In general, if the nth term of the series is designated by a_n, then $a_1 1+a_2+a_3+...+a_n=(n(a_1+a_n))/2$. If we try this out on 1+3+5+7+9, we get $5(1+9)/2=25$, which agrees with our answer from summing the terms in the usual way.

ARYABHATA THE ELDER (476-550)

Hindu mathematical astronomer

In a time and place where people believed certain distant stars, called "asuras," possessed malevolent powers capable of inflicting harm on Earth, Aryabhata the Elder took the first steps towards separating scientific explication from folklore and superstition. His *Aryabhatiya* was the first major book on Hindu mathematics, which summarized knowledge of his predecessors. While covering many aspects of **arithmetic**, **algebra**, and numerical notation, the majority of *The Aryabhatiya*

dealt with **trigonometric tables** and formulae for use in astronomy. Any astronomical observations Aryabhata made were most likely completely unaided. Although *The Aryabhatiya* contained errors, it was translated and reproduced as *Zij al–Arjabhar* by the Arabs. One of Aryabhata's methods, a solution to the indeterminate quadratic xy = ax + by + c, was rediscovered by **Leonhard Euler** in the 18th century.

Aryabhata the Elder was born near what is now the city of Patna in India. His year of birth is sometimes cited as 475 but also possibly 476. Aryabhata's hometown was called Kusumapura, or The City of Flowers. The two major centers in the area represented the two intellectual threads he contended with in his lifetime. At that time, Patna was a royal seat, and according to legend it was founded by a knight with magic powers in honor of his princess, and blessed by the Buddha. Further away in Ujjian, the study of science and astronomy began to flourish, and this knowledge was being disseminated in the form of rhymed, romantic stories. Various mathematical problems, similar to those used in textbooks today, were solved in verse form used for social amusements. The public challenge and the romantic forms are combined in the most famous quotation from *The Aryabhatiya*, as Aryabhata the Elder commands a "beautiful maiden" to answer a problem that requires inversion.

The Aryabhatiya was produced in the year 499. It described the Indian numerical system with nine symbols, and listed various rules for arithmetic and trigonometric calculations. It also made use of continued **fractions**, **square** and cube **roots**, and the **sine** function when needed. Solutions were given for linear and quadratic **equations** and **diophantine equations** of the first degree. These involved one of the first recorded uses of algebra and decimal place–value. Unlike the Greeks, the Hindus solved diophantines for all possible integral solutions, as they were more tolerant of negative, irrational, and other such numbers. For instance, one value for π given by Aryabhata is the **square root** of 10, generally called "the Hindu value."

The combination of correct and incorrect answers to the major questions of its time led one Arabic commentator, al–Biruni, to describe Hindu mathematics as a mixture of "common pebbles and costly crystals." Aryabhata gave an accurate approximation of π, although he overestimated the length of the year by 12 minutes and 30 seconds. He was singular in describing the orbits of the planets as ellipses. In the *Ganita*, a poem composed in 33 couplets, he correctly states the formulae for areas of a **triangle** and a **circle**. However, when Aryabhata attempts to extrapolate those to figure the volumes of three–dimensional shapes in the same couplets, he is not successful. Nonetheless, Aryabhata the Elder's commitment to general methods caused him to apply what is now nicknamed "the pulverizer." This rule finds the greatest common divisor of *a* and *b* by **division**, equivalent to Euler's later version of reducing *a* over *b* to a continued fraction. While "the pulverizer" has also been known as the Diophantine method, the fact that Diophantus himself never used it renders the term a misnomer.

Aryabhata died around 550, though it is not known where. As an astronomer he argued—against Vedic tradition—that the Earth was round and rotated daily. He correctly explained why equinoxes, solstices, and eclipses occur. These ideas were not accepted in Aryabhata's lifetime, but his mathematics had set the foundation for developments in the Eastern and Western worlds for centuries to come. In India particularly, Aryabhata the Elder marked the end of the sacred or "S'ulvasutra" period, during which mathematics was used primarily by priests for temple architecture. He ushered in the "astronomical period" that lasted until the year 1200.

The Aryabhatiya held the same stature in India that **Euclid**'s *Elements* did in ancient Greece. **Bhaskara** wrote a commentary on this work in 629. Aryabhata also influenced the work of Hindu astronomer **Brahmagupta**.

ASSOCIATIVE PROPERTY

In **algebra**, a binary operation is a rule for combining the elements of a set two at a time. In most important examples that combination is also another member of the same set. **Addition**, **subtraction**, **multiplication**, and **division** are familiar binary operations. A familiar example of a binary operation that is associative (obeys the associative principle) is addition (+) of **real numbers**. For example, the sum of 10, 2, and 35 is determined equally as well as $(10 + 2) + 35 = 12 + 35 = 47$, or $10 + (2 + 35) = 10 + 37 = 47$. The parentheses on either side of the defining equation indicate which two elements are to be combined first. Thus, the associative property states that combining a with b first, and then combining the result with c, is equivalent to combining b with c first, and then combining a with that result. A binary operation (*) defined on a set S obeys the associative property if $(a * b) * c = a * (b * c)$, for any three elements a, b, and c in S. Multiplication of real numbers is another associative operation, for example, $(5 \times 2) \times 3 = 10 \times 3 = 30$, and $5 \times (2 \times 3) = 5 \times 6 = 30$. However, not all binary operations are associative. Subtraction of real numbers is not associative since in general $(a - b) - c$ not equal $a - (b - c)$, for example $(35 - 2) - 6 = 33 - 6 = 27$, while $35 - (2 - 6) = 35 - (-4) = 39$. Division of real numbers is not associative either. When the associative property holds for all the members of a set, every combination of elements must result in another element of the same set.

ASYMPTOTE

An asymptote is a line a function will approach. The simplest varieties are either horizontal or vertical. A vertical asymptote occurs when the function at a certain value approaches **infinity** whether one comes at it from the positive or negative side. A horizontal asymptote occurs when the function approaches a certain value as it gets closer and closer to positive or negative infinity. The behavior of the function may oscillate around the asymptote, with gradually damping oscillations as it gets closer to infinity, or it may simply monotonically increase or decrease, depending on what type of function it is.

Functions which have a denominator of order one less than their numerator may have an oblique asymptote. An oblique asymptote is neither horizontal nor vertical, but rather is the portion of the function which approximates a straight line, appearing diagonal on the usual Cartesian axis. It clearly demonstrates the dominant terms in the function, those which determine the function's behavior far from the origin.

Asymptotic behavior can be useful in graphing a function. Its asymptotic behavior near **zero** and as the function approaches infinity can be combined to look very much like an exact graph of the function. Asymptotes can also be used in determining what approximations of a function may be appropriate. If the behavior of a function at a chosen limit is known, some series approximations may be ruled out for failure to match that behavior. While the asymptotic behavior of a function is never exact, it can nevertheless provide useful insight into how that function may be dealt with.

ATIYAH, MICHAEL FRANCIS (1929-)
English topologist

Michael Francis Atiyah has had a remarkably long and productive career that is based in **topology** but encompasses such diverse fields as **algebraic geometry**, **differential equations**, and theoretical physics. In recognition of three major theorems he developed during the first decade of his career, he was awarded the Fields Medal in 1966. Atiyah is also one of the pioneers of string theory, a new way of looking at the structure of matter. Claude LeBrun wrote in *American Scientist* that Atiyah has "played the role of the great unifier for a swath of mathematical subfields that had been developing into autonomous petty fiefdoms, oblivious to the outside world." He has also served in various capacities to help guide the scientific policy of the British government and has been an articulate spokesperson for the importance of theoretical research.

Atiyah was born in London on April 22, 1929. His father was Edward Selim Atiyah, originally from Lebanon, and his mother was the former Jean Levens, an English citizen. Atiyah was educated in Egypt, where his father was a broadcaster and commentator for the British Broadcasting Company. After completing his secondary education at Victoria College in Egypt, Atiyah enrolled at Manchester Grammar School, a preparatory school in England. After a year, he was accepted as a mathematics student at Trinity College, Cambridge.

During his second undergraduate year, Atiyah published his first paper, which dealt with higher-dimensional projective **geometry**. He earned his doctorate from Trinity in 1955, the same year he married Lily Brown, who would be the mother of his three sons. His first postdoctoral position was a fellowship at the Institute for Advanced Studies (IAS) in Princeton, New Jersey. He later told Glyn Jones in an interview for *New Scientist* that he went to the IAS (which had been founded by **Albert Einstein** and **John von Neumann**) to "get new ideas, meet new people, and open up new avenues." Over the next two decades, Atiyah held a variety of teaching and research positions on both sides of the Atlantic, at Cambridge University, Harvard University, Oxford University and the IAS. In 1972, he returned to Oxford, where he stayed as a Royal Society Research Professor for 18 years. In 1990, he was elected president of what is arguably the most prestigious scientific association in the world, the Royal Society of London. That same year, he was chosen for another post of significant influence: mastership of Trinity College, a 400-year-old institution that developed its mathematical reputation through such scholars as **Isaac Newton, Bertrand Russell, Alfred North Whitehead** and **Godfrey Hardy**.

Atiyah's major field of interest in mathematics has been topology, which is the study of properties do not change under continuous deformation (stretching and bending). It is sometimes referred as "rubber-sheet geometry." Because the principles of topology apply to such a wide variety of conditions, it has evolved as a fundamental field that unifies many other seemingly unrelated fields of mathematics. Topology has evolved from the geometrical side of mathematics. Atiyah titled his presidential address to the Mathematical Association "What Is Geometry?" In it, he said that "geometry is that part of mathematics in which visual thought is dominant whereas **algebra** is that part in which sequential thought is dominant.... Geometry is not so much a branch of mathematics as a way of thinking that permeates all branches."

The discovery of three theorems during his first decade of research was cited as the basis for Atiyah's 1966 Fields Medal. The first of these accomplishments was the development of a purely topological version of K-theory, which he derived with Friedrich Hirzebruch. The topic concerns systems of linear **equations** that depend continuously on auxiliary parameters. Such systems arise naturally in many different contexts, and K-theory has become a powerful and useful tool. Topological spaces are often characterized by the number of holes they exhibit; that number is **invariant** (it does not change when the object is deformed in a continuous way). K-theory revealed was a new topologically invariant quantity. Upon receiving the Feltrinelli Prize in 1984, Atiyah told the Accademia Nazionale dei Lincei that "[K-theory] led easily to the solution of many difficult problems, most notably that of the vector-field problem on spheres. This geometric problem, simple to state but difficult to prove, had long been regarded as a test case for new techniques, and it was finally solved by Frank Adams in 1962 using K-theory."

With the collaboration of Isadore Singer, Atiyah used K-theory to develop what is now known as the Atiyah-Singer Index Theorem. Elliptic differential equations describe various physical situations, but they are difficult to solve. In fact, it is even hard to tell how many independent solutions such an equation will have. However, if a pair of elliptic equations have n and m independent solutions, respectively, the index theorem provides a formula for calculating n-m (called the index of the pair). It turns out that the index can be used by physicists to determine the difference between the number of right-handed and left-handed particles in a system.

The index theorem led to another major result, which Atiyah developed with Raoul Bott. Fixed-point theorems deal with the number of points of a topological object that remain unchanged under a certain type of **transformation**, and they have many practical applications. Atiyah and Bott found a way to calculate the number of fixed points that exist under a transformation that preserves an elliptic system.

Beginning in the 1970s, Atiyah became interested in a new field of research known as string theory. Physicists have traditionally constructed theories about the nature of matter based on the assumption that the fundamental particles of matter can be thought of as discrete, dimensionless points. String theory adopts a radically new assumption that the fundamental units of matter do have a **dimension** (length) and can be thought of as stringlike objects. This approach to the study of matter has evolved out of mathematical theories than out of experimental observations. As a result, string theory tends to be both more complex and less easily interpreted in physical terms than traditional theories of matter. For example, one consequence of string theory is that matter has to be thought of in terms of many (often, more than a dozen) dimensions. Many scientists find it unnatural and difficult to speak about objects in, say, 14 dimensions. It is hardly surprising, then, that string theory has been received with something less than enthusiasm by many physicists, although it has become an influential and fruitful theory.

For his work in topology and string theory Atiyah has garnered a number of awards and honors. In addition to the 1966 Fields Medal, the highest honor in mathematics, he has received the Royal Medal and the Copley Medal of the Royal Society, the De Morgan Medal of the London Mathematical Society, the Feltrinelli Prize of the Accademia Nazionale dei Lincei, and the King Faisal International Prize for Science. He has receeived honorary doctorates from 20 universities, including those of Bonn, Dublin, Chicago, Helsinki, Rutgers, and Montreal. Atiyah was knighted by Queen Elizabeth II in 1983.

In an effort to make concise resources available to young mathematicians (particularly in China), in 1985 **Shiing-shen Chern** convinced Atiyah to compile and publish his collected papers. Atiyah commented in the preface on the appropriateness of publishing such a compilation prior to the author's death. With characteristic humor, he wrote, "there are several clear advantages to all parties: posterity is saved the trouble of undertaking the collection, while the author can add some personal touches in the way of a commentary. There are also disadvantages: the commentary will be biased, and the author may feel that he is being pensioned off." On the contrary, Atiyah has continued to generate important material in the decade since the collection was published.

Besides being a prolific researcher, Atiyah is a highly respected teacher. Even at times of his career when he does not conduct classes, he educates through the clear exposition of his books and articles. In his *Collected Works*, he wrote that he has often been asked to speak to a broad range of audiences: "In some cases I was talking to professional mathematics but in others I was almost the only mathematician in the room.

Giving such general talks... requires much greater thought on the material and presentation than for a normal seminar, but it is a worthwhile and important activity." It is one that Atiyah performs remarkably well, in print as well as in person. Two different reviewers described Atiyah's 1990 book *The Geometry and Physics of Knots* as "poetic," and his article "Geometry and Physics" is a very readable survey of highly theoretical topics.

For many years, Atiyah has also contributed to professional organizations in mathematics and science. He served as president of the London Mathematical Society and the Mathematical Association. From 1984 to 1989 he was a member of the British government's Science and Engineering Research Council. In 1991, Atiyah established the Isaac Newton Institute for Mathematical Studies in Cambridge, where researchers can work for six-months terms on their chosen topics.

AXIOM

Every mathematical theory is founded upon a set of axioms. Axioms are the basic **propositions** of a theory and are used as starting points. They are simply assumed to be true; all other true statements must be derived from them. By starting with these basic assumptions it is possible to construct a coherent, consistent, and unambiguous theory.

It helps to compare a mathematical theory to a house. The axioms are standardized building blocks; rules of **inference** connect the blocks to each other—stacking, gluing, nailing, etc. Every part of the house—the windows, the door, the roof—is constructed out of the same basic blocks, and held together by the same methods of connection.

Mathematical arguments usually have a simple structure. Begin with a set of premises, then use established rules of inference to manipulate and combine these premises until you reach the desired conclusion. Rules of inference are, for example, **modus ponens** and **modus tollens**. When used correctly, rules of inference produce conclusions that are certainly true if all the premises are true.

For the most part, scientists, mathematicians, and philosophers apply the rules of inference correctly in their arguments. When colleagues disagree, they rebut the arguments by showing that the set of premises was flawed. Some premises may be missing, or some may be plain false. Most often, the argument contains hidden premises, called warrants, that the arguer failed to state explicitly, but without which the argument would collapse. Quite often, these omissions can be fatal. Hidden assumptions often turn out to be hard to accept as true—in any case, they have to be justified by a separate argument. At times, the hidden premise presupposes the **truth** of the original conclusion. This is known as circular reasoning, an unacceptable error.

To avoid the pitfalls of hidden premises, mathematicians and logicians try to found their theories on small sets of basic, standardized premises: axioms. Ideally, axioms must be obviously true to any reasonable critic of the theory in question. They need never be proved by argument. Every true statement of the theory must be derived either directly from the axioms themselves, or from other theorems that were in turn derived legitimately. In this way, no hidden premises can ever enter an argument.

The axiomatic approach to logic and mathematics - starting with a few basic, unquestionable assumptions and constructing a theory from them—was first introduced by the Greek mathematician **Euclid** in his treatise on **geometry**, *The Elements*, around 300BC. Not only did Euclid write the definitive book on **plane geometry**, but established the very structure of Western science.

Once a mathematician defines a set of axioms, the question is still open as to whether the theory is complete. Could there be propositions that are true, but cannot be derived from the set of axioms? In fact, in his famous incompleteness **proof**, mathematician **Kurt Gödel** proved that theories of a certain complexity are always incomplete, even if infinitely many axioms are added. The theory of **arithmetic** is such an incomplete theory.

See also Euclid's axioms; Parallel postulate

B

BABBAGE, CHARLES (1792-1871)

English mathematician, inventor, and philosopher

Charles Babbage is considered the creator of modern computers. A mathematician and 19th-century British intellectual, he conceived of a steam driven Difference Engine that could automatically calculate and print error–free mathematical tables. He later developed the idea of the Analytical Engine, which could be programmed to make calculations and could store results in a memory unit. Babbage also invented the first automated typesetter for printing the results of computations. Although construction of his calculating machines was never completed in his lifetime, his concepts were used as the basis for the Harvard Mark I **Calculator**, the prototype of the modern digital computer, built by **Howard Aiken** in 1944.

Charles Babbage was born on December 26, 1792, in Teignmouth, Devon, England. He was the son of a London banker, Benjamin Babbage, from whom he inherited a sizable fortune, enabling him to devote his life to intellectual pursuits. As a child Babbage suffered several bouts of violent fever that interfered with his early education. He was placed in the care of a clergyman who ran a school in Devonshire, with instructions not to tax his health with too much knowledge. In his early teens Babbage attended a boarding school in London, where he developed a keen interest in **algebra**. He spent much of his leisure time studying mathematical works, and was especially influenced by Ward's *Young Mathematician's Guide*, a text he had found in his school library. In 1810 Babbage entered Trinity College at Cambridge University and soon discovered the knowledge of mathematics he had obtained through self–instruction exceeded that of his tutors. Together with a circle of friends Babbage founded the Analytical Society, whose purpose was to promote mathematics. A co–founder was to became a lifelong companion, John Herschel, a noted astronomer. Together, in 1813, they published a translation of LaCroix's *Differential and Integral Calculus*, accompanied by several volumes of mathematical

examples. Babbage graduated from Cambridge in 1814, and the following year wrote several papers on the **calculus** of **functions** for the Royal Society of London. In 1816 he was elected a Fellow of the Society.

While at Cambridge, Babbage had begun thinking about the possibility of building a machine that could compute **arithmetic** tables. In the early 19th century, actuaries, bankers, navigators, engineers, and others relied heavily on published numerical values of mathematical formulas and functions. Errors in calculation and transcription were so common that the tables were often accompanied not only by a list of corrections, but a list of corrections to the corrections. Mechanical calculators had existed since the time of **Blaise Pascal**, but they worked slowly and were only capable of performing single arithmetic calculations. Babbage designed a steam–driven machine he called the Difference Engine, which could rapidly calculate and automatically print the results of large numbers of mathematical operations. The first version of his Difference Engine was intended to compute values of squares and **quadratic functions**. It worked on the principle known as the method of finite differences, a technique that employs only **addition** to determine successive values for **polynomial** functions. In 1822 Babbage built a prototype that made accurate calculations up to five–place numbers. That same year, with backing from the Royal Society, he convinced the British government to provide funds for building a full–scale Difference Engine, with a capacity to work with numbers up to one million with 20 decimal places. The project, expected to take three years, was abandoned after a decade. Babbage's design called for a series of gear wheels on shafts that would be turned by cranks. The machine was to contain 25,000 die–cast pewter and precision gauged brass and steel parts. If finished, it would have weighed over two tons. Historians speculate its realization may have been beyond the engineering capacity of the era. The project was also hapered by Babbage's constant revisions of design. Construction of the Difference Engine was far from completion when financial arguments between Babbage and

Charles Babbage

his chief engineer brought the project to a halt. By then, Babbage had come upon a better idea.

Babbage's work on the Difference Engine led to the evolution of his new invention, the Analytical Engine. While the former machine was designed to work straight through a computational problem, the latter was designed to make calculations, store the results, analyze what to do next, then return to complete the problem. Babbage's design for the Analytical Engine had four key components: the mill, the card reader, the store, and the typesetter. The mill was the heart of the machine, where the four basic arithmetic operations could be performed with an accuracy of up to 50 decimal places. It received instructions and numerical data from punched cards that were deciphered by the card reader. Babbage borrowed the idea for this input system from the weaving industry, which in the mid–1700s had begun using punched cards as hand–held guides for creating different patterns in cloth. An automatic card reader that controlled a power loom had been invented in 1801 by the French carpet–maker Joseph Marie Jacquard. Babbage's Jacquard cards, as he referred to them, could provide instructions and data not only for the machine's mill, but for the store, a place where numbers were retained in memory for future use. The store consisted of a bank of one thousand registers, each of which could hold a 50–digit number. Finally, after mathematical operations had been per-

formed, the Jacquard cards could instruct the machine to type-set the results for printing.

In 1834 Babbage began an eight–year campaign to convince the government to fund construction of the Analytical Engine, but was unsuccessful. Britain had already spent seventeen thousand pounds on the Difference Engine to no avail. Babbage had contributed a comparable amount of money to the Difference Engine project, depleting most of his personal fortune. In 1848, Babbage drew up plans for a scaled–down version of the Analytical Engine, called the Difference Engine No. 2, but once again was unable to obtain funds for construction. The scaled–down version was finally built, nearly a century and a half later, in honor of Babbage's bicentenary, by the Science Museum of London. It weighed three tons and worked flawlessly.

Although the primary focus of Babbage's intellectual pursuits was his calculating engines, he devoted time to many other areas of scientific and practical interest. He published a paper with Herschel in 1825 on magnetization arising from **rotation**, and made contributions to the fields of geology, anthropology, and astronomy. In 1820 Babbage helped found the Royal Astronomical Society, in 1831 the British Association for the Advancement of Science, and in 1834 the Statistical Society of London. Babbage once descended into the crater of Mt. Vesuvius to research its volcanic activity. He studied glaciers and suggested a way of learning about past climatic conditions by measuring tree ring growth in fossilized wood. Babbage even designed a colored lighting system for theaters, an idea he reportedly dreamed up while bored during an opera performance. Concerned with economic efficiencies, in 1832 he wrote a pamphlet called "Economy of Manufactures and Machinery." Babbage advised the British postal service, consulted for the British rail system, and was the inventor of the "cowcatcher," the track clearing safety devise that protrudes from the front of a train engine. He ran twice, unsuccessfully, for a seat in Parliament. From 1828 until 1839, Babbage held the Lucasian Chair of Mathematics at Cambridge, although during his tenure he never taught or lived at the university.

Babbage was widely known in London's social circles, hosting regular Saturday night parties at his home at 1 Dorset Street. He was friends with naturalist Charles Darwin, German naturalist and statesman Alexander Humboldt, and Ada Byron, Countess of Lovelace, who published articles explaining Babbage's engines and authored the first computer program. Babbage is described as a man who in his final years was embittered and disappointed over the lack of support for his calculating engines. He was critical of the scientific establishment and of governmental funding policies, and published papers on what he described as the decline of science in England. Eventually, Babbage developed a reputation as an eccentric, launching a campaign to ban organ grinders as street nuisances. When he died in London on October 18, 1871, his *London Times* obituary commented that he lived to be almost eighty "in spite of organ–grinding persecutions."

Babbage married Georgiana Whitmore shortly after he graduated from Cambridge. Their eldest son was Herschel

Babbage. Another son, Henry Babbage, attempted to carry on in his father's tradition, presenting a paper on calculating engines to the British Association in 1888, and a year later editing a volume about his father's works.

BABBAGE'S ENGINE

The British scientist **Charles Babbage** (1791-1871) designed mechanical calculating machines many years before the computer age. Babbage lived during the Victorian era, when mathematicians compiled huge books of tables for use in **multiplication**, **division**, and finding **logarithms**, **square roots**, etc. The tables were expensive and tedious to prepare and inevitably contained mistakes. Babbage wanted to build a machine that would automatically calculate and print the values for such tables.

In 1821, he began work on his first computing machine, Difference Engine No. 1, which was designed to add, subtract, and solve polynomial **equations** using the method of finite differences. This method is best illustrated with an example. Consider the equation x^3. This is a third-order polynomial. The values of x^3 for $x=1$ to 5 are as follows:

x	x^3	1st Differences	2nd Differences	3rd Differences
1	1	7	12	6
2	8	19	18	6
3	27	37	24	
4	64	61		
5	125			

Taking the first set of differences between x^3 values (i.e., 8-1, 27-8, 64-27, and 125-64) provides 7, 19, 37, and 61. Taking the second set of differences (i.e., 19-7, 37-19, and 61-37) provides 12, 18, and 24. Taking the third set of differences (i.e., 18-12 and 24-18) provides 6 and 6. This set of differences is constant and defines the order of the polynomial equation. In other words, the nth set of differences that are constant defines the degree of the equation.

Table values of x^3 for very large values of x can be calculated by repeating the method. For example, the 3rd difference for $x=3$ is also 6, which means that the 2nd difference for $x=4$ is 6+24=30. Therefore, the 1st difference for $x=5$ is 30+61=91. Thus, the table value of x^3 for $x=6$ is 125+91=216. This is the concept upon which Babbage's Difference Engines were based.

The design for Difference Engine No. 1 called for approximately 25,000 precision-crafted parts of steel, bronze, and cast iron. Although Babbage employed expert craftsmen, the metal working technology of the time was not up to this mammoth task. After 11 years of work and the expenditure of a great deal of money from financial backers (including the British government), Babbage's team had only completed a small part of the engine, containing approximately 2,000 parts.

The completed portion, which did operate properly, contained three columns of figure wheels or cog wheels stacked one on top of another around an axis. Five of the figure wheels in the first column (the table column) could be set by hand to any digit from 0 to 9. The bottom-most of these figure wheels tracked digits in the ones place. The wheel above tracked digits in the tens place and so forth, up to the ten-thousands place. A number was set by hand on each column and then a handle was pulled. The machine added the number on the middle column (the 1st differences column) to the number on the Table column. Then, the number on the third column (the 2nd differences column) was added to the number on the 1st difference column to provide the 1st difference for the next computation. The final engine was to handle six orders of difference and compute answers containing up to 20 digits.

However, the huge expense and lack of significant progress caused the project to be abandoned. Although Babbage later drafted drawings for a simpler, more efficient Difference Engine No. 2, this machine was never completed during his life time due to lack of funding. Its design called for more than 4,000 parts in a cast-iron frame that stood seven feet tall and 11 feet long. An additional 4,000 parts were required for printing.

Even as work progressed on the Difference Engines, Babbage was designing a much more sophisticated machine called an Analytical Engine that could perform multiplication and division and algebraic **functions**. The Analytical Engine has many similarities to later computers in terms of its architecture and programmability, for example, it used punched cards for inputting data. This technique was borrowed from the textile industry, which used punched cards to "program" the Jacquard loom to weave intricate designs into cloth. The engine also contained a "mill" area, where arithmetical processing occurred (much like a central processing unit) and a "store" area where calculations in progress were held (much like memory chips).

Babbage wrote several computer programs for the Analytical Engine, but most were never published. In 1843, his friend Augusta Ada King (the daughter of Lord Byron) published a set of programming notes for the Analytical Engine that included a method for computing the Bernoulli Numbers. The engine was also capable of executing loops and conditional branching programming steps such as "if x, then y." Although some parts of the Analytical Engine were built during Babbage's lifetime, the project lacked funding and was far from complete at the time of his death in 1871.

In 1896, Babbage's son sent a small **model** of Difference Engine No. 1 to Harvard University. Professor **Howard Aiken** found the brass wheels in the attic of the Science Center in 1937 as he worked on his own proposal for a computing machine. Aiken realized the importance of Babbage's mechanism and collected a set of books from the inventor's grandson. Aiken created the Mark I, one of the first electromechanical computers, and credited Babbage for educating him about computers.

In 1990, the Science Museum in London used Babbage's original drawings to build a working version of Difference Engine No. 2 (without the printing mechanism) for the 1991 celebration of the 200-year anniversary of Babbage's birth. The Engine took a year to put together, cost approximately \$500,000, weighed three tons, and was 11 feet long and 7 feet tall. Most importantly, it worked. It was able to calculate successive values of seventh-order polynomial equations (e.g., 1^7=1, 2^7=128, etc.) containing up to 31 digits.

BABYLONIAN MATHEMATICS

The Babylonians, who lived in Mesopotamia from about 2000 B.C. until the last few centuries B.C., developed a substantial body of mathematical knowledge, well before most other civilizations. Their writings have been passed down to us on hundreds of clay tablets, on which they wrote in cuneiform (wedge-shaped) symbols. Most of the Babylonian mathematics that has come down to us on these tablets is from the period 1800-1600 B.C.; it is not clear whether mathematical understanding progressed significantly after that period.

Many of the mathematical tablets of the Babylonians consist of arithmetical tables. These tables, which are easier to translate than more complicated texts, proved invaluable to scholars who set out to decipher Babylonian writings. The tables made it clear that the Babylonians used a base-60 or sexagesimal system in place of the base-10 system in use today. Part of that system has been preserved even to the present in the way we measure time: sixty minutes to each hour, sixty seconds to each minute. In some ways, the base-60 system is much more cumbersome than the base-10 system. Instead of ten numerals (0, 1,..., 9), sixty are needed. This made learning the number system such a challenge for the Babylonians that it was generally reserved for the priesthood. On the other hand, 60 has a distinct advantage over 10, because it has many more divisors: 10 is only divisible by 2 and 5, while 60 is divisible by 2, 3, 4, 5, 6, 10, 12, 15, 20 and 30. This means that many **fractions** that have infinite decimal representations have finite sexagesimal representations. For example, the fraction 1/3 has the infinite decimal expansion 0.33333..., since 1/3=3/10+3/100+3/1000+.... But 60, unlike 10, is divisible by 3, and so 1/3=20/60, and 1/3 has a finite sexagesimal representation.

The Babylonians drew up tables of squares, **square roots**, cube roots, reciprocals, and some **powers**. They were adept at linear and quadratic **equations** and knew many geometric constructions. They had tables of **logarithms** and an approximation for the natural logarithm "**e**." They estimated **pi** and were aware of the **Pythagorean theorem**, as is shown in the following tablet inscription:

"4 is the length and 5 the diagonal. What is the breadth? Its size is not known. 4 times 4 is 16. 5 times 5 is 25. You take 16 from 25 and there remains 9. What times what shall I take in order to get 9? 3 times 3 is 9. 3 is the breadth."

This document shows that although the Babylonians had a good understanding of how to apply the **theorem**, they tended to think algorithmically; that is, in terms of a sequence of steps that could always be followed to arrive at the answer to a problem. They did not attempt any formal **proof** of the statement; that would not come until much later, when **Pythagoras** offered a proof in the 6th century B.C.

See also Egyptian mathematics

BACON, ROGER (CA. 1214-1292)
English philosopher

Roger Bacon, a medieval English philosopher and scholar, is believed to have been born near Ilchester, Somerset. His family was apparently wealthy. Following his studies at Oxford under Robert Grosseteste and in Paris under Peter Peregrinis, Bacon acquired a reputation for his research in philosophy, magic and alchemy. Because of these interests, Bacon was dubbed "doctor mirabilis," which translates as wonderful teacher. Bacon joined the Franciscan order sometime around 1257, but soon thereafter returned to Oxford University to pursue research in experimental science. One of his projects was an attempt to compile an encyclopedia of universal knowledge based on his three major works, but this project eventually proved to be too much of an undertaking for one man even during his lifetime. Between 1265 and 1268, Bacon was in favor with Pope Clement IV. But Bacon later earned a reputation for his outspokenness and his quarrelsome nature, and for refusing to compromise his opinions. As a result, his ideas met rejection and censorship. From 1257 onwards, he pursued studies independently of the University.

Bacon was eventually placed in a virtual prison in 1277 in Paris by the Franciscan Order for having abused all classes of society, in particular the Franciscan and Dominican orders for their educational practices. Fourteen years later, in 1292, he was released, subsequently returning to Oxford, where he died on June 11, 1292.

Although some have claimed that Bacon was the discoverer of the magnifying glass and gunpowder, they are without basis. He was however the author of several noteworthy speculations on aerial flight, mechanical air and sea transport, submarine exploration, global circumnavigation, and the construction of microscopes and telescopes.

Bacon believed that all branches of knowledge, which for him included languages, mathematics, optics, alchemy, metaphysics, and moral philosophy, were subordinate to theology. He considered knowledge to be of two types: inner knowledge of mystical origin, and practical knowledge acquired by mathematical observation and experimentation. For Bacon, these two realms of knowledge constituted experience. Bacon also considered alchemy to be the most valuable of the sciences because of its practical value.

Although Bacon never concerned himself with mathematical tables (of interest to many of his contemporaries in mathematics) in anything but a superficial way, he held to the idea that mathematics should be based on mathematical **proof** and that mathematics alone provides absolute certainty.

Roger Bacon

Nevertheless, his writings do not appear to contain a single proof; his preference for establishing veracity was to cite an authority such as Euclid. His writings on mathematics, philosophy, and logic were not widely appreciated in his time, their significance only becoming widely recognized some centuries later. Bacon's mathematics included astronomy and astrology, as well a geometrical theory of physical causation in optics. A believer that nothing in the world could be understood without the power of **geometry**, Bacon not surprisingly achieved some of his most fruitful research in the field of optics.

For Bacon, mathematics held potential applications to human, divine, ecclesiastical, and state affairs.

As a scientist, Bacon held that all speculation was pointless unless the facts had first been elucidated by experimentation. He demonstrated that the presence of air is necessary to bring about combustion. He experimented with lenses for improving vision. For these and other reasons, some historians regard Bacon as the first modern scientist.

BANACH SPACE

A Banach **space** is a vector space over the **field** of **real numbers**, or over the field of **complex numbers**, together with a norm. And in a Banach space the metric **topology** determined by the norm is complete. Thus the assertion that a certain object is a Banach space includes several different pieces of information.

We now describe a Banach space more carefully. We begin with a vector space X having either the real numbers or the complex numbers as its field of scalars, (see the article on abstract linear spaces.) By a *norm* on X we understand a func-

tion $\| \ \|: X \rightarrow [0, \infty)$ that satisfies the following three conditions:

- (1) $\|x\| = 0$ if and only if $x = 0$ is the **zero** vector in X,
- (2) $\|\alpha x\| = |\alpha| \ \|x\|$ for all scalars α and all vectors x,
- (3) $\|x + y\| \leq \|x\| + \|y\|$ for all pairs of vectors x and y.

The norm $\| \ \|$ on X allows us to define a closely related *metric* on X. We say that the **distance** from x to y is given by $\|x - y\|$. Alternatively, the function from $X \times X$ into $[0, \infty)$ defined by sending $(x, y) \rightarrow \|x - y\|$ is a metric (see the article on metric spaces.) This metric automatically induces a metric topology in the vector space X. If y is a point in X and $r > 0$ then we recall that the open ball centered at y and having radius r is the set

$$B_r(y) = \{x \in X: \|x - y\| < r\}.$$

More generally, a set in the metric topology is *open* if it is empty, or if it is a union of open balls. The pair $(X, \| \ \|)$ is often called a *normed linear space*. It consists of the vector space X over the field of real numbers or complex numbers, the norm function $\| \ \|$, and the metric topology in X which is induced by the norm. Thus the set X is both a vector space and a metric space.

Next we recall that a sequence $\{x_n\}$, where $n = 1, 2, 3,...$, in the metric space X is called a *Cauchy sequence* if for every $\varepsilon > 0$ there exists a positive integer N such that

$$\|x_n - x_m\| < \varepsilon \text{ whenever } N \leq m \text{ and } N \leq n.$$

If the sequence $\{x_n\}$ is convergent in X then it is easy to prove that it must be a Cauchy sequence. However, in general, a metric space may have Cauchy sequences which do not converge to a point in the space. We say that a metric space, or the metric itself, is *complete* if every Cauchy sequence does converge to a point in the space. We are now in position to give the **definition** of a Banach space: a Banach space is a normed **linear space** over the field of real numbers or over the field of complex number, such the metric determined by the norm is complete. Therefore, in a Banach space, every Cauchy sequence must converge to a point in the space.

Important examples of Banach spaces are the *finite dimensional* Banach spaces. For example, we could use the vector space $\Re^N$ of all column vectors

$$X = \begin{pmatrix} x_1 \\ x_2 \\ \vdots \\ x_N \end{pmatrix}$$

in which each coordinate x_n, $1 \leq n \leq N$, is an element from the field $\Re$ of real numbers. Next we need to specify a norm on $\Re^N$. This can be done in many ways, for example we could define the norm $\| \ \|_1$ by setting

$$\|x\|_1 = |x_1| + |x_2| + ... + |x_N|.$$

Then it is easy to check that $\| \ \|_1$ satisfies the three conditions of a norm, and the resulting metric topology in $\Re^N$ is in fact complete. Thus the pair $(\Re^N, \| \ \|_1)$ forms a Banach space.

Another example of a norm on $\mathfrak{R}^N$ is the *Euclidean norm*, which is often denoted by $\| \ \|_2$ and then defined by

$$\|x\|_2 = (|x_1|^2 + |x_2|^2 + \ldots + |x_N|^2)^{1/2}.$$

More generally, if p is a real number and $1 \le p < \infty$ then each of the **functions** $\| \ \|_p \colon \mathfrak{R}^N \to [0, \infty)$ defined by

$$\|x\|_p = \left(|x_1|^p + |x_2|^p + \cdots + |x_N|^p \right)^{\frac{1}{p}}$$

is a norm on $\mathfrak{R}^N$. There is also a norm denoted by $\| \ \|_\infty$ and then defined by

$$\|x\|_\infty = \max\{|x_1|, |x_2|, \ldots, |x_N|\}.$$

It can be shown that for each p with $1 \le p \le \infty$, the metric topology determined by $\| \ \|_p$ is complete. That is, for each p with $1 \le p \le \infty$ the pair $(\mathfrak{R}^N, \| \ \|_p)$ forms a Banach space.

Here is another example of a Banach space. Let $C[0,1]$ denote the set of all continuous, complex valued functions on the closed **interval** $[0,1]$. If α and β are complex numbers, if f and g are functions in the set $C[0,1]$, then the linear combination $\alpha f(x) + \beta g(x)$ turns out to be another function in $C[0,1]$. From this observation it is easy to see that $C[0,1]$ forms a vector space over the field of complex numbers. Now define $\| \ \|_\infty \colon C[0,1] \to [0, \infty)$ by

$$\|f\|_\infty = \sup\{|f(x)| \colon 0 \le x \le 1\}.$$

It turns out that $\| \ \|_\infty$ defines a norm on $C[0,1]$. And by using the fact that a continuous function on the compact interval $[0,1]$ is uniformly continuous, it can be shown that every sequence in $C[0,1]$ which is Cauchy with respect to the norm $\| \ \|_\infty$, converges to a function in $C[0,1]$. Thus the vector space $C[0,1]$ together with the norm $\| \ \|_\infty$ forms a Banach space.

Another method of constructing a norm on the vector space $C[0,1]$ is to define

$$\|f\|_1 = \int_0^1 |f(x)| \, dx.$$

Here we can use the Riemann integral from **calculus**, since the functions f in $C[0,1]$ are continuous and therefore Riemann integrable on the interval $[0,1]$. While $\| \ \|_1$ does satisfy the requirements of a norm on $C[0,1]$, there exist sequences of functions in $C[0,1]$ which are Cauchy with respect to this norm but which do not converge to a function in $C[0,1]$. Thus the pair $(C[0,1], \| \ \|_\infty)$ forms a normed linear space but it is *not* a Banach space. This example raises a natural question: is it possible to enlarge the vector space $C[0,1]$, and extend the definition of the norm $\| \ \|_1$ to the enlarged space, in such a way that we do get a Banach space? The answer to this question is yes. The enlarged vector space turns out to be the collection of **equivalence** classes of Lebesgue integrable functions on the interval $[0,1]$. The extended norm is defined in the same manner but now the **Lebesgue integral** must be used because, in general, the absolute value of the functions are no longer Riemann integrable.

See also Lebesgue integral

BANACH, STEFAN (1892-1945)
Polish analyst

In spite of his somewhat fragmented education (he never completed a formal doctoral program), Stefan Banach made important contributions to a number of fields of mathematics, including the theory of orthogonal series, **topology**, the theory of measure and integration, **set theory**, and the theory of linear spaces of an infinite number of dimensions. He is probably best remembered, however, for his work on functional **analysis**.

Stefan Banach was born on March 30, 1892, in Kraków, Poland. His father was named Greczek, a railway official from peasant background. He and Stefan's mother (whose name has been lost) abandoned their young child to a laundress almost immediately after his birth. The child took on his foster mother's surname of Banach, but almost nothing else is known about his early childhood. Banach apparently developed an interest in mathematics at an early age and taught himself the fundamentals of the subject. By the age of 15 he was supporting himself as a private teacher of mathematics. He also taught himself enough French to master Tannery's text on the theory of **functions**, *Introduction à la théorie des fonctions*. Banach attended lectures on mathematics at Jagellon University on an irregular basis before entering the Lwów Institute of Technology in the Ukraine in 1910. He did not, however, graduate from the institute.

In 1914, with the outbreak of World War I, Banach returned to Kraków. Two years later, a chance event was to change his life. While sitting on a park bench in Kraków talking with a friend about mathematics, he was overhead by the mathematician H. Steinhaus. Steinhaus later wrote that he was "so struck by the words 'the **Lebesgue integral**'" that he heard from the two that he came closer and introduced himself to the young men. As the group talked, the conversation turned to a problem on the **congruence** of a **Fourier series** on which Steinhaus had been working. "I was greatly surprised," Steinhaus went on to say, "when, after a few days, Banach brought me a negative answer with a reservation which resulted from his ignorance of [a technical point about which he did not know]." Banach and Steinhaus were later to collaborate on a number of mathematical studies.

Banach's natural gift for mathematics soon became more widely known, and at the conclusion of the war he was offered a position as mathematical assistant at the Lwów Institute of Technology by Antoni Lomnicki. For the first time in his life Banach had some degree of financial security and he married. Beginning in 1919, Banach was assigned to lecture on mathematics and **mechanics**. In the same year, he was awarded his doctoral degree although he had not completed the full program of courses expected for that degree.

The primary basis for Banach's degree was the paper he had written on integral **equations**, which had been published in *Fundamenta mathematicae* in 1922 as "Sur les opérations dans les ensembles abstraits et leur application aux équations intégrales." At about the same time he was made an instructor at the institute; in 1927 he was promoted to full professor. From

1939 to 1941 Banach also served as dean of the faculty at the institute.

The mathematical work for which Banach is best known is his book *Théorie des opérations linéaires,* which appeared in 1932 as the first volume in the Mathematical Monographs series, published in Warsaw. In this book, Banach developed a general theory for working with linear operations that proved to be a landmark in the field. Prior to this work, a number of individual, discrete methods had been developed for solving specific problems. But there was no comprehensive theory that could be applied to a great variety of problems. In his book, Banach introduced the concept of normed linear spaces, now known as Banach spaces, which, Steinhaus later wrote, can be used "to solve in a general way many problems which formerly called for special treatment and considerable ingenuity."

Banach's significance in the **history of mathematics** goes beyond his own research. He was also an effective teacher whose influence was spread throughout Europe and the United States by a number of brilliant students. In addition, he wrote an important popular textbook, *Differential and Integral Calculus* (1929–30) and was founder with Steinhaus of the journal *Studia mathematica.*

World War II was a personal disaster for Banach. After the German army occupied the city of Lwów, he was forced to work in a German laboratory studying infectious diseases. His job there was to feed the lice used in experiments. As degrading as this work was, Banach was able to continue teaching in underground schools and carry on his own research. By the time the war ended, however, his health had so badly deteriorated that he lived only a few more months. Banach died in Lwów on August 31, 1945. Among the honors accorded him during his lifetime were election as corresponding member of the Polish Academy of Sciences and of the Kiev Academy of Sciences. He also received the Prize of the City of Lwów in 1930 and the Prize of the Polish Academy in 1939. Upon his death, the city of Warsaw renamed one of its streets in his honor.

BARI, NINA (1901-1961)
Russian mathematician

Nina Bari's work focused on trigonometric series. She refined the constructive method of **proof** to prove results in function theory, and her work is regarded as the foundation of function and trigonometric series theory.

Nina Karlovna Bari was born in Moscow on November 19, 1901, the daughter of Olga and Karl Adolfovich Bari, a physician. In the Russia of her youth, education was segregated by gender and the best academic opportunities reserved for males only. Bari attended a private high school for girls, but in 1918 she defied convention and sat for—and passed—the examination for a boy's high school graduation certificate.

In 1917, Russia's political and social structure was shattered by the Russian Revolution. The power vacuum left the country at the mercy of the czarists, socialist revolutionaries, and Bolsheviks. While many of Russia's universities closed at

the beginning of the Revolution, the Faculty of Physics and Mathematics of Moscow State University reopened in 1918, and began accepting applications from women. Records show that Bari was the first woman to attend the university and was probably the first woman to graduate from it. Russia's educational institutions were in the same turmoil as the society around them. Graduation exams were scheduled on a catch-as-can basis, and Bari took advantage of the disorder to sit for her examinations early. She graduated from Moscow State in 1921—just three years after entering the university.

After graduation, Bari began her teaching career. She lectured at the Moscow Forestry Institute, the Moscow Polytechnic Institute, and the Sverdlov Communist Institute. Bari applied for and received the only paid research fellowship awarded by the newly created Research Institute of Mathematics and Mechanics. (Ten postgraduate students were accepted at the Research Institute; Bari won the stipend because her name appeared first on the alphabetically arranged list. According to a colleague, she shared the stipend with her fellow students.)

As a student, Bari was drawn to an elite group nicknamed the Luzitania—an informal academic and social organization. These scholars clustered around **Nikolai Nikolaevich Luzin**, a noted mathematician who rejected any area of mathematical study but function theory. With Luzin as her inspiration, Bari plunged into the study of trigonometric series and **functions**. She developed her thesis around the topic and presented the main results of her research to the Moscow Mathematical Society in 1922—the first woman to address the society. In 1926, she defended her thesis, and her work earned her the Glavnauk Prize.

In 1927, Bari took advantage of an opportunity to study in Paris at the Sorbonne and the College de France. She then attended the Polish Mathematical Congress in Lvov, Poland; a Rockefeller grant enabled her to return to Paris to continue her studies. Bari's decision to travel may have been influenced by the disintegration of the Luzitanians. Luzin's irascible, demanding personality had alienated many of the mathematicians who had gathered around him. By 1930, all traces of the Luzitania movement had vanished, and Luzin left Moscow State for the Academy of Science's Steklov Institute.

Bari returned to Moscow State in 1929 and in 1932 was made a full professor. In 1935, she was awarded the degree of Doctor of the Physical-Mathematical Sciences, a more prestigious research degree than the traditional Ph.D.

In 1936, during the dictatorship of Josef Stalin, Bari's mentor, Luzin, was charged with ideological sabotage. For some reason—possibly Stalin's preoccupation with more important enemies of the state—Luzin's trial was canceled. Luzin was officially reprimanded and withdrew from academia.

Bari managed to avoid the taint of association. She and D.E. Men'shov took charge of function theory work at Moscow State during the 1940s. In 1952, she published an important piece on primitive functions, and trigonometric series and their almost everywhere **convergence**. Bari also presented works at the 1956 Third All-Union Congress in

Moscow and the 1958 International Congress of Mathematicians in Edinburg.

Mathematics was the center of Bari's intellectual life, but she enjoyed literature and the arts. She was also a mountain hiking enthusiast and tackled the Caucasus, Altai, Lamir, and Tyan'shan' mountain ranges in Russia. Bari's interest in mountain hiking was inspired by her husband, Viktor Vladmirovich Nemytski, a Soviet mathematician, Moscow State professor, and an avid mountain explorer. There is no documentation of their marriage available, but contemporaries believe the two married later in life.

Bari's last work—her 55th publication—was a 900-page monograph on the state of the art of trigonometric series theory, which is recognized as a standard reference work for those specializing in function and trigonometric series theory.

Bari died July 15, 1961, when she fell in front of a train at the Moscow Metro. Colleagues, however, suspect her death was suicide; they speculate she was despondent over the death of Luzin in 1950, who some believe had been not only her mentor but her lover.

BARROW, ISAAC (1630-1677)
English geometer and theologian

Isaac Barrow is noted for his contributions to the field of optics. He is also remembered as the professor who served as inspiration and mentor to **Isaac Newton**.

Barrow was born in London. His father, Thomas, was a merchant who served as linen draper to King Charles I. Barrow's mother, Anne, died shortly after his birth. Barrow attended Charterhouse, a school noted for its emphasis on a classical education. He was a rowdy youngster, more interested in scrapping with other students than studying. Barrow was transferred to Felsted School in Essex, where it was hoped that schoolmaster Martin Holbeach's strict discipline would correct his bad habits and promote scholastic achievement.

Barrow stayed at Felsted for four years, thriving on the social and academic discipline it offered. He studied Latin, Greek, Hebrew, French, logic, and the classics. When Barrow's father suffered financial losses when a rebellion destroyed textile trade with Ireland, Barrow began tutoring Thomas Fairfax, fourth viscount Fairfax of Emely, Ireland. In 1646, Barrow finally secured a scholarship to Trinity College, Cambridge.

Barrow earned his baccalaureate in 1648 and was elected as a college fellow in 1649. He completed an M.A. in 1652 and was named a college lecturer and university examiner.

Barrow completed what would be his first published work in 1654, *Euclidis Elementorum libri XV*, a highly regard translation of **Euclid**. Designed as an undergraduate text, the work was reissued in 1657 and eventually reached a wide public in a pocket–sized edition. In 1655, Barrow was nominated for a prestigious professorship in Greek. But this was the decade of Cromwell, and Barrow had never bothered to hide his loyalty to the monarchy. The Regius Professorship went to

Isaac Barrow

Ralph Widdrington, a candidate who had the backing of the university's chancellor—Oliver Cromwell.

Frustrated and angry, Barrow sold his books, applied for and won a Trinity College traveling fellowship, and left England. He traveled to France, Italy, and Turkey, lingering abroad for nearly five years. It was during his travels that his interest in mathematics intensified, as he came into contact not only with classical scholars like himself, but scientists and mathematicians.

Barrow returned to England in 1660, the year King Charles II was restored to England's throne. He immediately was ordained in the Anglican Church and was promptly appointed to the Regius Professorship of Greek.

Barrow supplemented the modest pay of a professor of classics by accepting a professorship of **geometry** at Gresham College and filling in as a professor of astronomy. When he was named Lucasian professor of mathematics at Cambridge in 1663, however, the stipend attached to the professorship made it possible for him to give up extra teaching appointments. In that same year Barrow was the first to be named a fellow of the newly established Royal Society of London; it was in that capacity that in 1664 he served as a scholarship examiner for Isaac Newton. Apparently Barrow's wit and knowledge made an impression, and Newton began attending Barrow's lectures on optics.

Between 1663 and 1669, Barrow developed and presented the *Lectiones geometricae*, a series of lectures on geometry in which he defined time as the measure of **motion**, the properties of curves generated when moving points and lines are combined, the construction of tangents and the nature of quadrature.Although the combined lectures include elements key to the fundamentals of **calculus** theory, the work is not original. Barrow relied on contemporary mathematicians—including **René Descartes, John Wallis**, and **James Gregory**—for the information; his genius lay in combining these works into a comprehensible whole and relaying the results to a new generation of scholars.

Barrow is credited with developing the method of finding the point of refraction at a plane interface and the point construction of the diacuastic of a spherical interface. His work seems to have served as the starting point for Newton's ideas, although the most Newton admitted was that Barrow's lectures "might put me upon considering the generation of figures by motion, tho I no now remember it."

Barrow's work in opticswas quickly eclipsed by Newton's. Influenced both by Newton's genius and the tug of other interests, Barrow stepped down as Lucasian professor in favor of Newton in 1669.

Barrow devoted his energies to theology and served as royal chaplain in London. He returned to Trinity College in 1673 at the king's request and was appointed vice chancellor in 1675.

Barrow never married; he died in 1677, and contemporary accounts indicate his death at the age of 47 was the result of a drug overdose.

BAYES'S THEOREM

In the study of **probability**, two events A and B are said to be independent events if neither event influences or effects the other. For example, tossing a coin twice yields independent events because whether the coin shows heads or tails on the second toss is not influenced by what it showed on the first toss, or vice versa. Whenever two events, A and B, are independent, we can compute the probability of both A and B happening by simply multiplying the individual probabilities of A and B, i.e., P(A and B)=P(A)P(B), also written as $P(A \cap B) = P(A)P(B)$. The English mathematician **Thomas Bayes** (1702-1761) considered the situation in which events A and B are not independent. In this situation, the formula for $P(A \cap B)$ is not as simple as in the independent case. It first requires a notation for the probability of B happening given that A has happened. This notation is P(B|A) and this is known as "the conditional probability of B given A." Now if event B is influenced by event A, then the probability of both A and B happening is equal to the probability that A happens times the probability of B happening given that A happens, or, in symbols, $P(A \cap B) = P(A)P(B|A)$. To illustrate, suppose that we toss a fair die and ask for the probability that a number less than 4 turns up. If no further information is given, then the probability is P(1 or 2 or 3)= P(1)+P(2)+P(3)=1/6+1/6+1/6=1/2. But

suppose that the question is about the probability that a number less than 4 turns up given that the toss turned up an odd number. Then this is a conditional probability P(less than 4|odd) = P(less than 4 and odd)/P(odd) = (2/6)/(1/2)– 2/3. Note that the additional information that the upturned number was odd increased the probability from ½ to ⅔. Bayes would use this principle of the effect of increased **information** on probabilities to establish the theory of what is now called "Bayesian estimation." This program brought forth the possibility of using "subjective" probabilities as well as "objective" probabilities. Objective probabilities are those describing events such as coin tossing, dice throwing, card playing where the probabilities are objectively calculated by the underlying mathematics of the event. Subjective probabilities, on the other hand, can be human estimates of the likelihood that certain events will occur given some past history of similar events occurring. This use of subjective probability estimates is controversial in the community of classical statisticians, but it is used frequently in decision making when objective probabilities are unknown. Bayes was the first mathematician to suggest such an inductive approach to the calculation of probabilities.

Bayes's theorem first appeared in "An Essay Toward Solving a Problem in the Doctrine of Chances" published posthumously in 1763. The theorem addresses the following issue: Given that an event has occurred and that this event may have been the result of two or more causes, what is the probability that the event was the result of a particular cause. To start simply, let us assume that an event A may have been the result of one of the two causes: A_1 or A_2. Bayes asks the question: What is the probability that A_1 caused A? Or, in other words, what is the probability that A_1 occurred given that A occurred? Thus we are looking for $P(A_1 |A)$, which, from the discussion in the above paragraph, is equal to $P(A_1 \cap A)/P(A)$, which is equal to $P(A_1)P(A| A_1)/P(A)$. Now the "total" probability of A or $P(A)=P(A \cap A_1)+P(A \cap A_2)=P(A_1)P(A| A_1)+P(A_2)P(A| A_2)$. Putting all this together, we have Bayes's theorem:

$P(A_1|A)=P(A_1)P(A| A_1)/ (P(A_1)P(A| A_1)+ P(A_2)P(A| A_2))$

The theorem may be extended to any number of possible causes, in which case it is written as:

$P(A_k|A)=P(A_k)P(A| A_k)/\Sigma(P(A_k)P(A|A_k))$, where k takes on integer values from 1 to n, the total number of possible causes.

In the above formula, $P(A_k|A)$ is called the "posterior" probability of A_k given A. This is what we seek to calculate. The expressions $P(A|A_k)$ are called "prior" probabilities. These are often subjective and, therefore, controversial to some statisticians. The scientific, engineering, and business worlds seem to have made their peace with prior probabilities because these disciplines make extensive use of Bayesian **statistics** in decision making. Ironically, it has been in the latter half of the 20th century that this 18th century theorem has come into its own, largely due to the number-crunching capabilities of modern high speed computers. When the number of prior probabilities is large, the hand calculation of posterior probabilities by Bayes's theorem is practically impossible, but poses no problem for the lightning fast computers available at the end of the

20th century. In the 1990s, computing power and speed reached a level at which it was possible to bring Bayes's theorem to bear on complex decision making through the concept of "Bayesian networks." These are complex diagrams that organize a knowledge base in a given field by mapping out cause and effect relationships among key variables and assigning prior probabilities that represent the extent to which one variable is likely to affect another. When these Bayesian networks are programmed into computers, they can generate optimal predictions even in the absence of key pieces of information. The idea is that professionals in a field can make intuitive educated estimates about certain prior probabilities for which the precise values may be unknown. Furthermore, Bayesian networks allow practitioners to periodically update prior probabilities on the basis of new information. Microsoft Corporation was a pioneer in the research leading to the utilization of Bayesian networks, which it now includes in many of its software products. The German conglomorate Siemens has adopted Bayesian networks for application in industry. Hospitals are using Bayesian techniques to assist doctors in making diagnoses of diseases. General Electric has introduced Bayesian networks to take information from sensors attached to an engine and combine this information with data on past engine performance and expert opinion from its own engineers and scientists to predict the likelihood of various engine problems. The field of Artificial Intelligence has received a boost from the introduction of Bayesian networks that are able to "learn" automatically based on new knowledge. The essence of Bayes's message to the world is that decision making can be greatly improved when new information is allowed to be taken into account when calculating probabilities.

BAYES, THOMAS (1702-1761)

English probabilist and minister

Thomas Bayes, a Presbyterian minister, expressed a method of inductive **inference** in a precise and quantitative form, which lead to the development of Bayesian **statistics**, or Bayesian inference. His stature as a mathematician is based on only two short mathematical papers, both of which were published posthumously by the Royal Society of London. The first paper demonstrates what may be the first recognition of asymptotic behavior by series expansions. The second and far more important paper addresses a problem with continuing application in most areas of human endeavor. In this paper, Bayes discusses the estimation of future occurrences of an event, given knowledge of the history of the event—that it has occurred a number of times and failed a number of times. This work continues to spawn mathematical research, and provides the foundations for Bayesian statistical estimation, used today on such diverse problems as electoral polling or estimating time to failure of mechanical devices.

Little is known of Bayes' childhood. Some sources note that he was privately educated, while others state that he received a liberal education in preparation for the ministry. Bayes was the eldest of six children of Joshua and Ann Carpenter Bayes. His father was a Nonconformist minister, one of the first seven publicly ordained in England. Thomas' paternal grandfather, Joshua Bayes, had been a cutler and town collector in Sheffield. Thomas' place of birth is usually listed as London, but one biographer suggests that he was born in Hertfordshire, where his peripatetic father supposedly preached at the time of his birth. Unfortunately, the appropriate parish records of 1700–1706 have been lost. Thomas' epitaph in the family vault at Moorgate states his age at death as 59 in April 1761, placing his birth in 1701 or 1702.

Andrew I. Dale argues persuasively that Bayes was educated at Edinburgh University. Thomas Bayes' name appears in a 1719 catalogue of manuscripts in the Edinburgh University Library, and in a number of other records at the University over the period 1720–1722, including class lists and a list of theologues. The Bayes signature at Edinburgh matches closely that of the Royal Society records. Bayes received only licensure for the ministry at Edinburgh, but he was ordained during or before 1727, and is included in Dr. John Evans' 1727 list of "Approved Ministers of the Presbyterian Denomination." Bayes assisted his father at his ministry in Leather Lane for some years from 1728 before succeeding the Rev. John Archer as minister at Tunbridge Wells, Kent. He spent the remainder of his life in Tunbridge Wells as Presbyterian minister of the Mount Sion meeting house.

Two years after the publication of *The Analyst; or, a Discourse addressed to an Infidel Mathematician* (1734), **George Berkeley**'s famous attack of **Isaac Newton**'s work on **fluxions** (differentials), an anonymous tract was published that answered Berkeley and vigorously defended Newton's work. The tract, titled *An Introduction to the Doctrine of Fluxions and Defence of the Mathematicians against the Objections of the Author of the Analyst*, was widely attributed to Bayes and was probably the reason behind his election as a Fellow of the Royal Society in 1742. In it, he addresses the "business of the mathematician," and stated that "[he] is not inquiring how things are in matter of fact, but supposing things to be in a certain way, what are the consequences to be deduced from them; and all that is to be demanded of him is, that his suppositions be intelligible, and his inferences just from the suppositions he makes." The proposal for Bayes' election to the Royal Society read in part that he was "well skilled in **geometry** and all parts of mathematical and philosophical learning." It was signed by Eames, James Burrow, Cromwell Mortimer, Martin Folkes, and Earl Stanhope.

Bayes retired from the ministry around 1750. He died at Tunbridge Wells on April 17, 1761, leaving a fairly substantial estate, and was buried in the family vault at Bunhill Fields Burial Ground at Moorgate. Upon his death, Bayes' family asked his friend, the Unitarian Reverend Richard Price, to examine his papers. Among them Price found Bayes' work on probability.

In 1764, Bayes' paper on the Stirling–De Moivre **Theorem**, dealing with series expansions, was published in the *Philosophical Transactions of the Royal Society*. Price declared the problem in inductive reasoning stated by Bayes to be central "to the argument taken from final causes for the existence of the Deity." The same issue of *the Philosophical*

Transactions contains a second piece by Bayes, "An Essay towards Solving a Problem in the Doctrine of Chances," presented also by Price with his preface, footnotes and appendix. The problem posed in the essay, wrote Price, is "to find out a method by which we might judge concerning the probabilitythat an event has to happen, in given circumstances, upon supposition that we know nothing concerning it but that, under the same circumstances, it has happened a certain number of times, and failed a certain other number of times." The essay was followed in the next volume by "A Demonstration of the Second Rule in the Essay...," a continuation of Bayes' results which were further developed by Price.

A notebook belonging to Bayes has been preserved in the London records room of the Equitable Life Assurance Society, due to the action of Price and his nephew William Morgan, an actuary. Among other curiosities, the notebook includes the key to a system of shorthand, details of an electrifying machine, lists of English weights and measures, notes on topics in mathematics, natural philosophy and celestial **mechanics**, and a **proof** of one of the rules in the "Essay" that was published after Bayes' death.

BEAUTY IN MATHEMATICS

"Euclid alone has looked on Beauty bare," wrote the poet Edna St. Vincent Millay. And ever since Euclid's time, lovers of mathematics have marveled at its beauty, even if they have glimpsed it, as St. Vincent Millay wrote later in her sonnet, "once only and then from far away."

One of the beauties of mathematics is that it promises us eternal truths that do not depend on opinion or fashion. One plus one will never be three. But mathematics is not just about assertions of fact; the deeper **truth** usually lies in the *explanation* of the fact. A good definition of beauty in mathematics would be: *simplicity* that leads to *insight*. "Proof is beautiful," wrote the late Harvard mathematician Gian-Carlo Rota, "when it gives away the secret of the **theorem**, when it leads us to perceive the actual and not the logical inevitability of the statement that is proved."

Though the most beautiful solution is almost invariably the simplest one, it is not necessarily the most obvious. This counterintuitive kind of simplicity is hard won. It may come in a flash, but that flash is usually preceded by hours—or years—of hard mental labor. It reveals hidden patterns and deep meanings, and it often gives its discoverers the feeling that it was not the product of human intellect but must have been there all along. Surely this same sense of wonderment was felt by the twelfth-century Indian mathematician **Bhaskara**, who presented a famous **proof** of the **Pythagorean theorem** consisting of a single diagram with a one-word inscription: "Behold!"

Mathematical works, like symphonies or paintings, can be profoundly original. Often, this originality takes the form of a connection between two different problems or areas of mathematics that had not previously seemed to be related. **Leonhard Euler**, in his proof of the identity $1 + 1/2^2 + 1/3^2 + 1/4^2 +... = pi^2/6$, daringly treated the **sine** function (from the

world of **trigonometry**) as if it were a polynomial (from the world of **algebra**) and thereby discovered a connection that had never occurred to anyone. Quite apart from the beauty of Euler's proof, one could argue that the identity itself is beautiful, as it links two of the most organic concepts in mathematics: the series of **integers** 1, 2, 3, 4,... and the irrational number pi. It also contains a strong whiff of the unexpected: how did pi get there, and where did that 6 come from? To number theorists, Euler's formula is now "elementary," but it will never stop being elegant.

Visual attractiveness plays a role in mathematics and is increasingly revealed as graphic technology improves. Any mathematical portrait gallery should include the infinite regresses and labyrinths of **fractals** such as the **Mandelbrot set**; the holistic unity of the icosidodecahedron, which represents the molecular structure of carbon-60 or buckminsterfullerene; and the graceful sweep of minimal surfaces such as the helicoid, which is shaped like a DNA molecule. Most mathematicians would probably agree that the beauty of these objects was already immanent in the mathematics even before technology caught up to the task of portraying them. They are beautiful because they are symmetric, and mathematics places a high value on **symmetry**. (It is the motivating idea behind **group theory** and an important theme in many other branches of mathematics.)

The examples of buckminsterfullerene and DNA mentioned above bring up the point that mathematics and nature are inextricably intertwined. The ancient Pythagoreans taught that "all is number" and discovered the relationship between numbers and musical harmony: a lute string that is twice as long as another will sound one octave lower. The Fibonacci numbers can be found in the elegant spirals of a pine cone, a pineapple, or a sunflower. Physicists have used mathematical beauty as a compass in constructing new theories of the universe, such as the "eightfold way" of Murray Gell-Mann that led to the prediction of quarks. Galileo wrote, "The universe... cannot be understood unless one first learns to comprehend the language and interpret the alphabet in which it is written. It is written in the language of mathematics...." This connection between mathematics and the secrets and harmony of the universe may be, for many people, the best route to appreciating its beauty.

Nearly every subject in mathematics has been described as beautiful by someone at some time: public key encryption, **differential equations**, the **central limit theorem**, Maxwell's **equations**, fluid **mechanics**, graph theory. In mathematics, as in art, beauty is in the eye of the beholder. However, mathematicians—unlike some modern artists—have never turned their backs on beauty as a standard of merit. For them, the highest accolade that a new discovery can receive is not that it is logical or useful, but that it is elegant. "Beauty is the first test," wrote number theorist G. H. Hardy. "There is no permanent place in the world for ugly mathematics."

See also Mathematics and art; Mathematics and culture; Mathematics and literature; Mathematics and music; Mathematics and philosophy; Mathematics and physics; Symmetry; Truth

BELLOW, ALEXANDRA (1935-   )
Romanian-American mathematician

Ergodic theory, Alexandra Bellow's field of specialization, deals with the long term averages of the successive values of a function on a set when the set is mapped into itself, and whether these averages equal (converge to) a reasonable function on the set. The theory applies to probability and time series, and to the concept of entropy in physics and **information theory**. Bellow has proved significant results in this field.

Alexandra Bellow was born in Bucharest, Romania, on August 30, 1935. Her father, Dumitru Bagdasar, had studied medicine in the United States. He was a famous neurosurgeon who founded a school of neurosurgery in Romania in that year. Her mother, Florica Bagdasar, was a psychiatrist specializing in the treatment of mentally retarded children. The Romanian philosopher Nicolae Bagdasar was Bellow's uncle. After World War II, the politics of Romania were extremely unstable as the communists took control of the country. Bellow's parents supported the communists, and in 1946, her father was appointed minister of health in the Groza government. Bellow's father was soon accused of "defection," removed from the ministry, and imprisoned. He died in 1946, reportedly of cancer. Bellow's mother succeeded him as minister of health. According to the *New York Times*, she was the first woman to hold a ministerial position in Romania. However, by 1948 she was removed from that post. Bellow told Ruth Miller (a biographer of her second husband, Saul Bellow) that her mother was accused of "cosmopolitanism" and was prohibited from doing any work or practicing medicine.

In spite of these problems, Bellow studied mathematics at the University of Bucharest. In 1956, she married Cassius Ionescu Tulcea, a professor of mathematics at the university. She received a M.S. in mathematics in 1957. In that year, Bellow and her husband came to the United States to study at Yale University. He was a research associate at Yale while they were students. They both received Ph.D.s in mathematics from Yale in 1959. Bellow's thesis was titled "Ergodic Theory of a Random Sequence." After graduation, Bellow became a research associate at Yale from 1959 to 1961 and at the University of Pennsylvania from 1961 to 1962. She taught at the University of Illinois at Urbana as an assistant professor from 1962 to 1964, then as associate professor from 1964 to 1967. In 1967, Bellow went to Northwestern University as a professor. Her husband held positions at the same schools, although not exactly in the same years. Both remained at the mathematics department at Northwestern for the rest of their careers. Their book, *Topics in the Theory of Liftings*, appeared in 1969. They were divorced in that year.

Starting in 1971, Bellow published papers on ergodic theory. In 1974 she married the writer Saul Bellow, and in the following year they traveled to Israel, where she taught and worked with colleagues in mathematics at the University of Jerusalem. Saul Bellow won the Nobel Prize for literature in 1976. In the late 1970s Bellow published several papers on asymptotic martingales. Bellow edited, with D. Kolzow, the proceedings of a conference on Measure Theory held in

Oberwolfach in 1975. She was an editor for the *Transactions of the American Mathematical Society* from 1974 to 1977, associate editor of the *Annals of Probability* from 1979 to 1981, and associate editor for *Advances in Mathematics* since 1979. Bellow's 1979 paper with Harry Furstenberg on applying **number theory** to ergodic theory contains the Bellow-Furstenberg **Theorem**.

Bellow was a Fairchild scholar at CalTech in 1980, and also visited the University of California at Los Angeles. She was divorced from Saul Bellow in 1986. Bellow received an award from the Alexander von Humboldt Foundation, which sponsors visits to Germany by scholars, in 1987. In 1989, she and Roger L. Jones of De Paul University organized a conference on "Almost Everywhere Convergence in Probability and Ergodic Theory" at Northwestern University and edited the conference proceedings. In that same year, Bellow married **Alberto P. Calderon**, a distinguished mathematician and civil engineer retired from the University of Chicago, whose fields of research are partial **differential equations**, functional **analysis** and **harmonic analysis**.

In 1991, Bellow gave the **Emmy Noether** Lecture for the Association for Women in Mathematics. Bellow, Jones, and others have collaborated on eight recent papers that deal primarily with partial sequences of observations. This work consists of their attempts to identify when averages based on partial observations are probably valid for the whole population, when they are not valid, and by how much. A 1996 paper in this series had six authors, unusual for a paper in mathematics, which was featured in the *Mathematical Reviews*. It describes averages that behave very badly. Jones describes Bellow as an excellent collaborator who is extremely knowledgeable, "very clever and very careful," and, according to another colleague, "never seems to make a mistake." Jones also reports that Bellow enjoyed her teaching and was a very good instructor who supervised at least four Ph.D. students. A conference was scheduled at Northwestern in October 1997 on the occasion of Bellow's retirement. Bellow plans to continue her research in mathematics.

BERKELEY, GEORGE (1685-1753)
Irish Philosopher, Anglican cleric, and mathematician

Born in the same year as the great composers Johann Sebastian Bach, Georg Frideric Handel, and Domenico Scarlatti, Berkeley was one of the seminal figures in Western philosophy, his doctrines exerting a particularly significant influence on analytic philosophy. As a mathematician, George Berkeley is known for his thought–provoking critique of the mathematical theories of his time, particularly infinitesimal **calculus**.

Of English descent, Berkeley was born near Kilkenny, Ireland, and always considered himself an Irishman. Educated at Trinity College, Dublin, he studied mathematical logic, and philosophy. Berkeley graduated in 1704, publishing a short Latin work on mathematics in 1707. In 1710, the year he published his famous work *A Treatise Concerning the Principles of Human Knowledge*, he was ordained priest of the Church of England. After holding several academic appointments, he was named dean of Derry in 1724. Owing to his keen interest in education, Berkeley soon left for London, hoping to receive government funding for a college in Bermuda, where he intended to provide education for English and local youths. In 1728 he married Anne Forster, the well–educated daughter of a chief justice. Soon after the wedding, Berkeley and his new wife set sail, getting as far as Newport, Rhode Island, which he then decided was a better location for his school. However, the project was abandoned when promised funding failed to materialize, and the Berkeleys returned to London in 1731.

Berkeley was appointed bishop of Cloyne in 1734, and his home there became a social and cultural center, as well as a dispensary in times of epidemics. The Berkeleys eventually had six children, four sons (one became canon of Canterbury) and two daughters. They retired to Oxford in 1752, and after Berkeley's death in 1753, Anne Berkeley continued to defend her husband's philosophy. An indefatigable writer, polemicist, and researcher, Berkeley was also a clergyman, who took his pastoral duties very seriously, ministering to the needs of people far removed from the world of 18th–century philosophy and science.

Berkeley denied the existence of matter. The essence of his philosophy is expressed by the statement *esse est percipi* (to exist is to be perceived), which means that an object can be said to exist only insofar as it is perceived by a spirit—finite (a human being) or infinite (God). Certainly, Berkeley, being a very practical person, did not claim that the world of physical objects should be treated as an illusion, as some of his detractors naively assumed. In essence, Berkeley asserted that the **postulate** suggesting that matter existed independently of a percipient observer was based on illogical and unclear thinking. However, Berkeley does not deny the validity of general terms. As Copleston has written in his discussion of Berkeley's philosophy, "A proper name such as William, signifies a particular thing, while a general word signifies indifferently a plurality of things of a certain kind. Its universality is a matter of use or function. If we once understand this, we shall be saved from hunting for mysterious entities corresponding to general words. We can utter the term 'material substance', but it does not denote any abstract general idea; and if we suppose that because we can frame the term it must signify an entity apart from the objects of perception, we are misled by words.... 'Matter' is not a name in the way in which William is a name, though some philosophers seem to have thought mistakenly that it is."

Berkeley accepted mathematics as a practical science and pursuit, but adamantly rejected, in accordance with his criticism of meaningless concepts, the idea of number. As J. O. Urmson explains, to Berkeley, the term *ten* may denote the fact that there are ten *individual* entities in a group, but nothing more than that, and certainly not an abstract idea of *ten*, independent of any practical context. It is also important to note that Berkeley supplemented his purely philosophical critique of **infinitesimal** calculus with solid mathematical arguments. What Berkeley strenuously objected to was the

George Berkeley

practice, accepted by his contemporaries, of assuming that the quantity dx, being infinitesimally small, could simply be eliminated in mathematical derivations. Thus, though Berkeley's criticism may be irrelevant to modern applications of **infinitesimals**, he was nevertheless right in questioning the practice of treating an infinitesimal quantity as **zero**.

In the **history of mathematics**, Berkeley is best known for attacking the logical foundations of **Isaac Newton**'s calculus."Newton's theory," according to Tobias Dantzig, "dealt with continuous magnitudes and yet postulated the infinite divisibility of **space** and time; it spoke of a flow and yet dealt with this flow as if it were a succession of minute jumps." This theory of **fluxions** (Newton's term denoting the rate of change of a **variable**, such as length, speed, **area**, etc.) was open to criticism because it attempted to reconcile a smooth flow with a series of leaps. In *The Analyst* Berkeley asked: "And what are these fluxions? The velocities of evanescent increments. And what are these same evanescent increments? They are neither finite quantities, or quantities infinitely small, nor yet nothing. May we not call them the ghosts of departed quantities?" Although Berkeley felt that Newton's calculus yielded true results (even developing a clever explanation for these correct results), he nonetheless felt compelled to point out the logical fallacy on which he believed the calculus was based. In this way, he inspired other mathematicians to focus their attention on a logical clarification of calculus.

BERNAYS, PAUL (1888-1977)

English-born Swiss logician

Paul Bernays secured his reputation with a classic treatise on mathematical logic, the *Foundations of Mathematics,* and through his refinement and consolidation of **set theory** into the von Neumann-Bernays system. Bernays was a platonic mathematician—one who thought of the world of mathematics as separate from the world of material reality. Although Bernays's concept of mathematics as a mental product meant that no system could be designated as right or wrong, he believed that there were truths within the mathematical realm that allowed for a system to remain consistent and logical within itself.

Paul Isaac Bernays was born in London on October 17, 1888, to Julius Bernays, a Swiss businessman from a prominent Jewish family, and Sara Bernays. Shortly after his birth, the family moved to Paris and then to Berlin, where Bernays studied from 1895 to 1907. While studying engineering at the Technische Hochschule (Technical High School) in Charlottenburg, Germany, he developed an interest in pure mathematics. This led him to transfer to the University of Berlin where he studied for four semesters under a distinguished faculty that included philosopher Ernst Cassirer and physicist Max Planck. Bernays then attended the University of Göttingen where physicist Max Born and mathematician **David Hilbert** were among his professors. In 1912 Bernays received his doctorate degree from Göttingen under Hilbert.

In 1912 Bernays completed his postdoctoral thesis on modular elliptic **functions** at the University of Zurich in Switzerland under the German mathematician Ernst Zermelo. Bernays remained at Zurich until 1917 when he was invited by Hilbert to return to Göttingen to assist with a program on the foundations of mathematics. Bernays completed a second postdoctoral thesis at Göttingen in 1918 on propositional logic. In addition to serving as assistant to Hilbert, he gave lectures at Göttingen until the Nazi party's rise to power in 1933 when Bernays's right to lecture was withdrawn because of his Jewish background.

Bernays escaped to Zurich in 1934, eventually teaching at the Eidgenossische Technische Hochschule. In 1935 and 1936 he participated in the Institute for Advanced Study at Princeton University in New Jersey. Bernays published the first volume of his work on mathematical logic, *Foundations of Mathematics,* in 1934; the second volume was published in 1939. Research for this work was a collaborative effort between Bernays and Hilbert, but Bernays wrote both volumes singlehandedly. In this book Bernays and Hilbert created the mathematical discipline of **proof** theory, in which the correctness of a mathematical statement or **theorem** is demonstrated in terms of accepted axioms. E. Specker, a colleague of Bernays at the Eidgenossische Technische Hochschule, remarked in *Logic Colloquium '78* that the *Foundations of Mathematics* is unique because "it does not reduce mathemat-

ics to logic, or logic to mathematics—both are developed at the same time."

Over the years Bernays published a series of articles in the *Journal of Symbolic Logic* that was published collectively as *Axiomatic Set Theory* in 1958. Axiomatic set theory applies proof theory to set theory, the study of the properties and relationships of **sets**. Thus, axiomatic set theory involves the presentation of set theory in terms of fundamental axioms and logical rules of **inference**, rather than as a formalization of tabulated or intuitive knowledge. Classical set theory was largely established by **Zermelo** at the turn of the century and improved by the German mathematician Abraham Fraenkel in the 1920s. The **Zermelo-Fraenkel** (ZF) system was defined exclusively in terms of sets, but it could not address transfinite sets (for example, the set consisting of all possible sets). In the late 1920s the axioms of Hungarian mathematician **John von Neumann** accomplished many tasks previously left unsolved by the Zermelo-Fraenkel system. However, von Neumann's system was expressed in **symbolic logic** and was defined in terms of function rather than set, and it was less practical in both pure and **applied mathematics**.

Bernays's contribution to set theory both improved and simplified von Neumann's system. Bernays introduced a distinction between "sets" and "classes" to set theory. He did not view "classes" as mathematical objects in the normal sense. As G. H. Muller characterizes Bernays's distinction in *Mathematical Intelligencer,* a set is a collection of elements or members, a "multitude forming a proper thing." A class is a collection of objects that can be manipulated or extended, a "predicate regarded only with respect to its extension." For each set there was a corresponding class, but for each class there need not be a corresponding set. This idea created two axiomatic systems, one for sets and one for classes. The sets in the von Neumann-Bernays system operate similarly to those in the Zermelo-Fraenkel system, and thus a new system was created to allow for the construction of classes.

After World War II Bernays became Extraordinary Professor at the Eidgenossische Technische Hochschule. He also served as visiting professor at the University of Pennsylvania and at Princeton, where he was again a member of the Institute for Advanced Study in 1959–60. Bernays served as the president of the International Academy of the Philosophy of Science, as honorary chair of the German Society for Mathematical Logic and Foundation Research in the Exact Sciences, and as a corresponding member of the Academy of Science of Brussels and of Norway. He also served on the editorial boards of several journals, including *Dialectica, Journal of Symbolic Logic,* and *Archiv fur mathematische Logik und Grundlangenforschung.* Bernays received an honorary doctorate from the University of Munich in 1976 for his contributions to proof and set theory. Although he remained based in Zurich until his death from heart disease in 1977, Göttingen was always more of a home for Bernays. Bernays, who never married, lived most of his life with his mother and two sisters.

BERNOULLI, DANIEL (1700-1782)

Swiss mathematical physicist and scientist

Known as the discoverer of the Bernoulli principle, which applies the law of conservation of energy to fluids, Daniel Bernoulli was a true polymath, excelling in many fields of science, and contributing brilliant ideas and insights which not only shaped 18th century science, but also anticipated future discoveries. In essence, Bernoulli's scientific spirit easily transcended the boundaries set by scientific disciplines, surveying the landscape of science from a height where the idea of separate fields of study seems meaningless. Nevertheless, many of his insights stemmed from his profound understanding of mathematics, and from his extraordinary ability to illuminate physical phenomena by mathematical reasoning.

Born in Gröningen, Netherlands, Bernoulli was the second son of **Johann Bernoulli**, professor of mathematics at the University of Gröningen. In 1705, when Johann took over his brother **Jakob Bernoulli** 's chair of mathematics at the University of Basel, the family returned to Switzerland. A precocious student, young Daniel studied logic, philosophy, and mathematics, earning a master's degree at the age of 16. His father, who tried to keep him away from a scientific career, grudgingly allowed him to study medicine.

Bernoulli finished his medical studies in 1721, writing a dissertation on breathing entitled "De respiratione." Unable to obtain a teaching position, Bernoulli continued his medical studies in Italy, also keeping up with his mathematical studies. In fact, his most important accomplishment during his stay in Italy was a mathematical work entitled *Exercitationes quaedam mathematicae*, published in Venice in 1724. In this treatise, Bernoulli discussed a variety of subjects, including probability and fluid **motion**. Most notably, Bernoulli succeeded in integrating Jacopo Franceso Riccati's well–known differential equation by separating its variables. Bernoulli's *Exercitationes*so impressed the scientific community that he was immediately offered a teaching position at the St. Petersburg Academy (his older brother Nikolaus was also invited to teach mathematics). In 1725, having won his first prize from the French Académie Royale des Sciences (he would win another nine), Bernoulli moved to St. Petersburg.

Although officially a professor of mathematics, Bernoulli worked in several fields. For example, his medical publications include important papers on subjects such as muscular contraction and the optic nerve, and his writings on physics include a paper on oscillation. Nevertheless, Bernoulli's mathematical genius seemed to inform most of his endeavors, a case in point being research in the areas of probability and **statistics**. In St. Petersburg, Bernoulli further developed his ideas on probability, demonstrating, for example, the importance of probability for economic theory. Furthermore, he extended the relevance of probability to a seemingly unrelated field such as ethics, proposing a hypothesis which asserts that if a person's material fortune increases geometrically, his or her moral fortune will increase arithmetically. Bernoulli used the moral hypothesis in an attempt to solve the famous "Petersburg paradox," which he and his

Daniel Bernoulli

brother Nikolaus concocted. The paradox emerges from an imagined game of tossing a coin. When the first toss is head, B pays A one dollar. If the first toss is tail, A gets nothing, but if the second toss is head, he gets two dollars, and so on, the idea being that if head first appears on the nth toss, A should get 2^{n-1} dollars. The mathematical expectation of A's winnings is infinite, but when the French naturalist Georges–Louis–Leclerc, Comte de Buffon, sought an empirical verification of Bernoulli's paradox, he found that after 2,084 games, A would have received only $10,057, an average of less than $5 per game.

In 1726, Bernoulli discussed the parallelogram of forces in paper on **mechanics**; the following year, he started corresponding with his compatriot **Leonhard Euler**, one of the greatest mathematicians of that era, who was also interested in mechanics. This research in mechanics, begun in St. Petersburg, laid the foundations for Bernoulli's brilliant, and even prophetic, discoveries in the field of physics.

Although his work in St. Petersburg was highly successful, Bernoulli was so eager to obtain a post in Basel that he accepted a professorship in botany, a subject he did not particularly care for. Nevertheless, Bernoulli continued his research in mechanics, along with his multifarious scientific pursuits. Thus, for example, he delivered, in 1737, a historic lecture detailing the calculations needed to measure the work done by the heart.

In 1738, Bernoulli published his seminal book, *Hydrodynamica*, begun in St. Petersburg and completed in 1734. This may have inspired Bernoulli's father, Johann, who strongly resented his son's success, to publish his *Hydraulica*, predating the publication to 1732 in an effort to claim some of Bernoulli's discoveries for himself. However, *Hydrodynamica* established Bernoulli as one of the great scientists of his time. Among the important discoveries included in the book are Bernoulli's principle, one of his many insights into the behavior of fluids, which affirms that a fluid's pressure diminishes as its velocity increases. The application of Bernoulli's principle still underlies numerous industrial designs, including boats, automobiles, aircraft, and fluid conduits. Even more important, perhaps, is Bernoulli's explanation, presented in the tenth chapter, of the mechanics of gases. Gases, he posited, consist of fast and randomly moving particles. Placed in a **cylinder**, the particles exert constant pressure on the cylinder's walls. By applying outside pressure on a quantity of gas trapped in a cylinder by depressing a piston, Bernoulli confirmed Boyle's law, discovered in 1660 by **Robert Boyle**, which states that, in a closed container, gas pressure is inversely proportional to its **volume**, i.e., that the product PV (where P is pressure, and V is volume) remains constant if the temperature stays unchanged. As Sheldon Glashow pointed out in his *From Alchemy to Quarks*, "Bernoulli presented a mathematical **deduction** of Boyle's law from the principles of mechanics." Furthermore, Bernoulli also realized that gas pressure is in direct proportion to its temperature. Connecting the idea of pressure and particle motion, Bernoulli understood the correlation of temperature and particle speed: as the temperature rises, the particles accelerate. Bernoulli's discovery was one of the great moments in the history of science, for his extraordinary insights laid the foundations for the kinetic theory of gases, which in the 19th century fully validated his profound understanding of the nature of gases. "Once the temperature of a gas," Glashow wrote, "was recognized to be proportional to the mean kinetic energyof its constituent molecules, the transformation of mechanic energy into heat was demystified. It is the conversion of ordered motion into the disordered motion of molecules."

In 1743, Bernoulli became professor of physiology, a field which he much preferred to botany, bu it was not until 1750 that he obtained the chair of Natural Philosophy, or physics, an appointment which perfectly matched his scientific pursuits. An immensely popular lecturer, Bernoulli retained this post until his retirement in 1776.

As many scholars have asserted, it was in the field of physics that Bernoulli most brilliantly applied his mathematical genius. "Daniel Bernoulli," A. Wolf wrote in his *A History of Science, Technology, and Philosophy in the Eighteenth Century*, "applied himself especially to solving, with the aid of the new analysis, difficult mechanical problems of which the geometrical methods adhered to by Huygens, or by Newton in his *Principia*, offered no prospect of a successful solution. He

must be therefore regarded as one of the principal founders of that branch of science known as *Mathematical Physics*."

When Bernoulli tackled the phenomenon of kinetic energy, which was then called *vis viva*, or *living force*, he (going a step further than **Christiaan Huygens** and **Gottfried Wilhelm von Leibniz**, who studied kinetic energy in an uniform gravitational field) posited that the principle of the conservation of *vis viva* is valid for the entire universe. While he lacked the tools to verify his insight empirically, Bernoulli correctly anticipated the law of energy conservation, which was finally empirically confirmed by J. P. Joule in the 1840s. According to Donald E. Tilly and Walter Thumm in their textbook *College Physics*, since Joule's confirmation of Bernoulli's ideas, "the concept of energy, as a conserved quantity that can be transformed into many guises but never created or destroyed, has emerged perhaps as the most useful idea in all science."

Bernoulli proceeded to apply his mathematical insights to natural phenomena. While studying the nature of sound, Bernoulli discovered distinct mathematical regularities in the shape of sound waves. For example, he found that every particular body vibrates at certain natural, or *proper*, frequencies, the lowest natural frequency being the *fundamental frequency*, and the higher frequencies called the *overtones*. He also discovered that the number of *nodes*, or quiet spots where no **vibration** occurs, increases with higher frequency. Translating these insights into the language of mathematics, Bernoulli was able to calculate the natural frequency of a variety of musical instruments. In addition, he demonstrated that the intervals between overtones are not always harmonious.

Unlike his many predecessors, who often described music and mathematics as kindred arts forms from a purely philosophical point of view, Bernoulli demonstrated that these two arts indeed share certain fundamental traits. For example, his ideas about the nature of sound, particularly his remarkable insight that any small vibration consists of a series of natural frequencies, or modes, the higher frequency is superimposed on the lower, were expressed in the language of mathematics. It is significant that Bernoulli's seminal work in the field of acousticsdirectly anticipated such great discoveries of 19th–century mathematics as Jean–Baptiste–Joseph Fourier's **harmonic analysis**. What Bernoulli knew intuitively, namely that a sound can be represented as a trigonometric function, Fourier expressed in precise terms, as a series of **sine** waves. Fourier thus confirmed Bernoulli's understanding of the mathematical and physical nature of sound, particularly the extraordinary idea that each complex musical sound, such as the sound made by a musical instrument, consists of a series of simple sounds, such as those produced by a tuning fork.

Bernoulli died in Basel on March 17, 1782, five years after his retirement from the university.

BERNOULLI, JAKOB (1645-1705)
Swiss probabilist

Jakob Bernoulli, also known as Jacques, James, or Jakob I— to avoid any confusion with Jakob Bernoulli II—is one of the great names of 17th–century mathematics as well as the first member of the prodigiously mathematical Bernoulli family to attain international fame. Originally from the Spanish Netherlands, the Bernoullis moved to Basel, Switzerland, in 1583, to escape Spanish oppression; the first prominent Bernoulli was Nikolaus I, Jakob's father. A contemporary of the great German philosopher and mathematician **Gottfried Wilhelm von Leibniz**, with whom he maintained a correspondence, Bernoulli is known for his extraordinary contributions to **calculus** and the theory of probability.

Bernoulli earned a degree in theology in 1676, having studied mathematics and astronomy against his father's wishes. Employed as a tutor in Geneva in 1676, he later spent two years in France, where he studied the works of **René Descartes**. In 1681, Bernoulli traveled to the Netherlands and England, where he met **Robert Boyle**. During this period, Bernoulli wrote on a variety on scientific topics, including comets and gravity. Following his return to Basel 1682, Bernoulli founded a school for science and mathematics, where he lectured and conducted experiments in the field of **mechanics**. He also wrote articles for the two preeminent European scientific journals, *Journal des sçavans* and *Acta eruditorum.* In 1687, Bernoulli was named professor of mathematics at the University of Basel, a position he held until his death. Owing to his passion for learning, Bernoulli carefully studied the works of predecessors, such as Descartes, and contemporaries, such as Leibniz.

Leibniz's first discussion of the **differential calculus**, a six–page article, appeared in *Acta eruditorum* in 1684 (the presentation of the **integral calculus** was published two years later). While the term *calculus* refers to the process of calculation in general, Leibniz's calculus dealt with infinitesimals, quantities smaller than any definable finite quantity but larger than **zero**, and was therefore called **infinitesimal** calculus. It should be pointed out, however, that the term *infinitesimal* is no longer used in mathematical terminology, infinitesimal quantities being instead named *limit values*. Therefore, when mathematicians refer to calculus, the **predicate** *infinitesimal* is implied.

Calculus encompasses four distinct types. *Differential calculus* calculates derivatives. *Integral calculus* is the reverse of differential calculus—in other words, we use this type of calculus to determine a function when its derivative is known. *Calculus of variation* is used to find a function for which a given integral assumes a maximal or a minimal value. **Differential equations** are **equations** containing derivatives. Misunderstood by most of his colleagues, Leibniz's discovery nevertheless attracted a small following of mathematicians who realized the tremendous analytical power of calculus. Among Leibniz's followers, Bernoulli was among the first who completely grasped the essence of calculus, and he proceeded, in numerous contributions to *Acta eruditorum*, to develop the foundations of calculus.

In 1689, Bernoulli formulated the famous "Bernoulli inequality"—$(1 + x)^n > 1 + nx$, where x is real, $x > -1$, and $x \neq 0$, and $n > 1$—which had in fact already been presented in 1670 by **Isaac Barrow** in his *Lectiones geometriae*. Bernoulli

Jakob Bernoulli

also provided solutions for several famous mathematical problems. For example, he solved the **catenary** equation. When a flexible, non–elastic cable is suspended from two fixed points, the shape it assumes as a result of gravity is a catenary. Bernoulli formulated an equation to refute the traditional hypothesis among mathematicians that the catenary is a **parabola**. In his investigations of the isochrone, a plane curve enabling an object to fall with uniform velocity, Bernoulli found an equation which demonstrated that the necessary curve is a semicubical parabola. In fact, while working on the isochrone problem, Bernoulli used the term *integral,* in an 1690 article, to denote the inverse of the differential calculus. Leibniz later agreed that *calculus integralis* was a more concise term than the original *calculus summatorius*.

Bernoulli was among several mathematicians, including his younger brother **Johann Bernoulli** and Leibniz, who worked on the brachistochrone (the term was derived from the Greek words *brachystos*, meaning *the shortest* and *chronos*, meaning *time*) problem. Essentially, the problem challenged mathematicians to find a curve of quickest descent between two given points A and B, assuming that Bdoes not lie right beneath A. While his brother correctly assumed that the required curve is a cycloid but offered an incorrect **proof**, Bernoulli provided the correct proof. The competition between the two brothers became so intense that Johann appropriated the brachistochrone solutionas his own, which was just one episode in their bitter struggle for preeminence in mathematics.

Much of Bernoulli's work was devoted to the study of curves. The *lemniscate*, or figure eight curve (from the Latin term *lemniscus*, meaning *ribbon*), was named after him. However, he was totally fascinated by the logarithmic **spiral**, which in nature can be seen in a cross section of the shell of a chambered nautilus. Bernoulli noticed that the logarithmic spiral has several unique properties, including self–similarity, which means that any portion, if scaled up or down, is congruent to other parts of the curve. In fact, Bernoulli was so taken by this spiral, also called *spira mirabilis*, or *wonderful spiral*, that he requested that it be engraved on his tombstone, along with the Latin inscription *Eadem mutata resurgo* ("Though changed, I arise again the same").

Bernoulli's research on probabilityis documented in his treatise *Ars conjectandi* ("The Art of Conjecture"), published posthumously in 1713. Building on earlier writings on the subject, including **Girolamo Cardano**'s *Liber de ludo aleae* ("On Casting the Die"), published in 1663, the correspondence between **Pierre de Fermat** and **Blaise Pascal**, and **Christiaan Huygens**' 1656 book *De ratiociniis in ludo aleae* ("On Reasoning in Games of Chance"), Bernoulli created a work which scholars consider the substantial book on probability. In his book, to which he attached a treatise on **infinite series** (*Tractatus de seriebus infinitis*), Bernoulli presented his famous **theorem** which Siméon–Denis Poisson later named the "Law of Large Numbers." This law states that if a very large number of independent trials are made, then the observed proportion of successes for an event will, with probability close to 1, be very close to the theoretical probability of success for that event on each individual trial. Unfortunately, Bernoulli's treatise ends with his theorem, which he had hoped to use as the foundation of his project to apply the calculus of probability to a variety of fields, including demographics, politics, and economics.

BERNOULLI, JOHANN (1667-1748)
Swiss infinitesimal mathematician

Johann Bernoulli (also known as Jean Bernoulli) was born in Basel, Switzerland. He studied physics and medicine and wrote his doctoral dissertation on medicine. However, his primary interest was mathematics, and Bernoulli studied mathematics privately with his brother, **Jakob Bernoulli**, who was a professor of mathematics at the University of Basel. While still formally a medical student, Bernoulli became passionately involved in the emerging field of infinitesimal **calculus**, joining, along with Jakob, the select group of European mathematicians who fully grasped calculus. His first notable accomplishment in the field of mathematics is the solution to the **catenary** problem, which was posed by his brother Jakob. In fact, Bernoulli's doctoral dissertation, "De motu musculorum," although dealing with muscle contractions, is a mathe-

matical treatise, written under the influence of the Italian mathematician and physiologist Giovanni Alfonso Borelli.

Already regarded as a member of the European mathematical elite, Bernoulli traveled to Paris in 1691, quickly gaining access to the circle of France's elite scientists and philosophers, who were particularly impressed by his development of a formula for the radius of curvature of a curve, actually first given by Jakob. In Paris, Bernoulli became known as a partisan and representative of **Gottfried Wilhelm von Leibniz**'s **infinitesimal** calculus. In 1692, he began instructing a young mathematician, the Marquis **Guillaume–François–Antoine de L'Hospital**, in calculus. In addition to hiring Bernoulli as a teacher, L'Hospital also bought his mathematical discoveries. The method known as "L'Hospital's Rule" (see **L'Hôpital's rule**), for example, was among the discoveries that L'Hospital bought from his teacher. L'Hospital's Rule enabled mathematicians to determine the limiting value of a fraction in which both numerator and denominator tend toward **zero**. This idea, as well as other mathematical insights provided by Bernoulli, was eventually included in L'Hospital's book, *Analyse des infiniment petits* (1696), which was the first, and for almost a century the principal textbook of **differential calculus**. While the principal ideas used by L'Hospital came from Bernoulli, evidence suggests that the Marquis, a competent mathematician in his own right, contributed at least some ideas, such as the rectification of the logarithmic curve. That same year, in an anonymous contribution to the *Journal des sçavans*, Bernoulli used calculus to solve "Debeaune's problem." Posed by the French mathematician Florimond Debeaune, the problem consisted in determining a curve from a property of its tangent. In 1693, Bernoulli began a fruitful correspondence with Leibniz, vehemently defending the German philosopher as the true discoverer of calculus against any claims put forth by **Isaac Newton**'s partisans. During this period, Bernoulli also wrote extensively, publishing articles in *Acta eruditorum* and *Journal des sçavans*, Europe's leading scientific journals.

In 1695, thanks to a recomendation by **Christiaan Huygens**, Bernoulli obtained the chair of mathematics at the University of Groningen in The Netherlands. He eagerly accepted the position, knowing that a professorship at Basel was out of the question as long as his brother Jakob was a faculty member. Bernoulli felt intense animosity toward his brother, who insisted on portraying himself as the more competent mathematician. It is true that Bernoulli learned much from his older brother; however, the student had become his master's equal.

Wishing to challenge his colleagues, Bernoulli posed the famous brachistochrone problemin the June 1696 issue of *Acta Eruditorum*, also sending it, with a six–month deadline, to a select group of mathematicians. In essence, the challenge was to find the curve ensuring the quickest descent of an object between two points at different altitudes (but not along a vertical line). Leibniz, who solved the problem on the day he received it, correctly predicted that only five people would solve the problem. Indeed, only Leibniz, Newton, the

Bernoulli brothers, and L'Hospital found the solution—the cycloid curve.

Bernoulli was initially unable to provide a complete solution to the isoperimetric problem(involving the comparison of different **polygons** with equal perimeters). It was Jakob, realizing that the isoperimetric problem is essentially variational, who produced the solution, much to his brother's chagrin. However, Bernoulli used his brother's insight to further differential **geometry**, which uses differential and **integral calculus**, by discovering the variational properties of geodesic lines—curves describing the shortest **distance** between two points on a convex surface. As Morris Kline explains in his book *Mathematical Thought from Ancient to Modern Times*: "[Bernoulli] wrote to Leibnitz in 1698 to point out that the osculating plane at any given point of a geodesic is perpendicular to the surface at that point." The osculating plane is determined by the osculating circle, and a circle is determined by three points. Bernoulli published a complete solution of the isoperimetric problem in 1718, laying the foundations of the calculus of variation. In addition, Bernoulli studied exponential curves (curves whose formula is an exponential function, or an equation in which the unknown is an exponent), eventually establishing the basis for differentiating and integrating **exponential functions**.

When Jakob died in 1705, Bernoulli took over his position as professor of mathematics at the University of Basel. Having attained a more prominent position in the academic world, Bernoulli continued his efforts as a champion of Leibniz, who had no academic appointment. He criticized Newton's approach to calculus, focusing on the method of **fluxions**. While Newton viewed **variable** quantities as generated by the continuous **motion** of points, lines, and planes, actually calling them *fluents*, and defined their derivatives as *fluxions*, or rates of changes, Leibniz's system, which eventually prevailed, used the concept of differentials.

Bernoulli, whose authority as a mathematician (especially after Newton's death in 1727) seemed unrivaled, also furthered the field of **mechanics**. At this point in Bernoulli's career, there were two contending views in mechanics regarding the conservation of energy. While the Cartesians believed in the conservation of momentum, according to the Leibnizians, what was conserved was the *vis viva*, or living force. Although the Cartesian explanation was not incorrect, the Leibnizian concept prevailed, eventually having its original term replaced by the modern *kinetic energy*, defined as the energy a body possesses by virtue of its motion. As a Leibnizian, Bernoulli ardently supported the *vis viva*, fully accepting the idea put forth by Leibniz's student Christian Wolff, who postulated the universal validity of the conservation of the living force. Agreeing with Wolff, Bernoulli identified the living force as one of the fundamental principles in mechanics in his essay "De vera notione virium vivarum" in 1735. Scholars believe that Bernoulli may have been the first to realize the importance of the principle of conservation.

Johann Bernoulli

BESSEL, FRIEDRICH WILHELM (1784-1846)

German astronomer and mathematician

Unlike many men of achievement, Friedrich Wilhelm Bessel did little to give an indication of his forthcoming accomplishments during his early and limited education. He stumbled upon his genius by way of accepting an apprenticeship as a bookkeeper with a merchant house when he was 15 years old. Then, because Bessel wished to further his career and enter the world of foreign trade, he studied navigation as well as geography and foreign languages. This study of navigation soon blossomed into the study of astronomyand became an avocation that eventually led to a life's work affecting not only mathematics, but profoundly changing the world of astronomy. In his own lifetime, Bessel's discoveries were applauded as "inaugurating a new era of practical astronomy" and hailed as the "beginning of modern astronomy."

Bessel was born in Minden, Germany, on July 22, 1784. Not much is recorded about his family, except that his father was a government employee whose salary barely kept the nine children fed, and his mother came from Rheme, where her father was a minister. In 1812, Bessel married Johanna Hagen. The couple had five children, two boys and three girls, and the marriage was said to be successful except for the loss of both

Friedrich Wilhelm Bessel

sons when they were quite young. Bessel was fond of taking walks and, despite his small stature and less than robust health, enjoyed hunting. His only other form of relaxation was his correspondence with fellow astronomers and mathematicians.

Bessel's discoveries and achievements were not based on any sort of formal education; instead, he was almost self–taught. When Bessel was only 20 years old, he studied the 1607 observations of **Thomas Harriot** to calculate the **orbit** of Halley's Comet. After writing a paper on his calculations, Bessel sent it to astronomer Wilhelm Olbers, who was so impressed that he arranged for its publication in *Monatliche Corresondenz.* Olbers immediately recognized genius when he saw it and further prevailed on officials at the Lilienthal Observatory to make Bessel an assistant. During his four years at Lilienthal, Bessel attracted the attention of Friedrich Wilhelm III of Prussia, who was appointed him director of the Royal Observatory at Königsburg in 1810, where he spent the next 30 years.

While at the observatory, Bessel improved upon the work of past astronomers and paved the way for further discoveries for future cosmologists. Since the time of **Isaac Newton**, men of science had been attempting to calculate the **distance** of stars, and until Bessel they had failed. This failure can be attributed to two things—the lack of proper instruments

and the lack of a proper method for going about the measurement. In 1829, telescope maker Joseph Fraunhofer of Austria designed the equipment and Bessel designed the method that would finally provide the answer.

Bessel reasoned that the easiest and most reliable way to measure the distance of a star was to measure its annual parallax. This parallax, or shift in the apparent position of an object resulting from the Earth's orbit around the sun, had to be painstakingly measured every night for a year in order for Bessel's conjectures to be taken seriously by his peers. Using the speculations of astronomer James Bradley and his own intuition, Bessel accomplished what Bradley and all who had come before him could not and became the first person to accurately determine the distance of a star from Earth. Bessel's calculations vastly expanded astronomers' abstractions of **space**, and turned the conception that the Earth was part of a solar system into the realization that it is actually part of a universe. Bessel astounded his fellow astronomers when he proved that one of the nearest stars to Earth, 61 Cygni, was more than 60 trillion miles away. Bessel was also the first to use the term "light years" as a way of vividly explaining this distance. Traveling at the speed of **light**, 186,000 miles per second, it would take 10.3 years to reach 61 Cygni. Using his newly found method of computation, Bessel further compiled a catalog of the position of 75,000 stars.

In a paper written in 1840, Bessel recounted that Uranus displayed certain small but noticeable "irregularities" in its orbit. The planet, it seems, alternately slowed down and then ran ahead of its expected positions. Bessel surmised that this could only be due to the influence of an as yet unknown planet laying somewhere beyond it. Although Bessel did not live to see his hypothesis confirmed, that planet later proved to be Neptune. In 1842, Bessel calculated the mass of Jupiter by studying the orbital period of each of its major moons and established that while it had 388 times the mass of Earth, its overall density was only 1.35 times the density of water, bringing to light that this mammoth planet was essentially very light for its size.

The same sort of "irregularities" bothered Bessel when he studied the luminous stars Sirius and Procyon. Bessel proposed that the slight but significant erratic behavior of the two stars must be caused by "invisible companions." After Bessel's death, more powerful telescopes provided the evidence to prove him correct, and we now know those "invisible companions" as Sirius B and Procyon B.

Most of Bessel's gifts to the realm of mathematics materialized as a result of his astronomical breakthroughs, but that does not lessen the importance of his legacy to mathematical **functions** still used extensively today in such fields as geology, physics, engineering, and of course **applied mathematics**. These Bessel functions, also known as **cylinder** functions, were first devised to unlock the mystery of "planetary perturbation." This deviation of a planet from its regular orbit, usually caused by the presence of one or more bodies acting upon it, could be accurately calculated and anticipated due to Bessel's functions and later were used as solutions for a differential equation—intrinsic in the investigation of numerous problems in mathematical physics.

In addition to his duties at the Royal Observatory, Bessel was required to spend a considerable amount of time surveying. Like everything else he did, this task brought him acclaim and notoriety. Using his own idea for an improved measuring device, he commissioned the construction of an instrument that would accurately gauge base lines. Then, by using the methods of **Karl Friedrich Gauss**, he developed a new system of triangulation. These two accomplishments enabled Bessel to paint a mathematically accurate and, at that time, an astounding picture of the form and magnitude of the Earth. Bessel's work on triangulationsalso resulted in the eventual establishment of the International Bureau of Weights and Measures.

BESSEL FUNCTION

Bessel **functions** are solutions to Bessel's differential equation. The Bessel differential equation follows the form $x^2 y'' + xy' + (x^2 - n^2)y = 0$, where the prime notation is indicative of a derivative of y with respect to x. This equation results in an essential singularity at $x = 0$.

The Bessel functions are usually used in relation with the Bessel differential equation, but they also may be obtained from a generating function. If n is an integer, the Bessel function J_n can be obtained from the sum over s from 0 to **infinity** of the quantity $(x/2)^{n+2s} (-1)^s/s!(n + s)!$. This form is allows numerical evaluation of the Bessel functions, although a series solution of the differential equation (such as Frobenius's method) would allow for the same numerical inspection. The generating function can be manipulated to form several different recursion relations amongst Bessel functions. One of the most commonly used recursions is $J_{n-1}(x) - J_{n+1}(x) = 2J'_n(x)$. Another relation which does not involve derivatives is $J_{n-1}(x) + J_{n+1}(x) = (2n/x)J_n(x)$. These relations, combined with the orthogonality relations, are the basis of many useful proofs and manipulations of the Bessel **equations**; the differential equation itself and the generating function are almost never used directly in practical applications.

The Bessel differential equation has two **sets** of solutions, each of which forms an orthogonal set of functions (or **Hilbert space**). The Bessel functions, discussed above and denoted by J_n, are one of those sets of solutions. The other set is the Neumann functions, usually indicated with $N_n(x)$. (These are also called Bessel functions of the second kind.) The boundary or initial conditions of the specific case determine which set of functions are used. The major difference between Bessel and Neumann function is that Bessel functions are finite at **zero**, whereas Neumann functions approach negative infinity. Bessel and Neumann functions may be combined algebraically to form Hankel functions, which also solve the **differential equations** and fit some physical parameters better than either Bessel or Neumann functions would separately.

These functions are commonly used to solve problems with spherical **symmetry**. For example, the vibrations of a tympani drum when struck with a mallet can be expressed as Bessel functions of one type or another. The constants of the equation can be manipulated to fit a wide variety of physical conditions; the important common factor is spherical symmetry of the situation.

See also Differential equations

BETTI, ENRICO (1823-1892)

Italian mathematician

Enrico Betti, an Italian mathematician, was born near Pistoia, Italy on October 21, 1823. He died at Pisa, Italy on August 11, 1892.

With the loss of his father while still very young, Betti received his early education from his mother. He earned a degree in physical and mathematical sciences at the University of Pisa. While a university student, he championed ideas that would later propel him into the Italian war for independence.

In 1865, after a stint teaching mathematics at a high school in Pistoia, he was offered and accepted a professorship at the University of Pisa, where he remained until his death. In Pisa, he also served as rector of the university, and director of the teacher's college. He served as a member of parliament beginning in 1862, and as a senator starting in 1884.

For several months in 1874, Betti served as secretary of state for public education. His preference, however, was for academic life, with time alone or spent with close friends, such as **Bernhard Riemann** (1826-1866), the person for whom **Riemannian geometry** and Riemann surfaces are named. Betti had met Riemann in Italy, where the latter had gone to improve his health.

In mathematics, Betti is remembered for showing the relevance of the ideas of the Frenchman **Évariste Galois** (1812-1832) to the research of **Paolo Ruffini** (1765-1822) and the Norwegian **Niels Henrik Abel** (1802-1829). He made contributions to the solution of algebraic **equations** using a specific type of operation. He demonstrated, based on the theory of substitutions, the necessary and sufficient conditions for the solution of algebraic equations at a time when many believed that research related to Galois' was unproductive. Papers published in 1852 and 1855 make fundamental contributions to the development of abstract from classical **algebra**.

Betti also contributed to the theory of **functions**, and in particular, elliptic functions. He published his work in this area between 1860 and 1861. Fifteen years later, these ideas were further developed by **Karl Wilhem Theodor Weierstrass** (1815-1897).

By 1870, Betti had become interested in problems associated with the connectivity of higher-dimensional figures. Betti is especially remembered for his introduction of connectivity numbers. In one **dimension**, the connectivity number is the number of closed curves that can be drawn that do not divide a surface into separate regions. A two-dimensional connectivity number is the number of closed surfaces in a figure that do not, taken together, bound any three-dimensional regions. Higher-dimensional connectivity numbers are analogously defined. Betti succeeded in proving that the one-dimensional connectivity number is the same as the three-dimensional connectivity number. Betti's research in this area inspired the French mathematician **Henri Poincaré** (1854-1912) in the latter's own work, and resulted in the naming of "Betti numbers" in his honor.

Betti's mathematical ideas were either algebraic (related to Galois' research), or physicomathematic (influenced by the work of Riemann). Betti had a special fondness for theoretical physics, and devoted part of his research to further work on methods that had already been applied to electricity. In physicomathematics, Betti made contributions to the mathematical theories of elasticity and heat.

Betti was also devoted to the preservation of classical culture. He urged the return to the teaching of Euclid (whose work he felt exemplified the ideals of discipline and beauty) in secondary schools. A gifted teacher, Betti counted among his students individuals of later distinction, including U. Dini, L. Bianchi, and V. Volterra.

BÉZOUT, ÉTIENNE (1739-1783)
French algebraist and author

Étienne Bézout is best known as the author of one of the most widely used mathematics textbooks of his era, the six-volume *Cours de mathematique*. He is renowned for his work dealing with the usage of **determinants** in algebraic elimination. Born in Nemours, France, on March 31, 1739, Bézout was the second son of Pierre and Helene Jeanne Filz Bézout. His family was well established and respected in the region; his grandfather and father both served as district magistrates. Bézout, however, preferred numbers to politics and convinced his father to permit him to study mathematics instead of law. He was greatly influenced by the work of **Leonhard Euler**, and his abilities were quickly recognized by the Académie Royale des Sciences, which elected him an adjunct member at the age of 19 and a full member at age 29.

Little of Bézout's early life is documented, but it is recorded that he married at a young age. From the few reports available the marriage was a happy one, but it seems to have had a significant impact on his career. In 1763, at the age of 24, Bézout accepted a position as a teacher and examiner for aspiring naval officers who wished to serve the Duc de Choiseul in the Gardes du Pavillon et de la Marine. As a young father, his decision to accept the appointment was undoubtedly influenced by a need for a steady income.

His work as a teacher helped shape his approach to mathematics in general. Teaching mathematics to aspiring military officers (in 1768, he took on additional teaching duties for artillery officers) demanded a practical approach. In his courses, Bézout avoided detailed theory and shied away from intimidating terms. He introduced practical **geometry** before **algebra** to make his students more comfortable with the notion of mathematical reasoning and labored to make his lectures easy to understand.

Between 1764 and 1769 Bézout compiled the material he used in his lectures and published his six-volume *Cours de mathematiques: A l'usage des Gardes du Pavillon est de la Marine*. The work was criticized by some scholars as lacking academic rigor, but was received by educators with enthusiasm. *Cours de mathematiques* sold briskly and went into numerous reprints. It was translated into English in the early 1800s for use in schools in the United States and influenced American teaching methods well into the 19th century. John Farrar, one of the translators of *Cours de mathematiques*, based his Harvard University **calculus** course on Bézout's textbooks, and mathematics courses offered at West Point Military Academy were also based on Bézout's work. If academicians felt disdain for Bézout's straightforward approach, others owed a debt to him for his compilation of the works of such mathematicians as Euler and **Jean le Rond d'Alembert** in an easy-to-study format. Without such texts as Bézout's, general knowledge of important mathematical advances would have disseminated much more slowly.

Bézout's teaching duties left him little time for research. To make the most of the time he had, Bézout limited himself to the study of the theory **equations**; his early work probed integration, but he narrowed his investigations even further, and by 1762 he was absorbed by the questions posed by algebraic equations.

Bézout focused on the use of determinants in algebraic elimination. He developed artificial rules for solving simultaneous **linear equations** in unknowns. He is noted for developing an extension of this concept into a system of equations in one or more unknowns in which it is necessary to find the condition on coefficients for the equations to have a common solution.

In his paper *Sur plusieurs classes d'equations*, published in 1762, Bézout shifted his emphasis from solving the nth degree equation to elimination theory. Given n equations in n unknowns, Bézout searched for a resultant equation with the fewest possible number extraneous **roots**. In 1764, Bézout published *Sur le degre de equations resultantes de l'evanouissement des inconnues*. In this paper, he discussed Euler's methods for finding the equation that resulted from two equations in two unknowns, but rejected Euler's method as clumsy when applied to equations of high degree. Bézout identified determinants by listing permutation of coefficients in a table and used the table to determine if simultaneous linear equations could be solved.

In 1779, Bézout published *Theorie des ezuations algebriques*, his most important work on elimination theory, which includes the statement and **proof** of the following: The degree of the final equation resulting from any number of complete equations in the same number of unknowns and of any degrees, is equal to the product of the degrees of the equations. He was also the first to prove that two algebraic curves of degrees m and n intersect in general in mn points. Although this paper focused on algebraic theory, Bézout was compelled to point out his theory's geometric interpretation—a legacy of the practical concerns that dominated much of his professional life.

Some mathematical historians regard Bézout as a minor writer and mathematician. He was a product of his time and country; in 18th century France, most mathematicians were not affiliated with universities but with the Church or the military and the latter, at least, wanted practical applications. Bézout had practical concerns in his private life, as well. His family was well–regarded, but not wealthy, and his need to support his wife and children made his military teaching post his most professional priority. While Bézout's scholarly output was not large, his influence was significant. He instructed both **Gaspard Monge** and Lazare Carnot, and his textbooks had a profound influence on mathematics education in France and the United States.

Bézout died in 1783 in Basses–Loges, France, six years before the French Revolution rewove the country's social, academic, and intellectual fabric.

BÉZOUT'S THEOREM

Bézout's **theorem** is named after the French mathematician **Étienne Bézout**, who stated and gave a (partially correct) **proof** of the result in 1779. The statement itself is much much older, appearing in the works of **Jacques Bernoulli** (see **Jakob Bernoulli**), **Colin Maclaurin**, and others. The result deals with plane algebraic curves, which are subsets of the plane defined as the solution set of an equation f(x,y)=0, where f(x,y) is a **polynomial** in two variables. The total degree of the polynomial is called the degree of the curve. Bézout's theorem then states that two plane algebraic curves of degrees m and n intersect in at most mn points unless they have a common component; that is, unless there exists an algebraic curve which is a subset of both curves.

There is a more precise version of the theorem if one considers the complex projective plane, which is defined as follows. First of all, one allows the (x,y) coordinates to take **complex numbers** as values. Second, one introduces "points at infinity." These are ideal points, one for each direction, and are introduced in such a way that parallel lines meet at these points at **infinity**. Even after all these new points are introduced, it is still true that two-plane algebraic curves of degrees m and n intersect in at most mn points unless they have a common component. But even more can be established. If one counts each point of intersection with appropriate multiplicity (that is, if the curves are tangent to each other at a point, this point should count double or more) then there are exactly mn points of intersection, much as a polynomial in one **variable** and degree n will have exactly n complex **roots** counted with multiplicity, as in the fundamental theorem of **algebra**. For example, two conics will intersect in four points, since conics have degree two, and a conic and a cubic will intersect in six points, since cubics have degree three, and so on.

Bézout's theorem is a basic result in **Algebraic Geometry** and was one of the motivations for considering the complex projective plane. There are several modern extensions of Bézout's theorem in which one considers the intersection of two or more curves, surfaces, or higher dimensional

varieties inside higher dimensional projective **space** or even some other algebraic varieties. The subject dealing with these generalizations is called Intersection Theory. Solid foundations for Intersection Theory were only developed in the 20th century by B. L. van der Waerden, A. Weil, and others, who were trying to justify rigorously Schubert's enumerative **calculus**, as demanded by Hilbert's fifteenth problem.

See also Algebraic geometry

BHASKARA (1114-CA. 1185)
Indian mathematician

Bhaskara, or Bha-skara-cha-rya as he is sometimes known, was the leading mathematician of the 12th century. He applied the concept of **zero**, decimal notation, the use of letters to represent unknown quantities in **equations**, and he developed rules for equations for **trigonometry**.

Bhaskara was born in Biddur, in India, although his mathematical work was carried out whilst he was head of the astronomical observatory at Ujjain (where he eventually died). The three most important books he published were *Lilavati* (The Beautiful), which is about mathematics; *Bijaganita* (Seed Counting), which is about **algebra**; and an astronomical work, *Karanakutuhala* (The Calculation of Astronomical Wonders). *Lilavati* is the first known published work that uses the **decimal position system**.

Bhaskara spent much of his working life studyingdiophantine equations, and more specifically he studied what we now know as Pell's equation: $x^2 = 1 + py^2$ (in which p = 8, 11, 32, 61, and 67). For p = 61 he found the solutions x = 1776319049 and y = 22615390.

Bhaskara was a mathematician whose work predated much of what was to be achieved in Western mathematics by several centuries, and many of his principles are in wide use today.

See also Aryabhata the Elder; Brahmagupta

BINARY NUMBER SYSTEM

The binary number system, base two, uses only two symbols, 0 and 1. Two is the smallest whole number that can be used as the base of a number system. For many years, mathematicians saw base two as a primitive system and overlooked the potential of the binary system as a tool for developing computer science and many electrical devices.

Base two has several other names, including the binary positional numeration system and the dyadic system. Many civilizations have used the binary system in some form, including inhabitants of Australia, Polynesia, South America, and Africa. Ancient Egyptian **arithmetic** depended on the binary system. Records of Chinese mathematics trace the binary system back to the fifth century and possibly earlier. The Chinese were probably the first to appreciate the simplicity of noting **integers** as sums of **powers** of 2, with each coefficient being 0 or 1. For example, the number 10 would be written as 1010:

$$10 = 1 \times 2^3 + 0 \times 2^2 + 1 \times 2^1 + 0 \times 2^0$$

Users of the binary system face something of a trade-off. The two-digit system has a basic purity that makes it suitable for solving problems of modern technology. However, the process of writing out binary numbers and using them in mathematical computation is long and cumbersome, making it impractical to use binary numbers for everyday calculations. There are no shortcuts for converting a number from the commonly used denary scale (base ten) to the binary scale.

Over the years, several prominent mathematicians have recognized the potential of the binary system. Francis Bacon (1561-1626) invented a "bilateral alphabet code," a binary system that used the symbols A and B rather than 0 and 1. In his philosophical work, *The Advancement of Learning*, Bacon used his binary system to develop ciphers and codes. These studies laid the foundation for what was to become word processing in the late twentieth century. The American Standard Code for Information Interchange (ASCII), adopted in 1966, accomplishes the same purpose as Bacon's alphabet code. Bacon's discoveries were all the more remarkable because at the time Bacon was writing, Europeans had no information about the Chinese work on binary systems.

A German mathematician, **Gottfried Wilhelm von Leibniz** (1646-1716), learned of the binary system from Jesuit missionaries who had lived in China. Leibniz was quick to recognize the advantages of the binary system over the denary system, but he is also well known for his attempts to transfer binary thinking to theology. He speculated that the creation of the universe may have been based on a binary scale, where "God, represented by the number 1, created the Universe out of nothing, represented by 0." This widely quoted analogy rests on an error, in that it is not strictly correct to equate nothing with **zero**.

The English mathematician and logician **George Boole** (1815-1864) developed a system of Boolean logic that could be used to analyze any statement that could be broken down into binary form (for example, true/false, yes/no, male/female). Boole's work was ignored by mathematicians for 50 years, until a graduate student at the Massachusetts Institute of Technology realized that **Boolean algebra** could be applied to problems of electronic circuits. Boolean logic is one of the building blocks of computer science, and computer users apply binary principles every time they conduct an electronic search.

The binary system works well for computers because the mechanical and electronic relays recognize only two states of operation, such as on/off or closed/open.

Operational characters 1 and 0 stand for 1 = on = closed circuit = true 0 = off = open circuit = false

The telegraph system, which relies on binary code, demonstrates the ease with which binary numbers can be translated into electrical impulses. The binary system works well with electronic machines and can also aid in encrypting messages. Calculating machines using base two convert deci-

mal numbers to binary form, then take the process back again, from binary to decimal.

The binary system, once dismissed as primitive, is thus central to the development of computer science and many forms of electronics. Many important tools of communication, including the typewriter, cathode ray tube, telegraph, and transistor, could not have been developed without the work of Bacon and Boole. Contemporary applications of binary numerals include statistical investigations and probability studies. Mathematicians and everyday citizens use the binary system to explain strategy, prove mathematical theorems, and solve puzzles.

BINARY, OCTAL, AND HEXADECIMAL NUMERATION

A numeration system, or numeral system, is an orderly method in which each numeral of the system is associated with a unique number. Thomas Harriot (1560-1621) is recognized as having given the first generalized treatment of positional number systems. However, his work went unpublished, thus allowing **Gottfried Wilhelm von Leibniz** (1646-1716) to be credited for the achievement. Both men independently developed the theory of positional number systems.

The **decimal position system**, which uses ten as its base number, has become the dominant numeration system in the world. Three other important types of number systems are *binary, hexadecimal, and octal numeration*. The "binary number system" is a positional representation of numbers based on the number 2 and the **powers** of 2. The "octal number system" is a positional representation of numbers based on the number 8 and the powers of 8, and the "hexadecimal number system" is a positional representation of numbers based on the number 16 and the powers of 16. These number systems find their most extensive use in relation to digital computing.

Any real number x can be represented in a positional number system of base "b" by the expression $x = a_n b^n + a_{n-1} b^{n-1} + ... + a_1 b^1 + a_0 b^0 + a_{-1} b^{-1} + ... + a_{-(n-1)} b^{-(n-1)} + a_{-n} b^{-n}$. For the number systems considered here, b = 2 in the binary system, b = 8 in the octal system, and b = 16 in the hexadecimal system. The coefficients of the general equation for x given above, i.e., the values a_i, where "i" is a positive integer, may take on the numeric values of "0" through "b-1". So, for instance, the values of a_i in the **binary number system** may be 0 or 1, while a_i in the octal number system may take on any of the values 0 through 7.

The binary, or dyadic (dual), number system is based on the powers of the base 2, that is $(..., 128, 64, 32, 16, 8, 4, 2, 1, 1/2, 1/4,...) = (..., 2^7, 2^6, 2^5, 2^4, 2^3, 2^2, 2^1, 2^0, 2^{-1}, 2^{-2},...)$. The successive binary **integers** are 0, 1, 10, 11, 100, 101, 110, 111, 1000,... etc. The binary **fractions** are 0.1 (1/2), 0.01 (1/4), 0.001 (1/8), and so forth. In the binary system "b = 2" and the digits a_i are either "0" or "1". As an example of expressing a number in binary format, consider the number (in decimal notation) "83". In the binary system it is represented as "1010011". Using the binary numeric system described previously it is seen that $1010011 = (1 \times 2^6) + (0 \times 2^5) + (1 \times 2^4) + (0 \times 2^3) + (0 \times 2^2) + (1 \times 2^1) + (1 \times 2^0)$. In decimal notation this expression reads: $(1 \times 64) + (0 \times 32) + (1 \times 16) + (0 \times 8) + (0 \times 4) + (1 \times 2) + (1 \times 1) = 64 + 0 + 16 + 0 + 0 + 2 + 1 = 83$.

Of particular technical importance for the binary system is its common usage in digital computers. The binary system requires many more positions than the decimal system. However, with regards to **information theory** and computer technology, the binary system is much more convenient and reliable than the decimal system since the binary digits "0" and "1" can easily be converted to "off" and "on", "close" and "open", "negative" and "positive", to mention a few. Moreover, a number like 1,234 in base 10 appears more compact than the binary equivalent 10011000010. But since 10 states are required per digit in base 10, a 4-digit number requires $10 \times 10 \times 10 \times 10 = 10,000$ states, while the binary representation only requires $2^{11} = 2,048$ states.

The octal, or octenary, number system is based on the powers of the base 8, that is (..., 2097152, 262144, 32768, 4096, 512, 64, 8, 1, 1/8, 1/64,...) = (..., 8^7, 8^6, 8^5, 8^4, 8^3, 8^2, 8^1, 8^0, 8^{-1}, 8^{-2},...). In the octal system "b = 8" and the digits a_i have the values "0, 1, 2, 3, 4, 5, 6, 7". The decimal number "83" is written "123" in the octal system. Using the octal numeric system laid out previously we see that 123 (octal) = "$(1 \times 8^2) + (2 \times 8^1) + (3 \times 8^0)$". In decimal notation this expression reads: $(1 \times 64) + (2 \times 8) + (3 \times 1) = 64 + 16 + 3 = 83$.

The hexadecimal, or sexadecimal, number system is based on the number 16 and the powers of the base 16, that is (..., 1048576, 65536, 4096, 256, 16, 1, 1/16, 1/256,...) = (..., 6^5, 6^4, 6^3, 6^2, 6^1, 6^0, 6^{-1}, 6^{-2},...). In the hexadecimal system "b = 16" and the digits a_i have the values "0, 1, 2, 3, 4, 5, 6, 7, 8, 9, A, B, C, D, E, F". This notation is to be understood as follows: A represents ten, B represents eleven, C represents twelve, and so forth to the letter F, which represents fifteen. As an example of expressing a number in hexadecimal format, consider the number (in decimal form) "283.75". In the hexadecimal system it is represented as "11B.C". Using the hexadecimal numeric scheme described previously we see that 11B.C = $(1 \times 16^2) + (1 \times 16^1) + (B \times 16^0) + (C \times 16^{-1})$. In decimal notation this expression reads: $(1 \times 256) + (1 \times 16) + (11 \times 1) + (12 \times 1/16) = 256 + 16 + 11 + 0.75 = 283.75$.

In computer design, the hexadecimal numbering system is used as a convenient method for representing large binary numbers, where numbers often consist of long strings of zeros and ones. Thus, each hexadecimal digit stands for four binary digits. As an example, the number 101100010111110100 in binary notation can be converted to hexadecimal notation by dividing the number into groups of four binary digits, starting from the right, and replacing each group by the corresponding hexadecimal symbol. Where the left-hand group is incomplete, zeros are filled in as required. The result is $0010/1100/0101/1111/0100 = 2/C/5/F/4 = (2C5F4)_{16}$, where the subscript "16" next to the number denotes hexadecimal format.

See also Binary number system; Decimal position system; Gottfried Wilhelm von Leibniz

BINET, JACQUES-PHILIPPE-MARIE (1786-1856)

French mathematician and astronomer

The French mathematician and astronomer, Jacques-Philippe-Marie Binet, was born at Rennes, France on February, 2 1786. He died in Paris on May 12, 1856.

In 1806, following two years of study at the École Polytechnique, Binet received a government appointment as a student engineer in the department of bridges and roads. Finding himself attracted to the idea of teaching, he opted for a professorship of mathematics at the Lycée Napoléon. He later taught descriptive **geometry** and **mechanics** at the École Polytechnique.

In 1812, Binet stated his **multiplication theorem** for **determinants**, which holds that for nth-order determinants and the summation over the **integers** k, $c_{ij} = \Sigma\ a_{ik}b_{kj}$, the terms in the ith row and the jth column of the product is the sum of the products of the corresponding elements of the ith row of $|a_{ij}|$ and the jth column of $|b_{ij}|$. Note: Although Binet is credited with stating this theorem, it was actually his contemporary **Augustin-Louis Cauchy** (1789-1857) who was the first person to prove it, and to give the first systematic and modern treatment of determinants.

Binet was made a member of the Société Philomathique in 1812. Eleven years later, he was awarded the chair of astronomy at the Collège de France.

Binet was a loyal subject of Charles X, and as a result lost his position at the École Polytechnique when the French king was forced to abdicate during the insurrection of 1830. Binet was, however, allowed to retain his professorship at the Collège de France. In 1843, he was elected to the French Académie des Sciences, eventually becoming its president.

BINOMIAL THEOREM

The binomial **theorem** provides a simple method for determining the coefficients of each term in the expansion of a binomial with the general equation $(A + B)^n$. Developed by **Isaac Newton**, this theorem has been used extensively in the area of probability and **statistics**. The main argument in this theorem is the use of the combination formula to calculate the desired coefficients.

The question of expanding an equation with two unknown variables called a binomial was posed early in the **history of mathematics**. One solution, known as **Pascal's Triangle**, was determined in China as early as the thirteenth century by the mathematician Yang Hui. His solution was independently discovered in Europe 300 years later by **Blaise Pascal** whose name has been permanently associated with it since. The binomial theorem, a simpler and more efficient solution to the problem, was first suggested by Isaac Newton. He developed the theorem as an undergraduate at Cambridge and first published it in a letter written for Gottfried Leibniz, a German mathematician.

Expanding an equation like $(A + B)^n$ just means multiplying it out. By using standard **algebra** the equation $(A + B)^2$ can be expanded into the form $A^2 + 2AB + B^2$. Similarly, $(A + B)^4$ can be written $A^4 + 4A^3B + 6A^2B^2 + 4AB^3 + B^4$. Notice that the terms for A and B follow the general pattern A^nB^0, $A^{n-1}B^1$, $A^{n-2}B^2$, $A^{n-3}B^3$, **ellipse**, A^1B^{n-1}, A^0B^n. Also observe that as the value of n increases, the number of terms increases. This makes finding the coefficients for individual terms in an equation with a large n value tedious. For instance, it would be cumbersome to find the coefficient for the term A^4B^3 in the expansion of $(A + B)^7$ if we used this algebraic approach. The inconvenience of this method led to the development of other solution for the problem of expanding a binomial.

One solution, known as Pascal's Triangle, uses an array of numbers (shown below) to determine the coefficients of each term.

```
(A+B)⁰                    1
(A+B)¹                  1   1
(A+B)²                1   2   1
(A+B)³              1   3   3   1
(A+B)⁴            1   4   6   4   1
(A+B)⁵          1   5  10  10   5   1
```

This triangle of numbers is created by following a simple rule of **addition**. Numbers in one row are equal to the sum of two numbers in the row directly above it. In the fifth row the second term, 4 is equal to the sum of the two numbers above it, namely 3 + 1. Each row represents the terms for the expansion of the binomial on the left. For example, the terms for $(A+B)^3$ are $A^3 + 3A^2B + 3AB^2 + B^3$. Obviously, the coefficient for the terms A^3 and B^3 is 1. Pascal's Triangle works more efficiently than the algebraic approach, however, it also becomes tedious to create this triangle for binomials with a large n value.

The binomial theorem provides an easier and more efficient method for expanding binomials which have large n values. Using this theorem the coefficients for each term are found with the combination formula. The combination formula is $_nC_r = n! \div r!(n-r)!$

The notation n! is read "n factorial" and means multiplying n by every positive whole integer which is smaller than it. So, 4! would be equal to $4 \times 3 \times 2 \times 1 = 24$. Applying the combination formula to a binomial expansion $(A + B)^n$, n represents the power to which the formula is expanded, and r represents the power of B in each term. For example, for the term A^4B^3 in the expansion of $(A + B)^7$, n is equal to 7 and r is equal to 3. By substituting these values into the combination formula we get $7! / (3! \times 4!) = 35$ which is the coefficient for this term. The complete binomial theorem can be stated as the following

$$(A + B)^n = \Sigma_n C_r A^{n-r} B^r$$

The binomial theorem has been applied to many areas of science and mathematics. It is particularly useful for determining values for the **square** root of numbers. It is also extensively used in probability and statistics to describe the distribution of values about a **mean**.

BIRKHOFF, GEORGE DAVID (1884-1944)
American algebraist

George David Birkhoff's contributions as a theoretical mathematician, a teacher, and a member of the international scientific community rank him as one of the foremost mathematicians of the 20th century. He made extensive contributions to the area of **differential equations** and continued the work of the great French mathematician **Jules Henri Poincaré** on celestial **mechanics**. He is considered the founder of the modern theory of dynamical systems.

Born in Overisel, Michigan, on March 21, 1884, Birkhoff was the eldest of six children born to David Birkhoff, a physician, and Jane Gertrude Droppers. When Birkhoff was two years old, his family moved to Chicago, where he spent most of his childhood. From 1896 to 1902 Birkhoff studied at the Lewis Institute (now the Illinois Institute of Technology). Following a year at the University of Chicago as an undergraduate, Birkhoff transferred to Harvard University in 1903. In 1904, while still an undergraduate, he wrote his first mathematics paper, on **number theory**. He earned a bachelor's degree at Harvard in 1905 and a master's degree in 1906. Returning to the University of Chicago for his doctorate, Birkhoff wrote a dissertation on differential **equations** under the guidance of Eliakim Hastings Moore. He was awarded a doctorate *summa cum laude* in 1907.

Birkhoff taught mathematics at the University of Wisconsin from 1907 to 1909, when he took a position as assistant professor at Princeton University. He joined the faculty of Harvard University in 1912, teaching there until his death in 1944. Birkhoff, though not a great lecturer, was an inspiring teacher. Many of the influential American mathematicians of the mid-twentieth century, including Marston Morse and Marshall Stone, studied with Birkhoff at the doctoral or post-doctoral level. Six of his former students went on to become members of the National Academy of Sciences. From 1935 to 1939 Birkhoff also served as Dean of the Faculty of Arts and Science at Harvard.

The single most important influence on Birkhoff's mathematical research was that of Poincaré. The two never met, but Birkhoff studied Poincaré's work and adopted some of the problems in differential equations and celestial mechanics Poincaré left behind at his death in 1912. In 1913 Birkhoff first attracted international attention by proving a geometrical **theorem** that Poincaré had proposed but not proved in his last published paper. Birkhoff's accomplishment marked a special advance in solving the problem of three bodies. The three-body problem of celestial mechanics concerns trajectories and orbits of bodies moving in systems in such a way that each body affects the **motion** of the others.

From Poincaré's theorem, Birkhoff went on to consider the entire field of dynamical systems and made contributions that have become fundamental to this branch of mathematics. In his book *Dynamical Systems* (1927), Birkhoff wrote that "the final aim of the theory of motions of a dynamical system must be directed towards the qualitative determination of all possible types of motions and the interrelation of these motions." Using ideas developed by Poincaré, Birkhoff laid the foundations for the topological theory of dynamical systems by defining and classifying possible types of dynamic motions. Another signal achievement came in 1931, when Birkhoff offered further **proof** of the so-called ergodic theorem, which demonstrates the conditions needed for the behavior of a large dynamical system, such as a container of a gas, to reach equilibrium. This problem had baffled scientists for more than 50 years.

Birkhoff's journal articles and books reflect the breadth of his talent and the diversity of his interests. Among his published works is a basic **geometry** text that for many years formed the basis of high-school geometry curricula. Birkhoff also wrote extensively about the theory of relativity and **quantum mechanics**. Although his ideas in this field are not widely accepted, the mathematical tools that he developed for his approach play an important role in modern relativity theory. Birkhoff's other work encompasses number theory and point-set theory and the famous four-color problem, which is concerned with the possibility of coloring any map using only four colors. In 1933, his life-long passion for art, music, and poetry led him to write *Aesthetic Measure,* in which he attempted to create a general mathematical theory of the fine arts, starting out from the Pythagorean notion that beauty is mathematical in nature. He later extended the theory to ethics. Birkhoff was also the editor of several mathematical journals, including the *Annals of Mathematics, Transactions of the American Mathematical Society,* and the *American Journal of Mathematics.*

During his lifetime, Birkhoff received many honors and honorary degrees from universities worldwide. Among others, he was awarded the Querini-Stampalia prize of the Royal Institute of Science, Letters and Arts, Venice (1918); the Bôcher prize of the American Mathematical Society (1923) for his research in dynamics; the annual prize of the American Association for the Advancement of Science (1926); and the biennial prize of the Pontifical Academy of Sciences (1933) for his research on systems of differential equations. Birkhoff was elected to membership in the National Academy of Sciences (1918), the American Philosophical Society, and the American Academy of Arts and Sciences. He was made an officer of the French Legion of Honor in 1936, and was an honorary member of the Edinburgh Mathematical Society, the London Mathematical Society, the Peruvian Philosophic Society, and the Scientific Society of Argentina.

Birkhoff was fluent in French, the language of his famous mathematical predecessor, Poincaré, and presented several of his fundamental papers in that language. He traveled widely, promoting his belief in international fellowship among scientists. Because of his preeminence in research, he was able to represent mathematics in international scientific circles, and he played an important role in the creation of the mathematical institutes at Göttingen and Paris after World War I.

Birkhoff married Margaret Elizabeth Grafius of Chicago on September 2, 1908, and they had three children:

George David Birkhoff

Barbara, Garrett, and Rodney. Garrett went on to become a professor of mathematics at Harvard. Birkhoff died of a heart attack on November 12, 1944, in Cambridge, Massachusetts.

BLUM, LENORE (1943-   )
American algebraist, logician, and model theorist

Lenore Blum has played an integral role in increasing the participation of girls and women in mathematics. She was one of the founders of the Association for Women in Mathematics (AWM), acting as its president from 1975 to 1978. The AWM has membership totaling over 1,500 women and men. In addition to local, national and international meetings, the AWM sponsors the **Emmy Noether** Lecture series and has organized symposiums. It provides a list women who are available to speak at high schools and colleges and also contributes to the *Dictionary of Women in the Mathematical Sciences.*

Blum was born in 1943 and as a child, she enjoyed math, art and music. Finishing high school at 16, Blum applied to Massachusetts Institute of Technology (MIT), but was turned down, several times, in fact. After being turned down by MIT, Blum attended Carnegie Tech in Pittsburgh, Pennsylvania. She began studying architecture, then changed her major to mathematics. For her third year, Blum enrolled at

Simmons, a Boston area college for women. However, Blum found that she did not have to put forth much effort in the math classes. She then cross–registered at MIT, graduated from Simmons, and received her Ph.D. in mathematics from MIT in 1968. Blum continued her education as a postdoctorate student and lecturer at the University of California at Berkeley.

According to a biography written by Lisa Hayes, a student at Agnes Scott College, "Blum's research, from her early work in **model** theory, led to the formulation of her own theorems dealing with the patterns she found in trying to use new methods of logic to solve old problems in algebra." The work she did on this project became her doctoral thesis, which earned her a fellowship. Blum has also had the honor of reporting on work she did with **Stephen Smale** and Mike Shub in developing a theory of computation and complexity over **real numbers**.

Blum has written mathematical books with her husband, Manuel, a mathematician as well. They collaborated on a paper that proposed designing computers that had the ability to learn from example, much in the way young children learn. Blum has studied this project to discover why some computers learn the methods they do. Blum has been involved in other fields of research, in addition to working with her husband, which includes work in developing a new (homotopy) **algorithm** for **linear programming**.

When Blum was hired to teach **algebra** at Mills College, she was not happy with the program and sought a way to make the classes more interesting to the students and to the instructors. In 1973, she founded the Mills College Mathematics and Computer Science Department. Blum served as Head or Co–Head of the department for 13 years. While at Mills, Blum received the Letts–Villard Research professorship. Since 1988, she has been a research scientist in the Theory Group of the International Computer Science Institute (ICSI). In 1989, Blum was employed as an adjunct professor of Computer Science at Berkeley. During the 1980s, Blum became a research mathematician full–time, giving numerous talks at international conferences.

To further girls and women's participation in mathematics, Blum founded the Math/Science Network and its Expanding Your Horizons conferences. The Network began as an after–school problem–solving program. The aim of the program is to get high school girls interested in math and logic. The conference now travels nationwide. Blum served as its Co–Director from 1975–1981. Blum has written books and produced films, including *Count Me In*, *The Math/Science Connection*, and *Four Women in Science*, for the Network.

In addition to her work with the Math/Science Network, Blum is involved in the Mills College Summer Mathematics Institute for Undergraduate Women (SMI). The SMI is a six–week intensive mathematics program. Twenty–four undergraduate women are selected from across the nation to participate. According to the Mills College SMI page on the World Wide Web, the program aims "to increase the number of bright undergraduate women mathematics majors that continue on into graduate programs in the mathematical sciences and obtain advanced degrees."

Blum is an active member of several mathematical societies. She is a fellow of the American Association for the Advancement of Science and the American Mathematical Society (AMS), where she served as Vice President from 1990 to 1992. Blum represented the AMS at the Pan African Congress of Mathematics held in Nairobi, Kenya, in the summer of 1991. At that time she became dedicated to creating an electronic communication link between American and African mathematics communities. Blum also served as a member of the Mathematics Panel of Project 2061. The project was to determine how much a typical adult must know about science and technology to be prepared for the return of Halley's Comet. Blum also served as the first woman editor of the *International Journal of Algebra and Computation* from 1989 to 1991.

Blum served as the deputy director at the Mathematical Sciences Research Institute (MSRI) at U.C. Berkeley. She has participated in MSRI's Fermat Fest and has been an organizer of MSRI's "Conversations" between mathematics researchers and mathematics teachers.

BOHR, NIELS HENRIK DAVID (1885-1962)
Danish physicist

Niels Henrik David Bohr was born in Copenhagen, Denmark, on October 7, 1885, the second of three children born to Christian and Ellen Adler Bohr. Bohr's mother was a member of a wealthy Jewish banking family, and his father was a talented professor of physiology at the University of Copenhagen who conducted quantitative research on the physical processes underlying physiological functions. Bohr received his primary and secondary education at the Gammelholm School in Copenhagen.

In 1903, Bohr enrolled at the University of Copenhagen, where he majored in physics. His first research project was a precise measurement of the surface tension of water. He earned his bachelor of science degree in 1907, his master of science degree in 1909, and his doctorate in 1911, all at the University of Copenhagen. Bohr's thesis, which dealt with the electron theory of metals, convinced him that classical electromagnetic theory was inadequate for describing the quantitative properties of metals.

After completing his degree, Bohr joined J. J. Thomson, who had discovered the electron, at the Cavendish Laboratory at Cambridge University. Finding that Thomson was uninterested in his ideas, Bohr moved on in 1912 to spend three months at the University of Manchester working with **Ernest Rutherford**. Rutherford had the previous year published his nuclear **model** of the atom in which electrons **orbit** the atomic nucleus much like planets orbiting the sun. The chief shortcoming of Rutherford's model was that it could not explain why electrons in orbit about the nucleus did not lose energy and **spiral** into the nucleus. In July of 1912, Bohr returned to the University of Copenhagen to take a position as assistant professor of physics.

In 1913, Bohr published a series of papers describing his idea that electrons orbit the nucleus at fixed distances from the center of the atom, and that it requires the gain of loss of a discrete amount of energy for an electron to change its orbit. In Bohr's model, the electron's orbital momentum was quantized, i.e., it assumed only discrete values. Bohr explained atomic radiation as the emission of quantum amounts of radiation as electrons underwent transitions from one orbit to another. Unlike Rutherford's model, which predicted that the hydrogen spectrum should be continuous, Bohr's model accounted for the set of spectral lines that were actually observed. Although Bohr's work seemed to have a commonality with earlier work on quantized systems carried out by Max Planck and **Albert Einstein**, there really was no theoretical justification at the time for his ideas. But as a descriptive model of observed phenomena, Bohr's model seemed to work.

Bohr's ideas were slow to catch on with his contemporaries, many of whom were reluctant to accept the need for a break with classical physics that Bohr's theory seemed to imply. But as new spectral data were obtained that confirmed the predictions of the theory, the model, which was later shown to be a special solution of the Schrödinger equation, gained acceptance. The model was eventually found capable of explaining variations in properties of the elements and the chief features of X-ray and optical spectra exhibited by all of the elements.

In 1914, Bohr was appointed Reader in Theoretical Physics at Manchester University, where he remained for the next two years. In 1916, he returned to Copenhagen, and two years later he became the first director of the Institute for Theoretical Physics at the University of Copenhagen. This institute was to become one of the leading intellectual centers in Europe for young theoretical physicists. Bohr later characterized his early years at the Institute as a time of "unique co-operation of the whole generation of theoretical physicists from many countries." In 1922, he was awarded the Nobel Prize for physics for his work on the quantum mechanical model of the atom, which by that time had been accepted universally among physicists.

In 1927, in a lecture presented at the International physical Congress, Bohr introduced his notion of complementarity, which expressed the idea that it is impossible to make a sharp distinction between the behavior of physical objects and the objects' interaction with the instruments used to measure that behavior. This principle was brought to bear on the problem of wave-particle duality in which particles could be observed to behave both as waves and particles.

Also at this time, Bohr introduced his **correspondence** principle, which stated that, while quantum mechanical principles are required to described atomic phenomena, these principles must remain consistent with observations of the real world made at the macroscopic, or everyday, level.

In the 1930's, Bohr's interest turned to nuclear physics and the composition of the atomic nucleus. His chief contribution to nuclear structure theory was to point out that the atomic nucleus consisted of many protons and neutrons clus-

Niels Bohr

tered together like molecules in a drop of water. This idea was to heavily influence nuclear reaction theory.

Bohr also studied the physics of uranium fission, and the role of fast and slow neutrons. He showed, for example, that uranium-235 usually decayed by emitting slow neutrons, but uranium-238 usually absorbs neutrons without undergoing fission. Bohr's observation led to the awareness that one would have to use uranium-235 if one wanted to produce a fast chain reaction.

In 1943, when he learned that he was at risk of imprisonment in German-occupied Copenhagen for being a patriot, Bohr escaped with his family in fishing boats to Sweden. In October of that year, at the intervention of the British authorities, Bohr and his younger son, Aage, were flown to England. (Aage Bohr was himself later to win the Nobel Prize for physics in 1975 for his work on the structure of the atomic nucleus.)

Co-operative work in atomic energy had resumed between the United States and the United Kingdom by September of 1943, and arrangements were made to bring Bohr to the United States, where he spent time in Washington and at Los Alamos. At Los Alamos, Bohr contributed to the development of the atomic bomb, and also served as an elder statesman in the community of prominent scientists.

Bohr was particularly concerned with the implications that the atomic bomb had for future world history, and with keeping control of it out of the hands of the military and under civilian supervision. In 1944, he wrote to the United Kingdom's minister in charge of atomic energy research expressing his concern about the future control of these weapons.

Upon returning to Denmark after the war, Bohr continued to address these problems. In 1950, he sent a letter to the United Nations in which he made a plea for an "open world where each nation can assert itself solely by the extent to which it can contribute to the common culture and is able to help others with experience and resources."

In the 1950's, Bohr assumed a prominent role in the establishment of the European Centre for Nuclear Research, and served as a Council member of the center for the rest of his life. Although academic policy mandated his retirement at the age of 70 (in 1955) from the University of Copenhagen, Bohr retained his position as director of the Institute of Theoretical Physics until his death.

In his later years, Bohr attempted to apply the principle of complementary relationships to psychical phenomena, i.e., the relationship between human emotions as evidenced by behavior and consciousness of them; and to human society, i.e., the relationship between hereditary factors and tradition. He had thus come to view the principle of complementarity as a means for including in rational **analysis** any line of thought that provided useful results, and of maximizing the number of potential avenues of investigation.

Bohr was the first chairman of the Danish Atomic Energy Commission. He also served as president of the Royal Danish Academy of Sciences from 1939 until his death. He chaired a meeting of Academy just two days before his death in Copenhagen on November 18, 1962. His greatest contributions to physics were the formulation of the quantum theory of the hydrogen atom, and the explanation of the origin of the atomic spectra of hydrogen and helium.

BOLTZMANN, LUDWIG (1844-1906)

Austrian physicist and mathematician

Ludwig Boltzmann, an Austrian physicist and mathematician who won renown for his kinetic theory of gases and his work in statistical **mechanics**, was born in Vienna on February 20, 1844. A scientist of diverse interests, Boltzmann at one time or another in his career worked in the fields of physics, chemistry, mathematics, philosophy, and aeronautics.

Boltzmann studied at the University of Vienna, earning his doctorate there. He then went on to hold professorships in Graz (1869-73; theoretical physics); Vienna (1873-6; mathematics); Graz (1876-9; experimental physics); Munich (1889-93; theoretical physics); Vienna (1894-1900; theoretical physics); Leipzig (1900-2; theoretical physics); and Vienna (1902-6; theoretical physics). A popular lecturer, Boltzmann counted among his students Hermann Walther Nernst (1864-1941).

Ludwig Boltzmann

When Boltzmann began his career in 1866, major advances were already taking place in the three fields of theoretical physics that he himself would make major contributions to: **thermodynamics**, the theory of gases, and **electromagnetism**. It is of note that R. J. E. Clausius and Lord Kelvin has recently formulated the second law of thermodynamics based on the mechanical theory of heat; Clausius and **J. Clerk Maxwell** had laid the foundations of a kinetic theory of gases based on classical laws of probability and dynamics; and Maxwell had started work on his theory of electromagnetism.

In thermodynamics, Boltzmann provided the empirical law of black-body radiation first formulated by his teacher Joseph Stefan (1835-93) in Austria with a theoretical foundation (1884). Boltzmann's belief that thermodynamic phenomena are the macroscopic representation of microscopic atomic interactions that were governed by the laws of mechanics and probability was reflected in his work on his theory governing the equipartition of energy, and in his attempts to derive the second law of thermodynamics, these being based on mechanical principles and statistical methods and dating from 1866. In 1877, Boltzmann derived the equation named for him that relates entropy to the statistical distribution of molecular configurations, and which shows how entropy increases with increasing molecular randomness. Boltzmann was clearly a pioneer in the field of statistical mechanics, and although he never succeeded in establishing a statistical-mechanical derivation for the second law of thermodynamics, his treatments of the subjects of irreversibility and his correlations of entropy with probability paved the way for the later work in this area by Willard Gibbs and Max Planck.

In his work on the kinetic theory of gases between 1800 and 1802, Boltzmann made major contributions to the understanding of viscosity and diffusion. He also introduced the Maxwell-Boltzmann distribution function to describe the velocity distributions of colliding atoms. Much of his work on these subjects was published between 1896 and 1898.

Boltzmann was one of first scientists on the Continent to recognize the significance of Maxwell's work in electromagnetism. Besides lecturing and writing on the subject, Boltzmann in 1872 made measurements of the dielectric constant of a gas that led to the acceptance of Maxwell's theory throughout Europe.

Boltzmann's views came under attack by some of his contemporaries, notably the physicist Ernst Mach (1838-1916) and the chemist W. F. Ostwald (1853-1932), both of whom rejected Boltzmann's atomistic approach to science. Partly in response to the unpopularity of his views, Boltzmann lapsed into a severe depression beginning in 1900, which culminated with his suicide while on holiday in Duino, near Trieste, on September 5, 1906. In recognition of his contributions to science, the famous "Boltzmann equation" was later carved on his tombstone.

BOLYAI, JÁNOS (1802-1860)
Hungarian geometer

János Bolyai is remembered as the Hungarian mathematician whose work on **non–Euclidean geometry** was eclipsed by the work of **Nikolai Lobachevsky** in Russia and **Karl Gauss** in Germany. The son of Farkas (Wolfgang) and Susanna Arkos Bolyai, Bolyai was born in Kolosvar, Transylvania, Hungary, on December 15, 1802. Bolyai was education in Marosvasarhely at the Evangelical–Reformed College, where his father was a professor of mathematics, physics, and chemistry.

From childhood, Bolyai showed an aptitude for both **mathematics and music**; he was an accomplished violinist at an early age. His father—a notable mathematician in his own right who devoted considerable effort to trying to prove the Euclidean theory of parallels—wanted Bolyai to study mathematics in Germany with Gauss; instead, the young man enrolled in the imperial engineering academy in Vienna in 1818 and pursued a military education. However, he still shared his father's passion for proving the Euclidean **postulate** that there can be only one parallel to a line through a point outside it. The elder Bolyai, in despair about his own lack of success, wrote to his son: "For God's sake, I beseech you, give it up. Fear it no less than sensual passions because it, too, may take all your time, and deprive you of your health, peace of mind, and happiness in life."

While still a student in Vienna, Bolyai began to consider concepts that would alter the focus of studies. Inspired partly by his inability to prove Euclidean parallelismand partly by his exposure to other ideas, Bolyai began to explore the notion of a geometry constructed without the Euclidean **axiom**. He graduated from military college in 1822 and was commissioned as a sublieutenant. A flamboyant man, Bolyai developed a reputation as a competent swordsman and violinist. He readily accepted challenges to duel; in one account, he crossed swords with 13 consecutive opponents, stipulating only that he be permitted to pause between matches to play a violin selection. According to the story, he defeated all 13 swordsmen and was applauded for his musicianship. Bolyai still found time to pursue his mathematical inquiries, however. In 1823, he wrote to his father that he had made significant progress in his non–Euclidean geometry constructs, stating "from nothing I have created another entirely new world."

Bolyai's military duties took him to Temesvar, where he was stationed from 1823 to 1826. He managed a visit to his father in 1825 and took with him a manuscript that detailed his theory of absolute **space**. Although the father rejected the son's concept, he forwarded the manuscript to Gauss in 1831. In 1832, Gauss wrote to the elder Bolyai: "Now something about the work of your son.... The whole content of the paper, the paths that your son has taken, and the results to which he has been led, agree almost everywhere with my own meditations, which have occupied me in part already for 30–35 years... now I have been saved the trouble [of writing a paper]."

The younger Bolyai was distressed and humiliated; in either 1831 or 1832, Bolyai's work was published as an addendum to a longer work of his father's, entitled *Tentamen*. Bolyai was dismayed anew on publication—neither father or son had been aware that Lobachevsky had published a paper outlining the concepts of non–Euclidean geometry three years before their work was issued. Stung by the lack of recognition and his failure to establish his own priority, Bolyai virtually abandoned his scholarly efforts in mathematics. Plagued by poor health—he suffered from chronic fevers—he accepted an invalid's pension and left the military in 1833.

Bolyai returned to Marosvasarhely to live with his father, but the two strong–willed, emotional, and disappointed men were unable to peacefully share the same house. The younger Bolyai retired to a small family estate at Domald; in 1834 he contracted an "irregular marriage" (probably a live–in arrangement with no legal standing) with Rosalie von Orban, with whom he had three children.

In 1837 father and son tried again in vain to build a place for themselves in the world of mathematics. They entered the Jablonow Society prize contest, the subject of which was the geometric construction of imaginary quantities. Several mathematicians, Gauss included, were exploring the subject at the time. Unfortunately, the Bolyais' solutions to the problem were too complicated to win, and in fact, János' solution was similar to **William Rowan Hamilton**'s, who had already published a solution that was easier to understand.

Notes and letters indicate that Bolyai continued to dabble in mathematics. He was interested in absolute geometry, the relationship between absolute **trigonometry** and spherical trigonometry, and the **volume** of tetrahedronsin absolute space. Bolyai also dabbled in philosophy, outlining what he termed salvation theory, in which he examined the concepts of individual and universal happiness and the relationship of virtue to knowledge.

Bolyai's father died in 1856, at about the same time Bolyai's arrangement with Rosalie von Oraban ended. Bolyai continued to live at the family estate as a semi–recluse, and his occasional writings from 1856 to 1860 include a memorial to his mother and a lively appreciation of the ballet company at the Vienna Opera House. He died after a lengthy illness and was buried in Marosvasarhely.

BOLZANO, BERNHARD (1781-1848)

Czechoslovakian theologian, philosopher, and mathematician

Bernhard Placidus Johann Nepomuk Bolzano wrote and published pioneering works on **infinite series** and the **infinitesimal**.

Born in Prague, Bohemia in 1781, Bolzano entered the University of Prague at the age of 15 to study philosophy and mathematics. Four years later, he began theological studies at University, while simultaneously working on a doctoral thesis in **geometry**. In 1804, Bolzano was ordained a Roman Catholic priest, and received his doctorate in mathematics and an appointment to the chair of philosophy and religion at the University of Prague.

Over the next thirteen years, Bolzano published a number of mathematical works and papers. *Rein analytischer Beweis* (Pure Analytical **Proof**) published in 1817, contained a non-geometric, **arithmetic** proof of the location **theorem** in **algebra**. One of Bolzano's contemporaries, the French mathematician **Augustin Cauchy**, published a similar work several years later that received wider circulation and considerably more attention from the mathematical community. The similar mathematical concepts outlined by both men are thought to be a coincidental simultaneous discovery, as there is no written account of the two mathematicians ever meeting, nor of Cauchy having any familiarity with Bolzano's work.

Because of his pacifist views and his open criticism of the theology texts used at Prague, the Austrian government forced the suspension of Bolzano from his University position in 1819 under charges of heresy. Bolzano's work was largely censored in Austria from then on, but he managed to publish several more important mathematical papers over the next several decades. He was also active in the Royal Bohemian Society, and was named director in 1842.

Bolzano published *Wissenschaftslehre*, a comprehensive account of science and knowledge, in 1837. He continued his work with *Paradoxien des Unendlichen* (Paradoxes of the Infinite), an insightful work on infinite **sets** and theories of mathematical **infinity** that was published in 1850, several years after Bolzano's death. Bolzano left many unpublished manuscripts at the time of his death, some of which were published by students decades later. Although his ideas were revolutionary to the field of mathematics, they remained largely unnoticed until "rediscovered" by mathematicians and scholars that followed him. In some cases, Bolzano received posthumous recognition for his theories. The **Bolzano-Weierstrass theorem**, which states that a bounded set S containing infinitely many elements contains at least one limit point, was a mathematical concept originated by Bolzano, but rediscovered and reintroduced to the mathematical community by the University of Berlin's **Karl Weierstrass** some fifty years after Bolzano's original supposition.

BOLZANO-WEIERSTRASS THEOREM

The Bolzano-Weierstrass **Theorem** is an important result in point-set **topology**, the branch of mathematics that concerns the properties of **sets** and the points that comprise them. As an example, the real number line contains subsets of intervals, and the intervals contain **real numbers**, which are the points of this system. Thus we could consider the topology of the real number line to be the study of the properties of these intervals and the real numbers contained in them. We can also consider the topology of the Cartesian plane, of three-dimensional **space**, or of any higher dimensional spaces.

In order to understand the Bolzano-Weierstrass Theorem, it is necessary to understand four fundamental concepts: infinite sets, bounded sets, neighborhoods, and accumulation points. An infinite set is simply a set with an infinite number of points. The **interval** [0,1] is an infinite set in one **dimension**, as is a **circle** and its interior in the two-dimensional plane, as is a **sphere** and its interior in three dimensions. A bounded set is one for which there is both an upper bound, a value that is greater than or equal to every member of the set, and a lower bound, a value that is less than or equal to every member of the set. For example, the one-dimensional interval [0,1] is bounded above by 1 and below by 0. Note that [0,1] is also infinite, so it is clear that we may have sets that are both infinite and bounded. A neighborhood of a point p is essentially a small region completely surrounding p. On the one-dimensional real line, a neighborhood of the real number p would be any interval of real numbers that contained p. In two dimensions, a neighborhood of the point p would be a circle and its interior that contained p; in three dimensions a sphere and its interior that contained p. The fourth essential concept we need to understand before stating the Bolzano-Weierstrass Theorem is the accumulation point, also called a cluster point or limit point. A point p is called an accumulation point of a set S if every neighborhood of p contains a point in S that is different from p. Thus, in the Cartesian plane an accumulation point for a set S is a point such that if we draw smaller and smaller circles around p, no matter how small the radii of these circles become, we will always "capture" points of S inside the circles. In fact, we will always find an infinite number of points of S in these circular neighborhoods. On the real number line we use intervals of points as our neighborhoods and in three-dimensional space we use spherical neighborhoods.

Now we are in a position to state the Bolzano-Weierstrass Theorem: every bounded infinite set S has at least one accumulation point. An immediate consequence of Bolzano-Weierstrass is that every bounded infinite sequence of real numbers has a convergent subsequence. This fact is useful in proving many important results in the theory of infinite **sequences and series**. The **proof** of the Bolzano-Weierstrass Theorem can be found in most textbooks on topology or real **analysis**.

Bombelli, Rafaello (1526-1573)
Italian algebraist

Rafello Bombelli was the last of a long line of Italian algebraists who contributed to the theory of **equations** during the Renaissance. He was the first to develop a consistent theory of **imaginary numbers** which included the rules for the four operations on **complex numbers**. **Gottfried Wilhelm von Leibniz** complimented Bombelli years later referring to him as an "outstanding master of the analytical art."

Bombelli was born in 1526 in Bologna, Italy. His father, Antonio Bombelli, was a wool merchant. Bombelli choose not his father's profession but instead became an engineer. He never attended any university but instead was trained by Pier Francesco Clementi of Corinaldo, who as an engineer–architect was responsible for draining swamps. For the major part of his working life, Bombelli was employed as an engineer under the patronage of Monsignor Alessandro Rufini, who later became the Bishop of Melfi. Bombelli's engineering career included two major projects; the reclaiming of the Val di Chiana marshes from 1551 to 1560, and the failed attempt to repair the Ponte Santa Maria Bridge in Rome in 1561.

Bombelli wrote his *Algebra* text in 1560, just a few years after **Girolamo Cardano**'s *Ars magna* was published. Bombelli's *Algebra* was not published the first time until 1572 (and only a partial edition), and it became a very important work for several reasons. It featured 143 problems found originally in **Diophantus**' *Arithmetica*. Bombelli had found a copy of Diophantus' book in the Vatican Library and until that time the ancient Greek's Diophantus' work was mainly ignored. In *Algebra,* Bombelli made notable contributions to improvements in algebraic notation. He introduced an index notation for denoting **powers**, which he referred to by the term "dignita." Bombelli also introduced in this work a new symbol to indicate the root of an entire expression, was the predecessor of a modern bracket.

What Bombelli is best known for is his justification of conjugate imaginary **roots** for the "irreducible case" of **cubic equations**. Starting with the Cardano–**Tartaglia** formula, he developed a very detailed theory of imaginary numbers by arguing by analogy with the known rules for operating on **real numbers**. The irony of Bombelli's discovery was that the irreducible case actually results in three real roots, but the Cardano–Tartaglia formula produces two conjugate imaginary

roots involving the **square** root of a negative number. Bombelli knew that this type of cubic equation has three real roots and used the results of the Cardano–Tartaglia formula to demonstrate that real numbers can be the result of operations on complex numbers.

Bombelli never realized his great contribution, since he still referred to complex numbers as useless. Nonetheless, Bombelli's *Algebra* was widely read and respected. Leibniz used it to study cubic equations and **Leonhard Euler** quoted from it in his text, *Algebra*.

Boole, George (1815-1864)
English logician and algebraist

George Boole was the founder of the modern science of mathematical logic. He devised a system of binary **algebra** that today has broad applications in the design of computer circuits and telephone switching. He also made significant contributions to **probability theory**, the field of **differential equations**, and the **calculus** of finite differences.

George Boole was born in Lincoln, England, on November 2, 1815, into what was regarded at the time as a lower class family. His father, a tradesman, encouraged Boole to obtain an education in the classics, believing this might elevate him to a higher social rank. Because Latin and Greek were not offered at the National Schools, the only formal educational institutions then available for boys of Boole's social status, he studied the ancient languages on his own. At the age of 12 made a translation of a Latin ode by Horace, which his father submitted for publication in a local newspaper. When the translation appeared in print, a local Latin scholar argued that no 12–year–old could have been capable of such an accomplishment, and Boole was falsely accused of plagiarism. In addition to the classics, he was encouraged to learn mathematics. Boole's father had a personal interest in the subject and passed on what rudimentary knowledge he could from his own private study. Boole took a job as an assistant schoolteacher when he was 16, in order to help support his parents. He also began, in his spare time, to study for the clergy. In 1835, his parents' poverty worsened and Boole chose to abandon his religious pursuits. Boole opened his own elementary school and, in the course of preparing his students in mathematics, found that the available textbooks were inadequate. As he searched for better teaching materials, Boole began reading the classical mathematical works of the 17th and 18th century masters, developing an interest in algebra and calculus, and was influenced by the works of **Isaac Newton** and **Pierre Laplace**. He made a special study of *Mecanique analytique*, in which the author and French mathematician **Joseph–Louis Lagrange** set forth a purely analytical calculus of variations.

By the late 1830s, Boole was ready to make original contributions in the field of **analysis**. He established contact with Duncan Gregory, a Scottish mathematician who edited the newly founded *Cambridge Mathematical Journal*. This relationship was significant because, uncredentialed and lacking membership in a learned society, Boole had few options for presenting his ideas to a mathematical audience. Gregory, impressed with Boole's style and originality, began publishing his work. In 1841 Boole wrote a paper on the theory of algebraic invariants which greatly extended the work of Lagrange. On the basis of this paper, Boole is often credited with the discovery of invariants, a construct that has application in theoretical physics. In 1844 Boole published a paper on the calculus of operatorsin the *Philosophical Transactions of the Royal Society*. For his original contribution to this field, the Royal Society of London awarded him a gold medal. This honor led to Boole's correspondence with many of the prominent British mathematicians of his day. He was encouraged to take courses at Cambridge University, but finances never allowed. Boole continued teaching at his elementary school until 1849, when, owing to the reputation he had established, he was offered a university post. Despite his lack of formal training, Boole was welcomed as the Chair of Mathematics at the newly established Queens College, Cork, in Ireland. His professorship marked the end of his economic struggles, and allowed him to devote more time to his interests in both calculus and **symbolic logic**.

Boole's reputation as a prominent mathematician was secured by his work on the subject of mathematical logic. In 1847 he published a groundbreaking pamphlet, *The Mathematical Analysis of Logic*, in which he argued that logic was more closely allied with mathematics than philosophy. The pamphlet was written in reaction to an ongoing dispute between **William Hamilton**, professor of logic and metaphysics at the University of Edinburgh, and **Augustus De Morgan**, a logician and professor of mathematics at the University of London. Hamilton believed that the study of mathematics was a useless exercise, and that de Morgan would be a better philosopher if he were less of a mathematician. De Morgan, already recognized in Britain for his contributions to the study of logic, was also known for his satirical wit. He apparently made short–order of Hamilton's arguments and was in no need of additional aid from Boole. Nonetheless, Boole's pamphlet, which established that mathematical rules could be applied to logic, won de Morgan's admiration.

In 1854 Boole wrote *An Investigation of the Laws of Thought,* in which he greatly developed his ideas about logic. He reduced logical relationships to simple statements of equality, inequality, inclusion and exclusion, and expressed these statements symbolically, using a two digit or binary code. Boole then devised algebraic rules that governed the logical relationships. This bridge between mathematics and logic came to be known as **Boolean algebra**.

In the years following the publication of his treatise on logic, Boole turned his attention to calculus and differential **equations**. He wrote a standard textbook, *Treatise on Differential Equations*, in 1859, in which he investigated partial differential equations. Boole also advanced criteria for distinguishing between singular solutions and particular solutions of differential equations of the first order. In 1860 he published another textbook, *Treatise on the Calculus of Finite Differences*.Boole's two textbooks were in wide use at universities for many decades after his death.

In 1855 Boole married Mary Everest (Boole), niece of Sir George Everest, a Professor of Greek at Queen's College and the man for whom the world's highest mountain is named. They had five daughters, including the mathematician Alicia Boole Stott. Boole died prematurely at the age of 49. Attempting to get to a class on time, he walked two miles through a drenching rain. Soon after lecturing in his wet clothes he contracted pneumonia. The story is told that his wife, believing the remedy for an illness ought to bear resemblance to its cause, put him to bed and doused him with buckets of cold water. Boole died in Ballintemple on December 8, 1864, and is buried at St. Michael's Church in Blackrock, County Cork, Ireland.

Boole was honored in his lifetime with degrees from the universities of Dublin and Oxford. He was elected a fellow of the Royal Society of London in 1857. He was also made a member of the Royal Irish Academy. Although for many years the ideas of Boole's symbolic logicwere regarded mostly as philosophical curiosities, they eventually found important practical application. Much of modern computer processing is based on the binary system of Boolean algebra, as is the design of computer circuitry and telephone switching equipment.

BOOLEAN ALGEBRA

Boolean **algebra** is often referred to as the algebra of logic, because the English mathematician **George Boole**, who is largely responsible for its beginnings, was the first to apply algebraic techniques to logical methodology. Boole showed that logical **propositions** and their connectives could be expressed in the language of **set theory**. Thus, Boolean algebra is also the **algebra of sets**. Algebra, in general, is the language of mathematics, together with the rules for manipulating that language. Beginning with the members of a specific set (called the universal set), together with one or more binary operations defined on that set, procedures are derived for manipulating the members of the set using the defined operations, and combinations of those operations. Both the language and the rules of manipulation vary, depending on the properties of elements in the universal set. For instance, the algebra of **real numbers** differs from the algebra of **complex numbers**, because real numbers and complex numbers are defined differently, leading to differing definitions for the binary operations of **addition** and **multiplication**, and resulting in different rules for manipulating the two types of numbers. Boolean algebra consists of the rules for manipulating the subsets of any universal set, independent of the particular properties associated with individual members of that set. It depends, instead, on the properties of **sets**. The universal set may be any set, including the set of real numbers or the set of complex numbers, because the elements of interest, in Boolean algebra, are not the individual members of the universal set, but all possible subsets of the universal set.

A set is a collection of objects, called members or elements. The members of a set can be physical objects, such as

 •

people, stars, or red roses, or they can be abstract objects, such as ideas, numbers, or even other sets. A set is referred to as the universal set (usually called I) if it contains all the elements under consideration. A set, S, not equal to I, is called a proper subset of I, if every element of S is contained in I. This is written and read "S is contained in I."

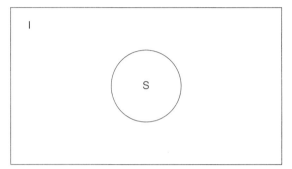

S is contained in I

Figure 1.

If S equals I, then S is called an improper subset of I, that is, I is an improper subset of itself (note that two sets are equal if and only if they both contain exactly the same elements). The special symbol is given to the set with no elements, called the empty set or null set. The null set is a subset of every set.

When dealing with sets there are three important operations. Two of these operations are binary (that is, they involve combining sets two at a time), and the third involves only one set at a time. The two binary operations are union and intersection. The third operation is complementation. The union of two sets S and T is the collection of those members that belong to either S or T or both.

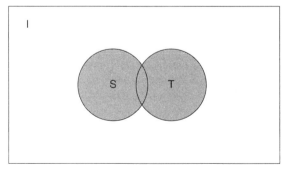

The union of S and T

Figure 2.

The intersection of the sets S and T is the collection of those members that belong to both S and T, and is written

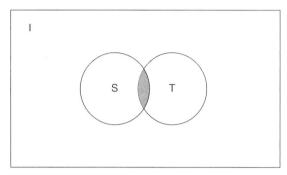

The intersection of S and T

Figure 3

The complement of a subset, S, is that part of I not contained in S, and is written S'.

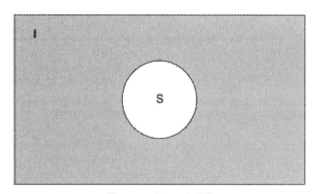

The complement of S

Figure 4.

The properties of Boolean algebra can be summarized in four basic rules.

(1) Both binary operations have the property of commutativity, that is, order doesn't matter.

S $\cap$ T = T $\cap$ S, and S $\cup$ T = T $\cup$ S.

(2) Each binary operation has an **identity element** associated with it. The universal set is the identity element for the operation of intersection, and the null set is the identity element for the operation of union.

S $\cap$ I = S, and S $\cup$ ø = S.

(3) each operation is distributive over the other.

S $\cup$ (T $\cap$ V) = (S $\cup$ T) $\cap$ (S $\cup$ V), and S $\cap$ (T $\cup$ V) = (S $\cap$ T) U (S $\cap$ V).

This differs from the algebra of real numbers, for which multiplication is distributive over addition, a(b+c) = ab + ac, but addition is not distributive over multiplication, a+(bc) not equal (a+b)(a+c).

(4) each element has associated with it a second element, such that the union or intersection of the two results in the identity element of the other operation.

A $\cup$ A' = I, and A $\cap$ A' = ø.

This also differs from the algebra of real numbers. Each real number has two others associated with it, such that its sum with one of them is the identity element for addition, and its product with the other is the identity element for multiplication. That is, a + (-a) = 0, and a(1/a) = 1.

The usefulness of Boolean algebra comes from the fact that its rules can be shown to apply to logical statements. A logical statement, or proposition, can either be true or false, just as an equation with real numbers can be true or false depending on the value of the **variable**. In Boolean algebra, however, variables do not represent the values that make a statement true, instead they represent the **truth** or falsity of the statement. That is, a Boolean variable can only have one of two values. In the context of **symbolic logic** these values are true and false. Boolean algebra is also extremely useful in the field of electrical engineering. In particular, by taking the variables to represent values of on and off (or 0 and 1), Boolean algebra is used to design and analyze digital switching circuitry, such as that found in personal computers, pocket calculators, cd players, cellular telephones, and a host of other electronic products.

BOREL, ÉMILE (1871-1956)

French number theorist

Émile Borel was one of the most powerful mathematicians of the twentieth century. He displayed great virtuosity in working in a number of different areas of mathematics, particularly **complex numbers** and **functions**, but in addition served visibly as a representative of the mathematical community to the general public. He brought the influence of the mathematical spirit to bear, thanks to his popular writings and to his participation in academic and national politics. His views on mathematics and how it should be supported affected French national policy on mathematics for many years after his death.

Émile Félix-Édouard-Justin Borel was born in Saint-Affrique in the Aveyron district of France, the son of Protestant pastor Honoré Borel and Émilie Teissié-Solier, the daughter of a successful merchant. He had two older sisters, but rapidly impressed the neighborhood with his intelligence. Born on January 7, 1871, he entered a France still in disarray after the setbacks of the Franco-Prussian war. His original education was at the hands of his father, from which he passed at the age of 11 to the lycée at Montauban, the nearest cathedral town. His merits were recognized quickly enough that he was sent to Paris to study at some of the leading preparatory schools for the university. His efforts were crowned by admission to the École Normale Supérieure, an institution that was at the summit of French scientific education and with which he was to be associated throughout the rest of his career.

Academically Borel made rapid progress through the next few years. He was welcome in the family circle of the eminent **geometry** specialist Gaston Darboux and married mathematician Paul Appell's daughter Marguerite in 1901. Although they had no children, they adopted one of Borel's nephews, Fernand, who was later killed during the First World War. After Borel graduated first in his class at the École Normale, he was offered a teaching position at the university at Lille, exceptionally early since he had not finished his doctoral thesis. He took his doctorate in 1894 and returned a couple of years later to the École Normale. Despite his busy schedule of research and publication, Borel always took his responsibilities at the École Normale seriously, paying attention both to his own teaching and to the curriculum. The École Normale rewarded his work by instituting a special chair in the theory of functions, of which Borel was the first occupant.

Borel started his career of research by publishing three brief articles at the age of 18. From there he went to a life in which writing always played a large part, as witnessed by his over 300 papers and books. In addition to his research articles, the dedicated teacher also devoted effort to the re-creation of French textbooks in mathematics. His activities as a writer were supplemented by his editing one of the best known series of expositions, known as the Borel collection. The first of these, written when he was still only 27, *Leçons sur la théorie des fonctions* (*Lessons on the Theory of Functions*), went through several editions and rapidly acquired the status of a classic, for it laid the foundation for the field of measure theory. The French publishing house of Gauthier-Villars owed its preeminence in mathematics to the work that Borel did on behalf of this series.

Borel's earliest mathematical work was already distinguished by its variety, but two of the areas in which his contributions attracted attention were the study of complex functions and of the **topology**, or properties of geometric configuration, of **real numbers**. Functions serve as a type of formula in which different numbers, known as variables, are plugged into a mathematical expression for a resulting value. Just as there are functions of real numbers that give out real numbers as values, so there are functions of complex numbers—numbers that include the imaginary unit i, the **square** root of –1—that give out complex numbers as values. There was a clear geometric significance attached to the existence of certain objects from **calculus** for real functions, but the geometry of curves representing functions of complex numbers was more complicated. Borel was able to use the information that complex functions had derivatives—the **limits** of rates of change—to prove what is known as Picard's **theorem**, concerning the number of possible values that a complex function can take. The theorem and the methods used to prove it are fundamental to the study of complex functions.

Borel's important results were not limited to complex functions, however. It might seem easy to talk about the **area** of a region in two dimensions, but as the region gets more complicated, it becomes harder to define exactly what is meant by area. Borel extended older ideas and results about areas to more general kinds of regions, including those defined by an infinite sequence of operations. Such regions or **sets** are said to be "Borel-measurable," and the more general notion of area introduced by Borel (called "measure" from the French "mesure") became the basis of the field called "measure theory." Although the most general notion of area that was taken up by the next generation of mathematicians came from Borel's younger colleague, Frenchman **Henri Lebesgue**, Borel influenced Lebesgue's work in many ways, including personal encouragement.

In his thesis, Borel articulated another idea that is now identified with his name. The idea of compactness has been a central one in twentieth-century mathematics and comes out of topology. The term is not easy to summarize, but perhaps the simplest way to consider compactness is that it describes the extent to which even an infinite set can have some of the simpler characteristics of a finite set. Borel showed that the set of real numbers includes the property that every closed and bounded **interval** (such as the real numbers between 1 and 2, inclusive) is compact. This theorem is known as the Heine-Borel theorem—a somewhat misleading name, as German mathematician **Heinrich Heine** observed but never articulated the significance of the idea.

One way of getting a perspective on Borel's central role in French mathematics from early in his career is to look at a debate about the role of the infinite that took place in the year 1905. As a result of the publication of a **proof** involving a controversial **axiom** in **set theory** called the axiom of choice (which allowed for an infinite number of choices from a set), Borel published an article expressing distrust of the general principle involved. There was a rapid exchange of letters between leading French mathematicians including Lebesgue, not all of whom agreed with Borel's restrictions. Nevertheless, the entire discussion was centered about Borel, who kept it going by forwarding letters he had received to those who might have more to say on the issue. There are many ways of describing the philosophy of mathematics that crystallized out of this discussion, and the views of Borel had been influenced by those of **Jules Henri Poincaré** of the previous generation, but Borel's role in raising the issue and trying to reach a modicum of agreement testifies to his central position.

Borel's interests had never been restricted entirely to mathematics. In 1906 he took the money that he had received for the Petit Prix d'Ormoy and used it to start a journal entitled *La revue du mois* (*Monthly Review*) that dealt with issues of general interest. Both he and his wife worked on the editing with success in the quality of the articles and the range of subscribers. It only ceased publication in 1920 as a result of the darker economic climate created by World War I.

The war also had the effect of moving Borel's interests in the direction of **applied mathematics**. Among the long-term consequences was a stream of publications in probability, which had become a standard subject in the mathematical curriculum. The original notions of finite probability had been developed in the seventeenth century, and the progress of calculus shortly thereafter allowed for the generalization of probabilistic notions to the real numbers as well. As in his work on measure, Borel was able to extend the notions of probability to more complicated sets of events. Some of his work in probability was fundamental to the new field called **game theory**, including his observation that there were situations where random behavior had its advantages. In addition to his strictly mathematical work on the subject, Borel wrote for a popular audience to make the subject of probability more widely accessible.

As an indication of the width of Borel's scientific interests, he wrote a volume in 1922 on **Albert Einstein**'s theory of relativity that was subsequently translated into English as *Space and Time*. The discussion proceeds almost entirely without **equations**, since Borel's concern once again was to make the subject comprehensible to the widest audience. The discussion includes both the special and general theories of relativity and is characterized by many references to familiar objects to remove the sense of strangeness. Although Borel did not contribute to the mathematical details of Einstein's theories, he took it as a public responsibility of the mathematical community to help make science comprehensible.

His sense of public responsibility extended to the political sphere as well. Although he remained a resident of Paris, he was elected mayor of Saint-Affrique, perhaps partly as a tribute to his scientific standing. He represented the area as a member of the Chamber of Deputies from 1924 to 1936 under the banner of the Radical-Socialist party, although it was neither radical nor socialist as those terms are customarily used. He also served as Minister of the Navy in 1925, but his most lasting accomplishments in the political arena concerned the funding of mathematical and scientific research. The Centre National de la Recherche Scientifique (CNRS) received funding under his aegis and has continued to be essential for the support of research in France. He helped plan the Institut Henri Poincaré and served as its director for many years, from its founding in 1928 until his death, thereby paying tribute to one of the great influences on his own approach to mathematics.

Borel was elected to the Académie des Sciences in 1921, rather later than one might have expected and perhaps out of a distrust of his political involvement. He served as president of the Académie in 1934, another expression of his stature in the wider scientific community, and received the gold medal of the CNRS on its first being awarded. He was decorated for his work during World War I, and again for having stood up to the German Gestapo during World War II. He had left his chair at the École Normale and had taken up a position at the Sorbonne (the University of Paris) instead, as his memories of the generation of French mathematicians dead on the battlefields of World War I were too painful in the familiar surroundings. He retired from the Sorbonne but remained mathematically active for the rest of his life, including regular attendance at international congresses. Part of the reason for his success in accomplishing so much was that he welcomed those who had something constructive to say and chose not to waste time on formalities and empty words. As a lecturer he was impressive, thanks to his tall, dignified manner and his air of distinction. His death on February 3, 1956, hastened by a fall on board ship returning from a conference in Brazil, was met with a sense of great loss by colleagues and pupils, but the heritage he left in his mathematics and in his writing testifies to a strong belief in the mathematician's obligation to serve the community at large.

BOSCOVIC, RUDJER (1711-1787)

Croatian mathematician and astronomer

Rudjer Boscovic, also known as Ruggero Giuseppe Boscovich, was a Jesuit mathematician and astronomer. He was born in Ragusa, Croatia (now Dubrovnik, Yugoslavia) on May 18, 1711, and died in Milan, Italy in February 13, 1787.

Boscovic's father was a Croatian merchant, and his mother was the daughter of a merchant originally from Italy. The family was of average means, but noted for its literary accomplishments. Boscovic began his education at the Jesuit college in Dubrovnik, and continued it in Rome, entering the novitiate of Sant'Andrea in 1725 at the age of fourteen.

Said to have been an excellent student, Boscovic first learned science through the independent study of mathematics, physics, astronomy, and geodesy.

He later studied **mathematics and physics** at the Collegium Romanum in Rome, where in 1740 he was appointed to the Chair in Mathematics in spite of the fact that he had not yet completed his theological studies. In 1754, he published a textbook, part of which was devoted to his theory of **conic sections**.

According to the custom of the times, Boscovic served on several practical commissions, including one investigating the origins of fissures in the cupola of St. Peter's, another involving the excavation of an ancient Roman villa, and another planning the draining of the Pontine marshes. The latter project led to his writing a series on hydraulic engineering. In 1758, he published a major work in the field on natural philosophy.

In 1759, Boscovic traveled to Paris, where he remained for six months, before moving on to London. In England, he met with representatives of the Church of England, and also met Benjamin Franklin. He was elected a fellow of the Royal Society in 1761.

Traveling to Istanbul to observe the 1671 transit of Venus, Boscovic eventually made his way to Holland, Germany, Bulgaria, Moldavia, Poland, Silesia, Austria, Venice, before reaching Rome in 1763 after an absence of four years.

In 1764, he became Professor of Mathematics at Pavia. Boscovic conducted research mainly in optics and in the improvement of telescopic lenses. He was especially interested in developing new methods to determine the orbits and rotational axes of the planets; he also investigated the shape if the Earth.

In 1770, he moved to the department of optics and astronomy at the Scuole Palatine in Milan. Soon Boscovic managed to provoke opposition from his colleagues at the observatory there. In 1772, the court in Vienna relieved Boscovic of responsibility for the observatory. Out of despair, he also resigned his professorship. And the next year, the pope suppressed the Jesuit order.

Now sixty-three years of age, Boscovic moved to Paris at the urging of friends where he was offered a directorship in optics for the navy. He remained in France until 1782, at which time he returned to Italy.

In 1785, the printing firm of the brothers Remondini brought out a five-volume set of Boscovic's writings. The preparation and proofreading of this work apparently exerted tremendous strain on Boscovic's health. Setting out again to travel, this time in Italy, Boscovic found that his mental faculties were declining. Some thought he was fortunate in that he died of a lung ailment before his mental decline had reached extreme severity. He was buried in the church of Santa Maria Podone in Milan.

Boscovic's work in instrumental science covered astronomy, optics, geodesy, mathematics, **mechanics**, and natural philosophy. As a theoretical scientist, he conducted investigations in mathematics, mechanics, the properties of and matter. Boscovic was also a strong advocate of Newton's theory of gravity, which he had began studying in 1735 while at the Collegium Romanum.

BOURBAKI, NICOLAS (1939-)
French mathematicians

Nicolas Bourbaki does not exist. This name is used as a pseudonym for a group of mainly French mathematicians who started publishing collectively in 1939 (the year of birth given above). The stated aim of the Bourbaki group is to produce a modern overview of mathematics relevant to university students.

The founding members of the Bourbaki group were all graduates from the Ecole Normale Superieure, in Paris. The original members were **Henri Cartan**, Claude Chevalley, Jean Dieudonne, Jean Deslarte, and **Andre Weil**. Their aim was to produce a modern text book for French University students to replace the classic work by Goursat. With their desire to keep the work up to date and fresh, a rule was adopted such that members upon reaching their 50th birthday would have to relinquish their membership in the group.

The first book in this series, *Elements of Mathematics*, was published in 1939 and it had as its aim the drawing together of all the different strands of modern mathematics. All of the works are concerned with the treatment and the relationship of the differing mathematical topics covered. There is no room for diversions such as the people involved in the discoveries of the various principles. The history and people involved have been looked at briefly in volume 36, *Elements d'historique des mathematiques*. To keep each volume fresh, the Bourbaki group would meet annually for one or two weeks. At these meetings all would contribute ideas and knowledge of the areas to be covered, after having spent the preceding year researching them from first principles. Then one member would produce the first draft which would be circulated to all other members for comments and corrections. At the next meeting the work in progress was greatly discussed and handed to another author to continue. This would continue until the work was judged to be completed by all. This method of composition with numerous rewrites ensured the work comprised the latest material, was not obviously by one author, and was not an encyclopaedia written by a panel of experts, each submitting a chapter on his own area of expertise.

Although the work of Bourbaki is highly regarded, the field of mathematics is so complex today that a book of this nature is no longer accessible to students embarking on their University mathematics course; and it is now recognized that the books are aimed at those who have at least a good working knowledge of the contents of the first year or two of a mathematics course. It is true to say that for many practical purposes the *Elements of Mathematics* is really only accessible to the graduate student and beyond. By 2000 over 45 volumes had been published in French with the majority being translated into English.

See also Henri Cartan; Andre Weil

BOYLE, ROBERT (1627-1691)

English chemist and physicist

Robert Boyle, often referred to as "the father of modern chemistry," was a revolutionary scientific figure in his belief that all scientific disciplines should be subjected to the rigors of scientific experimentation, and that science itself could be explained through mathematical laws.

Boyle was born in Lismore, Ireland in 1627, the fourteenth child of Lord and Lady Cork. Boyle's father, who was the Earl of Cork, was one of the wealthiest men in the country, owning a large amount of land throughout Ireland. This would have a significant impact on Boyle, as it would provide him with the means to pursue his scientific research unencumbered by financial concerns for the rest of his life.

At the age of 8, Boyle and his brother Francis, who was four years older than Robert, were sent to Eton College to begin their education. Boyle would spend three years at Eton before returning home in 1638 to receive private tutoring from a local parson.

In 1641, Boyle and his brother were sent away once again, this time to embark on a five year European tour. A tutor went with them, and the boys received instruction in Latin, French, religious studies, and mathematics. During the trip, Robert also learned Greek, Hebrew, and Italian. While in Italy, he studied the works of Galileo, which introduced him to the use of mathematics as a necessary method of scientific study. Towards the planned end of the journey, news of political strife and civil war back in Ireland reached the Boyle brothers. Francis returned to fight against the Irish rebels with his father, but Robert, who had fallen ill, stayed in Geneva. In 1643, he received news that his father had died, and Boyle was finally able to finance his return to England the following year after selling jewelry and borrowing from his tutor.

Once back in London, Boyle lived with his sister, Lady Katherine Ranelagh. Katherine was well connected with some of the leading political and academic minds of the day. Of particular interest to Boyle were the scholars of philosophy, astronomy, and other disciplines that would meet regularly at various places around town, including Ranelagh's home. Boyle called this group "the Invisible College," and he voraciously continued his studies in order to be able to participate in these discussions. The "College" eventually asked Boyle to join their group as a regular member, despite the fact that he was not affiliated with any university and was largely self-taught.

Robert Boyle

Boyle eventually returned to Stalbridge, the family estate that had been deeded to him by his father. Boyle arrived in 1646, and spent six years putting the estate, which had fallen into disarray, back into order while continuing his scientific and academic studies during his leisure time. His efforts at restoring Stalbridge did not go unrewarded, as the rents and other profits from the property brought him a considerable annual sum. Political changes in Europe meant that Boyle's family recovered much of the land they had previously lost to Irish rebels in Munster, increasing his wealth even further.

When Boyle returned to England in 1654, he settled down at lodgings in Oxford to set up his own laboratory in order to begin a series of scientific experiments. Boyle had the financial means to afford both the best equipment and the best laboratory assistants. At Oxford, Boyle began a series of famous experiments that used a vacuum pump, constructed in part by assistant **Robert Hooke**, to test the properties of a vacuum. In 1660, Boyle published the results of his experiments in a book entitled *New Experiments Physico-mechanical, touching the Spring of the Air and its Effects*. Among his many landmark findings were that a flame could not exist in a vacuum, that the pressure of air or atmosphere is what makes mercury in a barometer rise, that sound could not travel in a vacuum, and that air itself had weight.

In 1662 in response to doubts expressed by several scientists about his findings on barometric pressure, Boyle published a second edition of his first book which included the now-famous "Boyle's law." Boyle's law stated that doubling the pressure on a gas reduces its **volume** to one-half, as long as the temperature of the gas stays constant. This finding further strengthened Boyle's assertions that air had "springiness," or elasticity.

Boyle's reputation as a scientist grew tremendously after the publication of his book, and in 1960, Charles II asked Boyle for a private demonstration of his barometric experiment. Boyle was offered both a title and holy orders in recognition of his achievements, but refused both. However, he did use the King's esteem of his talents to set up a new scientific organization, known as the "Royal Society of London for Improving Natural Knowledge, to replace the now disbanded Invisible College. The King granted a charter for the group in 1662.

Boyle completed a significant work on chemistry in 1661, entitled *The Sceptical Chymist*, which sought to establish chemistry as a physical science based on mathematical principles. Up to that point, chemistry was not a science in and of itself, and was typically termed "alchemy," a craft that dealt with the separation of metals, the preparation of medicine, and spirit distillation. However, for the most part, there was no scientific discipline applied to the practice and much myth and mystery surrounding it. Boyle also argued against the Aristotelian idea that was still used in alchemy that everything in the world was composed of the four elements of earth, air, fire, and water. In Boyle's view, alchemy, or chemistry, was a necessary and useful science that needed to be subjected to the rigors of experimentation in order to place it on sound scientific footing.

Boyle returned to London in 1666 to be closer to the Royal Society. He moved back in with his sister Lady Ranelagh in 1668 when his health, which had never been good, began to fail. Boyle continued his experiments, and began to devote most of his laboratory time to work in chemistry. He conducted important work on the properties of phosphorus, acids, and alkalis here.

Boyle suffered a serious stroke in 1670 that left him paralyzed for nearly a year. He somehow managed an almost complete recovery and continued his work, although at a much slower pace. The Royal Society offered Boyle the presidency of that organization in 1680, but he declined. In 1691, Boyle passed away, one week after the death of one of his closest friends and allies, his sister.

See also Scientific method

BRAHE, TYCHO (1546-1601)

Dutch astronomer

Tycho Brahe made the most accurate observations ever in naked-eye astronomy, which led to the laws of planetary **motion** created by **Johannes Kepler**, one of his students.

Tycho was born December 14, 1546, in Knudstrup, Scania, a region of southern Sweden that was then part of Denmark, the survivor of a set of twin boys. Tycho lived with his parents (his father was of Swedish noble blood) until he was six, when he was kidnapped by his father's brother. Tycho remained in his uncle's house. From his uncle he learned Latin, and also studied philosophy and law, in preparation for a career in politics. But fate intervened when Tycho observed his first solar eclipse in 1560; he was fascinated, and decided to abandon politics for science and mathematics.

Tycho attended the University of Copenhagen and there studied science, math, ethics, music and philosophy, and later continued his studies in Germany.

A turning point came when he observed a close approach of Saturn and Jupiter and noted that it came a month earlier than was predicted in the tables he was using, which had been prepared under Alfonso X. Undoubtedly dissatisfied with this inaccuracy, he set about to create his own tables. He bought instruments for use in his observations. Among the instruments he used was a radius. On this instrument he slid a crosspiece along an arm, at the end of which was a fixed eyepiece. By aligning the fixed sight with one on the moveable crosspiece, he could measure the angular **distance** between any two objects.

Tycho's uncle died in 1565, and the young man spent the next few years traveling and studying (he graduated from the University of Rostock in 1566). Besides spending time casting horoscopes (as did many astronomers of the time) and dabbling in alchemy, during this time Tycho also lost part of his nose in a duel over a point of mathematical debate. From that time on, he wore a false nose made of silver.

Tycho's career began in earnest in 1572, when on November 11 he observed a star going nova in the constellation Cassiopeia. When a star goes nova, it explodes and increases tremendously in brightness; some stars that had been invisible to the naked eye become one of the brightest objects in the heavens. Tycho studied the star for seventeen months, using instruments of his own making, and published a fifty-two page book about it, *De Stella Nova*. He showed that the star was too far to measure, but was further away than the moon. Tycho not only gave us the name "nova" for an exploding star, but also proved that **Aristotle**'s belief that the heavens were unchanging was incorrect.

Tycho also refuted Aristotle's belief about comets when he studied the great comet of 1577. His studies suggested that the comet followed an elliptical **orbit** which took it across the "celestial spheres," one of Tycho's most fondly held beliefs. Despite his discoveries, he still held to Ptolemy's theory that the earth was the center of the solar system and the universe. In his 1583 book on the comet, he suggested that all planets but the earth revolved around the sun—a theory that was soundly ignored.

Although it may be easy to see this as a failing, the reluctance with which Tycho abandoned his beliefs was typ-

Tycho Brahe

ical of the era. However, his achievements in naked-eye astronomy cannot be ignored. His observations were accurate to within two minutes of arc. He calculated the length of the year so accurately that he was off by less than a second, and his work made it possible to create a more accurate **calendar**.

Through much of his career, Tycho enjoyed the patronage of King Frederick II, who in 1576 gave Tycho the island of Hven (now Ven), on which to build an observatory. The observatory, the first real astronomical observatory, drew scholars from all over Europe. But upon Frederick's death, Tycho lost his patronage. He moved to Germany (the very thing the King had hoped to prevent), in 1597, and settled in Prague (which is now in the Czech Republic). There, he took on a new assistant who would also make his mark on astronomy, Johannes Kepler.

Tycho gave Kepler his data, which would serve as the basis for **Kepler's laws** of planetary motion. Kepler is often considered the founder of modern astronomy.

In 1601, Tycho died after a short illness, and received a state funeral. Superseded by Galileo's telescope, his instruments were never used again, and they were destroyed during the Thirty Years War.

BRAHMAGUPTA (598-670)
Indian astronomer

Brahmagupta was an Indian astronomer and mathematician. He was the head of the astronomical observatory at Ujjain (his probable birthplace). His main, but not sole, achievements in the field of mathematics were the introduction of **zero** and **negative numbers**.

Brahmagupta wrote two main texts, both of which deal with **arithmetic** and astronomy. His first work in 628 was *Brahmasphuta siddhanta* (The Opening of the Universe), and in 665 he published *Khandakhadyaka*. Both of these texts are actually written in verse. Mathematically these works include the first known use of negative numbers and a figure for zero as well as a formula for finding the **area** of a cyclic quadrilateral based on its sides (this is now known as Brahmagupta's formula; it is a modified form of **Hero's formula**). Also in his first book Brahmagupta solved the **Chinese remainder theorem**, which looks at simultaneous linear congruences, by a method different to that used by the Chinese. Brahmagupta also proposed several algebraic rules for solving quadratic and simultaneous **equations**. Other work by Brahmagupta included arithmetic progression and theorems relating to right **angle** triangles. The astronomy included in these books deals with planetary movement and eclipses. Brahmagupta lived in a time when it was thought that the sun and other planets revolved around the earth, but he was still able to give an accurate figure for the length of a year, 365 days 6 hours 5 minutes and 19 seconds (which he later revised to 365 days 6 hours 12 minutes and 36 seconds). We now regard the length of a year as being 365 days 5 hours and 48 minutes. Some of the work of **Bhaskara**, which was published some 500 years later, shows little advancement form that produced by Brahmagupta. It is true to say that Brahmagupta was the leading mathematician of the seventh century and that his work has had a massive influence throughout the centuries. Sometime in the eighth century the work of Brahmagupta was brought to Baghdad where it was translated into Arabic and then subsequently it was translated into Latin, at which point it spread throughout the western world.

Like many Indian mathematicians of this and later periods, Brahmagupta was producing work that was many centuries ahead of the equivalent work being carried out in the western world.

See also Aryabhata the Elder; Bhaskara; Bourbaki

BRIANCHON, CHARLES-JULIEN (1783-1864)
French mathematician

The French mathematician Charles-Julien Brianchon was born in Sèvres, France on December 19, 1783. He died in Versailles, France on April 29, 1864.

Little is known of Brianchon's early life, except that he entered the École Polytechnique in 1804, where he studied under the geometer **Gaspard Monge** and read Carnot's work. As a student, he published his first paper in 1806, which contained a **theorem** that now bears his name. This theorem was an extension of a long-forgotten theorem that **Pascal** had proved in 1639, namely that if a hexagon is inscribed in a conic section, then the three points of intersection of the opposite sides always lie in a straight line. Brianchon's theorem stated that in any hexagon circumscribed about a conic section, the three diagonals cross each other at a single point. The theorems of Pascal and Brianchon later proved fundamental to the study of conics, and in special cases, to the study of pentagons, quadrilaterals, and triangles.

After graduating at the top of his class in 1808, Brianchon joined Napoleon's army as a lieutenant in the artillery. Serving in Spain and Portugal, he is said to have distinguished himself there. Military life took its toll on Brianchon's health, however, and after the end of the Napoleonic Wars in 1813, he applied for a teaching position. In 1818, he received an appointment as professor to the Artillery School of the Royal Guard.

Between 1816 and 1818, Brianchon published several works in **geometry**. In 1820, he co-authored an article with another graduate of the École Polytechnique, **Jean-Victor Poncelet** (1788-1867), that gave the first complete **proof** of the nine-point **circle** theorem, and made the first use of that term. By 1822, the focus of Brianchon's research had changed to include chemistry. Later, he ceased writing all together, and devoted his time entirely to teaching.

The theorems of Brianchon and Pascal formed the first clear instance of a pair of theorems that remain valid in **plane geometry** if the words point and line are interchanged. This notion was later exploited more fully by Poncelet.

BRIDGES OF KÖNIGSBERG

The bridges of Königsberg is a mathematics problem solved by **Leonhard Euler** that concerns whether there is a path through the city of Königsberg that traverses each one of its bridges exactly once. Although the problem was a very simple one to state, Euler's solution had a seminal effect on mathematics, introducing ideas that were to lead to the development of two new branches of mathematics: graph theory, which deals with questions about networks of points that are connected by lines, and **topology**, which is the study of those aspects of the shape of an object that don't depend on length measurements.

Königsberg, which in Euler's time was located in Prussia and is now part of Germany, is a town that is divided into several different parts by the Pregel River. There are two islands sandwiched in between the two banks of the river, and a total of seven bridges connecting the islands to each other

and the opposite banks. In Euler's time, it was a town tradition to go for a Sunday walk through the town, trying to cross each bridge exactly once. No one succeeded, however, and the question arose whether it was possible.

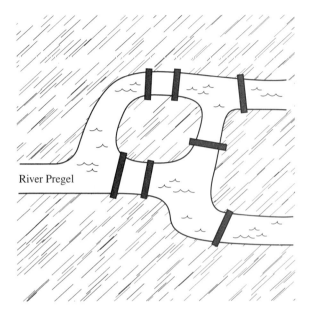

River Pregel

Euler heard of the question, and wrote,

"I was told that while some deny the possibility of doing this and others were in doubt, there were none who maintained that it was actually possible. On the basis of the above I formulated the following very general problem for myself: Given any configuration of the river and the branches into which it may divide, as well as any number of bridges, to determine whether or not it is possible to cross each bridge exactly once."

In 1736, Euler proved that it was not possible to cross the Königsberg bridges exactly once, in his paper "Solutio problematis ad geometriam situs pertinetis" (Solution of a problem relating to the **geometry** of position). His choice of title reflects the fact that the question of the Königsberg bridges was not a traditional geometry problem: the actual lengths of the bridges, and their precise geometric configuration, did not affect the answer to the problem. All that was important was how they were connected to each other. Euler's 1736 paper can be regarded as the birth of topology.

To attack the problem, Euler first simplified it by replacing each landmass with a single point, and each bridge by an arc connecting two of the points. Such a configuration of points and connecting arcs is now known as a graph; the points are more commonly referred to as vertices, and the arcs as edges. The question then becomes, is there a path along the edges of the graph that runs along each edge once, and only once? Such a path has since been named an Eulerian path.

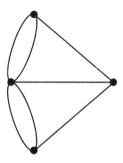

Euler's idea was a very simple one: he realized that the question was connected to the degree of the vertices. The degree of a vertex is the number of edges that come out of it. In the Königsberg graph, for example, three of the vertices have degree 3, and one has degree 5.

Euler realized that in a graph that did have an Eulerian path, there would be exactly two vertices whose degree was odd: the vertex where the path started, and the one where the path ended. The reason for this is elementary. At the starting vertex, there is an edge coming out along which the path starts. After that, every time the path returns to the starting vertex along some edge, there must be another edge to take the path out again (since the path eventually ends at the ending vertex). So the edges that come out of the starting vertex pair up, with one exception: the very first edge. This means that the degree of the starting vertex is an odd number. In a similar way, the degree of the ending vertex is odd. But at any other vertex, the edges pair up perfectly, since whenever the path comes in to the vertex along one edge, there must be another edge to carry the path out again: this means that the degree is even. Thus there are two odd-degree vertices, and all the other vertices have even degree.

The above argument leaves out one situation: that in which the path starts and ends at the same vertex. In that case, the same kind of reasoning as above shows that every vertex must have even degree.

Thus, a graph that admits a Eulerian path must be of one of two types: either it has exactly two odd-degree vertices, or it has no odd-degree vertices. But in the Königsberg graph, all four vertices have odd degree; therefore, it is impossible to have an Eulerian path along the Königsberg bridges.

See also Graph theory; Topology

BROGLIE, LOUIS VICTOR PIERRE RAYMOND DE (1892-1987)

French physicist

Louis Victor Pierre Raymond de Broglie, a twentieth century French theoretical physicist, established the foundations for wave **mechanics**, and as a result for much of modern physics. The principle underlying wave mechanics is that matter has both the characteristics of particles and waves, and the particular characteristic that is observed will depend on how the observation is made. These ideas fundamentally changed the way physicists view the world, and earned Louis de Broglie the 1929 Nobel Prize for physics.

Louis de Broglie was the youngest of five children. His early education was obtained at home. Born to Duc Victor and Pauline d'Armaille Broglie on August 15, 1892 in Dieppe, France, the young physicist-to-be was second in line to his elder (by 17 years) brother Maurice to inherit the titles of French Duc and German Prinz as the result of his family's past service to French and Austrian monarchies.

By the late nineteenth century, the de Broglie family had been supplying France with soldiers, diplomats, and politicians for centuries. Expected to follow in the family's traditional paths, elder brother Maurice was earmarked at an early age for a career as a diplomat or military officer. And so, with his paternal grandfather's consent, the young Maurice prepared for a naval career.

But by 1898, bolstered by further academic coursework, Maurice de Broglie had become increasingly interested in scientific work, and he approached his grandfather with the idea of resigning his naval commission and taking up a career in physics instead. Predictably, his grandfather would have none of it. The old man reportedly told de Broglie that science was like an old lady, content with the attentions of old men, and was an unsuitable career for a de Broglie.

As a compromise, Maurice set up a laboratory in one the rooms on the family estate, and meanwhile returned to his commission in the navy where he distinguished himself with scientific work. In 1904, after his grandfather's death, he took a military furlough, and eventually resigned his commission in 1908. At the same time, Maurice arranged for the younger Louis to study at the Lycée Janson de Sailly in Paris. Maurice, meanwhile, continued his studies in physics at the Collège de France, where he completed a thesis under the noted physicist, Paul Langevin.

In 1909, younger brother Louis graduated from the Sorbonne with baccalaureate degrees in philosophy and mathematics. But before settling on a career in physics, he also considered the fields of ancient history, paleography, and law. His decision to pursue scientific work was largely influenced by the writings of the French theoretical physicist, **Jules Henri Poincaré**.

Maurice, had already by this time begun experimental research on X rays and radioactivity. In 1912, with the discovery by Laue and the Braggs that X rays could produce diffraction patterns, Maurice suddenly began looking at X-ray spectra, which was to be the field of his greatest discoveries. His first achievement was to develop a crystal **rotation** method for X-ray analysis. Maurice was now in an ideal position to encourage his younger brother's budding interest in atomic theory. And in 1913, Louis obtained his diploma from the University of Paris' Faculté des Sciences.

During World War I, the de Broglies' scientific pursuits were put aside while Louis served in the French army, and Maurice in the navy. Louis' wartime service was spent with the French Engineers at the wireless (a field in which his brother had already distinguished himself) station under the Eiffel

Louis Victor Pierre Raymond de Broglie

Tower. In 1919, after six years of army service, Louis once again took up scientific work, this time in his brother's laboratory.

It was at this time that Louis began to look at a problem that had long been baffling the scientific community: why does **light** sometimes show the properties of particles, and at other time, the properties of waves? Based on work performed in his brother's laboratory, Louis wrote a doctoral thesis entitled "Investigations into the Quantum Theory," which he submitted to the Sorbonne in 1924.

Louis' thesis was that all elements of matter, including negatively charged electrons, behave as both as particles and waves, but that it is only at the atomic level that the wave characteristics show up. At ordinary dimensions, where the billiard-ball, i.e., particulate, properties of matter are observed. Louis argued that the wave and particle properties were actually one and the same; which one was observed depended only on the way the observation was made.

Louis' theory had the effect of making classical Newtonian mechanics a special case (e.g., one that applied to the everyday world) of a more general view of matter that included atomic effects. Louis argued that all matter has waves associated with it; the only reason that these waves are not usually observed is that they are very small compared to the dimensions of objects usually encoun-

tered. To formally express these ideas, he developed a mathematical formula, which is now known as the wave equation.

Because Louis' professors at the Sorbonne did not consider themselves competent to evaluate his thesis, they asked him to send a copy of his thesis to **Albert Einstein**. Einstein immediately recognized the significance of de Broglie's theory. Another early investigator in the field of **quantum mechanics**, Erwin Schrödinger, later based his own theory of wave mechanics on Louis' work.

De Broglie's theory remained unproven, however, until experiments by American and English physicists demonstrated conclusively that electrons can indeed be bent like waves. Other experiments followed demonstrating that protons, atoms, and molecules behave similarly. These discoveries paved the way for the invention of the electron microscope much later on.

In 1927, Louis attended the seventh Solvay Conference where the leading atomic physicists continued the debate about wave mechanics. At the conference, such theorists as **Werner Heisenberg**, **Niels Bohr**, and Max Born argued that matter waves provided only statistical information about the position of a particle (probabilistic theory), and nothing about the exact position. But Einstein, Louis de Broglie, and Schrödinger, being unwilling to believe that matter can act in a random way, supported the view that the waves determined exact positions and left nothing to chance (deterministic theory).

Unable to counter certain objections to his deterministic view of waves, Louis abandoned it, and thereafter taught his students probabilistic theory. In 1928, he accepted an appointment as professor of theoretical physics at the University of Paris' Faculty of Science. Despite his academic career, he established no significant record of guiding research students.

In 1929, the 37-year-old Louis was awarded the Nobel Prize for physics in recognition of his work on wave mechanics. In 1933, he moved to the Henri Poincaré Institute, where he was to remain for the next 29 years. While at the Institute, he established a center for the study of modern physical theories. Also in 1933, he was elected to the Académie of Sciences; he was later to use his influence as an academician to point out the harmful effects of nuclear explosions. In 1943, he established a center for applied mechanics at the Institute. A year later, he was elected to the Académie Francaise. In 1945, both he and his brother were appointed to the French High Commission of Atomic Energy.

Louis de Broglie was the author of over twenty books. His writings frequently dealt with the practical side of physics, and addressed such topics as **cybernetics**, atomic energy, particle accelerators, and optics.

With the death of his older brother in 1960, Louis inherited the title of titles of French duke and German prince. He died at the age of 95 on March 19, 1987, without ever having resolved all of the ambiguities of wave-particle duality.

Brouwer, Luitzen Egbertus Jan
(1881-1966)
Dutch logician

The Dutch mathematician Luitzen Egbertus Jan Brouwer made contributions in the fields of **topology** and logic. He founded the school of thought known as **intuitionism**, which is based on the notion that the only dependable basis of mathematics consists of proofs that can actually be constructed in the real world. In addition, he developed a fixed-point **theorem** and demonstrated the connection between two previously distinct fields of topology—point-set topology and combinatorial topology.

Brouwer was born on February 27, 1881, in Overschie, the Netherlands. His parents were Egbert Brouwer and Henderika Poutsma. Brouwer completed high school in the town of Hoorn at the age of fourteen and attended the Haarlem Gymnasium where he satisfied the Greek and Latin requirements needed to enter a Dutch university in 1897.

At the University of Amsterdam Brouwer easily moved through the traditional mathematics curriculum and began some original studies on four-dimensional **space** that were published by the Royal Academy of Science in 1904. A year later he published his first book, *Leven, Kunst, en Mystiek*, a philosophical treatise in which he considers the role of humans in society. In 1907 Brouwer presented his doctoral thesis and was granted his doctor of science degree by the University of Amsterdam. He began teaching at Amsterdam in 1909 and spent his entire academic career at the university.

His doctoral thesis, "On the Foundations of Mathematics," outlined a field of research that occupied Brouwer on and off for the rest of his life. In the early twentieth century the two primary schools of mathematics were logicism and **formalism**. Logicism is based on the premise that fundamental concepts in mathematics, such as lines and points, have an existence independent of the human mind. The job of mathematicians is to derive theorems from these concepts. Formalism is less concerned with the nature of fundamental concepts, but insists that those concepts be manipulated according to very strict rules.

Brouwer proposed a third concept of mathematics, later given the name intuitionism (also known as constructivism or finitism). The basic argument of intuitionism, according to Richard von Mises in *World of Mathematics,* is that "the simplest mathematical ideas are implied in the customary lines of thought of everyday life and all sciences make use of them; the mathematician is distinguished by the fact that he is conscious of these ideas, points them out clearly, and completes them. The only source of mathematical knowledge" in intuitionism, von Mises continues, is "the intuition that makes us recognize certain concepts and conclusions as absolutely evident, clear and indubitable."

Brouwer's intuitionist school was not particularly influential when it was first proposed in the 1910s. According to Victor M. Cassidy, in *Thinkers of the Twentieth Century,*

"Brouwer made few converts during his lifetime, and Intuitionism has only a tiny number of adherents today."

The 1910s were a period of intense activity in the field of topology, the mathematical discipline concerned with geometric point **sets**. In 1912 Brouwer announced perhaps his most famous theorem, the fixed-point theorem, also known as Brouwer's theorem. This theorem stated that during any **transformation** of all points in a **circle** or on a **sphere**, at least one point must remain unchanged. Brouwer was later able to extend this theorem to figures of more than three dimensions.

Brouwer's first appointment at the University of Amsterdam in 1909 was as a tutor. Three years later, he was promoted to professor of mathematics, a position he held for thirty-nine years. In 1951 he retired and was given the title of Professor Emeritus. He died in Blaricum, the Netherlands, on December 2, 1966. Brouwer had been married in 1904 to Reinharda Bernadina Frederica Elisabeth de Holl. She predeceased Brouwer in 1959. The couple had no children.

Brouwer received honorary doctorates from the universities of Oslo (1929) and Cambridge (1955) and was awarded a knighthood in the Order of the Dutch Lion in 1932. He had also been elected to membership in the Royal Dutch Academy of Sciences (1912), the German Academy of Science (1919), the American Philosophical Society (1943), and the Royal Society of London (1948).

Brouwer's fixed point theorem

The ball, B^n, is the subset of R^n (this is n-dimensional real **space**) equal to the set of all real n-tuples $(x_1,..,x_n)$ such that the **square** root of $(x_1^2 +... + x_n^2)$ is less than or equal to one. Brouwer's fixed point **theorem** states that if f is any continuous function from B^n to B^n then there is some point x in B^n with the property that f(x) = x. x is called a fixed point of f. For example, B^2 is just the unit disk in the plane. The function f((x,y)) = (x * cos (a) - y * sin (a), x* sin (a) + y* cos (a)) for some **angle** a, is the map which rotates the disk by the angle a. It has one fixed point at (0,0). The geometric shape of the ball is not really important; the theorem is true for any polygon or **polyhedron** or indeed any topological space homeomorphic to B^n.

Here is another example. Cut out a square piece of paper. Draw a grid on it and number each little square. Make a copy of it. Fold or crumple the copy without tearing and place it on top of the original. The theorem states that some little square on the original is directly below a little square on the copy that has the number as the square on the original. Of course, the two squares might not match up perfectly. The interesting thing is that it does not matter how small you make the little squares or how much the copy is crumpled.

Here is a **proof** sketch of Brouwer's theorem. Suppose that the theorem is false. Then there is a function, f say, from B^n to B^n with the property that for no point x of B^n is it true that f(x) = x. Then for each x in B^n draw the ray from f(x) to x. Let g(x) be the intersection point of this ray with S^{n-1}, which is the set of points $(x_1,...,x_n)$ such that the **square root** of $(x_1^2 +... + x_n^2) = 1$. S^{n-1} is called the n-1-dimensional **sphere** and it is the boundary of B^n. Notice if x is in S^{n-1} then g(x) = x. Hence g is a continuous map from B^n to S^{n-1} such that for every point x in S^{n-1}, g(x) = x. The fact is, such a function g cannot exist. The reason is that S^{n-1} is 'contractible' in B^n, i.e. it can be continuously deformed to a point. However, S^{n-1} is not contractible inside itself. For example, the boundary **circle** of the unit disk can be shrunk continuously to the origin. However, a circle cannot be continuously deformed to a point in such a way that the deformation takes place inside the circle itself. To be precise, a contraction is a continuous map C from S^{n-1} x [0,1] to B^n such that C(x,0) = x for all x in S^{n-1} and C(x,1) = C(y,1) for all x and y in S^{n-1}. If g exists, then the function H from S^{n-1} x [0,1] to S^{n-1} defined by H(x, t) = g(C(x, t)) is a contraction of S^{n-1} in S^{n-1}. The fact that no such contraction of S^{n-1} exists within S^{n-1}, can be proved by using a higher dimensional analogue of the fundamental group. Thus, H cannot exist. So g cannot exist either. This implies that the assumption that f exists is false. Hence Brouwer's theorem is not false. Therefore, it is true.

Brouwer rejected the validity of his proof for two reasons: first, it is nonconstructive because it does not show how to find any fixed points and second, it relies on the law of the excluded middle. This law states that a proposition is either true or false. The proof only shows that Brouwer's theorem is not false. According to the law of the excluded middle, the theorem must therefore be true. Brouwer, and the school of **intuitionism**, rejected the validity of that law and hence the validity of the above proof. No one has ever found a proof of Brouwer's theorem that avoids using the law of the excluded middle and many believe that so such proof exists.

Brouwer's theorem has inspired many other fixed point-theorems. These theorems are used to establish the existence of solutions to **differential equations**. Consider the differential equation dy/dx = F(x,y) for all x greater than or equal to **zero** and less than or equal to one with initial conditions y = 0 at x = 0. The solution y(x) satisfies y(x) = the integral from 0 to x of F(x, y(x)) with respect to x. We define an operation on continuous real **functions** of [0,1] called H by H(f) = the integral from 0 to x of F(x,f(x)) with respect to x. The solution to the differential equation is a fixed point of this operator, i.e. a function f such that H(f) = f. It is a fact that the space of continuous real functions on [0,1] satisfies the hypothesis of a fixed-point theorem similar to Brouwer's fixed point theorem so that the existence of a solution is guaranteed. Differential **equations** such as these are common in the **calculus** of variations and **hydrodynamics**.

See also Brouwer, Luitzen Egbertus Jan; Continuity; Differential equations; Intuitionism

C

CALCULATOR

The calculator is a computing machine. Its purpose is to do mathematics; basic calculators do the basic mathematical **functions** (**addition**, **subtraction**, **division**, and **multiplication**) while the more advanced ones, which are relatively new in the history of computing machines, do advanced calculations such as solving **polynomials**. The odometer, or mileage counter, in your car is a counting machine as is the calculator in your backpack and the computer on your desk. They may have different ability levels, but they all tally numbers.

Early history

Perhaps the earliest calculating machine was the Babylonian rod numerals. Not only was it a notational device, but administrators carried rods of bamboo, ivory, or iron in bags to help with their calculations. Rod numerals used nine digits.

The next invention in counting machines was the **abacus**. A counting board dating from approximately 500 BCE now resides in the National Museum in Athens. It is not an abacus, but the precursor of the abacus. Oriental cultures have documents discussing the abacus (the Chinese call it a *suan phan* and the Japanese the *soroban*) as early as the 1500s CE; however, they were using the abacus at least a thousand years earlier. While experts disagree on the origin of the name (from either the Semitic *abq*, or dust, or the Greek *abax*, or sand tray), they do agree the word is based upon the idea of a sand tray which was used for counting.

A simple abacus has rows (or wires) and each row has ten beads. Each row represents a unit ten greater than the previous. Thus, the first stands for units of one; the second, units of ten; the third, units of hundred; and so on. The appropriate number of beads are moved from left to right on the wire representing the unit. When all ten beads on a row have been moved, they are returned to their original place and one bead on the next row is moved. The soroban divides the wires into two unequal parts. The beads along the lower, or larger, part represent units, tens, hundreds, and so on. The bead at the top represent five, fifty, five hundred, and so on. These beads are stored away from the central divider, and as needed are moved toward it.

Finger reckoning must not be ignored as a basic calculator. Nicolaus Rhabda of Smyrna and the Venerable Bede (both of the 8th century CE) wrote, in detail, how this system using both hands could represent numbers up to one million. The numbers from 1 to 99 are created by the left and the numbers from 100 to 9900 by the right hand. By bending the fingers at their various joints and using the index finger and thumb to represent the multiples of ten, combinations of numbers can be represented. Similar systems were devised much earlier than the 8th century, probably by merchants and traders who could not speak each other's languages but needed a system to communicate: their fingers. Multiplication using the fingers came much later; well into the 15th century CE such complex issues, as multiplication, were left to university students, who were forced to learn a different finger reckoning system to accommodate multiplication.

Early calculators

Schickard, a German professor and Protestant minister, seems to have been the first to create an adding machine in the 1620s. It performed addition, subtraction, and carrying through the use of gears and preset multiplication tables. The machine only computed numbers up to six digits. Once the operator surpassed this limit, he was required to put a brass ring around his finger to remind him how many carries he had done. Schickard made sure to include a bell, so that the user would not forget to add his rings. His drawings of the machine were lost until the mid-1960s when a scrap of paper was located inside a friend's book which had a drawing of his machine. With this drawing and Schickard's letters, a reconstruction was made in 1971 to honor the adding machine's 350th anniversary.

Finished in 1642, **Blaise Pascal**'s calculating machine also automatically carried tens and was limited to six digit numbers. The digits from 0 to 9 were represented on dials; when a one was added to a nine the gear turned to show a **zero** and the next gear, representing the next higher tens unit, automatically turned. Over fifty of these machines were made. A few remain in existence.

Gottfried Wilhelm Leibniz, a German, created a machine in 1671 which did addition and multiplication. For over two hundred years this machine was lost; then it was discovered by some workmen in the attic of one of the Göttingen University's buildings.

Charles Xavier Thomas de Colmar, in 1820, devised a machine which added subtraction and division to a Leibniz type calculator. It was the first mass produced calculator and became a common sight in business offices.

Difference engine

The great English mathematician **Charles Babbage** tried to build a machine which would calculate mathematical tables to 26 significant figures; he called it the Difference Engine (see **Babbage's engine**). However, in the early 1820's his plans were stalled when the British government pulled its funding. His second attempt, in the 1830's, failed because some of the tools he needed had yet to be invented. Despite this, Babbage did complete part of this second machine, called the Analytical Engine, in the early 1840s. It is considered to be the first modern calculating machine. The difference between the Difference and Analytical Engines is that the first only performed a certain number of functions, which were built into the machine, while the second could be programmed to solve almost any algebraic equation.

By 1840 the first difference engine was finally built by the Swedish father and son team of George and Edvard Scheutz. They based their machine on Babbage's 1834 publication about his experiments. The Scheutz's three machines produced the first automatically created calculation tables. In one 80 hour experiment, the Scheutz's machine produced the **logarithms** of 1 to 10,000; this included time to reset the machine for the 20 polynomials needed to do the calculations.

Patents

The first patent for a calculating machine was granted to the American Frank Stephen Baldwin in 1875. Baldwin's machine did all four basic mathematical functions and did not need to be reset after each computation. The second patent was given in 1878 to Willgodt Theophile Odhner from Sweden for a machine of similar design to Baldwin's. The modern electronic calculators are based on Baldwin's design.

In 1910, Babbage's son, Henry P. Babbage, built the first hand held printing calculator based on his father's Analytical Engine. With this machine, he was able to calculate and then print multiples of π to 29 decimal places.

In 1936, a German student, Konrad Zuse, built the first automatic calculating machine. Without any knowledge of previous calculating machines, Zuse built the Z1 in his parents' living room. He theorized that the machine had to be able to do the mathematical fundamentals. To do this, he turned to binary mathematics, something no other scientist or mathematician had contemplated. (The mathematics we use in every day life, decimal, is based on 10 digits. Binary mathematics uses two digits: 0 and 1.) By using binary mathematics, the calculating machine became a series of switches rather than gears, because a switch has two options: on (closed or 1) or off (open or 0). He then connected these switches into logic gates, a combination of which can be selected to do addition or subtraction. Zuse's third **model** (finished in 1941) was not only programmable, but hand held; it added, subtracted, multiplied, divided, discovered **square roots**, and converted from decimal to binary and back again. However, to add it took a third of a second, and to subtract an additional three to five seconds. By changing his relays into vacuum tubes he believed he could speed his machine up 1000 times, but he could not get the funding from the Third Reich government to rebuild his machine.

Electronic predecessor to computer

In 1938, the International Business Machine Corporation (IBM) team of George Stibitz and S. B. Williams began building the Complex Number Calculator. It could add, subtract, multiply, and divide **complex numbers**. They completed the project in 1940 and until 1949 these calculators were used by Bell Laboratories. It was the first machine to use remote stations (terminals, or input units, not next to the computer) and to allow more than one terminal to be used. The operator typed the request onto a teletype machine and the response was sent back to that teletype. The relays inside the machine were basic telephone relays.

The IBM Automatic Sequence Controlled Calculator (ASCC) was based upon Babbage's ideas. This machine, completed in 1944, was built for the United States Navy by Harvard University and IBM. It weighed about 5 tons (10,000 kg) and was 51 ft (15.5 m) long and 8 ft (2.4 m) high. A second, more useful version, was completed in 1948.

The Electronic Numerical Integrator and Computer (ENIAC), also based on Babbage's concepts, was completed in 1945 at the University of Pennsylvania for the United States Army. It weighted over 30 tons (27,240 kg) and filled a 30 by 55 ft (9 by 17 m) room. In one second it could do 5000 additions, 357 multiplications, or 38 divisions. However, to reprogram ENIAC meant rewiring it, which caused up to two days in delays.

The first electronic calculator was suggested by the Hungarian turned American **John von Neumann**. Von Neumann introduced the idea of a stored memory for a computer which allowed the program and the data to be inputted to the machine. This resolved problems with rewiring computers and permitted the computer to move directly from one calculation to another. This type of machine architecture is called "von Neumann." The first completed computer to use von Neumann architecture was the English Electronic Delay Storage Automatic Calculator (EDSAC) finished in 1949. An added feature of the EDSAC was that it could be programmed in a type of shorthand, which it then converted into binary code, rather than its precursors which demanded the programmer actually write the program in binary code, a laborious process.

Inside calculators

The early counting machines, items like the car's odometer, work with a set of gears and wheel. A certain number of wheels are divided in ten equal parts on each of which one of the ten digits appears. (Windows are placed on top of these wheels so that only one digit appears at a time.) These wheels are then attached to gears which as they turn rotate the wheel so that the digit being displayed changes. When the right most wheel changes from 9 to 0, a mechanism is set in **motion** which turns the wheel to the left one unit, so that the digit it displays changes. This is the carry sequence. When this second wheel changes from 9 to 0, it too has a mechanism to carry to the next left wheel.

Electronic calculators have the four major units van Neumann created: input, processing, memory, and output. The input unit accepts the numbers keyed in, or sent through the reader in the case of the punch card, by the operator. The processing unit performs the calculations. When the processing unit encounters a complex calculation, it uses the memory unit to store intermediary results or to locate **arithmetic** instructions. At the completion of the calculation, the final answer is sent to the output unit which informs the operator of the result; this may be through a display, paper, or a combination. Since the calculator thinks in binary, the output unit must convert the result into decimal units.

Modern advances

At the end of 1947, the transistor was invented, eventually making the vacuum tube obsolete. This tiny creation, composed of semiconductors, were much faster and less energy consumptive than the tubes. Problems arose with the connections between the components with size, speed, and reliability as more complex machines needed more complex circuitry which in turn required more components soldered to more boards (the actual board to which the pieces were attached). The next breakthrough came with the invention of the integrated circuit (IC) in 1959 by Texas Instruments (TI) and Fairchild (a semiconductor manufacturing company). The integratedcircuit is akin to a solid mass of transistors, resistors, and capacitors. Again, the speed of computation increased (since the resistance in the circuit was reduced) and the energy required by the machine was decreased. Finally, a computer could fit on a spaceship (they were part of the Apollo computer) or missile. In 1959, an IC cost over $1,000 but by 1965 they were under $10.

Ted Hoff, an electrical engineer for Intel conceived of a radical new concept-the microprocessor. This incorporated the circuitry of the integrated circuit and the programs used in a computer onto a single chip, or piece of silicon. This model of this microprocessor was finished in 1970. The idea of a disposable piece of a calculator was revolutionary. Its compactness and speed changed the face of the computing industry.

The creation of ICs allowed calculators to become much faster and smaller. By the 1960s, they were hand held and affordable. By the late 1980s, calculators were found on watches. However, once again engineers are being blocked by the size factor. The **limits** are now the size of the chip which in turn limits the speed and programmability of the calculator, or computer.

CALCULUS

The invention of calculus by the mathematicians **Isaac Newton** (1642-1727) of England and **Gottfreid Wilhelm Leibniz** (1646-1716) of Germany, stands as one of the supreme intellectual achievements of the ages. The calculus, as applied by Newton to the study of **motion**, ushered in a revolution in science, the results of which are still being felt as we enter the 21st century. Newton had set for himself the task of describing the laws of gravity and the motion of the planets about the sun. The mathematics that existed before Newton was inadequate for this project. In order to bring his theories to life, Newton needed a more powerful mathematics that could deal with instantaneous rates of change instead of the average rates of change that could be calculated using **arithmetic** and **algebra**. In order to give a more accurate description of the motion of a body than had previously been given, Newton wanted to be able to calculate velocities and accelerations at any point along the path of motion. For this purpose, he introduced his "fluxions", which today are called derivatives. If the motion of an object is given by a mathematical function, then the derivative of that function at any point is the instantaneous velocity of the body at that point. The idea that Newton had was to imagine calculating average velocities at smaller and smaller intervals surrounding a point. Average velocity is simple to calculate: it is just the **distance** traveled divided by the time required to travel that distance. So Newton defined the instantaneous velocity as the limiting value of the average velocities as the time intervals became smaller and smaller. In a similar manner, he defined instantaneous acceleration at a point as the instantaneous rate of change of the velocity at that point. Thus was Newton able to define the position, velocity, and acceleration at any point on the path of a body in motion, thereby positing the most complete and accurate account of motion ever given. Newton introduced the scientific world to **differential equations**, which, in itself, changed forever the way scientists describe natural phenomena. A differential equation is just an equation that contains derivatives or instantaneous rates of change. Nature is full of changing phenomena, and the differential **equations** of Newton's calculus are tailor made to describe these phenomena. Today all of the laws of physics are stated in terms of differential equations. When the American astronaut Neil Armstrong became the first human to step on the surface of the moon in July of 1969, his "giant leap for mankind" was the most spectacular outcome of the scientific revolution that was born with the invention of the calculus.

It would be enough if the calculus of rates of change were the entire story, but it is not. The calculus also allows the mathematician to determine tangent lines to curves at particular points. This is useful in finding maximum and minimum values and points of inflection for a wide variety of **functions**. Tangent lines are also the best linear approximators to curves at particular points. As such they form the basis of numerical approximation algorithms such as Euler's method for numerically solving differential equations. Moreover, the differential calculus—the study of derivatives and differential equations—forms only one-half the subject matter of this immensely powerful discipline. The other half is called the **integral calculus** and deals with accumulations of rates of change, areas under curves, volumes of solids, lengths of curves, surface areas and much more. The **definite integral** is the centerpiece of the integral calculus. Historically, the integral developed from the attempt to calculate the **area** under a curve in the plane. This development can be traced as far back as the great **Archimedes** (287-212 BC) who developed a method for approximating areas under parabolas to any desired degree of accuracy. Archimedes' method was a precursor to the definite integral. In calculating the area under a curve defined by a function, one imagines the region under the curve being subdivided into rectangular strips extending from the x-axis to the curve. Now the area of a rectangle is easy to compute—it is just the base of the rectangle multiplied by its height. So if we add up the areas of all rectangles, we will have an approximation to the area under the curve. If we want a better approximation, we can take twice as many rectangles half as wide, which will do a better job of filling up the region under the curve. If we are still not satisfied, we can repeat the process again and again. The limiting value of such a process is defined to be the area under the curve and is one example of a definite integral. A more general **definition** of the definite integral is given in terms of so-called "Riemann Sums" named for the German mathematician Georg **Bernhard Riemann** (1826-1866). A key discovery of both Newton and Leibniz is now known as the "**Fundamental Theorem of Calculus**," which essentially states that differentiation and integration are inverse processes. Previously viewed as separate areas of study, differentiation and integration were brought together under the common umbrella of calculus by the Fundamental Theorem.

While Newton was busy developing the calculus as the basis for his revolution in physics, Leibniz was independently working on his own version of the calculus. Leibniz had a flair for elegant notation, which Newton lacked, and, by and large, it is Leibniz's notation that is most commonly seen in modern calculus textbooks. In particular, Leibniz gave us the dy/dx notation for derivatives and the elongated S for integrals. Although Newton wrote his "Method of Fluxions" in 1671, it was not published until 1736. Leibniz, on the other hand, did not begin to work on his version of the calculus until about 1673, but his first published work on the subject was in 1684. Thus, Leibniz's work on the calculus became more widely known sooner than Newton's version. This caused a controversy between the followers of Newton in England and Leibniz on the main European continent concerning who should be given credit for the invention of the calculus. Today it is commonly agreed that both men working independently invented the calculus.

Neither Newton nor Leibniz had concerned himself with the theoretical foundation upon which the calculus was built. The incredibly wide range of applicability to which the subject lent itself had mathematicians and physicists of the 1700s rushing to apply the calculus to a myriad of problems that had been intractable prior to its invention. Amid the excitement generated by the power and usefulness of this new discipline, a group of philosophers led by Bishop **George Berkeley** (1685-1783) pointed out some shaky theoretical underpinnings of the calculus. In particular, Berkeley attacked the notion of infinitesimals, the term used to denote the small changes in quantities that became closer and closer to **zero** as the derivative attempted to make small average rates of change into instantaneous rates of change. Berkeley, with no small amount of sarcasm, asked what these infinitesimals were. Did they, in fact, go to zero, and, if so how could you divide by them? Or did they remain ever so small, in which case you really did not have an instantaneous rate of change after all, did you? Were these infinitesimals, he asked, "the ghosts of departed quantities"? Berkeley's attack was not always kind, but it was important, for it led to a serious discussion among 19th-century mathematicians about the concepts upon which the calculus rested. To overcome Berkeley's objection to the rather fuzzy idea of infinitesimals, what was needed was an adequate and rigorous theory of what was meant by "getting closer and closer to zero." In 1821, the French mathematician **Augustin Louis Cauchy** (1789-1857) provided a theory of **limits** which answered this challenge. Cauchy's theory, with some additions by **Karl Weierstrass** (1815-1897), remains to this day the accepted foundation upon which the calculus is built. Cauchy defined all of the most fundamental ideas of the calculus—continuity, differentiability, **convergence**, and the definite integral—in terms of limits. Cauchy's definitions are essentially those encountered in modern textbooks on the calculus.

As the 21st century dawns with its emphasis on computers and all things digital, some say that the age of calculus is drawing to a close to be supplanted by the age of the **algorithm**. Whether or not that is true may be best judged by mathematicians and scientists alive at the beginning of the 22nd century. What is not in dispute is that the calculus stands as the most significant mathematical breakthrough in history and as one of humankind's grandest intellectual accomplishments.

CALDERÓN, ALBERTO P. (1920-1998)

Argentine American analyst

Calderón is known for his revolutionary work in the field of **analysis**, particularly in the area of partial differential equations. His influence helped turn the 1950s trend toward abstract mathematics back in a more practical direction. His award-winning research in the area of integral operators is an example of his impact on contemporary mathematical analysis. During his 45-year career, Calderón produced a number of seminal results and techniques. His accomplishments have been recognized by such prestigious awards as the National Medal of Science and the Wolf Prize in mathematics, and he was a member of the national academies of science of five countries as well as the Third World Academy of Sciences based in Italy. When Calderón was awarded the Steele Prize for a Fundamental Paper in mathematics, his work was described in the *Notices of the American Mathematical Society* as "a major progenitor of the modern theory of microlocal analysis."

Alberto Pedro Calderón was born on September 14, 1920, in Mendoza, Argentina, a small town at the foot of the Andes. His father, a descendant of notable 19th-century politicians and military officers, was a renowned medical doctor who helped organize the General Central Hospital of Mendoza. Calderón completed his secondary education in his hometown and in Zug, Switzerland. After graduating from the School of Engineering of the National University of Buenos Aires in 1947, he studied mathematics under Alberto Gonzalez Dominguez and Antoni Zygmund, a renowned mathematician who was a visiting professor in Buenos Aires in 1948.

A Rockefeller Foundation fellowship brought Calderón to the United States where he received his Ph.D. in mathematics at the University of Chicago (three years after earning his bachelor's degree in civil engineering). However, his teaching career had begun in 1948 when he was made an assistant to the chair of electric circuit theory in Buenos Aires. He went on to work as a visiting associate professor at Ohio State University from 1950 to 1953, moving to the Institute for Advanced Study (IAS) in Princeton in 1954. Calderón then spent four years at the Massachusetts Institute of Technology (MIT) as an associate professor. Returning to the University of Chicago, he served as professor of mathematics from 1959 to 1968, Louis Block professor of mathematics from 1968 to 1972, and chairperson of the mathematics department from 1970 to 1972.

By the early 1970s, Calderón's reputation was well established in scientific circles, and his research in collaboration with his longtime mentor Zygmund had already been dubbed "The Chicago School of Mathematics," also known today as "the Calderón-Zygmund School of Analysis." Their work had significant impact on contemporary mathematics by countering a predominant trend toward abstract mathematics and returning to basic questions of real and **complex analysis**. Their ideas in this regard became known as the Calderón-Zygmund Theory of Mathematics.

A landmark in Calderón's scientific career was his 1958 paper titled "Uniqueness of the Cauchy Problem for Partial Differential Equations." Two years later he used the same method to build a complete theory of hyperbolic partial **differential equations**. His theory of singular operators, which is used to estimate solutions to geometrical **equations**, contributed to linking together several different branches of mathematics. It also had practical applications in many areas, including physics and aerodynamic engineering. This theory has dominated contemporary mathematics and has made important inroads in other scientific fields, including quantum physics. Calderón had created what is now commonly known as pseudodifferential **calculus**.

Alberto P. Calderón

Calderón's article on the uniqueness of the Cauchy Problem had such a profound effect on the research community that in 1989 its legacy earned him both the Steele Prize from the American Mathematical Society and the Mathematics Prize of the Karl Wolf Foundation of Israel. In 1991, President George Bush awarded him the National Medal of Science for his career accomplishments.

The techniques devised by Calderón transformed contemporary mathematical analysis. In addition to his pseudodifferentials work, he also conducted fundamental studies in **interpolation** theory—the idea that any algebraic problem that has a rational solution is solvable. He was responsible (with R. Arens) for what is considered one of the best theorems in Banach algebras. Calderón also put forth an approach to energy estimates that has been of fundamental importance in dozens of subsequent investigations and has provided a **model** for general research in his field.

A summary of Calderón's work was written by three Yale University mathematicians for publication in the *Notices of the American Mathematical Society* in 1992. The authors assert that Calderón, working in collaboration with Zygmund, invented techniques that "have been absorbed as standard tools of **harmonic analysis** and are now propagating into nonlinear analysis, partial differential equations, complex analysis, and even signal processing and numerical analysis." They comment that after developing powerful methods in real analysis,

"Calderón was not reluctant to return to the methods of complex analysis... [which] had fallen out of favor, mostly due to the power of Calderón-Zygmund theory!" The resulting blend of complex and real analysis has been quite fruitful. In 1979 it earned him the Bôcher Memorial Prize, which is awarded once every five years for a notable research paper in analysis, in this case "Cauchy Integrals on Lipschitz Curves and Related Operators."

In the early 1970s, Calderón briefly returned to his home country to serve as a visiting lecturer and conduct mathematical Ph.D. dissertation studies at his alma mater, the National University of Buenos Aires. He has continued to encourage mathematics students from Latin America and the United States to pursue their doctoral degrees, in many instances directly sponsoring them. Some of his pupils, in turn, have become reputed mathematicians. For example, Robert T. Seeley applied Calderón-Zygmund techniques to singular integral operators on manifolds, laying the foundation for the Atiyah-Singer index **theorem**.

After his stay in Argentina, Calderón returned to MIT as a professor of mathematics and, in 1975, he became university professor of mathematics at the University of Chicago, a special position he held until his retirement in 1985. He subsequently served as a professor emeritus with a post-retirement appointment at that same institution. Beginning in 1975, he was also an honorary professor at the University of Buenos Aires. He lectured extensively around the world, and held occasional visiting professorships at several universities, including Cornell, Stanford, Bogotá, Madrid, Rome, Göttingen, and the Sorbonne.

An active member of the American Mathematical Society for more than 40 years, Calderón served as a member–at–large of its council in the mid–1960s and on several of its committees. He has also been an associate editor of various scientific publications, including the *Duke Mathematical Journal, Journal of Functional Analysis,* and *Advances in Mathematics.*

Calderón married **Alexandra Bellow**, a well-known mathematician in her own right and a professor at Northwestern University. Their daughter, Maria Josefina, holds a doctorate in French literature from the University of Chicago, and their son, Pablo Alberto, is a mathematician who has studied in Buenos Aires and New York. Calderón died on April 16, 1998, at Northwestern Memorial Hospital in Chicago at the age of 77.

CALENDARS

There are three units of time which have a direct basis in astronomy: the day, which is the period of time it takes for the Earth to make one **rotation** around its axis; the month, which is the period of time it takes for the Moon to revolve around the Earth; and the year, which is the period of time it takes for the Earth to make one revolution around the Sun.

The week has an indirect basis in astronomy—the seven days of the week probably were named for the seven objects which the ancients saw moving on the zodiac, which were the Sun, Moon, and the five planets that can be seen with the naked eye.

A calendar is a system for measuring long units of time, usually in terms of days, weeks, months, and years. The year is the most important time unit in most calendars, since the cycle of seasons, which are associated with change of climate in the Earth's temperate and frigid zones, repeat in a yearly cycle with the change in the Sun's apparent position on the ecliptic as the Earth revolves around the Sun.

There are three main types of calendars. One type of calendar is the Lunar Calendar, which is based on the month (Moon). A lunar calendar year is 12 synodic months long, where a synodic month is the time **interval** in which the phases of the Moon repeat (from one Full Moon to the next), and which averages 29.53 days. Thus, a lunar calendar year averages 354.37 days long. Since the Earth takes slightly longer than 365 days to revolve completely around the Sun, a lunar calendar soon gets out of phase with the seasons. Thus, most lunar calendars have died out over the centuries. The main exception is the Muslim Calendar, which is used in Islamic countries, most of which are in or near the Earth's torrid zone, where seasonal variation of climate is slight or non-existent, and the climate is usually consistently hot.

The second type of calendar is the Luni-Solar Calendar, in which most years are 12 synodic months long, but a thirteenth month is inserted every few years to keep the calendar in phase with the seasons. There are two important surviving luni-solar calendars: the Hebrew (Jewish) Calendar, which is used by the Jewish religion, and the Chinese Calendar, which is used extensively in eastern Asia.

The third type of calendar is the Solar Calendar, which is based on the length of the year. Our present calendar is of this type; however, it evolved from the ancient Roman calendar, which passed through the stage of being a luni-solar calendar. Let us summarize its early history.

In the first centuries after Rome was founded (753 B.C.), the Roman calendar consisted of ten synodic months; the year began near the start of spring with March and ended with December (the tenth month). The remaining 70 winter days were not counted in the calendar. Some centuries later two more months, January, named for Janus, the two-faced Roman god of gates and doorways, and February, named for the Roman festival of purification, were added between December and March. An occasional thirteenth month was later inserted into the calendar; at this stage the Roman calendar was luni-solar calendar. It was quite complicated and somewhat inaccurate even by the year 45 B.C.

That year, Julius Caesar (100-44 B.C.) commissioned the Greek astronomer Sosigenes (c. 50 B.C.) from Alexandria to plan a sweeping reform of the Roman Calendar. The calendar which Sosigenes devised and Caesar installed for the Roman Empire had the following main features.

The months January, March, May, July, August, October, and December each have 31 days. The months April, June, September, and November each have 30 days. February has 28 days in ordinary years, which have 365 days.

Every fourth year is a Leap year with 366 days. The 366th day appears in the calendar as February 29th.

The calendar year begins on January 1 instead of March 1. January 1 is set by the time of year when the Sun seems to set about half an hour later than its earliest setting seen in Rome, which occurs in early December.

This calendar was named the Julian calendar for Julius Caesar. He also had the month Quintilis (the fifth month) renamed July for himself. Augustus Caesar (63 B.C. - 14 A.D.) clarified the Julian calendar rule for leap year by decreeing that only years evenly divisible by four would be leap years. He also renamed the month Sextilis (the sixth month) August for himself.

The average length of the Julian calendar year over a century or more is 365.25 days. This time interval is between the lengths of two important astronomical years. The shorter one is the tropical year, or the year of the seasons; it is defined as the time interval between successive crossings of the Vernal Equinox by the Sun (which mark the beginning of spring in the Earth's northern hemisphere) and averages 365.2422 days long. The sidereal year, which is defined as the time interval needed for the Earth to make a complete 360° orbital revolution around the Sun, is slightly longer, being 365.25636 days long. The small difference between the lengths of the sidereal and tropical years arises because the Earth's rotation axis is not fixed in **space** but describes a cone around the line passing through the Earth's center that is perpendicular to the Earth's **orbit** plane (the ecliptic); the rotation axis describes a complete cone in 25,800 years. This phenomenon is called precession, and it causes the equinoxes (the intersections of the celestial equator and ecliptic) to shift westward on the ecliptic by 50."2 (0°0139) each year and also the celestial poles to describe small circles around the ecliptic poles. Since the Sun appears to move eastward on the ecliptic at an average rate of 0°.9856/day, the Sun moves only 359°.9861 eastward along the ecliptic in an average tropical year, whereas it moves 360° eastward in a sidereal year, making the tropical year about 20 minutes shorter than the sidereal year. Precession is caused by stronger gravitational pulls of the Sun and Moon on the closer parts of the Earth's equatorial bulge than on its more distant parts. This effect tries to turn the Earth's rotation axis towards the line perpendicular to the ecliptic, but since the Earth is rotating, it orbits like a rapidly spinning top, producing the effects described above.

An astronomer wants to make the average length of the calendar year equal to the length of the tropical year in order to keep the calendar in phase with the seasons. Sosigenes knew that precession of the equinoxes existed; it had been discovered by his predecessor **Hipparchus** (c. 166-125 B.C.). From his observations and records of earlier observations, Sosigenes allowed for precession of the equinoxes by making the average length of the Julian calendar year slightly shorter (0.00636 day, or about nine minutes) than the length of the sidereal year. But he did not know the physical cause of precession (a gravitational tidal effect), so he could not calculate what the annual rate of the precession of the equinoxes should be. The crude astronomical observations existing at that time

may have led Sosigenes to believe that the rate of precession of the equinoxes was about half its true value, and therefore, that the 365.25 day average length of the Julian calendar year was an adequate match to the length of the tropical year. Unfortunately, this is not true for a calendar intended for use over time intervals of many centuries.

The Sun appeared to reach the Vernal Equinox about March 25 in the years immediately after the Roman Empire adopted the Julian calendar. It continued to be the official Roman calendar for the rest of the empire's existence. The Roman Catholic Church adopted the Julian calendar as its official calendar at the Council of Nicaea in 325 A.D., soon after the conversion of the emperor Constantine I, who then made Christianity the Roman Empire's official religion. By that time, the Sun was reaching the Vernal equinox about March 21; the fact that the tropical year is 0.0078 day shorter than the average length of the Julian calendar year had accumulated a difference of three to four days from the time when the Julian calendar was first adopted. The Council of Nicaea also renumbered the calendar years; the numbers of the Roman years were replaced by a new numbering system in an effort to have had Christ's birth occur in the year 1 A.D. (Anno Domini). This effort was somewhat unsuccessful; the best historical evidence indicates that Christ probably was born sometime between 7 B.C. (before Christ) and 4 B.C. Another feature of this modified Julian calendar is that it has no year **zero**; the 1 B.C. is followed by the year 1 A.D.

This Julian calendar remained the official calendar of the Roman Catholic Church for the next 1,250 years. By the year 1575, the Sun was reaching the Vernal Equinox about March 11. This caused concern among both church and secular officials because, if this trend continued, by the year 11,690, Christmas would have become an early spring holiday instead of an early winter one, and would be occurring near Easter.

This prompted Pope Gregory XIII to commission the astronomer Clavius to reform the calendar. Clavius studied the problem, then he made several recommendations. The rate of the precession of the equinoxes was known much more precisely in the time of Clavius than it had been in the time of Sosigenes. The calendar which resulted from the study by Clavius is known as the Gregorian calendar; it was adopted in 1583 in predominantly Roman Catholic countries. It distinguished between century years, that is, years such as 1600, 1700, 1800, 1900, 2000, etc., and all other years, which are non-century years. The Gregorian calendar has the following main features.

All non-century years evenly divisible by four, such as 1988, 1992, and 1996 are leap years with February 29th as the 366th day. All other non-century years are ordinary years with 365 days.

Only century years evenly divisible by 400 are leap years; all other century years are ordinary years. Thus, 1600 was and 2000 will be leap years with 366 days, while 1700, 1800, and 1900 had only 365 days.

The Gregorian calendar was reset so that the Sun reaches the Vernal Equinox about March 21. To accomplish this, ten days were dropped from the Julian calendar; in the year 1582 in the Gregorian calendar, October 4 was followed by October 15.

The Gregorian calendar is the official calendar of the modern world. From the rules for the Gregorian calendar shown above, one finds that, in any 400-year interval, there are 97 leap years and 303 ordinary years, and the average length of the Gregorian calendar year is 365.2425 days. This is only 0.0003 day longer than the tropical year. This will lead to a discrepancy of a day in about the year 5000. Therefore, the Sun has usually reached the Vernal Equinox and northern hemisphere Spring has begun about March 21 according to the Gregorian calendar.

The Gregorian calendar was not immediately adopted beyond the Catholic countries. For example, the British Empire (including the American colonies) did not adopt the Gregorian calendar until 1752, when 11 days had to be dropped from the Julian calendar, and the conversion to the Gregorian calendar did not occur in Russia until 1917, when 13 days had to be dropped.

One feature of the Gregorian calendar is that February is the shortest month (with 28 or 29 days), while the summer months July and August have 31 days each. This disparity becomes understandable when one learns that the Earth's orbit is slightly elliptical with **eccentricity** 0.0167, and the Earth is closest to the Sun (at perihelion) in early January, while it is most distant from the Sun (at aphelion) in early July. It follows from **Kepler**'s Second Law (see **Kepler's Laws**) that the Earth, moving fastest in its orbit at perihelion and slowest at aphelion, causing the Sun to seem to move fastest on the ecliptic in January and slowest in July. The fact that the Gregorian calendar months January, February, and March have 89 or 90 days, while July, August, and September have 92 days makes some allowance for this.

Although the Gregorian calendar partially allows for the eccentricity of the Earth's orbit and for the dates of perihelion and aphelion, the shortness of February introduces slight inconveniences into daily life. An example is that a person usually pays the same rent for the 28 days of February as is paid for the 31 days of March. Also the same date falls on different days of the week in different years. These and other examples have led to several suggestions for calendar reform.

Perhaps the best suggestion for a new calendar is the World Calendar, recommended by the Association for World Calendar Reform. This calendar is divided into four equal quarters that are 91 days (13 weeks) long. Each quarter begins on a Sunday on January 1, April 1, July 1, and October 1. These four months are each 31 days long; the remaining eight months all have 30 days. The last day of the year, a World Holiday (W-Day), comes after Saturday December 30 and before January 1 (Sunday) of the next year; it is the 365th day of ordinary years and the 366th day of leap years. The extra day in leap years appears as a second World Holiday (Leap year or L-Day) between Saturday June 30 and Sunday July 1. The Gregorian calendar rules for ordinary, leap, century, and non-century years would remain unchanged for the foreseeable future.

One should mention a future Mars calendar for the human colonization of Mars in future centuries. There are about 668.6 sols (**mean** Martian solar days which average 24 hours 39 minutes 35.2 seconds of Mean Solar Time long) in a Martian sidereal year. At least one Martian calendar has been suggested, but much more must be done before an official Martian calendar is adopted.

This calendar should be mentioned because of the extensive use of the Julian date (J.D.) in astronomy, oceanography, and perhaps some other sciences. It must not be confused with the Julian civil calendar.

This calendar was devised in 1582 by Josephus Justus Scaliger; the Julian date for a given calendar date is the number of days which have elapsed for that date since noon (by Universal Time [U.T.]) on January 1, 4713 B.C. It is based on a time interval 7980 years long, which Scaliger called the Julian period. For example, noon (12:00 U.T.) on January 1, 1996 is Julian Day J.D. 2,450,084.0 = 1.5 January 1996 U.T.

CANTOR, GEORG (1845-1918)
Russian-born German algebraist and analyst

Georg Cantor, a German mathematician, developed a number of ideas that profoundly influenced 20th-century mathematics. Among other accomplishments, he introduced the idea of a completed **infinity**, an innovation that earned him recognition as the founder and creator of **set theory**. His revolutionary insights, however, were accepted only gradually and not without opposition during his lifetime. The praise for his work was best epitomized by the famous mathematician **David Hilbert**, who said that "Cantor has created a paradise from which no one shall expel us." Besides being the founder of set theory, Cantor also made significant contributions to classical **analysis**. In addition, he did innovative work on **real numbers** and was the first to define **irrational numbers** by sequences of **rational numbers**.

Georg Ferdinand Ludwig Philipp Cantor was born on March 3, 1845, in St. Petersburg, Russia, the first child of Georg Woldemar Cantor and Maria Böhm. The family moved to Frankfurt, Germany, in 1856 when the father became ill. His father, born in Copenhagen, had moved to St. Petersburg at a young age and had become a successful stockbroker there. His mother came from an artistic family. Cantor's brother Constantin was an accomplished piano player and his sister Sophie had drawing talents. Cantor himself sometimes expressed regret that he had not become a violinist. Of Jewish descent on both sides, Cantor was nevertheless raised in an intensely Christian atmosphere. The breadth and depth of his knowledge of the old masters, theologians, and philosophers was brought about by his religious upbringing and became evident in his more philosophical writings.

At a young age, while still in St. Petersburg, Cantor showed clear signs of mathematical talent. Though he wanted to become a mathematician, his father had charted out an engineering career for him. He attended several schools along the lines of his father's wishes, including the Gymnasium in

Wiesbaden and, from 1860, a Technical College in Darmstadt. Cantor finally received parental approval to study mathematics in 1862. He started his studies in the fall of that year in Zurich, but moved to Berlin after one semester. Cantor was a solid student. He spent a summer semester in Göttingen in 1866 and successfully defended a Ph.D. thesis in **number theory** on December 14, 1867, in Berlin. Cantor then moved to the University of Halle as a *Privatdozent,* becoming an associate there in 1872 and a professor in 1879. He remained at Halle for his entire career.

A friend of his sister's, Vally Guttmann, became his wife in 1874. During their honeymoon in Switzerland, the couple met **Richard Dedekind**, from then on a friend and mathematical confidant of Cantor's. Georg Cantor and Vally Guttmann had six children.

When Cantor arrived at Halle, the leading mathematician there was Heinrich Heine, under whose influence Cantor began to study **Fourier series**. His analysis of the **convergence** of these trigonometric series eventually led to far-reaching innovations. What started as a slight improvement of a **theorem** on the uniform convergence of Fourier series contained the first seeds of set theory. Cantor's first paper on set theory proper was published in 1874 under the title *Über eine Eigenschaft des Inbegriffes aller reellen algebraischen Zahlen* and dealt with **algebraic numbers**. An algebraic number is any real number that is a solution to an equation with integer coefficients. Cantor's paper contained the **proof** that the set of all algebraic numbers can be put in a one-to-one **correspondence** with the set of all positive **integers**. Moreover, Cantor proved that the set of all real numbers cannot be put into a one-to-one correspondence with the positive integers. As he later explained it, the set of positive integers has the same power (called *Mächtigkeit* in German) as the set of algebraic numbers, while the power of the set of real numbers is different from either. The 1874 paper was accepted for publication only after Dedekind's intercession.

The set of all algebraic numbers, containing, for example, the **square** root of 2, is properly larger than the set of all rational numbers (that is, quotients of integers). In turn, the set of rationals contains infinitely many more elements than the set of positive integers. In spite of that, Cantor showed that the three sets—the rationals, the algebraic numbers, and the positive integers—have the same power. **Sets** like the algebraic numbers or the rationals are said to be countable, and Cantor furnished proof that this is so. However, he discovered that the set of all real numbers is not countable. Encouraged by these successes, Cantor introduced the notion of equipollency of sets in his next paper, written in 1878. Two sets are equipollent if a one-to-one correspondence exists between them. Where he had previously shown that the set of algebraic numbers is equipollent with the set of positive integers, Cantor then proved that the set of points on any surface such as a plane is equipollent to the set of all real numbers. He finished the 1878 paper with the conjecture that every infinite subset of the set of real numbers is either countable or equipollent to the set of all real numbers. That conjecture became known as the continuum hypothesis. The possibility of a one-to-one correspon-

dence between an infinite set and one of its proper subsets had been observed earlier by scientists such as **Galileo** and **Gottfried Leibniz**. The novelty and courage of Cantor's contributions are in his refusal to consider this a contradiction and in using it to define infinite sets of equal power.

Cantor's next paper was published in six installments between 1879 and 1884. Where he had previously come to grips with the countable and had realized the gap between the countable and the continuum, his ideas about infinite sets in general had ripened. The paper broaches the idea of a proof of the continuum hypothesis. In 1882, Cantor defined another main concept of set theory, that of well-ordering. In 1883 he wrote that we may assume as a law of thought that every set can be well-ordered. Earlier, in 1878, Cantor had stated, without proof, "If two point-sets M and N are not equipollent then either M will be equipollent to a proper subset of N or N will be equipollent to a subset of N." This principle later became known as the trichotomy of cardinals. However, Cantor was not able to give a solution to the continuum hypothesis or a proof of the trichotomy of cardinals.

In 1884, Cantor had a nervous breakdown; several of such mental crises would follow. He had applied for a professorship in Berlin but was turned down, strongly opposed by **Leopold Kronecker**, a former teacher of his. In spite of his illness, Cantor remained active. He worked to institute the German Mathematical Society, founded in 1889, and was instrumental in establishing the first International Congress of Mathematicians in 1897 in Zurich. Between 1895 and 1897 Cantor published his last paper, *Beiträge zur Begründung der transfiniten Mengenlehre* ("Contributions to the Foundation of Transfinite Set Theory") in two parts. In these he defined the **transfinite numbers** that measure the magnitude of infinite sets.

While there was little enthusiasm for his discoveries within his own country of Germany at this time, Cantor's ideas were gaining support in the world mathematical community. The eventual recognition of sets as a notion underlying all of mathematics led to new fields like **topology**, measure theory, and set theory itself. Developments at the turn of the century reflected the importance of Cantor's work. At the Second International Congress of Mathematicians in Paris in 1900, the continuum hypothesis was first among 23 problems that Hilbert proposed as central to the development of twentieth-century mathematics. Not much later, in 1904, the mathematician Ernst Zermelo established that every set can be well-ordered, using the so-called **axiom** of choice. From it followed the trichotomy of cardinals. The earlier controversy between Kronecker and Cantor intensified into a new rage about what was and what was not permitted in mathematics. Much of the debate was later settled by the work of **Kurt Friedrich Gödel** and **Paul Cohen**. The first book on set theory was published in 1906 by William H. Young and **Grace C. Young**. These years also showed the beginnings of the study of topology. In 1911, **L. E. J. Brouwer** proved the topological invariance of **dimension**, at which Cantor himself had tried his hand earlier. In 1914, **Felix Hausdorff** published the first book on topology, entitled *Grundzüge der Mengenlehre* ("Principles of Set Theory").

Toward the end of his career Cantor's achievements were recognized with various honors. He became honorary member of the London Mathematical Society (1901) and the Mathematical Society of Kharkov and obtained honorary degrees at several universities abroad. A bust of Cantor was placed at the University of Halle in 1928. Perhaps more fittingly, one special subset of the real numbers that he introduced is now known under his name, the **Cantor set**. Cantor died on January 6, 1918, at the psychiatric hospital in Halle.

CANTOR SET

The Cantor set is an infinite set of numbers between 0 and 1, defined by an inductive process. To define this set, start with the closed **interval** [0,1]. Remove the middle third—the open interval (1/3,2/3). (That is, remove all the points between 1/3 and 2/3, except the points 1/3 and 2/3 themselves.) Next take each of the remaining two intervals, and remove *its* middle third. This leaves us with four intervals: [0,1/9], [2/9,1/3], [2/3,7/9], and [8/9,1]. Now remove the middle third of each of these intervals. This procedure is repeated *ad infinitum*. The resulting set is the Cantor set, named for **Georg Cantor**.

This set has many interesting properties. It is an example of a set of measure 0 (see Measurable and nonmeasurable). This is because the interval [0,1] has length 1, and if we add up the lengths of all the removed middle thirds, we also get 1. The Cantor set nevertheless contains infinitely many points— in fact, it contains many points. Moreover, it can be shown that all of these points are "cluster points"—that is, every point in the Cantor set has other points in the Cantor set that get arbitrarily close to it.

We saw above that every fraction whose denominator is a power of three is a member of the Cantor set. How can we tell whether the point 1/100 is in the Cantor set? We can, of course, simply follow the procedure for constructing the set, and check at each stage to see if 1/100 is in one of the intervals being removed. However, for some points, this can take a long time, and there is no way to know in advance how long it will take—the point in question may not be removed until, say, the billionth stage of the process.

There is an alternative method of viewing the Cantor set that makes it very easy to determine whether any given point is, in fact, in the Cantor set. To do this, we switch into base 3 notation. This means that we only use the numerals 0, 1 and 2—thus the number 3 is written as 10, because what we usually think of as the 10's column is now the 3's column. Then 4 is written 11, 5 is 12, and 6 becomes 20. The number 9 is written 100—just as in base 10, each column represents a power of 10, in base 3, each column represents a power of 3. We can also use base 3 notation to the right of the decimal point—the first column represents 1/3, the second 1/9, and so on. Thus, for example,.12 in base 3 is 1/3 + 2/9 = 5/9 (and so the "decimal" point is really now a "base-3" point).

Now consider the first stage in constructing the Cantor set, removing the interval (1/3,2/3). Written in base 3, we are removing the interval (.1,.2). Numbers in this interval have the

form.1..., i.e., they all have a 1 in the first column. So we have removed all numbers with a 1 in the first column, except for.1 itself.

In the second stage, we remove the intervals (1/9,2/9) and (7/9,8/9). In base 3 notation, these are written (.01,.02) and (.21,.22) respectively, and numbers in these intervals have the form.01... and.21.... So we are eliminating all numbers whose second digit to the right of the "decimal" point is 1, except for.01 and.21.

In the third stage, we eliminate the intervals (.001,.002), (.021,.022), (.201,.202) and (.221,.222). Thus we are eliminating all numbers whose third digit to the right of the decimal point is 1, except for the numbers.001,.021,.201, and.221.

Continuing this pattern, we can see that we eliminate all numbers that have a 1 *anywhere* in their base 3 expansion, with the exception of the numbers that have precisely one 1, as the final digit. Thus to test if any given number—say, 1/100— is in the Cantor set, we merely write it in base 3 (the procedure is the same as in base 10, i.e. long **division**) and check to see how many ones it has and where they are placed. In the case of 1/100, we get the base 3 expression.000210220022.... Since there is a 1 in this expression and it is not the final digit, we can see that the number 1/100 is not in the Cantor set.

Finally, one other way of looking at the Cantor set is geometrically. If we draw the Cantor set on a number line, we get a fractal: any small portion of it looks the same as the entire set. For example, the portion of the Cantor set in the interval [0,1/3] looks like the entire Cantor set: the middle third of this portion is missing, and the middle thirds of the two remaining segments are missing, and so on.

See also Georg Cantor; Fractals

CARDANO, GIROLAMO (1501-1576)
Italian algebraist and physician

It would be incomplete to simply list Girolamo Cardano as an Italian physician and mathematician. To reflect the true character of his life, one would have to add that he was the illegitimate son of a noted lawyer, a compulsive gambler, a popular astrologer, a one–time prisoner of the Inquisition, and the father of a convicted murderer. Despite the sordid nature of many aspects of his life, Cardano was also a brilliant physician and mathematician, as well as the author of more than 200 works on human medicine, mathematics, physics, philosophy, religion, natural science, and music. His contributions to mathematics were mainly in the area of **algebra** and included the first generalized solution to cubic (third degree polynomial) **equations** as well as the solution to certain **quartic equations** (fourth degree polynomial), even though some of the solutions had been borrowed from others. Cardano was also one of the first to recognize the existence of the **square** root of **negative numbers**, now called **imaginary numbers**, although he did not know how to deal with them.

Girolamo Cardano was born in Pavia, Italy, on September 24, 1501. His father was Fazio Cardano, a well–known lawyer and a friend of **Leonardo da Vinci**. His mother was Chiara Micheri. Cardano's parents were not married at the time he was born, and the stigma of being an illegitimate child followed him through life. As a youth, Cardano was often sick and mistreated. Although his parents did eventually marry, his father did not live with the family until Cardano was seven years old.

When Cardano became of age, his father encouraged him to study the classics, mathematics, and astrology. He entered the University of Pavia, where he completed his undergraduate studies. In 1524, his father died and left him a house and a small inheritance. Cardano returned to study in Pavia, where he earned a doctorate in medicine in 1526. Shortly thereafter he took up a medical practice in Saccolongo near Padua, where he married Lucia Bandareni in 1531. In time they had two sons and a daughter.

In 1534, friends of his father helped Cardano obtain a teaching position in mathematics at a school in Milan. He continued to practice medicine in addition to his teaching duties, and his success in treating several influential patients soon made him the most sought–after physician in Milan. By 1536, Cardano's thriving medical practice allowed him to leave his job as a teacher, although his interest in mathematics continued.

Cardano published his first book on mathematics, *Practica arithmetice et mensurandi singularis*, in 1539. In this book, he first demonstrated his superior mathematical skills in approaching problems in **algebra**. In one example, he was able to solve a specific cubic equation by manipulating the terms to reduce the problem to a second–degree quadratic equation that could be solved.

An algebraic solution to cubic and quartic equations had long eluded mathematicians, and Cardano's work in his *Practica arithmetice,* although limited to only certain **cubic equations**, was impressive. Cardano felt little satisfaction in his accomplishments, however, because he knew that fellow mathematician Nicolò Fontana, also known as **Tartaglia**, had achieved solutions to even more difficult cubic equations four years earlier, but had refused to divulge his methods. Cardano had repeatedly beseeched Tartaglia for his secret to no avail. Finally, just as Cardano's book was being published in 1539, Tartaglia agreed to share his methods, but only on the condition that Cardano take an oath that he would not reveal them until Tartaglia had written his own book.

At this point fate smiled on Cardano. A young man named Ludovico Ferrari came asking for a job as a servant. Cardano took him in and quickly discovered that his new servant possessed a brilliant mind. Eager to share his enthusiasm for mathematics, Cardano taught Ferrari algebra and eventually revealed Tartaglia's secret to him. One of the cubic equations Tartaglia had learned to solve was the so–called "depressed cubic," which lacked a second–power term. Working together, Cardano and Ferrari discovered a method to reduce any generalized cubic equation to a depressed cubic. Using Tartaglia's methods, they could then solve the equation. Their elation in making this monumental advancement in algebra was dampened by the knowledge that Cardano had given his oath not to reveal Tartaglia's methods. What was worse,

Girolamo Cardano

Tartaglia seemed in no hurry to publish his long–promised book, which would have freed Cardano from his oath.

In 1543, Cardano and Ferrari traveled to Bologna where they went over the papers of another mathematician, Scipione dal Ferro, who had died in 1526. They discovered that dal Ferro had solved the depressed cubic equation in 1515, but had kept it secret until just before his death, when he revealed it to his student, Antonio Fior. Fior had foolishly used his new–found knowledge to challenge Tartaglia to a contest, only to have Tartaglia discover the method for himself during the competition.

Armed with the knowledge that it was dal Ferro who had originally solved the depressed cubic equation, Cardano felt his obligation to Tartaglia was removed. Thus, in 1545, Cardano published his second book on mathematics entitled *Artis magnae sive de regulis algebraicis liber unus*, commonly called *Ars magna* ("Great Art"). In it, he revealed not only the solution to the generalized cubic equation, but also the solution to the biquadratic quartic equation, which had been developed by Ferrari.

Although Cardano gave full credit to dal Ferro, Fior, and Tartaglia for their work on cubic equations, Tartaglia was outraged. He claimed Cardano had broken his oath, and he began a long and vicious letter–writing campaign denouncing Cardano as a scoundrel. Despite this dispute, Cardano's book was widely acclaimed. Besides the solutions to cubic and quartic equations, Cardano presented many other new ideas in algebra that became the basis for the theory of algebraic equations.

In 1546, Cardano's wife died at the age of 31, leaving him with three children. His eldest son, Giambattista, was a promising scholar and it appeared he would become a successful physician like his father. Then, in 1557, Giambattista, poisoned his wife. Despite his father's appeals and influence, Giambattista was convicted of murder and was beheaded in 1560. In 1562, Cardano left Milan and took a position teaching medicine at the University of Bologna. Tragedy struck again in 1565, when his devoted student and collaborator Ferrari was poisoned.

In 1570, Cardano was accused of heresy for having cast the horoscope of Jesus Christ and for attributing the events of Christ's life to the influence of the stars. He was imprisoned by the Inquisition for several months before he was released on the condition that he abandon teaching. At the advice of friends, Cardano moved to Rome in 1571 and asked for protection from the Pope. Pope Pius V refused to help him because of his conviction for heresy, but Pope Gregory XIII was more lenient and granted Cardano a lifetime pension in 1573.

Cardano revised his *Ars magna* in 1570 to include a section dealing with solutions to the cubic equation that involve the square **roots** of negative numbers. Today, these numbers are known as imaginary numbers. Cardano had no such knowledge of imaginary numbers, but the fact that he recognized their existence led to further work by others. His work on solving cubic and quartic equations stimulated work by Thomas Harriot, **Leonhard Euler, René Descartes**, and others over the next several hundred years.

Cardano also wrote several books that presented new ideas in many other disciplines. He studied the physicsof projectiles in **motion** and correctly observed that their trajectories resembled a parabolic curve. Cardano was the first to deduce that the ratio of the distances of a projectile shot through air and water is inversely proportional to the ratio of the densities of the two mediums. In **hydrodynamics**, he observed and measured the flow of streams and stated that the velocity of the water was greater at the surface than near the bottom, contrary to what most people believed. Cardano's work in geology was also influential.

Cardano retired on his pension and lived quietly in Rome. He died on September 21, 1576, just three days short of his 75th birthday. A century after Cardano's death, **Gottfried Leibniz** summed up Cardano's turbulent life when he wrote "Cardano was a great man with all his faults; without them, he would have been incomparable."

CARDINAL NUMBERS

Cardinal numbers describe the size of a setor collection, for instance how many dollars or how many days there are. The word cardinal comes from the Latin *cardinalis* for "most

basic" or "most important," which underlines the significance people give to the idea of "how many."

When people count, they are actually using a combination of cardinal and **ordinal numbers**, two key mathematical concepts. Ordinal numbers also tell "how many" but in the context of "in what order"—for example, which day of the month it is. The Latin *ordinalis* means "order," as in a serial arrangement. Thus, when people count a collection of items, they are using cardinal numbers to tell how many, but the counting system itself is ordinal, because the numbers used for counting must be used in a specific sequence to be meaningful. It is possible for a set to contain only the number 0. **Zero** tells "how many" and is thus a cardinal number.

When people are trying to compare the size of two **sets** or collections, usually they will count "how many" in each collection and then compare the cardinal numbers. Cardinal numbers are an important part of **set theory**, which is a method of studying and explaining the basic properties of **arithmetic**. In set theory, a cardinal number is the number obtained by counting a set. If counting the set always yields the same result, then that result is a cardinal number. One could say that this is simply the theoretician's way of explaining "how many."

Although many math courses begin with an overview of cardinal numbers and set theory, theoreticians think set theory looks deceptively simple on the surface. They point to the impossibility of defining a cardinal number without first establishing many elaborate preconditions. **Gottlob Frege** (1848-1925) and **Bertrand Russell** (1872-1970) were among the first mathematicians to attempt such a **definition**. They said a cardinal number A consisted of all sets equipollent (of equal signification) to A. Again, one could boil this down to a consideration of "how many." Mathematicians disputed the definition of Frege and Russell on the grounds that it does not consider sets in the traditional way. Mathematics texts offer lengthy definitions and conditions for cardinal numbers, with variations such as finite and infinite cardinals, Dedekind-finite cardinals, successor cardinals, regular cardinals, and inaccessible cardinals.

The terms cardinal and ordinal are so closely linked that it is difficult to explain one without drawing on the other. In every society, in every language, the challenge is still unmet no one has found a way to explain cardinal and ordinal as independent concepts. A mathematician would say that all of the nonnegative **whole numbers** (0, 1, 2, 3, 4...) are simultaneously cardinal and ordinal.

CARDINALITY

Bernhard Bolzano defined cardinality in his book *Paradoxes of the Infinite* (1851). The cardinality of a set is, roughly speaking, its size. Precisely, the cardinality of X is said to be less than or equal to that of Y if there is a function f from X to Y with the property that if x and x' are elements of X with $f(x) = f(x')$ then $x = x'$. Such a function is said to be one to one or injective. Equivalently, the cardinality of X is said to be less than or equal to that of Y if there is a function f from Y to X with the property that if x is an element of X then there is an element y of Y with f(y) = x. Such a function is said to be onto or surjective. If the cardinality of X is less than or equal to Y and vice versa then X and Y are said to have the same cardinality and it can be proven that there is a bijective (i.e. both injective and surjective) function from X to Y. Such a function is also called a one-to-one **correspondence**. Accordingly, two finite **sets** have the same cardinality if and only if they have the same number of elements. Also, the set **N** of natural numbers {0,1,2,3,...} has the same cardinality as the set **Z** of **integers** {...,-3,-2,-1,0,1,2,3,...} because the function $f(x) = (-1)^x$ multiplied by the greatest integer less than or equal to one-half x is a bijection from **N** to **Z**. Any set that has the same cardinality as **N** is said to be countable or **denumerable**.

Georg Cantor (1845-1918) built on Bolzano's work and rigorously proved most of the theorems that draw on cardinality. In 1874, Cantor proved that the set **Q** of **rational numbers** has the same cardinality as **N** by his "diagonalization method." First, the rational numbers are arranged as follows:

0/1, 1/1, -1/1, 2/1, 2/1, 3/1,
0/2, 1/2, -1/2, 2/2, -2/2, 3/2,...
0/3, 1/3, -1/3, 2/3, -2/3, 3/3,...

Next, a zigzagging line is drawn which touches (in order) 0/1, 1/1, 0/2, 0/3, 1/2, -1/1, 2/1, -1/2, 1/3, and so on. The function f which assigns to each natural number n the (n+1)st number of the above sequence is onto (i.e., for every rational number r there is a natural number n with f(n) = r). The function g which assigns to each natural number n the rational number n/1 is one to one. Hence **Q** has the same cardinality as **N**.

For any set X, Cantor denoted by 2^X the set of all subsets of X, because if X is a finite set, 2^X has as many elements as 2 to the power of the number of elements of X. Cantor proved that any sct X is strictly smaller than 2^X. He reasoned thus: if f is a function from X to 2^X, then one can denote by Y the subset of X containing all elements x of X having the property "x is not an element of f(x)." Assuming that there is an element y in X with f(y) = Y results in a contradiction since by the very **definition** of Y, such a y cannot be in Y, but this implies that y is in Y. So f is not onto. Since f is arbitrary, this implies that there can be no bijection between X and 2^X. Cantor used this fact to show that the set **R** of **real numbers** is strictly larger than **N** by showing that **R** is bijective with the set 2^N. The same result can be demonstrated more easily as follows. Let f be a function from **N** to [0,1] (the set of real numbers between 0 and 1). Write the numbers f(0), f(1), f(2), f(3), and so on, in order and in binary form. For example, suppose

f(0) = 0.100101...
f(1) = 0.011100...
f(2) = 0.110001...

Let a(i) = the (i+1)st digit of f(i). Let x be the real number in [0,1] whose (i+1)st digit is equal to 1 minus a(i). In the example x = 0.001... By construction x is not in the image of f since if f(n) = x then the (n+1)st digit of f(n) is equal to one minus the (n+1)st digit of x, and this is a contradiction. Hence f cannot be onto. Since f is arbitrary, there are no onto **functions** from **N** to **R**.

Suppose that X is a set with cardinality strictly larger than that of **N** but smaller than or equal to that of **R**.

The continuum hypothesis is the assertion that the cardinality of X is equal to that of **R**. Cantor tried in vain to prove it. **David Hilbert** included it in his famous problem list at the beginning of the 20th century. In 1963, **Paul Cohen** of Stanford University proved that the continuum hypothesis is independent of the other well-accepted axioms of **set theory** (specifically, the Zermelo-Fraenkel axioms). In other words, it cannot be proved or disproved if one assumes only standard set theory.

Cantor's ideas were opposed by some prominent mathematicians of his time including **Leopold Kronecker**, Christian F. Klein, **Jules-Henri Poincaré**, and **Herman Weyl**, who doubted the soundness of Cantor's arguments and disliked the apparent paradoxes that arose from them. Today Cantor's theories are widely accepted and have applications to most areas of mathematical research. David Hilbert has said, "No one shall expel us from the paradise which Cantor created for us," and **Bertrand Russell** has said that Cantor's work is "probably the greatest of which the age can boast."

See also Continuum hypothesis; Correspondence; Denumerable; Hilbert's problems; Zermelo-Fraenkel set theory

CARLESON, LENNART AXEL EDVARD (1928-)

Swedish mathematician

Lennart Axel Edvard Carleson is Professor Emeritus of Mathematics at the Royal Institute of Technology (KTH) in Stockholm, Sweden. He is also a member of the faculties of the University of Uppsala, in Uppsala, Sweden, and the University of California at Los Angeles. Since his formal retirement at KTH in 1994, Carleson has held a senior professorship in the mathematics department there through an endowment provided by the Göran Gustafsson Foundation.

In 1966, Carleson proved the **theorem** that is named for him. The Carleson theorem states that the **Fourier series** for a square-integrable function converges almost everywhere, where terms are defined as follows:

- The Fourier series representation of a function is a representation of the function by a summation (over n) of the form $\Sigma a_n sin(n\pi x/\kappa)$. A practical example would be the representation of a complex musical waveform by the sum of many frequencies.

- A square-integrable function is a function f for which the integral over x from a to b of the absolute value of $(f(x))^2$ exists.

- The term "almost everywhere" describes a situation in which a pair of **functions** f and g differ for sufficiently few values of x that all integrals involving f have the same values as those involving g.

- **Convergence** means that the Fourier series of a given function f(x) actually converges to f(x) at all points over the **interval** of the expansion.

Between 1969 and 1984, Carleson served as director of the Mittag-Leffler Institute, a part of Swedish Royal Academy of Sciences. During his tenure there, Carleson transformed Mittag-Leffler into the institute for mathematics research that the founder had envisioned in 1916. Carleson was successful in attracting the financial backing required to attract foreign mathematicians to the institute with fellowships.

In 1984, Carleson received the American Mathematical Society's Leroy P. Steele Prize for his papers: "An interpolation problem for bounded analytic functions," *American Journal of Mathematics*, volume 80 (1958), pp. 921-930; "Interpolation by bounded analytic functions and the Corona problem," *Annals of Mathematics* (2), volume 76 (1962), pp. 547-559; and "On convergence and growth of partial sums of Fourier series," *Acta Mathematica*, volume 116 (1966), pp. 135-157.

In 1992, Carleson was awarded the Wolf Prize in recognition of his fundamental contributions to Fourier **analysis**, **complex analysis**, and quasi-conformal **mappings** and dynamical systems.

CARNAP, RUDOLF (1891-1970)

German philosopher

The German philosopher Rudolf Carnap was born on May 18, 1891, in Ronsdorf, Germany. He died on September 14, 1970, at Santa Monica, California.

Between 1910 and 1914, Carnap studied philosophy, physics and mathematics at the Universities of Jena and Freiburg. During those years, he was particularly interested in physics and the Kantian theory of **space**. By 1913, he was planning to write his dissertation on thermionic emission based on experimental work. But World War I intervened, and Carnap found himself assigned to the front. But in 1917, he was transferred to Berlin, where **Albert Einstein** was then professor of physics, and Carnap took advantage of his time in Berlin to study the theory of **relativity**.

In 1921, when Carnap finally did write his dissertation, he described a philosophical theory of space that showed the influence of Kantian philosophy. It is interesting that Carnap's first and last books both dealt with the philosophy of physics.

In 1926, Carnap relocated to Vienna to become an assistant professor at the University of Vienna. There he met Hans Hahn, Otto Neurath, **Kurt Gödel**, Ludwig Wittgenstein, and Karl Popper. Carnap became a leading member of the Vienna Circle and a strong proponent of logical positivism, the philosophy that holds that all valid philosophical problems are solvable by logical **analysis**.

In 1931, Carnap became professor of natural philosophy at the German University in Prague. With Adolf Hitler's rise to power in 1933, Carnap opted to emigrate and arrived in the United States two years later. He became an American citizen in 1941.

Carnap first became interested in semantics in the 1940s. Later, he also began to look at the structure of scientific theories. According to Carnap, any scientific theory must consist of

- a formal language that included both logical an non-logical terms;
- a set of logical-mathematical axioms and rules of inference;
- a set of non-logical axioms that express the empirical parts of the theory;
- a set of postulates that state the meaning of non-logical terms and formalize the analytical truths of the theory;
- a set of rules that provide an empirical interpretation of the theory.

Except for a short period as a visiting professor at Harvard University (1940-41), Carnap was a member of the faculty at the University of Chicago between 1936 and 1952. From 1952 to 1954, he was a professor at the Institute for Advanced Study at Princeton, and beginning in 1954, a professor at the University of California at Los Angeles.

CARROLL, LEWIS (1832-1898)

British mathematician and writer

Lewis Carroll is actually a pseudonym, the pen name taken by Charles Lutwidge Dodgson. Although best known for his children's books, Dodgson worked professionally as a mathematician, studying particularly recreational logic, **determinants**, **geometry**, and the mathematics behind tournaments and elections (see **game theory** and **voting paradox**).

Charles Lutwidge Dodgson was born in Daresbury, Cheshire, England, on January 27, 1832. He was the third child of 11 who would eventually be born to the clergyman Charles Dodgson and his wife, Frances Jane (Lutwidge). Charles, like his father and most of his siblings, suffered from a stutter his entire life.

Because of the relative isolation of the parsonage in which the family lived, young Charles created his own games. Biographer S.D. Collingwood wrote, "in this quiet home the boy invented the strangest diversions...he... numbered certain snails and toads among his intimate friends... He tried also to encourage civilised warfare among earthworms..."

Charles took his first lessons at home, under his father's tutelage. He learned mathematics, Latin, and literature with ease, and showed early promise at mathematics. In 1844, he went to the Richmond School, 10 miles from his father's posting at Croft-on-Tees in Yorkshire. James Tate, his headmaster, fostered Charles's interest in mathematics, even writing to his family that he had "a very uncommon share of genius."

He then attended the Rugby school, excelling in his studies. But he may not have enjoyed the boisterous atmosphere, and returned home in 1849, after four years, to get ready to attend Christ Church College, Oxford. He started there on January 24, 1851, but returned home almost immediately to attend his mother's funeral; she had died suddenly.

In 1854, he graduated from Christ Church College, and the following year was named lecturer in mathematics there. In addition, he worked as sub-librarian, and received a scholarship, which added an extra 55 pounds a year to his income. He would remain at Oxford his entire life.

He tutored pupils in mathematics, and was apparently quite a good teacher, able to provide clear explanations. However, he was not a stimulating lecturer, lacking charisma and communication skills (and he still had that stutter). Later in his life, he observed that nine students had attended his first lecture; only two attended his last one, which took place 25 years later.

In 1857 he received his master of arts from Christ Church College. In 1861, he was ordained as a deacon in the Church of England; records show that he performed several baptisms and funerals, and preached sermons, although public speaking must have been difficult for him.

Although he never married, and had few close friends, Dudgeon's life was not empty. He had always loved writing for the sheer joy of it, and his works amused people, which made him content. He never planned to publish, but several of his pieces appeared in British humor magazines, and out of some of these pieces grew *Alice in Wonderland* and *Through the Looking Glass,* two of the world's enduring classics of children's literature.

But a highlight of Dodgson's life was children. They accepted him—stuttering shyness and all—and he loved to tell them stories. He wrote his famous Alice books for the daugh-

Lewis Carroll

ters of Henry George Liddell. He published them under the name Lewis Carroll, under which he had also published light verse.

His scholarly writings are numerous, although not groundbreaking. He modestly described his work as "chiefly in the lower branches of Mathematics." Dodgson did not keep up with the latest developments in his field, nor did he concentrate on doing new mathematics. Instead, his focus was chiefly concerned with teaching, not with research—with providing the all-important underpinnings of knowledge that would allow students to pursue further work.

He wrote a textbook on geometry called *A Syllabus of Plane Algebraic Geometry,* which was published in 1860. Later, he turned his attention to determinants, first publishing a paper in the Proceedings of the Royal Society in 1866 and later expanding that paper into a book. However, *An Elementary Treatise on Determinants* was not well received, possibly because he created symbols to use solely in that book.

He distinguished himself particularly in his work in logic and on voting and tournaments. He published a pamphlet in 1873 regarding different ways of arriving at fair majority opinions; although he didn't review the existing literature on the topic, he improved on the ideas that he didn't even know existed.

One of his true loves was recreational logic. All his logic work was published under his pseudonym, indicating that he considered it an amusing pastime. He was particularly interested in the use of diagrams in logic, and he understood, used, and improved on James Venn's circular diagrams, making them **square** and using red and gray colored counters to indicate the contents of the **sets**.

Dodgson's clever mind is easily seen at work in the pages of the Alice books, which are riddled with logic puzzles. He also turned mathematics into comedy when he wrote the five-act play *Euclid and His Modern Rivals.* The play was widely read in English schools.

In addition to his work in mathematics and his writing, Dodgson was a gifted photographer, particularly of children. One of his favorite subjects as Alice Liddell. He has been called the most gifted photographer of children of his era.

Dodgson died on January 14, 1898, in Guildford, Surrey, England.

CARTAN, ÉLIE JOSEPH (1869-1951)

French number theorist, topologist, and geometer

Élie Joseph Cartan is one of the most important mathematical figures in the first half of the 20th century. Although recognition for his many accomplishments did not come until late in his career, his intellectual influence is still felt as modern mathematics develops. From the earliest point in his career, Cartan further developed Norwegian mathematician **Marius Sophus Lie**'s **group theory** which concerned continuous groups and symmetries within them. He interworked Lie's theory and the original means in which he studied its global properties with differential **geometry**, classical geometry, and **topology**. These combined areas are still a vital part of contemporary mathematics. Cartan also formulated many original theories based on his studies. Despite making such vital contributions to mathematics, Cartan was quite modest, good humored, and easy going. He was also a gifted and well–liked teacher who could break down his intricate theories for the consumption of an average student.

Cartan was born April 9, 1869, in Dolomieu Isère, France, a village in the Alps. Of peasant descent, he was the son of the village blacksmith, Joseph, and his wife, Anne (Cottaz) Cartan. Cartan was the second oldest of four children. An inspector of primary schools, Antonin Dubost, noticed Cartan's impressive scholastic aptitude while on a visit to Cartan's school. Though most children from poor families were not put on the track to attend university, Dubost helped Cartan to win a scholarship to attend lycée (secondary school). Cartan's success as a student inspired his youngest sister, Anna, to follow in his intellectual footsteps, and become a mathematics teacher. She also published several texts on geometry.

After attending three lycées, Cartan went on to the l'Ecole Normale Supérieure in Paris in 1888, obtaining his doctorate in 1894. In his graduate thesis, Cartan began the first phase of his life's research, that of Lie's theory of continuous

Élie Joseph Cartan

groups, a topic neglected by most of his contemporaries. In his thesis Cartan completed the classification of semisimple **algebra**s begun by Wilhelm Killing during the last two decades of the 19th century, and gave a rigorous foundation for the types of Lie algebras that Killing had shown to be possible.

Before pursuing a career as a mathematician, Cartan was drafted into the French Army for one year after his graduation, rising to the rank of sergeant before his discharge. Cartan then served as a lecturer at two successive institutions, the University of Montpellier (1894 to 1896), and the University of Lyons (1896 to 1903). While a lecturer, he continued his mathematical studies based on Lie's theory, and began bringing together the four disparate disciplines that became the hallmark of Cartan's work: differential geometry, classical geometry, topology, and Lie theory. Cartan began by exploring the structure for associative algebras, then moved onto semisimple **Lie groups** and their representations. In this period (roughly from 1894 to 1904), Cartan also helped create and develop the **calculus** of exterior differential forms. This calculus became an important tool Cartan used in his research and he applied it to various differential geometric problems.

In 1903, Cartan's life changed in two fundamental ways. He married Marie–Louise Bianconi in that year. With her he had four children (three sons and one daughter), and enjoyed a happy home life. His eldest son, **Henri Paul Cartan**, born in

1904, became a prominent mathematician in his own right. His daughter Hélène also became a mathematician who taught at lycées and published some mathematical papers. Cartan's two other sons died tragic deaths. Jean, a composer, died at 25 of tuberculosis; Louis, a physicist, was arrested and imprisoned by the Nazis for his activities in the French Resistance. Louis Cartan was executed by the Nazis in 1943, but his family did not learn of his death until 1945.

In 1903, Cartan also became a professor at the University of Nancy, where he worked until 1909. Cartan moved on to the University of Paris (the Sorbonne) in 1909, where he was a lecturer from 1909 to 1912 before becoming a full professor in 1912. He remained a professor there until he retired in 1940. In his first years at the Sorbonne, in 1913, Cartan made one of his most significant contributions to math when he discovered spinors. The spinors are complex vectors used to make two–dimensional representations out of three–dimensional rotations. The spinors were important in the development of **quantum mechanics**.

Though Cartan continued to intertwine the four subjects mentioned earlier, at the height of his career, he continued to look at them from different perspectives. For example, after 1916, Cartan's publications focused mainly on differential geometry. Cartan developed a moving frames theory and a generalization of Daboux's kinematical theory. In 1925, Cartan refocused his attention to the study of topology. In his paper "La géométrie des espaces de Riemann" ("The Geometry of Riemann Spaces"), Cartan came up with innovative ways to study Lie groups' global properties. He demonstrated, in topological terms, that a Euclidean **space** and a compact group can produce a connected Lie group. And, from 1926 to 1932, Cartan published treatises concerning his geometric theory of symmetric spaces.

Cartan's continual output of important work led to his appointment as a member of the Académie Royale des Sciences in 1931, one of many honors he received late in his career. About the same time, Cartan began collaborating with his son Henri on mathematical problems, building on his son's theorems. On Cartan's 70th birthday, in 1939, the Sorbonne honored him with a celebratory symposium which praised his many mathematical accomplishments. Although Cartan retired from the Sorbonne in 1940, he remained an honorary professor there until his death. He also continued to publish mathematical treatises and his love of classroom instruction led him to teach math at the l'Ecole Normale Supérieure for Girls. Cartan died in Paris on May 6, 1951, after a long illness.

CARTAN, HENRI PAUL (1904-   )

French algebraist

Henri Cartan has made monumental contributions in essentially every field of algebraic **topology**, including analytical **functions**, the theory of sheaves, homological theory, and potential theory. His most important works include *Homological Algebra* (1956) and *Elementary Theory of Analytic Functions of One or Several Complex Variables*

(1963). He also worked on a definitive **convergence** theorem on decreasing potentials of positive masses, which became a fundamental instrument for improving potential theory. Cartan is the recipient of the 1980 Wolf Prize in Mathematics.

Cartan was born in Nancy, France, in 1904. His father was **Élie Joseph Cartan**, a French mathematician who made significant contributions to the theory of subalgebras. Cartan was educated at the Ecole Normal Superieure in Paris, one of the finest schools for mathematics. Cartan attended the Ecole from 1923-1926. He was a protege of **Jules Henri Poincare**, who is considered the founder of algebraic topology and the theory of analytic functions of several complex variables. Along with **Albert Einstein** and **Hendrik Lorentz**, Poincare is considered a co-discoverer of the special theory of relativity. Cartan's studies with Poincare may explain his own interest in algebraic topology.

In 1935, Cartan was one of the founders of a group of young French mathematicians who had all graduated from the Ecole Normal Superieure. Other members included Claude Chevalley (or Chevallier), Jean Dieudonne, Jean Deslarte, and **Andre Weil.** All members were brilliant in their own right, however, Weil was the only "universalist" among them, accomplished in every area of mathematics. At the beginning, Cartan was not a universalist, although he did become more accomplished later in his career.

The group wrote under the pseudonym of **Nicolas Bourbaki**. In "Nicholas Bourbaki, Collective Mathematician," Claude Chevalley, as told to Denis Guedj, explains how Bourbaki got his name: "Weil had spent two years in India and for the thesis of one of his pupils he needed a result he couldn't find anywhere in the literature. He was convinced of its validity, but he was too lazy to write out the **proof**. His pupil, however, was content to put a note at the bottom of the page which referred to "Nicolas Bourbaki, of the Royal Academy of Poldavia'." The "real" Bourbaki liked to pretend he was in a secret society when he was young. From this, Weil tried to get the members of the group to stay anonymous. Members refused to answer questions about other members and of projects they were working, and even how the name of Bourbaki originated. The group did not remain anonymous for very long.

The Bourbaki group produced *Elements de mathematique*, a 30-volume textbook on **analysis** geared toward French university students, in 1939. It was intended to replace the class analysis textbook used in France written by Edouard Goursat. Bourbaki's book aimed to achieve the same high standards that Goursat set forth in his text, yet it also included mathematical advances. By 1968, there were 33 volumes in print. Early editions of *Elements* did not include credit to other contributors, so in 1960, *Elements d'historique des mathematiques* was published to rectify the situation.

The group sought the best way to produce the work together. They did not just want to assign topics to the person most qualified to write it, believing if they did, the book would turn out like an encyclopedia, which they did not want. Each member studied the same areas of mathematics, from the beginning. During the Bourbaki-congres, as their thrice yearly meetings were called, the group would discuss sections of the book, making numerous revisions. Bourbaki members had to retire from the group by the age of 50. This assurance was made to keep Bourbaki from "growing old," keeping him young in spirit. Women were not allowed to join the Bourbaki group. Through the Bourbaki members, a new kind of algebraic topology was born. Bourbaki played a key role in the rethinking of structural mathematics.

Cartan became president of the International Mathematical Union (IMU) in 1967, serving in that capacity until 1970. He then became the past-president, holding a seat on the executive committee for the following four years. The IMU is a scientific organization with the purpose of promoting international cooperation in mathematics and is part of the International Council of Scientific Unions. Founded in 1919, the IMU was disbanded in 1936, then reconstituted in 1951. Cartan was an invited speaker for a May 1995 IMU conference taking place at the Institut Henri Poincare.

Cartan's academic career began at the Lycee Caen in 1928, where he was a professor of mathematics. He left the Lycee in 1929, and was appointed a deputy professor at the University of Lille. From there, he went to the University of Strasbourg, which named him a professor of mathematics in 1931. From 1940 until 1969, Cartan was on the faculty at the University of Paris. After the University of Paris, Cartan taught at Orsay for about five years, until 1975.

In 1992, Cartan was part of a group of international scientists issuing a warning to humanity. The warning urged changes to protect the living world. According to the Worlds' Scientists Warning to Humanity, found on the World Wide Web (at various locations, e.g., http://deoxy.org/sciwarn.htm), "Human beings and the natural world are on a collision course.... If not checked, many of our current practices put at serious risk the future that we wish for human society and the plant and animal kingdoms, and may so alter the living world that it will be unable to sustain life in the manner that we know.... Fundamental changes are urgent if we are to avoid the collision our present course will bring about." The warning lists several environments in need of help. Cartan's name is among the 1,700 signatories, all of which are members of regional, national, and international science academies.

CARTESIAN COORDINATE SYSTEM

The Cartesian coordinate system is named after **René Descartes** (1596-1650), the noted French mathematician and philosopher, who was among the first to describe its properties. However, historical evidence shows that **Pierre de Fermat** (1601-1665), also a French mathematician and scholar, did more to develop the Cartesian system than did Descartes.

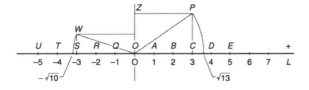

To best understand the nature of the Cartesian plane, it is desirable to start with the number line. Begin with line L and let L stand for a number axis (see figure 1). On L choose a point, O, and let this point designate the zeropoint or origin. Let the **distance** to the right of O be considered as positive; to the left as negative. Now choose another point, A, to the right of O on L. Let this point correspond to the number 1. We can use this distance between O and A to serve as a unit with which we can locate B, C, D,... to correspond to the $+1, +2, +3, +4,...$ Now we repeat this process to the left of O on L and call the points Q, R, S, T,... which can correspond to the numbers $-1, -2, -3, -4,...$ Thus the points A, B, C, D,..., Q, R, S, T,... correspond to the set of the **integers** (see figure 1). If we further subdivide the segment OA into d equal parts, the $1/d$ represents the length of each part. Also, if c is a positive integer, then c/d represents the length of c of these parts. In this way we can locate points to correspond to **rational numbers** between 0 and 1.

By constructing rectangles with their bases on the number line we are able to find points that correspond to some **irrational numbers**. For example, in Figure 1, rectangle OCPZ has a base of 3 and a segment of 2. Using the Theory of Pythagorus we know that the segment OP has a length equal to $\sqrt{13}$. Similarly, the length of segment OW is $\sqrt{10}$. The **real numbers** have the following property: to every real number there corresponds one and only one point on the number axis; and conversely, to every point on the number axis there corresponds one and only one real number.

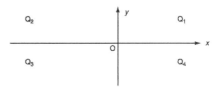

What happens when two number lines, one horizontal and the other vertical, are introduced into the plane? In the rectangular Cartesian plane, the position of a point is determined with reference to two perpendicular line called coordinate axes. The intersection of these axes is called the origin, and the four sections into which the axis divide a given plane are called quadrants. The vertical axis is real numbers usually referred to as the y axis or **functions** axis; the horizontal axis is usually known as the x axis or axis of the **independent variable**. The direction to the right of the y axis along the x axis is taken as positive; to the left is taken as negative. The direction above the x axis along the y axis is taken as positive; below as negative. Ordinarily, the unit of measure along the coordinate axes is the same for both axis, but sometimes it is convenient to use different measures for each axis.

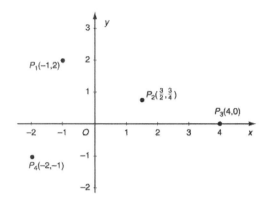

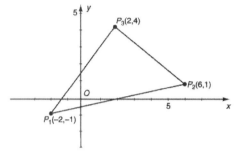

The symbol $P_1 (x_1, y_1)$ is used to denote the fixed point P_1. Here x_1 represents the x coordinate (abscissa) and is the perpendicular distance from the y axis to P_1; y_1 represents the y coordinate (ordinate) and is the perpendicular distance from the axis to P_1. In the symbol $P_1 (x_1, y_1)$, x_1 and y_1 are real numbers. No other kind of numbers would have meaning here. Thus, we observe that by means of a rectangular coordinate system we can show the **correspondence** between pairs of real numbers and points in a plane. For each pair of real numbers (x, y) there corresponds one and only one point (P), and conversely, to each point (P) there corresponds one and only pair of real numbers (x,y). We say there exists a one-to-one correspondence between the points in a plane and the pair of all real numbers.

The introduction of a rectangular coordinate system had many uses, chief of which was the concept of a graph. By the graph of an equation in two variables, say x and y, we mean the collection of all points whose coordinates satisfy the given equation. By the graph of the function f(x) we mean the graph of the equation y= f(x). To plot the graph of an equation we substitute admissible values for one **variable** and solve for the corresponding values of the other variable. Each such pair of values represents a point which we locate in the coordinate system. When we have located a sufficient number of such points, we join them with a smooth curve.

In general, to draw the graph of an equation we do not depend merely upon the plotted points we have at our disposal. An inspection of the equation itself yields certain properties which are useful in sketching the curve like **symmetry**, **asymptote**s, and intercepts.

CARTWRIGHT, MARY LUCY (1900-1998)
English mathematician

Dame Mary Lucy Cartwright's research and contributions to the field of mathematics have spanned more than seven decades. England claims her as one of its most brilliant citizens and has honored her for the past 50 years. As late as 1996, she was featured, along with two other women scientists, on British television, modestly answering questions about her impressive achievements.

Cartwright was born in Aynho, Northamptonshire, England, on December 17, 1900. Her father, William Digby Cartwright, served as a curate to his uncle, who was a rector at a church. Her mother was Lucy Harriette Maud Bury. The youngest of three children, Cartwright's two other brothers, John and Nigel, were both killed in World War I. When Cartwright was 11 years old, she was sent to live with her maternal uncle, Fred Bury, and his wife Annie, so she could attend Leamington High School, now known as the Kingsley School. There, she learned mathematics from a teacher she remembers as Miss Hancock, to whom she still pays tribute as "an excellent teacher of mathematics." In 1919, Cartwright entered St. Hughes College at Oxford and remained there until 1923. At the age of 23 she graduated from the University of Oxford, then taught first at the Alice Ottley School in Worcester, then at the Wycombie Abbey School for four years before returning to Oxford to complete her doctorate in mathematics. In 1925 Cartwright received the Hurry Prize for mathematics.

One of only five students, Cartwright began study and research with **Godfrey Hardy** in 1928 and it was under his supervision (as well as of E.C. Titchmarsh) that she pursued her degree. Initially, Hardy was Cartwright's advisor on her thesis, then Titchmarsh took over when Hardy left to spend some time at Princeton University. Hardy is best known for the Hardy-Weinberg law, which resolved the controversy over what **proportions** of dominant and recessive genetic traits would be reproduced in a large mixed population. Cartwright had great admiration for Hardy, both for his work and for the pains he took in instructing his students. Hardy also collaborated with John E. Littlewood on a series of papers that contributed fundamentally to the theory of Diophantine **analysis**, divergent series summation, and **Fourier series**.

Littlewood was Cartwright's examiner for her Ph.D. in 1930. She was the first woman to read for finals in mathematics at Oxford. The meeting was the first in a long series of conferences that would eventually lead to more than ten years of collaboration on a number of projects. Together, Cartwright and Littlewood published four papers on large parameters, **differential equations** combining topological and analytical methods, solutions for the Van der Pol equation, and **chaos theory**. They published several other papers, but the content was based on their collaborative work. Based on her own research, Cartwright also published a number of papers concerning classical analysis. Cartwright was a prolific writer; besides mathematical papers, she produced biographical essays on other women mathematicians such as Shelia Scott

Macintyre and **Grace Chisholm Young** for the London Mathematical Society's *Journal.*

Also in 1930 Cartwright became a Yarrow Research Fellow of Girton. During her fellowship she attended lectures given by Edward Collingwood concerning integral and metamorphic **functions**. She worked closely with Collingwood on cluster **sets** in the theory of functions of one complex **variable** and these papers were also published. Other publications by Cartwright include *Religion and the Scientist, Specialization in Education, The Mathematical Mind,* and *Integral Functions.*

In 1935, Cartwright was given a Faculty Assistant Lectureship at Cambridge University and her Yarrow Research Fellowship was extended. Cartwright contributes the appointments to a paper she published entitled "Mathematische Zeitschrift" in early 1935. An earlier version of the paper had been shown to Hardy and Littlewood, and Cartwright believes it was upon their joint recommendations that she was given her position at Cambridge.

During World War II, Cartwright volunteered and served with the British Red Cross Detachment from 1940 to 1944. Cartwright was elected to the Council of the London Mathematical Society for the first time in 1933 and served as a member until 1938. She was again elected in 1961 and served as president until 1963.

In 1949, she was appointed Mistress of Girton College and held that position until her retirement in 1968. During that time she received many honors, including an honorary Doctor of Laws degree from Edinburgh in 1953 and Doctor of Science degrees from Leeds in 1958 and from the University of Wales in 1967. Cartwright was elected to the Royal Society of London in 1955 and received the Sylvester Medal of the Royal Society in 1964. Succeeding her Yarrow Fellowships and Lectureship appointments, she became a Reader in Theory of Functions in 1959. At the same time, she additionally served as director of studies in mathematics. In 1968, she was awarded the De Morgan Medal of the London Mathematical Society.

Cartwright visited numerous countries around the world, sharing and acquiring knowledge. She was a consultant on the United States Navy Mathematical Research Project at Stanford and Princeton Universities in 1949. Cartwright returned to the United States in 1968 after her retirement and spent two years at Brown University in Providence, Rhode Island, as a Resident Fellow. While at Brown, she also lectured at Clairmont Graduate School and Case Western Reserve University. Cartwright's years in America made a lasting impression on her. She wrote extensively in her memoirs of the killing of three students at Kent State University and of the many college protests she witnessed, including those at the University of California at Berkeley and the Madison Army Research Centre.

In 1973, she received recognition for her lifetime achievements from the University of Jyvaskyla, Finland. The culmination of Cartwright's recognition came in 1969, when she was ordained Dame Commander of the British Empire by Queen Elizabeth II. Cartwright was henceforth known as Dame Cartwright.

CATASTROPHE THEORY

Catastrophe theory, or singularity theory, attempts to explain how discontinuous effects can arise from continuous causes. For example, a forest abruptly terminates and turns into a meadow, even though the weather conditions on each side of the boundary are almost identical. A small increase in tensions between two countries sometimes leads to violent conflict; other times it doesn't. Why?

Unfortunately, catastrophe theory is also known among mathematicians as a fad that failed to live up to its hype. In its heyday—the late 1960s and the 1970s—the advocates of catastrophe theory claimed that it was a new, "qualitative" breed of mathematics that would revolutionize biology or social science in the same way that **calculus** had revolutionized physics. But the revolution never happened. At best, the theory explained phenomena that people already knew, only in a more obscure and technical language. It had no predictive power, and in its more applied forms the theory died a quiet death.

However, the mathematical core of catastrophe theory remains solid and unimpeachable. This core is the classification of singularities of **functions** or mappings of several variables. To understand the concept of a singularity, imagine a smooth, partially translucent surface illuminated from above. Look at the shadow that it casts on the ground. In general, that shadow will have distinct regions, some of which will be darker than others because the **light** has to pass through more layers of the surface. The regions are separated by *folds,* which are the shadows of points where the surface becomes vertical, and *cusps,* which are the shadows of points where two folds come together. These are the two basic kinds of singularities that show up in the image of the surface.

It is possible to imagine other kinds of singularities. Such singularities occur, for example, in origami (Japanese paper folding), when one pinches several creases together at a point. But Hassler Whitney proved in 1955 that any such singularities are extremely non-random; they require intelligent design. The only kinds of random, or *generic,* singularities are folds and cusps.

In general, catastrophe theory deals with differentiable functions or processes that depend on a small number of control variables and have a small number of output, or *state,* variables. In the case of the translucent surface, the control variables are the coordinates of a point on the floor, and the state **variable** is the height above the floor of any point on the sculpture. If a fly crawls along the surface, when it reaches a fold point it cannot go any farther and falls off, causing a discontinuous change in the state variable. This is known as a catastrophe.

In the case of a country debating whether to go to war, the control variables might be the perceived amount of threat to the country and the cost of defense spending, and the state variable might be the amount of money actually spent on defense. Together, all the variables form the *control surface,* which is assumed to be smooth. Again, the control variables may reach a singularity, where the only way the country can stay on its control surface is to make a discontinuous leap in defense spending—as happens, for example, when a war breaks out (or, in the opposite direction, when peace is negotiated).

Remarkably, all simple catastrophes can be classified according to the number of control variables required to observe them. In the 1960s, John Mather proved that there are seven generic catastrophes that use fewer than five control variables. These are called:

- The fold, which requires 1 control variable and 1 state variable;
- The cusp, 2 control variables and 1 state variable;
- The swallowtail, 3 control variables and 1 state variable;
- The hyperbolic umbilic (or "purse"), 3 control and 2 state variables;
- The elliptic umbilic (or "pyramid"), 3 control and 2 state variable;
- The butterfly, 4 control variables and 1 state variable; The parabolic umbilic, 4 control and 2 state variables.

Now we can explain why catastrophe theory promised so much and delivered so little. The beauty of Mather's classification is that so little depends on the exact shape of the control surface. Once a generic singularity is formed, it won't go away, even if the shape changes a bit. This property is called *structural stability.* In many of the applications that were proposed for catastrophe theory, one could never hope to find an exact equation for the control surface. (What is the equation relating defense spending to cost and perceived threat? How can you even measure perceived threat?) But because of structural stability, it didn't matter. As long as you knew the qualitative shape of the surface, you could analyze the dynamics of the system it represented, and any quantitative information was simply superfluous.

In spite of the persuasiveness of the apostles of catastrophe theory (chiefly **René Thom**), such promises were too good to be true. Indeed, one could qualitatively predict that a large enough perceived threat would force a country to go to war, but that has been known for centuries. Without an equation for the control surface and numbers to plug into it, catastrophe theory could never predict *when* the catastrophe would occur, and thus had little practical value.

See also Differential calculus; Topology

CATENARY

A catenary is the shape assumed by a hanging chain. More precisely, it is the solution of the following mathematical problem: of all plane curves of a fixed length joining two fixed points, which has the least potential energy in a uniform gravitational **field**? (Or, alternatively, which curve has the lowest center of mass?) If the two fixed endpoints are at the same height, the catenary joining them looks roughly like a U or a **parabola**; however, it is not a parabola. While parabolas have the shape of the curve $y = x^2$, catenaries have the shape of the graph of the hyperbolic **cosine** function, $y = \cosh(x)$.

Curiously, catenaries also arise in an unrelated problem: if two circles lie in parallel planes, both centered on an axis perpendicular to the planes, what is the surface of least **area** joining them? The surface can be found experimentally by making the circles out of wire and dipping them in a soap solution. When the circles are removed, the soap will form a film with the two circles as boundary, and this film will have the least possible area because that is the configuration with the least surface energy. If the two circles are far apart, the soap will simply form two separate disks. But if the circles are close together, the soap will form a concave tube linking the two circles. This tube, called a catenoid, has the same shape as a catenary rotated about the central axis.

When a surface has less area than any other surface with the same boundary, it is called a *minimal surface.* Catenoids were among the first minimal surfaces to be known by an explicit formula, and they are the only minimal surfaces of revolution. The study of minimal surfaces has been an active area of mathematical research for more than 150 years, and many new minimal surfaces have been discovered in recent years thanks to computer graphics.

Catenaries also are useful in solving a more whimsical problem of recent vintage. What kind of road will allow a **square** wheel to roll smoothly? To accomplish this, the axle must move along a horizontal line even though the road itself is not level. The solution is a road made of cobblestones in the shape of upside-down catenaries, joined together at 90-degree angles. The side length of the square wheel is the same as the arc length of each catenary "cobblestone."

See also Hyperbolic functions

CAUCHY, AUGUSTIN–LOUIS (1789-1857)
French number theorist

Augustin–Louis Cauchy brought **formalism** to mathematics in the 19th century and defined the concepts of **limits, continuity,** and derivatives, familiar to modern–day students of **calculus.** He made contributions to **number theory,** developed ideas on the **convergence** of **infinite series,** and contributed to the field of astronomy. Cauchy's most significant contribution to mathematics, however, is as the co–founder, along with **Karl Gauss,** of complex **analysis.**

Less than two months after the fall of the Bastille, on August 21, 1789, Augustin–Louis Cauchy, the son of Louis–France Cauchy and Marie–Madeleine Desestre, was born in Paris. It was a time of great turmoil in France, and Cauchy's father, having served the recently overthrown monarchy as a parliamentary lawyer and a lieutenant of police, thought his new family would be safer if they moved to the countryside. Shortly after Cauchy's birth, they took up residence in the village of Arceuil. Cauchy spent the first 11 years

Augustin–Louis Cauchy

of his life there, enduring food scarcity and suffering from malnutrition. His frail condition kept him from being physically active and he spent much of his childhood studiously poring over books. Because schools had been shut down after the revolution, Cauchy was educated primarily by his father, whose lessons emphasized the study of religion and French and Latin verse. Cauchy's mathematical talents were recognized and encouraged by a family friend, one of the leading mathematicians of his day, **Pierre Laplace**. Laplace, too, had retreated to the countryside after the storming of the Bastille, and owned a large estate not far from the Cauchy's cottage.

In 1800, Cauchy's father accepted a position in the new French government as Secretary of the Senate. The family returned to Paris, where Cauchy soon became exposed to another leading mathematician, **Joseph–Louis Lagrange.** Impressed with the young boy's mathematical ability, Lagrange predicted that Cauchy would someday supplant both he and Laplace in mathematical prominence. Cauchy entered the Central School of the Pantheon at the age of 13, taking top honors. In 1805, he enrolled at the École Polytechnique, and two years later entered a college of civil engineering. By 1810 his superior academic performance had earned him a high ranking commission in Cherbourg as a military engineer.

Although Cauchy's first job was demanding, he still found time to begin an exhaustive study of mathematics and

astronomy. By 1813, when he moved back to Paris, he had already written one important treatise on symmetrical **functions** and another on polyhedra. In 1814, he published a manuscript on definite integrals with complex–number limits, independently deriving and more fully developing the theory of functions of a complex **variable** that had been established by Gauss only a few years earlier. This 1814 publication was to lay the foundation for a more lengthy treatment of **complex analysis** 12 years later.

In 1815, Cauchy proved a **theorem** of **Pierre de Fermat**'s that had defeated many of his distinguished predecessors, including **Leonhard Euler**, Lagrange, and **Adrien–Marie Legendre**. That same year he took a post at the École Polytechnique, lecturing on analysis. Then he turned himself to **applied mathematics**, winning the Grand Prize of the French Académie Royale des Sciences in 1816 for his work on the propagation of waves. Shortly afterward Cauchy was appointed a member of the Académie and was promoted to the position of professor at the Ecole Polytechnique. He also received an appointment at the College de France and the Sorbonne. Cauchy soon became known for the clarity of his mathematical presentations, especially on the subject of calculus. He was not yet 30 years old, and mathematicians from across Europe were traveling to Paris to attend his classes. In 1821, at the urging of Lagrange and others, Cauchy began publishing his lectures. His efforts resulted in three major textbooks, *Cours d'Analyse de l'Ecole Polytechnique* in1821, *Resume des Lecons sur le Calcul Infinitesimal* in 1823, and *Lecons sur les Applications de Calcul Infinitesimal a la Geometrie*, from 1826 to 1828. In these works, Cauchy established the definitions of limits and continuity and discussed conditions for convergence of infinite series. In his third book he also developed the field of complex analysis, which he had first written about in 1814. The material Cauchy set forth in his texts has stood the test of time and would be familiar to any student of calculus today. Through his definitions and proofs of theorems, Cauchy brought rigor to the field of calculus. Many believe these three volumes are the most important of Cauchy's writings.

During his mathematical career, Cauchy established a reputation not only for his lectures but for his prodigious output. He sometimes generated two full–length manuscripts a week. In 1826, he founded a personal journal, *Exercises de Mathematique*, superseded in 1830 by *Exercise d'Analyse Mathematique et de Physique*. In these publications Cauchy presented his expositions on various topics in **mathematics and physics**. His articles often ran several hundred pages in length. In 1835, when the Académie began publishing its weekly bulletin, the *Comptes Rendu*, Cauchy began flooding this publication with his memoirs as well. Eventually printing costs got out of hand and the Académie had to limit manuscripts to four pages. Over the course of his career, Cauchy wrote more mathematical works than anyone in the **history of mathematics**, other than Leonhard Euler.

In 1830, King Charles X was unseated from the throne in France and went into exile. Cauchy, who was loyal to the king, refused to take an oath of allegiance to the new government. He too went into exile, first to Switzerland, where he lived in a Jesuit community, then to Italy, where he was appointed professor of mathematical physics at the University of Turin. In 1833, Charles invited Cauchy, supposedly as a reward for his loyalty, to become the tutor to his 13–year–old grandson, the Duke of Bordeaux. Charles also bestowed upon Cauchy the title of Baron. Leaving Turin to tutor the Duke in Prague, Cauchy now entered a relatively unproductive phase of his mathematical career. His most important work during this time was a memoir in physics, pertaining to the **dispersion** of **light**. Cauchy returned to France in 1838 and resumed his post at the Académie in Paris, where a special dispensation from oaths of allegiance had, by now, been granted for its members. Cauchy's productivity soared. Over the remaining 19 years of his life, he was to publish approximately 500 manuscripts in areas ranging from pure mathematics to astronomy to **mechanics**.

Soon after Cauchy had returned to the Académie, a position at the College de France became available, which he was asked to fill. There was no special dispensation at the College regarding oaths of allegiance, so Cauchy turned down the post and went to work instead for the Bureau des Longitudes. There, the fact that he had not taken an oath of allegiance was temporarily overlooked. Over the next four years Cauchy made a variety of contributions to the field of astronomy, including a description of the **motion** of the asteroid Pallus. Eventually, the government attempted to oust him from the Bureau and Cauchy became embroiled in a political stand–off. The situation culminated in an open letter to the French citizenry regarding freedom of conscience and the government eventually retreated. Cauchy joined the Sorbonne faculty in 1848, where he remained until 1852.

On May 23, 1857, in the countryside near Paris, Cauchy died. He had left the city in order to recuperate from a bronchial condition, but developed a fever that eventually consumed him. Cauchy married Aloise de Bure, the daughter of a cultured family, in 1818. Together they had two children.

CAUCHY CONDITION

The Cauchy condition, or Cauchy criterion as it is sometimes called, describes the necessary and sufficient condition that needs to exist for a sequence to converge. It was developed by **Augustin Louis Cauchy**, a French mathematician, in the early part of the 19th century and hence bears his name. The Cauchy condition provides method for testing the **convergence** of a sequence of points.

Suppose we have a function that yields the points (x_n, y_n). Also suppose that as n increases the points (x_n, y_n) approach a single point whose coordinates are (a, b). This is what is defined in mathematics as convergence. When the points (x_n, y_n) approach the point (a, b) they get closer to each other. In order for the sequence of points to converge there must be an index N such that for $n \geq$ N and $m \geq$ N and any **distance** d, the distance between (x_n, y_n) and (x_m, y_m) is always less than d. This is called the Cauchy property and is

extremely important in testing convergence in mathematics. Any sequence that converges has the Cauchy property, and any sequence having the Cauchy property converges. If the Cauchy property is not met then the sequence is said to diverge. Although this test for convergence was developed in the early 19th century by Cauchy it can to this day still be found in any carefully written book on **calculus**.

CAUCHY'S INTEGRAL THEOREM

Cauchy's integral **theorem** allows for integration over a complex **variable**. It deals only with analytic **functions**, that is, those functions which have only a single value at each point in the region of interest. These functions have the same **limits** on a point no matter which direction that point is approached. The analyticity of a function can be tested with the Cauchy-Riemann **equations**, which relate the first derivatives of the real and imaginary components of a function in terms of the real and imaginary components of the variable. If a function is analytic, Cauchy's integral theorem states that the integral between two points is always the same regardless of what path is taken.

To state Cauchy's theorem another way, any closed curve in an analytic region integrates to **zero**. This means that a function can be integrated around a **circle**, a **square**, or a totally arbitrary closed curve and always get the same result: zero.

Cauchy's integral theorem also deals with functions which are not analytic, usually because of a discontinuity caused by attempting to divide by zero somewhere in the function's **range**. In this case, the result of the integration around the closed curve is 2 times **pi** times the **square root** of negative one (also denoted as i), times the residue of the function at the nonanalytic point. (The point of non-analyticity is sometimes called a pole.) The residue is calculated by decomposing the function into an analytic and a non-analytic part, and then taking the value of the analytic part at the pole.

This theorem may also cover situations where an entire half of the Cartesian plane—positive or negative values of the real or imaginary component—may be the integral surface. In that situation, all poles and discontinuities in the half-plane are accounted for, and the curve used must disappear at the appropriate (positive or negative) **infinity**.

This theorem is often used in physics and engineering applications. Most of the real-world functions scientists and engineers deal with work in at least two dimensions. Quantum **mechanics** especially requires many integrals in a two-dimensional coordinate system whose axes can be thought of as corresponding with real and **imaginary numbers**. Also, many quantum mechanical problems require an integration out to infinity to take into account all of **space** and time. In this case, Cauchy's integral theorem is one of the best ways to solve the problem, even if the original idea is formulated so that it's only one-dimensional. Taking an integral out to infinity and accounting for divisions by zero are the theorem's most useful features.

CAVALIERI, BONAVENTURA (1598-1647)

Italian geometer, physicist, and theologian

Bonaventura Cavalieri refined early Greek work on the concept of indivisibles. His work served as a stepping stone to the concept of infinitesimals and was the foundation of **Isaac Newton**'s development of the **calculus**. Cavalieri was born in Milan, Italy, in 1598. His actual birth date is uncertain; even his first name is unknown, because he entered a monastic order at an early age and adopted the first name Bonaventura as his religious name.

Cavalieri entered the Jesuatis (a Roman Catholic order founded on the rule of St. Augustine and not to be confused with the Jesuits or Society of Jesus) and took minor orders in 1615 at a monastery in Milan. In 1616, he transferred to a monastery in Pisa, where he met Benedetto Castelli, a former pupil of **Galileo**. Castelli introduced Cavalieri to the study of **geometry** and later introduced him to Galileo himself. From that point on, Cavalieri considered Galileo his mentor and teacher. During their long association, Cavalieri wrote more than a hundred letters to Galileo, which have survived in the national edition of the *Le opere di Galileo Galilei*.

Cavalieri's entry into religious life was typical of the route many aspiring scholars took in that era. Monasteries were centers of learning and operated the most comprehensive libraries; the Roman Catholic Church wielded tremendous power over education and the dissemination of information. (For instance, Galileo was forced by the Inquisition to repudiate his 1632 treatise that clarified the Copernican theory on the movement of the Earth around the sun. The Church did not officially lift the ban on treatises that treated as fact such celestial movement until 1828.) As a cleric, Cavalieri had access to the classic texts of **Euclid**, **Archimedes**, and **Apollonius**, and his status as a monk served as an introduction to the finest minds of his time.

In 1621, Cavalieri was ordained as a deacon to Cardinal Federigo Borromeo, whose regard for scholarship and appreciation of mathematics encouraged Cavalieri. The cardinal's esteem for Cavalieri's abilities made it possible for him to teach theology at the monastery of San Girolamo in Milan while still in his early twenties. It was during this time (1620 to 1623), that Cavalieri first began his work on the method of indivisibles. From 1623 to 1629 he served as prior of St. Peter's at Lodi and at the Jesuati monastery in Parma. On a trip to Milan, he fell ill with an attack of gout and was confined to Milan for several months. Perhaps the forced relaxation from his duties as prior gave him the opportunity to concentrate on his works in mathematics, because he told Galileo and Cardinal Borromeo that he had completed his *Geometria* in December of 1627.

In 1628, Cavalieri sought Galileo's help in securing a teaching post at the University of Bologna. Galileo wrote to a patron of the institution that "few, if any, since Archimedes, have delved as far and as deep into the science of geometry." Cavalieri won the academic appointment and was named the first chair in mathematics in 1629, a post he held until his death.

At the same time, his order appointed him prior of the Church of Santa Maria della Mascarella in the city, combining his responsibilities to the Church with convenient access to the university. Cavalieri was able to pursue his academic and theological ambitions, and this period was a fruitful one. While in Bologna, he published 11 books, including *Geometria* in 1635.

Cavalieri's inspiration was Archimedes, who first proposed the notion—unexamined for centuries—that indivisibles could be used to determine areas, volumes, and centers of gravity. Cavalieri developed a rational system that employed indivisibles to determine **area** and **volume**, a method that made calculations easier and quicker than the ancient Greek method of exhaustion. Cavalieri's **Theorem** states that if two solids have equal altitudes, and if sections made by planes parallel to the bases and at equal distances from the bases always has a given ratio, then volumes of the solids have the same ratio to each other. Cavalieri also refined a general **proof** of Guldin's theoremrelated to the area of a surface and the volume of rotating solids.

Cavalieri's work had tremendous impact on the works of his contemporaries. Torricelli, who expanded on Cavalieri's concepts, wrote "the geometry of indivisible was, indeed, the mathematical briar bush, the so–called royal road, and one that Cavalieri first opened and laid out for the public as a device of marvelous invention." **Pierre de Fermat, Blaise Pascal, Issac Barrow** and Newton were influenced by Cavalieri's work, which, according to Isaac Asimov, was "a stepping–stone toward... the development of the calculus by Newton, which is the dividing line between classical and modern mathematics."

CAYLEY ALGEBRA

Cayley **algebra** is the branch of the non-commutative algebras dealing with matrices. English mathematician **Arthur Cayley** developed it during 1840 to 1890. It is the only non-associative **division** algebra with real scalars. Division algebra is a type of algebra in which every nonzero element has a multiplicative inverse but where **multiplication** is non-commutative, that is $x * y \neq y * x$.

Arthur Cayley, considered a British mathematician although he practiced law the first 14 years of his professional career, was part of the movement in the 19th century by British mathematicians to study algebras. During these studies of various sorts of mathematical objects Cayley turned his attention to matrices and assorted operations that could be performed on them. This was a time when the scope of algebra was expanded and not limited to ordinary systems of numbers alone and one of the most important developments was the formulation of non-commutative algebras, those in which the operation of multiplication is not required to be commutative. **William Rowan Hamilton** was the first to develop such an algebra in 1843 with **quaternions**. Quaternions are an important example of a four-dimensional **vector space** that was employed by Einstein to develop his theory of special relativity. Two years before, Cayley had begun the first English contribution to the theory of **determinants**. During the late 1840s

Cayley went to a lecture Hamilton was giving on quaternions. He continued his development of **matrix** algebra and in 1853 Cayley published an article giving the inverse of a matrix. In the late 1850s Cayley formally introduced matrix algebras when he published *Memoir on the theory of matrices* in 1858 and noticed that quaternions could be represented by matrices. He went on to show that the coefficient arrays studied previously for quadratic forms and for linear transformations are special cases of his general concepts. In his book Cayley developed methods for carrying out **addition**, multiplication, scalar multiplication and inverses on matrices. He gave specifics of relating the construction of the inverse of a matrix in terms of the determinant of that matrix. He developed the theory of algebraic invariance and his work on matrices was the basis for **quantum mechanics** developed by **Werner Heisenberg** in 1925.

Cayley algebra is a non-associative algebra in that it does not satisfy $a(bc) = (ab)c$. It is a division algebra in which there is an eight-square identity and the elements are called Cayley numbers or octonions. Cayley numbers are typically of the form: $a + bi_0 + ci_1 + di_2 + ei_3 + fi_4 + gi_5 + hi_6$. Each of the triples (i_0, i_1, i_3), (i_1, i_2, i_4), (i_2, i_3, i_5), (i_3, i_4, i_6), (i_4, i_5, i_0), (i_5, i_6, i_1), (i_6, i_0, i_2) behaves like the quaternions (i, j, k). As mentioned before quaternions are a set of symbols that have the form: $a + bi + cj + dk$, where $a, b, c,$ and d are **real numbers**, and they multiply using the rules: $i^2 = j^2 = k^2 = -1$ and $ij = k$. These numbers used in Cayley algebra are not associative. They have been employed in the study of seven and eight-dimensional space. Cayley algebra has particularly important used in physical applications.

Arthur Cayley

CAYLEY, ARTHUR (1821-1895)

English algebraist and geometer

Arthur Cayley was one of the most prolific mathematicians of the 19th century. He was one of the individuals responsible for elevating English mathematics of the era to a position of visibility and authority. Although his best known for his work in **algebra**, Cayley was influential in generalizing **geometry**, particularly n–dimensional geometry and non–Euclidean geometry, in British mathematics and mathematical education.

Cayley was born on August 16, 1821, in Richmond, Surrey, England and spent his early childhood in Russia. His father, Henry Cayley, was a merchant who was based in St. Petersburg, Russia, and it has been claimed that Cayley's mother, Maria Antonia Doughty, was of Russian ancestry. On his father's side of the family, Cayley could trace his ancestry back to a Norman who had come to England with William the Conqueror in the 11th century.

Cayley's academic career was uniformly distinguished. At the age of 14 he entered King's College School and proceeded to Trinity College, Cambridge, when he was 17. In 1842, he was Senior Wrangler, placing first on the final university examinations in mathematics. So convincing was Cayley's performance that the examiners did not subject

him to the standard oral examination. At age 20 he was elected a fellow of Trinity, the youngest of any candidate in that century.

Cayley was reluctant to take holy orders, however, and in light of the regulations of the time he was required to give up his fellowship after a certain period. Cayley entered Lincoln's Inn, one of the Inns of Court for the training of prospective lawyers in 1846 and was admitted to the bar in 1849. For the duration of his legal career, Cayley limited his practice to conveyancing, and wrote about 300 mathematical papers during the 14 years of his legal practice. In 1852, Cayley was elected to the Royal Society.

The branch of mathematics to which Cayley devoted the most time was **invariant** theory. This involved looking at algebraic expressions, transforming them by classes of **functions**, and seeing what remained the same after the **transformation**. The subject lent itself to enormous amounts of calculation, an area in which Cayley excelled. In a series of writings dated from 1854 to 1878, Cayley not only performed a vast number of calculations but interpreted them for usage in other areas.

In 1863, Cayley was offered the Sadlerian chair of pure mathematics at Cambridge University, which he held until his death. It is a tribute to his sense of duty to the university that

Cayley put his knowledge of the law to help draft new statutes as necessary. That same year he married Susan Moline, by whom he had a son and a daughter.

Cayley's name is associated with some of the changes that were sweeping the mathematical community of the era. He was the first to use the term "n–dimensional geometry," recognizing the extent to which going beyond three dimensions was simply a matter of adding another **variable** to an algebraic expression. In geometry, perhaps Cayley's greatest contribution was developing a method for calculating distances in projective geometry. This quantity, called a cross–ratio, made it easier to consider projective geometry as a generalization of ordinary Euclidean geometry rather than as an entirely different field.

Cayley essentially created the theory of matrices, which has proved to be fundamental to **mathematics and physics** ever since. A **matrix** is a way of representing a linear transformation by an array of numbers, and Cayley developed an algebra of matrices to represent combinations of such transformations. Every matrix is associated by a **polynomial** whose **roots** reflected the behavior of the linear transformation that the matrix expressed. One of Cayley's striking results was that if one treated the matrix itself as an algebraic object, it was a solution of the equation derived by setting this characteristic polynomial to **zero**. From this **theorem** all sorts of matrix calculations can be simplified.

Cayley launched the new subject of group theory in a more abstract way than had been recognized previously. Aspects of **group theory** can be traced back to the French mathematician **Èvariste Galois**, but the groups Galois and his French successors considered were of particular kinds of objects. Cayley recognized that all types of objects could make up a group. A basic result which bears his name is that every group with a finite number of members has the same structure as a group of permutations. There was no need for such a theorem in the days when all groups were considered to be made up of permutations.

Cayley wrote on a variety of subjects that filled the 13 volumes of his collected works. He was quite a pure mathematician, but still pursued problems in dynamics as a source of problems in pure mathematics. Cayley received many awards and honors, including the De Morgan medal given by the London Mathematical Society and the Copley medal of the Royal Society in 1881. He was a guest lecturer at the Johns Hopkins University in the United States in 1881 and served as president of the British Association in 1883. Cayley was also an avid mountain climber and enjoyed painting in watercolors.

Cayley died on January 26, 1895. In light of his prolific composition, it is not surprising that there is some repetition in his papers. Nevertheless, the number of contributions that Cayley made to mathematics is staggering. His ability to reduce complicated expressions to manageable size made invariant theory the talk of the mathematical community and carried over to many other disciplines as well.

CENTER OF MASS

The center of mass is a fairly self-explanatory term: it is the location that is in the direct center of an object or system, in terms of the mass of that object or system. For uniform objects, the center of mass is a geometric center, but this concept can also take into account objects whose composition varies, changing the way the mass is distributed. It also covers systems which are only connected by forces: the Earth-moon system's center of mass can be calculated, taking into account the mass of both celestial bodies.

Center of mass can be calculated for discrete systems or for continuous systems. For discrete systems, the mass of each piece of the system times the vector position of that piece is summed over all the pieces of the system. Then the result is divided by the total mass of the system, giving a vector position that is defined as the center of mass. For continuous systems, this discrete sum is changed to a continuous integral, so one integrates over the mass function (as dependent upon position) times a position function, over the **limits** of the object or system.

The center of mass can be a useful physical or engineering concept, as single forces can be applied at the center of mass to move or contain an object without torque. If the center of mass calculations are not used, the forces to move or contain the object must be balanced. In the earth's gravitational well, center of mass is sometimes dealt with in the form of the center of gravity.

See also Force

CENTRAL LIMIT THEOREM

The central limit **theorem** is the name of a fundamental result in **probability theory** in which the so-called bell-shaped curve appears. Suppose, for example, that a fair coin is tossed N times. By a *fair* coin we understand an ideal coin in which the probability of getting heads is 1/2, and so the probability of getting tails is also 1/2. It is easy to compute the probability that after N tosses there will be exactly M heads, and so exactly N-M tails. When a fair coin is tossed N times we can record the outcomes as series of H's and T's. Thus after tossing a fair coin 10 times we might have an outcome such as T H H T T T H T H T. If we toss a coin N times the number of outcomes that contain exactly M heads is equal to the number of ways of selecting M positions among N possibilities in which the letter H will occur. This is given by the binomial coefficient

$$\binom{N}{M} = \frac{N!}{\{M!\}\{(N-M)!\}}.$$

Next we must divide this number by the total number of possible outcomes of any kind that can occur when we toss a fair coin N times. The total number of possible outcomes is clearly 2^N. Therefore the probability that there are exactly M heads when a fair coin is tossed N times is given by

$$2^{-N} \begin{pmatrix} N \\ M \end{pmatrix}.$$

Before we go further, consider the case in which the fair coin is tossed $N = 1000$ times. Since we are using a *fair* coin we might expect that the most likely number of heads to occur is 500. This is in fact the most likely number of heads, but what is the probability that exactly 500 heads will occur? By the previous formula the probability is only

$$2^{-1000} \begin{pmatrix} 1000 \\ 500 \end{pmatrix} = \frac{1000!}{2^{1000} \{500!\}^2} = 0.025225\dots.$$

Now suppose that we wish to determine an **interval** of **integers** close to 500 for which we can be much more confident that after 1000 tosses the number of heads will be in our chosen interval. How is this interval to be selected and what is the probability that the number of heads will be in the interval? For example, suppose that we select the interval [490, 510]. The probability that after 1000 tosses of a fair coin the number of heads will be greater than or equal to 490 and less than or equal to 510 is given by

$$2^{-1000} \sum_{M=490}^{510} \begin{pmatrix} 1000 \\ M \end{pmatrix} = 0.493340\dots.$$

The central limit theorem allows us to make approximate calculations of this sort when the number N tends to ∞.

Now let α and β be **real numbers** with $\alpha < \beta$. Assume as before that we toss a fair coin N times. We wish to determine the probability that the number of heads is in the interval $[\frac{1}{2}N + \frac{1}{2}\alpha\sqrt{N}, \frac{1}{2}N + \frac{1}{2}\beta\sqrt{N}]$. By our previous remarks, the probability that the number of heads is in this interval is exactly

$$2^{-N} \sum_{\frac{1}{2}N + \frac{1}{2}\alpha\sqrt{N} \le M \le \frac{1}{2}N + \frac{1}{2}\beta\sqrt{N}} \begin{pmatrix} N \\ M \end{pmatrix}$$

It turns out that as $N \to \infty$ this probability converges to a relatively simple expression that depends only on the numbers α and β. The precise limit is given by

$$\lim_{N \to \infty} 2^{-N} \sum_{\frac{1}{2}N + \frac{1}{2}\alpha\sqrt{N} \le M \le \frac{1}{2}N + \frac{1}{2}\beta\sqrt{N}} \begin{pmatrix} N \\ M \end{pmatrix} = \frac{1}{\sqrt{2\pi}} \int_{\alpha}^{\beta} e^{-x^2} \, dx.$$

This is the central limit theorem as it applies to tossing a fair coin. The graph of

$$y = \frac{1}{\sqrt{2\pi}} e^{-x^2}$$

is the so-called bell shaped curve. In probability theory it is known as the normal or Gaussian density function. The area under this curve is exactly one unit. The function

$$\Phi(x) = \frac{1}{\sqrt{2\pi}} \int_{\infty}^{x} e^{-t^2} \, dt$$

is the normal or Gaussian distribution function.

We have illustrated the central limit theorem with a simple example in which we have counted the number of heads that occur when a fair coin is tossed. But the central limit theorem applies also to much more general and complicated situations involving sums of independent random variables, and the Gaussian distribution function continues to appear in the limit.

CENTROIDS

The centroid, also called the **center of mass**, is one of the ways of defining the geometric center of a planar figure or a solid. If the planar figure was cut out of stiff paper, the centroid would be the point at which you could balance the object on a pin. The centroid of a solid is not as easy to visualize, but its coordinates also provide a center of balance. It is the point at which there is an equal amount of the figure in each direction, for as many degrees of freedom as are allowed.

To calculate each coordinate centroid, one takes the integral of the density function multiplied by the coordinate over the entire **area** or **volume** in consideration, divided by the total "mass" of the shape. For figures that consist of discrete points rather than a continuous **geometry**, the integral may be replaced by a sum over the individual points with mass weighting for each, divided by the sum of the masses. For equal masses at all points, the centroid is simply the average position of the figure. The centroid also is the point at which the total travel from all vertices of the figure is minimized. The centroid of a **triangle** is the point at which medians (lines from a vertex to the opposite **midpoint**) meet; they always meet at one point, although it is not always within the triangle. The centroid of a uniform quadrilateral is found at the point where the bimedians (lines joining opposite midpoints) intersect. Other geometric figures have well-defined **median** positions, but these are more complex calculations and are used far less often.

CHAIN RULE

The chain rule is a method of differentiating a function which is composed of two nested **functions**. That is, the chain rule provides a general case for taking the x derivative of a function f(g(x)). The chain rule states that the derivative of such a nested function is the product of the derivatives of each function, evaluated at appropriate points. That is, the derivative of such a function is the derivative of f, assuming that g is itself the **variable**, multiplied by the derivative of g, assuming that x is the variable.

For example, the function $\sin(x^2)$ can be evaluated using the chain rule. The derivative of $\sin(y)$ is $\cos(y)$, and the derivative of x^2 is $2x$. Therefore, the derivative of $\sin(x^2)$ is $2x \cos(x^2)$. The term that is particularly of interest is the argument of the **cosine** term. The only trick to the chain rule is remembering that the original argument of the outer function must be the argument of its derivative as well.

The chain rule is one of the basic rules of differentiation. It would be possible to step through the limit **definition** of a derivative every time a composite or nested function came up, but the chain rule provides a much more efficient method and is widely used wherever **calculus** is applied.

CHANDRASEKHAR, SUBRAHMANYAN

(1910-1995)

Indian-born American astrophysicist and applied mathematician

Subrahmanyan Chandrasekhar was an Indian-born American astrophysicist and applied mathematician whose work on the origins, structure, and dynamics of stars has secured him a prominent place in the annals of science. His most celebrated work concerns the radiation of energy from stars, particularly white dwarf stars, which are the dying fragments of stars. Chandrasekhar demonstrated that the radius of a white dwarf star is related to its mass: the greater its mass, the smaller its radius. Chandrasekhar made numerous other contributions to astrophysics. His expansive research and published papers and books include topics such as the system of energy transfer within stars, stellar evolution, stellar structure, and theories of planetary and stellar atmospheres. For nearly twenty years, he served as the editor-in-chief of the *Astrophysical Journal,* the leading publication of its kind in the world. For his immense contribution to science, Chandrasekhar received numerous awards and distinctions, most notably, the 1983 Nobel Prize for Physics for his research into the depths of aged stars.

Chandrasekhar, better known as Chandra, was born on October 19, 1910, in Lahore, British India (now Pakistan), the first son of C. Subrahmanyan Ayyar and Sitalakshmi née (Divan Bahadur) Balakrishnan. Chandra came from a large family; he had two older sisters, four younger sisters, and three younger brothers. As the firstborn son, Chandra inherited his paternal grandfather's name, Chandrasekhar. His uncle was the Nobel Prize-winning Indian physicist, Sir C. V. Raman.

Chandra received his early education at home, beginning when he was five. From his mother he learned Tamil, from his father, English and **arithmetic**. He set his sights upon becoming a scientist at an early age, and to this end, undertook at his own initiative some independent study of **calculus** and physics. The family moved north to Lucknow, in Uttar Pradesh, when Chandra was six. In 1918, the family moved again, this time south to Madras. Chandrasekhar was taught by private tutors until 1921, when he enrolled in the Hindu High School in Triplicane. With typical drive and motivation, he studied on his own and steamed ahead of the class, completing school by the age of fifteen.

After high school, Chandra attended Presidency College in Madras. For the first two years, he studied physics, chemistry, English, and Sanskrit. For his B.A. honors degree he wished to take pure mathematics but his father insisted that he take physics. Chandra resolved this conflict by registering as an honors physics student but attending mathematics lectures. Recognizing his brilliance, his lecturers went out of their way to accommodate Chandra. Chandra also took part in sporting activities and joined the debating team. A highlight of his college years was the publication of his paper, "The Compton Scattering and the New Statistics." These and other early successes while he was still an eighteen-year-old undergraduate only strengthened Chandra's resolve to pursue a career in scientific research, despite his father's wish that he join the Indian civil service. A meeting the following year with the German physicist **Werner Heisenberg**, whom Chandra, as the secretary of the student science association, had the honor of showing around Madras, and Chandra's attendance at the Indian Science Congress Association Meeting in early 1930, where his work was hailed, doubled his determination.

Upon graduating with a M.A. in 1930, Chandra set off for Trinity College, Cambridge, as a research student, courtesy of an Indian government scholarship created especially for him (with the stipulation that upon his return to India, he would serve for five years in the Madras government service). At Cambridge, Chandra turned to astrophysics, inspired by a theory of stellar evolution that had occurred to him as he made the long boat journey from India to Cambridge. It preoccupied him for the next ten years. He also worked on other aspects of astrophysics and published many papers.

In the summer of 1931, he worked with physicist Max Born at the Institut für Theoretische Physik at Göttingen in Germany. There, he studied **group theory** and **quantum mechanics** (the mathematical theory that relates matter and radiation) and produced work on the theory of stellar atmospheres. During this period, Chandra was often tempted to leave astrophysics for pure mathematics, his first love, or at least for physics. He was worried, though, that with less than a year to go before his thesis exam, a change might cost him his degree. Other factors influenced his decision to stay with astrophysics, most importantly, the encouragement shown him by astrophysicist Edward Arthur Milne. In August 1932, Chandra left Cambridge to continue his studies in Denmark under physicist **Niels Bohr**. In Copenhagen, he was able to devote more of his energies to pure physics. A series of Chandra's lectures on astrophysics given at the University of Liège, Belgium, in February 1933 received a warm reception. Before returning to Cambridge in May 1933 to sit for his doctoral exams, he went back to Copenhagen to work on his thesis.

Chandrasekhar's uncertainty about his future was assuaged when he was awarded a fellowship at Trinity College, Cambridge. During a four-week trip to Russia in 1934, where he met physicists Lev Davidovich Landau, B. P. Gerasimovic, and Viktor Ambartsumian, he returned to the work that had led him into astrophysics to begin with, white dwarfs. Upon returning to Cambridge, he took up research of white dwarfs again in earnest.

As a member of the Royal Astronomical Society since 1932, Chandra was entitled to present papers at its twice monthly meetings. It was at one of these that Chandra, in 1935, announced the results of the work that would later make his name. As stars evolve, he told the assembled audience, they emit energy generated by their conversion of hydrogen into helium and even heavier elements. As they reach the end of their life, stars have progressively less hydrogen left to convert and emit less energy in the form of radiation. They eventually reach a stage when they are no longer able to generate the pressure needed to sustain their size against their own gravitational pull and they begin to contract. As their density increases during the contraction process, stars build up sufficient internal energy to collapse their atomic structure into a degenerate state. They begin to collapse into themselves. Their electrons become so tightly packed that their normal activity is suppressed and they become white dwarfs, tiny objects of enormous density. The greater the mass of a white dwarf, the smaller its radius, according to Chandrasekhar. However, not all stars end their lives as stable white dwarfs. If the mass of evolving stars increases beyond a certain limit, eventually named the *Chandrasekhar limit* and calculated as 1.4 times the mass of the sun, evolving stars cannot become stable white dwarfs. A star with a mass above the limit has to either lose mass to become a white dwarf or take an alternative evolutionary path and become a supernova, which releases its excess energy in the form of an explosion. What mass remains after this spectacular event may become a white dwarf but more likely will form a neutron star. The neutron star has even greater density than a white dwarf and an average radius of about 0.18 m (0.15 km). It has since been independently proven that all white dwarf stars fall within Chandrasekhar's predicted limit, which has been revised to equal 1.2 solar masses.

Unfortunately, although his theory would later be vindicated, Chandra's ideas were unexpectedly undermined and ridiculed by no less a scientific figure than astronomer and physicist Sir Arthur Stanley Eddington, who dismissed as absurd Chandra's notion that stars can evolve into anything other than white dwarfs. Eddington's status and authority in the community of astronomers carried the day, and Chandra, as the junior, was not given the benefit of the doubt. Twenty years passed before his theory gained general acceptance among astrophysicists, although it was quickly recognized as valid by physicists as noteworthy as Wolfgang Pauli, **Niels Bohr**, Ralph H. Fowler, and **Paul Dirac**. Rather than continue sparring with Eddington at scientific meeting after meeting, Chandra collected his thoughts on the matter into his first book, *An Introduction to the Study of Stellar Structure,* and departed the fray to take up new research around stellar dynamics. An unfortunate result of the scientific quarrel, however, was to postpone the discovery of black holes and neutron stars by at least twenty years, and Chandra's receipt of a Nobel Prize for his white dwarf work by fifty years. Surprisingly,

Subrahmanyan Chandrasekhar

despite their scientific differences, he retained a close personal relationship with Eddington.

Chandra spent from December 1935 until March 1936 at Harvard University as a visiting lecturer in cosmic physics. While in the United States, he was offered a research associate position at Yerkes Observatory at Williams Bay, Wisconsin, starting in January 1937. Before taking up this post, Chandra returned home to India to marry the woman who had waited for him patiently for six years. He had known Lalitha Doraiswamy, daughter of Captain and Mrs. Savitri Doraiswamy, since they had been students together at Madras University. After graduation, she had undertaken a masters degree. At the time of their marriage, she was a headmistress. Although their marriage of love was unusual, neither of their families had any real objections. After a whirlwind courtship and wedding, the young bride and groom set out for the United States. They intended to stay no more than a few years, but, as luck would have it, it became their permanent home.

At the Yerkes Observatory, Chandra was charged with developing a graduate program in astronomy and astrophysics and with teaching some of the courses. His reputation as a teacher soon attracted top students to the observatory's graduate school. He also continued researching stellar evolution, stellar structure, and the transfer of energy within stars. In 1938, he was promoted to assistant professor of astrophysics. During this time Chandra revealed his conclusions regarding the life paths of stars.

During the World War II, Chandra was employed at the Aberdeen Proving Grounds in Maryland, working on ballistic tests, the theory of shock waves, the Mach effect, and transport problems related to neutron diffusion. In 1942, he was promoted to associate professor of astrophysics at the University of Chicago and in 1943, to professor. Around 1944, he switched his research from stellar dynamics to radiative transfer. Of all his research, the latter gave him, he recalled later, more fulfillment. That year, he also achieved a lifelong ambition when he was elected to the Royal Society of London. In 1946, he was elevated to Distinguished Service Professor. In 1952, he became Morton D. Hull Distinguished Service Professor of Astrophysics in the departments of astronomy and physics, as well as at the Institute for Nuclear Physics at the University of Chicago's Yerkes Observatory. Later the same year, he was appointed managing editor of the *Astrophysical Journal,* a position he held until 1971. He transformed the journal from a private publication of the University of Chicago to the national journal of the American Astronomical Society. The price he paid for his editorial impartiality, however, was isolation from the astrophysical community.

Chandra became a United States citizen in 1953. Despite receiving numerous offers from other universities, in the United States and overseas, Chandra never left the University of Chicago, although, owing to a disagreement with Bengt Strömgren, the head of Yerkes, he stopped teaching astrophysics and astronomy and began lecturing in mathematical physics at the University of Chicago campus. Chandra voluntarily retired from the University of Chicago in 1980, although he remained on as a post-retirement researcher. In 1983, he published a classic work on the mathematical theory of black holes. Following work on this book, he studied colliding waves and the Newtonian two-center problem in the framework of the general theory of relativity. His semi-retirement also left him with more time to pursue his hobbies and interests: literature and music, particularly orchestral, chamber, and South Indian. Chandrasekhar died on August 21, 1995, at the University of Chicago Hospitals.

During his long career, Chandrasekhar received many awards. In 1947, Cambridge University awarded him its Adams Prize. In 1952, he received the Bruce Medal of the Astronomical Society of the Pacific, and the following year, the Gold Medal of the Royal Astronomical Society. In 1955, Chandrasekhar became a Member of the National Academy of Sciences. The Royal Society of London bestowed upon him its Royal Medal seven years later. In 1962, he was also presented with the **Srinivasa Ramanujan** Medal of the Indian National Science Academy. The National Medal of Science of the United States was conferred upon Chandra in 1966; and the Padma Vibhushan Medal of India in 1968. Chandra

received the Henry Draper Medal of the National Academy of Sciences in 1971 and the Smoluchowski Medal of the Polish Physical Society in 1973. The American Physical Society gave him its Dannie Heineman Prize in 1974. The crowning glory of his career came nine years later when the Royal Swedish Academy awarded Chandrasekhar the Nobel Prize for Physics. ETH of Zurich gave the Indian astrophysicist its Dr. Tomalla Prize in 1984, while the Royal Society of London presented him with its Copley Prize later that year. Chandra also received the R. D. Birla Mcmorial Award of the Indian Physics Association in 1984. In 1985, the Vainu Bappu Memorial Award of the Indian National Science Academy was conferred upon Chandrasekhar. In May 1993, Chandra received the state of Illinois's highest honor, Lincoln Academy Award, for his outstanding contributions to science.

While his contribution to astrophysics was immense, Chandra always preferred to remain outside the mainstream of research. He described himself to his biographer, Kameshar C. Wali, as "a lonely wanderer in the byways of science." Throughout his life, Chandra strove to acquire knowledge and understanding, according to an autobiographical essay published with his Nobel lecture, motivated "principally by a quest after perspectives."

CHANG, SUN-YUNG ALICE (1948-)
American mathematician

Sun-Yung Alice Chang, working with Paul Yang, Tom Branson, and Matt Gursky, has produced what the American Mathematical Society has termed "deep contributions" to the study of partial **differential equations** in relation to **geometry** and **topology**.

Sun-Yung Alice Chang was born in Ci-an, China, on March 24, 1948, and studied for her bachelor's degree at the National University of Taiwan, which she received in 1970. Chang emigrated to the United States for graduate work and a series of teaching jobs. In 1974, Chang earned her Ph.D. from the University of California at Berkeley. She has served as assistant professor at SUNY-Buffalo, the University of California at Los Angeles, and the University of Maryland. Chang returned to UCLA as an associate professor, and she became a full professor in 1980. Her most visible performance was as a speaker at the International Congress of Mathematicians, held in Berkeley in 1986.

In 1995, Chang was awarded the third Ruth Lyttle Satter Prize at the American Mathematical Society's 101st annual meeting in San Francisco (the previous two recipients were Lai-Sung Young and Margaret McDuff). Young was on the selection committee that recommended Chang for that year's prize. In her acceptance speech, Chang acknowledged her debt to her collaborators and promised to "derive further geometric consequences" in various problems currently under study.

Reflecting momentarily on her own school years, Chang admitted that it had been important for her to have role models and female companionship. However, she stated, the deciding factor in the future will be to have more women proving theorems and contributing to mathematics as a whole. To that end, she has joined with a number of steering committees and advisory panels at the national level. After being a Sloan Fellow for the National Academy of Sciences in 1980, she returned ten years later as a member of their Board of Mathematical Sciences. Chang also advised the National Science Foundation and the Association for Women in Mathematics throughout the early 1990s.

Chang always finds time to involve her students in the sometimes arcane world of her specialty. At the University of Texas, she took part in their Distinguished Lecturer Series of 1996-97, a program that successfully targets an audience of young graduate students. She has been most active with the American Mathematical Society, working on a range of committees and speaking at a number of their meetings. Chang's most current professional positions include a three-year term on the Editorial Boards Committee of the AMS, expiring in 1998.

Chang was married in 1973 and has two children.

CHAOS THEORY

Chaos theory is the study of non-linear dynamic systems, that is, systems of activities (weather, turbulence in fluids, the stock market) that cannot be visualized in a graph with a straight line. Although dictionaries usually define "chaos" as "complete confusion," scientists who study chaos have discovered deep patterns that predict global stability in dynamic systems in spite of local instabilities.

Isaac Newton and the physicists of the 18th and 19th centuries who built upon his work showed that many natural phenomena could be accounted for in **equations** that would predict outcomes. If enough was known about the initial states of a dynamic system, then, all things being equal, the behavior of the system could be predicted with great accuracy for later periods, because small changes in initial states would result in small changes later on. For Newtonians, if a natural phenomenon seemed complex and chaotic, then it simply meant that scientists had to work harder to discover all the variables and the interconnected relationships involved in the physical behavior. Once these variables and their relationships were discovered, then the behavior of complex systems could be predicted.

But certain kinds of naturally occurring behaviors resisted the explanations of Newtonian science. The weather is the most famous of these natural occurrences, but there are many others. The **orbit** of the moon around the Earth is somewhat irregular, as is the orbit of the planet Pluto around the sun. Human heartbeats commonly exhibit minor irregularities, and the 24-hour human cycle of waking and sleeping is also irregular.

In 1961, Edward N. Lorenz discovered that one of the crucial assumptions of Newtonian science is unfounded. Small changes in initial states of some systems do not result in small changes later on. The contrary is sometimes true: small initial changes can result in large, completely random changes later. Lorenz's discovery is called the butterfly effect: a butterfly beating its wings in China creates small turbulences that eventually affect the weather in New York.

Lorenz, of MIT, made crucial discoveries in his research on the weather in the early 1960s. Lorenz had written a computer program to **model** the development of weather systems. He hoped to isolate variables that would allow him to forecast the weather. One day he introduced an extremely small change into the initial conditions of his weather prediction program: he changed one **variable** by one one-thousandth of a point. He found that his prediction program began to vary wildly in later stages for each tiny change in the initial state. This was the birth of the butterfly effect. Lorenz proved mathematically that long-term weather predictions based upon conditions at any one time would be impossible.

Mitchell Feigenbaum was one of several people who discovered order in chaos. He showed mathematically that many dynamic systems progress from order to chaos in a graduated series of steps known as scaling. In 1975 Feigenbaum discovered regularity even in orderly behavior so complex that it appeared to human senses as confused or chaotic. An example of this progression from order to chaos occurs if you drop pebbles in a calm pool of water. The first pebble that you drop makes a clear pattern of concentric circles. So do the second and third pebbles. But if the pool is bounded, then the waves bouncing back from the edge start overlapping and interfering with the waves created by the new pebbles that you drop in. Soon the clear concentric rings of waves created by dropping the first pebbles are replaced by a confusion of overlapping waves.

Feigenbaum and others located the order in chaos: apparently chaotic activities occur around some point, called an attractor because the activities seem attracted to it. This accounts for the clear, visual structure to what would otherwise appear to be disorder.

James Yorke applied the term "chaos" to non-linear dynamic systems in the early 1970s. But before Yorke gave non-linear dynamical systems their famous name, other scientists had been describing the phenomena now associated with chaos.

Chaos theory has a variety of applications. One of the most important of these is the stock market. Some researchers believe that they have found non-linear patterns in stock indexes, unemployment patterns, industrial production, and the price changes in Treasury bills. These researchers believe that they can reduce to six or seven the number of variables that determine some stock market trends. However, the researchers concede that if there are non-linear patterns in these financial areas, then anyone acting on those patterns to

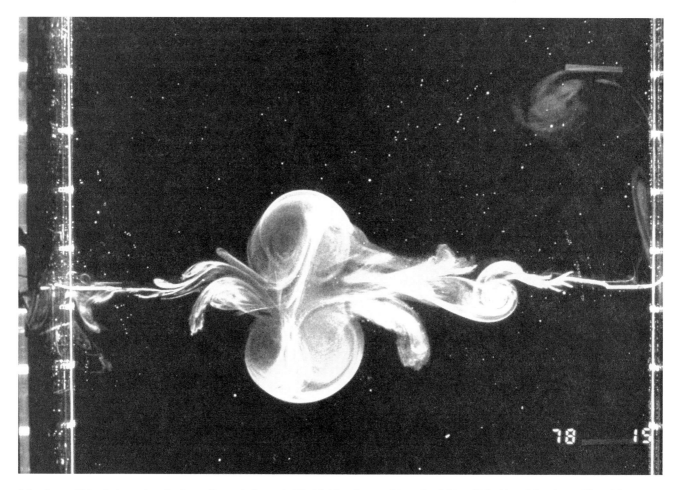

A head-on collision between two dipolar vortices entering a stratified fluid environment from the right and left sides of the picture. The original vortices have exchanged some of their substance to form two new mixed dipoles that are moving at roughly right angles to the original direction of travel (toward the top and bottom of the photo). Dipolar vortices are relevant to turbulence in large-scale geophysical systems like the Earth's atmosphere or oceans. Turbulence within a fluid is an example of a chaotic system.

profit will change the market and introduce new variables which will make the market unpredictable.

Population biology illustrates the deep structure that underlies the apparent confusion in the surface behavior of chaotic systems. Some animal populations exhibit a boom-and-bust pattern in their numbers over a period of years. In some years there is rapid growth in a population of animals, followed by a bust created when the population consumes all of its food supply and most members die from starvation. Soon the few remaining animals have an abundance of food because they have no competition. Since the food resources are so abundant, the few animals multiply rapidly, and some years later, the booming population turns bust again as the food supplies are exhausted from overfeeding. This pattern, however, can only be seen if many data have been gathered over many years. Yet this boom-and-bust pattern has been seen elsewhere, including disease epidemics. Large numbers of people may come down with measles, but in falling ill, they develop antibodies which protect them from future outbreaks. Thus, after years of rising cases of measles, the cases will sud-

denly decline sharply because so many people are naturally protected by their antibodies.. After a period of reduced cases of measles, the outbreaks will rise again and the cycle will start over, unless a program of inoculation is begun.

Chaos theory can also be applied to human biological rhythms. The human body is governed by the rhythmical movements of many dynamical systems: the beating heart, the regular cycle of inhaling and exhaling air that makes up breathing, the circadian rhythm of waking and sleeping, the saccadic (jumping) movements of the eye that allow us to focus and process images in the visual field, the regularities and irregularities in the brain waves of mentally healthy and mentally impaired people as represented on electroencephalo-grams. None of these dynamic systems is perfect all the time, and when a period of chaotic behavior occurs, it is not necessarily bad. Healthy hearts often exhibit brief chaotic fluctuations, and sick hearts can have regular rhythms. Applying chaos theory to these human dynamic systems provides information about how to reduce sleep disorders, heart disease, and mental disease.

CHÂTELET, GABRIELLE–ÉMILIE LE TONNELIER DE BRETEUIL DU (1706-1749)

French physicist, chemist, and translator

Gabrielle–Émilie Châtelet had a major role in the scientific revolution of the eighteenth century. By popularizing the theories of **Isaac Newton** she brought them more widespread acceptance in Europe where most people still followed the ideas of **René Descartes**. Châtelet's scientific work has been ignored and overshadowed by her relationship with the philosopher Voltaire.

Born Gabrielle–Émilie Le Tonnelier de Breteuil in Paris on December 17, 1706 into an aristocratic family, the marquise du Châtelet received an exceptional education at home, which included scientific, musical, and literary studies. In 1725, she married the marquis du Châtelet, who was also the count of Lomont. It was a marriage of convenience, but she nevertheless had three children with him. After spending some years with her husband, whose political and military career kept him away from Paris, Châtelet returned to the capital in 1730.

Initially leading a busy social life, Châtelet became the lover of the philosopher François–Marie Arouet de Voltaire in 1733. One of the greatest intellectual figures of 18th–century France, Voltaire recognized her exceptional talent for science, and encouraged her to develop her intellect. Châtelet consequently embarked on a study of mathematics, taking private lessons from the prominent French philosopher and scientist Pierre–Louis Moreau de Maupertuis. Both Voltaire and Maupertuis were enthusiastic supporters of Isaac Newton's scientific theories and world view, and it seems that the marquise was, as a result, immersed in Newtonian philosophy.

In 1734, Voltaire, who faced arrest because of his criticism of the monarchy, was offered sanctuary at Châtelet's chateau at Cirey, in Lorraine, where they spent many productive years. The two welcomed Europe's intellectual elite, thus creating a remarkable cultural center away from Paris. Châtelet was involved in a variety of literary and philosophical projects, eventually concentrating on the study of Newton's philosophy. She assisted Voltaire in the preparation of his 1738 book, *Elements of Newton's Philosophy*.

In 1737, Châtelet, like many other 18th–century scientists, attempted to explain the nature of combustion, submitting an essay entitled *Dissertation sur la nature et la propagation du feu*, as an entry for a contest organized by the Académie Royale des Sciences. Voltaire also participated in the contest, but was unaware of her work. When **Leonhard Euler** and two other scientists were declared the winners, Voltaire arranged that Châtelet's essay be published with the winning entries. In her study, she correctly argued that heat was not a substance, a view defended by the proponents of the phlogiston theory, which the great French chemist Antoine–Laurent Lavoisier empirically disproved in 1788. Furthermore, Châtelet put forth the original idea that **light** and heat were essentially the same substance.

Gabrielle–Émilie Châtelet

While writing her *Institutions de physique*, a work on Newtonian physics and **mechanics**, Châtelet became acquainted with the ideas of **Gottfried Leibniz**, particularly his conception of *forces vives*, which she accepted as true. While René Descartes described the physical world geometrically as extended matter, to which **force** can be applied as an external agent, Leibniz defined force as a distinctive quality of matter. In view of Châtelet's general Newtonian orientation as a scientist, her passionate interest in Leibnizian metaphysics, which essentially contradicts the Newtonian world view, may seem odd. However, as Margaret Alic argues, the marquise sought a synthesis of the two world views. "*Institutions*," Alic has written, "remained faithful to Newtonian physics, but Newton's purely scientific, materialistic philosophy did not completely satisfy the marquise. She believed that scientific theory demanded a foundation in metaphysics and this she found in Leibniz... She never doubted that Leibnizian metaphysics was reconcilable with Newtonian physics, as long as the implications of the Newtonian system were limited to empirical physical phenomena." Châtelet's acceptance of the metaphysical foundations of science was an implicit rejection of any mechanistic world view, Cartesian or Newtonian. Naturally, French scientists, most of whom tacitly accepted the Cartesian scientific paradigm, found the marquise's ideas offensive. For example, the eminent Cartesian physicist and mathematician Jean–Baptist Dortous de Mairan, whom she had singled out for criticism, responded sharply in 1741, representing a majority view which Châtelet was unable to refute alone.

Retreating from the philosophical war between the Cartesians and the Leibnizians, Châtelet focused on her Newtonian studies, particularly the huge task of translating Newton's *Principia mathematica* into French, an undertaking which she devoted the rest of her life. An excellent Latinist with a deep understanding of Newtonian physics, she was ideally suited for the project. Despite many obstacles, which included a busy social life and an unwanted pregnancy at the age of 42, Châtelet finished her translation. On September 4, 1749, she gave birth to a daughter, and died of puerperal fever shortly thereafter. Her translation of Newton's work remains one of the monuments of French scientific scholarship.

CHEBYSHEV, PAFUNTY LVOVICH (1821-1894)

Russian probabilist and analyst

Chebyshev has given his name to results in probability and **analysis**, one of the first Russian mathematicians by birth to be so recognized. His work reflected a great deal of mathematical sophistication, making connections between different areas and generalizing techniques. He played a primary role in establishing a viable mathematical curriculum at St. Petersburg University, which laid the foundations for subsequent achievements in Russian mathematics.

Pafnuty Lvovich Chebyshev was born on May 16, 1821 in Okatovo in the Kaluga region of Russia. His father, Lev Pavlovich Chebyshev, was a former army officer. In 1832, the family moved to Moscow, where Chebyshev was educated at home under the instruction of an author of popular arithmetics. As a result, he was well prepared when he enrolled in Moscow University in 1837. There, Chebyshev studied physics and mathematics.

In 1841, Chebyshev graduated with a bachelor's degree in mathematics and within two years he had passed his master's examination. Chebyshev's thesis, entitled "An Essay on an Elementary Analysis of the Theory of Probability," dealt with the derivation of a law of large numbers (one of a whole group of results that indicated the increasing reliability of experimental results the larger the number of trials) using the methods of analysis of which he had already shown himself a master. In general, Chebyshev was looking for derivations of the leading results of probability by methods that could not be faulted for rigor, but which were not dependent on mathematical ideas that seemed out of proportion to the depth of the subject.

Chebyshev could not find employment in Moscow. As in the German university system, it was necessary to produce a thesis to earn the privilege of teaching and Chebyshev's thesis examined integration by means of **logarithms**, a topic straight out of analysis. After its acceptance, Chebyshev joined the faculty at St. Petersburg University in 1847, lectur-

ing on **algebra**, **number theory**, and probability. In addition to his teaching, he helped prepare a new edition of **Leonhard Euler**'s papers on number theory. Not far removed from this task was the subject of the theory of congruences, which Chebyshev dealt with in his doctoral thesis. He defended the thesis in 1849 and received a prize from the Russian Academy of Sciences.

Between the years 1850 and 1860 Chebyshev spent much of his time working on questions of mechanical engineering. During this period he moved up the academic ladder and by 1859 he was a senior academician in the St. Petersburg Academy. The subject of hinges led Chebyshev to consider problems of best approximation to a function, one of the results of which was later known as the Chebyshev **polynomials**. In addition to his own polynomials, he studied other systems of what are called orthogonal polynomials as well (the orthogonality refers to an independence they have in being needed to represent a given function).

There had been work done in Russia on probability before Chebyshev, but the number and quality of the students picked up considerably as a result of his efforts. One contributing factor was his generosity with his time, which explains the impressive list of students who chose to study with him. He kept an open house for students and continued to do so even after his retirement. In his own work, Chebyshev preferred to use elementary methods, and that may have given his work an appearance of comprehensibility.

Among the subjects to which Chebyshev contributed was the distribution of **prime numbers**. There had been much work devoted to the question of whether the apparent irregularities in the distribution of prime numbers (any number only divisible by one and itself) disappeared as one looked at larger initial segments of the positive **whole numbers**. Chebyshev was able to get a decent approximation for the number of prime numbers less than a fixed number compared to known **functions** of that fixed number, but he did not prove that there was a limiting value. He did, however, demonstrate a conjecture that if n is greater than 3, there is always at least one prime between n and 2 *n*–2.

In addition to his own teaching, Chebyshev was also active in improving the quality of **mathematics and physics** teaching on a national basis. He built a calculating machine in the late1870s, although this was more as a demonstration of the potential usefulness of mechanical devices than as a genuine aid to calculation. He was perhaps the creator of the tradition in Russia of making probability a part of the general mathematical curriculum. In light of the profound contributions made to the subject by **Andrei Kolmogorov**, this was no slight step in the development of probability.

Chebyshev's virtuosity as an analyst enabled him to derive results in probability that had previously been used without being well understood. He was able to take advantage of material ndeveloped by the French probabilist I.J. Bienaymé to produce a convincing demonstration of the law of large numbers. Chebyshev was recognized at home and abroad, being the first Russian to be elected a foreign member of the Paris Académie des Sciences. At the time of his death on

December 8, 1894, the Russian mathematical tradition was stronger than ever before, thanks to his work and that of his students.

CHERN, SHIING-SHEN (1911-)
Chinese mathematician

Shiing-Shen Chern has specialized in differential **geometry** and he studied what are now known as the Chern characteristic classes in fibre spaces. This work has relevance in mathematics as well as mathematical physics. Chern also produced a **proof** of the Gauss-Bonnet formula during the mid 1940s.

Chern was born in what is now known as Jiaxing in Zhejiang in China. After initial studies at Nankai University, Chern undertook graduate studies at the Tsing Hua University of Beijing. Despite being offered a scholarship to the United States at the age of 23, he preferred to study at the University of Hamburg; this was due to a meeting two years previously with Hamburg mathematician Wilhelm Johann Eugen Blaschke, whose work he greatly admired. In 1936 Chern moved to Paris to continue working on differential geometry, this time under the guidance of **Elie-Joseph Cartan**. In 1937 Chern was returning to Tsing Hua University as a professor of mathematics when war broke out. This event thrust him into the American system where he produced one of his most famous pieces of work, a proof of the Gauss-Bonnet formula. At the end of the war he returned to China briefly until civil war broke out. To escape this event Chern returned to Princeton where he continued to work with **Oswald Veblen** and **Herman Weyl**. Chern was rapidly made professor of mathematics at the University of Chicago, where he remained for 11 years until, in 1960, he moved to the University of California, Berkeley.

The years following Chern's move to Berkley were characterized by a number of awards including the National Medal of Science (1975), the Wolf Prize (1983-1984), election as a Fellow of the Royal Society of London (1985), and honorary membership of the London Mathematical Society (1986). In 1979 a special Chern symposium was held in his honor. In a lifetime of achievement and awards the one fact that must give Chern the greatest pleasure must be his role in transforming differential geometry from a little studied area into a major topic area in mathematics.

See also Elie Joseph Cartan; Johann Carl Friedrich Gauss; Oswald Veblen; Herman Weyl

CHI SQUARE DISTRIBUTION

The chi square distribution is a **probability density function** used in **statistics** to determine the probability that numerical differences in data are significant and real (as opposed to just being due to random errors or chance). Chi is a Greek letter designated by χ. In practice, the χ^2 distribution is expressed as a **family of curves** or a table of critical χ^2 values that corre-

Shiing-Shen Chern

spond to a particular probability level. Tables of these "expected" or "theoretical" χ^2 values are published in reference books. During a χ^2 test, an "observed" χ^2 value is calculated for the data in question and compared to the table values to determine the probability that the numerical event in question is statistically significant.

The chi square distribution is closely related to another commonly used probability density function, the normal distribution, which shows the distribution of many random variables. Normal or gaussian curves are bell-shaped curves that are symmetrical about the **arithmetic mean** value and have two tails. The numerical values (e.g., measurements) are along the x-axis. The relative frequency of their occurrence is along the y-axis. Most of the values fall within the wide part of the curve near the mean. The frequency drops as values begin to deviate positively or negatively from the mean. Very high and very low values have a very low frequency of occurrence, and fall along the lower and upper tails of the **normal curve**, respectively.

Chi square distribution curves look much different, because they show the distribution of the variance of the data, not the data itself. Variance is calculated mathematically by squaring the **standard deviation**. Thus, χ^2 values can not be negative. The chi square distribution is a family of curves, because the curve shape varies depending on the number of degrees of freedom (d.f.) associated with the data in question. Degrees of freedom is an important concept in statistics and can be difficult to understand. Basically, it is a limit of the arbitrariness of a data set. For example, if three numbers sum to 50, defining two of these numbers limits the value of the other number. Thus, there are only two degree of freedom in this equation.

The chi square test can be used to test the homogeneity (or uniformity) within a data set or to determine the probability that there is a dependency relationship between two or more distinct data **sets** (e.g., in industry, the chi square test could be used to determine the probability that a particular machine in a group breaks down too often or that production increases when a raw material is changed).

A goodness-of-fit test based on χ^2 values determines the probability that a data set fits a defined pattern. For example, consider a large data set with a known frequency distribution. A small subset of it will have a different frequency distribution than the parent group. As the subset size increases, its frequency distribution more closely resembles that of the parent group. The chi square test can be used to determine what sample size will provide a reasonable approximation of the larger set.

The first step in any statistical test is to establish two hypotheses (educated guesses) about the data in question that can be accepted or rejected. The first hypothesis is called the null hypothesis and is designated H_0. For a goodness-of-fit test, the null hypothesis is that the data in question follow a particular pattern (e.g., H_0: The data set is uniform). The second hypothesis is called the alternative hypothesis and is designated by H_1. It is the opposite of the null hypothesis (e.g., H_1: The data set is not uniform).

Consider a simple example for testing data homogeneity. A die is rolled 120 times, and the number of occurrences of the values 1 through 6 is $x_1=17$, $x_2=19$, $x_3=16$, $x_4=18$, $x_5=16$, and $x_6=34$. Is this die equally balanced? The hypotheses are as follows:

- H_0: This is a uniform data set. Thus, variations are explained by chance and coincidence.
- H_1: This data set is not uniform, and variations are statistically significant.

The χ^2 value is calculated from: $\chi^2=\Sigma((Fo-Fe)^2/Fe)$ where Fo is the observed frequency and Fe is the expected frequency. The expected frequency for a uniform distribution is the mean (usually called xbar) of the data set or 120/6=20. Each value would have been expected to occur 20 times during 120 rolls. $(Fo-Fe)^2$ is the sum of the squares of the deviations from the mean or $\Sigma(x_i-x)^2=(17-20)^2+(19-20)^2+(16-20)^2+(18-20)^2+(16-20)^2+(34-20)^2=9+1+16+4+16+196=242$. Therefore, $\chi^2=242/20=12.1$. The degrees of freedom is one less than the number of categories or 6-1=5.

Compare this to a table of critical χ^2 values printed below. For a given d.f., the table provides the critical χ^2 value associated with a given probability level, i.e., the probability of obtaining that chi square value simply by chance. These probability levels are often called α values, referring to the **area** under the upper tail of the corresponding curve. In gen-

eral, a probability level of 0.05 is considered the threshold of significance in the scientific community. If the probability level associated with a calculated χ^2 value is less than or equal to 0.05, then the variance in the data is too great to occur by chance alone, and H_0 can be rejected with a high degree of confidence.

CRITICAL VALUES OF CHI SQUARE FOR A GIVEN PROBABILITY					
d.f.	0.900	0.750	0.250	0.050	0.010
1	0.016	0.102	1.323	3.841	6.635
2	0.211	0.575	2.773	5.991	9.210
3	0.584	1.213	4.108	7.815	11.345
4	1.064	1.823	5.385	9.488	13.277
5	1.610	2.675	8.626	11.071	15.086

From the chi square table, the probability of obtaining χ^2 greater than or equal to 12.1 (d.f.=5) just due to chance, if the null hypothesis is true, is less than 0.05. Therefore, the null hypothesis can be rejected with confidence. The variations in this data set are statistically significant. In other words, there is something "funny" about this die.

CHINESE REMAINDER THEOREM

We have an unknown number of things from which, when we count them by threes, we have two left over. If we count them by fives, we have three left over. If we count them by sevens, we have two left over. How many things are there?

The above problem is found in the "Sun Tzu Suan Ching," an early Chinese textbook on **arithmetic** dating from about the third century A.D. The book goes on to solve this particular problem and to give a general recipe to solve problems of this kind, which later acquired the name "Chinese remainder problems." Similar problems and solutions are also found in ancient Indian and Byzantine texts. These problems originated in astronomical questions, in which one is asked to predict the simultaneous occurrence of a number of periodic events with different periods.

More precisely, if one is given two or more **integers** as divisors, which are relatively prime in pairs (such as 3, 5, and 7 in the example from the "Sun Tzu Suan Ching"), and also remainders corresponding to each divisor (such as 2, 3, 2 in the example from the "Sun Tzu Suan Ching"), then the Chinese remainder **theorem** guarantees the existence of an integer which, when divided by the given divisors, leaves the corresponding remainders. Moreover, any two solutions to the problem will differ by a multiple of the product of all the divisors. In the example from the "Sun Tzu Suan Ching," one solution is 23, and any other solution differs from 23 by a multiple of 105.

The traditional way to find the solution is to express the question as a linear diophantine equation which can then be solved by the Euclidean **algorithm**. After **J. C. F. Gauss** introduced the notion of congruences, or **modular arithmetic**, the Chinese remainder theorem has often been stated as a theorem guaranteeing the existence of a solution to a system of congruences modulo pairwise relatively prime integers.

See also Euclid's algorithm

CHURCH, ALONZO (1903-1995)
American mathematician and logician

Alonzo Church was an American mathematician and logician who provided significant innovations in **number theory** and decision theory, the foundation of computer programming. His most important contributions focus on the degrees of decidability and solvability in logic and mathematics.

Church was born in Washington, D.C., on June 14, 1903, to Samuel Robbins Church and Mildred Hannah Letterman Church. He took his undergraduate degree from Princeton University in 1924. On August 25, 1925, he married Mary Julia Kuczinski. They had three children: Alonzo, Mary Ann, and Mildred Warner. Church completed his Ph.D. in mathematics at Princeton in 1927. After receiving his doctorate, he was a fellow at Harvard from 1927 to 1928. He studied in Europe from 1928 to 1929 at the University of Göttingen, a prestigious center for the study of **mathematics and physics**. He taught **mathematics and philosophy** at Princeton from 1929 to 1968. Among his Ph.D. students at Princeton was the British mathematician **Alan Turing**, who was to crack the German's World War II secret code, called Enigma, which significantly contributed to the Allied victory over Nazi Germany. In 1944, Church published his influential *Introduction to Mathematical Logic*. Upon his retirement, however, he became professor of philosophy and mathematics at the University of California, where he remained until his second retirement 1990. Church later moved to Hudson, Ohio, to be near his son. He died in Hudson on August 11, 1995. In addition to his work as a teacher, Church edited the *Journal of Symbolic Logic* from 1936 to 1979. His wife died in February 1976.

A very private person, Church led a quiet life. As Andrew Hodges said in his biography of Alan Turing (Church's famed student who killed himself in 1954 after being arrested for his homosexuality), Church "[is] a retiring man himself, not given to a great deal of discussion."

One of the key problems concerning foundations of mathematics was stated by the German mathematician **David Hilbert** (1862–1943) when he asked whether the **arithmetic** is consistent. The belief that formal systems, such as arithmetic, are consistent had been the cornerstone of mathematics for more than 2,000 years, and Hilbert devoted much energy to the task of showing the power of formal systems. The foundational idea of consistent formal systems, crucial to both mathematics and logic, was shattered in 1931, when **Kurt Gödel** published his epoch-making article "On the Formal Undecidability Thesis of *Principia Mathematica* and Related Systems." In essence, Gödel demonstrated that **proof** of consistency cannot be found within a formal system. Indeed, one can look for proof outside the system, perhaps within a larger system, but this still would not solve the problem of inconsis-

tency, because there would be no way of proving that the larger system is consistent. Influenced by Gödel's work, Church provided the proof, in 1936, that elementary quantification theory, the basic method that logicians use to express general statements, is not **decidable**. This means that in elementary quantification theory, there is no method, containing a finite number of steps, of proving a given statement.

For computer programs to run, programmers have to be able to reduce all problems to the kinds of simple binary logical (or on/off) statements that can be processed by the electronic circuits inside the computer. For a problem to be solvable by a computer, it must be possible to break it down into an operational set of rules and terms. Next it must be possible to apply these rules recursively—that is repeatedly—to the problem until it is solved in terms of the existing set of rules. In short, a computer's binary circuits can only solve a problem under three conditions: (1) if the problem can be expressed as a meaningful set of rules (i.e., meaningful to the computer); (2) if the result of each step is also meaningful in terms of the computer's predefined set of rules; (3) if the computer's set of rules can be applied repeatedly to the problem. For example, in a simple **addition** or **subtraction** computer program, it must be possible for a small number (e.g., 1) to be repeatedly added to or subtracted from a larger number (e.g., 100) to get some result, say 10 or 10,000. If any of these three conditions mentioned above is absent, then a computer program cannot solve the problem.

Church's contribution to the foundation of computer programming is that he discovered—as did Alan Turing and Emil Post simultaneously and independently—the importance of recursiveness in solving logical problems. That is, for calculations to take place, some actions (e.g., adding or subtracting) have to be repeated a certain number of times. In 1936, the same year he shook the foundations of logic, Church formulated the thesis that every intuitively calculable function is recursive. (which is often called the Church-Turing thesis) is that a function is computable or calculable if it is recursive. That is, the idea of recursiveness (repeatability) is tightly bound up with computability. **Church's thesis** is important because the repetition of a simple action can result in significant changes. It also means that one simple action can be useful over a broad range of problems, and at different levels of a problem.

Church's contributions to decidability theory earned him many honors, including **induction** into the National Academy of Science and the American Academy of Arts and Sciences. He received honorary doctorates from Case Western Reserve University in 1969, Princeton University in 1985, and the State University of New York at Buffalo in 1990.

CHURCH'S THESIS

Alonzo Church (1903-1995) was an American mathematician and logician who made groundbreaking contributions to mathematical logic, recursion theory, and computer science. Church's Thesis is a statement about the notion of "effective" or "mechanical" or "computable" procedures, so before stating the thesis itself, we shall first give an explanation of what "effective" or "mechanical" or "computable" procedures are. These three terms will be considered synonymous and for simplicity we shall use "effective" throughout this article. In essence, an effective procedure is a procedure which, in principle, could be carried out by a human being with nothing more than pencil and paper in a finite number of steps with no insight or ingenuity on the part of the human being. That is to say, the human being need only follow, without deviation, a finite fixed set of instructions to obtain the result of the procedure. Such a procedure is also called an **algorithm**. Very simple examples of effective procedures would be the algorithms for **addition**, **subtraction**, **multiplication**, and **division** of **whole numbers**. Such procedures have been of interest to mathematicians at least since the time of **Leibniz** (1646-1716), who thought that all reasoning could be ultimately broken down into steps simple enough for a machine to carry out. Leibniz's dream, unfulfilled even to this day, was to set up such a system that would eliminate the need for human beings to argue about logic and reasoning. Every question could be answered in one and only one way by simply following a prescribed sequence of instructions. Leibniz was, of course, foreshadowing the computers that would come to dominate life in the late twentieth and early twenty-first centuries. The modern quest to create "artificial intelligence" is a sophisticated version of Leibniz's desire to make reasoning and logic purely mechanical. And today's quest to build computers that "think" would not have reached its lofty, if incomplete, status were it not for the work of Church, **Kurt Gödel** (1906-1978), and **Alan Turing** (1912-1954).

Gödel introduced the notion of a **recursive function** into mathematical logic. A recursive function is a function which may be derived by a finite number of specified mechanical operations from one or more of the following **functions**: (1) the **zero** function Z for which Z(k)=0 for all k; (2) the successor function S, where S(k)=k+1 for all k; and (3) the identity function I for which I(k) = k for all k. Thus, the recursive functions may be computed by humans in a finite number of specified mechanical steps using nothing more than pencil and paper. Today, we would call such a mechanical procedure a "program" and we might have a computer execute it. The English mathematician, Alan Turing, in the 1930s, anticipated such a scenario, conceptualizing a machine he called a "logical computing machine" or "LCM." Such a machine is now universally known as a "Turing Machine." Turing dreamed up his machine in an attempt to answer a question posed by the great German mathematician, **David Hilbert** (1862-1943). Hibert's question was whether there exists some method by which all mathematical questions can be decided as either true or false. Turing thought that if such a method exists, it would have to consist of only effective (or mechanical or computable) procedures, i.e., procedures which, in principle, could be carried out by a human being with nothing more than pencil and paper in a finite number of steps with no insight or ingenuity on the part of the human being. Turing's imaginary machine consisted of a paper tape (theoretically infinite in

length) divided into squares; a finite set of symbols, namely 0 and 1; a reading head that could scan the tape, moving to the right or left one **square** at a time, and inscribe or erase symbols on the tape; and a finite set of states corresponding to various configurations of the reading head. The **Turing machine** could be given a finite set of instructions via symbols on the tape and the reading head could then carry out these instructions by scanning the tape, moving to the left or right inscribing or erasing or leaving symbols unchanged as instructed. Now Church and Turing were able to show that functions which could be computed by a Turing machine were precisely Gödel's recursive functions. Church's Thesis, then, is that "Any function which can be computed by an effective procedure is recursive." This can also be stated as "Any function which can be computed by an effective procedure can be computed by a Turing Machine." This is sometimes stated as "Whatever can be done effectively, can be done by a Turing machine." It is important to understand that this is a thesis, not a **theorem**. It has not been proved nor can it be proved in the mathematical sense of "proof." It is a statement which associates an informal human concept, the effective procedure, with a formal, mathematically defined concept, recursive functions or Turing Machines. Nevertheless, a great number of effectively computable functions have been investigated and all of them have been found to be computable by a Turing machine. Therefore, Church's Thesis is widely accepted as true among logicians.

Church's Thesis and the concept of the Turing Machine form the foundation of the "digital age" in which we now live. Although Turing's machine existed only in Turing's mind as a mathematical abstraction, it was essentially a blueprint for the first digital computers that would come into being during the decade after the 1936 paper in which Church's Thesis first appeared.

Chwistek, Leon (1884-1944)
Polish philosopher, logician, painter, aesthetician

The Polish philosopher, logician, painter, and aesthetician Leon Chwistek was born in Zakopane, Poland on January 13, 1884. His father was a doctor, and his mother a pianist. As a philosopher, Chwistek was particularly noted for his opposition to the metaphysical and idealistic modes of thought that were so prevalent in his lifetime.

Chwistek grew up in the pleasant town of Tatras where he was a friend of Stanislaw Witkiewicz. As a student in Cracow, he studied **mathematics and philosophy** at the Jagellonian University at the same time that he attended the Academy of Fine Arts.

After earning a doctorate in 1906, Chwistek became a teacher at the same gymnasium where he himself had studied. Upon being awarded a scholarship, he traveled abroad to study logic and mathematics, and attended the lectures of **David Hilbert** and **Henri Poincaré**.

After the First World War, Chwistek began lecturing in mathematics at the University of Cracow, becoming a qualified lecturer there in 1928. In 1921, he published his theory on the plurality of realities. In his theory, there are four concepts in reality: natural reality, physical reality, reality of sensation, and reality of images. He argued that each of these concepts has its own sphere of applications, and none should be confused with another. Chwistek applied his ideas to the study of movements and styles in art, which he felt should be evaluated based on form, rather than reality.

As a logician and mathematician, Chwistek believed that abstract concepts existed only as names, and were themselves without objective meaning. Starting with the views of **Bertrand Russell**, Henri Poincaré and David Hilbert, Chwistek developed the field of rational semantics, which he argued could be applied to problems in philosophy, science, social theory, and art.

In 1924, he proposed his theory of constructive types. In it, the members of a class are of a higher type than the class, and there is no highest type.

Between 1930 and the outbreak of the Second World War in 1940, Chwistek was a professor of logic at the University of Lvov. In 1940, being sympathetic to Marxism-Leninism, he took political refuge in the Soviet Union, where he involved himself in scientific research and political activity.

Chwistek died in Moscow on August 20th, 1944. He is today remembered for proposing the simple theory of types to eliminate the paradoxes of the theory of classes, for his criticism of the use of existence axioms in logic and mathematics.

CIRCLE

A circle is defined as the set of all points in a plane that are a given **distance** from a fixed point in the plane. The fixed point is referred to as the center of the circle. The fixed distance, which is the distance from the center of the circle to any point on the rim of the circle, is called the radius. The center of the circle is considered a point of **symmetry**. Any line through it is called a line a symmetry. A circle has infinite lines of symmetry and one point of symmetry. Circles that have the same center, but not the same radius are called concentric circles.

The circle belongs to the group of curves known as **conic sections**. The circle can be shown as the intersection of a plane that is perpendicular to the axis of the cone and a right circular cone.

The equation of a circle is $(x - h)^2 + (y - k)^2 = r^2$, where (h,k) is the center and r is the radius. For example, in the circle described by the equation $(x - 3)^2 + (y + 2)^2 = 100$, the center is $(3,-2)$ and the radius is 10.

A circle has many different terms associated with it. Any line or line segment that passes through the center of the circle is called a **diameter**. The fixed distance in the **definition**, or the radius, is half the length of the diameter. A chord is a segment that touches the circle in two places. It is a straight-line segment. An arc is part of the **perimeter** of the circle. It is the portion of the perimeter that lies between two points on the circle. A central **angle** is an angle formed by having its vertex at the center of the circle and its sides the length of the radius of the circle.

A very important symbol associated with the circle is π. The value, often rounded to 3.141592, represents the ratio of the **circumference** of the circle to the diameter of the circle. This symbol is used heavily in the study of physics and **trigonometry**. The Greek mathematician Archimedes described the value of π between 3 1/7 and 3 10/71. There are several very useful formulas associated with circles that utilize π in their calculations. The **area** of a circle, which represents the **space** inside the circle, is calculated by multiplying π times the radius squared. The circumference, which gives the distance around a circle, is found by multiplying π by the diameter of the circle.

CIRCUMFERENCE

The general definition of circumference is a line or external boundary of a closed curvilinear figure or object. The more common definition of circumference within mathematics is the measure of the outer boundary (commonly called the **perimeter**) of an elliptical **area**, especially a circular area. The letter "C" commonly denotes the length of a circle's circumference. The circumference of a **sphere** is defined as the length of any great **circle** on the sphere. A great circle is the circle on the surface of a sphere produced by a plane that passes through the center of the sphere.

One of the earliest references to circumference was by Heraclitus of Ephesos (535-475 B.C.) when he used the word "periphereia" to mean "The beginning and end join on the circumference of the circle." Circumference most likely is a combination of the Latin "circum" (around) and "ferre" (to carry) and a derivation of "circumferentia" (to carry around) that is a Latin translation of the earlier Greek term "periphereia."

A very simple way to directly measure the approximate circumference of a circle is to wrap a string around the figure, pull the string out straight, and then measure the string's length.

A historical approach to finding an approximation to the circumference of a circle involves an iterative calculation of the perimeters of regular **polygons** inscribed in the circle. Given a circle, let p_n be the perimeter of a regular polygon of n sides inscribed in the circle. Then as n gets larger and larger, the number p_n increases. (That is, the perimeter of each term in the sequence p_n is greater than the preceding term.) From **calculus**, if a sequence of numbers is increasing (that is, if each term in the sequence is greater than the preceding term), and if the sequence is bounded (that is, if there is a number that is equal to or greater than any term in the sequence), then the sequence has a limit. With the p_n sequence possessing those attributes, the circumference of a circle, therefore, is the limit of the sequence of perimeters p_n of the inscribed regular polygons, that is, $C = \lim_{n \to \infty} p_n$.

To calculate the circumference of a circle several common terms need to be defined. The point equidistant from all the points on the circumference of a circle is called the center of the circle. The straight-line segment (or **interval**) from the center of the circle to a point on the circumference is called a radius. A **diameter** is a straight line through the center of the circle with its two ends on the circumference and of length twice that of a radius of the circle. With regards to the radius, r, or diameter, d, of a circle, the equation to solve for circumference, C, is $C = 2\pi r = \pi d$, where **pi** is a constant number that is defined as the ratio of a circle's circumference to its diameter. As an example, if the radius of a circle is measured to be 20 centimeters (cm) and the approximate value of 3.14 is used for pi, then the circumference is calculated to be "2 x 3.14 x 20 cm", or C = 125.6 cm.

See also Circle; Diameter; Perimeter; Pi; Polygon; Radius

CLAIRAUT, ALEXIS–CLAUDE (1713-1765)
French geometer

A child prodigy, Alexis Claude Clairaut studied **calculus** at age 10, wrote mathematical papers at 13, and published a mathematical work on the gauche curve at age 18. Clairaut surpassed even **Isaac Newton** in his analysis of the effects of gravity and centrifugal **force** on a rotating body such as Earth, now known as Clairaut's **theorem**.

Clairaut was born in Paris, France, on May 7, 1713. His father, Jean–Baptiste, was a mathematics teacher who recognized his child's precociousness and guided his studies. Clairaut received his entire education at home; his father tutored him in **algebra** and **geometry**.

His mother, Catherine Petit, gave birth to 20 children, but most of them did not survive. Clairaut never married, but led an active social life. He and several other young mathematicians formed a society that served as a mathematical training ground for its members, and he often assisted his friends with their studies. He also visited and corresponded with most of the leading mathematicians of the age, including **Leonhard Euler** and **Johann Bernoulli**, and members of the Académie Royale des Sciences.

Clairaut's book, *Théorie de la figure de la terre*, which he published in 1743, was said to be responsible to a great degree for the acceptance of Isaac Newton's gravitational theories. The book was the result of Clairaut's journey to Lapland in 1736, where he assisted Pierre Louis Moreau de Maupertuis, director of the exploration, in measuring the curvature of the Earth inside the arctic circle. Their successful attempt at a meridian arc measurement proved Newton's theory that the Earth's shape was an oblate ellipsoid. Clairaut demonstrated, through an experiment which timed the swings of a pendulum, how the Earth's shape could be determined.

Clairaut was also interested in celestial **mechanics**. This field of study resulted in the first accurate determination of the size of the planet Venus (two–thirds the size of Earth). His studies of that planet also calculated its gravitational effects on the Earth as compared to the moon. Clairaut's work enabled him to determine a new figure for the size of the moon in relation to the Earth. He also was able to predict how close to the sun Halley's comet would come in its **orbit**, and correctly predicted the comet's return in 1759.

Clairaut was elected to the Académie Royale des Sciences in 1731, when he was 18 years old. The Académie awarded him a prize for his work on tides in 1740, and he became associate director of the Académie in 1743. He was also made a Fellow of the Royal Society of London in 1737, the Académie's counterpart in England.

A prolific writer, Clairaut published the first complete book on solid analytical geometry, *Recherches sur les courbes à double courbure*, in 1731 at age 18. In addition to *Théorie de la figure de la terre*, he also published books on motions of the moon (*Théorie de la lune*, 1752; *Tables de la lune*, 1754) and the comets (*Théorie du mouvement des cométes*, 1760). Clairaut died at age fifty–two on May 17, 1765, following a brief illness.

CLOSURE PROPERTY

"Closure" is a property which a set either has or lacks with respect to a given operation. A set is closed with respect to that operation if the operation can always be completed with elements in the set.

For example, the set of even natural numbers, 2, 4, 6, 8,..., is closed with respect to **addition** because the sum of any two of them is another even natural number. It is not closed with respect to **division** because the quotients 6/2 and 4/8, for instance, cannot be computed without using odd numbers or **fractions**.

Knowing the operations for which a given set is closed helps one understand the nature of the set. Thus one knows that the set of natural numbers is less versatile than the set of **integers** because the latter is closed with respect to **subtraction**, but the former is not. Similarly one knows that the set of **polynomials** is much like the set of integers because both **sets** are closed under addition, **multiplication**, negation, and subtraction, but are not closed under division.

Particularly interesting examples of closure are the positive and **negative numbers**. In mathematical structure these two sets are indistinguishable except for one property, closure with respect to multiplication. Once one decides that the product of two positive numbers is positive, the other rules for multiplying and dividing various combinations of positive and negative numbers follow. Then, for example, the product of two negative numbers must be positive, and so on.

The lack of closure is one reason for enlarging a set. For example, without augmenting the set of **rational numbers** with the irrationals, one cannot solve an equation such as x2 =2, which can arise from the use of the **Pythagorean theorem**. Without extending the set of **real numbers** to include **imaginary numbers**, one cannot solve an equation such as $x^2 + 1 = 0$, contrary to the **fundamental theorem of algebra**.

Closure can be associated with operations on single numbers as well as operations between two numbers. When the Pythagoreans discovered that the **square** root of 2 was not rational, they had discovered that the rationals were not closed with respect to taking **roots**.

Although closure is usually thought of as a property of sets of ordinary numbers, the concept can be applied to other kinds of mathematical elements. It can be applied to sets of rigid motions in the plane, to vectors, to matrices, and to other things. For example, one can say that the set of three-by-three matrices is closed with respect to addition.

Closure, or the lack of it, can be of practical concern, too. Inexpensive, four-function calculators rarely allow one to use negative numbers as inputs. Nevertheless, if one subtracts a larger number from a smaller number, the **calculator** will complete the operation and display the negative number which results. On the other hand, if one divides 1 by 3, the calculator will display .333333, which is close, but not exact. If an operation takes a calculator beyond the numbers it can use, the answer it displays will be wrong, perhaps significantly so.

COHEN, PAUL (1934-   )
American logician and analyst

Paul Cohen's reputation as a mathematician has been earned at least partly because of his ability to work successfully in a number of very different fields of mathematics. He received the highly regarded Bôcher Prize of the American Mathematical Society, for example, in 1964 for his research on the Littlewood problem. Two years later he was awarded perhaps the most prestigious prize in mathematics, the Fields Medal, for his research on one of **David Hilbert**'s "23 most important problems" in mathematics, proving the independence of the continuum hypothesis.

Paul Joseph Cohen was born in Long Branch, New Jersey, on April 2, 1934, but his childhood and adolescence were spent in Brooklyn, New York. His parents were Abraham Cohen and the former Minnie Kaplan. Both parents had immigrated to the United States from western Russia (now part of Poland) while they were still teenagers. Cohen's father became a successful grocery jobber in Brooklyn.

Cohen appears to have had a natural and precocious interest in mathematics from an early age. To a large extent, he was self-educated, depending on books that he could find in the public library or that his elder sister Sylvia was able to borrow for him from Brooklyn College. He told interviewers Donald J. Albers and Constance Reid for their book *More Mathematical People* that "by the time I was in the sixth grade I understood **algebra** and **geometry** fairly well. I knew the rudiments of **calculus** and a smattering of **number theory**."

For his secondary education, Cohen attended the Stuyvesant High School in lower Manhattan, widely regarded as one of the two (along with the Bronx High School of Science) best mathematics and science high schools in the United States. In 1950, having skipped "a few grades," as he told Albers and Reid, he graduated from Stuyvesant at the age of 16. He ranked sixth in his class and received one of the 40 national Westinghouse Science Talent Search awards given that year. He then enrolled at Brooklyn College, where he remained for two years. In 1952 he was offered a scholarship

at the University of Chicago, from which he received his M.S. in mathematics in 1954 and his Ph.D. in 1958.

Until he reached Chicago, Cohen had a relatively unstructured and diverse background in mathematics. He was fairly knowledgeable in some areas that interested him especially and that he had been able to teach himself. But he was still naive about some important areas of mathematics, such as logic, in which he had never had a formal course or even any informal training. Partly through the influence of one of his professors at Chicago, Antoni Zygmund, Cohen became interested in a classical problem in **harmonic analysis** commonly known as the Littlewood problem, named for the English mathematician John Edensor Littlewood. Cohen's solution to this problem won him the American Mathematical Society's Bôcher Prize in 1964.

On receiving his degree from Chicago, Cohen accepted a position as instructor of mathematics at the Massachusetts Institute of Technology (MIT). A year later he moved to the Institute for Advanced Studies at Princeton, New Jersey, where he was a fellow from 1959 to 1961. At MIT and Princeton Cohen continued to work on problems of analysis and seemed to have found a field to which he could devote his career. That illusion soon evaporated, however. As Cohen later told Albers and Reid, he has a restless mind and is constantly looking for new fields to conquer. "I [have been] told by many people that I should stick to one thing," he said, "but I have always been too restless."

An occasion for shifting gears presented itself to Cohen soon after he was appointed assistant professor at Stanford in 1961. At a departmental lunch, Cohen's colleagues were discussing the problems of developing a "consistency proof" in logic, first suggested by **Georg Cantor** in the late 19th century. The term *consistency* in mathematics refers to the condition that any mathematical **theorem** be free from contradiction. Developing a consistency **proof** had been listed as number one on David Hilbert's 1900 list of the 23 most important problems in mathematics for the twentieth century. Although he had no specific background in the field of logic, in which the consistency proof is particularly relevant, Cohen was intrigued by the challenge. He saw it as a way of providing convincing evidence "that **set theory** is based on some kind of truth," as he told Albers and Reid.

Cohen's work on the consistency proof went forward in fits and starts over the next two years. During one period he became so discouraged that he set the work aside and concentrated on other problems. He seems to have had a glimpse of the general approach for solving his problem during a vacation with his future wife to the Grand Canyon in late 1962. Still, it was another four months before the details of that approach were worked out and a solution produced. Two years later, Cohen received his second major award in mathematics, the International Mathematics Union's Field Prize, for his work on the consistency proof.

In 1964 Cohen was promoted to the post of professor of mathematics at Stanford, a position he has held since. He continues to work on a variety of problems, including those in the fields of analysis and logic. Cohen was married to Christina

Karls, a native of Sweden, in 1963. They have three sons, Steven, Charles, and Eric. In addition to the Bôcher Prize and the Fields Medal, Cohen was awarded the Research Corporation of America Award in 1964 and the National Medal of Science in 1967.

COMBINATORICS

Combinatorics is the study of combining objects by various rules to create new arrangements of objects. The objects can be anything from points and numbers to apples and oranges. Combinatorics, like **algebra**, **numerical analysis** and **topology**, is a important branch of mathematics. Examples of combinatorial questions are whether we can make a certain arrangement, how many arrangements can be made, and what the best arrangement for a set of objects is.

Combinatorics has grown rapidly in the last two decades making critical contributions to computer science, operations research, finite **probability theory** and cryptology. Computers and computer networks operate with finite data structures and algorithms which makes them perfect for enumeration and graph theory applications. Leading edge research in areas like neural networking rely on the contribution made by combinatorics.

Combinatorics can be grouped into two categories. Enumeration, which is the study of counting and arranging objects, and graph theory, or the study of graphs.

Leonhard Euler (1701-1783) was a Swiss mathematician who spent most of his life in Russia. He was responsible for making a number of the initial contributions to combinatorics both in graph theory and enumeration. One of these contributions was a paper he published in 1736. The people of an old town in Prussia called Königsberg (now Kaliningrad in Russia) brought to Euler's attention a stirring question about moving along bridges. Euler wrote a paper answering the question called "The Seven **Bridges of Königsberg**." The town was on an island in the Pregel river and had seven bridges. A frequently asked question there at the time was "Is it possible to take a walk through town, starting and ending at the same place, and cross each bridge exactly once?" Euler generalized the problem to points and lines where the island was represented by one point and the bridges were represented by lines. By abstracting the problem, Euler was able to answer the question. It was impossible to return to the same place by only crossing each bridge exactly once. The abstract picture he drew of lines and points was a graph, and the beginnings of graph theory. The study of molecules of hydrocarbons, a compound of hydrogen and carbon atoms, also spurred the development of graph theory.

To enumerate is to count. In combinatorics, it is the study of counting objects in different arrangements. The objects are counted and arranged by a set of rules called **equivalence** relations.

One way to count a set of objects is to ask, "how many different ways can the objects be arranged?" Each change in the original arrangement is called a permutation. For example,

changing the order of the songs to be played on a compact disc (CD) player would be a permutation of the regular order of the songs. If there were only two songs on the CD, there would be only two orders, playing the songs in the original order or in reverse order, song two and then song one. With three songs on the CD, there are more than just two ways to play the music. There is the original order, or songs one, two and three (123) and in reverse order, 321. There are two orders found by flipping the first two songs or the last two songs to get 213 or 132 respectively. There are another two orders, 312 and 231, found by rotating the songs to the right or left. This gives a total of six ways to order the music on a CD with three songs. By just trying different orders, it was intuitively seen how many combinations there were. If the CD had twelve or more songs on it, then this intuitive approach would not be very effective. Trying different arrangements would take a long time, and knowing if all arrangements were found would not be easy. Combinatorics formalizes the way arrangements are found by coming up with general formulas and methods that work for generic cases.

The power of combinatorics, as with all mathematics, is this ability to abstract to a point where complex problems can be solved which could not be solved intuitively. Combinatorics abstracts a problem of this nature in a recursive way. Take the CD example, with three songs. Instead of writing out all the arrangements to find out how many there are, think of the end arrangement and ask, "for the first song in the new arrangement, how many choices are there?" The answer is any three of the songs. There are then two choices for the second song because one of the songs has already been selected. There is only one choice for the last song. So three choices for the first song times two songs for the second choice gives six possibilities for a new arrangement of songs. Continuing in this way, the number of permutations for any size set of objects can be found.

Another example of a permutation is shuffling a deck of playing cards. There are 52 cards in a deck. How many ways can the cards be shuffled? After tearing off the plastic on a brand new deck of cards, the original order of the cards is seen. All the hearts, spades, clubs, and diamonds together and, in each suit, the cards are arranged in numerical order. To find out how many ways the cards can be shuffled, start by moving the first card to any of the 52 places in the deck. Of course, leaving the card in the first place is not moving it at all, which is always an option. This gives 52 shuffles only by moving the first card. Now consider the second card. It can go in 51 places because it can not go in the location of the first card. Again, the option of not moving the card at all is included. That gives a total of 52×51 shuffles which equals 2,652 already. The third card can be placed in 50 places and the fourth in 49 places. Continuing this way find to the last card gives a total of $52 \times 51 \times 50.... \times 4 \times 3 \times 2 \times 1$ possible shuffles which equals about 81 with sixty-six zeros behind it. A huge number of permutations. Once 51 cards were placed, the last card had only one place it could go, so the last number multiplied was one. Multiplying all the numbers from 52 down to one together is called 52 **factorial** and is written 52!.

The importance of binomial coefficients comes from another question that arises. How many subsets are contained in a set of objects? A set is just a collection of objects like the three songs on the CD. Subsets are the set itself, the empty set, or the set of nothing, and any smaller groupings of the set. So the first two songs alone would be a subset of the set of three songs. Intuitively, eight subsets could be found on a three song CD by writing out all possible subsets, including the set itself.

Unfortunately, the number of subsets also gets large quickly. The general way to find subsets is not as easily seen as finding the total number of permutations of a deck of cards. It has been found that the number of subsets of a set can be found by taking the number of elements of the set, and raising the number two to that power. So for a CD with three songs, the number of subsets is just two to the third power, $2 \times 2 \times 2$, or 8.

For a deck of 52 cards, the number of subsets comes to about 45 with fourteen zeros behind it. It would take a long time to write all of those subsets down. Binomial coefficients represent the number of subsets of a given size. Binomial coefficients are written C(r;c) and represent "the number of combinations of r things taken c at a time." Binomial coefficients can be calculated using factorials or with **Pascal's triangle** as seen below (only the first six rows are shown.) Each new row in Pascal's triangle is solved by taking the top two numbers and adding them together to get the number below.

$(A+B)^0$ 1
$(A+B)^1$ 1 1
$(A+B)^2$ 1 2 1
$(A+B)^3$ 1 3 3 1
$(A+B)^4$ 1 4 6 4 1
$(A+B)^5$ 1 5 10 10 5 1

The triangle always starts with one and has ones on the outside.

So for our three song CD, to find the number of two song subsets we want to find C(3,2) which is the third row and second column, or three. The subsets being songs one and two, two and three, and one and three. Binomial coefficients come up in many places in algebra and combinatorics and are very important when working with **polynomials**. The other formula for calculating C(r;c) is r! divided by c! × (r-c)!.

Equivalence relations is a very important concept in many branches of mathematics. An equivalence relation is a way to partition **sets** into subsets and equate elements with each other. The only requirements of an equivalence relation are that it must abide by the reflexive, symmetric and transitive laws.

Relating cards by suits in the deck of cards is one equivalence relation. Two cards are equivalent if they have the same suit. Card color, red or black, is another equivalence relation. In algebra, "equals," "greater than" and "less than" signs are examples of equivalence relations on numbers. These relations become important when we ask questions about subsets of a set of objects.

A powerful application of enumeration to computers and algorithms is the recurrence relation. A sequence of numbers can be generated using the previous numbers in a sequence by using a recurrence relation. This recurrence relation either adds to, or multiplies one or more previous elements of the sequence to generate the next sequence number. The factorial, n!, is solved using a recurrence relation since n! equals n × (n-1)! and (n-1)! equals (n-1) × (n-2)! and so on. Eventually one factorial is reached, which is just one. Pascal's triangle is also a recurrence relation. Computers, being based on algorithms, are designed to calculate and count numbers in this way.

Graphs are sets of objects which are studied based on their interconnectivity with each other. Graph theory began when people were seeking answers to questions about whether it was possible to travel from one point to another, or what the shortest **distance** between two points was.

A graph is composed of two sets, one of points or vertices, and the other of edges. The set of edges represents the vertices that are connected to each other. Combinatorally, graphs are just a set of objects (the vertex set) and a set of equivalence relations (the edge set) regarding the arrangement of the objects. For example, a triangle is a graph with three vertices and three edges. So the vertex set may be (x,y,z) and the edge set (xy,yz,zx). The actual labeling of the points is not as important as fundamental concepts which differentiate graphs.

Sometimes graphs are not easy to tell apart because there are a number of ways we can draw a graph. The graph (x,y,z) with edges (xy,yz,zx) can be drawn as a **circle** with three points on the **circumference**. The lines do not have to be straight. The vertex and edge sets are the only information defining the graph. So a circle with three distinct points on the circumference, and a triangle, are the same graph. Graphs with hundreds of vertices and edges are hard to tell apart. Are they the same?

One of a number of ways to tell graphs apart is to look at their connectivity and cycles, inherent properties of graphs.

A graph is connected if every vertex can be reached to every other vertex by traveling along an edge. The triangle is connected. A **square**, thought of as a graph with four vertices, (x,y,z,w) but with only two edges (xy,zw) is not connected. There is no way to travel from vertex x to vertex z. A graph has a cycle if there is a path from a vertex back to itself where no edge is passed over twice. The triangle has one cycle. The square, as defined above, has no cycles. Graphs can have many cycles and still not be connected. Ten disconnected triangles can be thought of as a graph with ten cycles. The two properties, connectivity and cycles, do not always allow for the differentiation of two graphs. Two graphs can be both connected and have the same number of cycles but still be different.

Another four properties for determining if two graphs are different is explained in a very nice introduction to the subject, *Introduction to Graph Theory* by Richard Trudeau.

Computer networks are excellent examples of a type of graph that demonstrates how important graphs are to the computer field. Networks are a type of graph that has directions and weights assigned to each edge. An example of a network problem is how to find the best way to send information over a national computer network. Should the information go from

Washington, D.C. through Pittsburgh, a high traffic point, and then to Detroit, or should the information be sent through Philadelphia and then through Toledo to Detroit? Is it faster to go through just one city even if there is more traffic through that city?

A similar issue involving networks is whether to have a plane make a stop at a city on the way to Los Angeles from Detroit, or should the trip be non-stop. Adding factors like cost, travel time, number of passengers, etc. along with the number of ways to travel to Los Angeles leads to an interesting network theory problem.

A traditional problem for the gasoline companies has been how to best determine their truck routes for refilling their gas stations. The gasoline trucks typically drive around an area, from gas station to gas station, refilling the tanks based on some route list, a graph. Driving to the nearest neighboring gas stations is often not the best route to drive. Choosing the cheapest path from vertex to vertex is known as the greedy **algorithm**. Choosing the shortest route based on distance between locations is often not the most cost effective route. Some tanks need refilling sooner than others because some street corners are much busier than others. Plus, like the Königsberg Bridge problem, traveling to the next closest station may lead to a dead end and a trucker may have to back track. The list of examples seems endless. Graph theory has applications to all professions.

Trees are yet another type of graph. Trees have all the properties of graphs except they must be connected with no cycles. A computer's hard drive directory structure is set up as a tree, with subdirectories branching out from a single root directory. Typically trees have a vertex labeled as the root vertex from which every other vertex can be reached from a unique path along the edges. Not all vertices can be a root vertex. Trees come into importance for devising searching algorithms.

COMBINED VARIATION

In order to set the stage for a **definition** of combined variation, we will first discuss some of the more basic types of variation. A quantity y is said to vary directly as x if $y=kx$, where k is a constant, called the constant of variation. In this case, we also say that y is directly proportional to x; so we also call k the constant of proportionality. This means that as x gets larger, so does y; and as x gets smaller, so does y. We may also have y varying directly as some power of x. For example, if $y=kx^2$, then we would say that y varies directly as the **square** of x. If $y=k/x$, where k is a constant, we say that y varies inversely as x, or that y is inversely proportional to x. If y varies inversely as x, then as x gets larger, y gets smaller, and vice versa. The equation $y=k/x^3$ would tell us that y varies inversely as the cube of x. If $y=kxz$, where k is a constant and x and z are variables, then y is said to vary jointly as x and z. If y varies jointly as the cube of x and the fourth power of z, we would write $y=kx^3y^4$. Finally, if y varies directly as one **variable** and inversely as another, we call this a combined variation. For

example, $y=kx/z$ is a combined variation; here, y varies directly as x and inversely as z. In the combined variation $y=kx^2/z^3$, y varies directly as the square of x and inversely as the cube of z. Some mathematicians consider joint variation to be a special type of combined variation.

Consider the following application of a combined variation from the field of electrical circuits: the electrical resistance R of a wire varies directly as the length l of the wire and inversely as the square of its **diameter** d. This can be written in equation form as $R=kl/d^2$, where, as usual, k is the constant of variation. Suppose that we know that a wire of length 50 feet and diameter 2 millimeters has a resistance of 8 ohms. What is the resistance of a 50 foot long wire made of the same metal if the wire has a diameter of 4 millimeters? We can answer this question by using the information given. First determine the constant k. Since $R=kl/d^2$, we have $8=50k/2^2$, which give k=0.64. Then we have $R=0.64(50)/4^2=2$ ohms. Here is another example of combined variation from physics. The time T, in hours, required for a satellite to complete a circular **orbit** around the earth varies directly as the radius r of the orbit measured from the center of the earth and inversely as the orbital velocity v in miles per hour. This could be expressed by the equation $T=kr/v$, where k is the constant of variation. So if we knew that a satellite traveling at 17000 mph completes an orbit 500 miles above the earth in 100 minutes, we could use this information to calculate the constant of variation. Thus, $5/3=4500k/17000$, where we have used an approximation of 4000 miles as the radius of the earth and have converted 100 minutes to 5/3 hours. This implies that k=6.296. Now if we want to know how long it would take a satellite to complete one orbit if it is circling 800 miles above the earth at 16000 mph, then use $T=(6.296)(4800)/16000=1.89$ hours or 113.3 minutes.

COMMUTATIVE PROPERTY

"Commutativity" is a property which an operation between two numbers (or other mathematical elements) may or may not have. The operation is commutative if it does not matter which element is named first.

For example, because **addition** is commutative, 5 + 7 has the same value as 7 + 5. **Subtraction**, on the other hand, is not commutative, and the difference 5 - 7 does not have the same value as 7 – 5.

Commutativity can be described more formally. If * stands for an operation and if A and B are elements from a given set, then * is commutative if, for all such elements A * B = B * A.

In ordinary **arithmetic** and **algebra**, the commutative operations are **multiplication** and addition. The non-commutative operations are subtraction, **division**, and exponentiation. For example, x + 3 is equal to 3 + x; xy is equal to yx; and (x + 7)(x - 2) is equal to (x - 2)(x + 7). On the other hand, 4 - 3x is not equal to 3x - 4; 16/4 is not equal to 4/16; and 5^2 is not equal to 2^5.

The commutative property can be associated with other mathematical elements and operations as well. For instance, one can think of a **translation** of axes in the coordinate plane as an "element," and following one translation by another as a "product." Then, if T_1 and T_2 are two such translations, T_1T_2 and T_2T_1 are equal. This operation is commutative. If the set of transformations includes both translations and rotations, however, then the operation loses its commutativity. A **rotation** of axes followed by a translation does not have the same effect on the ultimate position of the axes as the same translation followed by the same rotation.

When an operation is both commutative and associative (an operation is associative if for all A, B, and C, (A * B) * C = A * (B * C), the operation on a finite number of elements can be done in any order. This is particularly useful in simplifying an expression such as $x^2 + 5x + 8 + 2x^2 + x + 9$. One can combine the squared terms, the linear terms, and the constants without tediously and repeatedly using the associative and commutative properties to bring like terms together. In fact, because the terms of a sum can be combined in any order, the order need not be specified, and the expression can be written without parentheses. Because ordinary multiplication is both associative and commutative, this is true of products as well. The expression $5x^2y^3z$, with its seven factors, requires no parentheses.

COMPLETING THE SQUARE

Completing the **square** is a technique used in **algebra** to create an expression which is a perfect square where none existed before. The presence of a perfect square form in an expression can often simplify the steps in an algebraic process. Perhaps the most notable use of completing the square is in the solution of quadratic **equations** and ultimately in the derivation of the famous quadratic formula. As an example, consider the quadratic equation $x^2 + 6x - 17 = 0$. The expression on the left side of the equation is not a perfect square, but since $(x+3)^2 = x^2 + 6x + 9$, adding 26 to both sides of our quadratic equation will give $x^2 + 6x + 9 = 26$, which is equivalent to $(x+3)^2 = 26$. Now the advantage of having the perfect square on the left side of this last equation is that the equation may now be solved by taking square **roots** of both sides. This gives the two equations $x + 3 = \sqrt{26}$ and $x + 3 = -\sqrt{26}$, and the two solutions $-3 + \sqrt{26}$ and $-3 - \sqrt{26}$. These solutions in decimal form, accurate to the nearest thousandth, are 2.099 and -8.099. Note that to complete the square for the expression $x^2 + 6x$ we need the number 9 added to this expression. Notice that 9 is the square of 3 and 3 is half of 6, which is the coefficient of the x or "linear" term. This pattern is true for all cases of completing the square for quadratic expressions. For example, to complete the square for $x^2 + 8x$, we would need to add 4^2 or 16 because 4 is half of 8. Thus the expression $x^2 +8x + 16 = (x+4)^2$. Suppose we wanted to complete the square for the expression $x^* - 3x$. Then half of -3 is -3/2 and $(-3/2)^2 = 9/4$, so $x^2 - 3x + 9/4$ is the perfect square $(x-3/2)^2$.

It was mentioned above that the quadratic formula for finding solutions of **quadratic equations** is derived by com-

pleting the square. Here's how. The general quadratic equation has the form $ax^2 + bx + c = 0$. The technique of completing the square illustrated in the preceding paragraph only works when the coefficient of the x^2, or quadratic term, is 1; so our first move is to divide both sides by a. This gives $x^2 + (b/a)x + c/a = 0$, which is equivalent to $x^2 + (b/a)x = -c/a$. Now complete the square on the left side of this last equation: Half of b/a is b/(2a) and $(b/(2a))^2$ is $b^2/(4a^2)^2$. Now we must add this expression to both sides of our equation, which gives $x^2 + (b/a)x + b^2/(4a^2) = b^2/(4a^2) - c/a$, and this is equivalent to $(x + b/(2a))^2 = (b^2 - 4ac)/(2a)$. Taking square roots of both sides and subtracting b/(2a) from both sides gives $x = (-b + \sqrt{(b^2 - 4ac)})/(2a)$ and $x = (-b - \sqrt{(b^2 - 4ac)})/(2a)$. These two equations comprise what is usually called the quadratic formula. In essence, this process completes the square for every quadratic equation that exists, thus giving the solutions for every quadratic equation.

Another common use of the technique of completing the square is in converting certain quadratic relations given in standard form to special forms which reveal key properties of the graphs of these relations. For instance, the equation $y = x^2 - 10x + 15$ is the so-called "standard" form equation for a **parabola**. If we subtract 15 from each side of the equation, we get the equivalent $y - 15 = x^2 - 10x$. Completing the square on the right side of this equation is accomplished by adding 25, which must also be added to the left side, giving $y + 10 = x^2 - 10x + 25$, or $y + 10 = (x - 5)^2$. Now an equation of the form $y - k = a(x - h)^2$ is called the vertex form of the equation for a parabola because, written in this form, the vertex of the parabola (h,k) can be easily read from the equation. Thus in our specific case, we can see that the vertex of the parabola $y = x^2 - 10x + 15$ is (5,-10). As another example, the equation $x^2 + 2x + y^2 + 4y - 11 = 0$ can be changed, by completing the square in both the x and y expressions, to an equivalent form which reveals that it is the equation of a **circle** with center at (-1,-2) and radius 4. Here are the steps: first add 1 and 4 to both sides giving $x^2 + 2x + 1 + y^2 + 4y + 4 = 16$, which is equivalent to $(x+1)^2 + (y+2)^2 = 4^2$. In general, a circle with center (h,k) and radius r has equation $(x-h)^2 + (y-k)^2 = r^2$, which is the form of our equation with h = -1 and k = -2 and r = 4. Similarly, standard form equations for ellipses and hyperbolas, may be converted to more useful, but equivalent, forms by the process of completing the square.

One final example illustrates the use of completing the square in **calculus**. It is not necessary to understand all of the calculus terminology to see how completing the square is used. In calculus courses, one of the traditional challenges for students is to find a function that has a given function as its derivative. This can be anything from exceedingly simple to literally impossible. Suppose we wanted to find a function whose derivative is $1/(x^2 + 2x + 2)$. Note the quadratic expression in the denominator of this fraction. Many students might think that this was one of those impossible cases. However, a student who remembers the technique of completing the square, might notice that $x^2 + 2x + 2 = x^2 + 2x + 1 + 1 = (x + 1)^2 + 1$. Seen this way, the fraction has the form $1/(u^2 + 1)$, where u = x + 1, and it turns out that $1/(u^2 + 1)$ is the derivative of arctan(u), which implies that $1/((x + 1)^2 + 1)$ is the derivative of arctan(x+1). Again without understanding what a derivative is, one can, nevertheless, see the efficacy of completing the square in this process.

COMPLEX ANALYSIS

Complex **analysis** is the study of **functions** of a complex **variable**, and especially of those functions which are differentiable. Initially, complex analysis is concerned with generalizing the basic notions of **calculus**, such as **limits**, differentiation and integration, to complex valued functions defined on an open subset in the complex plane. Many of the first results proved in complex analysis are simple extensions of familiar facts from calculus. However, a function which is differentiable at each point of an open subset in the complex plane has many additional special properties that go far beyond merely being differentiable. If D is an open subset of the complex plane **C**, if $f:D \rightarrow$ **C** is a function, we say that f is differentiable at the point z in D if the limit

$$\lim_{h \to 0} \frac{f(z + h) - f(h)}{h} = f'(z)$$

exists. If the limit exists then the value of the limit defines the derivative $f'(z)$ at the point z. If f has a derivative at each point of the open set D we say that f is *analytic* or *holomorphic* on D. It is important to understand that in the limit defining the derivative of f the parameter h is complex. Complex analysis is the study of analytic functions.

It turns out that an analytic function $f:D \rightarrow$ **C** is also continuous, and has infinitely many continuous derivatives at each point of its **domain** D. Also, a function $f:D \rightarrow$ **C** is analytic on D if and only if it has a **power series** expansion about each point in its domain. More precisely, if f is analytic on D, if w is a point in D, if $r > 0$ is such that the disk $\Delta(w; r) = \{z \in$ **C**$: |z-w| < r\}$ is contained in D, then there exist **complex numbers** $c_0, c_1, c_2,...$ such that $f(z)$ is given by the absolutely convergent **infinite series**

$$f(z) = \sum_{n=0}^{\infty} c_n (z - w)^n,$$

at each point z in $\Delta(w; r)$. Moreover, the coefficients $c_0, c_1, c_2,...$ that occur in this series are given by the formula

$$c_n = \frac{f^{(n)}(w)}{n!}$$

for each nonnegative integer n.

As an example, a polynomial of degree at most N can be written as

$$p(z) = \sum_{n=0}^{N} c_n z^n,$$

for some complex numbers $c_0, c_1, c_2,... c_N$. Thus a polynomial is given by a finite power series expanded about the point $w = 0$, and so defines an analytic function at each point of the complex plane **C**. In this case the coefficients c_n are equal to 0 when $N < n$.

An example of an important analytic function that is not a polynomial is the exponential function. This is defined by the power series expansion

$$\exp z = \sum_{n=0}^{\infty} \frac{z^n}{n!},$$

which converges for all complex numbers z. Another common notation used to designate the exponential function is e^z, where $e = 2.71828182846...$ is the base for natural **logarithms** (see the entry on **e [number]**). The **trigonometric functions** sin z and cos z can also be defined by power series expansions. These are given by

$$\sin z = \sum_{n=0}^{\infty} (-1)^n \frac{z^{2n+1}}{(2n + 1)!}$$

and

$$\cos z = \sum_{n=0}^{\infty} (-1)^n \frac{z^{2n}}{(2n)!},$$

and again they converge for all complex numbers z. From these power series expansions it is easy to see that the functions satisfy the identity $\exp iz = \cos z + i \sin z$ for all complex numbers z. This connection between the three functions $\exp z$, $\cos z$ and $\sin z$ is not so apparent when they are viewed as functions of a real variable.

A further example of an analytic function is provided by the power series

$$\frac{1}{1 - z} = \sum_{n=0}^{\infty} z^n.$$

In this case the power series converges to $(1 - z)^{-1}$ at each point of the open disk $\Delta(0; 1)$. But if $1 < |z|$ then the series diverges and so does not define a function. However, it can be shown directly from the **definition** of the derivative that $(1 - z)^{-1}$ is analytic on the open set $D = \{z \in$ **C**$: z \neq 1\}$. Thus D is the natural domain of the analytic function $(1 - z)^{-1}$, but the power series expansion about the point 0 only defines the function in a proper subset of D. If we select a different complex number, say $3 + 4i$, we get the power series expansion

$$\frac{1}{1-z} = \sum_{n=0}^{\infty} (-2 - 4i)^{-n-1}(z - 3 - 4i)^n,$$

which converges in the disk $\Delta(3 + 4i; \sqrt{20}) = \{z \in \mathbf{C}: |z - 3 - 4i| < \sqrt{20}\}$. Notice that $\sqrt{20}$ is the **distance** from the point $3 + 4i$ to the point 1 where the function $(1 - z)^{-1}$ fails to be analytic.

COMPLEX NUMBERS

Complex numbers are those numbers that can be separated into both a real component and an imaginary component. Complex numbers are generally expressed in the form $a + bi$, where a represents any real number (rational or irrational) and b represents the real coefficient (rational or irrational) of the imaginary number bi. The symbol i represents the **square root** of -1. Since $a + 0i = a$, the set of all complex numbers wholly contains the set of purely **real numbers**. Similarly, since $0 + bi = bi$, the set of complex numbers wholly contains the set of purely **imaginary numbers**. Complex numbers, then, can be considered linear combinations of real and imaginary numbers.

The set of complex numbers is closed under both **addition** and **multiplication**, making it a **field** as well as a **vector space**. Addition and **substraction** with complex numbers are similar to addition and subtraction with real numbers, with the sums (or differences) of real components handled independently of imaginary components. For example, summing the two complex numbers $a + bi$ and $c + di$ becomes $(a + bi) + (c + di) = (a + c) + (b + d)i$. Finding the difference between these same numbers is simply $(a + bi) - (c + di) = (a - c) + (b - d)i$. It follows that the additive identity of a complex number is $0 + 0i = 0$, as $(a + bi) + (0 + 0i) = a + bi$.

Multiplication of complex numbers is similar to multiplying two first-order **polynomials**. Expressed generally, the product of two complex numbers is $(a + bi)(c + di) = ac + adi + bci + bdi^2 = (ac - bd) + (ac + bd)i$. It quickly follows that $1 + 0i = 1$ is the multiplicative identity of any complex number, as $(a + bi)(1 + 0i) = a + bi$.

Complex conjugate

Performing **division** with complex numbers gives rise to the extremely useful concept of the complex conjugate, the unique value that "cancels" the imaginary component of a complex number. For any complex number $a + bi$, its complex conjugate is $a - bi$. This allows one to simplify the quotient of two complex numbers $(a + bi)/(c + di) = (a + bi)/(c + di) \times (c - di)/(c - di) = [(ac - bd) + (ac + bd)i]/(c^2 + d^2)$.

Graphing Complex Numbers

Complex numbers can be plotted on an x-y graph as (a, b), where a represents the magnitude in the x-direction (or real component) and b represents the magnitude in the y-direction (or imaginary component). The **distance** from the origin to the point (a, b), using the **Pythagorean theorem**, is r = square root$(a^2 + b^2)$, which represents the magnitude of the complex number. The symbol θ is often used to represent the **angle** between the x-axis and the vector connecting the origin to the point (a, b), termed the phase. From **geometry** one finds $\tan \theta = b/a$.

Euler's Formula

Leonhard Euler (1707-1783) noted that any complex number $a + bi$ can be plotted along a **circle** of radius r = square root$(a^2 + b^2)$; the **circumference** of the circle represents the set of all complex numbers of magnitude r, with $0 \leq \theta \leq 2\pi$. For the unit circle (a circle of radius 1), Euler developed the formula $e^{i\theta} = \cos \theta + i\sin \theta$, recognizing that $\cos \theta$ represents the magnitude along the x-axis and $\sin \theta$ represents the magnitude along the y-axis. This formula links e to the complex number plane and gives rise to the elegant identity $e^{i\pi} = 1$.

See also Addition; Cosine; Distance; Division; e; Euler, Leonhard; Field; Imaginary numbers; Irrational numbers; Multiplication; Products and quotients; Rational numbers; Real numbers; Sets; Sine; Subtraction; Sums and differences; Square root; Vector spaces

CONGRUENCE

Two geometric figures are said to be congruent if they differ from each other only in their position in **space**. Perhaps the simplest examples of this are parallel line segments of the same length or parallel rays. Two triangles of the same size and shape are easily seen as congruent as well. In a more general sense, two geometric figures are congruent if they can be transformed into each other (as in being overlaid or made to coincide with one another) by some numeric **translation**, **rotation**, and/or reflection. This can easily be seen in the rectangular coordinate system using a linear equation such as $x_1 = 2y + 1$. If the **range** for y is from **zero** to two, this equation can be graphed as a line segment that starts at one on the x axis and ends at coordinates x=5, y=2. Given the same range for y, the equation $x_2 = 2y + 2$ graphs as a line segment parallel to and the same length as that created by the previous equation, but starting at x=2, y=0 and ending at x=6, y=2. This is a very simple example of congruence. A **transformation** equal to -1 is all that is required to transpose the first line segment into the other.

Two **square** matrices, A and B, are congruent if there is another **matrix** C that can be used to transform A into B or vice versa. C is said to be the transpose of A or B and contains numbers in each position in the matrix that provide the value to transform one matrix into the other.

See also Geometry; Graphs, domains, and ranges; Transformation

CONIC SECTIONS

A conic section is the plane curve formed by the intersection of a plane and a right-circular, two-napped cone. Such a cone is shown here.

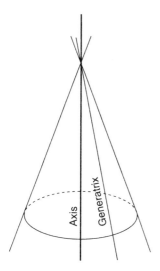

Figure 1.

The cone is the surface formed by all the lines passing through a **circle** and a point. The point must lie on a line, called the "axis," which is perpendicular to the plane of the circle at the circle's center. The point is called the "vertex," and each line on the cone is called a "generatrix." The two parts of the cone lying on either side of the vertex are called "nappes." When the intersecting plane is perpendicular to the axis, the conic section is a circle.

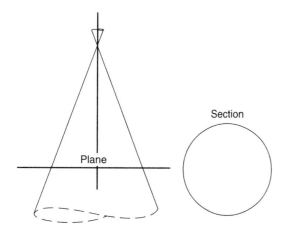

Figure 2.

When the intersecting plane is tilted and cuts completely across one of the nappes, the section is an oval called an **ellipse**.

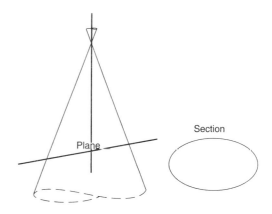

Figure 3.

When the intersecting plane is parallel to one of the generatrices, it cuts only one nappe. The section is an open curve called a **parabola**.

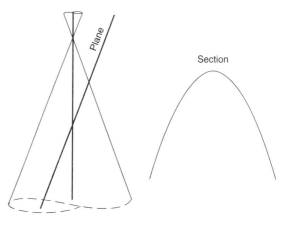

Figure 4.

When the intersecting plane cuts both nappes, the section is a **hyperbola**, a curve with two parts, called "branches."

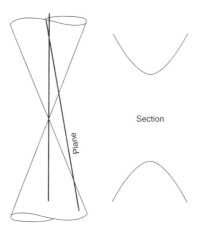

Figure 5.

All these sections are curved. If the intersecting plane passes through the vertex, however, the section will be a single point, a single line, of a pair of crossed lines. Such sections are of minor importance and are known as "degenerate" conic sections.

Since ancient times, mathematicians have known that conic sections can be defined in ways that have no obvious connection with conic sections. One set of ways is the following:

Ellipse: The set of points P such that $PF_1 + PF_2$ equals a constant and F_1 and F_2 are fixed points called the "foci."

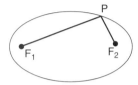

Figure 6.

Parabola: The set of points P such that PD = PF, where F is a fixed point called the "focus" and D is the foot of the perpendicular from P to a fixed line called the "directrix."

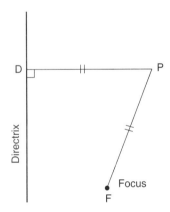

Figure 7.

Hyperbola: The set of points P such that $PF_1 - PF_2$ equals a constant and F_1 and F_2 are fixed points called the "foci."

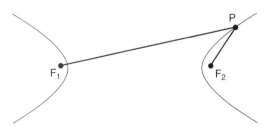

Figure 8.

If P, F, and D are shown as in Figure 7, then the set of points P satisfying the equation PF/PD = e where e is a constant, is a conic section. If $0 < e < 1$, then the section is an ellipse. If e = 1, then the section is a parabola. If e > 1, then the section is a hyperbola. The constant e is called the "eccentricity" of the conic section.

Because the ratio PF/PD is not changed by a change in the scale used to measure PF and PD, all conic sections having the same **eccentricity** are geometrically similar.

Conic sections can also be defined analytically, that is, as points (x,y) which satisfy a suitable equation.

An interesting way to accomplish this is to start with a suitably placed cone in coordinate **space**. A cone with its vertex at the origin and with its axis coinciding with the z-axis has the equation $x^2 + y^2 - kz^2 = 0$. The equation of a plane in space is $ax + by + cz + d = 0$. If one uses substitution to eliminate z from these **equations**, and combines like terms, the result is an equation of the form $Ax^2 + Bxy + Cy^2 + Dx + Ey + F = 0$ where at least one of the coefficients A, B, and C will be different from **zero**.

For example if the cone $x^2 + y^2 - z^2 = 0$ is cut by the plane $y + z - 2 = 0$, the points common to both must satisfy the equation $x^2 + 4y - 4 = 0$, which can be simplified by a **translation** of axes to $x^2 + 4y = 0$. Because, in this example, the plane is parallel to one of the generatrices of the cone, the section is a parabola.

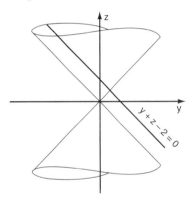

Figure 9.

One can follow this procedure with other intersecting planes. The plane z - 5 = 0 produces the circle $x^2 + y^2 - 25 = 0$. The planes $y + 2z - 2 = 0$ and $2y + z - 2 = 0$ produce the ellipse $12x^2 + 9y^2 - 16 = 0$ and the hyperbola $3x^2 - 9y^2 + 4 = 0$ respectively (after a simplifying translation of the axes). These planes, looking down the x-axis are shown here.

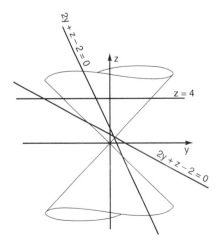

Figure 10.

As these examples illustrate, suitably placed conic sections have equations which can be put into the following forms:

Circle: $x^2 + y^2 = r^2$
Ellipse: $A^2x^2 + B^2y^2 = C^2$
Parabola: $y = Kx^2$
Hyperbola: $A^2x^2 - B^2y^2 = +C^2$

The equations above are "suitably placed." When the equation is not in one of the forms above, it can be hard to tell exactly what kind of conic section the equation represents. There is a simple test, however, which can do this. With the equation written $Ax^2 + Bxy + Cy^2 + Dx + Ey + F = 0$, the **discriminant** $B^2 - 4AC$ will identify which conic section it is. If the discriminant is positive, the section is a hyperbola; if it is negative, the section is an ellipse; if it is zero, the section is a parabola. The discriminant will not distinguish between a proper conic section and a degenerate one such as $x^2 - y^2 = 0$; it will not distinguish between an equation that has real **roots** and one, such as $x^2 + y^2 + 1 = 0$, that does not.

Students who are familiar with the quadratic formula

$ax^2 + bx + c = 0$

will recognize the discriminant, and with good reason. It has to do with finding the points where the conic section crosses the line at **infinity**. If the discriminant is negative, there will be no solution, which is consistent with the fact that both circles and ellipses lie entirely within the finite part of the plane. Parabolas lead to a single root and are tangent to the line at infinity. Hyperbolas lead to two roots and cross it in two places.

Conic sections can also be described with polar coordinates. To do this most easily, one uses the focus-directrix definitions, placing the focus at the origin and the directrix at $x = -k$ (in rectangular coordinates). Then the polar equation is $r = Ke/(1 - e \cos \theta)$ where e is the eccentricity.

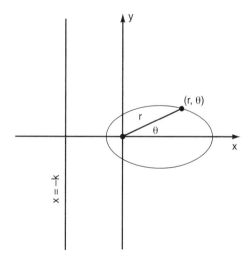

Figure 11.

The eccentricity in this equation is numerically equal to the eccentricity given by another ratio: the ratio CF/CV, where CF represents the **distance** from the geometric center of the conic section to the focus and CV the distance from the center to the vertex. In the case of a circle, the center and the foci are one and the same point; so CF and the eccentricity are both zero. In the case of the ellipse, the vertices are end points of the major axis, hence are farther from the center than the foci. CV is therefore bigger than CF, and the eccentricity is less than 1. In the case of the hyperbola, the vertices lie on the transverse axis, between the foci, hence the eccentricity is greater than 1. In the case of the parabola, the "center" is infinitely far from both the focus and the vertex; so (for those who have a good imagination) the ratio CF/CV is 1.

CONJUNCTION

Within propositional logic, the term conjunction is used in the following ways: (1) as a particular type of propositional statement that is the result of combining component **propositions**, and (2) as the logical operator "and" used to form a conjunctive statement.

Some background concerning propositional logic is necessary in order to develop a precise **definition** of conjunction. As stated above, conjunction is defined in terms of propositional statements. Within so-called classical **Aristotelian logic**, a proposition is defined as a linguistic formation used to communicate information within certain constraints. Those constraints are the "principle of the excluded middle," which states that a proposition must either be true or false, and the "principle of the excluded contradiction," which states that a proposition cannot be both true and false. It should be noted that the words proposition and statement are usually used interchangeably within propositional logic.

A further principle regarding propositions is that they may be combined according to certain rules to yield compound propositions. This is where conjunction comes in: it is used as a logical operator to combine two or more propositions to form a compound proposition. In propositional logic the word "and" is used synonymously with conjunction. In everyday conversation the word "and" is used in much the same way as it is defined for use in classical logic. Suppose a boy tells his mother before going off to school that "I made my bed and I fed the goldfish." If the boy is telling the **truth** in this statement then both component statements must be true; that is, he did make his bed (true) and he did feed the goldfish (true). However, if the boy lies about even one of the two tasks then his original conjunctive statement is false. Borrowing from these common sense ideas, it is then the task of propositional logic to precisely define the meaning of conjunction as a logical operator.

Though the ancient Greeks first developed a formal, propositional logic (carried out with ordinary conversational statements), it was only with German mathematician and philosopher **Gottfried Leibniz** (1646-1716) that the need for a totally "symbolic" logic was seriously considered. Starting with Leibniz's work, propositions and notation in logic would eventually be represented with symbols instead of just words. Hence, one may consider symbolic propositional logic where the symbol "$\wedge$" (and sometimes the dot "$\cdot$") is used to represent "and" or "conjunction." The conjunction of the two propositions A and B to form a third proposition may then be written completely symbolically as $A \wedge B$ or $A \cdot B$, where the statements A and B are called conjuncts. The truth or falsity (called the truth-value) of the conjunctive statement $A \wedge B$ is dependent solely upon the corresponding truth-values of statements A and B. This situation is best illustrated via the use of a **truth table**, as shown below:

	A	B	$A \wedge B$
1.	T	T	T
2.	T	F	F
3.	F	T	F
4.	F	F	F

This table demonstrates the four possible combinations of truth/falsity for statement A (first column), statement B (second column), and the resulting truth-value for the conjunction $A \wedge B$ (third column), where T = true and F = false. As demonstrated by the truth table, a conjunction is true only if its two component propositions are true. Although the truth table defines conjunction as a binary operation upon truth-values, it may be applied to "n" number of statements for integer values of n $\geq$ 2. For example, $A \wedge B \wedge C$ is a valid statement. To determine its truth-value evaluate the truth-value of one component conjunction, say $A \wedge B$, and in turn use its truth-value in a conjunction with statement C. This process could be represented symbolically as $(A \wedge B) \wedge C$, where the conjunctive statement in brackets $(A \wedge B)$ has its truth-value calculated first, and then the truth-value of the entire compound statement decided.

Of course conjunction is not the only logical operator. Conjunction is used along with other logical operators, like "or," "implication," "equivalent," and "not" to form compound statements. For example, "$\sim$" (not) operates on a statement to yield the opposite truth-value. That is to say, if statement A is true then $\sim A$ (read "not A") is false; likewise if A is false then $\sim A$ is a true statement. Hence, the logical operators "$\wedge$" and "$\sim$" can be used to form compound statements. The statement $A \wedge (\sim A)$ is an entirely permissible statement within symbolic propositional logic. Furthermore, it happens to be false for each possible truth-value of its component statements. That is, by the principles of the excluded middle and excluded contradiction, statement A must be either true or false (but not both true and false). If statement A is true then $\sim A$ is false and by the truth table for conjunction $A \wedge (\sim A)$ is false. On the other hand, if A is false then (according to the truth table) $A \wedge (\sim A)$ is false. One can imagine that by using these simple logical operators, large and complex compound statements can be formed.

The use of conjunction in this article has been repeatedly associated with propositional logic, the origins of which can be traced back to the ancient Greeks and especially to Greek philosopher **Aristotle** (384 B.C.-322 B.C.), which uses

statements, truth-values, and logical operators (like conjunction) to perform deductive reasoning. However, conjunction is used as a symbolic logical operator in other types of logic that may generally be called "non-Aristotelian" because of differences from classical propositional logic. For example, "m-valued" logics have been constructed that violate the principle of the excluded middle (i.e., two possible truth values for a statement are replaced with "m number" of truth-values, where m is an integer $\geq$ 3). Obviously those sorts of logical systems would dramatically alter the truth table for conjunction (if a conjunction operator was retained at all).

It should also be noted that the term conjunction is used in **Boolean algebra**, named after English mathematician and logician **George Boole** (1815-1864), a sort of "algebraic logic" involving notations and methods that represents relationships among classes or **sets**. The connective symbol $\oplus$ is called the conjunction and represents the additive operation.

See also Aristotelian logic; Aristotle; George Boole; Boolean algebra; Gottfried Wilhelm von Leibniz; Logical symbols; Propositions; Sets; Symbolic logic; Truth; Truth table

CONTINUITY

Continuity expresses the property of being uninterrupted. Intuitively, a continuous line or function is one that can be graphed without having to lift the pencil from the paper; there are no missing points, no skipped segments and no disconnections. This intuitive notion of continuity goes back to ancient Greece, where many mathematicians and philosophers believed that reality was a reflection of number. Thus, they thought, since numbers are infinitely divisible, **space** and time must also be infinitely divisible. In the fifth century B.C., however, the Greek mathematician **Zeno** pointed out that a number of logical inconsistencies arise when assuming that space is infinitely divisible, and stated his findings in the form of paradoxes. For example, in one paradox Zeno argued that the infinite divisibility of space actually meant that all **motion** was impossible. His argument went approximately as follows: before reaching any destination a traveler must first complete one-half of his journey, and before completing one-half he must complete one-fourth, and before completing one-fourth he must complete one-eighth, and so on indefinitely. Any trip requires an infinite number of steps, so ultimately, Zeno argued, no journey could ever begin, all motion was impossible. **Zeno's paradoxes** had a disturbing effect on Greek mathematicians, and the ultimate resolution of his paradoxes did not occur until the intuitive notion of continuity was finally dealt with logically.

The continuity of space or time, considered by Zeno and others, is represented in mathematics by the continuity of points on a line. As late as the seventeenth century, mathematicians continued to believe, as the ancient Greeks had, that this continuity of points was a simple result of density, meaning that between any two points, no matter how close together, there is always another. This is true, for example, of the

rational numbers. However, the rational numbers do not form a continuum, since **irrational numbers** like √2 are missing, leaving holes or discontinuities. The irrational numbers are required to complete the continuum. Together, the rational and irrational numbers do form a continuous set, the set of **real numbers**. Thus, the continuity of points on a line is ultimately linked to the continuity of the set of real numbers, by establishing a one-to-one **correspondence** between the two. This approach to continuity was first established in the 1820s, by **Augustin-Louis Cauchy**, who finally began to solve the problem of handling continuity logically. In Cauchy's view, any line corresponding to the graph of a function is continuous at a point, if the value of the function at x, denoted by f(x), gets arbitrarily close to f(p), when x gets close to a real number p. If f(x) is continuous for all real numbers x contained in a finite **interval**, then the function is continuous in that interval. If f(x) is continuous for every real number x, then the function is continuous everywhere.

Cauchy's **definition** of continuity is essentially the one we use today, though somewhat more refined versions were developed in the 1850s, and later in the nineteenth century. For example, the concept of continuity is often described in relation to **limits**. The condition for a function to be continuous, is equivalent to the requirement that the limit of the function at the point p be equal to f(p), that is:

$$\lim f(x) = f(p).$$

$$x \rightarrow p$$

In this version, there are two conditions that must be met for a function to be continuous at a point. First, the limit must exist at the point in question, and, second, it must be numerically equal to the value of the function at that point. For instance, polynomial **functions** are continuous everywhere, because the value of the function f(x) approaches f(p) smoothly, as x gets close to p, for all values of p.

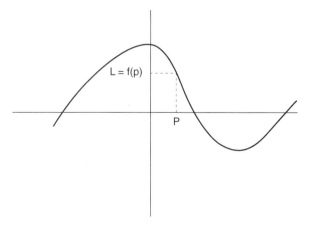

Figure 1.

However, a polynomial function with a single point redefined is not continuous at the point x = p if the limit of the function as x approaches p is L, and not f(p).

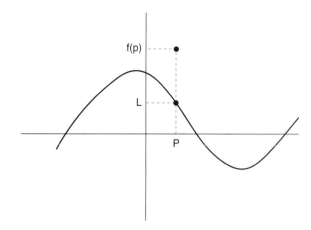

Figure 2.

This is a somewhat artificial example, but it makes the point that when the limit of f(x) as x approaches p is not f(p) then the function is not continuous at x = p. More realistic examples of discontinuous functions include the **square** wave,

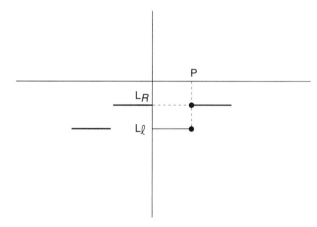

Figure 3.

which illustrates the existence of a right and left hand limits that differ; and functions with infinite discontinuities, that is, with limits that do not exist.

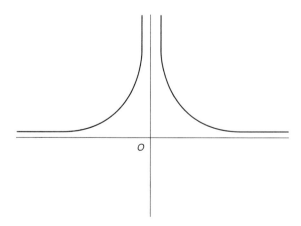

Figure 4.

These examples serve to illustrate the close connection between the limiting value of a function at a point, and continuity at a point.

There are two important properties of continuous functions. First, if a function is continuous in a closed interval, then the function has a maximum value and a minimum value in that interval. Since continuity implies that f(x) cannot be infinite for any x in the interval, the function must have both a maximum and a minimum value, though the two values may be equal. Second, the fact that there can be no holes in a continuous curve implies that a function, continuous on a closed interval [a,b], takes on every value between f(a) and f(b) at least once. The concept of continuity is central to isolating points for which the derivative of a function does not exist. The derivative of a function is equal to the **slope** of the tangent to the graph of the function. For some functions it is not possible to draw a unique tangent at a particular point on the graph, such as any endpoint of a step function segment. When this is the case, it is not possible to determine the value of the derivative at that point. Today, the meaning of continuity is settled within the mathematics community, though it continues to present problems for philosophers and physicists.

CONVERGENCE

Convergence can be defined for infinite sequences of numbers, **functions**, or more general objects such as points in topological spaces. The simplest case to consider is a sequence of numbers, such as (0.9, 0.99, 0.999, 0.9999,...). Such a sequence converges if the terms of the sequence approach a fixed number L, called the *limit* of the sequence, to any desired degree of accuracy. The sequence above, for example, converges to the number 1.

A sequence that does not converge is said to diverge. This can happen in various ways: the terms of the sequence can oscillate (1, 0, 1, 0,...) or they can increase without bound (1, 2, 4, 8, 16,...). In the latter case the limit is sometimes said to be infinite, but the sequence is still considered to be divergent.

Convergence can also be defined for sequences of functions $(f_1, f_2, f_3,...)$. Such a sequence has a limit function f if, for any fixed value of x, the limit of the corresponding sequence of numbers $(f_1(x), f_2(x), f_3(x),...)$ is $f(x)$. However, when dealing with functions, we can use other notions of convergence. The version just discussed, in which the functions converge point by point, is called *pointwise convergence*. For many applications in **calculus**, a stronger version called *uniform convergence* is required, in which the functions approach their limit "at a similar rate" for all values of x. An example to illustrate the distinction between these two types of convergence is the sequence of functions $(1, 1 + x, 1 + x + x^2, 1 + x + x^2 + x^3,...)$, which converges to the function $1/(1-x)$. Note, however, that the convergence is not uniform, and the pointwise convergence holds only for values of x between -1 and 1.

The concept of convergence can also be generalized to infinite sums (usually called **infinite series**) or products. The infinite sum $1 + x + x^2 2 + x^3 +...$ can be considered an abbreviation for the sequence at the end of the preceding paragraph. However, there are advantages to considering it simply as a sum. A number of practical tests (ratio test, root test, integral test) can be applied to determine whether a series converges.

For nineteenth-century mathematicians, convergence was generally applied to sequences of numbers or functions. Modern mathematicians have applied the concept to many more esoteric structures. One example of convergence that has attracted attention in recent years comes from **chaos theory**. The solutions to a dynamical system (for example, a set of **equations** describing the weather) may converge over time to an equilibrium state, or to a periodic "limit cycle," or to a quasiperiodic **orbit**, or to a *strange attractor*. Only in the case of an equilibrium state does the pattern obey the classical **definition** of convergence. But the other behaviors are interesting and important, and therefore the concept of convergence has been refined and generalized to include them. In the case of strange attractors it is difficult to determine the limiting structure precisely, but computer approximations suggest it is a complicated and often beautiful fractal.

See also Infinite series; Power series

COPERNICUS, NICOLAUS (1473-1543)
Polish astronomer

Nicolaus Copernicus, also known as Mikolaj Kopernik, was a groundbreaking astronomer who was the first to propose that the planets revolve around the Sun, not around Earth as was generally accepted by astronomers at the time. He published his findings in his seminal work, *De revolutionibus orbium coelestium*

Born in Torun, Poland, to a wealthy merchant family, Copernicus received the majority of his early education from his uncle, Lucas Watzenrode (1447-1512), the future Bishop of Olsztyn, who wished his nephew to become a church canon. From 1491-94, Copernicus studied astronomy and astrology at the University of Krakow then continued his studies for an additional four years at the University of Bologna. This Italian tenure was very important because he shared a house with the leading astronomer and astrologer of the city, Domenico Maria de Novara (1454-1504). Copernicus was introduced by de Novara to not only his methods of celestial observation, but also to the works of **Ptolemy** and **Regiomontanus**. Ptolemy provided Copernicus with an understanding of the prevailing geocentric view of the universe. Regiomontanus, however, provided unique planetary representations which may have provided a foundation for Copernicus' heliocentric theory.

Upon completion of his studies at Bologna, Copernicus went to the University of Padua to receive a degree in medicine. This may seem to be quite a dramatic shift from the study of the heavens, but for the Europeans of his time, the study of

Nicolaus Copernicus

astronomy, astrology, mathematics, and medicine were all closely related. In order to determine what was ailing an individual, medical doctor would observe, through mathematical techniques, which celestial bodies were adversely affecting his patient. Through this medicinal astrology, a remedy would be suggested. With this in mind, Copernicus' foray into medicine can be understood. However, in 1503, he did not receive a doctorate in medicine, but in canon law and subsequently returned to Poland where his uncle arranged a sinecure for him. Although the term sinecure suggests that Copernicus held a position of leisure, this was not the case. It was Copernicus' duty not only to offer medical care and advice to his uncle, but to also handle the administrative responsibilities of the church. This included a wide array of tasks such as collecting rent, organizing defenses, and managing food production. Considering that at this time the entire countryside was facing constant threats from the Teutonic knights, it is truly remarkable that Copernicus had the time to study the sciences.

Indeed, it was in Copernicus' spare time that he composed his important treatise *De revolutionibus orbium coelestium* (On the revolutions of the heavenly spheres). Here, Copernicus described his view of a heliostatic system of the planets which placed the sun, not the earth, at the center of the solar system. He retained the old Ptolemaic concept of orbits, but used a series of detailed mathematical techniques to alter the order of the planets. These techniques yielded the distances and orbital periods of the planets. *De*

revolutionibus orbium coelestium was publishd in 1543 at the urging of Copernicus' friend and former student, **Georg Rheticus**.

Orbits, also known as the sidereal periods, were easily calculated by observing the **orbit** relative to a fixed star. However, Copernicus realized that because the earth was in orbit around the sun and not the center of the planetary system, the old method of calculating orbits would be flawed. In order to compensate for this, he determined the orbits of the planets through a process called synodic periods. A synodic period was the length of time between the alignments of the Earth, Sun, and the planet under observation. Once the duration of the planetary orbit was calculated, Copernicus was able to determine the relative distances of the planets based upon astronomical units (AU). He had to develop the concept of the astronomical units (AU), where one AU was the **distance** from the earth to the Sun, because he did not know the true distance in miles or any other form of measurement.

Although the Copernican system was generally accepted when it was introduced to the academic community, it was gradually considered implausible and heretical within ecclesiastical circles. However, through the commitment of astronomers such as **Johannes Kepler** (1571-1630) and **Galileo Galilei** (1564-1642), the Copernican system eventually became the foundation by which all complex astronomical calculations and observations were made.

Copernicus would not live to see the repercussions of *De revolutionibus orbium coelestium*. He died in 1543 of a brain hemmorhage, shortly after the work was published.

CORRELATION

Correlation refers to the degree of **correspondence** or relationship between two variables. Correlated variables tend to change together. If one **variable** gets larger, the other one systematically becomes either larger or smaller. For example, we would expect to find such a relationship between scores on an **arithmetic** test taken three months apart. We could expect high scores on the first test to predict high scores on the second test, and low scores on the first test to predict low scores on the second test.

In the above example the scores on the first test are known as the independent or predictor variable (designated as "X") while the scores on the second test are known as the dependent or response variable (designated as "Y"). The relationship between the two variables X and Y is a positive relationship or positive correlation when high measures of X correspond with high measures of Y and low measures of X with low measures of Y. It is also possible for the relationship between variables X and Y to be an inverse relationship or negative correlation. This occurs when high measures of variable X are associated with low measures of variable Y and low measures on variable X are associated with high measures of variable Y. For example, if variable X is school attendance and variable Y is the score on an achievement test we could expect a negative correlation between X and Y. High measures of X (absence) would be associated with low measures of Y (achievement) and low measures of X with high measures of Y.

The correlation coefficient tells us that a relationship exists. The + or - sign indicates the direction of the relationship while the number indicates the magnitude of the relationship. This relationship should not be interpreted as a causal relationship. Variable X is related to variable Y, and may indeed be a good predictor of variable Y, but variable X does not cause variable Y although this is sometimes assumed. For example, there may be a positive correlation between head size and IQ or shoe size and IQ. Yet no one would say that the size of one's head or shoe size causes variations in intelligence. However, when two more likely variables show a positive or negative correlation, many interpret the change in the second variable to have been caused by the first.

CORRESPONDENCE

A correspondence, also called a one-to-one correspondence or a bijection, is a function from a set X to a set Y that has the property that for any element y in Y there is a unique element x in X such that f(x) = y. Most mathematical definitions of **equivalence** are correspondences that preserve some structure. For example, two groups are equivalent if there exists a correspondence between them that preserves their group structures. If there exists a correspondence between two **sets**, then they are said to have the same **cardinality**. A set X is said to be infinite if there is a subset Y of X such that Y is not equal to X, Y is not empty, and there is a correspondence between X and Y. The idea of correspondence originated with **Bernhard Bolzano** in the 1850s and was used extensively by **Georg Cantor** in his study of cardinality, infinite sets, and **ordinal numbers**. Cantor revolutionized the mathematical concept of **infinity**, but his contemporaries were hostile to his ideas and failed to realize their significance. Nowadays, Cantor's arguments are used without hesitation, and correspondences are used frequently in every area of mathematics.

See also Cardinality, Denumerable, Ordinal numbers, Nondenumerable infinite sets, Sets

COSINE

Cosine is a trigonometric function that provides information about the dimensions of an **angle** using a ratio. An angle can be illustrated geometrically as a slice of a **circle**. Consider angle AOB positioned within a unit circle centered at the origin (0,0) of the **Cartesian coordinate system**. A unit circle is defined by $x^2+y^2=1$, where 1 is the radius. The vertex of the angle (i.e., the point at which the legs intersect) is at the origin. The initial leg of the angle lies upon the positive x-axis and intersects the circle at point A with coordinates (1,0). The terminal leg intersects the circle at point B, which has the coordinates (x,y).

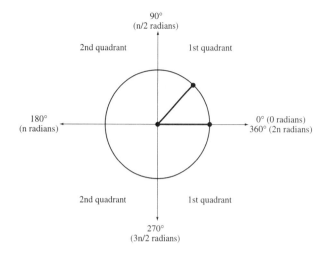

By definition, the cosine of angle AOB is cos AOB=x/r. This is true for any circle described by $x^2+y^2=r^2$. Because r=1 in this particular case, cos AOB=x (which is the **distance** from O to x and is sometimes called the abscissa).

The angle is said to be subtended by arc AB. The length of arc AB is a portion of the entire circle **circumference** $2\pi r$. The angle AOB is the same portion of the entire circle. Angle measurements are expressed in radians or degrees (i.e., **sexagesimal numeration**). A **radian** is defined as the angle size for which the subtended arc is equal in length to the radius of the circle. A degree is defined as 1/360 of a circle, because there are 360° in one complete revolution. The two measures are approximately related by 1 radian=57.29578° and 1°=0.01745 radian.

Assume that angle AOB is ⅛ of the circle. Because a circle is 2π radians, angle AOB is ⅛×2π=¼π radians or 0.7854 radians. Likewise, it is ⅛ of 360° or 45°. The cosine of a 45° angle to six decimal places is 0.707107. This value is available from **trigonometric tables** or can be calculated with a **calculator**.

If an angle θ is a real number expressed in radians, its cosine can be approximated using the first few terms of the following **infinite series**: cos θ=1-(θ^2/2!)+(θ^4/4!)-(θ^6/6!)+(θ^8/8!)—etc. Recall that a **factorial** is the product of all positive **integers** leading up to and including the indicated integer (e.g., 8!=1×2×3×4×5×6×7×8=40,320).

Therefore, the cosine of 45° (¼π or 0.7854 radians), can be approximated as follows: cos 0.7854=1-(0.6169/2)+(0.3805/24)-(0.2347/720)+(0.1448/40,320)=0.707082. This is a very good approximation.

Angle AOB is considered a positive angle, because its terminal leg rotates in a counterclockwise direction from its initial arm. If the terminal leg rotated in a clockwise direction, the angle would be negative. The cosine of any angle is always between -1 and 1. The cosine is positive for angles terminating in the first and fourth quadrants and negative for angles terminating in the second and third quadrants. For example, an angle of -100° terminates in the third quadrant, therefore its cosine is negative. However, the cosine of a negative angle is equivalent to the cosine of its positive value (e.g., cos -100°=cos 100°).

Because cosine is a periodic (repeating) function with a period of 2π, the cosine of any angle equals the cosine of the sum of that angle with any multiple of 360°. In other words, the terminal leg could be rotated many more times around the circle, and as long as it returns to the same position relative to the initial leg, the cosine of the angle would still be the same. Thus, cos 72°=cos (72+360°)=cos 432°.

The cosine and **sine** of an angle are closely related. By definition, the sine of example angle AOB is sin AOB=y/r. Because r=1 in this case, sin AOB=y (which is the distance from O to y and is sometimes called the ordinate). The sine and cosine of any angle θ are related by the following equation: cos θ=sin (90°-θ) in degrees and cos θ=sin (½π-θ) in radians. This equation is true even if the angle size exceeds 360° (2π radians) or if the angle is negative.

Cosine, and the other trigonometric **functions**, are very useful for solving triangles (i.e., finding a leg length or angle given enough information about other leg lengths and/or angles). Consider the triangles below comprised of three legs (a, b, and c) and containing three angles (A, B, and C). The **triangle** on the left is called an oblique triangle, because all of its angles are acute (i.e., less than 90°). The other triangle is a right triangle, because it contains a right angle of 90°.

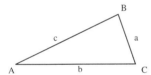 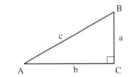

The law of cosines states the following for any triangle: (The **square** of any one leg length)=(The sum of the squares of the other 2 leg lengths)-(Their product times the cosine of the angle between them).

- $a^2=(b^2+c^2)-(2bc \cos A)$
- $b^2=(a^2+c^2)-(2ac \cos B)$
- $c^2=(a^2+b^2)-(2ab \cos C)$.

For the special case of the right triangle, leg c is the **hypotenuse** and angle C is 90°. Because cos 90°=0, the equation for leg c reduces to the **Pythagorean Theorem**: $c^2= a^2+b^2$.

In a right triangle, the cosine of each acute angle equals the length of the adjacent side divided by the length of the hypotenuse (i.e., cos A=b/c and cos B=a/c). This is a very useful formula that can be proven using the equation given earlier to describe angle AOB in a unit circle. Think of leg length c as the radius of the circle and leg length b as the x value. Then, cos AOB=x/r is equivalent to cos A=b/c.

COTES, ROGER (1682-1726)

English trigonometer and astronomer

Roger Cotes advanced the understanding of **trigonometric functions** through his original work on integration, **functions**, and the *nth* roots of unity. He is also remembered for his contributions to **Isaac Newton**'s work on universal gravitation.

Cotes was born in Burbage, England, the son of the Reverend Robert Cotes and his wife, Grace. He attended the Leicester School; his uncle, the Reverend John Smith, was so impressed by the boy's ability in mathematics that he decided to oversee Cotes' early education. When Cotes left Leicester School for St. Paul's School in London, he and his uncle regularly corresponded about science and mathematics. Cotes was admitted as a scholarship student to Trinity College in Cambridge in 1699. He completed his undergraduate work in 1702 and earned a master's degree in 1706.

Like many mathematicians of his era, Cotes was also an astronomer. The universe offered fascinating mathematical puzzles, and astronomical studies had an important practical benefit in an era when successful trade—especially for island nations like England—depended on sailing vessels that charted course by the stars.

In letters to Newton, Cotes described a heliostat telescope that used a clock mechanism to make an interior mirror revolve. He refined existing solar and planetary tables and made notes on the total solar eclipse in 1715.

Cotes was named as a fellow of Trinity College in 1705. In 1706—at the age of 24—he was appointed as the first Plumian professor of astronomy and natural philosophy at Cambridge. Cotes immediately began to solicit donations for an observatory at Trinity. Complete with living quarters, the observatory was built over King's Gate, and Cotes lived there with his cousin Robert Smith. The observatory itself was not completed in Cotes' lifetime and was demolished in 1797.

Cotes established a school of physical sciences at Trinity, and he and colleague William Whiston began a series of experiments in 1707. The details of these experiments were published after Cotes' death as *Hydrostatical and Pneumatical Lectures by Roger Cotes* in 1738.

In 1709, Cotes threw himself into the preparation of the second edition of Newton's *Philosophiae naturalis principia mathematica*, in which Newton refined his theory of universal gravitation. He and Newton worked closely for more than three years on the second edition. Cotes wrote a preface for the work that defended Newton's hypothesis against competing ideas. He argued that Newton's hypothesis, based on observation, was accurate and superior to theories based on description unaccompanied by rational explanation and, certainly, to theories based on superstition and the occult. Cotes' work on this edition was a labor of love—he received 12 copies of the book as payment for more than three years of work.

Cotes published only one paper on mathematics during his lifetime. His *Logometria* was released in 1714, and this paper is evidence not only of Cotes' brilliance, but his persistent and organized approach to the study of mathematics. Cotes calculated the natural base for a system of **logarithms** and introduced two inventive methods for calculating Briggsian logarithms for a number and interpolating intermediate values. He applied integration to problems related to quadratures, arc lengths areas of surfaces of revolution, and atmospheric density. In attempting to evaluate the surface **area** of an ellipsoid of revolution, Cotes identified not one, but two approaches to the problem, using both logarithms and arc sines to develop formulas to calculate area.

Logometria was included in a book of Cotes' papers compiled after his death by Robert Smith and published in 1722 as *Harmonia mensurarum*. The lengthiest section of the book includes Cotes' work in systematic integration. His work was based on a geometrical result involving *n* equally spaced points on a **circle**—now known as Cotes' theorem—that is equivalent to finding all the factors of $x^n - a^n$ when *n* is a positive integer. Another section of *Harmonia mensurarum* includes miscellaneous papers on estimating errors, Newton's differential method, the descent of heavy bodies, and cycloidal **motion**. Cotes' solution to determining the area under a curve, in its modern version, is known as the Newton–Cotes formula.

Cotes died from a fever at the age of 33. "Had Cotes lived," Newton mourned, "we might have known something."

COULOMB, CHARLES AUGUSTIN DE (1736-1806)

French physicist

The French physicist Charles Augustin de Coulomb was born in Angouleme to wealthy parents. During Coulomb's boyhood, the family moved to Paris.

Quarrelling with his mother over career plans, the young Coulomb joined his father in Montpellier, where the latter had become impoverished by his financial speculations. The younger Coulomb later returned to Paris, however, where he completed his studies at the École du Génie. Following his graduation, Coulomb undertook a lengthy period of service as a military engineer with the Corps du Génie in Martinique.

By 1779, Coulomb had returned to France. In that year, he published an analysis of friction in machinery in which he formulated the law governing the relationship between friction and normal pressure. He had already invented a torsion balance that would allow him to measure the forces of magnetic and electrical attraction which would lead to his formulation of the law (see below) that now bears his name.

During this period of his life, he held several public positions, but he had to give them up with the outbreak of the French Revolution in 1789. Forced to flee Paris, Coulomb settled in Blois where he conducted studies of magnetism, friction, and electricity.

Coulomb did not return to Paris until 1795, at which time he was elected a member of the new Institut de France. Between 1802 and 1806, he served as Inspector-General of Public Instruction. In the post-war years, he assisted the new government in establishing a **metric system** of weights and measures. The Système Internationale d'Unités (SI) unit of electrical charge is named after him.

Coulomb is today best known for his measurements of the forces of magnetic and electric attraction and for his formulation of the law now known as Coulomb's law. Coulomb's law states that the **force** between two small charged spheres is inversely proportional to the **square** of the **distance** between them.

Charles Augustin de Coulomb

COURANT, RICHARD (1888-1972)

German American algebraic geometer

Richard Courant received worldwide recognition as one of the foremost organizers of mathematical research and teaching in the twentieth century. Most of Courant's work was in variational **calculus** and its applications to physics, computer science, and other fields. He contributed significantly to the resurgence of **applied mathematics** in the twentieth century. While the Mathematics Institute in Göttingen, Germany, and the Courant Institute of Mathematical Sciences at New York University stand as monuments to his organizing and fundraising abilities, his numerous honorary degrees and awards, as well as the achievements of his students, testify to his noteworthy contributions to mathematics and other sciences.

Courant, the first of three sons, was born on January 8, 1888, in Lublinitz, a small town in Upper Silesia that was then German but later Polish. The family moved to Glatz when he was three; when he was nine they moved to the Silesian capital, Breslau (now Wroclaw). He was enrolled in Breslau's König Wilhelm Gymnasium, preparing to attend a university. At the age of 14, Courant felt a need to become self-supporting and started tutoring students for the high-school math finals, which he himself had not yet taken. He was asked to leave the school for this reason in 1905, and he began attending lectures in **mathematics and physics** at the local university. He passed the high-school finals later that year and became a

•

Richard Courant

full-time student at the University of Breslau. Unhappy with the lecture methods of his physics instructors, he began to concentrate on mathematics.

In 1907 Courant enrolled at the University of Göttingen to take courses with mathematician **David Hilbert**, a professor there. Soon Courant became an assistant to Hilbert, working principally on subjects in **analysis**, an area of mathematics with a close relationship to physics. Under Hilbert, Courant obtained his Ph.D. in 1910 for a dissertation in variational calculus.

In the fall of 1910, Courant was called up for a year of compulsory military service, during which he became a non-commissioned officer. After Courant completed his tour of duty, Hilbert encouraged him to come back to Göttingen for the *Habilitation,* an examination that qualified him for a license as a *privatdozent,* an unsalaried university lecturer or teacher remunerated directly by students' fees. In 1912 Courant received his license to teach at Göttingen.

Two years of teaching and other mathematical work in Göttingen came to an abrupt halt when Courant received his orders to serve in the Army on July 30, 1914. The Kaiser declared war the next day. Courant, like many others, thought the war would be over quickly and was eager to serve. He believed that Germany's cause was right and that his country would be victorious. After about a year of fighting in the

trenches, Courant was wounded and was subsequently deployed in the wireless communications department. Courant proposed that the use of mirrors to obtain visibility of what was going on above ground would help save lives. He also proposed the use of earth telegraphy, a means of communication that would use the earth as a conduit. Both ideas were utilized by the army.

About two weeks after the Armistice, which was signed November 11, 1918, in the midst of tremendous political turmoil, Courant managed to sign a contract with Ferdinand Springer to serve as editor for a series of mathematics books. Courant had made the original proposal for this series to Springer a year earlier. He envisioned timely mathematical treatises that would be especially pertinent to physics. These yellow-jacketed books became known worldwide as the *Yellow Series,* and their publication continued after Courant resigned as editor.

Courant returned to Göttingen in December of 1918 and resumed teaching as *Privatdozent* in the spring of 1919. During the summer of 1919, he completed a lengthy paper on the theory of eigen values of partial **differential equations** (of importance in **quantum mechanics**). After teaching at the University of Münster for a year, he returned as a professor to Göttingen in 1920; in addition to teaching, he was expected to take care of the informal administrative duties of the mathematics department. During the period from 1920 to 1925, Courant succeeded in making Göttingen an international center of theoretical and applied mathematics. Courant's emphasis on applied mathematics attracted physicists from all over the world, making the university a hub of research in quantum mechanics. His tireless efforts as a researcher, teacher, and organizer finally resulted in the creation of the Mathematics Institute of the University of Göttingen. Defining his vision of the future of mathematics, Courant said, "The ultimate justification of our institute rests in our belief in the indestructible vitality of mathematical scholarship. Everywhere there are signs to indicate that mathematics is on the threshold of a new breakthrough which may deepen its relationship with the other sciences and demand their mathematical penetration in a manner quite beyond our present understanding."

In the 1920s and 1930s, Courant worked with Hilbert on his most important publication, *Methoden der mathematischen Physik,* later translated into English as *Methods of Mathematical Physics.* The text was tremendously successful because it laid out the basic mathematical techniques that would play a role in the new quantum theory and nuclear physics. The Great Depression of the early 1930s created a need for university faculties to cut expenses, and there was an order to discharge most of the younger assistants. Courant successfully helped lead the fight of members of the mathematics and natural science faculty to pass a proposal that professors themselves pay the salaries of the assistants who were to be dismissed. He also helped students get scholarships and arranged for some to become part of his household to assist them financially. Those from wealthy families were encouraged to work without pay so that stipends could be available for needy students.

Courant took a leave of absence from Göttingen during the spring and summer of 1932 to lecture in the United States. His positions as professor and director of the Göttingen Mathematics Institute ended on May 5, 1933, when he and five other Göttingen professors received official word that they were on leave until further notice. The move reflected the National Socialist government's escalating campaign against German Jews as well as its displeasure with the university, which had become a locus of independent liberal thought. As Americans and other foreigners were leaving Göttingen at this time, Courant observed that the spirit of the institute had already been destroyed. During the 1933–1934 academic year he became a visiting lecturer at Cambridge in England. In January of 1934, he accepted an offer of a two-year contract with New York University.

In 1936, when his temporary position ended, he was appointed professor and head of New York University's mathematics department. In that position, he did for New York University what he had done for Göttingen by creating a center of mathematics and science of international importance. In recognition of his work, Courant was made director of the new mathematics institute at New York University, later named the Courant Institute of Mathematical Sciences. Courant's success as an organizer was largely due to his ability to attract promising young mathematicians. He was always available to them as a teacher, helped them to publish their work, and organized financial support for them if they needed it. His students often remained loyal to him for the rest of their lives and tended to stay in his orbit.

During World War II, Courant was a member of the Applied Mathematics Panel, which assisted scientists involved with military projects and contracted for specific research with universities throughout the country. While the group at New York University under contract with the panel made important contributions to the war effort, the contract in turn played a vital role in setting up a scientific center at New York University. Courant's mathematical work in **numerical analysis** and partial difference **equations** played a vital role in the development of computer applications to scientific work. Courant was also instrumental in getting the Atomic Energy Commission to place its experimental computer UNIVAC at New York University in 1953. Courant retired in 1958, the same year he was honored with the Navy Distinguished Public Service Award and the Knight-Commander's Cross and Star of the Order of Merit of the Federal Republic of Germany. In 1965, he received an award for distinguished service to mathematics from the Mathematical Association of America.

Married in 1912 to a woman he had tutored as an adolescent, Courant was divorced in 1916. In 1919 he married Nerina (Nina) Runge, and they had two sons and two daughters. He enjoyed skiing and hiking, and played the piano. In November 1971, Courant suffered a stroke. He died on January 27, 1972, a few weeks after his 84th birthday. According to a *New York Times* obituary by Harry Schwartz, Nobel laureate in physics **Niels Bohr** once remarked that

"every physicist is in Dr. Courant's debt for the vast insight he has given us into mathematical methods for comprehending nature and the physical world." At a memorial in Courant's honor, the mathematician Kurt O. Friedrichs said of him: "One cannot appreciate Courant's scientific achievements simply by enumerating his published work. To be sure, this work was original, significant, beautiful; but it had a very particular flavor: it never stood alone; it was always connected with problems and methods of other fields of science, drawing inspiration from them, and in turn inspiring them."

Cox, Gertrude Mary (1900-1978)
American statistician

Gertrude Cox organized and directed several agencies dedicated to research and teaching in **statistics**. "By her missionary zeal, her organizational ability and her appreciation of the need for a practical approach to the statistical needs of agricultural, biological and medical research workers she did much to counter the confused mass of theory emanating from mathematical statisticians, particularly in the United States, who had little contact with scientific research," eulogized Frank Yates in the *Journal of the Royal Statistical Society*.

Cox was born on a farm near Dayton, Iowa, on January 13, 1900, to John William Allen and Emmaline (Maddy) Cox. After graduating from Perry (Iowa) High School in 1918, she devoted several years to social service and training for the role of deaconess in the Methodist Episcopal Church. She spent part of that time caring for children in a Montana orphanage.

In 1925, in preparation for advancement to superintendent of the orphanage, Cox entered Iowa State College in Ames. Although she took courses in psychology, sociology, and other topics that would advance her intended career, she majored in mathematics because of her talent for it. She graduated in 1929 and registered for graduate work under the direction of Professor George Snedecor, a proponent of **Ronald A. Fisher**'s statistical methods. Cox and Fisher became friends when he worked at Iowa State during the summers of 1931 and 1936. In 1931, she earned Iowa State's first M.S. degree in statistics, and for the next two years Cox worked as a graduate assistant at the University of California, Berkeley, studying psychological statistics.

In 1933, Snedecor asked Cox to return to work at Iowa State's new Statistical Laboratory, where she built a reputation of expertise in experimental design. By 1939, Cox had become an assistant professor at Iowa State, although her teaching and consulting activities never allowed her time to write a doctoral dissertation. Eventually, in 1958, Cox was awarded an honorary Doctor of Science degree by Iowa State.

When Snedecor was asked to recommend nominees to head the new Department of Experimental Statistics being formed at North Carolina State (NCS) College's School of Agriculture, he showed his list to Cox, who asked why her name was not included. He then added a footnote to his letter: "Of course if you would consider a woman for this position I would recommend Gertrude Cox of my staff." She was hired in 1940, becoming the first woman to head a department at North Carolina State.

In 1944, Cox assumed additional duties as director of the NCS Institute of Statistics, which she had organized. By 1946, the University of North Carolina (UNC) joined the Institute, taking responsibility for teaching statistical theory while NCS provided courses in methodology. Cox saw the Institute's mission as developing strong statistical programs throughout the South, a vision described by colleagues as "spreading the gospel according to St. Gertrude." Always insistent that good statistical analysis depended on adequate computations, Cox unhesitatingly embraced the computer era. NCS became one of the first colleges in the country to install an IBM 650 computer, and statisticians in Cox's organizations developed powerful statistical software programs. Cox helped create the Biometric Society in 1947 and edited its journal *Biometrics* from 1947 to 1955. In 1949, she became the first female member of the International Statistical Institute, and was elected president of the American Statistical Association seven years later.

In 1950, Cox and her colleague William Cochran published *Experimental Designs* which was intended to be a reference book for research workers with little technical knowledge. In fact, it became a widely used textbook that Frank Yates described nearly 30 years later as "still the best practical book on the design and analysis of replicated experiments." In her own experimental design classes, Cox taught by focusing on specific examples gleaned from her years of consulting experience.

Although Cox made substantial contributions to the theory of statistics and experimental design, Richard Anderson wrote in *Biographical Memoirs* that her most valuable contribution to science was organizing and administering programs. She was exceptionally successful in generating financial support for research. For example, one large grant Cox obtained from the General Education Board established a revolving fund that supported fundamental statistical research for many years. Cox also played an integral role in planning what would become the Research Triangle Institute (RTI) for consulting and research, uniting the resources of NCS, UNC, and Duke University. In 1960, she retired from NCS and became the first director of RTI's Statistics Section.

Cox loved world travel, and during her lifetime she made 23 trips to various international destinations. After retiring a second time in 1965, she spent a year in Egypt establishing the University of Cairo's Institute of Statistics. On five different occasions, Cox worked on statistical assistance programs in Thailand. At the age of 76, she toured Alaska and the Yukon Territory by bus, train, and boat.

Although she received numerous honors, including her 1975 election to the National Academy of Sciences, Cox was particularly pleased with the dedication of the statistics building at North Carolina State University as "Cox Hall" in 1970 and the establishment by her former students of the $200,000 Gertrude M. Cox Fellowship Fund for outstanding students in statistics at NCS in 1977.

Cox died of leukemia on October 17, 1978, at Duke University Medical Center in Durham. During the preceding year, she had kept meticulous records of her treatment and response, making herself the subject of her final experiment.

CRAMER, GABRIEL (1704-1752)
Swiss geometer and probability theorist

Gabriel Cramer labored in the shadow of his more well–known mathematical contemporaries. Cramer added to mathematical knowledge in the areas of **analysis, determinants**, and **geometry**. Both Cramer's ruleand Cramer's paradox, discussed below, were not completely new ideas, but Cramer contributed significantly to both. Cramer also introduced the idea of utility to mathematics. Some of Cramer's most important work concerned the **history of mathematics**, as Cramer often served as the editor of other mathematicians' writings.

Cramer was born July 31, 1704, in Geneva, Switzerland, to physician Jean Isaac Cramer and his wife, Anne Mallet. Educated in Geneva, Cramer had two brothers. One, Jean–Antoine, also became a doctor; the other, Jean, was a law professor. All three brothers took an interest in local government and its inner workings. Cramer was a lifelong bachelor.

At the age of 18, Cramer defended his thesis with a topic about sound. At age 20, he was appointed co–chair of mathematics at the Académie de la Rive, and led the geometry and **mechanics** classes. His co–chair was Giovanni Ludovico Calandrini, a friend who taught **algebra** and astronomy. They also split the salary designated for one chair. As professors, Cramer and Calandrini broke with tradition and allowed recitations in French instead of the traditional Latin, so that students without a Latin background could participate.

The pair were encouraged to travel to enhance their scholarship. Cramer went to Basle for five months in 1727, where he met and befriended, among others, **Johann Bernoulli**. Cramer continued to travel until 1729, visiting London, Leiden, and Paris, before returning to Geneva. In 1734, Cramer became full chair of the department when Calandrini was appointed to a philosophy professorship. In the same year, Cramer's activity in local government was marked by his service on the Conseil des Deux–Cents.

Cramer's sense of community extended to his mathematical colleagues. He edited the collected works of fellow mathematician Johann Bernoulli in four volumes, which were published in 1742. Cramer's other editorial projects included the works of Johann's brother, **Jakob Bernoulli**, and Christian Wolff. His accomplishments as a mathematics editor were vital for the support and circulation of mathematical knowledge, and he was one of the first scholars of note to contribute to mathematics in this manner.

Cramer's most important mathematical work was published in 1750, the same year he was appointed professor of philosophy at Académie de la Rive when Caladrini left to work for the Swiss government. Within the four volumes of *Introduction à l'analyse des lignes courbes algebriques,*

CRAMER'S RULE

Cramer's rule gives a formula for the entries of a column vector x that satisfies Ax = b, where A is an "n x n" **matrix** with nonzero determinant and b is a column vector with n entries. A column vector in n x 1 matrix. Let $A_j(b)$ denote the n x n matrix which has the same entries as A except for the jth column which is equal to b. Cramer's rule is that the jth entry of x must be equal to the det($A_j(b)$)/det A. The notation det(M) means the determinant of the matrix M.

Here is the **proof** of Cramer's rule. Let e_j be the length 1 column vector that has jth entry equal to one and all other entries equal to **zero**. The identity matrix is $[e_1 \, e_2 ... \, e_n]$ = I. $AI_j(x) = [Ae_1 ... \, Ax ... \, Ae_n] = A_j$ (b). Because the determinant of a product of matrices is equal to the product of their **determinants**, det(A)*det($I_j(x)$)=det($A_j(b)$). Using cofactor expansion, it is easy to see that det($I_j(x)$) = the jth entry of x. Therefore the jth entry of x = det($A_j(b)$)/det(A).

Cramer's rule is not very practical as an **algorithm** for solving the equation Ax = b because too many computations are involved. Generally, however, there is some margin of error in figuring b. So, it is often useful to know all the solutions for Ax = b as b varies over its margin of error. In this case, Cramer's rule is helpful for understanding the set of solutions.

CRELLE, AUGUST LEOPOLD (1780-1855)

German editor, engineer, and mathematician

August Crelle was an engineer and amateur mathematician who founded one of the first European mathematical journals, *Journal für die reine und angewandte Mathematik*, also known as *Crelle's Journal*.

Born in 1780 in Eichwerder, Germany to a family of moderate means who could not afford private instruction or schooling for their son, Crelle was largely self-educated. He studied engineering, and instead of going on to University to pursue his passion—mathematics—Crelle secured a position with the Prussian Ministry of the Interior, where he worked on the planning and design of Germany's first railway line.

Established and well-compensated in his position, Crelle was finally able to devote his spare time to mathematical theory. At age 36, he completed a thesis entitled *De calculi variabilium in geometria et arte mechanica usu*, which he submitted to the University of Heidelberg. The thesis earned him a doctorate degree.

In 1826, Crelle used his financial and intellectual resources to found the *Journal für die reine und angewandte Mathematik*. It differed from other mathematical journals of the time in that it was devoted exclusively to applied mathematical theories. Crelle edited the journal himself, and encouraged both new and established mathematicians to submit their work for publication. He had an uncanny ability to discover promising mathematical minds, and was one of the first to recognize the importance of the work of Norwegian mathematician **Niels Abel** and his critical contributions to **group theory**.

Gabriel Cramer

Cramer delineated what came to be known as **Cramer's rule**, which provided a mechanism for linear equation solutions. Although deteminants were discovered by **Gottfried Wilhelm Leibniz** in 1693, it is Cramer who brought determinants and their uses to wide spread attention. Also in this text is Cramer's discussion of curve analysis which led to their rediscovery, which led to other mathematicians improving on his analysis.

Another important element featured in Cramer's text is what came to be known as Cramer's paradox. The paradox clarified a **theorem** first proposed by **Colin Maclaurin**, who pointed out that two separate cubic curves can meet at nine different points. Cramer added that a single cubic curve is itself defined by nine different points. Because Cramer's explanation for this phenomenon was lacking, other mathematicians fleshed it out. Cramer is also responsible for the concept of utility, which today links **probability theory** with mathematical economics.

In 1751, Cramer fell from a carriage and was confined to his bed for two months. His doctor recommended a rest in the south of France for his health because, in addition to the fall, he was overworked. Cramer died en route in Bagnoles, France, on January 4, 1752.

Abel was published extensively in early editions of the journal, as were **Peter Dirichlet**, **Karl Jacobi**, **Jakob Steiner**, and **Johann Gauss**. Crelle also acted as Abel's patron and mentor, introducing him to the leading mathematicians of Germany and France, and securing a teaching position for Abel at the University of Berlin. Unfortunately, the young mathematician died of tuberculosis several days after the appointment, but Crelle's assistance was instrumental in legitimizing Abel's work and in moving his ideas and theories into wide circulation in the mathematical community both before and after his death.

Crelle left his civil engineering post for a position with the Prussian Ministry of Education and Cultural Affairs in 1928. He advised the ministry on teaching techniques and policies for mathematical instruction in German schools, and spent considerable time studying the French teaching system. Crelle also published a number of mathematical textbooks and tables during this period.

In 1829, Crelle established a second mathematical journal, *Journal für die Bankunst*, which was devoted to practical mathematics (as opposed to the pure, **applied mathematics** of his original journal). Crelle published the journal until 1851, when it folded. Several years later, Crelle passed away in Berlin at age 75. However, his landmark journal, the *Journal für die reine und angewandte Mathematik* is still published today.

CUBIC EQUATIONS

A cubic equation is one of the form $ax^3+bx^2+cx+d = 0$ where a,b,c and d are **real numbers**. For example, $x^3-2x^2-5x+6 = 0$ and $x^3-3x^2+4x-2 = 0$ are cubic **equations**. The first one has the real solutions, or **roots**, -2, 1, and 3, and the second one has the real root 1 and the complex roots 1+i and 1-i.

Every cubic equation has either three real roots as in our first example or one real root and a pair of (conjugate) complex roots as in our second example.

There is a formula for finding the roots of a cubic equation that is similar to the one for the quadratic equation but much more complicated. It was first used by **Girolamo Cardano** in 1545, even though he had obtained the formula from **Niccolo Tartaglia** under the promise of secrecy.

CYBERNETICS

Cybernetics is a term that was originated by American mathematician **Norbert Wiener** (1894-1964) in the late 1940s. Based on common relationships between humans and machines, cybernetics is the study and analysis of control and communication systems. As Wiener explains in his 1948 book, *Cybernetics: or Control and Communication in the Animal and the Machine*, any machine that is "intelligent" must be able to modify its behavior in response to feedback from the environment.

This theory has particular relevance to the field of computer science. Within modern research, considerable attention is focused on creating computers that emulate the workings of the human mind, thus improving their performance. The goal of this research is the production of computers operating on a neural network. During the late 1990s work has progressed to the point that a neural network can be run, but, unfortunately, it is generally a computer software simulation that is run on a conventional computer. The eventual aim, and the continuing area of research in this field, is the production of a neural computer. With a neural computer, the architecture of the brain is reproduced. This system is brought about by transistors and resistors acting as neurons, axons, and dendrites. By 1998 a neural network had been produced on an integrated circuit, which contained 1024 artificial neurons. The advantage of these neural computers is that they are able to grow and adapt. They can learn from past experience and recognize patterns, allowing them to operate intuitively, at a faster rate, and in a predictive manner.

Another potential use of cybernetics is one much loved by science fiction authors, the replacement of ailing body parts with artificial structures and systems. If a structure, such as an organ, can take care of its own functioning, then it need not be plugged into the human nervous system, which is a very difficult operation. If the artificial organ can sense the environment around itself and act accordingly, it need only be attached to the appropriate part of the body for its correct functioning. An even more ambitious future for the cybernetics industry is the production of a fully autonomous life form, something akin to the robots often featured in popular science fiction offerings. Such an artificial life form with learning and deductive powers would be able to operate in areas that are inhospitable to human life. This could include long-term **space** travel or areas of high radioactivity.

CYCLOID

The cycloid is a planar curve, defined as the trajectory traveled by a marked point on the rim of a wheel as it rolls without slipping along a level surface. The cycloid resembles a series of arches, with cusps at the beginning and end of each arch. These cusps correspond to the locations where the marked point touches the ground. At that instant, the point's velocity is **zero**, thus providing the answer to an old conundrum, "What part of a rolling wheel is at rest?"

The cycloid has a distinguished history of turning up in geometric problems involving **motion**. The *brachistochrone problem* asks what shape of a ramp will enable an object to slide from a point A to another point B in the shortest possible time. Using the then newly discovered theory of **calculus**, **Gottfried Leibniz** and **Isaac Newton** proved in 1697 that the solution is an arc of an upside-down cycloid. In 1658, the Dutch physicist **Christiaan Huygens** discovered the isochronism of the cycloid: an object placed anywhere on an upside-down cycloid will slide to the bottom of the arch in the same amount of time. As a practical application, Huygens built pendulums that swung in a cycloidal arc, and therefore kept time at a constant rate even as their swings grew smaller. (Ordinary pendulums that swing in a circular arc are nearly, but not exactly, isochronous.)

The cycloid is part of a general class of curves called *roulettes* (curves produced by rolling) and was one of the first curves whose arc length was computed exactly. The length of each arc is eight times the radius of the wheel.

CYLINDER

A cylinder is a three-dimensional geometric figure generated by a straight line moving parallel to a fixed straight line along a plane curve. It is a body of roller-like form. It can also be thought of as a solid bounded by two parallel planes and as a surface having a closed curve, such as a **circle**, as a directrix or base. The name cylinder is derived from the Latin *cylindrus* which in turn is derived from the Greek *kulindros* from *kulindein*, meaning to roll.

There are two types of typical circular cylinders: right circular cylinders and oblique circular cylinders. A right circular cylinder is a body consisting of two circular bases of equal **area** that are in parallel planes and are connected by a lateral surface that intersects the boundaries of the circular bases at right angles. An oblique circular cylinder is similar to a right circular cylinder, but the lateral surface connecting the two circular bases intersects the boundaries of the bases at angles other than 90 and 70. The **volume** of a circular cylinder is equal to $\pi r^2 h$, where r is the radius of the bases and h is the height of the cylinder; that is, the **distance** between the two planes connecting the circular bases. The lateral surface area is equal to $2\pi rh$, and the total surface area of a right circular cylinder is equal to $2\pi r(r+h)$.

A cylinder does not necessarily have circular bases, nor must a cylinder form a closed lateral surface. A cylinder can have circular, ellipsoid, rectangular, **square**, or triangular directrixes. A cylindrical surface whose directrix is an **ellipse** is called a cylindroid. The lateral surface of a cylinder does not have to be a closed surface but can be composed of parallel lines that intersect the bases of the cylinder. The volume defined by these lines composes a cylindrical solid. If a cylinder is bent into a ring, the solid formed is known as a **torus**. A doughnut and the inner tube of a tire are examples of such forms.

The branch of **geometry** that deals with the measurement and properties of cylinders as well as other three-dimensional figures is called **solid geometry**. This field of geometry expands the theories of **plane geometry** to three-dimensional **space** and is the foundation for spherical **trigonometry**, solid analytical geometry, descriptive geometry, and other branches of mathematics concerned with three-dimensional space. Solid geometry has its origins in the demonstrative geometry of the Greeks that dealt with **polygons** and circles and with corresponding three-dimensional figures. In 300 B.C. the Greek mathematician **Euclid** wrote a set of thirteen books entitled *Elements* that covered the basic geometry of that time. The Greek scientist **Archimedes**, known for approximating the value of π, also devised a way to measure the surface area and volume of the cylinder.

D

D'ALEMBERT, JEAN LE ROND (1717-1783)

French calculist and physicist

Jean le Rond d'Alembert was a mathematician and physicist who applied his considerable genius to solving problems in **mechanics**. This is the branch of physics that deals with the effect of forces on matter, either at rest or in motion. His most important contribution was d'Alembert's principle, which states that the forces in an object that resist acceleration must be equal and opposite to the forces that produce the acceleration. D'Alembert was a pioneer in the development of **calculus**. He also served as science editor of Denis Diderot's *Encyclopédie*.

D'Alembert was born in Paris on November 17, 1717. He was the illegitimate son of Claudine–Alexandrine Guérin, Marquise de Tencin, an intelligent and unprincipled woman who broke her vows as a nun and became the mistress of many powerful men. D'Alembert's father was the Chevalier Louis–Camus Destouches, a military officer. D'Alembert's mother apparently regarded her pregnancy as an unwelcome accident and later abandoned her infant son on the steps of the church of Saint–Jean–le–Rond. The foundling was baptized with the name of the church, then sent to a foster home in Picardy.

After Destouches returned from military service, he brought his son back to Paris and arranged for him to be raised by one Madame Rousseau, the wife of a glazier. D'Alembert always regarded this woman as his real mother, and he continued to live in her home until he was 47 years old. Destouches provided for his son and paid for his education. When Destouches died in 1726, he left the boy a legacy that gave d'Alembert a modest lifetime income of 1,200 livres a year. D'Alembert always cherished the independence this income provided.

D'Alembert attended the Collège des Quatre–Nations (also called Mazarin College), a school run by Jansenists, members of a religious sect. While at college, he adopted the name Jean–Baptiste Daremberg, later shortened to d'Alembert. Despite the urging of teachers, however, d'Alembert rejected the religious life. After receiving a bachelor's degree in 1735, he went on to study law, receiving a license to practice in 1738. He also studied medicine for a year. Neither the law nor medicine held much lasting appeal for d'Alembert. He finally settled upon a career in mathematics, a vocation for which he had much natural talent.

In 1739, d'Alembert submitted his first paper to the French Académie Royale des Sciences, a critique of a mathematics book written by Father Charles Reyneau. Over the next two years, d'Alembert sent the Académie additional papers on such topics as fluid mechanics and the integration of **differential equations**. After several failed attempts to join the Académie, he was finally admitted in May 1741.

During the 1740s, d'Alembert became a fixture in French intellectual and social salons, where he was known for his gaiety and wit. He took his place among leading *philosophes*, thinkers of the Enlightenment, a philosophical movement marked by an emphasis on human reason and a rejection of traditional religious and political ideas. Although d'Alembert never married, he shared a close relationship for many years with Julie de Lespinasse, a popular salon hostess.

From 1741 through 1743, d'Alembert studied various problems in dynamics, the branch of mechanics that deals with the effect of forces on the motion of bodies. His writings were hastily collected into a book, *Traité de dynamique* (1743), that became his most important scientific work. This book introduced the famous principle that bears d'Alembert's name. The principle was actually an extension of **Isaac Newton**'s third law of motion, which states that for every force exerted on a static body, there is an equal and opposite force from that body. D'Alembert maintained that the law applied not only to bodies at rest, but to bodies in motion as well. This was the dawn of a new era in the science of mechanics. The next year, d'Alembert published *Traité de l'équilibre et du mouvement* this essay, he introduced such important concepts as the components of fluid velocity and acceleration. With *Élémens de musique théorique et pratique suivant les principes de M. Rameau* (1752), however, d'Alembert departed from science to indulge an interest in music. In this work, he described the new theory of musical structure advanced by French composer Jean–Philippe Rameau.

Much of the 1750s was devoted to work on Diderot's *Encyclopédie*. This monumental work was conceived by Diderot as a synthesis of all human knowledge, with an emphasis on new ideas and scientific discoveries. D'Alembert's first task was writing the *Discours préliminaire* (1751), an introduction that sought to show the links between disciplines and to trace the progress of thought, culminating in the philosophy of the Enlightenment. The discourse was widely praised. Its publication led to d'Alembert's acceptance into the French Académie in 1754. He later became very active in that organization, eventually being elected its permanent secretary.

D'Alembert wrote 1,500 articles for the *Encyclopédie*. While many of these articles discussed mathematics and science, others addressed philosophy and the arts. In fact, d'Alembert was increasingly drawn to nonscientific topics. Between 1753 and 1767, he published five volumes of *Mélanges de littérature et de philosophie*. This collection contained essays on music, law, and religion; a treatise on philosophy; translations of Tacitus; and a hodgepodge of other material.

In 1757, d'Alembert visited the French writer Voltaire, his closest friend among the *philosophes*. One result of the visit was an article on Geneva, Switzerland, which appeared in the seventh volume of the *Encyclopédie*. This article caused a furor with its depiction of Protestant ministers, managing to offend Roman Catholics and Calvinists alike. The uproar caused d'Alembert to resign as an editor of the project.

During the following decades, d'Alembert resumed scientific publication. From 1761 to 1780, he issued eight volumes of *Opuscules mathématiques*, which included essays on **hydrodynamics**, astronomy, and lenses. At this time, d'Alembert was almost alone in regarding the differential as the **limit** of a **function**, a key concept in modern calculus. However, he could never rise above the traditional focus on **geometry**, which prevented him from ever putting the idea of the limit into a purely algorithmic form.

In 1764, d'Alembert spent three months in the court of Frederick the Great. Although he was offered the presidency of the Prussian Academy, d'Alembert declined. He also turned down an offer from Russian Empress Catherine the Great to tutor her son for 100,000 livres a year. Above all, d'Alembert prized his financial independence and the intellectual freedom it afforded. D'Alembert published a work on religion the following year, in which he called for the suppression of both the Jesuits and their rivals, the Jansenists. This book, not one of d'Alembert's better efforts, was written at Voltaire's behest. It was issued anonymously, but the author's identity was known.

D'Alembert's final years were not easy. He became seriously ill in 1765 and moved into the home of Julie de

Jean le Rond d'Alembert

des fluides (1744), in which he applied his principle to the motion of fluids.

D'Alembert published *Réflexions sur la cause générale des vents* in 1747, a treatise on winds that won a prize from the Prussian Academy. This paper marked the first general use of partial differential equations in mathematical physics. That same year, d'Alembert published an article on the motion of vibrating strings. This paper is notable for the first use of a wave equation in physics. While d'Alembert pioneered both partial differential equations and wave equations, it was left to his Swiss contemporary **Leonhard Euler** to develop these concepts more fully.

Next, d'Alembert turned his attention to astronomy, applying calculus to celestial mechanics. In 1749 he issued *Recherches sur la précession des équinoxes et sur la nutation de la terre*. This book dealt with the precession of the equinoxes; that is, the slow, gradual westward motion of the equinoxes due to the movement of the Earth's axis. D'Alembert's research on astronomy continued in *Recherches sur différens points importants du systeme du monde*. This three–volume work, published in 1754–1756, dealt mainly with the motion of the moon.

D'Alembert issued one more scientific publication in the 1750s, returning to the subject of fluid mechanics in *Essai d'une nouvelle théorie de la résistance des fluides* (1752). In

Lespinasse, who nursed him back to health. He continued to live with Lespinasse until her death in 1776. After she died, d'Alembert discovered evidence of love affairs with other men among Lespinasse's effects, which made her loss doubly painful. He withdrew into a lonely, bitter retirement, living in a small apartment provided by the French Académie. D'Alembert died in Paris on October 29, 1783. He had outlived many of his fellow *philosophes*, but the scientific and philosophical legacy of this remarkable group of thinkers survives.

DA VINCI, LEONARDO (1452-1519)
Italian artist and scientist

Leonardo da Vinci was born in 1452, the illegitimate son of a Florentine notary, Ser Piero da Vinci, and an unmarried Tuscan peasant girl, Caterina. Raised by his father's grandparents and an uncle in the small countryside village of Vinci, the younger da Vinci moved to Florence at the age of 14 or 15 to pursue an apprenticeship and a career. As an illegitimate child, da Vinci was barred from pursuing an education at university, and from following in his father's footsteps as a notary or any position of similar stature in Florentine society (known as "the noble professions"). The only professions open to him were manual arts, such as military service, the arts, or literature. His father apprenticed him to local artist Andrea del Verrocchio, under whose tutelage he learned drawing, sculpting, engineering, and architecture techniques. After being accepted into the painters' guild in 1972 at age 20, da Vinci continued to work alongside Verrocchio as a master craftsman. Five years later, da Vinci opened his own studio in Florence.

From 1482 to 1500, da Vinci lived in Milan, painting such famous works of art as *The Last Supper*, and working as an engineer, architect, and military fortification consultant for the Duke of Milan. It is also during this time that his interest in **geometry** as a tool for art, architecture, and engineering began to grow. In 1496, da Vinci completed a series of illustrations for **Luca Pacioli**'s book, *De divina proportione*, a work on mathematical proportion. Leonardo believed in an intrinsic link between mathematics and his art, and stated in his *Quaderni D'Anatomia I-IV*, a treatise on the human anatomy, "Let no one read me who is not a mathematician."

Da Vinci and Pacioli both left Milan in 1500, and eventually Leonardo ended up back in Florence. Although his artistic talents were highly sought after, and he commanded a fair price for commissioned artwork, he gradually began to devote more and more time to mathematical and scientific studies, and less to painting. Over the next sixteen years, he spent his time in Florence, Milan, and Rome working on a variety of science-based projects, including the study of **hydraulics**, mirrors, flight, anatomical drawing, and cartography, and the design of military fortifications and canals. His surviving sketches also show that he constructed plans for elliptical, parabolic, and proportional compasses, perhaps one of his most significant contributions to the study of mathematics and geometry.

Leonardo da Vinci

In 1516, da Vinci was appointed to the position of first painter, architect, and mechanic of King Francis I of France. He moved to France, where he spent his time on a few final paintings and his scientific studies until his death in 1519 in Cloux, Amboise, France.

See also Luca Pacioli

DANTZIG, GEORGE BERNARD (1914-)
American applied mathematician

George Bernard Dantzig is a mathematician and the founder of **linear programming**, a mathematical technique that has had extensive scientific and technical applications in such areas as computer programming, logistics, and scheduling. Applicable to such endeavors as military research, industrial engineering, and business and managerial studies, linear programming is a method for formulating solutions to problems of how to optimally allocate resources among competitive activities. For example, linear programming could be used to develop a diet that contains all the necessary minimal quantities of dietary elements at a minimum cost by factoring in such variables as calories, protein, vitamins, and the prices of food. Dantzig also discovered the simplex method, an **algorithm** that was remarkably efficient for use in the linear programming of computers. It has been largely through Dantzig's vision that mathematical programming has become a field in which deep interactions between mathematics, computation, and application models are probed and developed. Dantzig is also coauthor of the book *Compact City,* which suggests improved approaches to urban development, including the use of computer programming that takes into account the "socioeconomic as well as physical aspects of complex urban systems."

Dantzig was born on November 8, 1914, in Portland, Oregon, to Tobias and Anja (Ourisson) Dantzig. Dantzig's father was born in Russia and participated in a failed revolution in 1905. After spending nine months in a Russian prison, Tobias Dantzig went to Paris and studied mathematics at the Sorbonne before immigrating to the United States in 1909. A well-known mathematician in his own right, Tobias Dantzig wrote the influential book, *Number, the Language of Science,* which focused on the concept of the evolution of numbers as related to the growth of the human mind.

Following in his father's footsteps, Dantzig attended the University of Maryland to study **mathematics and physics**. Upon graduation in 1936, Dantzig was appointed a Horace Rackham Scholar at the University of Michigan, where he earned his M.A. in mathematics in 1938. For the next two years, Dantzig worked as a junior statistician for the U.S. Bureau of Labor Statistics before enrolling in the mathematics doctoral program at the University of California, Berkeley; his studies were interrupted, however, when the United States entered World War II. Dantzig left Berkeley in 1941 to become chief of the combat analysis branch of the U.S. Air Force's statistical control headquarters. In 1944, he received the War Department Exceptional Civilian Service Medal for his efforts. In the meantime, Dantzig had returned to his doctoral studies at Berkeley, studying under **Jerzy Neyman**, a major contributor to modern mathematical statistics. He received his Ph.D. in mathematics in 1946.

Rapid advances in technology combined with the effects of World War II and urban development brought on a new era of large-scale planning tasks. With his valuable wartime military experience, Dantzig was asked to continue working for the Air Force and, in 1946, was appointed chief mathematical adviser on the staff of the Air Force Comptroller. At this time, the Air Force had begun Project SCOOP (Scientific Computation of Optimum Programs), which was designed to increase the mechanization and speed for planning and deploying military forces. Focusing primarily on the planning segment of the project, Dantzig discovered that linear programs could be used to solve a wide range of planning problems. Conceptually, this discovery was an important step toward a mathematical approach to many planning and management difficulties, but it was Dantzig's simultaneous discovery of the simplex method—an algorithm that could be efficiently used to solve programming problems—that revealed the enormous power of linear programming. Dantzig's discovery was facilitated by the fact that the modern era of computer research was also getting underway. The development of technology that could rapidly solve complicated equations—equations that otherwise could take years to complete—made linear computing programming a practical resource for use in such areas as industry and economics, which could now quickly compare the many factors involved in interdependent courses of action.

The key to linear programming and the simplex method is the use of a "best value" or set of best values for many variables involved in a certain problem. Linear programming works most efficiently when a quantity can be optimized, or made as perfect and functional as possible. This quantity, called the objective function, for example, could be the most economical way to produce and distribute a product taking into account various "system" factors, such as product composition, production scheduling, and distribution. A key to the programming's success is to develop proportional values for these factors, such as their linear interdependency, in which at least one linear combination of an element equals **zero** when the coefficients are taken from another given set and at least one of its coefficients is not equal to zero.

In 1952, Dantzig went to work with the RAND Corporation, one of the first private industries to use computer technology. As a research mathematician at RAND, Dantzig played a major role in developing the new discipline of operations research using linear programming, and became a pioneer in identifying its exhaustive uses. In Michael Olinick's book *An Introduction to Mathematical Models in the Social and Life Sciences,* Dantzig notes: "Industrial production, the flow of resources in the economy, the exertion of military effort in a war theater—all are complexes of numerous interrelated activities. Differences may exist in the goals to be achieved, the particular processes involved and the magnitude of effort. Nevertheless, it is possible to abstract the underlying similarities in the management of these seemingly disparate systems."

Over the years, Dantzig helped refine linear programming and contributed to establishing the field in both industry and academia. By the 1970s, decision-making software based on the principles of linear programming was being marketed for both technical and nontechnical users. As the growing importance of this field became apparent, universities began developing academic studies of operations research, also referred to as mathematical decision making, in such areas as business science, industrial engineering, and mathematical computing. In 1960, Dantzig left private industry and joined the University of California, Berkeley, as chairman of the Operations Research Center. Located in the heart of the "silicon valley," home of the computer programming and software industry, Dantzig was ideally situated to continue his studies. In 1963, he published the highly influential book *Linear Programming and Extensions,* which includes discussions of the origins of linear programming; according to Dantzig, the theories of linear programming dates back to **Jean Baptiste Joseph Fourier**, a French mathematician known for his research into numerical **equations** and the conduction of heat. In the book, Dantzig also delves into how the field was developed both on a theoretical basis and by real-life problems presented in the military and economics. His book has become a classic in the field.

In 1966, Dantzig became a professor of operations research and computer science at Stanford and served as acting chairman of the Operations Research Department from 1969 to 1970. He contributed to the development of such major areas of mathematical programming and operations research as quadratic programming, complementary pivot theory, nonlinear equations, convex programming, integer programming, **stochastic** programming, dynamic programming, **game theory**, and optimal control theory. With the mathematician Philip Wolfe, he also originated the decomposition principle, a method for solving large systems by exploiting the special characteristics of their block-diagonal structure. This procedure was successfully used in 1971 to solve an equation containing 282,468 variables and 50,215 equations in just 2.5 hours; such an equation would have otherwise required 37 years to complete. Throughout his career, Dantzig consulted on the development of large-scale management planning models and created mathematical models of chemical and biological processes. He also utilized computers as a fundamental aspect of mathematical programming—for example, he participated in the development of a computer language and compiler which was designed to facilitate experimentation on mathematical programming algorithms.

One of Dantzig's primary interests was in the development of analytical models of transportation systems; in 1974, he was the recipient of an endowed chair at Stanford, the C. A. Criley Chair of Transportation. Dantzig's interest in transportation and the efficient and most economical use of resources through mathematical programming led him to write *Compact City, A Plan for a Livable Urban Environment* with Thomas L. Saaty. Dantzig and Saaty set out to learn more about city planning by consulting with a range of experts, including engineers, economists, social workers, sociologists, seismologists, waste-removal engineers, and environmentalists. Concerned with such urban crises as the shortage of energy, growth of slums, congestion, and pollution, the book focuses on finding more advanced ways of developing urban areas while increasing the standard of living and minimizing the consumption of nonrenewable resources.

Compact City describes a new concept of living in which as many as two million people could live in ideal weather in spacious homes and gardens and walk to work within a few minutes. An integral part of the planning process was to transform urban development from "flat, predominantly two-dimensional cities to four-dimensional cities in which vertical **space** and time are exploited." In addition to simplifying transportation systems and alleviating the burden on energy consumption, Dantzig and Saaty were also concerned with "bringing the community together" and offering new opportunities for the underprivileged. Computers and linear programming played an integral part in the planning by taking into account not only the physical aspects but also the socioeconomic aspects of urban development.

As the conceptual developer of linear programming, Dantzig was invited to lecture around the world. He went on a one-year sabbatical in 1974 as head of the methodology group at the International Institute for Applied Systems Analysis, in Laxenburg, Austria, and received an honorary degree of doctor of science from the Israel Institute of Technology in 1973. During his career, Dantzig also received honorary degrees from the University of Linköping in Sweden, the University of Maryland, and Yale University. In 1975, U.S. President Gerald Ford awarded him the National Medal of Science in recognition of his inventing linear programming, developing methods that allowed it to be applied widely in industry and science, and using computers to incorporate mathematical theory. On November 1, 1976, California passed State Resolution No. 1748, honoring Dantzig's contributions to applied science.

In addition to his many honors, including being elected to the National Academy of Sciences in 1971, Dantzig served in many scientific societies and was the founder of the Mathematical Programming Society. Dantzig married Anne S. Shmumer on August 23, 1936. They have three children, David Franklin, Jessica Rose, and Paul Michael.

DAUBECHIES, INGRID (1954-   )
Belgian-American applied mathematician

Ingrid Daubechies was born August 17, 1954, in Houthalen, Belgium. Her father, Marcel Daubechies, is a retired civil engineer and her mother, Simone, is a retired criminologist. Daubechies credits her parents with giving her a love of learning and her mother with teaching her by example to be her own person. Her father always encouraged her to pursue her interest in science. She has one brother.

As a small child, Daubechies displayed an insatiable interest in how things worked and in making things with her hands. She took up the hobbies of weaving and pottery at a young age and continues to produce *objets d'art* in both crafts. At the age of eight or nine Daubechies' favorite hobby was to sew clothes for her dolls because it fascinated her that flat pieces of material could be worked into curved surfaces that fit the angles of the doll's body. But she also fascinated with machinery and mathematical axioms. Daubechies used to lie in bed and compute the **powers** of two, or test the mathematical law that any number divisible by nine produces another number divisible by nine when the digits are added together. Reading has been a lifelong hobby.

Daubechies spent her entire childhood and school years in Belgium. She was educated at the Free University Brussels, earning a B.S. degree in 1975 and a Ph.D. in 1980, both in physics. Her thesis was entitled "Representation of Quantum Mechanical Operators by Kernels on Hilbert Spaces of Analytic Functions." Between 1978 and 1980 Daubechies wrote ten articles based on her own original research. While pursuing her own studies, she taught at the Free University Brussels a total of 12 years. Daubechies first visited the United

States in 1981, staying for two years, then returned to Belgium believing she would not come back to America.

In 1984, Daubechies was the recipient of the Louis Empain Prize for physics. The prize is given every five years to a Belgian scientist for scientific contributions done before the age of 29. She returned to the United States in 1987 and joined AT&T Bell Laboratories, where she was a technical staff member for the Mathematics Research Center. During her employment with AT&T, she concurrently took leaves of absences to teach at the University of Michigan and later at Rutgers University. In 1993, Daubechies became a full professor at Princeton University in the Mathematics Department and Program in Applied and Computational Mathematics, where she has remained to date. Daubechies is the first woman to obtain this position at Princeton. Her responsibilities include teaching both undergraduate and graduate courses, directing Ph.D. students in thesis work, and collaborating with postdoctoral fellows in research. She has also devoted much time to creating mathematics curriculums for grades kindergarten through 12th grade that reflect present-day applications of mathematics.

Daubechies' original intent was to become a physicist (particularly in the field of engineering). But she involved in mathematical work which was very theoretical in nature. She soon found herself caught up in mathematical applications. Her designation as a mathematician was sealed through her brilliant and innovative work in wavelet theory.

In 1987, Daubechies made one of the biggest breakthroughs in wave analysis in the past two hundred years. Prior to the development of Daubechies' **theorem**, signal processing was accomplished by using French mathematician **Jean-Baptiste Fourier**'s series of **trigonometric functions**, breaking down the signal into combinations of **sine** waves. Sine waves can measure the amplitude and frequency of a signal, but they can't measure both at the same time. Daubechies changed all that when she discovered a way to break signals down into wavelets instead of breaking them down into their components; a task thought by most mathematicians to be impossible.

This discovery has changed the image-processing techniques used by the Federal Bureau of Investigation for transmitting and retrieving the information contained in their massive database of fingerprints. With more than 200 million fingerprints on file, the technique also allows for data compression without loss of information, and eliminates extraneous data that slows or clutters the procedure. Of more significance to Daubechies is the application of her discovery to the field of biomedicine. She likens a wavelets transform to "a musical score which tells the musician which note to play at what time," and this is of particular importance to medical science. Through the analysis of signals used in electrocardiograms, electroencephalograms, and other processes used in medical imaging, the medical world hopes to employ Daubechies' development to detect disease and abnormalities in patients much sooner than is presently possible. The development and implementation of wavelet imagery in medicine would improve the ability of an ECG from a simple recording of a heartbeat to a digitized record of complete heart function.

Other applications for wavelets still in the research stage include video and speech compression, sound enhancement, statistical analysis, and partial **differential equations** involving shock waves and turbulence, to name only a few.

Daubechies' work has not gone unnoticed by her peers. She has been a fellow of the John D. and Catherine T. MacArthur Foundation from 1992 to 1997 and an elected member of the American Academy of Arts and Sciences since 1993. She was the recipient of the American Mathematical Steele Prize for Exposition for her "Ten Lectures on Wavelets" in 1994, and received the Ruth Lyttle Satter Prize in 1997. Daubechies is also a member of the American Mathematical Society, the Mathematical Association of America, the Society for Industrial and **Applied Mathematics**, and the Institute of Electrical and Electronics Engineers.

Daubechies has written more than 70 articles and papers during her career, more than 20 of them dealing with the nature, application, and interdisciplinary use of wavelets. She has held memberships in more than 17 professional organizations and committees, including her current memberships with the United States National Committee on Mathematics and the European Mathematical Society's Commission on the Applications of Mathematics. Daubechies has been a guest editor or member of the editorial board for ten professional journals and has served as editor-in-chief for the publication *Applied and Computation Harmonic Analysis*.

Daubechies married A. Robert Calderbank, a mathematician, in 1987 and has two children.

DE MOIVRE, ABRAHAM (1667-1754)

French mathematician

Abraham de Moivre, a French mathematician chiefly remembered for the **theorem** in **trigonometry** that now bears his name, was born in Champagne, France on may 26, 1667. He died in London on November 27, 1754.

De Moivre was the son of a surgeon. Being a Huguenot, he chose to leave France when the Edict of Nantes was revoked in 1685. In London, he supported himself as a lecturer and private teacher in mathematics and natural science, while he continued his studies in mathematics. There he became a close friend of **Isaac Newton**, and is said to have torn the latter's *Principia* into sheets that he could carry with him and study in his spare time.

Although de Moivre aspired to a university position in mathematics, he never succeeded in securing one. Most of de Moivre's life was spent in poverty, and he managed to outlive most of his friends. He did achieve recognition for his work during his lifetime, which include a demonstration of the process of finding a root of an **infinite series**, and was admitted to the Royal Society (1697), and to the Berlin and Paris Academies.

In 1711, de Moivre published a long memoir about the laws of chance in the *Philosophical Transactions*. His first book, entitled *Doctrine of Chances*, dealt with questions about games of chance (such as dice) and was published in 1718. The book, which described methods for approximating function of large numbers, later gave him the idea of the normal distribution curve.

In 1725, he published *Annuities on Lives*, which was perhaps the first important mathematical treatment of that subject. In the book he adopted as the rule the idea that annuities can be computed on the assumption that that the number of persons in a given group that die will be the same during each year.

In 1730, De Moivre published in *Miscellanea Analytica* the formula for the factorials of large numbers that is now ironically known as Stirling's approximation.

De Moivre was one of the first mathematicians to make use of **complex numbers** in trigonometry. The theorem that he is most remembered (de Moivre's Theorem) for is as follows:

$$(\cos(\theta) + i\sin(\theta))^n = \cos n(\theta) + i\sin n(\theta)$$

Twenty years after de Moivre derived this formula, Euler enlarged upon it to shift trigonometry from the **domain** of **geometry** into mathematical **analysis**.

Newton, impressed with the significance of this result, is said to have referred those who came to him with questions about mathematics to de Moivre, with the statement "He knows these things better than I."

When de Moivre was elected to the Royal Society in 1712, it was at the proposal of Jean Bernoulli, with whom de Moivre had been in extensive correspondence since 1704. In the same year, the Royal Society appointed de Moivre a partisan commissioner to evaluate the claims of Newton and **Leibniz** concerning the invention of the **calculus**.

Augustus De Morgan

DE MORGAN, AUGUSTUS (1806-1871)

English algebraist and logician

Augustus De Morgan entered the English mathematical scene during a period of inactivity and by the time of his death it had regained the stature it had since the time of **Isaac Newton**. Although De Morgan did not devote himself wholeheartedly to the pursuit of mathematics, he is credited for promoting its study by his publications and his teaching. He worked outside the established universities and was able to appeal to a wider audience than merely the mathematics graduates of Oxford and Cambridge.

De Morgan was born in June 1806 in Madurai, a picturesque town in southern India. Shortly after his birth he lost the sight of his right eye. The De Morgans moved back to England when he was seven months old. His father, a colonel in the Indian army, continued to spend time in India and died on St. Helena in 1816. De Morgan was educated at a series of private schools before enrolling at Trinity College, Cambridge, in 1823 and graduated as fourth wrangler.

After graduation, De Morgan entered Lincoln's Inn, one of the Inns of Court intended to prepare students for a legal career. He may have been discouraged by some aspects of the mathematics curriculum at Cambridge at the time, which was

still recovering from the inertia of the 18th century. Partly in response to the quarrel between **Isaac Newton** and **Gottfried Leibniz** over the invention of **calculus**, the English mathematical community began to isolate itself from the continent in the 18th century, idolizing Newton and reluctant to change. The result was a petrification of both the foundations of the calculus and of its notation, an area in which Leibniz's version was clearly an improvement over Newton's. Cambridge was drifting along serenely, unaware of the progress of mathematics outside its walls until the arrival of a group of students who were known as the Analytical Society. They devoted themselves to the reform of mathematical education.

De Morgan soon found the legal profession unappealing and applied for the chair of mathematics at University College, London. He was offered the chair in 1828 on the strength of recommendations from his former tutors, who included some members of the Analytical Society. As a teacher, De Morgan was devoted to the presentation of ideas and principles rather than techniques, and his pupils included some of the most distinguished British mathematicians of the next generation. In addition, he also produced a series of textbooks on **arithmetic**, **algebra**, **trigonometry**, calculus, **complex numbers**, **probability**, and logic. These were written clearly and with attention to giving an intuitive understanding as well as one based on calculation.

In his own mathematical work, De Morgan made major contributions in the area of logic. The Aristotelian tradition of logic had become fossilized during the Middle Ages, and instruction in logic frequently resorted to memorizing a few lines of low Latin and a little caution about the misuse of rhetoric. If reasoning about mathematics was being carried out in ordinary language, there was not much advantage to studying mathematics in approaching questions of logic.

One of the main interests of the English mathematical community at that time was the status of the laws of algebra. It was clear that some of the laws applied to all the systems of numbers then known, but other laws did not apply beyond a restricted **domain**. The **quaternions** discovered by Sir **William Rowan Hamilton**, for example, (which are related to vectors in three dimensions) had an operation of **multiplication** which was not commutative (that is, a x b was not equal to b x a). An obvious question was that of which laws are automatically satisfied by any objects whatever, which could be considered laws of logic.

The crucial respect in which De Morgan sought to improve on the traditional logic of **Aristotle** was in the treatment of the logic of relations. In **Aristotelian logic**, all statements had to be analyzed into the form "A is (or is not) B," with the possible inclusion of "all" and "some." It was not clear how this could be used to handle statements like "A is taller than B" or "A is closer to B than C." De Morgan's work on the logic of relations did not become part of the mainstream, due to the shortcomings of his notation. More successful in the reform of logic was **George Boole**, author of *The Mathematical Analysis of Logic* and creator of a superior notation. Boole acknowledged his debt to De Morgan, whose name remains attached to two laws of **Boolean algebra** involving the negations of compound expressions.

De Morgan also wrote on probability. As far as the interpretation of probabilities was concerned, De Morgan fell in the "subjectivist" rather than the "frequentist" school. In other words, probability statements were reflections of the degree of belief attached to **propositions** rather than features of the natural world itself.

In 1831 De Morgan resigned his chair at University College. By 1836, following the death of his successor, however, De Morgan returned to his position. The next year he married Sophia Elizabeth Frend, who wrote De Morgan's biography after his death. Although De Morgan found his family life a comfort after some of the controversies in the academic world, his later years were saddened by the deaths of a couple of his children.

De Morgan wrote many articles for the popular press. This type of writing did not command much respect, and it is worth noting that De Morgan never became a Fellow of the Royal Society, of whom he criticized for being too open to social influence, as indicated by the proportion of nobility among its members.

Much more to De Morgan's taste was the London Mathematical Society, which he cofounded and served as first president. This type of mathematical organization was more fitting for someone who had contributed 850 articles for one reference work alone.

De Morgan resigned a second time from University College in 1866 and on this occasion could not be tempted to return. He died in London on March 18, 1871. Not so much by his research as by his pedagogical efforts had De Morgan transformed the mathematical community in England into a setting for the discussion of the current topics of interest in mathematics both English and European.

DE MORGAN'S LAWS

Augustus De Morgan (1806-1871), the British mathematician and logician, contributed what we now call De Morgan's Laws to the science of logic. These laws allow the substitution of one statement for another logically equivalent statement. There are two pairs of statements that operate according to De Morgan's law, and each statement in a pair means, logically speaking, the same thing as its partner. These laws are helpful in making valid inferences in proofs of deductive arguments, and the laws are based on the concept of logical **equivalence** and the logical **functions** of the connectives "not," "and," and "or." There are other statements that are logically equivalent and used in deductions, but De Morgan's laws employ and clarify quantificational relationships moreso than do other equivalence rules.

To assert that two statements are logically equivalent means that they share the same **truth** values in every possible instance. The concept of logical equivalence between two or more statements relies, then, on the concept of truth-values. A truth-value is the truth or falsity of a given statement. Every statement is a sentence that is either true or false. Thus, "The cat is on the mat," has a truth-value of true or false depending upon whether it is true or false that the cat is in fact on the mat. When two statements are compared with each other, there are four possible truth values which exhaust all the permutations of possible truth values. For example, when comparing the statements, "The cat is on the mat," and "The bug is under the rug," there will be one instance in which both statements are true at the same time, one in which they are false at the same time, but in the other two instances the values are not equal. Consequently, the statements are not logically equivalent.

De Morgan's laws of replacement are based on the equivalent truth-values of the two statements. In other words, De Morgan's laws assert that there are two pairs of statments, each pair of which shares the same truth-values in every possible instance. Consequently, one of the two statements may be inferred from, or replaced by, another in each pair. The main operator functioning in each pair is a negation, which sets De Morgan's statements apart from other logically equivalent statements.

De Morgan's laws assert that the following pair of statement forms is logically equivalent: "Not both" and "Either not one or not the other." Additionally, the following pair of statement forms is also logically equivalent: "Neither one nor the other," and "Not one and not the other." It is easier to recognize these statement forms when they are symbolized. Lower case letters p and q are stand-ins for any statement, like the mathematical variables x, y, and z. The negation "~", **conjunction** "·", and "v" symbols are the operators that join the statements together to form more complex statements.

In symbolic form, then, the first pair of logically equivalent statements is the negation of a conjunction: $\sim(p \cdot q)$ and the negation of both sides of a disjunction: $\sim p$ v $\sim q$. The second pair of logically equivalent statements is the negation of a disjunction: $\sim(p$ v $q)$ and the negation of both sides of a conjunction: $\sim p \cdot \sim q$. Thus, $\sim(p \cdot q)$ is logically equivalent to $\sim p$ v $\sim q$, and vice versa, and $\sim(p$ v $q)$ is logically equivalent to $\sim p \cdot \sim q$, and vice versa. Considered in terms of their truth-values, each statement in a pair shares the same truth-values as its partner in each instance, and so the statements in each pair are logically equivalent.

The statement, $\sim(p \cdot q)$ asserts the following: "It is not the case that both p and q obtain," or "Not both p and q obtain," and its logical equivalence, $\sim p$ v $\sim q$, asserts that "either p does not obtain or q does not obtain." The statement $\sim(p$ v $q)$ asserts the following: "Neither p nor q obtains, which is logically equivalent to $\sim p \cdot \sim q$.

When we claim, for example, "Not both Kim and Reggie are in class," we are claiming that "Either Kim is not in class or Reggie is not in class," we just do not know which of the two is absent. Similarly, when we claim, for example, "Neither Latoya nor Quentin is in class," we are claiming that "Latoya is not in class and Quentin is not in class." We can see that "Not both" does not make the same assertion as "Both are not." Likewise, "Neither/nor" does not make the same assertion as "either one is not or the other is not." However, "Not both," logically speaking, means the same thing as "either one is not or the other is not," and "Both are not," means the same thing, logically speaking, as "Neither/nor." Such distinctions help to ensure valid reasoning from one proposition to another.

DECIDABLE

The word "decidable" has two distinct meanings in mathematics. The first meaning came from logic. A set of axioms is a set of statements that are regarded as true in the system for which the axioms hold. If a statement in the language of the system can be proven true or false within the system itself, then it is said to be decidable. For example, the statement "$1 + 1 = 2$" can be proven true in Zermelo-Fraenkel **set theory**. So, this statement is decidable in **Zermelo-Fraenkel set theory**. On the other hand, the continuum hypothesis is undecidable in Zermelo-Fraenkel set theory. **Kurt Gödel**'s incompleteness **theorem** states that for most systems (those whose axioms are defined recursively), the consistency of the system is undecidable within itself. This means that it cannot be proven, by

using only the axioms within the system, that internal contradictions do not occur, i.e. that no statement can be proven to be both true and false. This theorem effectively stopped a fifty year long effort to prove the consistency of various branches of mathematics.

The second meaning of decidable comes from the study of algorithms. A class of **propositions** is called "decidable" if an **algorithm** exists that can be applied to each proposition in the class to determine whether or not the proposition is true. For example, Wang tiles are 1-by-1 squares that have colored edges. A tile is allowed to butt up against another tile only when the edges that touch are the same color. The tiling problem is: does there exist an algorithm that can be applied to any finite set of Wang tiles to determine whether those tiles could be used to tile a room of any given (rectangular) size? Robert Berger proved the answer is no in 1966. It is still unknown what the answer is to the analogous question in Lobachevsky's plane.

This idea of decidability has been around at least since 1920s. But there was no **definition** of an algorithm that was rigorous enough to allow one to prove that none existed when this was the case. This changed in 1936 when **Alan Turing** (the "father of computer science") published "On Computable Numbers, with an application to the Entscheidungsproblem". In it he introduced what has since been the most useful **model** of computers to this date, the **Turing machine**. The tiling problem, for example, is undecidable because if the tiles are encoded as numbers and fed into a Turing machine, the machine's output will not correspond to whether or not the given set of tiles can tile the plane.

Other famous decidability questions include the word problem, the knot problem, and the single tile problem. The word problem is this: given a finite presentation for a group, is there an algorithm that decides whether any given word in the generators is equal to the **identity element**? This answer is no. Novikov gave an explicit example of a finite group presentation without this property. A knot is loop of string whose ends have been glued together. Two knots are equivalent if one can be stretched or compressed and moved around (without cutting or tearing) to look like the other knot. The knot problem is this: does there exist an algorithm for determining whether two given knots are equivalent? The answer is yes! **Wolfgang Haken** wrote the algorithm in 1961, but it is so complicated that it has never been implemented. The single tile problem is this: does there exist an algorithm for deciding whether a single polygon can tile the infinite plane? This problem is still open.

See also Algorithm; Continuum hypothesis; Gödel's theorem; Turing, Alan; Turing machines; Zermelo-Fraenkel set theory

DECIMAL FRACTIONS

All common **fractions** can be converted to decimal fractions by dividing the numerator by the denominator. For example:
- $\frac{1}{2} = 0.5$

Ancient mathematical records—Hindu-Arabic, Greek, Egyptian, and Babylonian—show many different methods for noting fractions. However, Islamic scholars were the first to convert fractions to decimal analogues. During the Middle Ages, scholars used the Babylonian system of sexagesimal (base 60) fractions for computation, particularly when calculating **square roots** and cube roots. In the fifteenth century, the astronomer al-Kashi invented a method for converting sexagesimal fractions to decimal fractions. Al-Kashi also discovered a way to approximate the value of 2π in both sexagesimal and decimal fractions.

Despite al-Kashi's discovery, the use of decimal fractions evolved very slowly. **Johannes Regiomontanus** (1436-1476) compiled tables of sines and cotangents that were very close to decimal fractions. In 1585, **Simon Stevin** (1548-1620) was the first to offer conceptual details of decimal fractions and appreciate their significance. Stevin's system was cumbersome because he used a number in a **circle** to indicate each decimal place: 0 for the units place, 1 for the tens place, and so forth. The 1616 English translation of the **logarithms** of **John Napier** (1550-1617) marked the first time the decimal point appeared in print. In the late seventeenth century, some mathematicians were signifying a decimal by using the left half of a pair of brackets or other such notations. Decimal fractions still not used universally until the eighteenth century; until then, about half of the **arithmetic** texts mentioned decimals and half did not.

The International System of Units (SI) favors the use of the decimal comma as a decimal marker and most countries follow this form. Many countries used a decimal comma even before the SI recommendation. Most English-speaking countries use a decimal point. In the British press, sometimes a dot raised halfway above the line is used as a decimal marker. This notation causes difficulty because it can be mistaken for a **multiplication** sign. Many English-speaking countries have started calling the entire number a decimal rather than a decimal fraction. Strictly speaking, a decimal is the part that follows the decimal marker. When reading a decimal fraction out loud, "point" is a common abbreviation for "decimal point." For clarity, decimal fractions whose absolute value is less than one are written with the unit **zero** in front of the decimal point—for example, 0.5.

In long decimal fractions, digits are often divided into groups of three, working both ways from the decimal marker:

- 123 000.000 321

In scientific tables, by contrast, the digits are often shown in groups of five.

DECIMAL POSITION SYSTEM

The decimal position system, also called the Hindu-Arabic or Arabic system, is a numeral system in which all derived units are based on the number 10 and the **powers** of 10. The decimal system in nearly universal use today requires ten different symbols, or digits (0, 1, 2, 3, 4, 5, 6, 7, 8, 9) and a decimal point (dot) to represent numbers. In this scheme, the numerals used in denoting a number take different values depending upon position. A number is written as a row of digits, with each position in the row corresponding to a certain power of 10. A decimal point in the row divides the row into those powers of 10 equal to or greater than **zero** and those less than zero. Positions farther to the left of the decimal point correspond to increasing positive powers of 10 and those farther to the right to increasing negative powers.

As previously stated, the decimal number system is based on the powers of the base 10, that is (..., 1000, 100, 10, 1, 1/10, 1/100,...) = (..., 10^3, 10^2, 10^1, 10^0, 10^{-1}, 10^{-2},...). Any real number can be represented in a variety of positional number systems, each of a unique base "b". Since the decimal number system has 10 as its base (i.e., b = 10), any real number x can be represented as: $x = a_n10^n + a_{n-1}10^{n-1} +... + a_110^1 + a_010^0 + a_{-1}10^{-1} +... + a_{-(n-1)}10^{-(n-1)} + a_{-n}10^{-n}$. The coefficients of this equation (i.e., the values a_i, where i = {0, 1, 2,..., n}) may take on the numeric values of 0 through 9; or stated another way, $0 \leq a_i$ b. As a particular example of the general equation just given, the decimal notation "647.25" represents the sum $(6 \times 10^2) + (4 \times 10^1) + (7 \times 10^0) + (2 \times 10^{-1}) + (5 \times 10^{-2}) = (600 + 40 + 7 + 2/10 + 5/100) = (600 + 40 +7) + (0.20 + 0.05)$.

The decimal system is derived from the Arabic system and can be traced back to ancient Egyptian, Babylonian (Sumerian), and Chinese roots. This Near East (Arabic digits) system was first developed by the Hindus and was in use in India in the third century B.C.. At that time the numerals 1, 4, and 6 were written in substantially the same form that is used today. The Hindu numeral system was probably introduced into the Arab world about the seventh or eighth century. The early Egyptians used a base-10 system that had different symbols for each power of 10 up to 10^6, and lacked a place-value notation and an explicit number zero. The Chinese, Cretans, Greeks, Hebrews, and Romans also used similar decimal systems that lacked the features of our modern decimal system, but helped to formulate that system. The important innovation in the Arabic system was the use of positional notation, in which individual number symbols assume different values according to their position in the written number system. Positional notation was made possible beginning around AD 800 by the use of a symbol for zero. The creation of the symbol "0" made it possible to differentiate between 11, 101, and 1001, for instance, without the use of additional symbols. Therefore, all numbers could be expressed in terms of only ten symbols, the numerals from 1 to 9 plus 0. Positional notation also greatly simplified all forms of written numerical calculation. The majority of the credit for the origin of the modern base 10 system goes to the Hindu-Arabic mathematicians between the eighth and eleventh centuries. The first recorded use of the system in Europe was in AD 976. The notation of the modern-day decimal system is credited to the work of Leonardo of Pisa (**Fibonacci**) in AD 1202. Thomas Harriot (1560-1621) and **Gottfried Wilhelm von Leibniz** (1646-1716) both independently developed the generalized treatment of positional number systems that included the decimal number system.

Decimal is derived from Latin as "decem" (10) and Greek "deka" (10). It is probable, but not certain, that the special position occupied by 10 is biologically connected to the number of human fingers. In fact, the word "digit" is from the Latin *digitus* and means "finger". The base 10 system is still evident in modern usage not only in the logical structure of the decimal system, but also in the English names for numbers. Thus, eleven comes from Old English *endleofan* that literally means "[ten and] one left [over]", and twelve from *twelf* that is defined as "two left". In addition, the endings "-teen" and "-ty" both refer to 10, and the word "hundred" comes originally from a pre-Greek term meaning "ten times [ten]". The powers of ten commonly possess their own names such as (within the United States) 1 million for $10^6 = 1,000,000$; 1 billion for $10^9 = 1,000,000,000$; and 1 trillion for $10^{12} = 1,000,000,000,000$.

In the base-10 number system, the number 983.75, for instance, is a compact way of writing $(900 + 80 + 3 + 7/10 + 5/100) = (9 \times 10^2) + (8 \times 10^1) + (3 \times 10^0) + (7 \times 10^{-1}) + (5 \times 10^{-2})$. For comparison, the number 983.75 is expressed in base 2 as 1111010111.11 (that is, $(1 \times 2^9) + (1 \times 2^8) + (1 \times 2^7) + (1 \times 2^6) + (0 \times 2^5) + (1 \times 2^4) + (0 \times 2^3) + (1 \times 2^2) + (1 \times 2^1) + (1 \times 2^0) + (1 \times 2^{-1}) + (1 \times 2^{-2})$ and in base 8 as 1727.6 (that is, $(1 \times 8^3) + (7 \times 8^2) + (2 \times 8^1) + (7 \times 8^0) + (6 \times 8^{-1})$).

The decimal system is widely used in various systems employing numbers. The **metric system** of weights and measures, used in most of the world, is based on the decimal system, as are most systems of national currency. In the course of history, the decimal system finally overshadowed all other positional number systems, and it is now found in all technologically advanced nations.

See also Babylonian mathematics; Binary, octal, and hexadecimal numeration; Decimal fractions; Egyptian mathematics; Numbers and numerals; Zero

DEDEKIND, (JULIUS WILHELM) RICHARD (1831-1916)
German number theorist

Richard Dedekind is best known for his work in **number theory**. He redefined **irrational numbers**, proposing that rational and irrational numbers form a continuum in which **real numbers** are located by "cuts" in the realm of **rational numbers**. He also introduced the notion of an ideal (for example, the collection of all integer multiples of a given integer), which allowed wider application of factorizationand is fundamental to modern ring theory. In addition to Dedekind cuts, about a dozen mathematical concepts carry his name. Although he was among the most capable and original mathematicians of his day, Dedekind was a modest man, spending most of his professional life as a teacher at the technical high school in his hometown of Brunswick. He was a gifted teacher, and his teaching was an integral part of his mathematical thinking.

Dedeking was born Julius Wilhelm Richard Dedekind on October 6, 1831, in Brunswick (Braunschweig), now Germany, the last of four children. As an adult he dropped his first two names. His father, Julius Levin Ulrich Dedekind, was a lawyer and professor at Caroline College in Brunswick and the son of a physician and chemist. His mother, Caroline Marie Hanriette Emperius Dedekind, was a daughter of a professor at the College and a granddaughter of an imperial postmaster. Dedekind's brother, Adolf, became a district court president; his sister, Julie, became a novelist.

From age seven to age sixteen, Dedekind studied at the Gymnasium in Brunswick. At first, he concentrated on physics and chemistry, considering mathematics merely a scientific tool; however, he eventually became enthralled with the logic of mathematics. From 1848 to 1859, Dedekind attended Caroline College where he studied **analytic geometry**, advanced **algebra**, the **calculus**, and higher **mechanics**, and gave private lessons. In 1850, at age nineteen, he entered the University of Göttingen, where he became the last doctoral student trained by **Karl Gauss**. At Göttingen, Dedekind studied calculus, elements of higher **arithmetic**, least squares, higher geodesy, and experimental physics. Dedekind's doctoral thesis on Eulerian integrals, completed after only four semesters at Göttingen, was a solid but uninspired piece of work. However, Gauss praised his knowledge and independence, and predicted future success for him.

Dedekind continued to study and attend lectures, and in 1854 he was appointed lecturer *(privatdozent)* at Göttingen, just a few weeks after his friend **Georg Riemann**, who also studied under Gauss, received a similar appointment. Some time later Dedekind and Riemann traveled together to Berlin to meet with the mathematical community there. When Gauss died in 1855, Dedekind served as a pallbearer at the funeral service. When **Peter Gustav Lejeune Dirichlet** came to Göttingen from the University of Berlin to take Gauss' place, he and Dedekind became close friends and colleagues. Dedekind attended Dirichlet's lectures, and their discussions inspired Dedekind's investigations in new directions, making a "new man" of him. Dedekind was among the first to recognize the application of Galois groups in algebra and arithmetic, and in 1857–58 gave a course to two students on **Évariste Galois'** theory of **equations** (see **Galois therory**).

In 1858, Dedekind was invited to succeed Joseph Ludwig Raabe at the Polytechnic School in Zürich. In recommending him for the position, Dirichlet described Dedekind as "an exceptional pedagogue." A position in Zürich was traditionally a first step toward a professorship in Germany. However, after five years in Zürich, in 1862 Dedekind succeeded Wilhelm Julius Uhde as professor of higher mathematics at the technical high school in Brunswick. He stayed there the remainder of his life; Dedekind directed the school from 1872 to 1875, and was named professor emeritus in 1894.

In assuming the position at the technical school, which had been created under the auspices of Carolina College, Dedekind was following in his father's administrative footsteps. In Brunswick, he lived in close association with his family and did not aspire to a greater position. He was an accomplished cellist and pianist, and composed a chamber opera for his brother's libretto. Dedekind lived with his sister Julie until her death in 1914. He died in Brunswick on February 12, 1916.

Dedekind's work focused almost totally on the area of numbers. As a result of attempts to answer questions about real numbers, several ideas had been put forth, all involving infinite **sets** or sequences. The simplest idea was Dedekind's. He defined a real number to be a partition, or "cut," of the rational numbers into two sets, so that each member of one set is less than all numbers of the other. These **Dedekind cuts**— which he said occurred to him on November 24, 1858—gave a precise **model** for the continuous number line, since they filled all the gaps in the rationals. Other formulations followed from his **definition**. In 1872, Dedekind defined an infinite set in his paper *Stetigkeit und irrationale Zahlen*. In 1888, he expanded these ideas in a book, *Was sind und was sollen die Zahlen*? With his 1872 paper he had joined **Karl Weierstrauss** and **Georg Cantor** in defining a new mathematical area.

In 1879, Dedekind published *Über die Theorie der ganzen algebraischen Zahlen*, in which he introduced the notion of an ideal, which is fundamental to ring theory. Dedekind formulated his theory in the ring of **integers** of an algebraic number **field**. His idea was later extended by **David Hilbert** and **Emmy Noether**. Dedekind also collected, explained, extended, and published the works of those mathematicians who had influenced him: Gauss (1863), Riemann (1876), and Dirichlet (1863, 1871). His work on Dirichlet's lectures led him to a theory of generalized **complex numbers** and forms that can be resolved into linear factors. His work was characterized by exceptional clarity, and he has been credited with creating a "style" of mathematics.

Dedekind was a corresponding member of the Göttingen Academy (1862), the Berlin Academy (1880), and the Paris Académie des Sciences (1910). He was a member of the Leopoldino–Carolina Naturae Curiosorum Academia and the Academy of Rome. Dedekind also received many honors, including honorary doctorates from Brunswick and the University of Oslo. On one occasion, his death was listed on a *Calendar for Mathematicians* as September 4, 1899; an amused Dedekind wrote the publisher that he had spent that day talking with his friend Georg Cantor.

DEDEKIND CUT

Real **analysis**, which studies the theoretical foundations of the **calculus** of real-valued **functions**, depends upon an accurately defined real number concept. Such a precise **definition** eluded mathematicians from the time of **Pythagoras** (c.560-c.480BC) to the latter half of the nineteenth century. The problem was not with all **real numbers**, but with that subset of the reals known as the **irrational numbers**. Pythagoras, it is said, was so horrified by the appearance of the irrational number $\sqrt{2}$, i.e., the positive **square root** of 2, in his calculation of the **hypotenuse** of a right **triangle** that he swore his followers to secrecy on the matter. The Pythagoreans called such numbers "incommensurable" because they could not be obtained by multiplying two **rational numbers** together. A rational number can always be expressed as the quotient of two integers—that's the definition of a rational number—but these

"incommensurables" cannot be expressed in this way; hence the name "irrational." Nevertheless, since the real number line is a continuum with no "gaps" or "holes," there must be a location on the real line for numbers such as $\sqrt{2}$. Another way of looking at it is that if one takes the hypotenuse of an isosceles right triangle whose sides have length 1 and places it along the real number line with one end at 0, then clearly there is a point on the line at which the other end rests. But where, exactly, does $\sqrt{2}$ reside on the real line. We can determine that it lies somewhere between the rational numbers 1.4 and 1.5. We can pin it down even more accurately between 1.41 and 1.42, or, more accurately still, between 1.414 and 1.415, and so on. Even so, with each new upper and lower bound for $\sqrt{2}$, we have still not reached the precise point on the real number line where $\sqrt{2}$ should be located. The problem is that we can approximate the positive square root of 2 by rational numbers to any desired number of decimal places. But the square root of 2 has a non-ending and non-repeating decimal expansion; so how can we ever define the "exact" value of the positive square root of 2. Enter **Richard Dedekind** (1831-1916), who saw in this method of approximating irrational numbers by rational numbers a potential definition for the irrational number $\sqrt{2}$, and indeed for all irrational numbers. His idea was to define a "cut" of the rational numbers corresponding to each irrational number. Consider the case of the cut corresponding to $\sqrt{2}$. He defined this cut as the set of rational numbers which are less than $\sqrt{2}$ but which contains no largest rational number. Thus the cut contains an infinite set of rational numbers which progress to the right on the number line but which are always less than $\sqrt{2}$. Dedekind then actually defines $\sqrt{2}$ as this cut. More generally, Dedekind gives the following definition of a cut.

A set K of rational numbers is said to be a cut if (i) K contains at least one rational number, but not every rational number; (ii) if p is in K and q is a rational number less than p, then q is in K; and (iii) K contains no largest rational number.

It can be shown that each rational number is also associated with its own cut. It is then possible to define a real number as a cut, which is what Dedekind did. All the familiar properties of **arithmetic**, in particular, thefield properties, hold for Dedekind cuts. That is, the arithmetic of cuts obeys the same operational rules that govern the rational numbers. Because of this, Dedekind's work in this area has sometimes been called the "arithmetization" of the irrational numbers or, more broadly, the "arithmetization" of real analysis. Although **Georg Cantor** (1845-1918) put forth an alternative definition of the real numbers based on infinite sequences, Dedekind's characterization of real numbers as cuts of rational numbers is today the standard approach taken in most textbooks on real analysis and **number theory**.

DEDUCTION

Deduction is the process of deriving conclusions from logical premises without resort to empirical evidence. Deductive reasoning is the primary method of reasoning used in mathematical **proof**, whereas, inductive reasoning, or reasoning from specific empirical facts to more general conclusions, is the method most often practiced in the natural sciences. In a deductive mathematical system, certain undefined terms, definitions, and "self-evident" assumptions called axioms or postulates are stated, after which theorems or **propositions** of the system are derived, based only upon what has been previously assumed or proved. The Greek philosopher, **Aristotle** (384-322 B.C.) laid the foundation for deductive argument with his method of the syllogism. Syllogistic reasoning, in which conclusions are derived from stated premises, dominated deductive logic for nearly 2000 years after Aristotle. Modern logicians give less emphasis to syllogistic reasoning, but the **symbolic logic** developed during the 20th century remains deductive in requiring rigorous arguments from a bare minimum of assumptions. Most modern mathematicians believe that the essence of mathematics is that all of its branches can be developed axiomatically, that is, deductively, from a few basic assumptions. In this view, mathematics is a purely logical discipline unaffected by empirical evidence. This position allows for the **modeling** of "real world" processes with mathematics, but considers the mathematics itself to be in the realm of Plato's ideas or "pure forms." The mathematical Platonist believes that all of mathematics can be "discovered" deductively with no reference to natural phenomena. If the mathematics so derived happens to lend itself well to the needs of physical science and engineering, so much the better, but this is not the pure mathematicians's primary concern. The pure mathematician desires a solid foundation of indisputable definitions and axioms from which her entire mathematical system may be constructed by proving theorems using nothing more than these axioms. This is the essence of axiomatic or deductive mathematics.

Historically, the most famous example of a deductive mathematical system is the axiomatic development of **geometry** set forth in the *Elements* of **Euclid** (c. 300 BC). In this work, Euclid proposes to deductively derive the entire body of mathematics as it then existed from a few undefined terms, definitions, and five postulates. Starting with this bare minimum of assumptions, Euclid then stated and proved "propositions" or what we today call "theorems." Each proposition had to be proved either from one or more of the five postulates or from one or more propositions that had already been proved. In this way, the mathematical structure so developed could have absolute certainty, since every step in the deductive process led ultimately back to the five postulates, which were chosen for their own indubitability. If this process were carried out in what we would today call a "rigorous" manner, then the resulting system of mathematics could be doubted only if one or more of the five postulates were doubted. Through the ages, following Euclid's writing of the *Elements*, many mathematicians and philosophers called into question the claim of the "self-evidence" of the fifth **postulate**, also called the "parallel postulate," because of Euclid's rather lengthy wording of that postulate. Many thought that Euclid could have proved the fifth postulate from the first four, thus eliminating the system's dependence on what some believed

to be a too elaborate statement. In the end, Euclid was exonerated as 18th-century mathematicians showed that the **parallel postulate** could not be proved from the other four. It had to be an independent fifth postulate just as Euclid had said. In any case, the *Elements* became the paradigm for the development of deductive mathematical systems. The geometry course still taught in most American high schools is usually called "Euclidean" geometry because it is based on modern variations of the deductive process introduced by Euclid more than two-thousand years ago.

In the late nineteenth century, the Italian mathematician, **Giuseppe Peano** (1858-1932) followed Euclid's example and established his own system of five postulates, or axioms, of **arithmetic** that became the foundation upon which 20th century deductive mathematics would be built. Peano's axioms, together with the work of **Cantor**, **Dedekind**, **Frege**, **Russell**, and many other 20th-century mathematicians and logicians, made possible the deductive construction of the real number system and the mathematics which is based upon it. Although the logician **Kurt Gödel**, in 1930, proved some results that brought into question the limits of what we can know through deduction alone, it is safe to say that most theoretical mathematicians of the 21st century remain comitted to the deductive method in practice.

DEFINITE INTEGRAL

The definite integral of a function f on the **interval** from a to b, where $a < b$, is a unique number I that is greater than or equal to all the lower sums and less than or equal to all of the upper sums of that function. The lower and upper sums of a function f for a partition P of $[a, b]$ are defined as $L_f(P) = m_1\delta x_1 + m_2\delta x_2 + ... m_n\delta x_n$ and $U_f(P) = M_1\delta x_1 + M_2\delta x_2 + ... M_n\delta x_n$, where for any integer within a subinterval the ms and Ms are the minimum and maximum values of f on that particular subinterval. A definite integral is usually written as $\int_a^b f(x)dx$. The $\int$ is called the integral sign, the numbers a and b are referred to as the **limits** of integration, and the function f appearing in the integral is called the integrand. The main difference between a definite integral and an indefinite integral is that a definite integral is defined on a definite interval, a and b, and is equal to a unique number whereas an indefinite integral is not. The definite integral is a number depending only on f, a and b. The **variable** x is a dummy variable in that it may be replaced by any other variable not already in use.

The definite integral of a function f is equal to the **area** bounded by f on one side, the x-axis, on the left by the line $x = a$ and on the right by the line $x = b$. This region bounded by this **space** is called the region between the graph of f and the x-axis on $[a, b]$. Although in some situations it is possible to calculate the definite integral of f on the interval $[a, b]$ by calculating formulas for the lower and upper sums some **functions**, such as x^2 and $\cos x$ is tedious at least and sometimes very difficult.

Although we defined the definite integral of the function f on an interval where $a < b$ we can relate the situations where the limits of integration are reversed, that is $\int^a_b f(x)dx$. This definite integral is related to the one we spoke about above as $\int^a_b f(x)dx = -\int^b_a f(x)dx$. We should also mention that the area of a line segment is equal to 0. That is $\int^a_a f(x)dx = 0$.

The upper and lower sums are also known as Riemann sums of f on $[a, b]$. An important attribute of a Riemann sum is that it approximates the definite integral of a function. **Georg Bernhard Riemann**, for whom the sums are named, was the German mathematician who in the 19th century clarified the concept of the integral while employing such sums. The first formal **definition** of the integral is attributed to him.

DEFINITION

Definition is the description of the meaning of a word, phrase, or concept. There are various kinds of definition, including truth-functional, recursive (or inductive), lexical, stipulative, ostensive, definition by abstraction, and definition by genus and difference. The truth-functional definitions and definitions by abstraction are the most commonly used in logic and mathematics.

A lexical definition is the report of a meaning a word already has. This is mainly what a dictionary definition is. A stipulative definition is one which assigns a new meaning to a word. Stipulative definitions are often used by theoreticians attempting to use a word in a way that will capture an idea in a way not thought previously. An ostensive definition is an example of the word to be defined. Pointing to an apple would be an ostensive definition of the word, "apple." The definiendum (that which is defined in a definition) in a definition by abstraction is a class term, and it is defined by the properties a thing must have in order to be a member of that class. For example, whatever properties are necessary for calling something a human constitute a definition by abstraction. A definition by genus and difference, like the definition by abstraction, is determined by classes. The genus is the larger class, and the species is the smaller class. For example, human being is a species of the genus animal. What differentiates humans from other species in the genus is a peculiar form of rationality. The genus is the class, and the 'difference' is the property or properties that distinguish a species of a genus. So, human being is defined by its genus, animal and its difference, rationality, from other species.

The definition of a logical operator is called a truth-functional definition, or simply a truth-function. This definition is determined by the components of the statement which are connected by the operator. For example, the definition for the **conjunction** (or "and") operator is determined by the **truth** of the statement's components so that the conjunction is true if and only if both sides of the conjunct are true. If one side of the conjunct is false, the whole statement is false. Likewise, if both sides are false, the whole statement is false. The statement, "Tom and Jane are in class," is true if and only if Tom is in class and Jane is in class. The definition of conjunction applies to any statement employing a conjunction, as illustrated by the Tom and Jane example.

A recursive, or inductive definition applies a definition repeatedly to various instances, like a formula. For example, the series of natural numbers can be defined recursively. 1 is the first in the series, then 2, 3...n, $n+1$, and so on, where n is some successive number in the series, and the series is defined by applying the formula $n+1$ to any number in order to generate the next in the series.

Definition is important to mathematics because it provides clarity to concepts that are meant to have universal application. **Analytic geometry**, for example, first presented by **Descartes** in the 17th century, is defined by the relationship of points on an axis. The "**Cartesian coordinates**," as they are now known, show us that all **quadratic equations** become geometrical figures when graphed as connected points on an axis of intersecting lines.

DELIGNÉ, PIERRE RENÉ (1944-)
Belgian algebraic geometer and number theorist

Pierre Deligné is a research mathematician who has excelled at making connections between various fields of mathematics. His research has led to several important discoveries, the most critical of which is the **proof** of three famous conjectures made by the mathematician **André Weil.** For this work, Deligné received both the Fields Medal, the highest honor in mathematics, and the Crafoord Prize. In recognition of his reception of the Fields Medal, David Mumford and John Tate, both of the Harvard University Department of Mathematics, wrote in *Science* magazine that "There are few [mathematical] subjects that [Deligné's] questions and comments do not clarify, for he combines powerful technique, broad knowledge, daring imagination, and unfailing instinct for the key idea."

Pierre René Deligné was born on October 3, 1944, in Brussels, Belgium, where he and his parents, Albert and Renee Bodart Deligné, lived throughout his childhood. The young Deligné showed an early affinity for mathematics, and his interest was encouraged by M. J. Nijs, his high school teacher. Nijs loaned Deligné several books by **Nicolas Bourbaki** that introduced concepts of modern mathematics, such as **topology**, long before discussing the topics traditionally studied first. Despite the unfamiliar and complicated terminology, Deligné's understanding of mathematics flourished, and after completing high school he enrolled at the University of Brussels. Deligné obtained his degree in mathematics there in 1966 and remained for graduate study.

Deligné's adviser at the University of Brussels, group theorist Jaques Tits, suggested in 1965 that Deligné travel to Paris. Since Deligné was interested in **algebraic geometry**, Tits felt that he should study where some of the most important researchers in that field were teaching and researching—mathematicians such as **Jean-Pierre Serre** and **Alexander Grothendieck.** Deligné went and met both Serre and Grothendieck; his association with them would strongly influence his career. After returning to Brussels to complete work on his dissertation, he received his Ph.D. in 1968.

Following completion of his doctorate, Deligné took up residence in Bures-sur-Yvette, a small community south of Paris, where the Institut des Hautes Etudes Scientifiques (Institute for Advanced Scientific Study—IHES) is located. He had been appointed a visiting member of this organization so that he could continue his research with Grothendieck; he became a permanent member of the IHES in 1970. For several years, Grothendieck had been working to generalize and update the field of algebraic **geometry** by making it more compatible with recent abstract mathematical theories. Deligné admired and learned from Grothendieck's work, although he followed a different approach. Whereas Grothendieck tried to connect algebraic geometry with all other fields by creating new theories or rules, Deligné instead worked to uncover connections already implied by previous work in these fields. Contrasting the two men's styles, Mumford and Tate observed, "One could say that Grothendieck liked to cross a valley by filling it in, Deligné by building a suspension bridge."

A prime example of Deligné's methods is his work on the Weil conjectures. Proposed in 1949 by the mathematician André Weil, these three conjectures state that it should be possible to determine the number of solutions for certain systems of **equations** by predicting the shapes of the graphs of the solutions. In other words, by using certain topological concepts, algebraic results can be obtained. As explained by **Michael Atiyah** in his 1975 Bakerian Lecture, this amounted to finding an algebraic technique for identifying holes in the **manifold** of complex solutions of an equation. Although Weil felt certain that he was correct, he was never able to prove his conjectures. Over a period of several years, Deligné whittled away at the conjectures. Combining the new theory of étale cohomology (a branch of topology), which had been developed by Grothendieck, and a related conjecture by the Indian number theorist **S. I. Ramanujan,** he completed the final proof in 1973.

Deligné's work has been valued not only because he solved an important problem in mathematics, but also because he proved that seemingly disparate subjects can be connected. Referring to Deligné's use of a 1939 paper on Ramanujan's conjecture along with the new étale cohomology, Mumford and Tate wrote, "It is hard to imagine two mathematical schools more different in spirit and outlook than were those of the British analytic number theorists in the 1930s and of the French algebraic geometers in the 1960s. That Deligné's proof is a blend of ideas from both is an indication of the universality of his mathematical taste and understanding." For this reason, as much as for actually proving the Weil conjectures, the International Mathematics Union in 1978 awarded Deligné its highest honor, the Fields Medal, noting that his work "did much to unify algebraic geometry and algebraic number theory."

Deligné continued to study the Weil conjectures even after his initial success, attempting to use automorphic forms (equations involving multiple **functions**) and **prime numbers** to determine more and more exact solutions. He worked on several problems proposed by the American mathematician **Robert Langlands,** who was leading a major research program in the area of automorphic forms at Princeton's Institute for Advanced Studies (IAS). At Langlands' invitation, Deligné traveled to the United States in 1977 to help organize a conference at Oregon State University.

Also, in the late 1970s, Deligné gave a series of lectures in étale cohomology with the help of Grothendieck and others in the field of algebraic geometry. These lectures were considered definitive in describing this relatively new field, but Deligné's contributions were not limited to lecturing. He added significantly to the content by his work with Shimura varieties and by his proofs of some conjectures proposed by William V. D. Hodge. In 1988, the Royal Swedish Academy of Sciences awarded Deligné and Grothendieck the Crafoord Prize for Mathematics for their work in defining étale cohomology and applying it to algebraic geometry (Grothendieck declined his prize).

In 1980, Deligné married Elena Vladimirovna Alexeeva, who would become the mother of his two children. He enjoys simple pleasures such as vegetable gardening, bicycle riding, and hiking. He brought his young family to the United States in 1984 to continue his mathematical research at the IAS, where he has remained. In 1993, Deligné and G. Daniel Mostow coauthored a book titled *Commensurabilities Among Lattices in PU(1,n).* Reviewing this book for the *Bulletin of the American Mathematical Society,* P. Beazely Cohen and F. Hirzebruch wrote that the authors "extract the best aspects of the previous techniques of algebraic and differential geometry... together with function theory, giving an overall coherent presentation yielding new results." A quarter of a century after cracking the Weil conjectures, Deligné has not lost his powerful touch for creating mathematics.

DEMOCRITUS OF ABDERA (CA. 460 B.C.-CA. 370 B.C.)
Greek mathematician and philosopher

Democritus of Abdera is best known for his atomic theory of the universe, but also made significant contributions to the study of **geometry**. Not much is known of his life, but rather extensive knowledge remains about his philosophies. He wrote a large amount of works on a variety of subjects including ethics, language, literature, logic, mathematics, music, and physics. Some of the titles relevant to mathematics are: *On Numbers, On Geometry, On Tangencies, On Mappings,* and *On Irrationals.*

Born to a wealthy family in Abdera (Avdhira, Greece), an ancient port on the coast of Thrace, Democritus used his inheritance for the sole purpose of acquiring knowledge. He traveled extensively (by ancient standards) and visited Egypt, Ethiopia, Persia, and India. He may have even visited Athens to study under **Anaxagoras of Clazomenae** (499-428 B.C.), the great mathematician and philosopher.

The foundation for all of Democritus' mathematics was based upon the theory of Atomism. An elaboration of his mentor Leucippus' theory (c. 480-420 B.C.), Atomism explains that the universe is composed of a void, or vacuum, and an infinite number of atoms. These atoms are *atomon* (the Greek word for

Democritus

indivisible), impassable (completely filling the **space** they occupy), and eternal. Although each is indivisible, they possess the ability to link with others to form larger objects—the visible entities of reality. And through variations in the shapes and arrangement of these atoms, coupled with the degree of void within the substance, individual objects are created. From this concept, Democritus could explain all facets of everything in existence within the physical world. For example, Democritus claimed that the atoms of iron and water were identical. Their inherent differences come from the fact that water atoms are glossy spheres that are unable to hook onto one another. Thus, the water atoms continually roll over one another creating a liquid form. On the other hand, the atoms of iron are jagged and rough, and therefore cling together to form a solid mass.

Since all things were predicated upon the theory of Atomism, Democritus' mathematics focused primarily on **infinitesimal** problems and the concept of the geometrical atom. For example, Democritus had the idea of a solid being the sum of many parallel planes, and he may have used this concept while calculating the line segment, **area**, or **volume** of a cone and pyramid.

See also Infinitesimal

DENUMERABLE

A denumerable (or countable) set is one for which there is a **correspondence** from the set to **N**, the set of natural numbers. To put it in another way: the **cardinality** of the set is equal to that of the natural numbers. Every infinite subset S of the natural numbers is countable since the function that takes the least element in S to 1 and the next to least element to 2 and the next to 3 and so on, is a correspondence. This also shows that no infinite set can be "smaller" than the natural numbers. Moreover any infinite set that can be put into correspondence with a denumerable set is denumerable. Hence the **integers** are denumerable since the function that takes x to $2*X^2 + x + 1$ is a correspondence from the integers to a subset of the natural numbers. Also if X and Y are denumerable **sets** then the set XxY of all ordered pairs (x, y) where x is in X and y is in Y is a denumerable set. To prove this, notice that since both X and Y are denumerable, it suffices to prove that **NxN** is denumerable. The function which takes (x, y) in **NxN** to $x*(x-1)/2 + y$ is a one-to-one correspondence. Another way to show the **NxN** is denumerable is to write out the elements of **NxN** in sequence like so:

(1,1) (1,2) (1,3)...
(2,1) (2,2) (2,3)...
(3,1) (3,2) (3,3)...

and so on. Then draw a zigzagging line which starts at (1,1) and proceeds across successive diagonals so that the sequence of elements it crosses is (1,1), (1,2), (2,2), (3,1), (2,2), (1,3), (1,4), etc. Then the function which maps 1 to the first element in the sequence just given and 2 to the next element and so on, is a correspondence from **N** to **NxN**. Any fraction, such as p/q, expressed in lowest terms, can be identified with the ordered pair (p, q). From this identification, the **rational numbers** are a subset of **ZxN** where **Z** stands for the set of integers. Since both **Z** and **N** are denumerable, so is the set of rationals.

A root of a polynomial P is a number n such that P(n) = 0. If the coefficients of P are rational numbers, then n is called an algebraic number. The set of **algebraic numbers** is denumerable. To see this, notice that a denumerable union of denumerable sets is denumerable. What this means is that if we have a denumerable "number" of denumerable sets then the total "number" of elements in the sets is denumerable. To prove this, notice that since each set in the union is denumerable, each can be identified with a copy of **N**. Then since the number of sets is also denumerable, each set can be numbered with a natural number. Therefore the union can be put into correspondence with **NxN**. Now given any number d, the number of **polynomials** of degree less than or equal to d with rational coefficients can be put into correspondence with Q^d (where **Q** denotes the rational numbers) since each polynomials can be identified with the ordered set of its coefficients. The union of all such polynomials is therefore denumerable. Since each polynomial has a finite number of **roots**, the algebraic numbers are countable.

See also Cardinality; Non-denumerably infinite sets; Sets

DEPENDENT VARIABLE

A dependent **variable** is the variable which expresses the results of a function. In a normal algebraic function such as y = 3x or y = x² + 6, "y" is the dependent variable. In the usual method of graphing a function, the dependent variable will be the one that is indicated as the height of the graph. The shape of the function thus indicates the behavior of the dependent variable. The use of inverse **functions** or algebraic manipulation often can easily make the dependent variable in one equation look like the **independent variable** in a different way of writing the same relationship. However, there are significant differences in how the dependent variable is treated in experimental and functional situations. The dependent variable is the one whose derivative is taken in **calculus**, for example, and the independent variable is the one with respect to which the derivative is taken. Similar actions upon functions are taken upon dependent variables with respect to independent ones.

In scientific experiments, the dependent variable is the "experimental" variable. The independent variable is the "control" variable. For example, the current sent through a circuit (the independent variable) could determine the voltage output of that circuit (the dependent variable). The dependent variable is thus the important variable for interesting, sometimes surprising, results in any experiment.

See also Functions; Independent variables

DERIVATIVES AND DIFFERENTIALS

The derivative of a function represents the limiting value of the ratio of the change in that function to the corresponding change in its **independent variable**. This describes the instantaneous rate of change of a function with respect to a **variable**. In practical terms the derivative of a function represents the **slope** of the graph of that equation at a given point. The derivative of a function $f(x)$ is defined mathematically as: $f'(x) = \lim_{h \to 0} (f(x + h) - f(x))/h$, where h is a defined **interval** on x. The differential of a function $f(x)$ is the derivative of that function with respect to x multiplied by the increment interval, usually denoted as dx. $df = f'(x)dx$, where df is the differential of f at x with increment dx. From this equation, along with the fact that $dx = h$, it is clear to see the relationship between the derivative, $f'(x)$, and the differential, df. The differential is just the numerator in the limit defining the derivative of the function: $df = f(x + h) - f(x)$.

Derivatives must conform to specific rules that help us standardize the process for determining them. First, the derivative is mathematically defined as the limit as $h \to 0$ and so both the limit as $h \to 0+$ and the limit as $h \to 0-$ must exist for the overall limit to exist. This means in effect that the function is continuous. This condition is a necessity for a function to be differentiable but it may not be the only condition. There are several important rules for computing derivatives. The product and quotient rules are employed for determining derivatives when **functions** are multiplied and divided by one another. The

power rule is used to take the derivative of a function that contains a variable with an exponent, raised to a power. Other rules that are very useful for finding derivatives include knowing that the derivatives of sums of functions are equal to the sum of the derivatives of those functions and the **chain rule**, which is useful in determining derivatives of composite functions. The **trigonometric functions** have derivatives that are specific to each function. This is also true for functions containing logarithmic terms. Determining the derivative of a function may be a difficult task.

The derivative of a function, sometimes called a fluxion, has several notations that are commonly used. Two of the most commonly used notations for the derivative are: $f'(x)$ and df/dx. There are other common notations that express special attribute derivatives. Newton's overdot notation is used when derivatives are taken specifically with respect to time: $dx/dt = \dot{x}$. The derivative of a function can be taken over and over and a special notation results: $d^n f/dx^n$, with n being the number of times the derivative was taken. Newton's overdot notation can be extended to multiple derivatives by just placing the equivalent number of dots over the variable but again this notation is only for derivatives taken with respect to time.

As well as having several notations to represent derivatives in general there are several types of derivatives. There is something called a partial derivative that is obtained when the function depends on more than one variable. The partial derivative of a function $f(x, y,...)$ is denoted: $\partial f/\partial x$ for a derivative of the function taken with respect to x or: $\partial_2 f/\partial x \partial y$ for a derivative of the function taken twice with respect to more than one variable. The partial derivative is obtained by holding all but one variable constant while the derivative is taken with respect to the one remaining variable. A directional derivative is the generalization of the derivative with respect to an arbitrary direction in three-dimensional **space**. Instead of yielding a line tangent to the graph of the function at a point it would yield the plane tangent to a surface at a point. The term vector derivative refers to the derivative of a vector function and so has a particular direction associated with it.

DESARGUES, GIRARD (1591-1661)
French geometer and engineer

Little regarded in his own time, Girard Desargues developed a treatise and **theorem** that formed the basis for the development of projective **geometry**, breaking with the straight Euclidian traditions that had informed geometry since the Hellenistic age. However, it was not until almost two centuries after his death that Desargues was rediscovered and his work on conic sectionsgained its rightful place in mathematics. Although there is scant knowledge about Desargues' personal life, it is recorded that he was, as an amateur mathematician, a member of a renowned group of Parisian mathematicians, and that his work influenced the young **Blaise Pascal**. An engineer and inventor, Desargues also developed a pump device and wrote on such practical subjects as stonecutting, the production of sundials, and music composition.

Desargues was born on February 21, 1591 in Lyons, France, one of nine children of Girard Desargues and Jeanne Croppet. The elder Desargues was a tithe collector, and the family owned several large houses in Lyons as well as a chateau and vineyards near the city. Not much is recorded of his early education, but it would appear that the family was sufficiently well off that Desargues could indulge in various pursuits, including the sciences. An early report places him in Paris in 1626, proposing to the municipality that it raise the level of the Seine by the use of machines so as to be able to pump water through the city. As an engineer, Desargues is also reported as having participated in the 1628 siege of La Rochelle, where he met the philosopher and mathematician **René Descartes**. He is also supposed to have been an engineer and technical advisor for the French government under Cardinal Richelieu.

By about 1630, Desargues was spending more time in Paris and became part of the intellectual circle of **Marin Mersenne**, Étienne Pascal and his son Blaise Pascal, and René Descartes. Desargues published two papers in 1636, one a musical treatise on harmony, and the other entitled *Traite de la section perspective* ("Treatise on the Perspective Section"), which laid out his universal method of perspective and contained the initial ideas that led to the theorem named after him. This theorem was later published by Desargues's friend and confidante, Abraham Bosse, and has since been attributed directly to Desargues. The theorem, which holds for either two or three dimensions, states that two triangles may be positioned so that the three lines joining corresponding vertices meet in a point if and only if the three lines containing pairs of corresponding sides intersect in three collinear points. It was this theorem that the French mathematician **Jean–Victor Poncelet** rediscovered in the 19th century and used to help modify Euclidian geometry into projective geometry. With this universal method, Desargues hoped to develop a completely geometric study of perspective, and therefore stands between the artist, **Albrecht Dürer**, and the French geometer, **Gaspard Monge**, in the development of a graphical representation of perspective.

As with all the work in his lifetime, this early theorem of Desargues's received little attention. Such neglect, however, did not keep Desargues from his studies, and in 1639 he privately printed 50 copies of what is his most famous work, *Brouillon project d'une atteinte aux evenemens des rencontres d'une cone avec un plan* ("Proposed Draft of an Attempt to Deal with the Events of the Meeting of a Cone with a Plane"), a treatise which developed new practices in projective geometry, especially as applied to **conic sections**. As René Taton wrote in *Dictionary of Scientific Biography,* the treatise was "a daring projective presentation of the theory of conic sections.... But the use of an original vocabulary and the refusal to resort to Cartesian symbolism make the reading of his essay rather difficult and partially explains its meager success." Julian Lowell Coolidge, in his *A History of Geometrical Methods*, is more to the point: "The style and nomenclature are weird beyond imagining." Desargues employed botanical terms for his mathematical concepts, including among others, "tree" to denote a straight line with three pairs of points of an involution, and a "stump" to refer to the mate of an infinite point. Such

nomenclature makes the essay incredibly difficult to understand, but beneath this arcane language Desargues had done no less than present a unified theory of conics. It took another two centuries for this legacy to be accepted, however.

Possessing an inquisitive mind, Desargues did not simply focus on geometry for his life's work. He was also an accomplished architect and applied his spatial capacities to creating useful designs. Desargues is particularly noted for his spiraling staircases, the construction of which was aided by his theories of **projection** and perspective. About 1645 he began designing houses in Paris, and then returning to his birthplace of Lyons, he supervised the construction of several more houses and mansions, both private and public, between 1650 and 1657, after which time he returned to Paris.

As an engineer, Desargues is especially noteworthy for his early use of a cycloidal or epicycloidal teeth for gear wheels. These wheels were used in a system for raising water at the chateau of Beaulieu near Paris, one that Desargues installed in the 1630s. Desargues also extended the pragmatic uses of his method of perspective to the craft of stonecutting and developed principles to simplify the construction of sundials.

Desargues was not, however, free from criticism in his day. Regarding his optimism that graphical representation alone was sufficient to breathe new life into geometry, Descartes took the mathematician to task. A strong proponent of **algebraic geometry**, Descartes doubted such an assertion. The mathematician Jean de Beaugrand was a more ardent critic, asserting that many of Desargues's supposedly original ideas on conic sections were taken directly from the ancient geometer, **Apollonius**. The two exchanged broadsides, in the form of critical essays, for several years.

It was not only his theoretical work that drew criticism, however. Desargues's methods applied to stonecutting brought him into direct conflict with the trade guilds, which had responsibility for such matters. Also, his methods outraged practitioners of older methods of perspective. Such attacks eventually took their toll on Desargues, and increasingly he kept out of the public eye, entrusting the dissemination of his works and ideas to his disciple and friend, the engraver Bosse. It was several of Bosse's publications which kept Desargues's name alive; that, and a manuscript copy of Desargues's *Brouillon project* which another colleague, Phillipe de la Hire, made. It was not until 1951 that an original copy of that treatise was uncovered. When Desargues died in October of 1661, there was little to indicate that history would grant him immortality. As Taton noted in *Dictionary of Scientific Biography,* "Desargues's work was rediscovered and fully appreciated by the geometers of the nineteenth century. Thus, like that of all precursors, his work revealed its fruitfulness much more by its remote extensions than by its immediate repercussions."

DESCARTES, RENÉ (1596-1650)

French geometer, algebraist, and philosopher

René Descartes was an analytical genius. He conceived and articulated ideas about the nature of knowledge that were essential to the Enlightenment and created the philosophical underpinnings for the development of modern science, which included the idea that laws of nature are constant and are sufficient to explain natural phenomena. Descartes felt that **truth** was clear and accessible to the ordinary human intellect, if the search for truth was directed properly. Two of his writings, *Rules for the Direction of the Mind* and *Discourse on the Method of Rightly Conducting the Reason* defined ways of obtaining knowledge. The latter work contained *Geometry,* that introduced the **Cartesian coordinate system** and marked the birth of **analytic geometry**, in which geometric relationships are investigated by means of **algebra**. Descartes also contributed to areas of music theory, **mechanics**, physics, optics, anatomy, and physiology.

René du Perron Descartes was born on March 31, 1596 in La Haye (now Descartes), in the province of Touraine, France. He was born into the gentry, a well–to–do class of landowners between the nobility and the bourgeoisie. His father, Joachim, was a councilor to the high court at Rennes in Brittany. From his mother, Jeanne Brochard, Descartes received the property that gave him his financial independence. Descartes was her third and last surviving child. She died in childbirth in 1597 and he and his older brother and sister were brought up by their maternal grandmother, Jeanne Sain. In 1600, Descartes' father remarried and moved to Chételleraut. Descartes seems not to have had enduring relationships with his father or siblings; however, the elder Descartes early on recognized his youngest child's curiosity, referring to him as "my little philosopher."

In 1606, Descartes was sent to La Fléche, the Jesuit school at Anjou. Descartes' health was considered delicate and the rector, Father Charlet, allowed him to spend mornings in bed in contemplation, a habit he continued throughout most of his life. Descartes spent nine years at La Fléche, where he perfected his Latin, studied humanities, philosophy, and mathematics, and was introduced to new developments in opticsin astronomy. Although Descartes expressed high regard for his education, it was at La Fléche he realized that, with the exception of mathematics and **geometry**, he had learned nothing that was absolute truth. He first saw mathematics only as the servant of mechanics, but was struck "by the certainty of its proofs and the evidence of its reasonings" and was surprised that nothing loftier had been erected upon its foundations.

Descartes moved to a house outside Paris in 1614, where he shut himself off from others. Although he was self–assured and expected admiration from others, scholars have suggested that he suffered from depression. He spent the year 1615–1616 at the University of Poitiers, where he earned a law degree.

The law did not interest Descartes; he chose instead to become a gentleman soldier. He had resolved "to seek no knowledge other than that which could be found in myself or else in the great book of the world," and in the summer of 1618 he traveled to Holland, where he joined the army of Prince Maurice of Nassau as an unpaid volunteer. In Breda he met Isaac Beeckman, the Dutch philosopher, doctor and physicist. Descartes' discussions with Beeckman rekindled his

René Descartes

interest in applying mathematical reasoning to problems in physics. Descartes' first work, *Compendium Musicae*, an arithmetical account of sound, was dedicated to Beeckman and given to him as a New Year's gift in 1619. At this time, Descartes also worked on problems in falling bodies, hydrostatics, a proportional compass, and a theory of proportional magnitudes.

Descartes resigned from Maurice's army, and traveled to Bavaria to join the Bavarian Army. Stationed at Ulm in Neuburg, he met the Rosicrucian and mathematician Johannes Faulhaber. On November 10, 1619, in a "stove–heated room," Descartes had the mystical experience that set his life's course. He had been searching for a method of obtaining knowledge, and in a state of delirium had three vivid dreams in succession. Much has been made of the dreams (even the suggestion that they were symptomatic of migraine headaches), but their result was to convince Descartes of his divine mission to found a new philosophical system, in which he would reduce physics to geometry and connect all sciences through a chain of mathematical logic.

Descartes subsequently gave up the military life and traveled widely for several years, visiting Italy, Germany, and Holland, where along the way he studied glaciers, made meteorological observations, and computed the heights of mountains. From 1625 to 1628, he lived in Paris and became friends with **Marin Mersenne**, a Franciscan friar who had also attended La Fléche. In Paris, Descartes produced *Regulae* ("Rules for the Direction of the Mind"), which was published in 1701, after his death.

In 1629 Descartes retired to Holland, where he devoted the next 20 years to studies of science and philosophy. During this time he made three trips back to Paris, where Mersenne acted as his editor and agent. The tolerant Protestant climate of Holland protected Descartes from academic and theological disputes, at least in the beginning, and he moved frequently to avoid visitors. He was not a recluse, however; he visited universities and talked with mathematicians, philosophers, and physicians. Descartes studied anatomy and frequently visited butcher shops to obtain animal carcasses for dissection. In 1633, he completed *Le monde* ("Of the World"), which included his theories in physiology, perception, and a heliocentric cosmology. When Descartes learned that **Galileo** had been condemned by the Inquisition for embracing **Copernicus**' ideas, he withheld *Le monde* from publication. He modified information from *Le monde* for use in his 1637 masterpiece, *A Discourse on the Method of rightly conducting the Reason and seeking Truth in the Sciences. Further, the Dioptric, Meteors, and Geometry, essays in this Method*. The *Meteors* was the first attempt to give a scientific theory of the weather. The *Dioptric* explained rainbows, and contained the law of refraction, describing the behavior of **light** rays transmitted from one medium to another.

Descartes' *Geometry,* essentially an appendix to the *Discourse on Method*, revolutionized mathematics and provided the foundation for what is now known as analytic geometry. It enabled the use of algebra, a relatively new branch of mathematics, for the discovery and investigation of geometrical theorems. He introduced the use of coordinates, by which is possible to begin with **equations** of any degree of complexity and interpret their algebraic and analytic properties geometrically. In the *Geometry,* Descartes introduced algebraic notation that is still in use today, dealt with the problem of **Pappus**, and provided a systematic **definition** of curves.

In Amsterdam, Descartes had formed a liaison with his serving girl, Héléne, who bore him a daughter, Francine, on July 19, 1635. Héléne and Francine came to live with him in Santpoort, and he made arrangements for Francine to be educated in France. Unfortunately, she died in 1640, probably of scarlet fever.

Descartes published his major metaphysical work, *Meditations on First Philosophy*, in which the Existence of God and the Distinction between Mind and Body are Demonstrated in 1641. Although he quickly published an edition containing solicited objections and his replies to them, he was particularly criticized and attacked by the president of the University of Utrecht, Gisbert Voet, and published his lengthy defense as *Episula at Voetium*.

In 1643 Descartes began a long–lasting correspondence with 24–year–old Princess Elizabeth of Bohemia, who lived in exile in Holland. In his letters, Descartes discussed his philos-

ophy of the mind and its relation to the body, and the relationship between reason and the passions. He dedicated his 1644 *Principles of Philosophy*, which contains a naturalistic theory of the solar system, to Princess Elizabeth.

Descartes accepted an invitation to tutor 20–year–old Queen Christina of Sweden in 1649. After much hesitation he left for Sweden on September 1, where the energetic Queen put him to work writing verses and a pastoral comedy, and planning a Swedish academy of science. She insisted that he meet with her at five in the morning when her mind was most active. The lessons began in mid–January 1650, but the early hours and the record cold winter quickly took their toll on Descartes. On February 1, he contracted pneumonia. He refused to see the royal physician and instead relied on his own remedy, wine flavored with tobacco. He died in Stockholm on February 11, 1650.

In 1666, Descartes' remains were exhumed and returned to France, where they were moved several times before being permanently placed in the chapel of the Sacré Coeur in the church of St. Germain–des–Prés in 1819. At the time of the original exhumation, the French ambassador was given permission to cut off Descartes' right forefinger. Descartes' skull was said to have been removed by a guard and it was sold several times, coming into the possession of Georges Cuvier in 1821. Although it has not been authenticated, the skull is on display at the Musée de l'Homme in the Palais de Chaillot.

DESCARTES' RULE OF SIGNS

Descartes' rule of signs, first published by **René Descartes** in 1637, is a method for determining the largest number of positive or negative **roots** that a polynomial equation may have. If the terms of the polynomial $p(x)$ are written in the customary fashion—that is, with the terms given in decreasing order of the exponent of x—then the number of positive roots of the polynomial cannot be greater than the number of sign changes in the coefficients. (A sign change occurs whenever one coefficient is positive and the next one is negative, or vice versa. If any coefficients are **zero**, they are simply ignored.) A test for the number of negative roots can be created by replacing x with $-x$ and counting the sign changes in the new polynomial.

A few examples make the application of the rule easy to understand. In the polynomial $p(xx) = x^4 - 12x^3 + 39x^2 - 8x - 60$, there are three sign changes (because the signs of the coefficients are +, -, +, -, - respectively) Therefore Descartes' rule guarantees that there are no more than three positive roots. If we substitute $-x$ for x, we get $p(-xx) = x^4 + 12x^3 + 39x^2 + 8x - 60$, with signs +, +, +, +, -. Since there is only one sign change, there is at most one negative root of $p(x)$. In this example, the estimates happen to be precise; the polynomial $p(x)$ factors as $(x+1)(x-2)(x-5)(x-6)$ and therefore has three positive roots (2, 5, and 6) and one negative root (-1).

As another example, Descartes' rule predicts that the polynomial $q(x) = x^4 - 7x^3 + 21x^2 - 30x + 18$ should have no more than four positive roots (because there are four sign changes). In fact, it has none. This polynomial factors as $(x^2 - 3x + 3)(x^2 - 4x + 6)$, and one can check with the quadratic formula that neither of the quadratic factors has any real roots, let alone any positive real roots. Hence, Descartes' rule only gives an upper bound for the number of positive roots—not the precise number. **Carl Friedrich Gauss** proved that the discrepancy between the number of sign changes and the actual number of positive roots is always even (in this case, 4). This fact is often stated as part of Descartes' rule, even though Descartes never proved it.

In Descartes' day, the distinction between positive and **negative numbers** was quite important; Descartes himself used the terms "true roots" and "false roots." Mathematicians no longer consider negative numbers to be "false." However, in practical applications, where the number x is likely to represent a physical measurement of some sort, positive solutions may be the only ones that make sense in the context of the problem. In such cases, applying Descartes' rule of signs can still be a useful preliminary step before one actually tries to solve the equation.

See also Algebra

DETERMINANTS

A determinant, signified by two straight lines ||, is a **square** array of numbers or symbols that has a specific value. For a square **matrix**, say, A, there exists an associated determinant, |A|, which has elements identical with the corresponding elements of the matrix. When matrices are not square, they do not possess corresponding determinants.

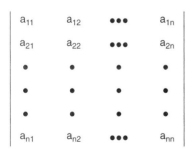

Figure 1.

In general, determinants are expressed as shown in Figure 1, in which a_{ij}s are called elements of the determinant, and the horizontal and vertical lines of elements are called rows and columns, respectively. The sloping line consisting of a_{ii} elements is called the principal diagonal of the determinant. Sometimes, determinants can be written in a short form, $|a_{ij}|$. The n value, which reflects how many n^2 quantities are enclosed in ||, determines the order of a determinant.

For determinants of third order, that is, n = 3, or three rows of elements, we can evaluate them as illustrated in Figure 2.

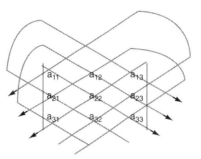

$$= a_{11}a_{22}a_{33} + a_{12}a_{23}a_{31} + a_{13}a_{21}a_{32} - a_{13}a_{22}a_{31} - a_{12}a_{21}a_{33} - a_{11}a_{23}a_{32}$$

Figure 2.

By summing the products of terms as indicated by the arrows pointing towards the right-hand side and subtracting the products of terms as indicated by the arrows pointing towards the left-hand side, we can obtain the value of this determinant. The determinant can also be evaluated in terms of second-order determinants (two rows of elements), as in Figures 3(a) or 3(b).

(a)
$$a_{11} \begin{vmatrix} a_{22} & a_{23} \\ a_{32} & a_{33} \end{vmatrix} - a_{21} \begin{vmatrix} a_{12} & a_{13} \\ a_{32} & a_{33} \end{vmatrix} + a_{31} \begin{vmatrix} a_{12} & a_{13} \\ a_{22} & a_{23} \end{vmatrix}$$

or

(b)
$$-a_{12} \begin{vmatrix} a_{21} & a_{23} \\ a_{31} & a_{33} \end{vmatrix} + a_{22} \begin{vmatrix} a_{11} & a_{13} \\ a_{31} & a_{33} \end{vmatrix} - a_{32} \begin{vmatrix} a_{11} & a_{13} \\ a_{21} & a_{23} \end{vmatrix}$$

Figure 3.

Each of these second-order determinants, multiplied by an element a_{ij}, is obtained by deleting the ith row and the jth column of elements in the original third-order determinant, and it is called the "minor" of the element a_{ij}. The minor is further multiplied by $(-1)^{i+j}$, which is exactly the way we determine either the "+" or "-" sign for each determinant included in Figures 3 as shown, to become the "cofactor", C_{ij}, of the corresponding element.

Determinants have a variety of applications in engineering mathematics. Now, let's consider the system of two linear **equations** with two unknowns x_1 and x_2: $a_{11}x_1 + a_{12}x_2 = b_1$ and $a_{21}x_1 + a_{22}x_2 = b_2$.

We can multiply these two equations by a_{22} and $-a_{12}$, respectively, and add them together. This yields $(a_{11}a_{22} - a_{12}a_{21})x_1 = b_1a_{22} - b_2a_{12}$, i.e., $x_1 = (b_1a_{22} - b_2a_{12})/(a_{11}a_{22} - a_{12}a_{21})$. Similarly, $x_2 = (b_1a_{21} - b_2a_{11})/(a_{12}a_{21} - a_{11}a_{21})$ can be obtained by adding together the first equation multiplied by a_{21} and second equation multiplied by $-a_{11}$. These results can be written in determinant form as in Figure 4.

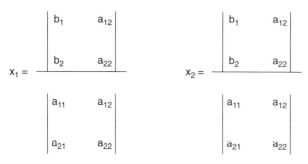

$$x_1 = \frac{\begin{vmatrix} b_1 & a_{12} \\ b_2 & a_{22} \end{vmatrix}}{\begin{vmatrix} a_{11} & a_{12} \\ a_{21} & a_{22} \end{vmatrix}} \qquad x_2 = \frac{\begin{vmatrix} b_1 & a_{12} \\ b_2 & a_{22} \end{vmatrix}}{\begin{vmatrix} a_{11} & a_{12} \\ a_{21} & a_{22} \end{vmatrix}}$$

Figure 4.

This is generally called **Cramer's rule**. Notice that in Figure 4, elements of the determinant in the denominator are the same as the coefficients of x_1 and x_2 in the two equations. To solve for x_1 (or x_2), we then replace the elements that correspond to the coefficients of x_1 (or x_2) of the determinant in the numerator with two constant terms, b_1 and b_2. When b_1 and b_2 both are equal to **zero**, the system defined by the two equations is said to be homogeneous. In this case, it will have either only the trivial solution $x_1 = 0$ and $x_2 = 0$ or additional solutions if the determinant in the denominator in figure 5 is zero. When at least b_1 or b_2 is not zero (that is, a nonhomogeneous system) and the denominator has a value other than zero, the solution to the system is then obtained from figure 4. Cramer's rule is also applicable to systems of three **linear equations**. Therefore, determinants, along with matrices, have been used for solving simultaneous linear and **differential equations** involved in various systems, such as reactions in chemical reactors, stiffness of spring-connected masses, and currents in an electric network.

DIAMETER

A diameter is a straight-line segment that passes through the center of a **circle** or **sphere** and whose two end points lie on the **circumference** of the surface. The center is a point of **symmetry** within a circle or sphere, with any diameter passing through that center denoted as an axis of symmetry. An infinite number of diameters are possible for any given circle or sphere, and are the longest line segments that terminate on the circumference. With one revolution around a circle or sphere equal to 360 degrees, the two end-points of a diameter are 180 degrees away from one another.

Diameter is an appropriate name for the measure across a circle or sphere, as the word is the union of the Greek roots "dia" (across) and "metros" (to measure). One of the earliest uses for the word diameter came from **Euclid of Alexandria**, a fourth-to-third century B.C. Greek mathematician, who used the word with relationship to the line bisecting a circle and also to mean the diagonal of a **square**.

If r is the radius of a circle or a sphere, then the diameter, d, equals d = 2r. For example, if the radius of a circle has a length of ten centimeters (cm), then the circle's diameter is "d = 2 x 10 cm" or "d = 20 cm". The diameter is also defined as d = C / π, where C equals the circle's circumference and **pi**, π, is approximately valued at 3.14. In this example, if the circumference of the circle equals a length of thirty centimeters, then the circle's diameter is "d = 30 cm / 3.14", or the diameter is approximately equal to 9.55 cm. An interesting take on the formula "d = C / π" is that if C is given as, say, an integer, then that circle's diameter, d, must be irrational (it cannot be expressed by the ratio of two **integers**) and transcendental (it satisfies no algebraic equation with integer coefficients) because pi also possesses those properties.

The history of diameter is rooted in the attempts by the ancient Greeks to find an exact ratio for a value of pi, that of circumference over diameter (C / d). The quest of the Greeks took the form of an effort to "square the circle"; that is, to construct with straight edge and compass both a square and a circle of identical areas. A very early approximation for pi was the ancient Greek value of "22 / 7". Greek mathematician, physicist, and inventor **Archimedes of Syracuse**, in the third century B.C., is recorded to have made the first scientific effort to compute pi. Archimedes computed pi to a value between 3-10/71 and 3-1/7, or about 3.14. Four centuries later, at around AD 200, pi was valued to 3.1416. By the early sixth century Chinese and Indian mathematicians had independently confirmed or improved on the number of decimal places that pi was calculated past the decimal point. By the end of the seventeenth century in Europe, new methods of mathematical **analysis** provided various ways of calculating pi. In 1767 **Johann Heinrich Lambert** showed that the ratio C/d is irrational, and in 1882 **Carl Louis Friedrich von Lindemann** proved that an exact value for pi is impossible. Early in the twentieth century the Indian mathematician **Srinivasa Ramanujan** developed ways to calculate pi so efficiently that they were eventually incorporated into computer algorithms, permitting expressions of pi at, presently, billions of digits. Throughout the many centuries of gaining more precise values of pi, diameter played an important role in this development.

See also Archimedes of Syracuse; Circle; Euclid of Alexandria; Johann Heinrich Lambert; Carl Louis Friedrich von Lindemann; Pi; Radius; Srinivasa Ramanujan; Sphere

DIFFERENTIAL CALCULUS

Elementary **calculus** is usually divided into two branches, differential calculus and **integral calculus**. Differential calculus is the branch of calculus that is based on the determination of the limit of a certain ratio whereas integral calculus is based on the determination of the limit of a certain sum. Differential calculus is the portion of calculus that deals with derivatives. The derivative of a function is representative of an **infinitesimal** change in the function with respect to its parameters. The derivative of a function f with respect to x is denoted either as $f'(x)$ or df/dx. **Leibniz** developed this notation in 1684. All applications of differential calculus are concerned with interpretations of the derivative as the **slope** of the line tangent to the curve at a specific point or as the rate of change of the dependent **variable** with respect to the **independent variable**. Employing differential calculus provides a method for determining the slope of a line tangent to a curve, rates of change, points moving on a straight line or other curve, and absolute maxima and minima. It is used by the physical and biological sciences as well as statistical analysis used in business and social studies.

In order to describe differential calculus as the determination of the limit of a certain ratio we must first understand the ratio. Let $f(x) = y$ where y is the dependent **variable** that is a function of the independent variable x. If x_0 is a value of x defined in the **domain** identified for x, then $y_0 = f(x_0)$ is the corresponding value of y. Let h and k be **real numbers**, and $y_0 + k = f(x_0 + h)$. So $k = f(x_0 + h) - f(x_0)$ and $k/h = [f(x_0 + h) - f(x_0)]/h$. This ratio is the difference quotient and equals the tangent of the curve drawn between the points (x_0, y_0) and $(x_0 + h, y_0 + k)$. The difference quotient can be thought of as the average rate of change of $y = f(x)$ with respect to x within the defined **interval**. If the limit of this ratio k/h exists as h approaches 0 then this limit is called the derivative of y with respect to x, evaluated at $x = x_0$. The limit is written as $\lim_{h \to 0} k/h$. This derivative can be interpreted as the slope of the curve $y = f(x)$ at $x = x_0$. It can also be interpreted as the instantaneous rate of change of y with respect to x at x_0. Rules and methods developed by this limit process enabled mathematicians to formulate various **equations** that provide for rapid calculation of the derivatives of various **functions**. The derivative of a function $f(x)$ of x is usually denoted as $f'(x)$. When the derivative of $f(x)$ is found for all values of x a new function is obtained which is itself a function of x. The derivative of $f'(x)$ can be found and is called the second order derivative of y with respect to x and is usually denoted as $f''(x)$, $(d^2y)/(dx^2)$ or $(d^2f(x))/(dx^2)$. Higher order derivatives are expressed in the same manner.

Isaac Newton and Gottfried Wilhelm Leibniz are attributed with inventing calculus in the 17th century. Although these two are credited with inventing calculus isolated results concerning its fundamental problems had been known for thousands of years. The Egyptians used calculus to determine the **volume** of a pyramid and the **area** of a **circle**. Ancient Greek scientists employed calculus to study the stars and planetary **motion** long before calculus was formally recognized. By the early 17th century mathematicians had developed methods for determining the areas and volumes of a wide variety of shapes and forms. In about 1720 **Isaac Barrow** published a book stating geometrically the inverse relationship between problems of finding tangents and areas. This relationship is now known as the **fundamental theorem of calculus**. In 1665-1666 Newton's discovery of combining infinite sums or **infinite series**, the **binomial theorem** for fractional **exponents**, and the algebraic expression of the inverse relationship between tangents and areas led his to develop methods used in calculus. He was reluctant to publish these results and so Leibniz published his discovery of differential calculus in 1684. Although Leibniz developed his methods of calculus independently of Newton and so much later than Newton he is still recognized as the codiscoverer of calculus since he published his results. Leibniz also replaced Newton's symbols in differ-

ential calculus with those that are used today. Calculus is used in the physical sciences to study the speed of a falling object, the rates of change in a chemical reaction, and the rate of decay of a radioactive material. The biological sciences employ calculus to study problems such as the rate of growth of a colony of bacteria as a function of time. The social sciences use calculus to study problems concerned with **statistics** and probability, such as population.

DIFFERENTIAL EQUATIONS

A differential equation is a mathematical relationship which contains one or more derivatives of a function. The solution of the differential equation is the original function. These **equations** are often used in scientific applications, particularly in physics and engineering, because they express rates of change in quantities of interest such as position, temperature, and other physical parameters. They also can be considered an extension of the basic study of **calculus**.

Differential equations are classified by several features. The order of a differential equation is given by the largest number of derivatives of a function featured in that equation— thus, an equation featuring third-derivatives of a function would be a third-order differential equation, but an equation featuring both third- and fourth-derivatives would be a fourth-order differential equation. Most differential equations used in practical applications are first- or second-order differential equations, as there are very few third-order rates of change that are useful in physical settings (see **First-order ordinary differential equations** and **Second-order ordinary differential equations**). However, the higher order equations do appear and may be useful for obscure and theoretical studies.

Differential equations can also be divided into ordinary and partial categories. Ordinary differential equations only feature ordinary derivatives, whereas partial differential equations involve partial derivatives. In the case of partial differential equations, the partial derivatives may be in more than one **variable**. Often, differential equations will appear with partial derivatives in all spatial directions and time.

Linearity is another way to classify differential equations. If a differential equation is linear, it does not feature the product or any other non-additive combination of different orders of derivative—that is, a linear differential equation will not feature a term like yy', but y + y' is acceptably linear. Nonlinear equations are generally much harder to solve because they cannot be separated into differential and non-differential components as easily.

Systems of differential equations with more than one unknown function can occur. These contain more than one unknown function in differential form, at any level of derivative. As with **systems of linear equations**, systems of differential equations must be solved as an entire unit, combining and eliminating **functions** instead of algebraic variables.

Methods of solving differential equations vary according to the type of equation being solved. Some simple differential equations may be separated into a differential and a non-differential component. Both sides of the equation are then integrated to find the solution. For some differential equations, a simpler related equation can be solved with a correction term for the more complex case. All differential equations may be solved with a polynomial series. The polynomial solution is known as Frobenius' Method. For other differential equations that are to be used in a practical settings, numerical solutions may be provided, in which the behavior of the equation is examined in small incremental steps. The advent of the personal computer in the workplace makes this solution method much more feasible than it was when all calculations were done by humans directly. The numerical calculations are often used when the equation has been separated and the resultant integral is intractable to solve in closed form.

To get exact solutions to differential equations, one must have some information about the equation's specific behavior. Usually there is a family of solutions which would solve the differential equation, varying by one or several constants, depending upon the order of the equation. Boundary value problems tell the person solving the equation what behavior the solution function will have at more than one variable location. Initial value problems tell the solver the behavior of all the derivatives of the function at one variable location. In either case, the number of specified parameters must be equal to the order of the equation. The family of equations is then examined algebraically to determine the values of the constants in the solution.

There are many famous differential equations whose solution is already known. In this case, it only remains to recognize the equation one has as a form of the known equation. The **motion** of waves in a medium, simple oscillatory motion, and the motion of a struck drumhead are all expressed in familiar, solved differential equations, and the application of the equations to these situations only requires a knowledge of boundary or initial value conditions. Most physics problems involve the solution of a differential equation at some point. They are essential to classical and quantum physics alike, as well as to some chemical rate reactions and many formulas used in chemical, electrical, mechanical, and environmental engineering.

See also Differential equations

DIFFERENTIATING ALGEBRAIC FUNCTIONS

Differentiation of algebraic **functions** works by one of the simplest rules in **calculus**. Each component of the algebraic function is differentiated separately, and then the differentials are added together. If the components of the algebraic function are the **variable** taken to different **powers** and multiplied by coefficients, the process is even easier.

First, the power of the variable is decreased by one, so that if the original power is x^3, the result contains a power of x^2. Then, the original power of the variable is multiplied by the coefficient of the term for the new coefficient. In this manner,

$5x^4$ becomes $20x^3$ when differentiated once. Constant terms, of course, disappear.

Each of these rules has a specific name, within calculus. The rule that a constant's derivative is **zero** is, of course, the constant derivative rule. The power rule for positive integer powers of x states that, when taking a derivative of a function of a positive integer power of x, the power of the variable is decreased by one and the function is multiplied by the original power. The constant multiple rule simply affirms that the derivative of a constant times a function is the same as the constant times the derivative of the function. Finally, the sum and difference rule states that the derivatives of **sums and differences** of algebraic terms are the same as the sums and differences of the derivatives.

The differentiation of algebraic functions is probably one of the most used skills in calculus. Many rate of change problems occur at linear rates, and still more can be expressed in terms of algebraic functions. Since calculus is largely the study of rates of change, this skill and information become particularly useful.

DIFFERENTIATING DISTANCE AND VELOCITY

The kinematic quantities of **distance** and velocity are some of the most basic concepts in introductory physics. The differential of each of these quantities is very important in kinematics **equations**, used to calculate the **motion** of traveling bodies.

Differentiating these quantities has some of the most conceptually useful results in basic physics. The derivative of distance over time is simply velocity, since the **definition** of velocity is the rate of change of distance (or position) over time. Differentiating velocity, or taking the second derivative of position, results in acceleration, defined as the rate of change of velocity over time.

The differentiated form of these terms can be either a function or a specific value. If it is a function, the instantaneous acceleration or velocity can be obtained by substituting a specific time value into the function.

The derivatives of distance and velocity may give insight into the behavior of the system. For example, if the system is under constant acceleration over time, the velocity term will be a linear term and the distance will contain a quadratic term, so that the derivatives come out to a constant. Further, taking the second derivative of a periodic motion function will indicate that this function's acceleration and position are greatest at the same time, while velocity is precisely out of phase with them. This kind of information can be used to figure out the behavior of a system with these known parameters.

Kinematics equations are often the first lesson learned in introductory physics classes. Studying them in more mathematically sophisticated terms of differentials of distance and velocity can provide a deeper insight into the physics of the problem.

DIFFERENTIATING TRANSCENDENTAL FUNCTIONS

One of the surprises of **calculus** is the fact that the derivatives and integrals of some **functions** look very different from the original functions. This is usually the case for inverse trigonometric and logarithm functions. These functions are "transcendental," in the sense that they cannot be written as **polynomials**, quotients of polynomials, or **roots** of polynomials. Yet their derivatives are rational functions or roots of rational functions. The simplest illustrations are:

- the natural logarithm of x, whose derivative is $1/x$;
- the inverse tangent of x, whose derivative is $1/(1 + x^2)$;
- and the inverse **sine** of x, whose derivative is $1/((1-x^2)$.

A common feature of all of these examples is the fact that the transcendental function is the inverse of a function that satisfies a simple first-order differential equation. The natural logarithm, for example, is the inverse of the exponential function, which satisfies the equation $dy/dx = y$. In each case, the formula for the derivative of the **inverse function** follows easily from the corresponding differential equation and from the inverse function **theorem**.

For the student of calculus, it is also important to realize that each of these derivative formulas leads to a corresponding integral formula, and hence that the integral of an innocuous-appearing rational function may be a transcendental function.

DIMENSION

One of the most misunderstood concepts in mathematics, dimension is a central concept in modern **geometry**, **topology**, **algebra**, and the theory of **fractals**. To make matters even more complicated, each of these branches of mathematics has its own version of dimension (or even several versions).

In simplest terms, the dimension of a mathematical object is the number of independent parameters required to describe that object. It is well known that the dimension of a line is 1, the dimension of a plane is 2, and the dimension of **space** is 3. For example, any point in space can be described by three coordinates; a box in space is described by three parameters, its height, width and depth.

However, most people have difficulty imagining dimensions larger than 3; it is sometimes asserted that "The fourth dimension is time." As a statement about our physical universe, this may be correct; as a mathematical statement, it is false. Mathematical structures can have any number of dimensions, and none of these need to have any connection with time.

Here are four different ways that mathematicians approach the concept of dimension.

- The vector space approach. An n-dimensional Euclidean vector space is described as the set of all points with coordinates $(x_1, x_2,..., x_n)$, where the numbers $x_1, x_2,..., x_n$ can take any real values. More generally, **vector spaces**

can be defined with coordinates in any **field**, such as the **complex numbers**, and the dimension equals the number of coordinates.

- The differential geometric approach. Differential geometry deals primarily with the properties of smooth manifolds, which are generalizations of curves and surfaces. On a small scale, any **manifold** "looks like" an n-dimensional Euclidean vector space. For example, any small piece of a **sphere** looks like a flat plane; therefore the sphere is 2-dimensional.

- The topological approach. Topologists use the "Lebesgue covering dimension" or "topological dimension," an ingenious method that makes no mention of **real numbers** or Euclidean space, and hence might be considered more fundamental. Roughly speaking, a topological space has dimension n if it can be covered by open **sets** that overlap no more than $(n+1)$ at a time, and if these coverings can be made arbitrarily fine. For example, a plane can be covered by disks that intersect no more than 3 at a time, and the disks may be made as small as desired. Hence the plane is two-dimensional.

- The fractal approach. The various approaches to fractal dimension all take into account the scaling of an object. They take the viewpoint that the dimension of a box is 3 not because it resides in Euclidean 3-dimensional space, but because its **volume** scales as the third power of its linear size. Remarkably, for fractals the "volume"—measured, for example, by the number of disks of a given (fixed) size needed to cover the fractal—need not scale as an integer power. For example, the **snowflake curve** has a fractal dimension of 1.26..., because when its size is tripled the number of disks required to cover it increases by a **factor** of four, or $3^{(1.26...)}$. Yet the topological dimension of the snowflake curve is 1. The fractal dimension, in fact, is the only concept of dimension that allows fractional values; this motivated the name "fractal."

See also Fractals; Geometry; Hausdorff dimension; Manifolds; Space; Space-time; Topology; Vector space

DIOPHANTINE EQUATIONS

A Diophantine equation is characterized not by the shape of the equation but by the fact that one is interested only in solutions in **integers** (**whole numbers**) or **rational numbers**. Typically, it will have more than one **variable**. For example, xy = x + y + 2 can be looked at as a Diophantine equation and it has only the solutions (x,y)=(2,4),(4,2),(0,-2),(-2,0) in integers. The equation 2x+4y = 1 is also a Diophantine equation, and in this case it has no integral solutions, since the left hand side is even if x and y are integers and 1 is not even.

Historically, the first record of the study of Diophantine **equations** is a Babylonian tablet, dating from before 1600 B.C., which lists integer solutions of the Pythagorean equation $x^2+y^2=z^2$, which leads to right-angled triangles with integral sides. Diophantine equations were also studied by the Greeks, in particular **Diophantus of Alexandria** (hence the name) around 200 AD because they preferred their problems to have integer or rational solutions. The great French mathematician **Pierre de Fermat** became interested in Diophantine equations by reading a (then recent) translation of Diophantus's book *Arithmetica* into Latin, in the seventeenth century. He developed the subject extensively, creating the method of infinite descent which is still basic for the subject and by raising a number of interesting questions, some of which he said he could solve but did not divulge the solutions. This was meant as a challenge for his contemporaries such as **Blaise Pascal** and **John Wallis**, but it not have the desired effect. The challenge was taken up only much later by other mathematicians, such as **Leonhard Euler** who proved many of Fermat's assertions such as that any prime number of the form $4n+1$ is a sum of two perfect squares, and **Joseph-Louis Lagrange** who proved that every integer is a sum of four perfect squares. Eventually, the only assertion made by Fermat left unproved was what became known as **Fermat's last theorem,** which was only proved in 1993, by **Andrew Wiles**, almost 350 years after it was posed.

There is still much interest today in Diophantine equations, since our knowledge is very much incomplete. Study of Diophantine equations has been a driving force in the development of mathematics. For instance, the theory of **algebraic numbers** was developed by E. Kummer partly as an attempt to prove Fermat's last **theorem**.

The simplest type of Diophantine equations are the linear ones. The basic case of two variables, that of equations of the type $ax+by=c$, can be solved by the Euclidean **algorithm**, as has been known since classical times. The so-called Pell equation (the name is due to a historical inaccuracy of Euler's) $x^2-dy^2=1$ was studied by the Indian mathematicians **Brahmagupta** in the sixth century AD and **Bhaskara** in the eleventh century AD, who had a method for finding the solutions. This method was rediscovered by Fermat and first published by Euler. There is a vast general theory for **quadratic equations** in many variables one of whose central results is the Hasse-Minkowski theorem that gives necessary and sufficient conditions for the solvability of homogeneous quadratic equations. **Cubic equations** in two variables fall into the theory of Elliptic curves which is a very developed theory but still an important topic of current research. A lot is known also about equations in two variables in higher degrees, specially since the German mathematician G. Faltings proved the Mordell conjecture in 1983, which ensures that equations that define the so-called curves of genus bigger than one have only finitely many rational solutions. Faltings received the Fields medal in 1986 for this work. For equations with more than three variables and degree at least three, very little is known. An important modern theme in the theory of Diophantine equations is the use of methods of **algebraic geometry**. One of the early proponents of this approach was Weil and the subject of Diophantine **geometry** reached maturity with the work of **Alexander Grothendieck** who unified **number theory** and algebraic geometry with his theory of schemes.

Hilbert's tenth problem asked for an algorithm that would decide whether any given Diophantine equation had a solution or not. The Russian mathematician Yuri Matiyasevich proved in 1970, building on work of H. Putnam, M. Davis and J. Robinson, that such algorithm cannot possibly exist.

See also Fermat's last theorem; Algebraic Geometry; Number Theory

DIOPHANTUS (3RD CENTURY)
Greek mathematician

Little is known about the life of the Greek mathematician Diophantus. However, his work led to one of the greatest mathematical challenges of all time, **Fermat's last theorem**.

Diophantus was born and lived in Alexandria, now in Egypt, which was at the time a great center of culture and learning in the Greek world. We have no record of the date of his birth or death, but we do have two pieces of evidence regarding when and how long he lived. One is a letter written in the 11th century, that tells that the bishop of Laodicea, Anatolius, dedicated his work on Egyptian computation to Diophantus, who was a good friend; since we know that Anatolius was bishop around 270 a.d., it would make sense that Diophantus was his contemporary.

A bit more sleuthing is required to learn a few details of Diophantus's life, which were left behind in a mathematical puzzle called "Diophantus's riddle." If it is to be believed, Diophantus married at age 33, had a son who died at age 42, and the mathematician himself died four years later, at age 84.

Where little is known about his personal life, Diophantus's work in mathematics is documented in his *Arithmetika*, which was supposed to have consisted of 13 books, but only six have survived. (It is possible that the six books were the only ones completed.) His works show that he may have studied with Babylonian teachers, for he was influenced by both Greek and Babylonian practices.

Diophantus devised an early form of **algebra**. He worked with **equations** that are solved in terms of **integers**. Some of his equations had no single solution; such equations, with more than one or even an infinite number of solutions, are called Diophantine, or indeterminate, equations today. Although they can't be solved down to one answer, they can be generalized to fit a number of solutions.

Diophantus used a symbol to indicate the unknown quantities in his equations, which was a major innovation. He was also the first Greek mathematician to treat **fractions** as numbers.

Before Diophantus, most Greek mathematics were concerned with practical problems drawn from everyday concerns, such as agriculture and finance; the math was either computational or geometric. However, by devising indeterminate equations, Diophantus opened the door to **number theory**.

The eighth problem of the second book of *Arithmetika* led to a puzzle that stumped generations of mathematicians. When looking at the problem in 1637, **Pierre de Fermat** wrote:

"On the other hand it is impossible to separate a cube into the sum of two cubes... or any power except a **square** into the sum of two **powers** with the same exponent. I have discovered a truly marvelous **proof** of this, which however the margin is not large enough to contain." For over 300 years, Fermat's last **theorem** went unsolved. It was a riddle that Diophantus surely would have appreciated greatly.

DIRAC, PAUL ADRIEN MAURICE (1902-1984)

British physicist

Paul Adrien Maurice Dirac was one of the leading theoretical physicists of the twentieth century. He made significant contributions to the early development of **quantum mechanics**, to theories of atomic structure and properties, and to quantum electrodynamics (the study of electrical interactions between atomic particles). Dirac successfully predicted the existence of the positron, i.e., a positively charged electron, and established the theoretical foundations for later discoveries related to antimatter. Dirac shared the Nobel Prize for Physics with Erwin Schrödinger in 1933 for his "discovery of new fertile forms of the theory of atoms and for its applications." Few of Dirac's theories were simple to grasp, and for that reason he had few students during his career.

Dirac was born in Bristol, England on August 8, 1902 to Charles Adrien Ladislas Dirac, a Swiss immigrant, and Florence Hannah (Holten) Dirac, a native of Britain. Dirac's father taught French at Merchant Venturer's Technical College, where Paul received his early education. As a youth, Dirac is said to have preferred to spend his time alone, taking long walks, and enjoying nature and gardening in preference to social activities. At school, he excelled in science and mathematics, but showed little interest in humanistic studies. Following his graduation in 1918 from Merchant Venturer's, Dirac enrolled in Bristol University, where he received a bachelor's degree in electrical engineering in 1921. At the time of his graduation, Britain was still in a postwar economic depression, so the young Dirac decided to return to school; this led to his accepting a two-year scholarship from the department of mathematics at Bristol University.

When Dirac's scholarship ran out at Bristol University, he enrolled as a research student in mathematics in St. John's College at Cambridge University. There, Dirac discovered in class and through his reading the work of many of the then leading atomic theorists, including **Niels Bohr**, Max Born, Arnold Summerfeld, and **Werner Heisenberg**, some of whom paid visits to Cambridge.

Sometime around 1924, Dirac made the statement at an academic tea party that any really interesting mathematical theory should have an application in the physical world. This idea was to evolve into Dirac's methodology for doing physics: find a beautiful mathematical theory and try to connect it to the physical world.

Dirac published his first research paper in 1925, and earned his Ph.D. in physics (at Cambridge University, theoret-

Paul Dirac

ical physics was a discipline with the mathematics department) one year later by writing a thesis that elaborated quantum mechanic concepts originally developed by Werner Heisenberg.

In 1926, Dirac traveled to Copenhagen, where he met with Niels Bohr. One year later, he moved to Göttingen, where Max Born, J. Robert Oppenheimer, James Franck, and Igor E. Tamm were conducting research. When he returned to Cambridge, he was elected a fellow at St. John's College.

During the winter of 1927-8 at Cambridge, Dirac attempted to make improvements in Schrödinger's wave equation. Schrödinger's theory attempted to explain the behavior of an electron in an atom at the expense of ignoring relativistic effects, and Dirac set out to more completely describe the electron's behavior based only on that particle's mass and charge. Dirac succeeded in accurately predicting the spin angular momentum of the electron, its magnetic moment, and other properties as confirmed by later experimental measurements.

While working with his **equations** for the electron, Dirac discovered that his theory actually allowed for the existence of a positively charged electron, besides the already known negatively charged one. He also provided a road map for experimentalists to help them search for this particle, predicting that it would always be produced in connection with a negatively charged electron and in such a way that the positive and negative electrons would annihilate each other, producing chargeless photons. This particle, later named the positron, was discovered in 1932 by an American physicist, C.D. Anderson.

By extending this theory, Dirac was also able to predict the existence of other types of antimatter, including the antiproton and antineutron, which were also later observed experimentally. Dirac's ideas were to prove essential to later investigations in particle physics.

In 1929, Dirac was appointed university lecturer and praelector at St. John's. Because this post had few regular duties associated with it, Dirac was able to devote his time to research, writing, and travel. In the same year, he spent brief periods teaching at the University of Michigan and the University of Wisconsin. Returning to England by way of Japan and Siberia, Dirac reached Cambridge in 1930, and found himself elected to the Royal Society. His classic text, *The Principles of Quantum Mechanics*, was published the same year. In 1932, he was appointed Lucasian Professor of Mathematics at Cambridge, a post previously held in the seventeenth century by **Isaac Newton**.

In 1934, Dirac once again visited the United States, this time spending most of the 1934-5 academic year at the Institute for Advanced Studies at Princeton University. There, he met his future wife, Margit Wigner (the sister of physicist Eugene Wigner), whom he married in London three years later.

Unlike many of his colleagues, Dirac was not directly engaged in war-related work during World War II. Remaining at Cambridge during the war, Dirac did participate in government projects related to atomic energy, however. A particular interest of his was the separation of uranium isotopes, but none of his ideas ended up being used in later weapons development programs.

In 1969, Dirac retired from his academic position at Cambridge, and relocated to the United States. After a brief stay at the Center for Theoretical Studies at the University of Miami, he accepted an appointment as professor of physics at Florida State University in Tallahassee in 1972. For the next decade, he continued to travel, write, and lecture. His health began to deteriorate in 1982, and he died in Tallahassee on October 20, 1984.

DIRECT VARIATION

If one quantity increases (or decreases) each time another quantity increases (or decreases), the two quantities are said to vary together. The most common form of this is direct variation in which the ratio of the two amounts is always the same. For example, speed and **distance** traveled vary directly for a given time. If you travel at 4 mph (6.5 kph) per hour for three hours, you go 12 mi (19.5 km), but at 6 mph (9.5 kph) you go 18 mi (28.5 km) in three hours. The ratio of distance to speed is always 3 in this case.

The common ratio is often written as a constant in an equation. For example, if s is speed and d is distance, the relation between them is direct variation for d = ks, where k is the constant. In the example above, k = 3, so the equation becomes d = 3s. For a different time **interval**, a different k would be used.

Often, one quantity varies with respect to a power of the other. For example, of $y = kx^2$, then y varies directly with the **square** of x. More than two variables may be involved in a direct variation. Thus if z = kxy, we say that z is a joint (direct) variation of z with x and y. Similarly, if $z = kx^2/y$, we say that z varies directly with x^2 and inversely with y.

DIRICHLET, JOHANN PETER GUSTAV LEJEUNE (1805-1859)
German number theorist and analyst

Johann Peter Gustav Lejeune Dirichlet was born in 1805, the son of the town postmaster of Düren (then part of the French empire). He was initially educated at public schools, then at a private school which stressed Latin. Interest in mathematics surfaced early for Dirichlet and by age 12 he was purchasing mathematics textbooks. Dirichlet enrolled at Bonn's Gymnasium in 1817, where he showed great interest in mathematics and history.

Two years later, Dirichlet's parents sent him to a Jesuit college in Cologne. He was a student of physicist George Simon Ohm, under whom he received thorough training in theoretical physics. At the young age of 16, Dirichlet completed his Abitur examination. His parents wanted him to study law, but he was already well on his way in the field of mathematics.

Other than **Karl Gauss,** Germany, at that time, had no notable mathematicians. Paris, on the other hand, boasted such luminaries as **Pierre Simon Laplace, Adrien–Marie Legendre**, and **Jean Baptiste Jospeh Fourier**. In 1822, Dirichlet visited Paris. He was not there long before he caught a mild case of smallpox, but it was not severe enough to keep him from continuing classes at the College de France, and the Faculte des Sciences. In 1823, he was appointed to a well–paid position as a tutor to General Maximilian Fay's children. Fay was a national hero of the Napoleonic wars and a liberal opposition leader in the Chamber of Deputies. Dirichlet was treated as a member of the family, thus meeting prominent French intellectuals, including mathematician Joseph Fourier. Fourier's ideas influenced Dirichlet's later works on trigonometric seriesand mathematical physics.

Dirichlet's main interest was **number theory**, which was first ignited through an early study of Gauss' *Disquistiones arthmeticae* (1801). In June 1825, Dirichlet presented his first paper on mathematics, "Memoire sur l'impossibilite de quelques equations indeterminees du cinquieme degre," to the French Académie Royale des Sciences. The paper explored **Diophantine equations** of the form $x^5 + y^5 = Az^5$ using algebraic number theory, Dirichlet's favorite area of study.

Johann Peter Gustav Lejeune Dirichlet

Legendre extended these results to give a **proof** of **Fermat's last theorem** for n=5.

With General Fay's death in 1825, Dirichlet returned to Germany. Fellow German scientist Alexander von Humboldt strongly supported Dirichlet's return, as Germany needed strengthening in the natural sciences. Although Dirichlet did not have the required doctorate, he was permitted to qualify at the University of Breslau for the habilitation required to teach at a German university.

Breslau was not an inspiring environment for scientific work, so in 1828, again with Humbolt's assistance, Dirichlet moved to Berlin and began teaching mathematics at the military academy. At age 23, he was first appointed to a temporary position at the University of Berlin and in 1831 became a member of the Berlin Academy of Sciences. That same year Dirichlet married Rebecca Mendelssohn–Bartholdy, granddaughter of philosopher Moses Mendelssohn, and sister to composer Felix Mendelssohn. In 1832, he published a proof of **Pierre de Fermat**'s last **theorem** for n=14.

During his 27 years as professor in Berlin, Dirichlet influenced the development of German mathematics through his lectures, pupils, and scientific papers. He taught with great clarity and his published scientific papers were of the highest quality.

Dirichlet was a shy, modest man, who rarely made public appearances, or spoke at meetings. His lifelong friend was mathematician **Karl Jacobi**. Both mathematicians influenced each other's work, particularly in number theory. In 1843, Jacobi moved to Rome for health reasons. This prompted Dirichlet to request a leave of absence to move his family to Rome with Jacobi. Dirichlet remained in Italy for a year and a half, visited Sicily, and spent the winter in Florence.

At a meeting of the Academy of Science, held on July 27, 1837, Dirichlet presented a paper on analytic number theory. The paper offers proof of the fundamental theorem that bears his name: any **arithmetic** sequence of **integers**: an + b, n = 0, 1, 2..., where a and b are relatively prime, must include an infinite number of primes. This paper was followed in 1838 and 1839 by a two–part paper on analytic number theory, "Recherches sur diverses applications de l'analyse infinitesimale a la theorie des nombres." After publication of his fundamental papers, the importance of his number theory work declined, but Dirichlet continued to publish papers in other areas.

Dirichlet is best known for his work on trigonometric series and mathematical physics. In an 1828 paper in *Crelle's Journal*, he gave the first rigorous proof of sufficient conditions for the **convergence** of the **Fourier series** for a function. His investigations of equilibrium of systems and potential theory gave rise to what is now called the Dirichlet problem about formulating and solving a class of partial **differential equations** that arise from the flow of heat, electricity, and fluids subject to given boundary conditions.

In 1837 Dirichlet proposed the modern **definition** of a function: if a **variable** y is so related to a variable x that whenever a numerical value is assigned to x, there is a rule according to which a unique value of y is determined, then y is said to be a function of the **independent variable** x. In another 1837 paper, Dirichlet proved that in an absolutely convergent series, one may rearrange the order in which terms are added in whatever way one wishes and not change the sum of the series. While this is immediate for a finite sum, the fact that an infinite sum of numbers could always be rearranged without changing the value was surprising. He also gave examples of conditionally convergent series in which the sum *was* altered by rearrangement of the terms. Almost 20 years later, **Georg Riemann** proved that the terms of a conditionally convergent series could be rearranged to yield a sum of any desired value.

At the golden jubilee celebration of his doctorate, Dirichlet's teacher Karl Gauss tried to light his pipe with a piece of his original manuscript *Disquisitiones arithmeticae*. Dirichlet was overcome by the sacrilege of such an action. He rescued the piece of paper from Gauss' fire. Dirichlet treasured the paper for his remaining years, and his editors found it among his papers after his death.

With Gauss's death in 1855, the University of Göttingen sought a successor of great distinction and chose Dirichlet. His current position at the military academy was unappealing and lacked scientific stimulation and he was required to lecture 13 times a week. Dirichlet accepted the university's offer.

Dirichlet moved to Göttingen in 1855, where he purchased a house with a garden. He enjoyed a quiet life there with excellent students and the time available for research. But his new contentment did not last long. During a speech in Montreaux, Switzerland, he suffered a heart attack and barely made it home. During his illness, his wife died of a stroke, and Dirichlet subsequently died the following spring in 1863.

After his death, Dirichlet's pupil and friend, **Julius Dedekind**, published Dirichlet's *Vorlesungen über Zahlentheorie*, adding several supplements of his own investigations on algebraic number theory. The addenda are regarded as one of the most important sources for the creation of the theory of ideals, and are the core of algebraic number theory.

DISCRIMINANT

A discriminant is a number, usually **invariant** under suitable transformations, which characterizes some properties of the **roots** of a certain quantity.

For instance, the most common application of the concept of discriminant is in the second-degree equation $ax^2 + bx + c = 0$.

In this context, the discriminant is the quantity $D = b^2 - 4ac$. According to the sign of D, such an equation can have either two distinct roots (if D is greater than **zero**), or only one root (in the case $D=0$), or no root (if D is less than zero).

The concept of discriminant is also used for **polynomials**. In the case of a polynomial with real coefficients, the discriminant is defined as the product of the squares of the differences of the complex roots. Since the non-real roots of such a polynomial are in conjugated pairs, the discriminant is a real number. Moreover, the **definition** of discriminant in the case of the second degree equation (as given above) coincides with the one of the second degree polynomial up to a multiplicative positive quantity, which does not change the sign of the discriminant.

It is possible to define a discriminant also for **conic sections**. In this context, the sign of the discriminant depends on whether the conic is a **parabola**, a **hyperbola**, or an **ellipse**.

Given a general quadratic curve $Ax^2 + Bxy + Cy^2 + Dx + Ey + F = 0$, the quantity D is known as the discriminant, where $D = B^2 - 4AC$. The discriminant defined in this way is invariant under **rotation** and **translation**.

This invariant quantity provides a useful shortcut to determining the shape represented by a quadratic curve. If D is negative, the equation represents an ellipse, a **circle** (degenerate ellipse), a point (degenerate circle), or has no graph.

If D is positive, the equation represents either a hyperbola or a pair of intersecting lines (degenerate hyperbola).

If $D=0$, the equation represents a parabola, a line (degenerate parabola), a pair of parallel lines (degenerate parabola), or has no graph.

A very useful application of the concept of discriminant is the so-called "second derivative test." Suppose that f is a twice-differentiable function of two variables and that a is a stationary point for f. In this case, the discriminant D of f at the point a is defined as the determinant of the **matrix** of the second derivatives evaluated at the point a (this matrix is sometimes called a Hessian matrix and its determinant is called Hessian determinant).

If D is greater than zero, f_{xx} at a is greater than zero, and $f_{xx} + f_{yy}$ at a is greater than zero, then the point a is a relative minimum.

If D is greater than zero, f_{xx} at a is less than zero, and $f_{xx} + f_{yy}$ at a is less than zero, then the point a is a relative maximum.

If D is less than zero, the point a is a saddle point.

If $D=0$, higher order tests must be used to determine the local behavior of the function f.

The concept of discriminant can also be used for: binary quadratic forms, elliptic curves, metrics, modules, quadratic fields, and quadratic forms. The interested reader may look at the following books: H. Cohn, *Advanced Number Theory* (New York, Dover, 1980); D. Hilbert and S. Cohn-Vossen, *Geometry and the Imagination* (New York, Chelsea, 1999); J. Silverman, *The Arithmetic of Elliptic Curves* (New York, Springer-Verlag, 1986).

DISPERSION

In **statistics**, the scattering of values in a distribution of data from an average value is called dispersion. The term dispersion generally means the spread of a series of values, usually about some central point such as the mean, also called the average, or the **median** of the values. The measure of dispersion normally used is the **standard deviation**, but the **mean** deviation and the mean difference are also used.

A well known example of what might be called physical dispersion is the refraction of **light** into its component colors when it passes through a material in which the velocity of the light waves varies with the wavelength. In a vacuum, light may be thought of as being composed of superimposed trains of light waves of many wavelengths. When light passes through some materials, like a glass prism, it is bent and spread out into its component colors because the different component wave trains travel at different speeds through the glass resulting in a spectrum. An example of such a spectrum in nature is the rainbow. When split into its colors, the light is said to be dispersed. Given appropriate media through which to send them, other wavelengths along the electromagnetic spectrum can be dispersed in a similar way. Each wave changes its length as it moves through the material, at each point having a length determined by the time of arrival of the waves traveling the point of origin.

As defined above, dispersion is the separation of a complex wave into components. The term is also applied to the property of an optical device or medium giving rise to the phenomenon, or the numerical value of this property.

See also Electromagnetism

DISTANCE

Distance has two different meanings. It is a number used to characterize the shortest length between two geometric figures, and it is the total length of a path. In the first case, the distance between two points is the simplest instance.

The absolute distance between two points, sometimes called the displacement, can only be a positive number. It can never be a negative number, and can only be **zero** when the two points are identical. Only one straight line exists between any two points P_1 and P_2. The length of this line is the shortest distance between P_1 and P_2.

In the case of parallel lines, the distance between the two lines is the length of a perpendicular segment connecting them. If two figures such as line segments, triangles, circles, cubes, etc. do not intersect, then the distance between them is the shortest distance between any pair of points, one of which lies on one figure, one of which lies on the other.

To determine the distance between two points, we must first consider a coordinate system. An xy coordinate system consists of a horizontal axis (x) and vertical axis (y). Both axes are infinite for positive and negative values. The crossing point of the lines is the origin (O), at which both x and y values are zero.

We define the coordinates of point P_1 as (x_1, y_1), and point P_2 by (x_2 and y_2). The distance, the length of the connecting straight line ($P_1 P_2$) which is the shortest distance between the two points, is given by $d=\sqrt{(x_1-x_2)^2 + (y_1-y_2)^2}$ in many types of physics and engineering problems, for example, in tracking the trajectory of an atomic particle, or in determining the lateral **motion** of a suspension bridge in the presence of high winds.

The other meaning of distance is the length of a path. This is easily understood if the path consists entirely of line segments, such the **perimeter** of a pentagon. The distance is the sum of the lengths of the line segments that make up the perimeter. For curves that are not line segments, a continuous path can usually be approximated by a sequence of line segments. Using shorter line segments produces a better approximation. The limiting case, when the lengths of the line segments go to zero, is the distance. A common example would be the **circumference** of a **circle**, which is a distance.

DISTRIBUTIVE PROPERTY

The distributive property states that the **multiplication** "distributes" over **addition**. Thus $a \times (b + c) = a \times b + a \times c$ and $(b + c) \times a = b \times a + c \times a$ for all real or **complex numbers** a, b, and c.

The distributive property is behind the common multiplication **algorithm**. For example, 27×4 means $4 \times (2$ tens + 7 ones). To complete the multiplication, you use the distributive property: $4 \times (20 + 7) = (4 \times 20) + (4 \times 7) = 80 + 28 = 108$.

We use the distributive property more than once in carrying out such computations as $(3x + 4)(x + 2)$. Thus $(3x + 4)(x + 2) = (3x + 4)x + (3x + 4)2$ where $3x + 4$ is "distributed over"

x + 2 and then $(3x + 4)x + (3x + 4)2 = 3x^2 + 4x) + (6x + 8) = 3x^2 + 10x + 8$ where x and 2 are "distributed" over $3x + 4$.

DIVISION

Situations in which a collection of objects is separated into several smaller and equal groups predate recorded history. With the advent of **arithmetic** and number systems, this sort of 'dividing' process was generalized and given precision as an arithmetic operation. That is, **division** can be defined as an operation performed to determine how many times one quantity is contained in another quantity. For example, the number "2" is contained three times within the number "6". The number of times one number must be added together to yield another number is symbolized by the arithmetic expression "a / b"; with the number "a" called the "dividend", the symbol "b" called the "divisor", and the result called the quotient. So "6 / 2" is a quotient whose value is 3; that is, 2 must be added three times to produce 6. The word quotient is from the Latin "quotiens" for "how many times". In addition to the symbol "/" (diagonal or solidus), the symbol "÷" (obelus) is frequently used to denote the operation of division. Either symbol is read as "divided by" (i.e., a divided by b).

There are four basic arithmetic operations: addition, **subtraction**, **multiplication**, and division. The fundamental idea of an operation is that of a procedure which for any pair of numbers assigns a third unique number. With regards to **whole numbers**, **integers**, **rational numbers**, and **real numbers**, the operations of addition and multiplication are closed. That is to say, the addition or multiplication of two whole numbers is another whole number, with the same holding true for real, integer, and rational numbers. The other two fundamental operations of arithmetic, subtraction and division, are considered the inverse operations to addition and multiplication, respectively. These two operations are not closed with regards to some of the number systems described above. For example, the expression "2 - 5" has no solution in the set of whole numbers (there are no negative whole numbers). Likewise, for the operation of division, "5 / 2" has no solution in the set of whole numbers or the set of integers. However, "5 / 2" is a meaningful expression with regards to rational and real numbers. It was a blessing for the advancement of mathematics that ancient civilizations were faced with the task of finding solutions to expressions like "2 - 5" and "5 / 2" because it eventually led to the extension of their number systems to more advanced number systems, like the rational numbers.

For the rational and real number systems, division may be defined as the inverse operation of multiplication, where "a / b = a · (1 / b)". For instance, the quotient "6 / 2" is represented by the number "6" (the dividend of the quotient) multiplied by "1/2" (the reciprocal of the divisor); or written as an equation: "6 / 2 = 6 · (1 / 2)".

Besides the closure law, there are other familiar rules that are valid for addition and multiplication, like the property of commutativity "a + b = b + a" and "a · b = b · a", which are not necessarily valid for their corresponding inverse operations ("$(5 - 2) \neq (2 - 5)$" and "$(5 / 2) \neq (2 / 5)$"). For real or rational numbers (a, b, c) the following rules are applicable for the operation of division:

- Closure law: the quotient "a / b" is also a real number (for $b \neq 0$).

- Commutative law: "a / b = b / a" is not a valid rule for division.

- Associative law: "(a / b) / c = a / (b / c)" is not a valid rule for division.

- Division by **zero** is undefined.

- "0 / a = 0" is valid for all $a \neq 0$.

- Dividing a by b is equivalent to multiplying a by 1 / b, where "$(1 / b) \cdot b = b \cdot (1 / b) = 1$" where $b \neq 0$.

- Monotonic law: for $c \neq 0$, "(a / c) < (b / c)" is valid, where $a < b$ for positive numbers and $a > b$ for negative numbers.

- Distributive law: unlike multiplication, the distribution of division over addition, "a / (b + c) = (a / b) + (a / c)", is not a valid rule.

In addition to rational and real numbers, division is a valid operation on the set of **complex numbers** and **functions**. However, division is undefined as an operation on vectors and matrices.

The division of **fractions** is aided by use of the following procedure: the 'bottom' fraction is inverted, multiplied by the 'top' fraction, and the result is reduced to its lowest terms; that is "$(a / b) / (c / d) = (a / b) \cdot (d / c) = (a \cdot d) / (b \cdot c)$", where b, c, and d are not equal to zero. For example "$(2 / 3) / (5 / 6) = (2 / 3) \cdot (6 / 5) = 12 / 15 = 4 / 5$".

Complex numbers are expressed with a real part followed by an imaginary part (i.e., for the complex number "a + ib", "a" is a real number and "ib" an imaginary number). The operation of division is defined by multiplying both the dividend and the divisor by the complex conjugate of the divisor. That is, dividing "a + ib" by "c + id" results in "(a + ib) / (c +id) = ((a + ib) \cdot (c + id)*) / ((c + id) \cdot (c + id)*)", where the complex conjugate "(c + id)*" is equal to "c - id". Therefore, substituting that equality results in "((a + ib) \cdot (c - id)) / ((c + id) \cdot (c - id)) = ((ac + bd) + i(bc - ad)) / (c^2 + d^2)", where $(i \cdot i) = -1$.

The division of functions is denoted "f / g", and is defined as $(f / g)(x) = f(x) / g(x)$. For example if $f(x) = x + 5$ and $g(x) = x^2 - 3x + 4$, then "$f(x) / g(x) = (x + 5) / (x^2 - 3x + 4)$". If $x = 2$, then "$f(2) / g(2) = (2 + 5) / (4 - 6 + 4) = 7 / 2$".

See also Addition; Arithmetic; Complex numbers; Fractions; Functions; Integers; Matrix; Multiplication; Negative numbers; Numbers and numerals; Products and quotients; Rational numbers; Real numbers; Subtraction; Vector analysis; Whole numbers

DOMAIN

The domain of a relation is the set that contains all the first elements, x, from the ordered pairs (x,y) that make up the relation. In mathematics, a relation is defined as a set of ordered pairs (x,y) for which each y depends on x in a predetermined way. If x represents an element from the set X, and y represents an element from the set Y, the Cartesian product of X and Y is the set of all possible ordered pairs (x,y) that can be formed with an element of X being first. A relation between the **sets** X and Y is a subset of their Cartesian product, so the domain of the relation is a subset of the set X. For example, suppose that X is the set of all men and Y is the set of all women. The Cartesian product of X and Y is the set of all ordered pairs having a man first and women second. One of the many possible relations between these two sets is the set of all ordered pairs (x,y) such that x and y are married. The set of all married men is the domain of this relation, and is a subset of X. The set of all second elements from the ordered pairs of a relation is called the **range** of the relation, so the set of all married women is the range of this relation, and is a subset of Y. The **variable** associated with the domain of the relation is called the **independent variable**. The variable associated with the range of a relation is called the **dependent variable**.

Many important relations in science, engineering, business and economics can be expressed as **functions** of **real numbers**. A function is a special type of relation in which none of the ordered pairs share the same first element. A real-valued function is a function between two sets X and Y, both of which correspond to the set of real numbers. The Cartesian product of these two sets is the familiar **Cartesian coordinate system**, with the set X associated with the x-axis and the set Y associated with the y-axis. The graph of a real-valued function consists of the set of points in the plane that are contained in the function, and thus represents a subset of the Cartesian plane. The x-axis, or some portion of it, corresponds to the domain of the function. Since, by definition, every set is a subset of itself, the domain of a function may correspond to the entire x-axis. In other cases the domain is limited to a portion of the x-axis, either explicitly or implicitly.

Example 1. Let X and Y equal the set of real numbers. Let the function, f, be defined by the equation $y = 3x^2 + 2$. Then the variable x may range over the entire set of real numbers. That is, the domain of f is given by the set $D = \{x \mid -\infty \leq x \leq \infty\}$, read "D equals the set of all x such that negative **infinity** is less than or equal to x and x is less than or equal to infinity." Example 2. Let X and Y equal the set of real numbers. Let the function f represent the location of a falling body during the second 5 seconds of descent. Then, letting t represent time, the location of the body, at any time between 5 and 10 seconds after descent begins, is given by $f(t) = 1/2gt^2$. In this example, the domain is explicitly limited to values of t between 5 and 10, that is, $D = \{t \mid 5 \leq t \leq 10\}$.

Example 3. Let X and Y equal the set of real numbers. Consider the function defined by $y = \pi x^2$, where y is the **area** of a **circle** and x is its radius. Since the radius of a circle cannot be negative, the domain, D, of this function is the set of all real numbers greater than or equal to **zero**, $D = \{x \mid x \geq 0\}$. In this example, the domain is limited implicitly by the physical circumstances.

Example 4. Let X and Y equal the set of real numbers. Consider the function given by $y = 1/x$. The variable x can take on any real number value but zero, because **division** by zero is undefined. Hence the domain of this function is the set D{x | x ≠ 0}. Variations of this function exist, in which values of x other than zero make the denominator zero. The function defined by y = 1/ 2-x is an example; x=2 makes the denominator zero. In these examples the domain is again limited implicitly.

DONALDSON, SIMON K. (1957-   )
English geometer and topologist

Simon Donaldson shocked the mathematical world during the 1980s with a series of papers on the structure of four-dimensional spaces. Researchers had produced a collection of results during the previous decade that outlined a general understanding of the properties of spaces of five or more dimensions, and of course, the cases of one- and two-dimensional spaces were well known. Ironically, three- and four-dimensional spaces were the hardest to interpret, even though they are the most applicable to physical **space** (if time is considered to be the fourth **dimension**). Great progress was made in three-dimensions by William Thurston,who received a Fields Medal in 1982 for his efforts. That same year, Donaldson published his most remarkable result: four-dimensional space has highly unusual properties that are found in no other dimension. Speaking on the occasion of Donaldson's presentation with the 1986 Fields Medal, **Michael Atiyah** commented, "When Donaldson produced his first few results on four-manifolds [four-dimensional topological surfaces], the ideas were so new and foreign to geometers and topologists that they merely gazed in bewildered admiration... Donaldson has opened up an entirely new area; unexpected and mysterious phenomena about the **geometry** of four dimensions have been discovered."

Simon K. Donaldson was born on August 20, 1957, in Cambridge, England. He attended Pembroke College in Cambridge University and received his B.A. degree in 1979. During his second year of graduate studies at Worcester College in Oxford University, Donaldson made the spectacular discovery of "exotic" or nonstandard differential structures of four-dimensional Euclidean space. In other words, he found that there were different ways of orienting a mathematical structure in ordinary space with the addition of a fourth dimension. Because the standard differential structure is the only one possible in all other dimensions, the mathematical community was amazed at the exceptions created by the addition of the fourth dimension. After completing his doctorate in 1984, Donaldson spent a year at Princeton University's Institute for Advanced Study (IAS) and was a visiting scholar at Harvard University during the spring of 1985. He then returned to England where he holds an appointment at the Mathematics Institute in Oxford.

In 1954, Chen Ning Yang and Robert Mills collaborated on derivations of mathematical formulas that combined the branch of mathematics known as **topology** (the study of the ways in which coordinate structures attach at a point) and the branch of physics called quantum electrodynamics (the study of electromagnetic phenomena under the rules of **quantum mechanics**). In doing so, they built upon the work of **James Clerk Maxwell**, who had introduced **equations** in the 19th century to describe the behavior of electromagnetic waves. The Yang-Mills equations generalize **Maxwell's equations** to more complex spaces. Since they are nonlinear partial **differential equations**, the Yang-Mills equations are very difficult to solve, even for specific cases. Work by Atiyah and others during the 1970s led to important connections between the equations and techniques from differential and **algebraic geometry**. (See **Yang-Mills existence and mass gap**.)

While others were working on methods for solving the Yang-Mills equations, Donaldson approached the topic from a fresh viewpoint. As described by John D. S. Jones in *Nature*, "Donaldson argues as follows: if we know something about the solutions of the Yang-Mills equations then we must be able to extract information about the underlying space... Donaldson starts by treating the solutions of the equations as, in some sense, the known quantity."

It is common for mathematicians to look to theoretical physics for problems to investigate. In this unusual reversal, Donaldson used the tools of physics to explore purely mathematical ideas. In four-dimensional Euclidean space, the absolute minimum solutions (under certain boundary conditions at **infinity**) of the Yang-Mills equations are called "instantons." Donaldson's inspiration was to look at the nonlinear space of parameters for these instantons as a lens through which he could examine the space on which the equations are defined.

One of the basic goals of topology is to classify multidimensional spaces into categories that have the same basic structure. Jones described this as being similar to the taxonomic classification system in biology. In mathematics, spaces can be classified in terms of their topology (connectedness) or their smoothness (lack of corners, as shown by **continuity** of derivatives). For example, the surface of a **sphere** belongs to a different topological class than that of a **torus** (doughnut–like shape) because of the existence of the hole. In three dimensions, there is no difference between the results of classifying spaces topologically or smoothly. In five or more dimensions, there are relatively minor differences that are well understood. **Michael Freedman** obtained clear results for topological classification of four-dimensional spaces about the same time that Donaldson was establishing very different results using smoothness criteria. In the words of Atiyah, this "shows that the differentiable and topological situations are totally different"—a situation that occurs only in spaces of four dimensions.

Another way of describing Donaldson's results was offered by John Baez in *This Week's Finds in Mathematical Physics*. He stated the basic question as being whether n-dimensional Euclidean space allows any smooth structure other than the usual one. He concluded, "The answer is no—EXCEPT if n=4, where there are uncountably many smooth structures!" This unexpected result generated great excitement within the mathematical community and earned Donaldson the

Fields Medal, the most prestigious international mathematics award, in 1986.

Donaldson used intersection matrices as a tool for exploring four–dimensional spaces. Any four–dimensional space can be characterized by a **matrix** of integersin a way that describes how two–dimensional spaces intersect within it. This symmetric, **invariant** matrix will be the same for all topologically equivalent spaces. This means that if one finds two spaces with nonequivalent intersection matrices, those spaces will be topologically distinct. Conversely, Freedman showed that at most two topologically distinct spaces can be represented by equivalent intersection matrices. Examining the situation from a smoothness perspective, Donaldson found that there are unlimited possibilities for distinct spaces with equivalent intersection matrices.

It was in calculating these intersection matrices that Donaldson used the Yang–Mills equations. According to Jones, the instanton solutions are "concentrated in a very small ball and they behave like particles placed at the centre of the ball. This gives a way of recovering the points of the space from the solutions of the equations." With continued work, Donaldson was able to identify other invariants capable of distinguishing between two smoothly different, topologically equivalent manifolds.

One of the dramatic byproducts of Donaldson's discoveries was the description of exotic four–spaces. In Atiyah's description, these remarkable spaces "contain compact **sets** which cannot be contained inside any differentiably embedded 3–sphere!" In addition to producing startling results about the mathematics of four-dimensional spaces, Donaldson's work has also generated useful information for physicists. For instance, the earliest link between topology and quantum theory was **Paul Dirac**'s idea that the electric charge of a particle (which is an integral multiple of the charge of a single electron) has a basis in magnetism. He described this in terms of a hypothetical "magnetic monopole"—a basic particle that radiates a magnetic **field** just as a charged particle radiates an electrical field. Donaldson established a direct link between the parameter space of monopoles having magnetic charge k and the space of rational **functions** of a complex **variable** of degree k.

In addition to continuing his research, Donaldson currently serves as one of nine voting members of the executive committee of the International Mathematical Union (IMU). He is a fellow of the Royal Society of London, from which he received the Royal Medal in 1992. In 1994 he shared with**Shing-Tung Yau** the Swedish Academy of Science's Crafoord Prize, which is presented once every six years in the field of mathematics to provide financial support for research in an area of "particular interest and considerable activity."

DOUADY, ADRIEN (1935-   )

French mathematician

Adrien Douady is Professor of Mathematics at the University of Paris-Sud Orsay. He was born September 25, 1935. He

became a correspondent of the French Academy of Sciences on March 3, 1997.

Douady is best known for his studies of the **Mandelbrot set** and its role in the description of dynamics of iterative processes. Mandelbrot **sets** describe the behavior of **functions** of the form $f(z)=z*z+c$ as c varies over the complex plane. Iterative processes are those in which each consecutive term in a sequence is obtained by applying a given function its predecessor. For example, given the simple **parabola** defined by function $f(x) = x^2$ with x real, it is possible to start with a point x_0 and then form the following sequence by **iteration**: x_0, $x_1 = f(x_0)$, $x_2 = f(x_1)$, $x_3 = f(x_2)$, etc.

According to a fundamental **theorem** of mathematics, **polynomials** should be studied in the complex plane rather than on the real lines. Douady's study of simple second-degree polynomials in the complex plain provided new insight into the behavior of more general iterative processes.

Fundamental contributions to the understanding of iterative processes had been made by Gaston Julia (1893-1978) and Pierre Fatou (1878-1929) in the early part of the twentieth century. But their work went largely forgotten by mathematicians until **Benoit Mandelbrot**'s discovery of **fractals** (Mandelbrot coined the term in 1975). Mandelbrot's discovery led to a revival of interest in the properties of iterations, which proved to be essential for the theory of fractals. Mandelbrot drew the first of what Douady very quickly named Mandelbrot sets in 1980.

DUODECIMAL SYSTEM

The duodecimal system is a mathematical plan that uses a base of twelve rather than the standard base ten. Under base twelve, the place value changes from 10 to 12. Quantities are explained in terms of twelves, such as dozens, grosses, and great-grosses, rather than tens, hundreds, and thousands. In the duodecimal system, there are new symbols for 10 (X or *dek*) and 11 (E or *el*). Dozen is called *do*, and dozenal is a synonym for duodecimal.

The rationale for base twelve lies in many duodecimal models. Some of the best known examples include 12 months in a year, 12 hours on a clock face, 12 items to a dozen, and 12 inches to a foot. The system of 12 hours in a day came from the Egyptians. Shopkeepers favored selling by the dozen because it was easy to store goods on a shelf in three rows of four. The world **calendar**, easily divided into 12 months and four quarters, is already suited for duodecimal use.

Supporters of the duodecimal system point out the limitations of base ten: it can only be evenly divided by 2 and 5. However, base twelve offers many more options, as it is evenly divisible by 2, 3, 4, and 6. Supporters of base twelve say **fractions** are much easier to work with, since there are more divisors. Thus, fractions can readily be reduced to their simplest form.

To indicate temperature in base twelve, a Do-Metric system is used, with the boiling and freezing points of water separated by 100 degrees. The duodecimal **abacus** has 11 or

12 beads, as opposed to nine or 10 beads on a base ten abacus. **Zero** is used the same way in duodecimal and decimal systems, as a placeholder.

Other mathematical bases have been considered: two, eight, and sixteen. These numbers are easily divisible by 2 but they are not divisible by 3, so they do not hold much of an advantage over base ten. However, there are some remnants of base sixteen in the system of 16 ounces to a pound.

The duodecimal system was first introduced by Babylonians around 2400 B.C., but they ended up working with several different bases. The Babylonian sexagesimal system was based on 60 units, but they found they needed a smaller unit for counting and they chose base ten. Many observers trace the popularity of base ten to ten human fingers.

The Romans spread the duodecimal system in the second century B.C.. The term duodecimal is based on the Latin *duodecim* for 12. The Latin word for one-twelfth, *unus*, means unit. The English words ounce and inch are both derived from this root. The Latin *pes* means foot. The popularity of the Latin terminology is borne out in the system of 12 inches to a foot that is still widely used.

Napoleon reiterated the acceptance of the foot as a unit of measurement when the decimal system was first introduced in France. In Napoleon's view, base twelve was more logical because he could understand one-twelfth of a foot, but he found it hard to grasp the concept of one-thousandth of a meter. John Quincy Adams declared his preference for base twelve in 1821. He criticized the decimal system as being unnatural, commenting that there was no **model** for base ten in nature. At the end of the nineteenth century, some countries tried to change from a system using a base of 12 hours to a system using a base of 10 hours, so the hours of the day would be consistent with other metric systems already in use. However, 12-hour timekeeping was very firmly ingrained. The British tried to incorporate base ten and base twelve in their monetary system. Until 1971, the British used a duodecimal system combined with a vigesimal system (base twenty, for 10 fingers and 10 toes). The British pound sterling was divided into 20 shillings of 10 pence each.

During the 1950s, there was renewed enthusiasm for base twelve and a Duodecimal Society was formed in the United States. The society published one of the few books devoted to base ten and devised tables for **logarithms** and **trigonometry**. Musicians and artists have proposed base twelve systems for musical notation and color values, but these have never been widely used. Despite the vestiges of base twelve in many activities of daily life, the duodecimal system has consistently been overshadowed by the success of the decimal system.

DUPLICATION OF THE CUBE

Duplication of the cube-one of the fundamental problems of Greek **geometry**, together with squaring the **circle** and trisecting the **angle**. It asks whether, given a cube of a certain size, it

is possible to construct a cube of double the original size, using only a compass and a straightedge. Although the Greeks came up with several ingenious methods for doubling the cube, they were never able to do so without resorting to additional instruments. It is, in fact, impossible to double the cube with only a compass and a straightedge, but a **proof** of this fact did not come until more than two thousand years after the problem was first posed.

There are several different stories about the origin of the cube-doubling problem. One story, mythological in origin, says that King Minos of Crete ordered a tomb built for his son Glaucus, who died by drowning in a jar of honey. But when the tomb was finished, Minos said, "Too small is the tomb you have marked out as the royal resting place. Let it be twice as large. Without spoiling the form, quickly double each side of the tomb." The poet who related this story may have been a fine wordsmith, but he was far from an able mathematician, since doubling each **dimension** of the tomb would produce a tomb that is 8 times as large as the original, not 2 times.

According to another story, when the Delians were suffering from a terrible plague, the oracle told them that they would be spared if they constructed an altar that was double the size of the existing one. **Eratosthenes**, as quoted by Theon of Smyrna, wrote that "Their craftsmen fell into a great perplexity in their efforts to discover how a solid could be made the double of a similar solid; they therefore went to ask **Plato** about it, and he replied that the oracle meant, not that the god wanted an altar of double the size, but that he wished, in setting them the task, to shame the Greeks for their neglect of mathematics and their contempt of geometry."

The first genuine progress in solving the problem was made by **Hippocrates**, who realized, using purely geometric considerations, that the cube-doubling problem could be solved if it were possible to find two **mean** proportionals between 1 and 2. Following Hippocrates' insight, many of the best Greek mathematicians came up with clever mechanical solutions to the problem. Eratosthenes was one of those who devised a mechanical method, and in fact he built a column at Alexandria for King Ptolemy, with his cube-doubling method inscribed on it. He suggested that Ptolemy always adopt his method, and warned him against his rivals' methods: "Do not thou seek to do the difficult business of **Archytas**' cylinders, or to cut the cone in the triads of **Menaechmus**, or to compass such a curved form of lines as is described by the god-fearing **Eudoxus**.... Thus may it be, and let any one who sees this offering say 'This is the gift of Eratosthenes of Cyrene.'"

The Greeks probably suspected that the problem of doubling the cube using compass and straightedge was impossible. But this was not proved until 1837, when Pierre Wantzel used algebraic techniques to prove the impossibility not only of doubling the cube, but also of **squaring the circle** and **trisecting the angle**.

The key idea in this algebraic method was to turn the problem into a question about numbers, by asking what kind of lengths can be constructed using straightedge and compass, starting with a unit length. For example, 2 is a constructible number, since a line segment of length 2 can be constructed simply by building two unit-segments that abut each other. Once mathematicians had made the **definition** of constructible numbers, they could ask what kind of **equations** a constructible number could satisfy. They discovered a remarkable fact: no constructible number could satisfy a cubic polynomial equation, like $x^3-5=0$ (unless the polynomial was of a very simple type called reducible).

If you start with a cube whose dimensions each measure one unit in length, then doubling the cube amounts to building a cube whose dimensions have length equal to the cube root of 2 (so that their product, the **volume** of the cube, is equal to 2). So the question of doubling the cube reduces to the question, is the cube root of 2 a constructible length? The cube root of 2 is a solution to the irreducible cubic equation $x^3-2=0$, so by the algebraic **theorem**, it cannot be a constructible number! Hence it is impossible to double the cube with only a compass and straightedge.

See also Squaring the circle; Trisecting the angle

DÜRER, ALBRECHT (1471-1528)
Hungarian German artist and mathematician

Albrecht Dürer was a world-renown German artist who incorporated the use of **geometry** and other mathematical theories into his woodcuts, illustrations, and other graphic artwork. Dürer also published a series of treatises on the applications of mathematics to art.

Dürer was born in 1471 in Nuremberg, the third of eighteen children of Albrecht Dürer the Elder and his wife Barbara. Dürer the Elder was a master goldsmith and artisan who served as Nuremberg's official assayer of gold. Dürer the Younger attended school for several years as a child, until leaving at age 10 or 12 to learn goldsmithing and jewelry-making from his father. As an apprentice, Dürer mastered drawing and metal working techniques, and his innate talents as an artist were apparent from an early age. A self-portrait made by Dürer at age 13 that is still in existence today shows his remarkably realistic and finely-detailed drawing skills.

At age 15, Dürer began a second apprenticeship with Michael Wolgemut, a successful Nuremberg painter who taught him drafting skills and woodcut illustration. At the advice of Wolgemut and his father, who both recognized Dürer's immense talents, he toured Germany to further develop his craft and study and meet with other German artists. When he returned home in 1494, he entered into an arranged marriage with Agnes Frey, daughter of a wealthy Nuremberg artisan, and used the dowry money to finance his first trip to Italy. It was in Italy that Dürer hoped not only to broaden his artistic horizons, but to learn more about the study of mathematics in order to raise his art and the study of art in general to new levels.

While in Italy, Dürer studied the classical Italian art forms. While there are no records of Dürer meeting personally with any of the famous Italian mathematical figures of the

Albrecht Dürer

time, there are indications that he followed their work while in Italy and discussed their theories with several Italian artists during his stay. Dürer also did not meet with **Leonardo da Vinci**, another outspoken proponent of the use mathematics in art, although there is later evidence that he knew of da Vinci's theories and had met at least one of da Vinci's confidants.

When he returned to Nuremberg in 1495, Dürer began an intensive study of the mathematical theories of **Euclid**, Vitruvius, and the Italians **Leone Alberti** and **Luca Pacioli**. Dürer's close friend Willibald Pirckheimer, a well-connected humanist translator and writer fluent in Italian, Greek, and Latin, provided Dürer with ready access to classic mathematical texts and Italian Renaissance literature, and introduced him to many leading figures of both Nuremberg and Italian society.

Dürer set up his own artists workshop in 1496, and began to incorporate many of the mathematical principles he had studied into his woodcuts, sketches, and other works. The influence of Pacioli, in particular, surfaced in many of his works as he incorporated the concepts of perspective and proportion into his artwork. He used compass and ruler to construct proportionate human figures, and geometrical patterns and shapes began to figure prominently in his artwork. Dürer's mathematical influences are also clear in his later works. His application of mathematics to art would continue throughout his career. Dürer's famous 1514 engraving "Melancholia," incorporates the figure of a **magic square**, which is believed to be the first representation of that mathematical figure seen in Europe.

After a number of commissions for portraits by wealthy, highly-ranked members of German society, Dürer's reputation as a master artist began to grow throughout Germany. By 1500, he was acknowledged as both a well-respected and well-compensated member of the artistic community. Between 1505 and 1507, Dürer visited Italy once again to consult and study with Italian mathematicians. During this trip he is thought to have met with Pacioli and mathematician Jacopo de Barbari. Upon his return to Nuremberg, Dürer began work on a series of books that would explain the mathematical principles of proportion and their applications to art. The books, which formed a treatise

entitled *Unterweisung der Messung mit dem Zirkel und Richtscheit*, (or "Investigation of the Measurement with Circles and Straight Lines of Plane and Solid Figures"), were a blend of artistic instruction and classical and contemporary mathematical theory. They dealt with the construction of curves, **polygons**, and other geometrical elements; the nature of pyramids, cylinders, and other solids; and the theory of perspective. Dürer published them himself in 1525. Three years later, Dürer was in the final stages of preparing his newest work *Four Books on Human Proportion* for publication when he died. At age 57, Albrecht Dürer had become one of the world's most famous artists and an important figure in Renaissance mathematics and geometrical theory.

See also Geometry; Luca Pacioli; Magic (and other) squares; Proportions

E

E (NUMBER)

The number e, like the number **pi**, is a useful mathematical constant that is the basis of the system of natural **logarithms**. Its value correct to nine places is 2.718281828... The number e is used in complex **equations** to describe a process of growth or decay. It is therefore utilized in the biology, business, demographics, physics, and engineering fields.

The number e is widely used as the base in the exponential function $y = Ce^{kx}$. There are extensive tables for e^x, and scientific calculators usually include an e^x key. In **calculus** one finds that the **slope** of the graph of e^x at any point is equal to e^x itself, and that the integral of e^x is also e^x plus a constant.

Exponential functions based on e are also closely related to sines, cosines, hyperbolic sines, and hyperbolic cosines: $e^{ix} = \cos x + i\sin x$; and $e^x = \cosh x + \sinh x$. Here i is the imaginary number $\sqrt{-1}$. From the first of these relationships one can obtain the curious equation $e^{i\pi} + 1 = 0$, which combines five of the most important constants in mathematics.

The constant e appears in many other formulae in **statistics**, science, and elsewhere. It is the base for natural (as opposed to common) logarithms. That is, if $e^x = y$, then $x = \ln y$. ($\ln x$ is the symbol for the natural logarithm of x.) $\ln x$ and e^x are therefore inverse **functions**.

The expression $(1 + 1/n)^n$ approaches the number e more and more closely as n is replaced with larger and larger values. For example, when n is replaced in turn with the values 1, 10, 100, and 1000, the expression takes on the values 2, 2.59..., 2.70..., and 2.717....

Calculating a decimal approximation for e by means of the this **definition** requires one to use very large values of n, and the equations can become quite complex. A much easier way is to use the Maclaurin series for e^x : $e^x = 1 + x/1! + x^2/2! + x^3/3! + x^4/4! + ...$ By letting x equal 1 in this series one gets $e = 1 + 1/1 + 1/2 + 1/6 + 1/24 + 1/120 + ...$ The first seven terms will yield a three-place approximation; the first twelve will yield nine places.

ECCENTRICITY

One of the dictionary definitions of the word eccentric is "not a **circle**." Another is "not at the geometric center." Both definitions help in understanding the meaning of eccentricity. A **circle** and other symmetrical shapes can be formed by "cutting" or passing a plane surface through a right circular cone. This creates a cross-sections of the cone. When the cut or plane is perpendicular (at 90 degrees) to the axis of the cone, a circle results. If the plane is positioned at some **angle** less than 90 degrees but not quite parallel to the axis, the resulting **conic section** is an **ellipse**. Within an ellipse can be found two points, called foci. The sum of the **distance** from each focus to any point on the ellipse is the same for all points. This is also a property of a circle, but in this special type of ellipse, the two foci are superimposed at the same place, which is the center of the circle.

Mathematically, the eccentricity of an ellipse is found by using the equation e=c/a where e is the eccentricity, c is the distance from the center of the ellipse to one focus and a is the semi-major axis of the ellipse. This last term is one-half the largest **dimension** of the ellipse, which is called the major axis. The first definition above implies that something that is circular has no eccentricity. According to e=c/a, this definition applies numerically to circles because both foci lie at the center of the circle. Therefore, for circles, c=0. This means that for a circle e=0, so a circle can be thought of as not being eccentric. An ellipse is similar to a circle except for the fact that it is not perfectly round. The second definition mentioned above refers to the foci of an ellipse which are not at the center. Values for eccentricity range from e=0 for circles to e<1 for very long, narrow ellipses. If e were to equal 1, the distance between a focus and the center and the length of the semi-major axis would have to be the same, so a true ellipse can never quite reach this value. If it did, it would look like a line or, rather, a line segment.

The grandest examples of ellipses in nature are the orbits of planets and moons. Man-made satellites also circle the earth in elliptical orbits. Of all the planets, Venus has the least (e=0.0068), and Pluto the most eccentric **orbit** (e=0.253). But Nereid, one of the moons of Neptune, has an eccentricity of 0.76).

See also Circle; Ellipse; Elliptic geometry

EGYPTIAN MATHEMATICS

The earliest records of advanced, organized mathematics (beyond just counting) date back to the ancient Mesopotamian country of Babylonia and to Egypt of the fourth millennium B.C.. The Egyptian civilization, one of the world's oldest, developed in the valley of the Nile. Its study and use of mathematics developed only in response to the practical needs within agriculture, business, and industry. Specifically, Egyptian mathematics was primarily **arithmetic**, with an emphasis on measurement, surveying, and calculation in **geometry**. Its use was elementary, with no trace of the necessity of later abstract mathematical concepts such as proofs and axioms. This mathematics was generally arrived at by trial and error as the way to obtain desired results. However, no records exist that show how the Egyptians reached their conclusions. It is true, though, that they possessed advanced mathematical knowledge, for without it their accomplishments in engineering, astronomy, and administration would not have been possible.

Some of our knowledge of ancient Egyptian mathematics first appeared from the sacred carvings in the temples called hieroglyphics. Occurring as far back as 3400 B.C., these inscriptions of a simple grouping system were made primarily on stone. Later a more complex number system called hieratic evolved from hieroglyphics. Hieratic developed on papyrus, wood, and pottery as a method to more rapidly write. From hieratic developed the demotic numeral system that was eventually adopted for general use. Another important source of mathematical knowledge about the Egyptians came from texts that were composed about 1800 B.C. and written by a class of literate professionals, called scribes. Because of their reading, writing, and mathematical skills, the scribes performed the duties of modern-day civil servants: that of tax accounting, record keeping, management of public works, and overseeing military payrolls and supplies. However, only a few of these original documents survived. Two papyrus documents that once served as scribal textbooks were discovered intact and contain collections of mathematical problems along with their solutions and a decimal numeration system with separate symbols for the successive **powers** of 10 (1, 10, 100, etc.). The first text, the *Rhind Mathematical Papyrus* (also called the **Ahmes** Papyrus or **Rhind Papyrus**), was named after Scottish Egyptologist A. Henry Rhind who purchased it in Luxor in 1858. It was written around 1650 B.C. by the scribe Ahmes from another text that was two centuries older. The Rhind discussed unit **fractions**, presented solutions of 84 specific problems in arithmetic and geometry, and contained a table of representa-

tions of "2/n" as a sum of distinct unit fractions for odd "n" between 5 and 101. The second text was called the *Moscow Mathematical Papyrus*, also known as the Golenishchev Papyrus, and was purchased by V. S. Golenishchev. Its origin was from the nineteenth century B.C. and presented 25 problems involving the distribution of beer and bread as wages, and how to calculate the areas of fields and the volumes of pyramids and other solids.

In Egyptian mathematics the symbol for 1 "vertical stroke" was written five times to represent the number 5; the symbol for 10 "heal bone (or 'hobble' for cattle)" was written seven times to represent the number 70; and the symbol for 100 "snare (or coil of rope)" was written four times to represent the number 400. Together, these three groups of symbols (smallest symbols on the right and largest symbols on the left) were brought together to represent the number 475. Some of the remaining symbols, and what it is believed the symbols represent, are 1,000 "lotus flower"; 10,000 "bent finger"=; 100,000 "burbot fish (or frog)"; and 1,000,000 "kneeling god with arms raised above his head".

As indicated above, the Egyptians possessed an early (hieroglyphics) grouping system that separated the units (10s, 100s, etc.) in the numbers to be added. Thus, the number 475 is represented as (4 snares) + (7 lotus flowers) + (5 vertical strokes). The number 263 is represented as (2 snares) + (6 lotus flowers) + (3 vertical strokes), and 475 + 263 is represented as (6 snares) + (13 lotus flowers, or 1 snare and 3 lotus flowers) + (8 vertical strokes) = (7 snares) + (3 lotus flower) + (8 vertical strokes).

Multiplication was basically binary arithmetic, based on successive doublings. For instance, multiplying 47 by 27 is solved by associating $1 = 2^0$ with 47, $2 = 2^1$ with 94 (47 x 2), $4 = 2^2$ with 188 (47 x 4), $8 = 2^3$ with 376 (47 x 8), and $16 = 2^4$ with 752 (47 x 16). Since 32 > 27, there is no need to go beyond the 16 entry. Now, go through a number of subtractions (27 - 16 = 11, 11 - 8 = 3, 3 - 2 = 1, 1 - 1 = 0) to show that 27 = 16 + 8 + 2 + 1. Next, add together the numbers associated with the right-hand numbers. Therefore, 47 x 27 = 47 x (16 + 8 + 2 + 1) = 47 x ($2^4 + 2^3 + 2^1 + 2^0$). Substitute into this equation the previously calculated values for the powers of 2 multiplied by 47 to obtain: 47 x 27 = 752 + 376 + 94 + 47 = 1,269.

Division is based on the reverse process of multiplication. Using the previous example, dividing 1,269 by 47 is solved by consecutive subtractions of "1,269 - 752 = 517; 517 - 376 = 141; 141 - 94 = 47; 47 - 47 = 0". Therefore, 1,269 /47 = 16 + 8 + 2 + 1 = 27.

The Egyptians used sums of unit fractions (1/n), supplemented by the fraction 2/3 that was not represented by a sum of unit fractions, to express all other fractions. For example, if 5 loaves of bread needed to be divided among 12 men, the fraction 5/12 became the sum of the unit fractions 1/3 and 1/12 where 1/3 + 1/12 = 4/12 + 1/12 = 5/12. Therefore, each man received one-third and one-twelfth of a loaf. Using this system, the Egyptians were able to solve all problems of arithmetic that involved fractions, as well as some elementary problems in **algebra**. However, the procedure can sometimes become quite complicated for large numbers and several combinations of unit fractions are possible for the solution.

Historians view the Egyptian achievement in mathematics as modest. But, their methods persisted for nearly three thousand years. Egyptian mathematics was most important in its effect on the emerging Greek mathematics between the sixth and fourth centuries B.C..

See also Babylonian mathematics; Binary number system

EINSTEIN, ALBERT (1879-1955)

German-born American physicist

Albert Einstein ranks as one of the most remarkable theoreticians in the history of science. During a single year, 1905, he produced three papers that are among the most important in 20th-century physics, and perhaps in all of the recorded history of science, for they revolutionized the way scientists looked at the nature of **space**, time, and matter. These papers dealt with the nature of particle movement known as Brownian **motion**, the quantum nature of electromagnetic radiation as demonstrated by the photoelectric effect, and the special theory of **relativity**. Although Einstein is probably best known for the last of these works, it was for his quantum explanation of the photoelectric effect that he was awarded the 1921 Nobel Prize in physics. In 1915, Einstein extended his special theory of relativity to include certain cases of accelerated motion, resulting in the more general theory of relativity.

Einstein was born in Ulm, Germany, on March 14, 1879, the only son of Hermann and Pauline Koch Einstein. Both sides of his family had long-established roots in southern Germany, and, at the time of Einstein's birth, his father and uncle Jakob owned a small electrical equipment plant. When that business failed around 1880, Hermann Einstein moved his family to Munich to make a new beginning. A year after their arrival in Munich, Einstein's only sister, Maja, was born.

Although his family was Jewish, Einstein was sent to a Catholic elementary school from 1884 to 1889. He was then enrolled at the Luitpold Gymnasium in Munich. During these years, Einstein began to develop some of his earliest interests in science and mathematics, but he gave little outward indication of any special aptitude in these fields. Indeed, he did not begin to talk until the age of three and, by the age of nine, was still not fluent in his native language. His parents were actually concerned that he might be somewhat mentally retarded.

In 1894, Hermann Einstein's business failed again, and the family moved once more, this time to Pavia, near Milan, Italy. Einstein was left behind in Munich to allow him to finish school. Such was not to be the case, however, since he left the *gymnasium* after only six more months. Einstein's biographer, Philipp Frank, explains that Einstein so thoroughly despised formal schooling that he devised a scheme by which he received a medical excuse from school on the basis of a potential nervous breakdown. He then convinced a mathematics teacher to certify that he was adequately prepared to begin his college studies without a high school diploma. Other biographies, however, say

Albert Einstein

that Einstein was expelled from the *gymnasium* on the grounds that he was a disruptive influence at the school.

In any case, Einstein then rejoined his family in Italy. One of his first acts upon reaching Pavia was to give up his German citizenship. He was so unhappy with his native land that he wanted to sever all formal connections with it; in addition, by renouncing his citizenship, he could later return to Germany without being arrested as a draft dodger. As a result, Einstein remained without an official citizenship until he became a Swiss citizen at the age of 21. For most of his first year in Italy, Einstein spent his time traveling, relaxing, and teaching himself **calculus** and higher mathematics. In 1895, he thought himself ready to take the entrance examination for the Eidgenössiche Technische Hochschule (the ETH, Swiss Federal Polytechnic School, or Swiss Federal Institute of Technology), where he planned to major in electrical engineering. When he failed that examination, Einstein enrolled at a Swiss cantonal high school in Aarau. He found the more democratic style of instruction at Aarau much more enjoyable than his experience in Munich and soon began to make rapid progress. He took the entrance examination for the ETH a second time in 1896, passed, and was admitted to the school. (In *Einstein,* however, Jeremy Bernstein writes that Einstein was admitted without examination on the basis of his diploma from Aarau.)

The program at ETH had nearly as little appeal for Einstein as had his schooling in Munich, however. He apparently hated studying for examinations and was not especially interested in attending classes on a regular basis. He devoted much of this time to reading on his own, specializing in the works of Gustav Kirchhoff, Heinrich Hertz, **James Clerk Maxwell**, Ernst Mach, and other classical physicists. When Einstein graduated with a teaching degree in 1900, he was unable to find a regular teaching job. Instead, he supported himself as a tutor in a private school in Schaffhausen. In 1901, Einstein also published his first scientific paper, "Consequences of Capillary Phenomena."

In February, 1902, Einstein moved to Bern and applied for a job with the Swiss Patent Office. He was given a probationary appointment to begin in June of that year and was promoted to the position of technical expert, third class, a few months later. The seven years Einstein spent at the Patent Office were the most productive years of his life. The demands of his work were relatively modest and he was able to devote a great deal of time to his own research.

The promise of a steady income at the Patent Office also made it possible for Einstein to marry. Mileva Marić (also given as Maritsch) was a fellow student in physics at ETH, and Einstein had fallen in love with her even though his parents strongly objected to the match. Marić had originally come from Hungary and was of Serbian and Greek Orthodox heritage. The couple married on January 6, 1903, and later had two sons, Hans Albert and Edward. A previous child, Liserl, was born in 1902 at the home of Marić's parents in Hungary, but there is no further mention or trace of her after 1903 since she was given up for adoption.

In 1905, Einstein published a series of papers, any one of which would have assured his fame in history. One, "On the Movement of Small Particles Suspended in a Stationary Liquid Demanded by the Molecular-Kinetic Theory of Heat," dealt with a phenomenon first observed by the Scottish botanist Robert Brown in 1827. Brown had reported that tiny particles, such as dust particles, move about with a rapid and random zigzag motion when suspended in a liquid.

Einstein hypothesized that the visible motion of particles was caused by the random movement of molecules that make up the liquid. He derived a mathematical formula that predicted the **distance** traveled by particles and their relative speed. This formula was confirmed experimentally by the French physicist Jean Baptiste Perrin in 1908. Einstein's work on the Brownian movement is generally regarded as the first direct experimental evidence of the existence of molecules.

A second paper, "On a Heuristic Viewpoint concerning the Production and Transformation of Light," dealt with another puzzle in physics, the photoelectric effect. First observed by Heinrich Hertz in 1888, the photoelectric effect involves the release of electrons from a metal that occurs when **light** is shined on the metal. The puzzling aspect of the photoelectric effect was that the number of electrons released is not a function of the light's intensity, but of the color (that is, the wavelength) of the light.

To solve this problem, Einstein made use of a concept developed only a few years before, in 1900, by the German physicist Max Planck, the quantum hypothesis. Einstein assumed that light travels in tiny discrete bundles, or "quanta," of energy. The energy of any given light quantum (later renamed the photon), Einstein said, is a function of its wavelength. Thus, when light falls on a metal, electrons in the metal absorb specific quanta of energy, giving them enough energy to escape from the surface of the metal. But the number of electrons released will be determined not by the number of quanta (that is, the intensity) of the light, but by its energy (that is, its wavelength). Einstein's hypothesis was confirmed by several experiments and laid the foundation for the fields of quantitative photoelectric chemistry and quantum **mechanics**. As recognition for this work, Einstein was awarded the 1921 Nobel Prize in physics.

A third 1905 paper by Einstein, almost certainly the one for which he became best known, details his special theory of relativity. In essence, "On the Electrodynamics of Moving Bodies" discusses the relationship between measurements made by observers in two separate systems moving at constant velocity with respect to each other.

Einstein's work on relativity was by no means the first in the field. The French mathematician and physicist **Jules Henri Poincaré**, the Irish physicist George Francis FitzGerald, and the Dutch physicist **Hendrik Lorentz** had already analyzed in some detail the problem attacked by Einstein in his 1905 paper. Each had developed mathematical formulas that described the effect of motion on various types of measurement. Indeed, the record of pre-Einsteinian thought on relativity is so extensive that one historian of science once wrote a two-volume work on the subject that devoted only a single sentence to Einstein's work. Still, there is little question that Einstein provided the most complete analysis of this subject. He began by making two assumptions. First, he said that the laws of physics are the same in all frames of reference. Second, he declared that the velocity of light is always the same, regardless of the conditions under which it is measured.

Using only these two assumptions, Einstein proceeded to uncover an unexpectedly extensive description of the properties of bodies that are in uniform motion. For example, he showed that the length and mass of an object are dependent upon their movement relative to an observer. He derived a mathematical relationship between the length of an object and its velocity that had previously been suggested by both FitzGerald and Lorentz. Einstein's theory was revolutionary, for previously scientists had believed that basic quantities of measurement such as time, mass, and length were absolute and unchanging. Einstein's work established the opposite—that these measurements could change, depending on the relative motion of the observer.

In addition to his masterpieces on the photoelectric effect, Brownian movement, and relativity, Einstein wrote two more papers in 1905. One, "Does the Inertia of a Body Depend on Its Energy Content?," dealt with an extension of his earlier work on relativity. He came to the conclusion in this paper that the energy and mass of a body are closely interrelated. Two years later he specifically stated that relationship in a formula,

E=mc² (energy equals mass times the speed of light squared), that is now familiar to both scientists and non-scientists alike. His final paper, the most modest of the five, was "A New Determination of Molecular Dimensions." It was this paper that Einstein submitted as his doctoral dissertation, for which the University of Zurich awarded him a Ph.D. in 1905.

Fame did not come to Einstein immediately as a result of his five 1905 papers. Indeed, he submitted his paper on relativity to the University of Bern in support of his application to become a *privatdozent,* or unsalaried instructor, but the paper and application were rejected. His work was too important to be long ignored, however, and a second application three years later was accepted. Einstein spent only a year at Bern, however, before taking a job as professor of physics at the University of Zurich in 1909. He then went on to the German University of Prague for a year and a half before returning to Zurich and a position at ETH in 1912. A year later Einstein was made director of scientific research at the Kaiser Wilhelm Institute for Physics in Berlin, a post he held from 1914 to 1933.

In recent years, the role of Mileva Einstein-Marić in her husband's early work has been the subject of some controversy. The more traditional view among Einstein's biographers is that of A. P. French in his "Condensed Biography" in *Einstein: A Centenary Volume.* French argues that although "little is recorded about his [Einstein's] domestic life, it certainly did not inhibit his scientific activity." In perhaps the most substantial of all Einstein biographies, Philipp Frank writes that "For Einstein life with her was not always a source of peace and happiness. When he wanted to discuss with her his ideas, which came to him in great abundance, her response was so slight that he was often unable to decide whether or not she was interested."

A quite different view of the relationship between Einstein and Marić is presented in a 1990 paper by Senta Troemel-Ploetz in *Women's Studies International Forum.* Based on a biography of Marić originally published in Yugoslavia, Troemel-Ploetz argues that Marić gave to her husband "her companionship, her diligence, her endurance, her mathematical genius, and her mathematical devotion." Indeed, Troemel-Ploetz builds a case that it was Marić who did a significant portion of the mathematical calculations involved in much of Einstein's early work. She begins by repeating a famous remark by Einstein himself to the effect that "My wife solves all my mathematical problems." In addition, Troemel-Ploetz cites many of Einstein's own letters of 1900 and 1901 (available in *Collected Papers*) that allude to Marić's role in the development of "our papers," including one letter to Marić in which Einstein noted: "How happy and proud I will be when both of us together will have brought our work on relative motion to a successful end." The author also points out the somewhat unexpected fact that Einstein gave the money he received from the 1921 Nobel Prize to Marić, although the two had been divorced two years earlier. Nevertheless, Einstein never publicly acknowledged any contributions by his wife to his work.

Any mathematical efforts Mileva Einstein-Marić may have contributed to Einstein's work greatly decreased after the birth of their second son in 1910. Einstein was increasingly occupied with his career and his wife with managing their household; upon moving to Berlin in 1914, the couple grew even more distant. With the outbreak of World War I, Einstein's wife and two children returned to Zurich. The two were never reconciled; in 1919, they were formally divorced. Towards the end of the war, Einstein became very ill and was nursed back to health by his cousin Elsa. Not long after Einstein's divorce from Marić, he was married to Elsa, a widow. The two had no children of their own, although Elsa brought two daughters, Ilse and Margot, to the marriage.

The war years also marked the culmination of Einstein's attempt to extend his 1905 theory of relativity to a broader context, specifically to systems with non-zero acceleration. Under the general theory of relativity, motions no longer had to be uniform and relative velocities no longer constant. Einstein was able to write mathematical expressions that describe the relationships between measurements made in *any* two systems in motion relative to each other, even if the motion is accelerated in one or both. One of the fundamental features of the general theory is the concept of a **space-time** continuum in which space is curved. That concept means that a body affects the shape of the space that surrounds it so that a second body moving near the first body will travel in a curved path.

Einstein's new theory was too radical to be immediately accepted, for not only were the mathematics behind it extremely complex, it replaced Newton's theory of gravitation that had been accepted for two centuries. So, Einstein offered three proofs for his theory that could be tested: first, that relativity would cause Mercury's perihelion, or point of **orbit** closest to the sun, to advance slightly more than was predicted by **Newton's laws**. Second, Einstein predicted that light from a star will be bent as it passes close to a massive body, such as the sun. Last, the physicist suggested that relativity would also affect light by changing its wavelength, a phenomenon known as the redshift effect. Observations of the planet Mercury bore out Einstein's hypothesis and calculations, but astronomers and physicists had yet to test the other two proofs.

Einstein had calculated that the amount of light bent by the sun would amount to 1.7 seconds of an arc, a small but detectable effect. In 1919, during an eclipse of the sun, English astronomer Arthur Eddington measured the deflection of starlight and found it to be 1.61 seconds of an arc, well within experimental error. The publication of this **proof** made Einstein an instant celebrity and made "relativity" a household word, although it was not until 1924 that Eddington proved the final hypothesis concerning redshift with a spectral analysis of the star Sirius B. This phenomenon, that light would be shifted to a longer wavelength in the presence of a strong gravitational field, became known as the "Einstein shift."

Einstein's publication of his general theory in 1916, the *Foundation of the General Theory of Relativity,* essentially brought to a close the revolutionary period of his scientific career. In many ways, Einstein had begun to fall out of phase with the rapid changes taking place in physics during the 1920s. Even though Einstein's own work on the photoelectric effect helped set the stage for the development of quantum theory, he was never able to accept some of its concepts, particularly the uncertainty principle. In one of the most-quoted comments in the history of science, he claimed that **quantum mechanics**, which could only calculate the probabilities of physical events, could not be correct because "God does not play dice." Instead, Einstein devoted his efforts for the remaining years of his life to the search for a unified field theory, a single theory that would encompass all physical fields, particularly gravitation and **electromagnetism**.

Since the outbreak of World War I, Einstein had been opposed to war, and used his notoriety to lecture against it during the 1920s and 1930s. With the rise of National Socialism in Germany in the early 1930s, Einstein's position became difficult. Although he had renewed his German citizenship, he was suspect as both a Jew and a pacifist. In addition, his writings about relativity were in conflict with the absolutist teachings of the Nazi party. Fortunately, by 1930, Einstein had become internationally famous and had traveled widely throughout the world. A number of institutions were eager to add his name to their faculties.

In early 1933, Einstein made a decision. He was out of Germany when Hitler rose to power, and he decided not to return. Instead he accepted an appointment at the Institute for Advanced Studies in Princeton, New Jersey, where he spent the rest of his life. In addition to his continued work on unified field theory, Einstein was in demand as a speaker and wrote extensively on many topics, especially peace. The growing fascism and anti-Semitism of Hitler's regime, however, convinced him in 1939 to sign his name to a letter written by American physicist Leo Szilard informing President Franklin D. Roosevelt of the possibility of an atomic bomb. This letter led to the formation of the Manhattan Project for the construction of the world's first nuclear weapons. Although Einstein's work on relativity, particularly his formulation of the equation E=mc², was essential to the development of the atomic bomb, Einstein himself did not participate in the project. He was considered a security risk, although he had renounced his German citizenship and become a U.S. citizen in 1940.

After World War II and the bombing of Japan, Einstein became an ardent supporter of nuclear disarmament. He also lent his support to the efforts to establish a world government and to the Zionist movement to establish a Jewish state. In 1952, after the death of Israel's first president, Chaim Weizmann, Einstein was invited to succeed him as president; he declined the offer. Among the many other honors given to Einstein were the Barnard Medal of Columbia University in 1920, the Copley Medal of the Royal Society in 1925, the Gold Medal of the Royal Astronomical Society in 1926, the Max Planck Medal of the German Physical Society in 1929, and the Franklin Medal of the Franklin Institute in 1935. Einstein died at his home in Princeton on April 18, 1955, after suffering an aortic aneurysm. At the time of his death, he was the world's most widely admired scientist and his name was synonymous with genius. Yet Einstein declined to become enamored of the admiration of others. He wrote in his book *The World as I See It:* "Let every man be respected as an individual and no man idolized. It is an irony of fate that I myself

have been the recipient of excessive admiration and respect from my fellows through no fault, and no merit, of my own. The cause of this may well be the desire, unattainable for many, to understand the one or two ideas to which I have with my feeble powers attained through ceaseless struggle."

ELECTROMAGNETISM

Electromagnetism is a branch of physical science that involves all the phenomena in which electricity and magnetism interact. This field is especially important to electronics because a magnetic field is created by an electric current. The rules of electromagnetism are responsible for the way charged particles of atoms interact.

Some of the rules of *electrostatics*, the study of electric charges at rest, were first noted by the ancient Romans, who observed the way a brushed comb would attract particles. It is now known that electric charges occur in two different types, called positive and negative. Like types repel each other, and differing types attract.

The **force** that attract positive charges to negative charges weakens with **distance**, but is intrinsically very strong. The fact that unlike types attract means that most of this force is normally neutralized and not seen in full strength. The negative charge is generally carried by the atom's electrons, while the positive resides with the protons inside the atomic nucleus. There are other less well known particles that can also carry charge. When the electrons of a material are not tightly bound to the atom's nucleus, they can move from atom to atom and the substance, called a conductor, can conduct electricity. On the contrary, when the electron binding is strong, the material is called an insulator.

When electrons are weakly bound to the atomic nucleus, the result is a semiconductor, often used in the electronics industry. It was not initially known if the electric current carriers were positive or negative, and this initial ignorance gave rise to the convention that current flows from the positive terminal to the negative. In reality we now know that the electrons actually run from the negative to the positive.

Electromagnetism is the theory of a unified expression of an underlying force, the so-called electromagnetic force. This is seen in the movement of electric charge, which gives rise to magnetism (the electric current in a wire being found to deflect a compass needle), and it was a Scotsman, **James Clerk Maxwell**, who in 1865 published the theory unifying electricity and magnetism. The theory arose from former specialized work by **Gauss**, **Coulomb**, Ampère, Faraday, Franklin, and Ohm. However, one factor that did not contradict the experiments was added to the **equations** by Maxwell so as to ensure the conservation of charge. This was done on the theoretical grounds that charge should be a conserved quantity, and this addition led to the prediction of a wave phenomena with a certain anticipated velocity. **Light**, which has the expected velocity, was found to be an example of this electromagnetic radiation.

Light had formerly been thought of as consisting of particles (photons) by **Newton**, but the theory of light as particles was unable to explain the wave nature of light (diffraction and the alike). In reality, light displays both wave *and* particle properties. The resolution to this duality lies in quantum theory, where light is neither particles or wave, but both. It propagates as a wave without the need of a media and interacts in the manner of a particle. This is the basic nature of quantum theory.

Classical electromagnetism, useful as it is, contains contradictions (acausality) that make it incomplete and drive one to consider its extension to the area of quantum physics, where electromagnetism, of all the fundamental forces of nature, it is perhaps the best understood.

There is much **symmetry** between electricity and magnetism. It is possible for electricity to give rise to magnetism, and symmetrically for magnetism to give rise to electricity (as in the exchanges within an electric transformer). It is an exchange of just this kind that constitutes electromagnetic waves. These waves, although they don't need a medium of propagation, are slowed when traveling through a transparent substance.

Electromagnetic waves differ from each other only in amplitude, frequency and orientation (polarization). Laser beams are particular in being very coherent, that is, the radiation is of one frequency, and the waves coordinated in **motion** and direction. This permits a highly concentrated beam that is used not only for its cutting abilities, but also in electronic data storage, such as in CD-ROMs.

The differing frequency forms are given a variety of names, from radio waves at very low frequencies through light itself, to the high frequency X and gamma rays.

Many a miracle depends upon the broad span of the electromagnetic spectrum. The ability to communicate across long distances despite intervening obstacles, such as the walls of buildings, is possible using the radio and television frequencies. X rays can see into the human body without opening it. These things, which would once have been labeled magic, are now ordinary ways we use the electromagneticspectrum.

The unification of electricity and magnetism has led to a deeper understanding of physical science, and much effort has been put into further unifying the four forces of nature. The remaining known forces are the so called weak, strong, and gravitational forces. The weak force has now been unified with electromagnetism, called the electroweak force. There are proposals to include the strong force in a grand unified theory, but the inclusion of gravity remains an open problem.

The fundamental role of special relativity in electromagnetism

Maxwell's theory is in fact in contradiction with Newtonian **mechanics**, and in trying to find the resolution to this conflict, **Einstein** was lead to his theory of special **relativity**. **Maxwell's equations** withstood the conflict, but it was Newtonian mechanics that were corrected by relativistic mechanics. These corrections are most necessary at velocities, close to the speed of light. The many strange predictions about **space** and time that follow from special relativity are found to be a part of the real world.

Paradoxically, magnetism is a counter example to the frequent claims that relativistic effects are not noticeable for low

velocities. The moving charges that compose an electric current in a wire might typically only be traveling at several feet per second (walking speed), and the resulting Lorentz contraction of special relativity is indeed minute. However, the electrostatic forces at balance in the wire are of such great magnitude, that this small contraction of the moving (negative) charges exposes a residue force of real world magnitude, namely the magnetic force. It is in exactly this way that the magnetic force derives from the electric. Special relativity is indeed hidden in Maxwell's equations, which were known before special relativity was understood or separately formulated by Einstein.

Technological uses of electromagnetism

Before the advent of technology, electromagnetism was perhaps most strongly experienced in the form of lightning, and electromagnetic radiation in the form of light. Ancient man kindled fires which he thought were kept alive in trees struck by lightning.

Much of the magic of nature has been put to work by man, but not always for his betterment or that of his surroundings. Electricity at high voltages can carry energy across extended distances with little loss. Magnetism derived from that electricity can then power vast motors. But electromagnetism can also be employed in a more delicate fashion as a means of communication, either with wires (as in the telephone), or without them (as in radio communication). It also drives our electronics devices (as in computers).

Magnetism has long been employed for navigation in the compass. This works because the Earth is itself a huge magnet, thought to have arisen from the great heat driven convection currents of molten iron in its center. In fact, it is known that the Earth's magnetic poles have exchanged positions in the past.

ELEMENTARY FUNCTIONS

Simple algebraic relationships that involve only two variables such as x and y are called elementary **functions**. In general, these can be denoted as y=f(x), where f symbolizes constants and operators on x. Basic examples of such a function would be y=2x or y=5x+1. It can be seen that these are easily evaluated given a value for x or y. Exponential functions that use a function of x as an **exponent** are also considered to be elementary functions. These take the form $y=e^{f(x)}$ where f(x) is composed of the same type of constants and operations upon x as before. Again, knowing a value for x or y allows a solution to the equation to be determined. In addition, the inverse of **exponential functions**, the logarithmic functions, are included. Since trigonometric functions can be expressed as **equations** that include exponential functions and their inverse, such as $2\cos x = e^{ix} + e^{-ix}$, the **trigonometric functions** must also be included as elementary functions. Any combinations of the above functions, complex as those combinations might seem are still elementary functions.

The **derivative** of an elementary function is always an elementary function, but the inverse is not always true. Not all elementary functions can be obtained by derivation. Some examples of non-elementary functions include the **Bessel function**, the **gamma function**, and others.

See also Inverse function; Logarithms

ELLIPSE

An ellipse is a kind of oval. It is the oval formed by the intersection of a plane and a right circular cone-one of the four types of **conic sections**. The other three are the **circle**, the **hyperbola**, and the **parabola**. The ellipse is symmetrical along two lines, called *axes*. The *major axis* runs through the longest part of the ellipse and its center, and the *minor axis* is perpendicular to the major axis through the ellipse's center.

Other definitions of an ellipse

Ellipses are described in several ways, each way having its own advantages and limitations:

1. The set of points, the sum of whose distances from two fixed points (the foci, which lie on the major axis) is constant. That is, $P: PF_1 + PF_2 = $ constant.

2. The set of points whose distances from a fixed point (the focus) and fixed line (the directrix) are in a constant ratio less than 1. That is, $P: PF/PD = e$, where 0 |less than| e |less than| 1. The constant, e, is the **eccentricity** of the ellipse.

3. The set of points (x,y) in a Cartesian plane satisfying an equation of the form $x^2/25 + y^2/16 = 1$. The equation of an ellipse can have other forms, but this one, with the center at the origin and the major axis coinciding with one of the coordinate axes, is the simplest.

4. The set of points (x,y) in a Cartesian plane satisfying the parametric **equations** $x = a \cos t$ and $y = b \sin t$, where a and b are constants and t is a **variable**. Other **parametric equations** are possible, but these are the simplest.

Features

In working with ellipses it is useful to identify several special points, chords, measurements, and properties:

The major axis: The longest chord in an ellipse that passes through the foci. It is equal in length to the constant sum in Definition 1 above. In Definitions 3 and 4 the larger of the constants a or b is equal to the semimajor axis.

The center: The **midpoint**, C, of the major axis.

The vertices: The end points of the major axis.

The minor axis: The chord which is perpendicular to the major axis at the center. It is the shortest chord which passes through the center. In Definitions 3 and 4 the smaller of a or b is the semiminor axis.

The foci: The fixes points in Definitions 1 and 2. In any ellipse, these points lie on the major axis and are at a **distance** c on either side of the center. If a and b are the semimajor and semiminor axes respectively, then $a^2 = b^2 + c^2$. In the examples in Definitions 3 and 4, the foci are 3 units from the center.

The eccentricity: A measure of the relative elongation of an ellipse. It is the ratio e in Definition 2, or the ratio FC/VC (center-to-focus divided by center-to-vertex). These two definitions are mathematically equivalent. When the eccentricity is close to **zero**, the ellipse is almost circular; when it is close to 1, the ellipse is almost a parabola. All ellipses having the same eccentricity are geometrically similar figures.

The **angle** measure of eccentricity: Another measure of eccentricity. It is the acute angle formed by the major axis and a line passing through one focus and an end point of the minor axis. This angle is the arc **cosine** of the eccentricity.

The **area**: The area of an ellipse is given by the simple formula πab, where a and b are the semimajor and semiminor axes.

The **perimeter**: There is no simple formula for the perimeter of an ellipse. The formula is an elliptic integral which can be evaluated only by approximation.

The reflective property of an ellipse: If an ellipse is thought of as a mirror, any ray which passes through one focus and strikes the ellipse will be reflected through the other focus. This is the principle behind rooms designed so that a small sound made at one location can be easily heard at another, but not elsewhere in the room. The two locations are the foci of an ellipse.

Drawing ellipses

There are mechanical devices, called ellipsographs, based on Definition 4 for drawing ellipses precisely, but lacking such a device, one can use simple equipment and the definitions above to draw ellipses which are accurate enough for most practical purposes.

To draw large ellipses one can use the pin-and-string method based on Definition 1: Stick pins into the drawing board at the two foci and at one end of the minor axis. Tie a loop of string snugly around the three pins. Replace the pin at the end of the minor axis with a pencil and, keeping the loop taut, draw the ellipse. If string which does not stretch is used, the resulting ellipse will be quite accurate.

To draw small and medium sized ellipses a technique based on Definition 4 can be used: Draw two concentric circles whose radii are equal to the semimajor axis and the semiminor axis respectively. Draw a ray from the center, intersecting the inner circle at y and outer circle at x. From y draw a short horizontal line and from x a short vertical line. Where these lines intersect is a point on the ellipse. Continue this procedure with many different rays until points all around the ellipse have been located. Connect these points with a smooth curve. If this is done carefully, using ordinary drafting equipment, the resulting ellipse will be quite accurate.

Uses

Ellipses are found in both natural and artificial objects. The paths of the planets and some comets around the Sun are approximately elliptical, with the sun at one of the foci. The seam where two cylindrical pipes are joined is an ellipse. Artists drawing circular objects such as the tops of vases use ellipses to render them in proper perspective. In Salt Lake City the roof of the Mormon Tabernacle has the shape of an ellipse rotated around its major axis, and its reflective properties give the auditorium its unusual acoustical properties. (A pin dropped at one focus can be heard clearly by a person standing at the other focus.) An ellipsoidal reflector in a lamp such as those dentists use will, if the **light** source is placed at its focus, concentrate the light at the other focus.

Because the ellipse is a particularly graceful sort of oval, it is widely used for esthetic purposes, in the design of formal gardens, in table tops, in mirrors, in picture frames, and in other decorative uses.

EQUATIONS

Equations are statements that use numbers and symbols to demonstrate that two groups of mathematical data are equal. The **roots** or solutions to an equation are numbers that replace the variables (unknown quantities) to make the equation true.

The first records of equations come from early Greek scholars, 540 B.C.-250 B.C. **Euclid**, **Pythagoras**, and his Pythagorean followers used algebraic problems with geometric proofs. This method of mixing **algebra** and **geometry** led to complex constructions. The Greek system of mathematics used layers of numbers, letters, and punctuation marks piled on top of each other. This approach probably favored interpretations of geometry rather than **arithmetic** calculations. Other early mathematical records show that Egyptians, Indians, Muslims, and Greeks have been using **cubic equations** for more than 2000 years.

Some of the best known categories of equations include:

- Algebraic equations: The variables appear with coefficients that can have **addition**, **subtraction**, **multiplication**, or **division** performed on them. Some algebraic equations have non-algebraic coefficients.

- Root equations: The **variable** is a radicands (a number under a radical sign).

- Transcendental equations: These equations are not algebraic, but they may have algebraic coefficients. They often deal with relationships between non-algebraic numbers and quantities.

- Exponential equations: These are a form of transcendental equation that is usually solvable by algebra. The variables often appear as **exponents** but can appear as bases or coefficients.

- Logarithmic equations: These are also a form of transcendental equation and are difficult to distinguish from exponential equations.

- Trigonometric equations: These are also a form of transcendental equation and have trigonometic terms with algebraic coefficients. These equations are often solvable by algebra.

- Identity equations: These equations use a statement of equality that is true for all values in a universal set.

- Conditional equations: These equations are true only for certain values of variables.

- **Linear equations**: In these equations, the unknown variable is expressed in first-degree terms.

- Polynomial equations: There are two kinds of polynomial equations: linear equations with integer coefficients, and fractional equations.

- Equations with absolute values: In these equations, a first-degree equation with one variable usually has one solution. However, first-degree equations with absolute-value expressions have more than one solution.

- **Quadratic equations**: These are defined as a polynomial equations of the second degree.

- Cubic equations: These require third-degree equations. Mathematical records show that although cubic equations were written for many centuries, Italian mathematicians of the sixteenth century were the first to actually solve such an equation.

- **Quartic equations**: These are regarded as very cumbersome and can be solved with rational operations and rational expressions.

- Simultaneous equations: In this form of problem, all equations must be true at the same time.

- **Diophantine equations**: These are named for the mathematician **Diophantus of Alexandria** (c. A.D. 220). The **definition** of Diophantine equations have evolved over time and today this term refers to equations that can be solved with integers.

EQUIVALENCE

Within propositional logic, the term equivalence is used in either of the following two ways: (1) as a particular type of propositional statement or (2) as the logical operator used to form an "equivalence" statement. In order to develop a precise **definition** of equivalence, some background regarding propositional logic is necessary. The Greek philosopher **Aristotle** (384 B.C.-322 B.C.) provided the first systematized approach to propositional logic. Keep in mind that logic at that time was expressed entirely in ordinary language and not at all symbolically. Nevertheless, "**Aristotelian logic**" was, and remains, the starting point to formal logic. Within the Aristotelian system the concepts of "proposition" and "truth" are of paramount importance. A proposition (also called a statement) is defined as a linguistic formation used to communicate information that may be labeled as "true" or "false". Moreover, there exist several principles related to **propositions**: the *principle of identity* states that every proposition is equal to itself; the *principle of the excluded middle* means that every proposition is either true or false; and the *principle of excluded contradiction* states that no proposition is both true and false. An "equivalence statement", then, is simply a type of proposition (as stated at the beginning in definition #1) that may be evaluated as being true or false.

More than 2000 years after Aristotle, linguistic logic gave way to a purely **symbolic logic** (also called mathematical logic). In this symbolic form of propositional logic, statements are often denoted by capital letters (*A*, *B*, *C*, etc.) and "logical operators" are used to form compound statements. Equivalence (as stated at the beginning in definition #2) is used along with other logical operators, like "∧" (and), "~" (not), "→" (**implication**), etc. to form new compound statements. Thus, two propositions, say A and B, and the symbol "↔" are used to form the equivalence statement $A ↔ B$, where ↔ denotes "equivalence". The **truth** or falsity (called the truth-value) of the equivalence statement $A ↔ B$ is dependent solely upon the corresponding truth-values of statements A and B. This situation is best illustrated via the use of a **truth table**, as shown below:

	A	B	A↔B
1.	T	T	T
2.	T	F	F
3.	F	T	F
4.	F	F	T

This truth table demonstrates the four possible combinations of truth/falsity for statement A (the first column), statement B (the second column), and the resulting truth-value for the equivalence $A ↔ B$ (the third column), where T = true and F = false. As demonstrated by the truth table, $A ↔ B$ is true if A and B are both true, or A and B are both false. If A and B have different truth-values (one is true and the other false) then the proposition $A ↔ B$ is false.

For a valid (i.e., true) equivalence statement the truth-value of one component statement ensures the same truth-value for the other. (See rows 1 and 4 in the above truth table.) This relationship can be expressed by saying that A is true if and only if B is true, and likewise A is false if and only if B is false. The phrase "if and only if" can be replaced by the statement "A is a necessary and sufficient condition for B", which means that for a valid equivalence statement the truth of statement A is a "necessary and sufficient condition" for the truth of statement B (likewise, the falsity of A is a "necessary and sufficient condition" for the falsity of B). The condition of "if and only if" does not hold between the statements A and B in the case of invalid (i.e., false) equivalence statements.

An interesting and important **theorem** within propositional logic is the tautology $((A → B) ∧ (B → A)) ↔ (A ↔ B)$. A tautology is a statement that is true for every possible truth-value of its component statements. The truth table below demonstrates that the statement in column (g) is true for every possible truth-value of the component statements A and B, and therefore the statement $((A → B) ∧ (B → A)) ↔ (A ↔ B)$ is indeed a tautology.

	(a)	(b)	(c)	(d)	(e)
	A	B	A→B	A→B	(A→B) ∧ (B→A)
1.	T	T	T	T	T
2.	T	F	F	T	F
3.	F	T	T	F	F
4.	F	F	T	T	T

The four possible truth-values of statement A and statement B appear in columns (a) and (b) of the above truth table. The propositions $A → B$ and $B → A$ within columns (c) and (d), respectively, are called implications, or conditionals. An implication, in general, can be read as "If A, then B" or as "A implies B". The implication $A → B$ is false only for A = true and B = false, as shown in cell (2c). It is true for all other truth-value combinations of A and B, as shown within the table. The same conditions apply to the implication $B → A$ (i.e., false only for B = true and A = false).

The statement in column (e), which is the **conjunction** of two implications, is called a biconditional, or double implication. The fundamental purpose of the preceding truth table is to demonstrate that, given statements A and B, the double implication of statements A and B (i.e., $(A → B) ∧ (B → A)$) is true if and only if the equivalence $A ↔ B$ is true. This relationship between the double implication of two statements and their corresponding equivalence can often be of use in mathematical proofs. Specifically, one can demonstrate rigorously that "$A ↔ B$" is true using the following procedure: one proves that the implication $A → B$ is true, and further that the implication $B → A$ is true, then one can say that both $A → B$ and $B → A$ are true statements, or symbolically $(A → B) ∧ (B → A)$ is true. This last statement is the double implication for statements A and B; then according to the preceding truth table the truth of the equivalence $A ↔ B$ has been proven.

See also Aristotle; Logical symbols; Propositions; Symbolic logic; Tautologies; Truth; Truth table

ERATOSTHENES OF CYRENE (CA. 285 B.C.-CA. 205 B.C.)

Greek mathematician, astronomer, geographer, philosopher, and poet

Eratosthenes was one of the great thinkers of ancient Greece whose accomplishments included accurately measuring the Earth's **circumference** and discovering a simple method for finding **prime numbers**. In addition to his contributions in mathematics, geography, and astronomy, Eratosthenes was a poet, philosopher, and educator whose wide ranging interests served him well as director of the great ancient library at Alexandria. Eratosthenes was nicknamed "Beta" (Greek for two) by some of his contemporaries. This name may refer to his status as second only to **Plato** in knowledge, or it may also have been a derogatory term used by detractors to indicate that Eratosthenes had not achieved top standing in any discipline of knowledge.

Eratosthenes was born circa 285 B.C. in Cyrene, which is now Libya, in northern Africa. His father's name was Aglaus, but little else is known about his family or its station in Greek society. At that time, Cyrene was a prosperous Hellenistic Greek city–state known for its culture and as a successful commercial harbor on the Mediterranean coast. Eratosthenes is believed to have studied under Lysanias, a renowned grammarian, and Callimachus, a poet. He developed an early interest in literature and philosophy and, in his teens, traveled to Athens, where he probably studied in the famed Academy and Lyceum.

After completing his studies, Eratosthenes turned his attention to writing, especially poetry. None of his poems from this period survives, but references to them include the poem *Hermes*, which describes the astronomical "heavens" and is a homage to the Greek god, and *Ergion*, about a young woman's suicide. It is a testament to Eratosthenes intellect that he achieved wide renown as a poet before his scientific accomplishments. His literary abilities likely attracted the attention of Ptolemy III, who invited Eratosthenes to Alexandria to tutor his son, the crown prince. At the age of 40, he also garnered the prestigious appointment as director of the library at Alexandria. He then embarked on a career unparalleled at that time for its breadth of scientific, artistic, and academic contributions.

Eratosthenes was a consummate scholar with wide ranging interests, but his most important contributions, including those in geography and astronomy, were based on his mathematical acumen. In the field of mathematics, Eratosthenes' most famous achievement is his simple but foolproof method, or **algorithm**, for finding prime numbers. Known as the **sieve of Eratosthenes**, the approach allowed him to "sift" out prime numbers (divisible only by one and the number itself) from a composite list of natural numbers arranged in order. Eratosthenes' discovery has led to centuries of **number theory** research.

In the area of **geometry**, Eratosthenes is credited with developing a practical method for finding an infinite number of **mean** proportionals between two given straight lines. He developed the mean–finder, or mesolabe, which led to solving the Delian problem, otherwise known as doubling the cube. A letter to Ptolemy III, which is repeated in the commentary of Eutocius on **Archimedes**, is the only remaining writing on geometry by Eratosthenes. In the letter, Eratosthenes describes the mesolabe, a mechanical device constructed of three equal rectangular frames, each with a diagonal that creates two triangles within each frame. These frames could be pushed along grooves back and forth, and over and under each other. With this device, Eratosthenes was able to demonstrate a mathematical method for doubling the cube. This unique approach to solving problems of **proportions** was vitally important to the Greeks, since they used proportions to solve geometrical problems. Eratosthenes was so proud of this accomplishment that he gave a bronze model of his invention with an inscription of how it worked as a gift to the king and the people of Alexandria.

Eratosthenes also wrote several books or treatises on mathematics, although none of them survive today. In one book, he wrote on the fundamentals of mathematics in relation to the philosophy of Plato, including definitions in geometry and **arithmetic** and theories of proportions. His most famous and comprehensive mathematical work is called *On Means*, which was considered important enough to be included by Pappus in his *Treasury of Analysis*. However, the actual content of *On Means* is unknown. Perhaps the greatest testament

to Eratosthenes' mathematical abilities was made by Archimedes, who sent him a manuscript of his *Method*, stating that his treatise, which explains the methods he used to make his mathematical discoveries, was written especially for Eratosthenes.

Eratosthenes' accurate measurement of the Earth's circumference is his most renowned achievement. It marks the first important step in the science of geodesy, a branch of mathematics concerned with determining the Earth's size and shape and location of points on its surface. Prior to this, the Earth's circumference was estimated to be approximately 300,000 stades. Incorporating two geometrical **propositions** set forth by **Euclid**, Eratosthenes first measured the **distance** from Alexandria to Syene, Egypt (located south of Alexandria) at 5,000 stades. He then used calculations of the position of the sun during the summer solstice over both cities to estimate the distance from the two cities to represent 1/50th of the Earth's surface. With some minor adjustments to correct errors in his readings, Eratosthenes came up with his final estimate of 252,000 stades (24,662 miles), which is within 1 percent of modern estimates of the Earth's circumference.

Eratosthenes' most famous book on geography, *On the Measurement of the Earth*, probably included a detailed explanation of how he measured the Earth's circumference. The lost book is believed to have focused on mathematical geography and included calculations of the distance of the Earth from the moon and sun, the distance between the tropic circles, and the size of the sun. His three–volume *Geographica* included a history of geography to his time. It then discussed specifics of mathematical geography based on the belief that the Earth was round or spherical. It was largely through these books that Eratosthenes established the science of geodesy. Eratosthenes used this newly found science to make some of the most accurate maps of the world in his time, including the first maps to incorporate the system of latitude and longitude.

In addition to his commitment to precise measurements in geography and astronomy, Eratosthenes also established the first scientific chronology for recording history in Greece. As a scholar, he wanted to accurately record his country's past based solely on documented records and without the inclusion of legends. In his two books on the subject, *Ilympionikai* and *Chronographiai* (Chronological Tables), he develops and details his method, which is based largely on using existing records of the Olympiads (the precursor of the modern day Olympic Games) to set an event in time.

Among Eratosthenes' other notable scientific accomplishments is his accurate measurement of the tilt of the Earth's axis and a comprehensive catalogue of 675 stars. He is also believed to have created the improved **calendar** adopted by Julius Caesar. Similar to the modern–day calendar, Eratosthenes' calendar incorporated the addition of one day every four years, which is now known as Leap Year.

A man of boundless intellectual curiosity, Eratosthenes also wrote a number of literary subjects, including books *Good and Evil* and *Comedy*, which discussed authors of that genre. His last known written work, *Biography of Arsinoe III*, revealed some of his intimate knowledge of the royal court.

Befitting his versatility and genius, Eratosthenes was referred to as a philologist, or lover of learning. Whether Eratosthenes himself created this title or it was bestowed upon him by others is unknown. Eratosthenes went blind in his later years. For a man committed to knowledge and learning, his loss of sight must have been a terrible blow, barring him from ever again using the greatest library in the world. Eratosthenes chose a path of voluntary starvation and died at the age of 80.

Erdös, Paul (1913-1996)

Hungarian number theorist

For Paul Erdös, mathematics was life. **Number theory**, **combinatorics** (a branch of mathematics concerning the arrangement of finite **sets**), and discrete mathematics were his consuming passions. Everything else was of no interest: property, money, clothes, intimate relationships, social pleasantries—all were looked on as encumbrances to his mathematical pursuits. A genius in the true sense of the word, Erdös traveled the world, living out of a suitcase, to problem solve—and problem pose—with his mathematical peers. A small, hyperactive man, he would arrive at a university or research center confident of his welcome. While he was their guest, it was a host's task to lodge him, feed him, do his laundry, make sure he caught his plane to the next meeting, and sometimes even do his income taxes. Cosseted by his mother and by household servants, he was not brought up to fend for himself. Gina Bari Kolata, writing in *Science* magazine, reports that Erdös said he "never even buttered his own bread until he was 21 years old."

Yet this man, whom Paul Hoffman called "probably the most eccentric mathematician in the world" in the *Atlantic Monthly*, more than repaid his colleagues' care of him by giving them a wealth of new and challenging problems—and brilliant methods for solving them. Erdös laid the foundation of computer science by establishing the field of discrete mathematics. A number theorist from the beginning, he was just 20 years old when he discovered a **proof** for **Chebyshev**'s theorem, which says that for each integer greater than one, there is always at least one prime number between it and its double.

Erdös was born in Budapest, Hungary, on March 26, 1913. His parents, Lajos and Anna Erdös, were high school mathematics teachers. His two older sisters died of scarlet fever when he was a newborn baby, leaving him an only child with a very protective mother. Erdös was schooled at home by his parents and a governess, and his gift for mathematics was recognized at an early age. It is said that Erdös could multiply three-digit numbers in his head at age three, and discovered the fact of **negative numbers** when he was four. He received his higher education from the University of Budapest, entering at age 17 and graduating four years later with a Ph. D. in mathematics. He completed a postdoctoral fellowship in Manchester, England, leaving Hungary in the midst of political unrest in 1934. As a Jew, Hungary was then a dangerous place for him to be. During the ensuing Nazi reign of power,

four of Erdös's relatives were murdered, and his father died of a heart attack in 1942.

In 1938, Erdös came to the United States, but because of the political situation in Hungary, he had difficulty receiving permission from the U. S. government to come and go freely between America and Europe. He settled in Israel and did not return to the United States until the 1960s. While in the country, he attended mathematical conferences, met with top U. S. mathematicians such as Ronald Graham, Ernst Straus and **Stanislaw Ulam**, and lectured at prestigious universities. His appearances were irregular, owing to the fact that he had no formal arrangements with any of the schools he visited. He would come for a few months, receive payment for his work, and move on. He was known to fly to as many as fifteen places in one month—remarking that he was unaffected by jet lag. Because he never renounced his Hungarian citizenship, he was able to receive a small salary from the Hungarian Academy of Sciences.

So esteemed was Erdös by his colleagues that they invented the term "Erdös number" to describe their close connections with him. For example, if someone had coauthored a paper with Erdös, they were said to have an Erdös number of one. If someone had worked with another who had worked with Erdös, their Erdös number was two, and so on. According to his obituary in the *New York Times*, 458 persons had an Erdös number of one; an additional 4,500 could claim an Erdös number of two. It is said that **Albert Einstein** had an Erdös number of two. Ronald Graham, director of information sciences at AT&T Laboratories, once said that research was done to determine the highest Erdös number, which was thought to be 12. As Graham recalled, "It's hard to get a large Erdös number, because you keep coming back to Erdös." This "claim to fame" exercise underscores Erdös's monumental publishing output of more than 1,500 papers, and is not only a tribute to his genius but also to his widespread mathematical network.

Throughout his career, Erdös sought out younger mathematicians, encouraging them to work on problems he had not solved. He created an awards system as an incentive, paying amounts from $10 to $3,000 for solutions. He also established prizes in Hungary and Israel to recognize outstanding young mathematicians. In 1983, Erdös was awarded the renowned Wolf Prize in Mathematics. Much of the $50,000 prize money he received endowed scholarships made in the name of his parents. He also helped to establish an endowed lectureship, called the Turan Memorial Lectureship, in Hungary.

Erdös's mathematical interests were vast and varied, although his great love remained number theory. He was fascinated with solving problems that looked—but were not—deceptively simple. Difficult problems involving number relationships were Erdös's special forte. He was convinced that discovery, not invention, was the way to mathematical **truth**. He often spoke in jest of "God's Great Mathematics Book in the Sky," which contained the proofs to all mathematical problems. Hoffman in the *Atlantic Monthly* says "The strongest compliment Erdös can give to a colleague's work is to say, 'It's straight from the Book.'"

Erdös's mother was an important figure in his life. When she was 84 years old, she began traveling with him, even though she disliked traveling and did not speak English. When she died of complications from a bleeding ulcer in 1971, Erdös became extremely depressed and began taking amphetamines. This habit would continue for many years, and some of his extreme actions and his hyperactivity were attributed to his addiction. Graham and others worried about his habit and prevailed upon him to quit, apparently with little result. Even though Erdös would say, "there is plenty of time to rest in the grave," he often talked about death. In the eccentric and personal language he liked to use, God was known as S. F. (Supreme Fascist). His idea of the perfect death was to "fall over dead" during a lecture on mathematics.

Erdös's "perfect death" almost happened. He died of a heart attack in Warsaw, Poland, on September 20, 1996, while attending a mathematics meeting. As news of his death began to reach the world's mathematicians, the accolades began. Ronald Graham, who had assumed a primary role in looking after Erdös after his mother's death, said he received many electronic-mail messages from all over the world saying, "Tell me it isn't so." Erdös's colleagues considered him one of the 20th century's greatest mathematicians. Ulam remarked that it was said "You are not a real mathematician if you don't know Paul Erdös." Straus, who had worked with Einstein as well as Erdös, called him "the prince of problem solvers and the absolute monarch of problem posers," and compared him with the great 18th-century mathematician **Leonhard Euler**. And, Graham remarked, "He died with his boots on, in hand-to-hand combat with one more problem. It was the way he wanted to go."

ERLANG, AGNER KRARUP (1878-1929)

Danish mathematician

Erlang is regarded as the founder of queuing theory and of operations research. His formulas, designed and published in 1917, enabled early telephone switching systems to become operational. These formulas give the probability that a user will encounter a busy signal instead of a dial tone, or the length of waiting time for a system that can hold calls. They are used to calculate the number of circuits needed to give a specified level of service. Erlang's formulas may be applied to any system with a limited number of servers and customers that arrive at random times.

Agner Krarup Erlang was born on January 1, 1878, at Lonborg, near Tarm in Jutland, the mainland of Denmark. His parents were Hans Nielson Erlang, the parish clerk and schoolmaster, and Magdalene Krarup Erlang. Many of his mother's family were clergymen. Erlang had an older brother, Frederik, and two younger sisters, Marie and Ingeborg. Erlang studied at his father's school, then was tutored at home by his father and the assistant school teacher, P.J. Pedersen, for his preliminary examination. Erlang passed with distinction, although at age 14 he needed special permission to take the examination. Erlang served as assistant teacher at his father's school for two years,

then stayed for two years with M. Funch, in Hillerod, to prepare for his university examination at the Frederikborg Grammarschool. In 1896, Erlang passed this examination and attended the University of Copenhagen. He studied mathematics, astronomy, physics and chemistry, and completed an M.A. in 1901.

After graduation, Erlang taught at a number of schools. He joined a Christian students' association where he met his friend H.C. Nybolle, who later became a professor of **statistics** at Copenhagen University. Erlang also was a member of the Mathematics Association. In 1904, he won a distinction for his answer to the University mathematics prize question about Christiaan Huygen's methods of solving **infinitesimal** problems. At the Mathematics Association, Erlang met J.L.W.V. Jensen of the Copenhagen Telephone Company. Jensen introduced Erlang to Fr. Johannsen, his managing director, who hired Erlang in 1908 as scientific collaborator and leader of the laboratory.

Johannsen had already published two essays on the barred access and waiting time problems inherent in telephone systems. He suggested that Erlang study these problems further. Erlang demonstrated in a 1909 paper that the number of calls to arrive during a period of time follows a **Poisson distribution**, and treated the problem of waiting time when holding times are constant, for the simplest case of one circuit. In 1917, Erlang's most important paper was published, in which he gives his B-formula for the probability of barred access, or a busy signal, for a group of circuits, and formulas for waiting time. His **proof** of the B-formula is based on the idea of statistical equilibrium, that transitions between pairs of states are in balance. In 1922 and 1926 Erlang published lectures on the contents of his earlier papers; the 1922 paper containing a new interconnection formula. In 1924 he wrote about a principal of K. Moe for deciding whether to add circuits to large or small groups of circuits. Erlang often presented his results as tables as well as formulas. He wrote several papers about producing accurate tables and published tables produced by his methods.

Erlang also wrote about cables, the **induction** coil in a telephone, and about a device for measuring transmission in cables. Also interested in more theoretical matters, Erlang used the idea of statistical equilibrium to prove Maxwell's Law in the kinetic theory of gases, and wrote short papers on **geometry** and on proportional representation in voting.

Erlang, who never married, lived with his sister Ingeborg in Copenhagen. He had a large collection of books on science and mathematics. Erlang's sister founded a home for mentally ill women, which he supported generously. Erlang died after a brief abdominal illness on February 3, 1929, at 51 years of age.

In addition to his widely used formulas, Erlang's name is attached to the gamma probability distribution. In 1946, the C.C.I.F. (La comite consultatif des communications telephoniques a grande **distance**) adopted the Erlang as the unit of telephone traffic. The average traffic in Erlangs is the sum of the lengths of calls originating during an **interval** of time, divided by the length of the time interval. ERLANG is also the name of a programming language which was developed at Ericsson and Ellemtel Computer Science Laboratories for programming telecommunications switching systems.

ERLANGER PROGRAM

The term Erlanger program comes from an article written by the great mathematician **Felix Klein** in 1872, in which he outlined his vision of the direction in which the study of **geometry** should proceed in the years that followed. Klein's key idea was that geometric structures on spaces should be understood in terms of the group of transformations that preserve the structures. Klein's insight, which seemed very abstract to his contemporaries, was a source of inspiration to the mathematicians that followed him, notably **Henri Poincaré**, and has become fundamental to our current understanding of geometry.

In the years preceding Klein's Erlanger program, the world of geometry had been changed by the discovery of **non-Euclidean geometries**, in which Klein was instrumental. Mathematicians realized that the same **space** could admit more than one geometric structure, and it became important to make a precise **definition** of a geometric structure on a space. Two complementary pictures evolved. One was the idea of **Bernhard Riemann** that a geometric structure was given by a local definition of lengths and angles; this gave rise to the field of differential geometry. The other was Klein's Erlanger program, which argued that a geometric structure could be defined in terms of the transformations that left it **invariant**.

A **transformation** of a space is a **mapping** of the space onto itself (usually required to be continuous and have a continuous inverse) that never maps two points to the same point. For example, in the plane, a **rotation** (say, a quarter-turn) about a fixed point is a transformation, since it maps the plane onto itself, and no two different points get rotated to the same point. Another transformation is the 'stretch' map that leaves the origin fixed, and sends every other point to the point twice as far from the origin, along the same line (in symbols, this would be the function $f(x,y)=(2x,2y)$). Like the rotation, the stretch map sends the plane onto itself, and no two points get mapped to the same point. On the other hand, a map that sends the plane onto the upper-half plane by folding it along the x-axis like a sheet of paper is not a transformation, since it does not map the plane onto the entire plane, and two different points can get mapped to the same point.

A transformation is said to preserve a geometric structure if it leaves all the measurements of that structure (lengths, angles, etc.) unchanged. Thus, the rotation preserves the Euclidean structure on the plane, since rotation does not change **angle** or length measurements. On the other hand, the stretch map does not preserve the Euclidean structure on the plane; it preserves angle measurements, but not length measurements.

According to Klein, a geometric structure consists of a space together with a particular group of transformations of the space. The transformations that preserve Euclidean space are the rotations, translations (maps that shift the plane in some fixed direction, by some fixed amount), reflections across a line, and glide reflections (maps that consist of a reflection across a line, followed by a shift in the direction of the line). Klein would say that two-dimensional Euclidean geometry is the two-dimensional plane together with the group of rotations, translations, reflections and glide reflections of the plane. **Hyperbolic geometry** can be modeled by the upper-half plane, together with the following transformations: horizontal translations, reflections across vertical lines, stretch maps with center on the x-axis, inversions with center on the x-axis, and combinations of those maps (an inversion is a map that leaves a given **circle** fixed, and sends every other point to a point on the same line through the center, according to a precisely defined rule).

As mathematicians began to have a better understanding of what constitutes a geometric structure, they realized that the coordinate line, plane and higher-dimensional spaces were far from the only spaces that could admit geometric structures. The notion of a **manifold** evolved: a manifold is a space for which, around every point, there is a neighborhood that looks like a bent or distorted piece of the coordinate space of the appropriate **dimension**. Thus, the surface of the earth is a 2-dimensional manifold, since every point has a small neighborhood that looks like a slightly curved piece of the coordinate plane.

Riemann's and Klein's ideas generalize to give definitions of what it means to have a geometric structure on a manifold. Riemann would say that a geometric structure consists of a definition of lengths and angles on the manifold. But a geometric structure can also be defined in terms of transformations. When we say that every point in a manifold has a neighborhood that "looks like" coordinate space, what we mean is that there is a function that maps the neighborhood onto a piece of coordinate space. But there is not just one such function; each point has many neighborhoods that look like many different pieces of coordinate space. Hence there are many different pieces of coordinate space associated to the same piece of the manifold, and there are transformations that take one of the corresponding pieces of coordinate space to another; these transformations are called the transition maps. We can put a geometric structure on the manifold by requiring that the transition maps satisfy geometric requirements. A manifold is said to have a Euclidean structure if the transition maps preserve Euclidean length and angle measurements. A manifold has a hyperbolic structure if the transition maps preserve hyperbolic measurements. This type of definition is not limited to geometric structures: a manifold is said to have a differentiable structure if the transition maps and their inverses are differentiable. The question of what kinds of additional structures different manifolds can support is one of the most active fields of mathematical research today.

See also Geometry; Hyperbolic geometry; Manifold

ERRORS, APPROXIMATION, AND ROUNDING

People prefer to work with round numbers because they are easy to remember and easy to use. For example, most people would rather multiply 4×10 than 3.5×9.5. It is important to know when a round number can be used for convenience. Sometimes the way a number is rounded will affect the accuracy of a computation, introducing a degree of error. If a person is using a blend of exact and rounded numbers in one computation, it may be very difficult to know the accuracy of the answer. There are some basic ground rules for approximation and rounding.

There are actually two forms of approximation: truncation and rounding.

A truncated number is cut off. The truncation is shown by three dots following a series of decimals. All of the digits are valid up to the point of truncation. The truncated value is always lower than exact value.

When a number is rounded, it is replaced by another number with fewer decimal digits or fewer non-zero digits if it is an integer. A number can be rounded down or can be rounded up. Something at the halfway point, 5, can be rounded down or up; 4 would be rounded down and 6 would be rounded up. The International Standard Organization (ISO) offers rounding rules and encourages scientific rounding to reduce risk of errors. The organization emphasizes the importance of doing the rounding all in one step to avoid errors, and of using consistent principles of rounding.

An approximate value might be sufficient for a narrow purpose, but an approximate value is not exact. Under the principles of numerical **analysis**, an approximation should result in the intended value and should reduce the number of **arithmetic** calculations.

Why would an approximate value need to be used? Sometimes an **irrational number** can only be represented as a nonperiodic infinite decimal fraction. Nonperiodic infinite **decimal fractions** can only be represented as approximate numbers. A nonperiodic infinite decimal fraction can be made more precise with a greater number of decimal digits, but it can never be written as an exact number. Periodic infinite decimal **fractions**, also called repeating infinite decimal fractions, have an interminable number of decimals and are also approximations. However, they can be written as common fractions in exact numbers. Finite decimal fractions are often shown with a limited sequence of decimals—this is an approximation. Sometimes a formula will generate a sum with an infinite number of terms. There is no way of showing an infinite number, so an approximation is a necessity as well as a convenience.

ESCHER, MAURITS CORNELIS (1898-1970)

Dutch artist

The Dutch artist M. C. Escher was born on June 17, 1898, in Leeuwarden, the Netherlands. Trained as a graphic artist, Escher became well known for his drawings of scenes that seem to play visual logic against realistic impossibility.

Escher was the youngest of three sons born to George Arnold Escher and Sarah Gleichman Escher. As a youth, Escher was encouraged by his father, a civil engineer, to learn carpentry and other craft skills. He attended elementary and secondary school in Arnhem and in the seaside town of Zandvoort, a place that was selected for the benefit of his poor health. Failing his final exam, Escher never officially graduated from secondary school.

Nevertheless, in 1918, Escher began architectural studies at the Higher Technology School in Delft. Later he moved to Haarlem to study at the School for Architecture and Decorative Arts. But by 1919 Escher had discovered that his real interest lay in graphic arts rather than architecture.

For the next two years, therefore, Escher pursued studies in art school, gaining a mastery of graphics and wood cutting techniques. Following his graduation, he traveled in southern Europe in search of inspiration for his work.

In 1924, he held his first one-man art show in Holland. Later that year, he married Jetta Umiker in Viareggio, Italy. The late 1920s saw Escher living in Rome with his wife and child. In 1929, he held five shows in Holland and Switzerland. Some of Escher's most striking Italian landscapes date from this period.

Escher's early work tended to be realistic portrayals of the landscape and architecture that he saw during his travels. Visiting the Alhambra in Granada, he was intrigued by the Moorish designs that covered the walls. These designs had a profound influence on his work after 1937.

In the 1930s, Escher moved away from the true-to-life themes inspired by southern Italy, and began to look instead toward Switzerland, Belgium, and Holland for ideas.

With the invasion of Holland and Belgium by the Nazi army in 1940, Escher and wife relocated to Baarn, Holland. By the late 1940s, besides creating woodcuts, lithographs, and occasional mezzotints, Escher was involved in the design of a tapestry and even a ceiling decoration.

By the early 1950s, Escher had also achieved popularity as a lecturer, and was in demand as a speaker to groups of artists and scientists alike.

Beginning in 1956, Escher tried to find ways to express the notion of **infinity** within the bounds of a finite print. The objects in most of his works prior to 1958 appear to be shrinking toward the center of the print, but after that time, they give the appearance of shrinking toward the outer edges.

This change in Escher's focus reportedly came in response to his reading an article by the mathematician H. M. S. Coxeter of Ottawa that included an illustration of a system for reducing a plane-filling motif as the **distance** from the center of a **circle** increases. At least six of Escher's major works appear to be based on Coxeter's system.

In 1959, Escher met Professor Caroline MacGillavry. MacGillavry arranged for Escher present a lecture on **symmetry** to an international meeting of crystallographers in England in 1960. (The lecture was later repeated in Canada and the US.) In the same year, Escher received a copy of an article co-authored by **Roger Penrose** that described that mathematician's notion of impossible objects. Penrose's article referred to Escher's early works, and actually inspired others.

Escher's work with periodic repetitions had achieved considerable popularity in the late 1950s and 1960s, and in 1961 he authorized the International Union of Crystallography to publish a book about them. The book, Symmetry Aspects of M.C. Escher's Periodic Drawings, was published in 1965.

In Escher's most famous work, patterns are repeated in regularly divided planes to create impossible constructions, and the illusion of infinite **space**. His later work drew heavily on his intuitive understanding of mathematical concepts.

EUCLID OF ALEXANDRIA (ca. 300 B.C.)
Greek geometer

Euclid was a preeminent Greek mathematician who is often to referred to as the founder of **geometry**. His historical stature as one of the most influential mathematicians of all time stems from his authorship of the *Elements*, a textbook of elementary geometry and logic. The accumulated knowledge set forth in the *Elements* was so exhaustive that the book has been a standard text on geometry throughout the centuries.

Little is known about Euclid's life. Most of what can be inferred concerning Euclid stems from a brief summary by Proclus, who wrote a commentary on the *Elements*. It is generally agreed that Euclid lived about 300 B.C. Although ancient Arab authors claim that Euclid was Greek and born in Tyre, most historians are certain that Euclid was either Greek or Egyptian. He is believed to have been a student at Plato's Academy in Athens, where most of the accomplished geometers of Euclid's time studied. Little is known of Euclid's personality or beliefs other than he was completely devoted to the study of mathematics and was considered a wise, patient, and kind teacher. His acknowledgment that he derives most of the knowledge in the *Elements* from his predecessors reveals Euclid's sense of fair play and respect for the mathematicians who came before him.

Perhaps because of civic unrest or at the request of Ptolemy I, Euclid eventually traveled to Alexandria, Egypt, around 322 B.C. and established a famous school of mathematics. Built by Alexander the Great between two arms of the Nile River, Alexandria became an intellectual center of education and learning unparalleled in the Hellenistic Age. Not long after Euclid established his school, Ptolemy founded a museum that became the first national university. It included the most comprehensive library in the ancient world, housing more than 600,000 "books," actually papyrus rolls. Euclid was the museum's first teacher of mathematics.

When it came to mathematics, however, Euclid was unyielding in his insistence on the careful study of the discipline and its worth. According to one story, Ptolemy was observing one of Euclid's geometry classes and, afterwards, asked Euclid if there was a "shorter road to its mastery." Euclid replied, "There is no royal road to geometry." Another story relates the tale of a student who, upon learning a new proposition asked his master "... what will I get by learning these difficult things?" Euclid called for his slave and said, "Give this man an *obol* [a Greek coin], for he must make gain from what he learns."

The genius and monumental impact of Euclid's *Elements* is unquestioned. Not only was it immediately recognized as the definitive book on geometry in Euclid's own time, but it has also been a standard mathematical text for more than 2,000 years, forming the foundation of all geometry taught until the beginning of the 20th century. To write the *Elements*, Euclid drew upon three centuries of mathematical thought, including the work of **Pythagoras, Hippocrates**, and **Menaechmus**. As the book's reputation grew and handwritten copies spread throughout the ancient world, the *Elements* stimulated mathematical thought and discovery in ancient Greece, Egypt, Persia, Arabia, and India. The first printed edition appeared in 1482, and the first complete English edition in 1570.

Euclid himself is attributed with only developing a few of the theorems in the *Elements*. Composed of thirteen books, the genius of the *Elements* stems from its systematic presentation of the basic principles of geometry and the accompanying statements and **theorem** proofs. With astounding clarity, Euclid states his theorems and problems and then logically proceeds to present rigorous **propositions**.

Containing more than 450 propositions, the 13 books of the *Elements* begins with definitions, postulates, and axioms. In Book I, Euclid **sets** forth the definitions of points, lines, planes, angles, circles, triangles, quadrilaterals, and parallel lines. Book II focuses primarily on rectangles and squares and contains many problems now treated algebraically. Book II considers the geometry of the **circle**, including the relationship between circles that intersect or touch. Book IV continues with Euclid's examination of the circle and of **polygons**. Book V sets forth a theory of proportion and areas, and Book VI applies this theory to **plane geometry**. Books VII, VIII, and IX focus on **arithmetic, irrational numbers**, and the theory of **rational numbers**. Book XI, XII, and XIII deals with three dimensional, or solid, geometry.

In addition to containing the accumulated knowledge in mathematics up to Euclid's time, these 13 books include some original analyses and proofs. For example, Euclid developed a new **proof** for the **Pythagorean theorem** and, in the process, proved the existence of irrational numbers. Euclid also devised an ingenious proof showing that the number of primes is infinite. His development of the methods of exhaustion for measuring areas and volumes were later used by **Archimedes**.

One of the most important contributions made by Euclid in the *Elements* is the five postulates, which embody the distinctive principles of Euclidean geometry. Postulates one through three focus on construction concerning the straight line and the circle. **Postulate** four states that all right angles are equal. However, it is the Fifth Postulate, which Euclid most certainly invented, that stands out as evidence of both Euclid's mathematical genius and the *Elements* greatest stumbling block.

The Fifth Postulate, also known as the "parallel postulate," has remained unchanged for 23 centuries. It states: If a straight line falling on two straight lines make the interior angles on the same side less than two right angles, the two straight lines, if extended indefinitely, meet on that side on which the angles less than two right angles are. It is the Fifth Postulate that lays the foundation for the theory of parallel lines, including the theory that only one parallel to a line can be drawn through any point external to the line.

Ironically, Euclid and the mathematicians who followed him recognized that this postulate had no proof from the other four postulates. As a result, Euclid himself avoided using it as much as possible. In the following centuries, several mathematicians developed alternatives to the Fifth Postulate. One of the most famous is the **axiom** developed by Proclus stating: "Through a given point in a plane, one and only one line can be drawn parallel to a given line." This postulate became known as Playfair's Axiomin 1795, when John Playfair proposed that it replace the Fifth Postulate.

Numerous other attempts have been made to prove the Fifth Postulate, primarily from the other four postulates. While some of these proofs were accepted for periods of time, they eventually were found to contain mistakes and be untrue. However, like Playfair's Axiom, some of the substitutes have merit.

Ironically, the problems associated with the Fifth Postulate led to the creation of **non–Euclidean geometry**, which is based on the assumption that the Fifth Postulate is not true. First publicly proposed by **Nikolai I. Lobachevsky** in 1826 and then by Eugenio Beltrami in 1868, non–Euclidean geometry has contributed greatly to the study of physics and relativity.

The *Elements* was so comprehensive and successful that Euclid's other works have been nearly forgotten. However, a few of these works have survived the ages. *Data*, which is most closely aligned with the *Elements*, deals with plane geometry and contains 94 propositions. An Arabic translation of Euclid's work in pure geometry, called *On Division*, also survives and was published in 1851. Euclid's other surviving works belong to **applied mathematics**, including the *Phenomena*, which focuses on spherical geometry. *Optics*, an elementary treatise on perspectives, may have been used to warn against paradoxical astronomical theories like the belief that the stars and other heavenly bodies were actually the size that they appear to be to the human eye.

Unfortunately, several of Euclid's works have been lost. The *Pseudaria*, or as Proclus called it, "The Book of Fallacies," presented fallacies in some of the elemental geometry of Euclid's day side by side with valid theorems. The loss of *Porisms*is believed to be the greatest loss of all Euclid's books. Focusing on higher geometry, the *Porisms* was reported to be a comprehensive and difficult work containing types of propositions intermediate between theorems and problems. Another work, the *Conics*, was reported to consist of four books, with **Apollonius** adding another four books. The *Surface–Loci* was a treatise in two books that focused on the relationships of points on surfaces.

Many works have also been falsely attributed to Euclid, including two books added to the *Elements*, the so–called Books XIV and XV. Others include the *Elements of Music*, the *Sectio Canonis*, and the *Introduction to Harmony*. Ancient Arab scholars also attributed several works on **mechanics** to Euclid, including the "Book of Heavy and Light." However, there is little corroborating evidence that Euclid ever wrote on mechanics.

Euclid

As the author of the most successful textbook ever written, Euclid remains the most influential figure in the history of geometry and, perhaps, all mathematics. Euclid's impact on science included influencing such luminaries as **Galileo** and Sir **Isaac Newton**. **Eratosthenes** also used Euclid's theorems to make accurate measurements of the Earth's **circumference**, confirming its shape as spheroid, or round. From the very beginning, the *Elements* inspired Hellenistic mathematicians to write numerous commentaries, both praising it and pointing out mistakes and inconsistencies. Still, Euclid achieved the status of hero or god in his own time.

EUCLID'S ALGORITHM

Euclid's **algorithm** is a technique used to find the greatest common divisor of two numbers. The greatest common divisor of two numbers is the greatest number that is a divisor or **factor**, a number that divides another number, of those two numbers. Euclid's algorithm for **rational numbers** was given in Book VII of *Elements*, a book written by **Euclid**, the famous Greek mathematician. The corresponding algorithm for **real numbers** appeared in Book X of the same publication. It is the earliest example of an integer relation algorithm which are sometimes referred to as Euclidean algorithms or multidimensional continued fraction algorithms.

Euclid's algorithm is an example of a P-problem, a polynomial time problem whose number of steps is bounded by a polynomial. In this particular case the time complexity is bounded by a quadratic function of length equal to the input values. In Euclid's Book VII of *Elements* he uses the algorithm to find the greatest common divisor of two numbers that are not prime. For this problem, called "Proposition 2" in his book, we let $a = bm + r$ and seek to find a number u which divides both a and b. ($a = su$ and $b = tu$) The number u also will divide r since:

$r = a - bm = su - mtu = (s - mt)u.$

Similarly we can find a number v which divides b and r. ($b = s'v$ and $r = t'v$) In this case the number v will also divide a since:

$a = bm + r = s'vm + t'v = (s'm + t')v.$

This means that every common divisor of a and b is a common divisor of b and r. The procedure can be iterated to find the greatest common divisor of a and b:

$m_1 = a/b$ $a = bm_1 + r_1$ $m_2 = b/r_1$ $b = m_2r_1 + r_2$ $m_3 = r_1/r_2$ $r_1 = m_3r_2 + r_3$ $m_n = (r_{n-2})/(r_{n-1})$ $r_{n-2} = m_nr_{n-1} + r_n$ $m_{n+1} = (r_{n-1})/(r_n)$ $r_{n-1} = m_{n+1}$ $rn + 0.$

When a and b are **integers** Euclid's algorithm terminates when m_{n+1} divides r_{n-1} exactly. At this point r_n is equal to the greatest common divisor of a and b. When a and b are real numbers, Euclid's algorithm gives either an infinite sequence of approximate relations or an exact relation.

Lame's **theorem** deals with the number of steps in the **iteration** of Euclid's algorithm needed to find the greatest common divisor of two numbers. If the two numbers are less than n then the number of steps is:

steps $\leq 4.785\log_{10}n + 1.6723.$

Lame's theorem shows that the worse case, the most number of iterations required, exists when Euclid's algorithm is applied to two consecutive Fibonacci numbers.

Euclid's algorithm was first published in Book VII of Euclid's *Elements* sometime between 330 BC and 265 BC. This book was a compilation of mathematical knowledge that was known and accepted at that time. It became the central teaching book for mathematics for 2000 years. It is thought that no results in *Elements* were first proven by Euclid, but he is attributed with the organization of the material and its exposition. More than 1000 editions of *Elements* have been published since it was first printed in 1482. Over the years various attempts were made to generalize Euclid's algorithm to find integer relations between $n \geq 3$ variables. A generalized form of the algorithm for these types of systems would make obtaining solutions much easier. These attempts failed until 1999 when Ferguson and coworkers discovered the Ferguson-Forcade algorithm. Since that time several other integer relation algorithms have been found.

See also Euclid of Alexandria

EUCLID'S AXIOMS

Euclid's axioms are five postulates about the behavior of geometric objects; they constitute the foundation upon which **Euclid** built the entire edifice of **geometry** that is known today as Euclidean geometry. Before Euclid wrote his famous book *The Elements* around 300 BC, many geometric ideas were well understood, but in a disorganized way that obscured their logical structure. It was often unclear which facts depended on which others for their proofs, and this vagueness opened the door to circular reasoning and other logical errors. To systematize the study of geometry, Euclid formulated five axioms, statements so simple he considered them self-evident, and then attempted to prove all other geometric facts using only these five axioms and the principles of logical reasoning. Euclid's analysis was so definitive and far-reaching that it laid the foundation for the study of geometry for the next 2000 years. Still more, it took the groundbreaking step of subjecting mathematics to the rigors of logic, making *The Elements* one of the milestones in the history of human thought.

Euclid realized that it is impossible to prove anything without starting with a few basic assumptions, or axioms; the ideal in mathematics is to start with as few and simple axioms as possible, and prove as many statements from the axioms as possible. Euclid limited himself to the following five axioms:

- First, any two points can be connected by one and only one straight line.
- Second, any line segment is contained in a full (infinitely long) line.
- Third, given a point and a line segment starting at the point, there is a **circle** that has the given point as its center and the given line segment as a radius.
- Fourth, all right angles are equal to each other (Euclid defines a right **angle** to be the angle formed when two lines intersect each other perpendicularly, that is, forming equal angles on both sides of the intersection).
- Fifth (known as Euclid's **parallel postulate**), given a line and a point P that is not on the line, there is one and only one line through P that never meets the original line.

The fifth is the most controversial of his assumptions, and it has been framed in many different ways; first by the mathematician Proclus in the 5th century.

Using these five axioms, Euclid was able to prove, for example, that two triangles that have all equal side-lengths are congruent; that two triangles that have all equal angles are similar; that a tangent line to a circle is perpendicular to the **diameter** that it intersects; and that in a right **triangle**, the sum of the squares of the lengths of the two legs is equal to the **square** of the length of the **hypotenuse**, the famous **Pythagorean theorem**.

Of Euclid's five axioms, the fifth is conspicuously more complicated than the others. Through the centuries many believed, therefore, that the fifth **axiom** was not fundamental enough to be one of the basic axioms, and should instead be provable using the other four. Euclid himself avoided any use of the fifth axiom in the proofs of the first 28 **propositions** of

The Elements. This evasion has led some historians to the conclusion that even Euclid felt the fifth axiom to be less natural than the others. Many mathematicians, first the Greeks, then the Arabs in the Middle Ages, then the Europeans in the Renaissance, tried to show that the fifth axiom was a consequence of the others, but with no success. In the middle of the 19th century, the mathematicians **Nicolay Lobachevsky**, **János Bolyai** and Eugenio Beltrami independently made a further attempt, using a technique that mathematicians call *reductio ad absurdum* (see **proof by reductio ad absurdum**): to prove that something is true, assume the opposite and follow a logical line of arguments until you reach something ridiculous. So they assumed that, given a line and a point P not on the line, it is possible to have more than one line through P that never meets the original line. With this assumption, they built up a framework of logical consequences, waiting for an absurd conclusion to emerge. There was just one problem: the absurd conclusion failed to appear. Using the strict reasoning of pure logic, they constructed an alternative geometry in which Euclid's first four axioms are true, but the fifth is not. Thus, the quest to prove that the fifth axiom was a consequence of the first four led instead to a discovery that astonished the mathematical world: an entirely new, counterintuitive geometry known as **hyperbolic geometry**. The geometry of Euclid's five axioms is now called Euclidean geometry.

The existence of hyperbolic geometry in no way undermines Euclid's geometry. Euclidean geometry remains as consistent as it ever was; it is simply not the only consistent geometry, as was originally believed. This consistency is not quite perfect, however. Mathematicians have long been aware that Euclid's systemization contains many small flaws—uses of vaguely defined terms and hidden propositions that had not yet been proven. In 1899 the famous mathematician **David Hilbert** set out to correct these gaps with a more complete system of axioms; his system has the defect, however, that it has many axioms and is very complex, making the proofs of even the simplest propositions extremely cumbersome. Euclid's system remains unrivaled in its simplicity and power. The elegance and logical strength of his arguments have led many to regard *The Elements* as the pinnacle of pure reasoning.

See also Euclid of Alexandria; Euclidean geometries; Geometry; Hyperbolic geometry; Non-Euclidean geometries

EUCLIDEAN CONSTRUCTION

The ancient Greeks felt that the line and the **circle** are the basic figures and the straightedge and compass are their physical analogues. Hence they were interested in what can be done with a straightedge and compass, i.e. what figures are Euclidean constructions. For example, given a line l and a point P on l, it is possible to construct the perpendicular to l though P using only a straightedge and compass as follows: with the compass draw a circle C0 with center P. Let A and B be the intersection points of l with C0. Next draw the circles with centers A and B which pass through B and A respectively.

These two circles intersect in two points E and F say. The line segment EF is perpendicular to l.

If one is given segments AB and CD of lengths x and y respectively then one can construct segments of length x + y and x/y as follows. Let's assume for simplicity that x is less than or equal to y. Put the compass point on point A and the tip on point B. Now move the compass point to C and draw the circle of radius x. Draw the line through CD. This lines intersects the circle in two points say E and F. Then F is interior to CD then segment ED has length x + y and segment CF has length y - x. Now draw the circle of radius 1 with center at D. Also draw the perpendiculars to CD which pass through points F and D. The perpendicular through F intersects the circle of radius one in a point G. The segment GC intersects the perpendicular through F in a point H. Since **triangle** CFH is similar to triangle CDG, the length of FH is equal to y/x.

The Greeks knew that if one draws a **square** with side length x then the diagonal has side length equal to x times the **square root** of 2 which is the side length of a square with twice the **area** of the given square. So they wondered if it was possible to construct a segment of length equal to x times the positive cube root of 2 since this is the side length of a cube with twice the **volume** of a cube with side length x. That problem is called "doubling the cube". They also asked if it was possible to trisect any given **angle** (bisecting is easy; see **Trisecting an angle**) and if given any circle can one construct a square with the same area. The latter is called "**squaring the circle**". They could do all these things with other instruments (for example there is a linkage that can be used to trisect any angle).

It was not until 1837 that it was proven (by Wantzel and Gauss) that all three of the above problems are impossible. Today, students learn how to prove their impossibility using **Galois theory**. Most of the works of **Evariste Galois** (1811-32) were lost before anyone else read them. Before he died in a duel at age 21, he summarized his theories in a note to a friend, August Chevalier. Today his ideas are frequently applied in the fields of **number theory** and **algebraic geometry**.

Most theorems about constructions can be derived from a study of K the set of constructible numbers which is a subset of C the **complex numbers**. C can be identified with two dimensional real **space** by the map which assigns the number a + bi (where a and b are real and i is a square root of -1) to the point with coordinates (a,b). The **definition** of K is that it is the set of all points in C which can be obtained from straightedge and compass constructions from the initial point 0 and 1. It can be shown that if x and y are in K and y is nonzero then x + y, x-y, x*y, x/y and the square root of x (or y) is in K. So K contains the **rational numbers** and i. If points A, B, C and D have been constructed and X and Y are either circles or lines constructed directly from A, B, C and D then it is not hard to show that the points in the intersection of X with Y satisfy a quadratic or linear equation with coefficients that are derived from A, B, C and D by a combination of **multiplication**, **division**, **addition**, **subtraction**, and taking square **roots**. Hence K is the smallest subset of C containing 0 and 1 which has the property that if x and y are in K and y is nonzero then x + y, x-y, x*y, x/y and the square root of x (or y) is in K.

Doubling the cube of side length one requires constructing the cube root of 2. Squaring the circle of radius one requires constructing the square root of **Pi**. It can be shown using elementary **trigonometry** that if A is any angle then cos A = 4*cos(A/3) - 3*cos(A/3). Hence to trisect a sixty degree angle, one must construct the roots of the equation 8*x³ - 6*x - 1. Using Galois theory (and the fact that Pi is a **transcendental number**) it is possible to show that the cube root of 2, the square root of Pi, and the roots of 8*x³ - 6*x - 1 are not in K. Hence the famous three problems are impossible.

See also Algebraic geometry; Galois theory

EUCLIDEAN GEOMETRIES

Euclidean **geometry**, sometimes called parabolic geometry, is a geometry that follows a set of **propositions** that are based on Euclid's five postulates (see **Euclid's axioms**), as defined in his book *The Elements*. More specifically, Euclidean geometry is different from other types of geometry in that the fifth **postulate**, sometimes called the **parallel postulate**, holds to be true. Non-Euclidean geometry replaces this fifth postulate with one of two alternative postulates and leads to **hyperbolic geometry** or elliptic geometry. There are two types of Euclidean geometry: **plane geometry**, which is two-dimensional Euclidean geometry, and **solid geometry**, which is three-dimensional Euclidean geometry.

Euclid's five postulates can be stated as follows:

1. It is possible to draw a straight line segment joining any two points.

2. It is possible to indefinitely extend any straight line segment continuously in a straight line.

3. Given any straight line segment, it is possible to draw a **circle** having the segment as a radius and one endpoint as its center.

4. All right angles are equal to each other or congruent.

5. If two lines are drawn so that they intersect a third in such a way that the sum of the interior angles on one side is less than two right angles, then those two lines, if extended far enough, must intersect each other on that particular side.

The fifth postulate is equivalent to what is known as the parallel postulate. The parallel postulate states that given any straight line segment and a point not on that line segment, there exists one and only one straight line which passes through that point and never intersects the first line, no matter how far the line segments are extended. Although Euclid's fifth postulate cannot be proven as a **theorem**, over the years many purported proofs were published. Much effort was devoted to formulating a theorem to this postulate since it was needed to prove important results and it did not seem as intuitive as the other postulates. After more than two millennia of study the fifth postulate was found to be independent of the other four. It is this fifth postulate that must hold for geometry to be considered Euclidean. In 1826 **Nikolay Lobachevsky**, and, in 1832, **János Bolyai**, independently developed entirely self-consistent **non-Euclidean geometries** in which the fifth

postulate did not hold. **Johann Carl Friedrich Gauss** had previously discovered this but had not published his results. Euclid tried to avoid using this fifth postulate and was successful in the first 28 propositions of *The Elements,* but for the 29th proposition he needed it. That part of geometry that can be derived using only the first four of Euclid's postulates came to be known as absolute geometry. As stated above, the fifth postulate and therefore the parallel postulate describe Euclidean geometry. If part of the parallel postulate is replaced by "no line exists which passes through that point" then elliptic or spherical geometry is described. If part of the parallel postulate is replaced by "at least two lines exist that pass through that point" then hyperbolic geometry is described.

As was stated above, the two types of Euclidean geometry, plane geometry and solid geometry, are distinctly different. Plane geometry is that portion of geometry in two-dimensional **space** that deals with figures in a plane, such as the circle, line and polygon. Solid geometry is that portion of geometry in three-dimensional space that deals with solids, such as polyhedra, spheres, and lines and planes. In both types of Euclidean geometry Euclid's fifth postulate holds—but each describes figures in different types of space. Euclidean space is the space of all n-tuples of **real numbers** and is denoted as R^n. This space is a vector space and has topological **dimension** (Lebesgue covering dimension) n (see **topology**). Contravariant and covariant quantities are equivalent in Euclidean space. R^1 is the real line, that is a line with a fixed scale on which numbers correspond to unique points on the line. The generalization of the real line in two-dimensional space is called the Euclidean plane and denoted R^2.

See also Euclid of Alexandria; Geometry; Non-Euclidean geometries

EUDOXUS OF CNIDUS (408 B.C.-355 B.C.)
Greek mathematician

Eudoxus was a Greek mathematician from Cnidus, in what is now Turkey. He is famous for his work on calculating the **area** bounded by a curve as well as the ladder of Eudoxus. It is also believed that some of the work of **Euclid of Alexandria** is derived from lost work of Eudoxus.

Eudoxus was given a good mathematical education: one of his main teachers was **Archytas of Tarentum**, who was in turn a follower of **Pythagoras**. In his education Eudoxus also studied medicine with Philiston of Sicily, and Theomedon. While in Athens Eudoxus also attended discussions with **Plato** and other philosophers. After this period, presumably on the advice of Plato, Eudoxus spent nearly two years in Egypt studying astronomy. After this Eudoxus felt his basic apprenticeship had been completed, and on his return to Asia Minor he opened up a school that proved very successful. In 386 B.C. Eudoxus and a number of mathematicians from the school visited Athens and Plato. Finally Eudoxus returned to his beloved Cnidus where he was pressed into service as a congressman. This work in running the city still allowed him to continue his other pursuits of mathematics, astronomy, and philosophy. Eventually Eudoxus died in 355 B.C. carrying on the above mentioned work right until the end.

We know that Eudoxus wrote several books covering such topics as mathematics, astronomy, and geography; unfortunately no copies of his work exist. What records we do have of the work of Eudoxus are in the acknowledgements of other, later authors. These later authors include **Hipparchus**, writing on astronomy, specifically the rising and setting of constellations and planetary movement (Eudoxus postulated a series of concentric spheres holding the planets in their paths). Euclid of Alexandria quoted Eudoxus when writing on relationships between **ratios** (the foundation of **real numbers**, which moved away from the Pythagorean love affair with **irrational numbers**). It is possible that Eudoxus also calculated an accurate length for the solar year, which was a precursor to the later **calendar** reforms of Julius Caesar. More recent mathematicians have also acknowledged the inspiration that has been given by Eudoxus; these include such people as the early 20th-century mathematician **Julius Wilhelm Richard Dedekind**.

See also Democritus of Adbera; Diophantus of Alexandra; Eratosthenes of Cyrene; Euclid of Alexandria; Hero of Alexandria; Hipparchus; Hippocrates of Chios; Hypatia of Alexandria; Menaechmus; Nicomachos of Gerasa; Pappus of Alexandria; Plato; Claudius Ptolemy; Pythagoras; Thales of Miletus; Zeno

EULER CHARACTERISTIC

The Euler characteristic is a number that can be associated to a **polyhedron** that gives an indication of its topological qualities (qualities of an object that don't change when the object is stretched or distorted, without tearing). The characteristic number was first studied by the great mathematician **Leonhard Euler**, who wrote his famous formula for a polyhedron in a letter in 1750 to **Christian Goldbach**: v-e+f=2, where v is the number of vertices (corners) of the polyhedron, e is the number of edges, and f is the number of faces (flat sides). Thus, for example, a cube has 8 vertices, 12 edges, and 6 faces, and 8-12+6=2. Although polyhedra had been studied quite intensely before the time of Euler by such great mathematicians as **Archimedes** and **Descartes**, historians believe that none of them noticed this simple relationship. Euler's work was something of a departure from the usual study of polyhedra, since it did not depend on the usual length measurements of **geometry**, but only on properties of polyhedra that are unchanged under distortion—this was a new focus that was to lead to the field of mathematics now known as **topology**.

In Euler's papers, he assumed that the polyhedra were convex (an object is convex if, for every pair of points in the object, the line connecting them is also in the object). This at first does not appear to be a substantial restriction, since the cube, tetrahedron, octahedron, and the other familiar polyhedra are all convex. However, many polyhedra are non-convex, and in 1813 Antoine-Jean L'huilier discovered than when a

polyhedron is not convex, Euler's formula is not always true. For example, a polyhedron that is shaped like a box with a tunnel through the middle (so that it looks something like a distorted donut) satisfies the relationship v-e+f=0. In fact, any polyhedron that is shaped like a distorted donut (in mathematical language, a **torus**) satisfies v-e+f=0. L'huilier discovered further that when a polyhedron was shaped like a distorted donut with g holes, then v-e+f=2-2g.

With L'huilier's work it became clear that the number v-e+f was a measurement not so much of a particular polyhedron, but more of its topological shape. All of the polyhedra studied by Euler—the cube, the octahedron, and so forth—are topologically equivalent to the **sphere**, since by poking them here, pulling them there, you can make each one look spherical without ever tearing it. It can be proven that any polyhedron that is topologically equivalent to the sphere satisfies v-e+f=2; in mathematical language, the sphere has a Euler characteristic equal to 2. In the same way, the torus has an Euler characteristic equal to 0.

It is not hard to show that continuous deformations of a surface (distortions that never tear the surface or glue parts of it together) never change the Euler characteristic. Thus, the Euler characteristic is, historically, the first example of what is known as a topological **invariant**. It can be used to distinguish between topologically distinct surfaces—if two surfaces have different Euler characteristics, then they must be topologically different: the only way to distort one of them into the other is through tearing and gluing.

When the two surfaces are the sphere and the torus, it is easy just by looking at them to see that they are topologically different, without calculating their Euler characteristics. However, mathematicians often wish to study a surface that is given not by a concrete picture but instead by a collection of maps of the surface (this is rather like our understanding of the earth. We never see the entire globe at one time, but, through looking at maps and knowing how the maps fit together, we can form a good impression of the shape of the earth.) If a surface is given by a collection of maps, it is not always easy to tell by inspection what kind of surface it is; but a simple Euler characteristic calculation will determine whether it is a sphere, a torus, or a torus with more than one hole (we are excluding from this discussion non-orientable surfaces, which add a few technicalities. For more information, see entries on the **Möbius strip** and the **Klein bottle**).

The Euler characteristic has analogues in higher dimensions. A three-dimensional analogue of a polyhedron is a shape made up of solid polyhedra that meet each other along their faces, edges, and vertices. The Euler characteristic of such a shape is v-e+f-s, where v is the number of vertices, e the number of edges, f the number of 2-dimensional faces, and s the number of solid pieces. In higher dimensions, a Euler characteristic can also be defined in a similar fashion. Once again it can be shown that the Euler characteristic is a topological invariant—it does not change when the object is distorted. In high dimensions it becomes much more difficult to visualize objects than in **dimension** 2, and hence the Euler characteristic increases in value, because it can be calculated even in situations where the actual object cannot be visualized. The Euler characteristic belongs to the branch of topology now called homology theory, which attempts to distinguish between different topological spaces by examining polyhedra that are contained in them.

See also Manifold; Topology

EULER, LEONHARD (1707-1783)

Swiss geometer and number theorist

Leonhard Euler advanced every known field of mathematics in his day. A prolific author, among his greatest writings are treatises on **analytic geometry**, differential and integral **calculus**, and the calculus of variations. Euler developed spherical **trigonometry**, demonstrated the importance of **convergence** in algebraic series, proved important assertions in **number theory**, and made contributions to **hydrodynamics**, celestial **mechanics**, and optics. Euler also brought into common usage such mathematical notations "**e**" for the base of the natural logarithm, "i" for the **square** root of negative 1, and $f(x)$ for a function of x. A variety of mathematical concepts bear his name, including the **Euler characteristic** in **topology**, Euler's triangle in **geometry**, Euler's **polynomials**, Euler's integrals, and Euler's constant. His accomplishments are especially remarkable in that many were made during the last quarter of his life, when he was totally blind.

Euler was born in Basel, Switzerland, on April 17, 1707 to Paul Euler and Marguerite Brucker. In 1708, his family moved to the nearby village of Reichen, where his father, a Calvinist pastor, had taken a parish. Before joining the clergy, Euler's father had studied mathematics under the tutelage of **Jacob Bernoulli**. Following in his father's footsteps, Euler also took his formal education in religion and mathematics, studying theology and Hebrew at the University of Basel, and taking weekly mathematics lessons from **Johann Bernoulli**, Jacob Bernoulli's younger brother. The Bernoullis recognized Euler's talent and when he received his master's degree from the University of Basel at age 17, they advised him to pursue a mathematical career. The advice was met with resistance from Euler's father, who wished for his son to inherit the pastorship in Reichen. Euler was to remain a devout Calvinist throughout his life, but the Bernoullis eventually convinced his father that Euler's true destiny was not with the church.

When Euler was 19 years old, he produced his first mathematical work, entering a contest sponsored by the French Académie Royale des Sciences. The object of the contest was to solve a problem related to the optimum placement of masts on sailing ships. Euler received an honorable mention for his effort, his solution suffering primarily in the area of practicality. Having not yet traveled outside of Switzerland, he had never seen a ship. Over the course of his career, Euler would eventually receive a total of twelve prizes from the French Académie for his mathematical solutions.

Around the time that Euler was attempting to solve the ship mast problem, he was also trying, unsuccessfully, to obtain a post as a professor of mathematics at the University of Basel. Determined to hold an academic position, he corresponded with friends **Daniel Bernoulli** and his cousin, Nicolaus, who were members of the newly established St. Petersburg Academy of Sciences. They wrote to him about a post available in the medical section of the Academy, and, hoping to qualify, Euler immediately began studying physiology. Within three months he was considered sufficiently prepared for the medical post, and in 1727 he traveled to St. Petersburg to join the Academy. Euler's arrival in Russia coincided with the death of Catherine, the wife of Peter the Great. A period of political oppression ensued, lasting several decades, and in the initial turmoil Euler slipped quietly into the Academy's mathematical section. Academic and political freedoms eventually became so stifled that, in 1733, **Daniel Bernoulli** decided to leave Russia and return to Switzerland. Euler, then 26 years old, inherited Bernoulli's post, the top mathematical position in St. Petersburg. Two years later, he lost the vision in his right eye. According to some historians, Euler developed an eye infection while solving an astronomical problem that had been put forth by the French Académie. It is possible that he injured his eye by staring into the sun while working on the problem. Euler derived a solution in the course of only three days, and won the Académie's contest.

In 1736, Euler wrote a paper on the solution of the Königsburg Bridge Problem, a puzzle concerning attempts to cross seven different bridges in one journey (see **Bridges of Königsburg**). This work led to the development of the modern field of graph theory. Between 1736 and 1737, Euler wrote *Mechanica*, in which he demonstrated that mathematical **analysis** could be applied to Newtonian dynamics. This treatise and the wealth of articles he had already published secured his mathematical prominence. By the end of the 1730s, he had also established a reputation as a gifted educator, having written both elementary and advanced mathematical textbooks for the Russian schools. As a member of the Russian Academy, Euler was called upon to solve many practical problems for the benefit of the Russian government. He created a test for determining the accuracy of scales, developed a system of weights and measures, and supervised the government's department of geography. Although political oppression in Russia continued, Euler was never restricted in the pursuit of his own mathematical interests. However, he was growing increasingly weary of the injustices that surrounded him. In 1740, Euler accepted an invitation from Frederick the Great, the Prussian king, to join the Berlin Academy. He left Russia on sufficiently good terms, however, that throughout his tenure in Berlin the St. Petersburg Academy provided part of his salary. He was to remain at his Berlin post for the next 24 years.

Frederick the Great, while lacking in his own mathematical ability, did appreciate the utility of mathematics. He directed Euler to work on calculations related to diverse practical matters, including pension plans, navigation, water supply systems, and the national coinage. In Berlin, Euler accomplished what many consider his most important work. He wrote *Methodus inveniendi lineas curvas maximi minimive proprietate gaudentes*, on the calculus of variations in 1744, and its publication led, in 1746, to his election as a Fellow of the Royal Society of London. This masterpiece was followed by several texts on calculus that were to become instant classics. These included *Introductio in analysin infinitorum*, in 1748, and *Institutiones calculi differentialis*, in 1755.

While in Berlin, Euler corresponded with many of his mathematical contemporaries, including **Johann Lagrange** and **Jean d'Alembert**. He was introduced to the field of number theory by **Christian Goldbach**, who presented him with the various challenges of **Pierre de Fermat**. Euler proved many assertions in the field of number theory and was the first mathematician to make serious progress in solving **Fermat's Last Theorem**. In a letter to Goldbach in 1753 he described a partial **proof**, ultimately shown to contain a fallacy, but which laid the foundation for its eventual solution.

In 1766, at the age of 59, Euler returned to Russia. He had fallen into gradual disfavor in King Frederick's court because of the positions he took in metaphysical arguments with contemporaries such as Voltaire. The king eventually concluded that Euler was unsophisticated, and took to calling him a "mathematical Cyclops," in reference to his partial loss of vision. When Euler went back to St. Petersburg, he was greeted with much greater esteem. Catherine the Great, now in power, provided him with a large estate and one of her personal cooks.

Not long after resettling in Russia, Euler developed a cataract in his left eye and totally lost his vision. He nonetheless entered one of the most prolific periods of his career. Nearly half of the 886 books and manuscripts Euler wrote were composed during this second tenure at the St. Petersburg Academy. From 1768 to 1770, he drafted a classical treatise on **integral calculus**, *Institutiones calculi integralis*. He went on to tackle the lunar theory problem, researching the phases of the moon and the tidal fluctuations on Earth. His calculations relating to the gravitational interactions among the moon, sun, and Earth won him a 300 pound prize from the British government.

Euler was known for his remarkable memory. As a boy, he had memorized the entire text of Virgil's *Aeneid*, and 50 years later could still recite it. His ability to perform complex calculations in his head was also renown. Once, when two of his students disagreed on the answer to a problem that required they sum a complicated convergent series to 17 terms, Euler settled the matter using only mental **arithmetic**. His memory and mental calculation skills undoubtedly allowed him to cope with the blindness during the latter part of his life.

Euler was married in 1733 to Catharina Gsell, the daughter of the Swiss painter Gsell, that Peter the Great had brought to Russia. They had 13 children, of whom only three sons and two daughters survived beyond their early years. Catharina died in 1776, and a year later Euler married her aunt and half–sister, Salome Abigail Gsell. He was known as a kind and generous man. Euler was especially fond of children, often writing mathematical treatises with a child on his lap. On September 18, 1783, while playing with his grandson, he suffered a stroke and died. Just before his death he had calculated the **orbit** of the newly discovered planet Uranus.

EULER-LAGRANGE EQUATION

The Euler-Lagrange equation is a fundamental equation of the **calculus** of variations that is used to determine if a function has a stationary value. In physical terms a stationary value corresponds to a minimum or maximum of a function. The Euler-Lagrange equation is usually written as: $(\partial f/\partial y) - (d/dt)(\partial f/\partial y) = 0$. One of the fundamental **equations** of calculus of variations states that for a function P defined by an integral of the form $P = \int f(x, y, y)dx$, where $y = dy/dt$, there is a stationary value if the corresponding Euler-Lagrange equation for that function is satisfied. When the time derivative notation is replaced by space **variable** notation often times, in physical problems, the partial derivative of f with respect to x is equal to 0. When this is the case the Euler-Lagrange equation is reduced to a simplified form known as the Beltrami identity: $f - y_x(\partial f/\partial y_x) = c$, where c is a constant. The Euler-Lagrange equation plays a pivotal role in the calculus of variations since this generalization of calculus focuses on finding the path, curve, surface, etc., for which a given function has a stationary value. Solving an appropriate Euler-Lagrange equation often lends itself to solutions of problems in the calculus of variations.

The Euler-Lagrange equation arose out of studies conducted by Swiss mathematician **Leonhard Euler** and Italian mathematician **Joseph-Louis Lagrange** in the mid 1700s. During the early 1750s Lagrange began studying the tautochrone, the curve on which a weighted particle always arrives at a fixed point in the same time regardless of its initial position. His discoveries concerning this curve were substantial to the calculus of variations. In 1755 Lagrange sent Euler his results on the tautochrone which contained his ideas for methods determining maxima and minima. A year later Lagrange sent Euler the results of applying the calculus of variations to **mechanics**. These results were a generalized form of results Euler had already obtained himself. From these communications rose the Euler-Lagrange equation. Lagrange, using this equation, went on to publish works in the foundations of dynamics which were based on the principle of least action and on kinetic energy. In 1766 Euler officially named these studies calculus of variations.

The Euler-Lagrange equation and calculus of variations are well known in classical mechanics since they are often used to determine minimum and/or maximum points of **functions**. Studying critical behavior such as in surface tension phenomena often employs Euler-Lagrange equations and calculus of variations. A generalized form of calculus of variations, called Morse theory, employs nonlinear techniques in studying variational problems. This particular theory relies on the thought that there is a relationship between the stationary points of a smooth, real-valued function on a **manifold** and the global **topology** of that manifold. Morse theory was related to quantum **field** theory in 1982 by a paper published by **Edward Witten**.

See also Calculus

EULER'S LAWS OF MOTION

Swiss mathematician **Leonhard Euler** formulated two laws to describe the **motion** of a rigid body relative to an inertial reference frame. The first law describes how applied forces change the velocity of the **center of mass** of a rigid body. The second law describes how the change in angular momentum of a rigid body is influenced by the moment of the applied **force** and any attachments to other bodies an object may have. These laws were written for bodies of fixed matter. That is the body must be composed of a fixed amount and type of matter that does not change.

The first of Euler's two laws is written as $\Sigma F = d/dt(G)$ where F is each individual force applied to the body and G is the linear momentum of the body. This form is written relative to an inertial reference frame. The points of the inertial reference frame in this case have no acceleration, neither translational nor rotational. For a particle with mass m its linear momentum G is given by its mass times its velocity, $G = mv$. For a body composed of many particles its linear momentum is assumed to be the sum of the linear momentum of its particles. $G = \Sigma m_i v_i$, where m_i is the individual mass of a particle and v_i is the velocity of that particle. So for a continuous body the summation can be replaced by $G = \int_m v\,dm$ The location of the center of mass of a body composed of several particles relative to an inertial frame of reference can be found from $r_{CM} = 1/m \int_m r\,dm$, where r_{CM} is the position of the center of mass relative to the inertial frame and r is the position of a particle of the body relative to the inertial frame. When the time derivative of this equation is taken we obtain a form that contains velocity of the center of mass relative to the inertial frame: $mv_{CM} = \int_m v\,dm$, where v_{CM} is the velocity of the center of mass with respect to the reference frame and m is the total mass of the body. So the linear momentum of a body is given by: $G = mv_{CM}$. This by substitution yields an alternate form of Euler's first law of motion for a rigid body: $\Sigma F = d/dt(mv_{CM})$. Since mass is constant in Newtonian **mechanics** this becomes $\Sigma F = ma_{CM}$, where a_{CM} is the acceleration of the center of mass. For a single particle the location of the center of mass and the location of the particle are the same and so Euler's first law of motion is the same as Newton's second law.

Euler's second law describing how the moment of an applied force changes the angular momentum of a rigid body is written as $\Sigma M_0 = d/dt\,(H_0)$, where M_0 is the moment of an applied force of the body relative to the frame of reference and H_0 is the angular momentum of the body relative to the reference frame. This inertial frame of reference can be removed if the center of mass is used so that the second law is rewritten as $\Sigma M_{CM} = d/dt\,(H_{CM})$. This form of Euler's second law is especially useful if the body is accelerating since the calculation can be performed relative to the body's center of mass as opposed to some inertial reference point.

Euler used these laws to develop an understanding of the motion of the Moon around the Earth. At that time, the motion of the Moon around the Earth was not seen as governed by the same laws that governed the motion of the planets around the Sun. **Newton** developed ideas concerning the law of universal gravitation and his three laws of motion and used these as the basis for the *Principia*. These laws were used to explain planetary motion around the Sun but Newton thought the three-bodied problem too complex to be solved. Even if the Earth-Moon system were considered a two-bodied problem the orbits would not be simple ellipses since neither of them is a perfect **sphere** and so does not behave as a point mass. Euler developed methods of integrating linear **differential equations** in 1739 and this led him to draw up lunar tables in 1744 demonstrating the gravitational attraction of the Earth, Moon, Sun system. In the 1750s a small perturbation in the precession of the Earth's axis of **rotation** caused by the gravitational **field** of the Moon led Euler to develop his mechanics of rigid bodies and formulate his laws of motion. From 1760 onwards Euler was the first to study the general three-bodied problem under mutual gravitation. The Paris Academy Prize of 1772 was won jointly by Lagrange and Euler for their work on the **orbit** of the Moon.

EXISTENCE

The term "existence" is a good example of how something familiar can take on a slightly different technical meaning in a mathematical context. If a book exists, for example, it has been written, and usually published. There is a physical manifestation of that book somewhere. But if a mathematical solution to a system exists, it is not automatically implied that there is any example of that solution which is physically observable. The existence of that solution simply means that it is a possibility, logically and mathematically, for that solution to happen. According to the formalist **definition**, something which exists is free from internal contradictions (see **Formalism**).

If it is mathematically impossible for something to exist, a logical contradiction must result from that item's existence: true = false, say, or 0 = 4. For it to be physically impossible for something to exist, on the other hand, some law of the universe must be broken by that item's existence. For example, a particle which had mass and traveled at the speed of **light** would be in violation of our current knowledge of **relativity**. However, there is nothing inherently and ultimately logical about relativity. It is easy to imagine an internally consistent system with different parameters than ours has. Such a system is thus not mathematically impossible to create. It is merely physically impossible to create. However, all things that physically exist must also mathematically exist.

This existence difference is reflected in the way one resolves the existence of something apparently impossible. If one is looking at something which seems logically or mathematically impossible, the proper course of action is to find where the definitions had gone wrong, since usually the problem is that two different logical descriptions or definitions have been used, or the situation has been misunderstood. If a physically impossible event is occurring, and it does not look to be a logical contradiction, the proper course of action at that point is to look for where human understanding of the laws of nature has gone wrong. Much scientific progress occurs due to events that are not supposed to have physical existence at all. Seemingly mathematically impossible events, on the other hand, are simply human mistakes requiring clarification.

Unfortunately, this formalist definition of existence, while generally functionally acceptable, runs into contradictions in generalized situations. Specifically, **Gödel**'s incompleteness **theorem** shows that it is impossible to construct an entirely complete and self-consistent set of postulates from which the rest of mathematics may be derived. So while the freedom from contradictions seems to be necessary for something to mathematically exist, it is not sufficient, and a system of mathematics cannot be wholly derived from noncontradiction. Some mathematicians, called intuitionists, insist that any **proof** must demonstrate the existence of something by constructing an example of it, rather than by showing that it contains no internal inconsistencies. Of course, this is somewhat more difficult.

See also Necessary condition

EXPONENT

Exponents are numerals that indicate an operation on a number or **variable**. The interpretation of this operation is based upon exponents that symbolize natural numbers (also known as positive **integers**). Natural-number exponents are used to indicate that **multiplication** of a number or variable is to be repeated. For instance, $5 \times 5 \times 5$ is written in exponential notation as 5^3 (read as any of "5 cubed," "5 raised to the exponent 3," or "5 raised to the power 3," or just "5 to the third power"), and $x \times x \times x \times x$ is written x^4. The number that is to be multiplied repeatedly is called the base. The number of times that the base appears in the product is the number represented by the exponent. In the previous examples, 5 and x are the bases, and 3 and 4 are the exponents. The process of repeated multiplication is often referred to as raising a number to a power. Thus the entire expression 5^3 is the power.

Exponents have a number of useful properties:

- 1) $x^a \times x^b = x^{a+b}$
- 2) $x^a \div x^b = x^{a-b}$
- 3) $x^{-a} = 1/x^a$
- 4) $x^a \times y^a = (xy)^a$
- 5) $(x^a)^b + x^{(ab)}$
- 6) $x^{a/b} = (x^a)^{1/b} = {}^b\sqrt{x^a} = ({}^b\sqrt{x})^a$

Any of the properties of exponents are easily verified for natural-number exponents by expanding the exponential notation in the form of a product. For example, property number (1) is easily verified for the example $x^3 x^2$ as follows:

$$x^3 \times x^2 = (x \times x \times x) \times (x \times x) = (x \times x \times x \times x \times x) = x^5 = x^{(3+2)}$$

Property (5) is verified for the specific case $x^2 y^2$ in the same fashion:

$$x^2 \times y^2 = (x \times x) \times (y \times y) = (x \times y \times x \times y) = (x \times y) \times (x \times y) = (xy)^2$$

Exponents are not limited to the natural numbers. For example, property (3) shows that a base raised to a negative exponent is the same as the multiplicative inverse of (1 over)

the base raised to the positive value of the same exponent. Thus $2^{-2} = 1/2^2 = 1/4$.

Property (6) shows how the operation of exponentiation is extended to the **rational numbers**. Note that unit-fraction exponents, such as 1/3 or 1/2, are simply **roots**; that is, 125 to the 1/3 power is the same as the cube root of 125, while 49 to 1/2 power is the same as the **square** root of 49.

By keeping properties (1) through (6) as central, the operation is extended to all real-number exponents and even to complex-number exponents. For a given base, the real-number exponents map into a continuous curve.

EXPONENTIAL FUNCTIONS

Exponential **functions** have the form $f(x) = ab^x$ where a is a non-zero real number, b is a positive real number not equal to 1, and x can take on all real number values. Examples of exponential functions include $f(x) = 2^x$, $g(x) = 3(0.5)^x$, and $h(t) = 1300(1.02)^t$. In general, the graph of $f(x) = ab^x$ has a y-intercept at (0,a) and is asymptotic to the x-axis from above if a is positive and from below if a is negative. If b is greater than 1, then the graph is asymptotic to the x-axis as x decreases without bound and increases without bound as x increases without bound. If b is less than 1, then the graph is asymptotic to the x-axis as x increases without bound and increases without bound as x decreases without bound.

Exponential functions have a wide range of applications in finance, economics, biology, ecology, physics, and other sciences. A simple example is a bank account in which a certain initial amount of money is deposited at a given interest rate and we wish to track the growth of the money in this account over some specified time period. For example, if $1000 is deposited in an account earning 5% interest compounded annually, then the exponential function $f(t) = 1000(1.05)^t$ gives the amount of money in the account after t years. So at the end of 10 years we would calculate $1000(1.05)^{10} = \$1628.89$. Scientists from various fields use exponential functions as models for growth and decay phenomena in which a quantity is assumed to grow or decay at a rate which is proportional to the amount of the quantity at any given time. This assumption leads to a differential equation of the form $dq/dt = kq(t)$ where $q(t)$ is the amount of the quantity present at time t. In **calculus** courses it is shown that this differential equation has a solution of the form $q(t) = q(0)e^{kt}$ where e is the base of the so-called natural logarithm system and is approximately 2.71828. The equation $q(t) = q(0)e^{kt}$ is, of course exponential. When k is positive, the equation models growth; when k is negative, it models decay. As an example, suppose that a biologist is studying the growth of a certain colony of bacteria. She estimates that 1000 bacteria are present at her initial observation, and that an hour later the population has tripled. The equation she can use to **model** this growth has the form $q(t) = 1000e^{kt}$ where k can be determined by using the fact that the population tripled in one hour. In this case $k = \ln(3)$, the natural logarithm of 3, and the simplest form of the exponential model is $q(t) = 1000(3)^t$. So the biologist could predict that after 5 hours the population would be approximately $1000(3)^5 = 243,000$. Physicists use exponential functions to study the decay of radioactive substances. Such functions can be used to approximate the age of ancient archeological organisms by estimating the amount of, say, radioactive carbon-14 present today in the object, knowing the amount of C-14 which would have been in the object when it was alive, and using the exponential model to solve for the number of years since the organism was alive.

The power of exponential growth can be illustrated by the following simple example. Suppose you are taking on a certain job for one month and you are given the choice of being paid a fee of $20,000 for the month or being paid 1 penny on the first day, 2 pennies on the second day, 4 pennies on the third day, 8 pennies on the fourth day, and so on with the number of pennies earned doubling each day through the end of the month. A person who does not understand the nature of exponential growth might opt for the flat $20,000 fee. However, the exponential function $y = 2^{x-1}$ will give the number of pennies earned on day x. So on the 31st day of the month, you would be paid 2^{30} pennies or $10,737,418.24! That's right, more than ten million dollars! Note that this is just the amount paid on the last day of the month and does not include the total from the previous 30 days! In a more practical example, suppose you invest $1000 in a stock market fund which has historically earned 12% per year, about average for the stock market. In 30 years, you would have $1000(1.12)^{30} = $29,959.92 and that's if you added nothing else to your account during the 30 years.

Exponential functions and modifications of exponential functions are currently being used to study such phenomena as the growth of the internet, the spread of AIDS, the projected growth or decay of the national debt, and much more. Wherever growth and decay are studied, exponential models are sure to be a fundamental part of the study.

EXTRANEOUS SOLUTIONS

Extraneous solutions are those solutions encountered when solving an equation or system of **equations** that at first glance appear reasonable, but do not actually satisfy the original conditions of the problem. Extraneous solutions can occur with any type of equation—including differential equations—but most often occur when a problem involves applying the **square**, quartic, or any other even-numbered power to both sides of an equation. Extraneous solutions also sometimes occur when the problem involves a radical term (that is, a term with a non-integer exponent) or trigonometric function, or when the restrictions on the domains of terms in the equation(s) are not well-considered. Extraneous solutions are also quite common when a mathematical **model** is applied to a real-world event, scientific experiment, or engineering problem, as models generally apply only to a well-defined **domain**, such as the time between the start of a race and the time that the last competitor crosses the finish line. Extraneous solutions are particularly common with a quadratic equation.

How an extraneous solution is created in a just few simple steps is easily demonstrated. Consider, for example, the equation $x = 1$. Squaring both sides of the equation, which is

a perfectly legitimate operation, results in the equation $x^2 = 1$. However, the second equation ($x^2 = 1$) has two solutions—$x = 1$ and $x = -1$! Clearly, $x = -1$ is not a valid solution of the first equation (that is, $x = 1$ (-1)). The solution $x = -1$ is extraneous; although the solution $x = -1$ was derived as the result of applying a perfectly legitimate operation to an equation, the outcome of the operation led to two results—one that was correct and one that was incorrect. Failing to isolate and then eliminate extraneous solutions is a common failing in mathematical problem-solving, which can lead to false assumptions and operating parameters. Such missteps are particularly troublesome in computer programming, as a computer program may generate and utilize many mathematical solutions without a human in the loop to review the reasonableness (that is, correctness) of the solution. Establishing tight parameters in which a computer-generated solution will be accepted and then used in a subsequent part of the computer program is an important element in locating and correcting errors.

One of the most common mathematical models that results in an extraneous solution is the path of an object through free **space** (usually considered a vacuum or frictionless air, which is physically unattainable). Using the **Cartesian coordinate system** and parametric equations—as well as assuming the object is moving through a frictionless medium—the path of an object can be modeled by the equations $x - x_0 = (v_0 * \cos\theta_0) * t$ and $y - y_0 = (v_0 * \sin\theta_0) * t - 1/2\, gt^2$, where x_0 is the initial displacement in the x-direction, y_0 is the initial displacement in the y-direction, v_0 is the initial displacement of the object, t is time, g is the gravitational constant, and θ_0 is the initial **angle** of **projection** for the object. (Note that x_0, y_0, v_0, g, and θ_0 are all constants, so they do not vary with time.)

In the example, suppose that $x_0 = 0$ (that is, there is no displacement x-direction at $t = 0$, such as the starting line for a shotput competition), that $y_0 = 5$ feet (the shoulder height of a shotput athlete), that $z_0 = 30$ degrees, that the initial velocity v_0 is 40 feet per second, and g, the gravitational constant, is approximately equal to 16 feet per second squared. For the y-variable, which represents the height of the shotput as it travels through the air, the equation becomes $y - 5 = (40 * \sin 30°) * t - 1/2 * 16 * t^2$, or $y = -8t^2 + 20t + 5$. If one solves this equation for $y = 0$ (that is, the height of the ground), one finds t is approximately equal to -0.23 seconds and 2.73 seconds. The first solution, however, is not possible, as it does not account for the fact that the shotput started its parabolic path of travel at the athlete's shoulder height, not the ground. The second solution is the amount of time elapsed before the shotput reaches the ground after it is tossed. In the first case, the solution represents a solution in which the model of the object's path is not appropriate and is therefore an extraneous solution.

See also Cartesian coordinate system; Domain; Equations; Exponents; Graphs, domains, and ranges; Modeling; Necessary conditions; Parametric equations; Powers; Power functions; Quartic equations; Trigonometric equations; Trigonometric functions

EXTREMA AND CRITICAL POINTS

The concepts of extrema and critical points of **functions** is an extremely useful concept in mathematics, particularly the areas of **calculus** and **differential equations**. Extrema may be absolute, such as the maximum (or highest) value or the minimum (or lowest) value attained by the function over its defined **domain**, or extrema may be local, such as the maximum or minimum value attained by a function in a particular, localized **interval** of the domain. Critical points are any interior points of the domain at which the derivative of the function is either **zero** or undefined (that is, does not exist). Importantly, on all closed intervals (that is, finite intervals that include boundary points), continuous functions will take both an absolute maximum and an absolute minimum; it is impossible to construct a continuous function over a closed interval that does not take on absolute maximum and minimum values.

Defining Absolute and Local Extrema

The definitions of absolute maximum and minimum values, as well as local maximum and minimum values, is exactly as expected:

- The absolute maximum is the maximum value of the function f across its entire domain D (that is, $f(c)$ is the absolute maximum if $f(c) \geq f(x)$ for all x in D)
- The absolute minimum us the minimum value of the function across its entire domain (that is, $f(c)$ is the absolute minimum if $f(c) < = f(x)$ for all x in D)
- A local maximum is the maximum value of the function in a particular interval (that is, $f(c)$ is a local maximum if $f(c) \geq f(x)$ for all x in a defined interval, which must contain c)
- A local minimum is the minimum value of the function in a particular interval (that is, $f(c)$ is a local minimum if $f(c) \leq f(x)$ for all x in a defined interval, which must contain c)

Finding Absolute Extrema on Closed Intervals

To find an absolute extreme value of a function f on a closed interval, it is necessary to first find the critical points by solving $f = 0$ (that is, setting the first derivative of the function to zero and finding all solutions to this problem). Then, one evaluates the function f at all critical points, as well as the endpoints (that is, boundary points) of the closed interval. The absolute maximum of the function on the interval is the largest of all these values, and the absolute minimum of the function is the smallest of all these values.

Using the First Derivative to Find Local Extrema

Since the first derivative of a function provides the **slope**, or instantaneous rate of change, of the function at any given point, evaluating and comparing the signs of the first derivative on both "sides" of a critical point establishes whether a critical point is a local extreme value. If f, the first derivative is positive, then the function is *increasing*. If f is negative, then the function is *decreasing*. A local maximum occurs when f changes from positive to negative at the critical value c (that is, $f^{(x)} > 0$ for $x < c$ and $f^{(x)} < 0$ for $x > c$). A local minimum occurs when f changes from negative to positive at the critical value c (that is, $f^{(x)} < 0$ for $x < c$ and $f^{(x)} > 0$ for $x > c$). If f does not change sign at the critical point c, then no local maximum or local minimum occurs at this point. This practice is sometimes known as the first derivative test.

Using the Second Derivative to Find Local Extrema

Interestingly enough, the second derivative can also be used to find local extreme values. The second derivative of a function, of course, is the derivative of the first derivative of the function. The second derivative of a function is often represented by the mathematical symbol f'. The second derivative of a function provides insight into its concavity—over intervals where $f' < 0$, the function is *concave down*, and over intervals where $f' > 0$, the function is *concave up*. When the concavity of a function changes at a critical point, either a local maximum or local minimum is observed. More specifically, if $f^{(c)} = 0$ and $f^{(c)} < 0$, the function has a local maximum at $x = c$. Similarly, $f^{(c)} = 0$ and $f^{(c)} > 0$, the function has a local minimum at $x = c$. This evaluation is sometimes called the second derivative test.

See also Calculus; Continuity; Derivatives and differentials; Differential calculus; Differential equations; Domain; Functions; Instantaneous velocity; Interval; Mathematical symbols; Negative numbers; Slope; Zero

F

FACTOR

In mathematics, to factor a number or algebraic expression is to find parts whose product is the original number or expression. For instance, 12 can be factored into the product 6×2, or 3×4. The expression $(x^2 - 4)$ can be factored into the product $(x + 2)(x - 2)$. Factor is also the name given to the parts. We say that 2 and 6 are factors of 12, and $(x-2)$ is a factor of $(x^2 - 4)$. Thus we refer to the factors of a product and the product of factors.

The fundamental **theorem** of **arithmetic** states that every positive integer can be expressed as the product of prime factors in essentially a single way. A prime number is a number whose only factors are itself and 1 (the first few **prime numbers** are 1, 2, 3, 5, 7, 11, 13). **Integers** that are not prime are called composite. The number 99 is composite because it can be factored into the product 9×11. It can be factored further by noting that 9 is the product 3×3. Thus, 99 can be factored into the product $3 \times 3 \times 11$, all of which are prime. By saying "in essentially one way," it is meant that although the factors of 99 could be arranged into $3 \times 11 \times 3$ or $11 \times 3 \times 3$, there is no factoring of 99 that includes any primes other than 3 used twice and 11.

Factoring large numbers was once mainly of interest to mathematicians, but today factoring is the basis of the security codes used by computers in military codes and in protecting financial transactions. High-powered computers can factor numbers with 50 digits, so these codes must be based on numbers with a hundred or more digits to keep the data secure.

In **algebra**, it is often useful to factor polynomial expressions (expressions of the type $9x^3 + 3x^2$ or $x^4 -27xy + 32$). For example $x^2 + 4x + 4$ is a polynomial that can be factored into $(x + 2)(x + 2)$. That this is true can be verified by multiplying the factors together. The degree of a polynomial is equal to the largest exponent that appears in it. Every polynomial of degree n has at most n polynomial factors (though some may contain **complex numbers**). For example, the third degree polynomial

$x^3 + 6x^2 + 11x + 6$ can be factored into $(x + 3)$ $(x^2 + 3x + 2)$, and the second factor can be factored again into $(x + 2)(x + 1)$, so that the original polynomial has three factors. This is a form of (or corollary to) the fundamental theorem of algebra.

In general, factoring can be rather difficult. There are some special cases and helpful hints, though, that often make the job easier. For instance, a common factor in each term is immediately factorable; certain common situations occur often and one learns to recognize them, such as $x^3 + 3x^2 + xy = x(x^2 + 3x + y)$. The difference of two squares is a good example: $a^2 - b^2 = (a + b)(a - b)$. Another common pattern consists of perfect squares of binomial expressions, such as $(x + b)^2$. Any squared binomial has the form $x^2 + 2bx + b^2$. The important things to note are: (1) the coefficient of x^2 is always one (2) the coefficient of x in the middle term is always twice the **square** root of the last term. Thus $x^2 + 10x + 25 = (x+5)^2$, $x^2 - 6x + 9 = (x-3)^2$, and so on.

Many practical problems of interest involve polynomial **equations**. A polynomial equation of the form $ax^2 + bx + c = 0$ can be solved if the polynomial can be factored. For instance, the equation $x^2 + x - 2 = 0$ can be written $(x + 2)(x - 1) = 0$, by factoring the polynomial. Whenever the product of two numbers or expressions is **zero**, one or the other must be zero. Thus either $x + 2 = 0$ or $x - 1 = 0$, meaning that $x = -2$ and $x = 1$ are solutions of the equation.

FACTOR ANALYSIS

The process of transforming statistical data, such as some sort of measurements, into **linear equations** that usually combine a number of independent **variables** is called factor **analysis**. The techniques for determining or estimating various parameters in such algebraic equations or the values of the variables therein are also included in the **definition** of this term. Many techniques for performing factor analysis exist. Because they are the most likely to provide direct numerical results, the most accepted by mathematical theoreticians require laborious computations to be carried out. For centuries, as mathematics developed, this fact limited the accuracy and amount of analysis that any particular mathematician was able or willing to attempt. But with the advent of sophisticated **calculators** and computers, such computations are much easier and faster to perform with greater accuracy than ever before, allowing factor analysis to be completed much more quickly even with extremely complex **equations**.

The goal of factor analysis is to account for a set of observed results in terms of variables in the least complicated equation or set of equations possible. In addition, an attempt is made to identify **factors** within the equation(s) that are common to or independent of other factors. If one type of factor has an influence on all of another type, it is known as a general factor, if only on some of the other type, it is called a group factor. A common factor with both positive and negative effects on others is called bipolar. Identification of the types and values of factors is vital to factor analysis.

See also Mathematics and computers; Statistics

FACTORIAL

The number n! is the product $1 \times 2 \times 3 \times 4 \times... \times n$, that is, the product of all the natural numbers from 1 up to n, including n itself where 1 is a natural number. It is called either "n factorial" or "factorial n." Thus 5! is the number $1 \times 2 \times 3 \times 4 \times 5$, or 120.

Older books sometimes used the symbol In for n factorial, but the numeral followed by an exclamation point is currently the standard symbol.

Factorials show up in many formulas of **statistics**, probability, **combinatorics**, **calculus**, **algebra**, and elsewhere. For example, the formula for the number of permutations of n things, taken n at a time, is simply n!. If a singer chooses eight songs for his or her concert, these songs can be presented in 8!, or 40,320 different orders. Similarly the number of combinations of n things r at a time is n! divided by the product r!(n - r)!. Thus the number of different bridge hands that can be dealt is 52! divided by 13!39!. This happens to be a *very* large number.

When used in conjunction with other operations, as in the formula for combinations, the factorial function takes precedence over **addition**, **subtraction**, negation, **multiplication**, and **division** unless parentheses are used to indicate otherwise. Thus in the expression r!(n - r)!, the subtraction is done first because of the parentheses; then r! and (r - n)! are computed; then the results are multiplied.

As n! has been defined, 0! makes no sense. However, in many formulas, such as the one above, 0! can occur. If one uses this formula to compute the number of combinations of 6 things 6 at a time, the formula gives 6! divided by 6!0!. To make formulas like this work, mathematicians have decided to give 0! the value 1. When this is done, one gets 6!/6!, or 1, which is, of course, exactly the number of ways in which one can choose all six things.

As one substitutes increasingly large values for n, the value of n! increases very fast. Ten factorial is more than three million, and 70! is beyond the capacity of even those calculators which can represent numbers in **scientific notation**.

This is not necessarily a disadvantage. In the series representation of **sine** x, which is $x/1! - x^3/3! + x^5/5! -...$, the denominators get large so fast that very few terms of the series are needed to compute a good decimal approximation for a particular value of sine x.

FAMILY OF CURVES

A family of curves is a collection of curves which share many of the same properties. For example, although there are infinitely many parabolas, all of them share the same basic shape. All parabolas consist of points which are equally distant from a fixed point, called the focus, and a fixed line, called the directrix. All have a vertex that is halfway between the focus and directrix. In fact, all parabolas can be seen as "descending" from a common "parent," the **parabola** whose algebraic representation is $y=x^2$. This simplest of all parabolas has its vertex at the origin, its focus at the point (0,1/4) and its directrix with equation y=-1/4. Any other parabola can be "generated" from the $y=x^2$ "parent" by one or more mathematical transformations. As an example, the parabola with equation $(y-3)=(x-4)^2$ can be created by translating the $y=x^2$ parabola 4 units to the right and 3 units vertically upward. Thus, all parabolas do share a kind of "family" resemblance, and the same can be said of other types of curves. This is important to the mathematician because it allows her to study the parent curve and draw conclusions about the entire family of curves, without having to investigate each curve individually. Many of the key properties of a curve will be "inherited" from its parent.

Families of curves play an important role in the study of **differential equations**. For example, it is known that 2x is the derivative of x^2, but it is also the derivative of x^2+1 and x^2+13 and $x^2+0.5$, among infinitely many others. In fact, 2x is the derivative of any member of the family of curves with **equations** of the form x^2+C, where C can be any constant. This means that the differential equation y'=2x has a family of solutions of the form x^2+C. This form is also called the general solution of the differential equation. If some specific condition is given, then it is possible to pick out the one member of the family of solutions that satisfies this condition. This member is called a particular solution of the differential equation for which the given condition is true. For instance, if it is known that the point (2,7) is on the graph, then we can say that the particular solution of y'=2x which satisfies this condition is $y=x^2+3$. Thus, there is an entire family of curves that have the property that y'=2x, but only one member of that family will pass through the point (2,7). Such a family of curves may be illustrated by means of a "slope field" in which short segments of tangent lines are drawn at each point on a grid. Such a picture gives one a sense of the "flow" of the family of solutions curves for a differential equation.

FERMAT NUMBERS

Fermat numbers, named after the French mathematician **Pierre de Fermat**, are numbers of the form $F_n = 2^{2n}+1$, where n is some non-negative integer.

The first few Fermat numbers are $F_0=3$, $F_1=5$, $F_2=17$, $F_3=257$, $F_4=65537$ and $F_5=4294967297$ and were computed by Fermat himself, who claimed that they were all **prime numbers**. On this basis, he conjectured that the Fermat numbers were always prime but this turned out to be false. The Swiss mathematician **Leonhard Euler** showed that F_5 was divisible by 641 and therefore not prime, contrary to what Fermat had claimed.

To date, no other Fermat number was found to be prime and it is known that all Fermat numbers F_n with n between 5 and 30 (and several other values of n) are composite. The current guess is that no other Fermat number beyond the first five, is prime.

A test devised by Pepin in 1877 allows the primality of a Fermat number to be tested relatively quickly on a computer, even though these numbers get huge very quickly. However, Pepin's test does not give the factors of a Fermat number when it is composite and it remains a challenge to **factor** the larger composite Fermat numbers.

See also Mersenne numbers

FERMAT, PIERRE DE (1601-1665)

French number theorist

Pierre de Fermat, a lawyer and jurist by profession, made major contributions to every field of mathematics that existed in the seventeenth century. He developed the principles of **analytic geometry** independently from his contemporary **René Descartes**. Fermat is regarded by some mathematicians as the inventor of differential **calculus**. He created modern **number theory**, and, together with **Blaise Pascal**, developed the theory of probability. Fermat's wide range of accomplishments is especially remarkable considering that mathematics was for him only a hobby.

Pierre Fermat was baptized and most likely born on August 20, 1601, in Beaumont–de–Lomagne. His father, Dominique Fermat, was bourgeois second consul of Beaumont–de–Lomagne and a prosperous leather merchant. His mother, Claire de Long, came from a prominent family of parliamentary lawyers. Fermat had three siblings, a brother, Clement, and two sisters, Louise and Marie. His early education was most likely received at the Franciscan monastery in Beaumont. During the 1620s he attended the University of Toulouse, then moved for a few years to Bordeaux, where he undertook an informal study of mathematics and made contact with students of **François Viète**, the noted French algebraist. Fermat then entered the University of Orleans, where he received a Bachelor of Civil Law. In 1631 he began practicing law, purchasing positions as councilor of the Parliament of Toulouse and Commissioner of

Pierre de Fermat

Requests. It was then that he added "de" to his name, an indication of his social standing afforded by his offices. In the same year Fermat married his fourth cousin, Louise de Long. Together they had five children, Clement–Samuel, Jean, Claire, Catherine and Louise. In Fermat's profession as a jurist, he rose gradually through the ranks, being named to the criminal court in 1638 and promoted in 1648 to a King's councilorship. He retained the latter post until his death in 1665.

Within a few years of taking his first parliamentary post, Fermat began corresponding with several prominent Parisian mathematicians, including **Marin Mersenne, Gille Personne de Roberval**, and Étienne Pascal, father of probability theorist Blaise Pascal. In his first communications he described his work on geostatics. He proposed that **Galileo** had been incorrect in stating that a freely falling cannonball should follow a semicircular path. In the course of proving the path would be **spiral**, Fermat developed a new method of quadrature for curves. He also posed several analytical problems to the Parisian group, which he claimed to have already solved on his own. Roberval and Mersenne found those and subsequent problems difficult and eventually requested that Fermat describe the techniques by which he had derived their solution. In response, Fermat sent a paper called *Method for*

Determining Maxima and Minima and Tangents to Curves Lines, along with his restoration of the Greek mathematician **Apollonius**'s *Plane Loci*. These papers, received in Paris in 1636, set out the fundamentals of **differential calculus**. In 1637 Fermat wrote a manuscript which the Parisian mathematicians circulated called *An Introduction to Plane and Solid Loci*. At about the same time René Descartes had sent Mersenne the galley proofs of his *Discourse on the Method* and accompanying *Essays*. The group in Paris quickly realized that Fermat and Descartes had independently developed the principles of analytic **geometry**, deriving the same basic technique for treating geometric **locus** problems algebraically. While Fermat's treatise arrived first in Paris, Descartes had laid his foundation earlier. A bitter and protracted argument arose between the two men, ignited by issues of priority, but eventually focusing on the subject of maxima and minima. The dispute engaged most of the mathematicians in Paris and forced Fermat to prove the generality of his methods. Descartes finally admitted defeat regarding the mathematical controversy, but Fermat's reputation was damaged. While he had established himself as one of the best mathematicians in Europe, he was resented by some of his colleagues for communicating his work in piecemeal and sending problems in the form of challenges. He was even accused of posing problems that had no solutions, in order to expose his rivals.

In the 1640s Fermat began his most important work, the development of modern number theory. Civic duties, however, prevented him from any meaningful mathematical correspondence between 1643 and 1654. In 1648 Fermat was occupied with the Fronde, a civil war, and in 1649, the Spanish raid on Languedoc. In 1651 the plague struck Toulouse, and Fermat became so ill that in 1653 his death was mistakenly reported. During a decade of isolation from the mathematical community he focused attention on developing a method of determining whether numbers were prime, and if not, on finding their divisors. The culmination of this work is today known as Fermat's **Theorem**. It states that if n is any whole number and p any prime, then $n^p - n$ is divisible by p. Fermat then went on to explore the concept of decomposition of primes of various forms into sums of their squares.

In the spring of 1654, Fermat received a letter from Blaise Pascal, asking for advice on a problem involving the consecutive throws of a die. The question to be resolved was how to divide the stakes in a dice game between two players, when their game was prematurely interrupted. The exchange that ensued between Pascal and Fermat over the course of just a few months helped lay the foundations of **probability theory**. By August 1654 Fermat was trying to divert Pascal's attention to his work on the theory of numbers. But Pascal, like most of his French contemporaries, saw little importance in this topic. In 1656 Fermat began to correspond about number theory with the Dutch physicist and astronomer **Christiaan Huygens**, expositor of the wave theory of **light** and inventor of the pendulum clock. Fermat described to him his method of infinite descent or "reduction analysis," in which larger problems are broken down into groups of problems more readily solvable.

Though Huygens admired the seminal contributions of Fermat to calculus and analytical geometry, he believed Fermat's number theory had no practical application and that Fermat had become out of touch with important mathematical questions. Fermat tried to engage the two English mathematicians **John Wallis** and William Brouncker in discussion of his new theories, also to little avail. Finally, between 1658 and 1662 Fermat turned to the topic of optics. Using the method of maxima and minima, he investigated the laws of reflection and refraction. In the course of his work he discovered that light travels by the path of least duration, a concept now known as Fermat's Principle.

After Fermat's bout with the plague of 1651 he suffered from frequent illnesses, and in 1662 he ended all scientific and mathematical correspondence. He died on January 12, 1665, and was buried in the Church of St. Dominique in Castres.

During his lifetime Fermat had frequently been asked by his mathematical contemporaries to publish descriptions of his work. He refused, expressing in his correspondence the desire to remain anonymous. His colleagues had resorted to including descriptions of Fermat's theories in their own writings, and circulating Fermat's letters and manuscripts by hand. Fermat's eldest son Clement–Samuel, published much of Fermat's work posthumously. In the process of collating the work, he examined his father's copy of the Latin translation of **Diophantus of Alexandria**'s *Arithmetic*. In a page margin Fermat had written a theorem accompanied by a brief note which read, "I have discovered a truly remarkable **proof** which this margin is too small to contain." Known as **Fermat's Last Theorem**, it stated that for the equation $x^n + y^n = z^n$ there are no positive integer solutions for x, y, and z, when n is greater than two. The proof of the theorem was to become one of the most famous mathematical challenges of the last three centuries. In 1994, an English–born American mathematician, **Andrew J. Wiles**, was credited with providing the first acceptable proof. The complexity of the problem was such that some historians question whether indeed Fermat had ever actually derived the correct proof himself.

FERMAT'S LAST THEOREM

Fermat's last **theorem** is one of the most famous theorems of mathematics. It was first stated by the brilliant amateur mathematician **Pierre de Fermat**, but it wasn't proved until more than 350 years later. Over the centuries it provided a tantalizing challenge to great mathematicians and amateurs alike, but it did not yield to **proof** until 1994, when a proof by Princeton mathematician **Andrew Wiles** was accepted by the mathematical community as correct.

Although it turned out to be one of the most challenging problems in mathematics, Fermat's last theorem is exceptionally easy to state. It says that whenever n is a whole number bigger than two, the equation $x^n + y^n = z^n$ has no nonzero solutions x, y, and z among the **whole numbers**.

Fermat's statement was found by his son as a marginal note that the great mathematician had made in his volume of

Diophantus's *Arithmetica*. In the margin Fermat wrote the above statement, then words that have become notorious among mathematicians: "I have discovered a truly remarkable proof which this margin is too small to contain." Since that time, mathematicians have speculated whether Fermat was really able to prove the theorem, or whether he was mistaken or bluffing. Mathematicians agree that the techniques employed by Wiles in his proof are so modern that they could not possibly have been conceived of by Fermat. To this day there are mathematicians who try to find a proof that uses only techniques that could have been known to Fermat.

In the first two centuries that followed Fermat's statement, many eminent mathematicians made progress toward proving Fermat's theorem, but in agonizingly small steps. Fermat's theorem is a statement about all possible **exponents** n, and all possible solutions x, y, and z. Therefore, it is not possible to check the statement by brute force: even if mathematicians were to test all the numbers x, y, and z less than a billion, say, with all the exponents less than a trillion, and none of them satisfied the equation, there could still be a solution involving larger numbers. Over the years, computers were used to test larger and larger numbers, and their evidence greatly strengthened mathematicians' belief in the **truth** of the theorem, but it did not blind them to the fact that no amount of checking of specific numbers can constitute a proof.

The first progress towards a proof was made by **Leonhard Euler**, who claimed in a letter to **Christian Goldbach** in 1753 that he had proved the theorem for the exponent n=3. The proof that he gave in his book *Algebra* in 1770 has a fallacy that is not easy to correct, but that is possible to correct using other methods from his book.

The next step towards a proof was made by **Sophie Germain**, who proved that when n and 2n+1 are both **prime numbers**, potential solutions x, y, and z could only be of a few narrowly restricted types. Using Germain's idea, mathematicians were able to prove that there are no solutions for several more exponents n.

Germain's discovery led to a flurry of new interest in Fermat's theorem, and the French Academy of Sciences offered a series of prizes to anyone who could come up with a correct proof of the theorem. On March 1, 1847 the academy had one of its most exciting meetings ever, when Gabriel Lame, a highly respected member, informed the academy that he was on the verge of proving Fermat's theorem, and gave a rough outline of his plan of attack. No sooner had he finished than the great mathematician **Augustin Louis Cauchy** rose to announce that he himself was also near to a proof of the theorem, following somewhat similar lines to those of Lame. A race was on between the two mathematicians, and three weeks later each of them delivered a sealed envelope to the academy detailing their techniques; this was a common proceeding at the time, by which mathematicians could establish priority for a result without having to make their techniques public before they were ready. But on May 24, the hopes of the two mathematicians and the entire academy were dashed when they received a letter from the German mathematician Ernst Kummer. He wrote to point out that both Lame and Cauchy

were in error, because they were relying on unique factorization, the fact that any number can be written as a product of primes in a unique way (for example, 18=2x3x3). While this is true of whole numbers, Kummer realized that it was not true of **complex numbers**, numbers that involve the **square root** of -1. Lame and Cauchy had been assuming unique factorization of certain kinds of complex numbers, so neither of their proofs was correct.

In 1908, the amateur mathematician Paul Wolfskehl, who had always been fascinated by Fermat's last theorem, bequeathed a prize of one hundred thousand marks to whoever could prove the theorem. This generous prize greatly increased public awareness of the problem, with the result that the University of Göttingen, which was to administer the prize, was deluged with attempts at proofs. Eventually, **Edmund Landau**, the head of the mathematics department, resorted to sending printed cards acknowledging submissions and stating on which page and line the first error occurred, as it unfailingly did. Even after World War I, when inflation severely damaged the value of the prize, countless seekers of glory tried to prove the theorem.

Substantial progress occurred in 1955 when the Japanese mathematicians Yutaka Taniyama and Goro Shimura did seminal work on elliptic curves, solutions to **equations** of the form $y^2 = x^3 + ax + b$, where a and b are constants. Although their work was halted tragically by Taniyama's suicide in 1958, they produced a conjecture that was later shown by Gerhard Frey to imply Fermat's theorem. In other words, if someone could prove the extremely difficult Taniyama-Shimura conjecture, Fermat's theorem would be proved as well.

The person who succeeded in proving the Taniyama-Shimura conjecture was Andrew Wiles, who had been fascinated by Fermat's last theorem since childhood. In order to have complete concentration and not to be scooped by competitors, Wiles shut himself up in an office inside his home and worked steadily on the theorem for seven years, without telling his colleagues what he was doing. Finally in 1993, amid rumors that something dramatic was to occur, Wiles gave a series of lectures on his work at the **Isaac Newton** Institute. At the end of the last lecture Wiles wrote Fermat's Last Theorem on the blackboard as a consequence of his work, saying "I think I'll stop here," to tumultuous applause.

The voluminous proof still had to be subjected to peer review, and during that process an error was discovered in the proof, which Wiles was at first unable to correct. He began to collaborate with Richard Taylor to try to fill the gap, and after an anxious year, Wiles realized that he could correct the error using an extension of a technique of Matheus Flach that Taylor had brought to his attention. With that insight he was in fact able to simplify the proof, and the current proof as it stands is generally accepted as correct, bringing to a close one of the most dramatic stories of mathematics.

See also Pierre de Fermat

FERMAT'S SPIRAL

Fermat's **spiral** is a special kind of spiral shape having the polar equation: $r = a\Theta^{1/2}$. **Pierre de Fermat**, a French lawyer who studied mathematics in his spare time, developed this shape in 1636. The equation for Fermat's spiral shows that for any given value of Θ there are two corresponding values of r. One value is negative and the other corresponding value is positive. Because of this the resulting spiral is symmetric about the line $y = -x$ as well as the origin. Fermat's spiral is sometimes also referred to as the parabolic spiral.

The general form of a spiral is given by the polar equation $r = a\Theta^{1/n}$, where r is the radial **distance**, Θ is the **angle** in polar coordinates, and n is a constant determining how tightly the spiral is wound. There are special names given to spirals having $n = -2, -1, 1$, and 2. As can be seen in the equation given in the paragraph above for $n = 2$, the spiral is called Fermat's spiral. If $n = -2$ the spiral is called the lituus and is equal to the inverse curve of Fermat's spiral with the origin as the inversion center. "Lituus" is translated as "crook" in the sense of a bishop's crosier. The lituus curve originated with Cotes in 1722, well after Fermat developed the spiral that bears his name. The curvature of Fermat's spiral is given by K where: $K(\Theta) = (((3a^2)/(4\Theta)) + a^2\Theta)/(((a^2/4\Theta) + a^2\Theta)^{3/2})$.

FEYNMAN, RICHARD PHILLIPS (1918-1988)

American physicist

Richard P. Feynman was a joint winner of the 1965 Nobel Prize for physics for the theory of quantum electrodynamics; he also helped develop the atom bomb.

Richard P. Feynman made significant advances in the understanding of superfluidity, weak nuclear interactions, and quarks, and shared the Nobel Prize for physics in 1965 for his contributions to the theory of quantum electrodynamics. In early 1986, Feynman served on the presidential commission that investigated the **space** shuttle *Challenger* incident, demonstrating to the nation that defective O-rings reacted too slowly to hot gases in cold temperatures, causing the shuttle to explode.

Richard Phillips Feynman was born in New York City on May 11, 1918, to Melville Arthur and Lucille Phillips Feynman. His father worked a number of jobs, but spent most of his years as a sales manager for a uniform manufacturer. Feynman had a younger sister, Joan. A brother, Henry, died in infancy. His youth was spent reading assorted mathematics books and the *Encyclopaedia Britannica*, as well as conducting experiments and fixing radios. As a student, Feynman excelled in math and science. He participated in the physics club and, as head of his high school's **algebra** team, placed first in the New York City math team competition when he was a senior. He was so adept at mathematics that his high school **geometry** teacher let Feynman teach the class.

Feynman applied to Columbia University but was not accepted, in part because he lacked in other subjects what he made up for in math and science. Accepted by the Massachusetts Institute of Technology (MIT), he enrolled as a mathematics major in 1936 but became disillusioned with its prospect as a viable career. He eventually leap-frogged from electrical engineering to physics, wherein he became fascinated with nuclear physics, a field that had only just begun to flourish after the discovery of the neutron in 1932. When MIT gave a graduate course on nuclear physics in the spring of 1938, Feynman jumped at the chance to take it, even though he was only a junior. At MIT he was regarded by several of his professors as one of the best students they had had in years. When he was a senior he competed in the nation's most prestigious mathematics contest, the Putnam competition. As James Gleick wrote in *Genius: The Life and Science of Richard Feynman*, "One of Feynman's fraternity brothers was surprised to see him return home while the examination was still going on. Feynman learned later that the scorers had been astounded by the gap between his result and the next four." Feynman's senior thesis, "Forces and Stresses in Molecules," caused a stir among MIT's physics faculty because it presented a much simpler way to calculate the electrostatic **force** in a solid crystal. He received his bachelor's degree in physics from MIT in 1939 and later published a shorter version of his thesis in *Physical Review*.

Feynman's admittance to Princeton's graduate school was almost hampered by the fact that he was Jewish. Though, unlike many institutions of the time, Princeton did not set a quota on the number of Jews they admitted, they were reluctant to accept very many because they were not easily placed among American industries strongly prejudiced against them. Also, Feynman's test scores in history, literature, and the fine arts were very weak. Nonetheless, he was accepted to graduate studies at Princeton in 1939 and became the teaching assistant for American physicist John Archibald Wheeler, who coined the term "black hole." In February 1941, Feynman gave his first professional presentation at the physics department seminar. In attendance were some of the most notable figures of early twentieth-century physics, including Swiss-born American physicist Wolfgang Pauli, Hungarian American mathematical physicist Eugene Wigner, and Hungarian American mathematician **John von Neumann**; American physicist **Albert Einstein** was also present. Feynman was so scared that he later remembered little of his presentation.

In anticipation of war, the U. S. government began recruiting physicists in a push to develop instrumentation and weaponry that would eventually change the face of war forever. In the spring of 1941, Feynman turned down a long-awaited job offer from Bell Laboratories, opting instead to spend that summer working at the Frankford Arsenal in Philadelphia. Returning to Princeton, he was pulled into the Manhattan Project by American physicist Robert Wilson. Like many major universities and development companies throughout the United States, Princeton was employed in the effort to isolate the lighter, or radioactive, isotopes of uranium. Wilson sought Feynman's help in calculating the speed at which uranium atoms of varying weight would separate in the magnetic field produced by Wilson's isotron, or isotope separator. While working on this problem, Feynman completed his dissertation on "The Principle of Least Action in Quantum Mechanics," and received his Ph.D. in theoretical physics in June 1942. Shortly after graduation, Feynman married his high school sweetheart, Arline Greenbaum, who had been diagnosed with tuberculosis a year earlier. After a short wedding on Staten Island, which neither family attended, Feynman drove his wife to a charity hospital in New Jersey.

In 1942, Wilson's isotron was shut down so isolation efforts could be concentrated on Berkeley's seemingly more productive Calutron. Early in 1943, American physicist J. Robert Oppenheimer recruited Feynman and other members of the Princeton team to develop the atomic bomb in the secluded desert surroundings of Los Alamos, New Mexico. On March 28, Feynman and Arline boarded a train for Albuquerque, where Oppenheimer had arranged for Arline to stay in a sanitarium. Feynman would hitchhike or borrow a car to visit his wife on many weekends. At Los Alamos, Feynman worked with some of the world's top physicists and mathematicians, including Oppenheimer, American physicist Edward Teller, and von Neumann. One of the most important of these colleagues was group leader and future Nobel Prize-winner Hans Bethe, a nuclear physicist and professor from Cornell University. While at Los Alamos, Feynman worked on computing how fast neutrons would diffuse through a critical mass of uranium or plutonium as it approached the reaction that leads to a nuclear explosion. He also developed calculating techniques to do many of the complex mathematical computations involved in designing a nuclear bomb, and he determined the **limits** of how many radioactive materials could be stored together safely without starting an explosive chain reaction. Tuberculosis wasted Arline Feynman away. She died on June 16, 1945. A month later to the day, at 5:30 in the morning, her husband stood twenty miles from the steel tower that held the first atomic bomb and watched it explode. Less than a month later, Hiroshima and Nagasaki were both destroyed by nuclear bombs.

Feynman left Los Alamos in October 1945 and, at Bethe's invitation, accepted a position as associate professor of physics at Cornell. Feynman began to think more about problems in the theory of quantum electrodynamics, problems that he had worked on while at MIT and Princeton, and even sporadically at Los Alamos. Quantum electrodynamics is the study of electromagnetic radiation and how electrically charged particles, such as atoms and their electrons, interact. It also investigates what happens when electrons, positrons, and photons collide with each other. The main principles of quantum electrodynamics had been formulated in the mid 1920s by English physicist **Paul Dirac**, German physicist **Werner Heisenberg**, and Pauli. Although quantum electrodynamics explained how **light** could be made of waves and particles, problems in the theory soon developed. The chief problems were the need for electrons to have infinite masses and infinite electric charges so that other aspects of the theory would work. American physicist Willis E. Lamb also demonstrated, in a discovery known as "the Lamb shift," that the hydrogen atom gave off energy levels that were not predicted by earlier theories.

Richard P. Feynman

Feynman was inspired one day in the Cornell cafeteria. He noticed a student spinning a cafeteria plate in the air. As the plate spun around, it wobbled, though the two motions were not synchronized. Thus a mathematical description of the **motion** of the plate had to account for two motions. Feynman's computations concerning such motions further advanced previous theories of quantum electrodynamics by eliminating the need for the idea of **infinity** and explaining the Lamb shift. His theory also made quantum electrodynamics vastly more accurate in predicting subatomic events. Subsequent experiments have confirmed the great accuracy of Feynman's reformulation of quantum electrodynamics. He also introduced a graphic aid that came to be known as a "Feynman diagram," which shows how subatomic particles, represented by arrows and lines, interact through space and time. Feynman's diagrams were simple, elegant, and powerful expressions of fundamental subatomic actions. They were also very popular; by the early 1950s, Feynman's diagrams began to appear frequently in academic articles about quantum electrodynamics.

Two other physicists were also reformulating quantum electrodynamics. American physicist Julian Schwinger was no stranger to Feynman; both hailed from New York and were the same age. Japanese physicist Sin-Itiro Tomonaga had developed his theories by 1943, but was unable to publish them in the West until after the War. Both men had independently developed different mathematical approaches that led to results similar to Feynman's. The three men shared the Nobel Prize for physics in 1965. In 1950, Feynman left Cornell for the California Institute of Technology near Los Angeles. The offer included an immediate sabbatical year which he spent in Brazil, performing sambas as a musician while theorizing the role of mesons (fundamental particles made up of a quark and an antiquark) in the cohesion of the atomic nucleus. When he returned in June 1952, he married Mary Louise Bell, whom he divorced four years later. In 1954, Feynman became the third person to win the Albert Einstein Award, after American mathematician **Kurt Gödel** and Schwinger.

While at Cal Tech, Feynman worked on a variety of other problems. One of these was a theory of superfluidity, which he developed to explain why liquid helium behaves in such strange ways (e.g., defying gravity, losing heat instead of gaining it when it flows). An interest in the weak nuclear force, that is, the force that makes the process of radioactive decay possible, led Feynman and American physicist Murray Gell-Mann to the supposition that the emission of beta-particles from radioactive nuclei acts as the chief agitator in the decay process. As Gleick explained, Feynman also contributed to "a theory of partons, hypothetical hard particles inside the atom's nucleus, that helped produce the modern understanding of quarks." Quarks are paired elementary particles, one of which has a charge of +2/3, and the other, +1/3. In 1973, Feynman received the Niels Bohr International Gold Medal.

In 1958, Feynman met Gweneth Howarth while in Switzerland. Gweneth, a domestic servant, turned down Feynman's first offer to come to the United States and work for him as a maid. After his long-distance pursuit of her by mail, she was finally won over and arrived in California in the summer of 1959. They were married on September 24, 1960. Their first child, Carl, was born in 1962. They adopted a daughter, Michelle, six years later. In October 1978, Feynman was diagnosed with myxoid liposarcoma, a rare cancer that affects the soft tissues of the body. The tumor from the cancer weighed six pounds and was located in the back of his abdomen, where it had destroyed his left kidney. His long-term chances were not good: less than fifty percent of patients survived for five years, and the chances for lasting ten years were close to **zero**.

The space ship *Challenger* exploded above Cape Kennedy in Florida on January 28, 1986. NASA's acting chief administrator, William R. Graham, asked Feynman to serve on the presidential commission that would investigate the accident. Graham's wife had suggested Feynman because she remembered some of his lectures. The appointment to the commission was not well timed, as Feynman had been recently diagnosed with a second rare form of bone cancer, Waldenström's macroglobulinemia. When the presidential commission began its public hearings on the accident in early February 1986, the discussion quickly turned toward the effects of cold temperatures on O-rings. These rubber rings seal the joints of the solid rocket boosters on either side of the large external tank that holds the liquid oxygen and hydrogen

fuel for the shuttle. Feynman demonstrated to the public the cause of the disaster. He bent a piece of O-ring in a clamp and immersed it in a glass of ice water, then released the piece of O-ring from the clamp to reveal how slowly it regained its original shape when it was cold. Feynman thus showed that the slow reaction time of the O-ring had allowed hot gases to escape, erode the O-ring, and burn a hole in the side of the right solid rocket booster, ultimately causing the destruction of the spacecraft and the deaths of the seven astronauts.

In Appendix F to the commission's final report, "Personal Observations on the Reliability of the Shuttle," Feynman contended that the managers of the NASA shuttle project came up with wildly unreasonable numbers to show that the spacecraft would be safe. He concluded the report with these words: "For a successful technology, reality must take precedence over public relations, for nature cannot be fooled." Feynman, diagnosed with yet another cancerous abdominal tumor in October 1987, died of complications on February 15, 1988.

FIBONACCI, LEONARDO PISANO (1170-1250)

Italian number theorist

Leonardo Pisano Fibonacci is considered one of the most talented mathematicians of the Middle Ages. He is credited with introducing the Hindu–Arabic numbering system into western European culture at the beginning of the 13th century. His series of books on mathematical subjects helped revive the tradition of ancient mathematics and laid the foundation for the development of **number theory**. While Fibonacci's introduction of the modern day numbering system had a profound impact on the subsequent **history of mathematics**, his fame as a mathematician is perhaps more often associated with his development of the **Fibonacci sequence**, a series of numbers that he derived in order to solve a riddle about the reproduction of rabbits.

Also known as Leonardo of Pisa, Fibonacci, was born sometime during the latter half of the 12th century in Pisa, Italy. His father, William Bonacci, was a merchant and a government representative of Pisa, then an independent city–state. The name "Fibonacci" is believed to have been derived from the contraction of *Filiorum Bonacci*, or possibly, *Filius Bonacci,* meaning respectively, "of the family of Bonacci" and "Bonacci's son." What is known of Fibonacci's life has been gleaned mostly from the brief autobiographical notes he included in the introduction to his first mathematical treatise, *Liber Abaci*, in 1202, from the dedication in his *Liber Quadratorum*, written in 1225, and from his only surviving letter, written to the Emperor Frederick II's philosopher, Magister Theodoris. Based on incidents described in these writings, Fibonacci's year of birth is approximated at 1170.

During the 12th century, Pisa had about 10,000 inhabitants, and was an important center of commerce. Its merchants traded throughout the Mediterranean region and maintained warehouses in the coastal cities. When Fibonacci was a boy, his father was appointed as the head of a warehouse in the city of Bugia, on the North African coast. Fibonacci traveled there to join his father and to receive a business education. He studied **arithmetic** under the tutelage of a Moorish schoolmaster, and learned to make calculations using the Hindu–Arabic numerals 0 through 9. Through Italy and western Europe the seven Roman symbols, I, V, X, L, C, D, and M were still used in their various combinations to express all possible **integers**. These symbols represented respectively: one, five, ten, fifty, one hundred, five hundred, and one thousand. Because the Roman numeral system lacked the concepts of **zero** and place–value, **multiplication** and **division** were cumbersome and virtually impossible operations. Performing such calculations required the use of an **abacus**, a mechanical device which allowed for no written verification of the result.

Fibonacci was undoubtedly taught the merits of the Hindu–Arabic numbering system and the methods of using it to perform various mathematical operations. It is not clear how much more advanced his studies were, or how long Fibonacci remained in Bugia. He returned to Italy in about 1200, after having traveled extensively throughout the Mediterranean. Fibonacci's travels brought him in contact with leading scholars of the day and exposed him to the monetary systems of Egypt, Syria, Constantinople, Greece, Sicily, and France.

In 1202, Fibonacci wrote *Liber abaci* ("Book of Calculations"). Its intent was to introduce Hindu–Arabic numerals to Western culture and to explain the utility of the new numbering system in business transactions. Divided into 15 chapters, Fibonacci's book covered a broad range of topics. There was a chapter on how to read and write the numerals, as well as separate chapters on how the numerals could be used practically to make additions, subtractions, multiplications, and divisions. The concepts of **fractions** and of squared and cubic **roots** were explained. A series of chapters dealt with such business practices as pricing, bartering, and partnership. His final and perhaps most important chapter dealt with the more sophisticated topics of **geometry** and **algebra**. Throughout the text, Fibonacci posed and showed solutions for various mathematical puzzles and riddles.

Fibonacci wrote *Practica geometriae* in 1220. This manuscript dealt with practical problems in geometry and the measurement of objects. It also covered algebraic and trigonometric operations and the use of **square** and cubic roots. These writings revealed Fibonacci's familiarity with the works of **Euclid** and other mathematicians of antiquity.

Fibonacci's reputation as an influential mathematician became widespread. The Holy Roman Emperor Frederick II had read and was impressed with *Liber Abaci*, and in 1225 he traveled to Pisa to conduct a mathematical tournament as a test of Fibonacci's skills. Johannes of Palermo, a member of the emperor's staff, composed three tournament questions, sent in advance to Fibonacci and several competitors. At the emperor's court in Pisa, Fibonacci demonstrated his mathematical ability by deriving correct answers to each of the questions. His competitors withdrew, unable to provide any of the solutions.

The first question posed by Johannes of Palermo was a second–degree problem, that is, one involving squares.

Leonardo Pisano Fibonacci

Specifically, the contestants were asked to determine values of x and y, such that $x^2 + 5 = y^2$, when $x^2 - 5 = y^2$. The next question was one of the third degree, involving cubes. The third question, a first–degree problem, was posed in the form of a riddle. Three men owned, respectively a half, a third and a sixth of an unknown quantity of money. Each man took an unspecified amount of the money, leaving none left. Then each man, respectively, returned a half, a third, and a sixth of what he had first taken. The returned money was divided into equal thirds and redistribute to the men. This resulted in each man acquiring his fair share. The contestants were asked to determine the quantity of money owned by each man.

Having displayed his skills as a mathematician before the emperor, Fibonacci went on that same year to write his *Liber quadratorum*. The manuscript was dedicated to the emperor and in it Fibonacci described both their meeting and the particulars of the contest. *Liber quadratorum* contained a collection of theorems about indeterminate **analysis**, specifically relating to second degree **equations**, and its introduction included a description of the second–degree contest problem. The first and third–degree contest problems were described in a separate manuscript of Fibonacci's, entitled *Flos*. The originality of his work and the power of his methods for solution have caused Fibonacci to be ranked among the most important seminal figures in the field of number theory.

Fibonacci revised *Liber abaci* in 1228, and it is this text that eventually became most widely distributed in Europe. It is dedicated to Michael Scot, a friend of Fibonacci's who wrote science texts and served as chief astrologer to Emperor Frederick II. As in the first edition, Fibonacci argues strongly for the adoption of the Hindu–Arabic numbering system.

Tourists to Pisa, Italy, are today most frequently drawn to the city by its famous leaning tower, designed and partially built by a contemporary of Fibonacci's, Bonnano Pisano. Across the Arno River from the leaning tower is a lesser known monument, a statue representing Fibonacci. Since no drawings of Fibonacci has survived, the statue was created in the likeness of a "generic" Pisan of the 12th century. While the potential lack of resemblance might strike Fibonacci as an odd legacy, the enduring association of his name with a sequence of numbers he generated, in order to solve a puzzle about rabbits, might seem even more peculiar. In his *Liber abaci*, he described his solution to a well–known mathematical problem of his time. A pair of rabbits is kept in an enclosure and begins producing offspring at the rate of one pair per month, beginning in the second month. Each new pair reproduces at that same rate after its second month. Assuming there is no mortality, how many rabbits will exist at the end of a year? The numbers in the Fibonacci sequence represent the quantity of rabbit pairs at the end of each month, namely 1, 1, 2, 3, 5, 8, 13, 21, 34, 55, and so on. Fibonacci was aware that the sequence was recursive, that is, one in which the relationship between successive terms can be expressed with a formula. Various modern–day mathematical societies bearing Fibonacci's name, have devoted themselves to exploring the interesting features of this sequence.

Fibonacci is believed to have died around 1250, when Pisa was defeated by Genoa in a naval battle.

FIBONACCI SEQUENCE

The Fibonacci sequence is a series of numbers in which each succeeding number (after the second) is the sum of the previous two. The most famous Fibonacci sequence is 1, 1, 2, 3, 5, 8, 13, 21, 34, 55, 89.... This sequence expresses many naturally occurring relationships in the plant world.

History

The Fibonacci sequence was invented by the Italian Leonardo Pisano Bigollo (1180-1250), who is known in mathematical history by several names: Leonardo of Pisa (Pisano means "from Pisa") and **Fibonacci** (which means "son of Bonacci").

Fibonacci, the son of an Italian businessman from the city of Pisa, grew up in a trading colony in North Africa during the Middle Ages. Italians were some of the western world's most proficient traders and merchants during the Middle Ages, and they needed **arithmetic** to keep track of their commercial transactions. Mathematical calculations were made using the Roman numeral system (I, II, III, IV, V, VI, etc.), but that system made it hard to do the **addition, subtrac-**

tion, multiplication, and **division** that merchants needed to keep track of their transactions.

While growing up in North Africa, Fibonacci learned the more efficient Hindu-Arabic system of arithmetical notation (1, 2, 3, 4...) from an Arab teacher. In 1202, he published his knowledge in a famous book called the *Liber Abaci* (which means the "book of the abacus," even though it had nothing to do with the **abacus**). The *Liber Abaci* showed how superior the Hindu-Arabic arithmetic system was to the Roman numeral system, and it showed how the Hindu-Arabic system of arithmetic could be applied to benefit Italian merchants.

The Fibonacci sequence was the outcome of a mathematical problem about rabbit breeding that was posed in the *Liber Abaci*. The problem was this: Beginning with a single pair of rabbits (one male and one female), how many pairs of rabbits will be born in a year, assuming that every month each male and female rabbit gives birth to a new pair of rabbits, and the new pair of rabbits itself starts giving birth to additional pairs of rabbits after the first month of their birth?

Other Fibonacci sequences

Although the most famous Fibonacci sequence is 1, 1, 2, 3, 5, 8, 13, 21, 34, 55..., a Fibonacci sequence may be *any*-series of numbers in which each succeeding number (after the second) is the sum of the previous two. That means that the specific numbers in a Fibonacci series depend upon the initial numbers. Thus, if a series begins with 3, then the subsequent series would be as follows: 3, 3, 6, 9, 15, 24, 39, 63, 102, and so on.

A Fibonacci series can also be based on something other than an integer (a whole number). For example, the series 0.1, 0.1, 0.2, 0.3, 0.5, 0.8, 1.3, 2.1, 3.4, 5.5, and so on, is also a Fibonacci sequence.

The Fibonacci sequence in nature

The Fibonacci sequence appears in unexpected places such as in the growth of plants, especially in the number of petals on flowers, in the arrangement of leaves on a plant stem, and in the number of rows of seeds in a sunflower.

For example, although there are thousands of kinds of flowers, there are relatively few consistent **sets** of numbers of petals on flowers. Some flowers have 3 petals; others have 5 petals; still others have 8 petals; and others have 13, 21, 34, 55, or 89 petals. There are exceptions and variations in these patterns, but they are comparatively few. All of these numbers observed in the flower petals—3, 5, 8, 13, 21, 34, 55, 89—appear in the Fibonacci series.

Similarly, the configurations of seeds in a giant sunflower and the configuration of rigid, spiny scales in pine cones also conform with the Fibonacci series. The corkscrew spirals of seeds that radiate outward from the center of a sunflower are most often 34 and 55 rows of seeds in opposite directions, or 55 and 89 rows of seeds in opposite directions, or even 89 and 144 rows of seeds in opposite directions. The number of rows of the scales in the spirals that radiate upwards in opposite directions from the base in a pine cone are almost always the lower numbers in the Fibonacci sequence—3, 5, and 8.

Why are Fibonacci numbers in plant growth so common? One clue appears in Fibonacci's original ideas about the rate of increase in rabbit populations. Given his time frame and growth cycle, Fibonacci's sequence represented the most efficient rate of breeding that the rabbits could have if other conditions were ideal. The same conditions may also apply to the propagation of seeds or petals in flowers. That is, these phenomena may be an expression of nature's efficiency. As each row of seeds in a sunflower or each row of scales in a pine cone grows radially away from the center, it tries to grow the maximum number of seeds (or scales) in the smallest **space**. The Fibonacci sequence may simply express the most efficient packing of the seeds (or scales) in the space available.

FIELD

A "field" is the name given to a pair of numbers and a set of operations which together satisfy several specific laws. A familiar example of a field is the set of **rational numbers** and the operations **addition** and **multiplication**. An example of a set of numbers that is not a field is the set of **integers**. It is an "integral domain." It is not a field because it lacks multiplicative inverses. Without multiplicative inverses, **division** may be impossible.

The elements of a field obey the following laws:

1. Closure laws: $a + b$ and ab are unique elements in the field.
2. Commutative laws: $a + b = b + a$ and $ab = ba$.
3. Associative laws: $a + (b + c) = (a + b) + c$ and $a(bc) = (ab)c$.
4. Identity laws: there exist elements 0 and 1 such that $a + 0 = a$ and $a \times 1 = a$.
5. Inverse laws: for every a there exists an element $-a$ such that $a + (-a) = 0$, and for every $a \neq 0$ there exists an element a^{-1} such that $a \times a^{-1} = 1$.
6. Distributive law: $a(b + c) = ab + ac$.

Rational numbers (which are numbers that can be expressed as the ratio a/b of an integer a and a natural number b) obey all these laws. They obey closure because the rules for adding and multiplying **fractions**, $a/b + c/d = (ad + cb)/bd$ and $(a/b)(c/d) = (ac)/(bd)$, convert these operations into adding and multiplying integers which are closed. They are commutative and associative because integers are commutative and associative. The ratio $0/1$ is an additive identity, and the ratio $1/1$ is a multiplicative identity. The **ratios** a/b and $-a/b$ are additive inverses, and a/b and b/a ($a, b \neq 0$) are multiplicative inverses. The rules for adding and multiplying fractions, together with the distributive law for integers, make the distributive law hold for rational numbers as well. Because the rational numbers obey all the laws, they form a field.

The rational numbers constitute the most widely used field, but there are others. The set of **real numbers** is a field. The set of **complex numbers** (numbers of the form $a + bi$, where a and b are real numbers, and $i^2 = -1$) is also a field.

Although all the fields named above have an infinite number of elements in them, a set with only a finite number of elements can, under the right circumstances, be a field. For example, the set {0,1} constitutes a field when addition and multiplication are defined by these tables:

+	0	1		×	0	1
0	0	1		0	0	0
1	1	0		1	0	1

With such a small number of elements, one can check that all the laws are obeyed by simply running down all the possibilities. For instance, the **symmetry** of the tables show that the commutative laws are obeyed. Verifying associativity and distributivity is a little tedious, but it can be done. The identity laws can be verified by looking at the tables. Where things become interesting is in finding inverses, since the addition table has no negative elements in it, and the multiplication table, no fractions. Two additive inverses have to add up to 0. According to the addition table $1 + 1$ is 0; so 1, curiously, is its own additive inverse. The multiplication table is less remarkable. **Zero** never has a multiplicative inverse, and even in ordinary **arithmetic**, 1 is its own multiplicative inverse, as it is here.

This example is not as outlandish as one might think. If one replaces 0 with "even" and 1 with "odd," the resulting tables are the familiar parity tables for catching mistakes in arithmetic.

One interesting situation arises where an algebraic number such as $=\sqrt{2}$ is used. (An algebraic number is one which is the root of a polynomial equation.) If one creates the set of numbers of the form $a + b\sqrt{2}$, where a and b are rational, this set constitutes a field. Every sum, product, difference, or quotient (except, of course, $(a + b\sqrt{2})/0$) can be expressed as a number in that form. In fact, when one learns to rationalize the denominator in an expression such as $1/(1 - \sqrt{2})$ that is what is going on. The set of such elements therefore form another field which is called an "algebraic extension" of the original field.

FIGURATIVE NUMBERS

Figurative numbers are numbers which can be represented by dots arranged in various geometric patterns. For example, triangular numbers are represented by the patterns shown in Figure 1.

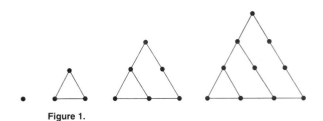

Figure 1.

The numbers they represent are 1, 3, 6, 10, and so on.

Figurative numbers were first studied by the mathematician **Pythagoras** in the sixth century B.C. and by the Pythagoreans, who were his followers. These numbers were studied, as were many kinds of numbers, for the sake of their supposed mystical properties rather than for their practical value. The study of figurative numbers continues to be a source of interest to both amateur and professional mathematicians.

Figurative numbers also include the **square** numbers which can be represented by square arrays of dots, as shown here.

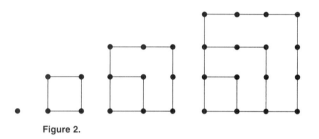

Figure 2.

The first few square numbers are 1, 4, 9, 16, 25, etc.

There are pentagonal numbers based on pentagonal arrays. Figure 3 shows the fourth pentagonal number, 22.

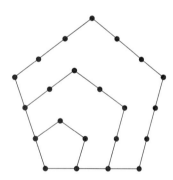

Figure 3.

Other pentagonal numbers are 1, 5, 12, 22, 35, and so on.

There is, of course, no limit to the number of polygonal arrays into which dots may be fitted. There are hexagonal numbers, heptagonal numbers, octagonal numbers, and, in general, n-gonal numbers, where n can be any number greater than 2.

One reason that figurative numbers have the appeal they do is that they can be studied both algebraically and geometrically. Properties that might be hard to discover by algebraic techniques alone are often revealed by simply looking at the figures, and this is what we shall do, first with triangular numbers.

If we denote by T_n the n-th **triangle** number, Figure 1 shows us that $T_1 = 1$ $T_2 = T_1 + 2$ $T_3 = T_2 + 3$ $T_4 = T_3 + 4$.

$$T_n = T_{n-1} + n$$

These formulas are *recursive*. To compute T_{10} one must compute T_9. To compute T_9 one has to compute T_8, and so on.

For small values of n this is not hard to do: $T_{10} = 10 + (9 + (8 + (7 + (6 + (5 + (4 + (3 + (2 + 1)))))))) = 55$.

For larger values of n, or for general values, a formula that gives T_n directly would be useful. Here the use of the figures themselves comes into play: From Figure 4 one can see that $2T_3 = (3)(4)$; so $T_3 = 12/2$ or 6.

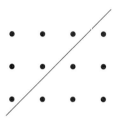

Figure 4.

The same trick can be applied to any triangular number: $2T_n = n(n + 1)$ $T_n = n(n+ 1)/2$ When n = 10, $T_{10} = (10)(11)/2$, or 55, as before.

In the case of square numbers, the general formula for the n-th square number is easy to derive, It is simply n^2. In fact, the very name given to n^2 reflects the fact that it is a square number. The recursive formulas for the various square numbers are a little less obvious. To derive them, one can turn again to the figures themselves. Since each square can be obtained from its predecessor by adding a row of dots across the top and a column of dots down the side, including one dot in the corner, one gets the recursive pattern $S_1 = 1$ $S_2 = S_1 + 3$ $S_3 = S_2 + 5$ $S_4 = S_3 + 7...$

$$S_n = S_{n-1} + 2n - 1$$

or, alternatively

$$S_{n+1} = S_n + 2n + 1$$

Thus $S_8 = 15 + (13 + (11 + (9 + (7 + (5 + (3 + 1))))))$ or 64.

Because humans are so fond of arranging things in rows and columns, including themselves, square numbers in use are not hard to find. Tic-tac-toe has S_3 squares; chess and checkers have S_8. S_{19} has been found to be the ideal number of points, says a text on the oriental game of "go," on which to play the game.

One of the less obvious places where square numbers-or more correctly, the recursive formulas for generating them-show up is in one of the algorithms for computing **square roots**. This is the **algorithm** based on the formula $(a + b)^2 = a^2 + 2ab + b^2$. When this formula is used, b is not 1. In fact, at each stage of the computing process the size of b is systematically decreased. The *process* however parallels that of finding S_{n+1} recursively from S_n. This becomes apparent when n and 1 are substituted for a and b in the formula: $(n + 1)^2 = n^2 + 2n + 1$. This translates into $S_{n+1} = S_n + 2n + 1$, which is the formula given earlier.

Formulas for pentagonal numbers are trickier to discover, both the general formula and the recursive formulas. But again the availability of geometric figures helps. By examining the array in Figure 5

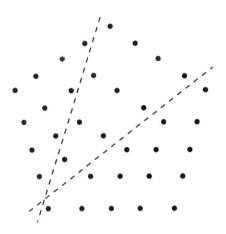

Figure 5.

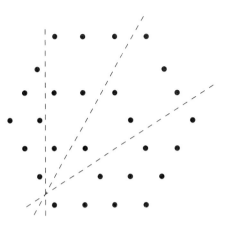

Figure 7.

one can come up with the following recursive formulas, where P_n represents the n-th pentagonal number: $P_1 = 1$ $P_2 = P_1 + 4$ $P_3 = P_2 + 7$ $P_4 = P_3 + 10$. One can guess the formula for P_n: $P_n = P_{n-1} + 3n - 2$ and this is correct. At each stage in going from P_{n-1} to P_n one adds three sides of n dots each to the existing pentagon, but two of those dots are common to two sides and should be discounted.

To compute P_7 recursively we have $19 + (16 + (13 + (10 + (7 + (4 + 1)))))$, which adds up to 70.

To find a general formula, we can pull another trick. We can cut up the array in Figure 5 along the dotted lines. When we do this we have $P_5 = T_5 + 2T_4$ or more generally $P_n = T_n + 2T_{n-1}$.

If we substitute algebraic expressions for the triangular numbers and simplify the result we come up with $P_n = (3n^2 - n)/2$.

The fact that pentagonal numbers can be cut into triangular numbers makes one wonder if other polygonal numbers can, too. Square numbers can, and Figure 6 shows an example.

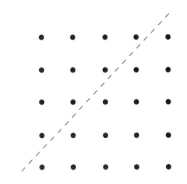

Figure 6.

It yields $T_5 + T_4$, or in general $S_n = T_n + T_{n-1}$. Arranging these formulas for triangular dissections suggests a pattern: $T_n = T_n$ $S_n = T_n + T_{n-1}$ $P_n = T_n + 2T_{n-1}$ $H_n = T_n + 3T_{n-1}$ where H_n is the n-th hexagonal number. If we check this for H_4 in Figure 7, we find it to be correct.

In general, if N_k represents the k-th polygonal number for a polygon of N sides (an N-gon) $N_k = T_k + (N - 3)T_{k-1}$.

The Pythagoreans were also concerned with "oblong" arrays (here "oblong" has a more limited meaning than usual), having one column more than its rows.

Figure 8.

In this case the concern was not with the total number of dots but with the ratio of dots per row to dots per column. In the smallest array, this ratio is 2:1. In the next array it is 3:2; and the third, 4:3. If the arrays were further enlarged, the **ratios** would change in a regular way: 5:4, 6:5, and so on.

These ratios were related, by Pythagoreans, to music. If, on a stringed instrument, two notes were played whose frequencies were in the ratio of 2:1, those notes would be an octave apart and would sound harmonious when played together. (Actually the Pythagoreans went by the lengths of the strings, but if the string lengths were in the ratio 1:2, the frequencies would be in the ratio 2:1.) If the ratio of the frequencies were 3:2, the notes would be a perfect fifth apart and would also sound harmonious (in fact, violinists use the harmony of perfect fifths to tune the strings of the violin, which sound a perfect fifth apart). Notes in the ratio 4:3 were also considered harmonious. Other ratios were thought to result in discordant notes.

The Pythagoreans went well beyond this in developing their musical theories, but what is particularly interesting is that they based a really fundamental musical idea on an array of dots. Of course, for reasons having little to do with figurative numbers, they got it right.

Figurative numbers are not confined to those associated with plane figures. One can have pyramidal numbers based on figures made up of layers of dots, for example a tetrahedron made up of layers of triangular arrays. Such a tetrahedron would have T_n dots in the first layer, T_{n-1} dots in the next, and so on. If there were four such layers, the tetrahedral number it represents would be $1 + 3 + 6 + 10$, or 20.

A general formula applicable to a pyramid whose base is an N-gon with N_k points in it is $(k + 1)(2N_k + k)/6$. In the example above N = 3 and k = 4. Using these values in the formula gives $5(20 + 4)/6 = 20$.

If the base is a square with 5 points on a side, N = 4 and k = 5. Then the total number of points is $6(50 + 5)/6$, or 55. To arrange cannon balls in a pyramidal stack with a square array of 25 balls for a base, one would need 55 balls.

FINCKE, THOMAS (1561-1656)

Danish mathematician

In 1583 Thomas Fincke produced a mathematical text book entitled *Geometria Rotunda*, which brought the world the terms tangent and secant. This work also includes several trigonometric **equations** relating to these **functions**.

Thomas Fincke was born in Flensburg in what was then part of Denmark (Flensburg is now a part of Germany). His university studies covered such places as Strasbourg, Jena, Wittenberg, Heidelberg, Leipzig, Basel, Padua, and Pisa. After all of this study he qualified as a medical doctor, a role which he took from 1587 to 1591 in Flensburg. Fincke left this position to become professor of mathematics at Copenhagen (after a personal invitation by the chancellor, Niels Kaas), which he relinquished in 1602 to become professor of rhetoric. This chair lasted for only one year as in 1603 he became professor of medicine at Copenhagen. On five occasions between 1598 and 1633 Fincke served as rector of the university. As well as his mathematical text book *Geometria Rotunda,* Fincke also produced several works on astronomy and astrology. He died in Copenhagen at the age of 95. For the last 16 years of his life he was still professor of medicine, though he appointed a deputy to carry out the majority of his duties.

The generally accepted form of spelling of his name is Thomas Fincke, though some texts may refer to the archaic Danish spelling of Finke or indeed Finck.

FINITE SEQUENCES AND SERIES

Mathematicians define a sequence as a function which maps some subset of the positive **integers** onto some subset of the **real numbers**. In simplest terms, a sequence is an ordered list of elements. In this article, we will take these elements to be real numbers to simplify the discussion. A finite sequence, then, is a finite ordered list of real numbers: $a_1, a_2, a_3, ..., a_n$. Thus 1,2,4,8,16,32 is a finite sequence, as is -1,0,1,-1,0,1, and 1,3,5,7,9,...,99. When a finite sequence has more terms than we care to write down, we use three dots (...) to designate the numbers we are not listing individually. The three sequences given above all have a clearly discernable pattern, but that is not a requirement for sequences in general. It does happen that

most of the sequences that are important in mathematics do have some kind of pattern that allows one to find a general formula for the nth term, or, at least, a recursive formula that allows one to link each term to the one preceding it in the sequence. For example, in the sequence 1,2,4,8,16,32, we can see that each term is a power of 2. The first term is $2^0 = 1$; the second term is $2^1 = 2$; the third term is 2^2; and so on. Therefore we can say that, in general, the nth term of this sequence is given by 2^{n-1}. We can also describe this sequence by a recursive rule that relates each term of the sequence to the one before it. Since we obtain each term after the first by multiplying the preceding term by 2, our recursive **definition** could be written as $a_1 = 1$ and $a_{k+1} = 2a_k$. Note that when writing the recursive rule we must give the value of the first term and the relationship of each term to the one preceding it in the sequence. For the sequence 1,3,5,7,9,...,99, each term is an odd number that is 2 greater than the term before it. The general term can be written as $a_n = 2n - 1$. A recursive rule would be $a_1 = 1$ and $a_{k+1} = a_k + 2$. The idea is that if you know the first term of a sequence and the rule which tells you how to get any term from the one preceding it, then you know everything about the sequence. In principle, you could write down the first term, use it to get the second term, use the second term to get the third, and so on. A recursive rule is very effective in programming a computer or **calculator** to generate hundreds or thousands of terms of a sequence, as is sometimes done in computer simulations of some natural processes in which each step in the process depends on the preceding step.

Two of the most important types of sequences in mathematics are geometric sequences and **arithmetic** sequences. We have seen an example of each kind in the preceding paragraph. The sequence 1,2,4,8,16,32 is geometric, also called exponential because of its relationship to **exponential functions**. In fact, each member of this sequence is a value of the exponential function $f(x) = 2^x$. In general, a finite geometric sequence looks like $a, ar, ar^2, ar^3, ..., ar^{n-1}$. Using this notation, a is the first term of the geometric sequence and r is called the "common ratio." Essentially, one obtains each term of this sequence by multiplying the preceding term by r, hence the name common ratio. In 1,2,4,8,16,32, the first term is 1 and the common ratio is 2. The sequence 1,3,5,7,9,...,99 is arithmetic. A general notation for finite arithmetic sequences is a, a+d, a+2d, a+3d,..., a+(n-1)d. Again a is the first term, as in the geometric sequence, but the d is called the "common difference." Here one obtains a term of the sequence by adding d to term before it. In the sequence 1,3,5,7,9,..., 99, the first term is 1 and the common difference is 2.

Once we have a basic understanding of finite sequences, we can give the definition of a finite series: A finite series is the sum of the terms of a finite sequence. So, whereas, 1,2,4,8,16,32 is a finite sequence, $1 + 2 + 4 + 8 + 16 + 32$ is a finite series. In fact, it is called a finite **geometric series** because it is the sum of the terms of a finite geometric sequence. Likewise, $1 + 3 + 5 + 7 + 9 +... + 99$ is a finite **arithmetic series**. Thus, given any finite sequence, we can create a corresponding finite series by simply adding together all the terms of the sequence. An understanding of finite **sequences**

and series forms the necessary background for the study of infinite sequences and series, which are studied extensively in **calculus** and other advanced mathematics courses.

FIRST-ORDER ORDINARY DIFFERENTIAL EQUATIONS

Ordinary **differential equations**, sometimes abbreviated as ODEs, are **equations** comprised of a function containing an unknown and the derivatives of that function. Since the order of an ordinary differential equation is the order of the highest-order derivative of the function appearing in the equation a first-order ordinary differential equation is one that contains the first derivative of the function. They commonly have the form: $dy/dx = f(x, y)$, where $f(x, y)$ is a function of x and y, and dy/dx is the first derivative of that function with respect to x. A solution to a first-order ordinary differential equation is any function y that satisfies that differential equation. First-order ordinary differential equations have one linearly independent solution. They can describe the change in the size of a population, the **motion** of a falling body, the flow of current in an electric circuit, and the motion of a pendulum just to mention a few. First-order ordinary differential equations have a wide variety of uses and as such are used in a variety of fields such as chemistry, physics, economics, engineering, and electronics.

There are different types of first-order ordinary differential equations described as linear, exact, separable, homogeneous or cross multiple. The methods of finding solutions depend upon the classification of the specific equation. If a first-order ordinary differential equation has the form: $dy/dx + P(x)y = Q(x)$ where P and Q are continuous then it is said to be a linear first-order differential equation. This type of equation can be solved by the method of finding an integrating **factor**. This method involves finding a function by which the original ordinary differential equation can be multiplied to make it integrable. When $Q(x) = 0$ the linear first-order differential equation is said to be homogeneous as well. If the linear first-order ordinary differential equation is nonhomogeneous, if $Q(x) \neq 0$, then another method called variation of parameters can be employed to find the solution.

If a first-order differential equation is of the form: $p(x, y)dx + q(x, y)dy = 0$ and $\partial p/\partial y = \partial q/\partial x$ then it is said to be exact and means that a conservative **field** exists and that a scalar potential can be defined. The solution to a first-order ordinary differential equation that is exact is given by:

$$\int_{(x_0,y_0)}^{(x,y)} p(x,y)dx + q(x,y)dy = 0$$

If $\partial p/\partial y \neq \partial q/\partial x$ then the first-order ordinary differential equation is said to be inexact and may be solved by defining an integrating factor so that the resulting equation becomes exact and can be solved as above.

If a first-order ordinary differential equation can be expressed as: $dy/dx = X(x)Y(y)$ it is called a separable equation and can be solved using a technique called separation of variables. This involves rearranging the above form and integrating both sides to yield the solution: $\int dy/Y(y) = \int X(x)dx$.

Although there are many methods for solving the different classes of first-order ordinary differential equations the only practical solution method for very complicated equations is to use numerical methods. All solutions to first-order ordinary differential equations satisfy existence and uniqueness properties. That is if one solution can be found for a problem then it is said that a solution exists and that it is the only solution. The methods for solving first-order ordinary differential equations are very important because every higher order ordinary differential equation can be expressed as a system of first-order differential equations. In some cases it is helpful to break a higher order differential equation down into a system of first-order equations and to solve them.

FISHER, RONALD AYLMER (1890-1962)

English statistician and geneticist

Sir Ronald Aylmer Fisher formalized and extended the field of **statistics** and revolutionized the concept of experimental design. He worked for 14 years as a research statistician and later held professorships in genetics, another field to which he made significant contributions. Fisher wrote about 300 papers and seven books during his prodigious career.

The youngest of seven children, Ronald Fisher was born on February 17, 1890, in a northern suburb of London. His father, George Fisher, was a partner in a fine arts auction firm. Because of poor eyesight, the young Fisher was not allowed to read or write under artificial **light**. Consequently, he rarely took notes at lectures he attended, and he preferred to solve problems mentally rather than on paper. He developed a facility for visualizing complex geometrical relationships in his mind. This ability later proved fruitful, as his geometrical interpretation of statistics led him to previously unattainable results.

In 1909, Fisher earned a scholarship to attend Gonville and Caius College in Cambridge, where he specialized in mathematics and theoretical physics while also studying genetics. He graduated in 1912. For the next six years, Fisher searched for the right type of occupation, even working briefly as a farm laborer in Canada. Primarily, however, he worked as a statistician for the Mercantile and General Investment Company in London (1913 to 1915) and as a public school teacher (1915 to 1919). He was unhappy and, apparently, ineffective as a teacher—throughout his career, he was recognized as a brilliant thinker who had difficulty explaining his ideas to others. In 1917, he married Ruth Eileen Guinness; they had eight children and eventually separated.

Even though his jobs did not support research, Fisher published several papers during this period. One of his earliest accomplishments in statistics was determining the distribution of the **correlation** coefficient in normal samples, a problem he solved by formulating it in terms of n–dimensional Euclidean **space** with n representing the sample size. He also wrote two

Ronald Aylmer Fisher

papers on eugenics (the long-discredited science of improving the human race through selective mating); he was concerned that the lower classes produced offspring at a faster rate than the upper classes (this bigotry influenced his choice to have a large family).

Because of his growing reputation as a mathematician, in 1919 Fisher was offered a job analyzing a 66–year accumulation of statistical data at the Rothamsted Experimental Station, an agricultural research laboratory about 25 miles north of London. For the next 14 years, Fisher took advantage of the huge data resources at Rothamsted to derive new analysis techniques as well as agricultural results. He formulated the analysis of variance, which is now a fundamental tool of statistical analysis; it isolates the effects of several variables in an experiment, showing what contribution each made to the results. Consequently, he advocated **factorial** experimentation (in which several factors can vary simultaneously) rather than attempting to vary only one **factor** at a time. This not only speeds results by gathering information on the effects of several variables at one time, but it also allows investigation of interactions of variables that differ from any of the variables acting alone.

In another innovation of experimental design, Fisher advocated random arrangement of samples receiving different treatments. Traditional agricultural experiments arranged

samples according to elaborate placement schemes on checkerboard plots in an attempt to avoid bias from extraneous factors such as variations in soil and exposure to weather. Fisher demonstrated that by assigning these positions randomly, rather than according to a systematic pattern, facilitated statistical analysis of the results. His 1925 textbook, *Statistical Methods for Research Workers,* is considered a landmark work in the field. Unfortunately, it is so difficult to read that Fisher's friend and colleague M. G. Kendall wrote in *Studies in the History of Statistics and Probability,* "Somebody once said that no student should attempt to read it unless he had read it before."

During the course of his career, Fisher's theoretical work also included improvements to the Helmert–Pearson 2 distribution (including the **addition** of degrees of freedom) and the Student's t distribution, and development of what would eventually be called the F–distribution in his honor. He introduced the concept of the "null hypothesis" and developed procedures for decision-making based on the percentage difference between experimental results and the null hypothesis. He derived the distributions of numerous statistical **functions** including partial and multiple correlation coefficients and the regression coefficient, and he clarified the concept of the maximum likelihood estimate.

Fisher became a Fellow of the Royal Society in 1929, the same year he published a paper on sampling moments that would provide the foundation for future development of that topic. During the 1930s, he wrote several substantial papers on the logic of inductive **inference**.

Fisher left Rothamsted in 1933 to occupy the Galton Chair of Eugenics at University College in London, only to return during World War II when his department was evacuated to the Experimental Station. In 1935, he established a blood–typing department in the Galton Laboratory, which developed important information on Rh factor inheritance. That same year, he published *Design of Experiments*, another landmark text in statistical science. The following year, he published his first paper on **discriminant** analysis, which is now used in such areas as weather forecasting, medical research, and educational testing. During summer lectureships at Iowa State College's agricultural research center at Ames in 1931 and 1936, Fisher established contacts that helped popularize his techniques among American educators and psychologists as well as agriculturalists.

In 1943, Fisher joined the University of Cambridge as Balfour Professor of Genetics. He was knighted in 1952, and served as president of the Royal Society from 1952 until 1954. Both the Royal Society and the Royal Statistical Society awarded him several prestigious medals during his tenure at the University of Cambridge. In 1950, Fisher published *Contributions to Mathematical Statistics*, an annotated collection of 43 of his most significant papers, many of which had originally appeared in rather obscure journals. He formally retired in 1957, but continued working until a successor was found in 1959.

When Fisher left Cambridge in 1959, he moved to Adelaide, Australia, to join several of his former students and work as a statistical researcher for the Commonwealth Scientific and Industrial Research Organization. He died on July 29, 1962, as a result of an embolism following an intestinal disorder.

FLÜGGE-LOTZ, IRMGARD (1903-1974)
German-born American applied mathematician

Flügge-Lotz conducted pioneering studies of aircraft wing lift distribution and made significant contributions to modern aeronautic design. She served as an advisor to the National Aeronautics and Space Administration (NASA) as well as to German and French research institutes. During an era when there were few women engineers, Flügge-Lotz was named the first female professor in Stanford University's College of Engineering. Describing her 20-year career at Stanford, John R. Spreiter, and Wilhelm Flügge wrote in *Women of Mathematics,* "her work in fluid **mechanics** was directed toward developing numerical methods for the accurate solution of problems in compressible boundary-layer theory. She pioneered the use of finite-difference methods for such purposes and was quick to employ the emerging capability of computers to deal with the large computations inherent in the use of these methods... She applied these methods to solve a series of important and previously unsolved problems in compressible boundary-layer theory."

Flügge-Lotz was born on July 16, 1903, in Hameln, Germany. Her father, Oskar Lotz, was a journalist and amateur mathematician. Her mother, Dora (Grupe) Lotz, came from a family that had been in the construction business for generations. Visiting the firm's building sites, in addition to watching Count von Zeppelin conduct airship tests, fueled her interest in engineering. When she was a teenager, Flügge-Lotz's father was drafted into the German army during World War I. To help support her mother and younger sister, she began tutoring fellow students in mathematics and Latin. She continued this work after her father returned from the war in ill health. After graduating from high school in 1923, she studied **applied mathematics** and engineering at the Technical University of Hanover. In 1929, she earned a doctorate in engineering. Her dissertation explored the mathematical theory of circular cylinders and heat conduction.

Opportunities for women in engineering were limited in the 1930s, and Flügge-Lotz had difficulty finding a level of employment that was commensurate with her education. When she began working at the Aerodynamische Veruchsanstalt (AVA) research institute in Göttingen, she spent half of her time as a cataloguer. Perceptions of Flügge-Lotz's abilities changed dramatically after she solved a problem that had stymied Ludwig Prandtl and Albert Betz, two leading aerodynamicists at the AVA (Betz was the director of the institute). A decade earlier, Prandtl had developed a differential equation for his theory about the lift distribution of an airplane wing. However, he had made little progress in solving the equation. Flügge-Lotz tackled the problem and solved the equation for the general case. Continuing to work with the

equation, she developed it so that it had widespread practical applications. Her cataloguing days ended, and she was named supervisor of a group of engineers who researched theoretical aerodynamics within the AVA.

In 1931, Flügge-Lotz published a technique she had developed for calculating the lift distribution on aircraft wings. Later dubbed the "Lotz method,"it is still used today. Continuing to delve into wing theory, she added to the knowledge base of the effects of control surfaces, propeller slipstream, and wind-tunnel wall interference.

The course of Flügge-Lotz's career changed in 1938 when she married Wilhelm Flügge, a civil engineer from Göttingen. The husband and wife team went to work at Berlin's Deutsche Versuchsansalt für Luftfahrt (DVL), a German agency similar to NASA. Beginning work as a consultant in flight and aerodynamics, Flügge-Lotz conducted groundbreaking research in automatic control theory, especially pertaining to on-off controls. Subsequently, these controls came into widespread use because they were reliable and inexpensive to build.

In 1944 Flügge-Lotz and her husband moved to the small town of Saulgau, where they continued to work for the DVL. After Germany's defeat in World War II, the town and its surrounding area became part of France. The Flügges joined the staff of the French National Office for Aeronautical Research (ONERA) in Paris. Flügge-Lotz headed a research group in theoretical aerodynamics and continued her work in automatic control theory. In 1948 the couple accepted positions at Stanford University in California.

Flügge was hired as a professor at Stanford. Nepotism regulations prohibited his wife from holding a similar position, so she was hired as a lecturer. Flügge-Lotz developed graduate and undergraduate courses in mathematical hydro- and aerodynamics and automatic control theory. She also designed a weekly seminar in fluid mechanics that was attended not only by Stanford students but also by young engineers from the National Advisory Committee for Aeronautics—NASA's predecessor. The seminar has continued to serve as an important forum for faculty and students of varying specializations to share their findings. The Flügges had no children, but they frequently invited their students to their home for dinner parties with other faculty members and visitors.

In addition to teaching, Flügge-Lotz also continued her research while at Stanford. She applied numerical methods and analog computer simulations to boundary layer problems in fluid dynamics and continued studying automatic control theory. Her 1953 book, *Discontinuous Automatic Control,* presented the theoretical foundation for using discontinuous controls for flight paths of missiles. Specifically, the problem concerns a missile in flight encountering some **force** (such as a wind gust) that causes it to begin oscillating. An automatic control mechanism could be used to counteract the **vibration**. Flügge-Lotz was particularly interested in discontinuous controls, which could be activated only when needed. A simple analogy is a furnace thermostat; rather than continuously adjusting the amount of heat generated by a furnace, the thermostat simply turns the furnace on when needed and off when the proper temperature is achieved. The fact that discontinuous automatic controls were simpler and cheaper to make than continuous-control mechanisms made them particularly desirable for one-time use in missiles that would crash after flight.

Flügge-Lotz's second book, published in 1968, presented techniques for optimizing discontinuous automatic controls for objects such as airplanes, rockets, and satellites. In the preface to *Discontinuous and Optimal Control*, she noted that optimal control theory literature was difficult for practicing engineers to read because it was so mathematically rigorous. "The purpose of this book," she wrote, "is to acquaint the reader with the problem of discontinuous control by presenting the essential phenomena in simple examples before guiding him to an understanding of systems of higher order."

Flügge-Lotz attended the First Congress of the International Federation of Automatic Control in Moscow as the only female delegate from the United States. Upon her return from that congress, Stanford University finally offered her a professorship in 1961. In fact, she was granted that rank in two departments—aeronautics and astronautics, and engineering mechanics. She retired from teaching seven years later. Flügge-Lotz continued her research activities, however, studying heat transfer and the control of satellites. Her retirement years were physically difficult, due to progressive arthritis.

The Society of Women Engineers gave Flügge-Lotz its Achievement Award in 1970. The American Institute of Aeronautics and Astronautics (AIAA) chose her to deliver the prestigious annual von Kármán Lecture in 1971 and elected her as its first woman fellow. The University of Maryland awarded her an honorary degree, citing "contributions [that] have spanned a lifetime during which she demonstrated, in a field dominated by men, the value and quality of a woman's intuitive approach in searching for and discovering solutions to complex engineering problems." Flügge-Lotz died a year later, on May 22, 1974.

FLUXIONS

In 1665 **Isaac Newton**, an English mathematician, introduced the concept of a fluxion. Fluxion was the early term for what we now call a derivative, which is just the instantaneous rate of change of a function with respect to a **variable**.

While Newton was at Cambridge the university was closed because of the plague of 1665. During this time Newton commenced his studies to revolutionize mathematics, optics, physics and astronomy. He laid the foundations for differential and integral **calculus** several years before its independent discovery by **Leibniz**. It was during this time that he created his method of fluxions that was based on his important insight that the integral of a function is simply the result of the inverse procedure applied to a derivative. Newton visualized a curve as an entity being traced by a particle moving according to two moving lines that were the coordinates. The velocity of the particle was defined as the derivative of the *x*-line and the

derivative of the *y*-line. These derivatives were the fluxions of *x* and *y* coordinates connected with the flux of time. *x* and *y* were the flowing quantities themselves. Using the fluxion notation that he developed he described the tangent to the function $f(x,y) = 0$ as y'/x', where y' and x' are the derivatives. Using the fluxion he developed simple analytical methods that unified the separate techniques employed to find areas, tangents, the lengths of curves and the maxima and minima of **functions**. Newton considered integration a process by which fluents, or functions in modern terms, were determined for a given fluxion, or derivative in modern terms. His method implied that integration and differentiation were inverse procedures. Newton wrote *Method of fluxions and infinite series* detailing the development of his methods in 1671. The book was eventually published in English in 1736. It was later, after G. W. Leibniz independently developed his methods of differential and **integral calculus**, that Newton's notations and terms of fluxions were discarded and replaced by Leibniz's terms of **derivatives and differentials**.

FORCE

The concept of force is vital to **Isaac Newton**'s formulation of classical physics. The force is the first derivative of momentum with respect to time. That is, it's the rate of change of momentum with respect to time. While in many cases the mass is constant, leaving the force with only one term, the product of mass and acceleration, this can by no means be assumed, as there are also many cases of changing mass in the physical world. An alternate, though equivalent, view of force, is as a position-respective directional derivative (or gradient) of the potential energy. Using these two views of force, many physical problems can be solved which would be otherwise intractable. In most introductory physical problems, the force equation is the easiest way to find a solution. Newton's Second Law, $F = dp/dt$, indicates that the sum of the forces on an object equals the derivative of its momentum over time. It should be noted that force is a vector quantity, existing in three dimensions, rather than a scalar. While this may make energy considerations (which are scalars) easier to deal with in more complex problems, eventually anything which deals with force has to be translated back into three-dimensions, usually using the gradient.

Sometimes fictitious forces may arise—that is, forces which appear due to inertia and the choice of the coordinate system rather than due to any actual action or reaction. These include the centrifugal force, the Coriolis force, the azimuthal force, and the translational force. Centrifugal force, Coriolis force, and azimuthal force are due to **rotation** of a coordinate system and can relate to anything from falling off a merry-go-round to aiming ballistics in the Northern hemisphere of the Earth versus in the Southern hemisphere. Translational forces appear when a coordinate system is in accelerated **motion** itself. An example of these forces is when a pendulum in a moving car swings when the car is brought to a halt: no actual force acts upon the pendulum, but inertia makes a fictitious force show up to the observer's eye.

The physical world features four known basic forces: strong, weak, electromagnetic, and gravitational, in order of decreasing strength. The electromagnetic and weak forces can be unified at high energies to be the electroweak force. While there has been much talk of unifying the other forces with the electroweak and with each other, this has not yet been accomplished to the satisfaction of the vast majority of physicists, and it is not yet known whether it will ever be possible or whether these forces are fundamentally different. All other forces—friction, spring compression forces, and others commonly used in physics and engineering problems—are at some level an expression of the four fundamental forces.

See also Isaac Newton

FORMALISM

Formalism is the mathematical school of thought which holds that mathematics consists of symbols, rules for combining those symbols, some minimal number of assumptions or axioms, and certain agreed upon rules of **inference**. Formalism was introduced in the early twentieth century by the great German mathematician, **David Hilbert** (1862-1943), in response to a certain uneasiness that had arisen among some mathematicians concerning the logical foundations of mathematics. The German logician **Gottlob Frege** (1848-1925), had made an attempt to derive all of the laws of **arithmetic** from logic alone, but the young British logician **Bertrand Russell** (1872-1970) discovered a paradox in Frege's system which doomed it to failure. Russell, with **Alfred North Whitehead** (1861-1947), made modifications in Frege's work to eliminate Russell's paradox and others that had been discovered by other mathematicians. Russell and Whitehead produced the massive three-volume work *Principia Mathematica*, which, they claimed, did the job Frege had set out to do. Nevertheless, there remained an undercurrent of opinion among mathematicians that Russell and Whitehead had made too many concessions to eliminate the paradoxes. Rather than arguing the fine points of logic, Hilbert proposed a different way of looking at the foundations of mathematics. Hilbert's formalistic program essentially reduces mathematics to a game, albeit an important game, in which a very strict set of rules must be followed if one is to be allowed to play the game. The equipment needed for playing the game consists of the symbols of the mathematical system, stripped of meaning, except the meaning they are given by the rules of syntax by which they may be combined. The rules for playing the game are the rules of logical inference. The goal of the game is to prove, consistently, i.e., without contradiction, all the theorems that may be stated within the system. The players, the mathematicians, are governed by the very strict rules of the game, the axioms. In this view, Euclidean **geometry** is just a game and the **non-Euclidean geometries** of **Riemann** and **Lobachevsky** are just different games played using different rules. So long as the rules are

applied consistently, the interpretation of the mathematics to any physical reality is irrelevant. In Hilbert's formalism, mathematics is not about symbols which stand for idealized physical objccts as in Platonist mathcmatics. For the formalist, mathematics is about the symbols themselves, which are devoid of any outside meaning. The symbols are strung together according to very strict syntactical rules to make formulas that may then be used in the proving of theorems. The proving of theorems, by established rules of inference, is the point of mathematics for the formalist. Under Hilbert's plan, theorem-proving would become a mechanical procedure.

Hilbert required that a formal mathematical system should meet three standards: consistency, completeness, and decidability. Consistency means that no contradictions will be found in the system, i.e., a **theorem** and its negation cannot both be true. Completeness implies that every theorem which is true in the system can be proved within the system. Decidability requires that a finite mechanical procedure exists that determines whether any given claim made by the system may be proved within the system. In 1931, the brilliant logician **Kurt Gödel** (1906-1978) published a paper showing that any formal system containing the natural numbers cannot be both complete and consistent. If a system is complete, it will contain a contradiction. If it is consistent, then it will not include all possible theorems of the system. Gödel's result doomed **Hilbert's program**, as well as Russell and Whitehead's, to failure. Nevertheless, the influence of Hibert's formalism did not end with Gödel's publication. Although formalism was unsuited for a philosophy of mathematics, the idea of a mechanical process for computation was adopted by the British logician, **Alan Turing** (1912-1954), and became the basis for the digital computer which would spur the information revolution of the late 20th century. Turing proposed a machine that could be instructed or programmed to make any computation that could be carried out by a human being with pencil and paper and a sufficient amount of time. The set of instructions given to such a machine would come to be known as a program. Modern computer programs are written in formal languages which can be traced back to Hilbert's desire to formalize the entire discipline of mathematics. Gödel's work showed that this desire was untenable as a foundation upon which to build mathematics, but the position of software developers as among the largest corporations in the world demonstrates that formalism was by no means a dead end.

FOUR-COLOR MAP THEOREM

The four-color map **theorem** states that, at most, four colors are needed to color any map, if two adjoining countries are always to be assigned different colors. The statement of the theorem is deceptively simple: it says that every imaginable map—regardless of how complicated the countries are and how many countries adjoin any given country—can be colored using only four colors. For the purposes of this problem, two countries are considered to adjoin each other if the boundary that they share has non-zero length, so that two countries that only meet each other in a corner can be given the same color (this fits in with commonsense principles of map-coloring, since there is no danger of confusion when two countries that only meet at a corner have the same color). So for example, a chessboard can be colored using only two colors, black and white.

Although the four-color theorem can be described in language that a child can understand, its **proof** is so difficult that more than one hundred years passed between the first statement of the problem and its successful proof. During that **interval**, many of the finest mathematicians worked on the problem, and one invalid proof was accepted as correct for more than a decade. The correct proof, discovered in 1976, was no less controversial, for it made unprecedented use of computers to do calculations that mathematicians could not check by hand. This brought into question the feasibility of examining the proof for errors, a key part of the acceptance of a proof by the mathematical community.

The first person to pose the map-coloring question appears to have been Francis Guthrie, a student at University College London. In 1852, Guthrie passed the question on through his brother to **Augustus De Morgan**, who immediately wrote to **William Rowan Hamilton**,

"A student of mine asked me today to give him a reason for a fact which I did not know was a fact—and do not yet. He says that if a figure be anyhow divided and the compartments differently coloured so that figures with any portion of common boundary line are differently coloured—four colours may be wanted, but not more.... If you retort with some very simple case which makes me out a stupid animal, I think I must do as the Sphinx did...."

Hamilton's reply was quick and terse: "I am not likely to attempt your quaternion of colour very soon."

De Morgan continued to publicize the new problem, and soon it had attracted a good deal of attention. In 1879 Alfred Bray Kempe announced that he had proven the four-color theorem. His argument hinged on a method to color the countries one at a time, sometimes adjusting the color of previously colored countries to allow one of the colors to be used for a new country. Kempe's proof was hailed by the mathematical community, and Kempe was elected a fellow of the Royal Society. However, in 1890 Percy John Heawood found errors in the proof that Kempe was unable to correct, and the four-color theorem was returned to the status of conjecture.

In the decades that followed, mathematicians tried to shed light on the problem by relating it to the ideas of graph theory. A graph is a collection of points, called vertices, that are connected to each other by curves, called edges. A map can be turned into a graph by placing a vertex inside each country, and then connecting two vertices by an edge if the two countries share a boundary. A graph of this type will be a planar graph, that is, a graph whose edges never cross each other. The four color conjecture turns into the question, Is it possible to color each vertex of the graph using four colors, if two vertices that are connected by an edge are required to have different colors?

This translation of the problem, although elementary, transformed it from a problem about maps, which could have arbitrarily complicated boundary curves, to a question about graphs, which could be represented by a few points and edges. Mathematicians began to ask whether planar graphs could be put into a finite number of categories, so that each type of configuration could be analyzed to see if it was four-colorable.

In 1976 the mathematicians **Kenneth Appel** and **Wolfgang Haken** finally proved the four-color theorem, breaking down the different kinds of graphs into more than 1500 different configurations. Then, using a combination of ingenuity and computing power, they showed that each of those configurations could be colored with four colors. To perform the calculations, they used more than 1000 hours of computing time on three high-speed computers. Recently, the mathematicians Robertson, Sanders, Seymour, and Thomas have simplified the proof, which now relies on 633 configurations.

The reliance of the proof on computers made many mathematicians uneasy, although the proof is generally regarded as correct. The great mathematician **Paul Erdös** said, "I assume the proof is true. However, it's not beautiful. I'd prefer to see a proof that gives insight into why four colors are sufficient."

FOURIER, JEAN BAPTISTE JOSEPH (1768-1830)
French mathematician

Jean Baptiste Joseph Fourier is best known for the trigonometric technique that we now know as the **Fourier series**. His life was not without excitement outside the mathematical world. During his life he was a political activist caught up in the French Revolution, imprisoned on two separate occasions, a campaigner with Napoleon, and a monarchist!

Jean Baptiste Joseph Fourier was born in the French town of Auxerre in 1768, the twelfth of fifteen children. After showing an initial aptitude for mathematics at the Ecole Royale Militaire Fourier's life changed direction in 1787. At this time he entered a Benedictine Abbey at St. Benoit-sur-Loire to train to become a priest. He still loved mathematics and he was unsure if he was making the right decision. Fourier submitted a paper in **algebra** to Montucla in Paris and held a long-running correspondence with Professor Bonard of Auxerre. Fourier subsequently left St. Benoit in 1789 without taking his final vows. After reading a paper on algebra at the Royal Academy of Sciences in Paris Fourier became a teacher in Auxerre. Fourier was caught up in the French Revolution, and he even went to prison expecting to be guillotined. At the end of the revolution Fourier was released and he returned to Paris to undertake formal teacher training, where he was taught by **Joseph-Louis Lagrange**, **Pierre-Simon Laplace**, and **Gaspard Monge**. He taught in Paris from 1795 to 1798, eventually replacing Lagrange in the chair of **analysis** and **mechanics**. It was during this period that he was briefly imprisoned

Jean Baptiste Joseph Fourier

again due to his earlier political views. When he left, it was to join Napoleon I in Egypt as scientific adviser. Whilst in Egypt Fourier was one of the founders of the Cairo Institute where he was one of the members of the mathematics division. In 1802, on his return to France (after the collapse of the French occupation of Egypt), Fourier published extensively on Egyptian antiquities, and he was the Prefect of the Department of Isere. This administrative position in Grenoble took Fourier away from his beloved research; but as it was a direct request from Napoleon I he could not refuse. In 1810 Fourier published *The Description of Egypt,* which was extensively rewritten by Napoleon (the second edition removed all references to Napoleon). Much to Fourier's delight he found he was able to continue mathematical research, and in 1807 he published *On the Propagation of Heat in Solid Bodies.* This was the first appearance of a trigonometric series that we now know as the Fourier series. This publication won the Paris Institutes mathematics prize in 1811. In 1816 he was elected to the Academy of Sciences and in 1827 the French Academy. In 1822 Fourier was elected to the position of Secretary to the mathematical section of the Academy of Sciences. In the same year the Academy published *The Analytical Theory of Heat.* Fourier was created a Baron by Napoleon in 1808. Fourier eventually died in Paris in 1830 at the age of 62.

FOURIER SERIES

A Fourier series is the expansion of any other function in terms of the Fourier **functions** of **sine** (nx) and **cosine** (nx). (These can also be expressed as a combination of positive and negative exponentials.) Since these functions compose a **Hilbert space** or a basis set, any continuous function can be composed of a combination of them as long as it has a finite number of finite discontinuities and a finite number of extrema in the **interval** over which the Fourier series is summed, usually 0 to 2 **Pi**. That is, as long as there's a set number of places where the function is not continuous, and as long as there are not an infinite number of local maxima and minima, the Fourier series can express the function. Each function will have a unique Fourier series expansion.

The Fourier coefficients, that is, the numbers multiplied by each sine or cosine term, may be found by taking the integral of the function times the desired trigonometric term. In many cases, these integrals will be very similar from term to term, providing a generalized way of expressing the coefficients. The theory that allows this formula to work is called Sturm-Liouville Theory. The interval in which the series holds can be changed by simply dividing the original denominator in the sine or cosine expression by the length of the new interval.

At a discontinuity, the Fourier series will converge on the **arithmetic mean** of the terms on either side of the discontinuity. The discontinuity must be finite for this formulation (or the series itself) to work: a jump from one to **zero** is fine, but from one to **infinity** is not.

Fourier series yield a continuous function as long as they are expressions of continuous functions. In this case, they will also be uniformly convergent. On the average of any given function, even discontinuous functions will yield convergent Fourier series results.

Fourier series expansions are used in most applications dealing with wave **mechanics**, including optics and acoustics. Since any periodic function meeting the above conditions can be described as the superposition of waves in a Fourier series, sometimes one will speak of "Fourier decomposing" a function, or breaking it up into its component Fourier series parts. Fourier **analysis** is also common for filtering in optics, although this generally uses Fourier transformations rather than Fourier series. They are also used in electronics applications when dealing with a periodic signal. With a wide variety of physical applications, Fourier series are one of the most commonly used mathematical tools of physicists and engineers.

FRACTAL

A fractal is a geometric figure, often characterized as being self-similar; that is, irregular, fractured, fragmented, or loosely connected in appearance. **Benoit Mandelbrot** coined the term fractal to describe such figures, deriving the word from the Latin "fractus" meaning broken, fragmented, or irregular. He also pointed out amazing similarities in appearance between some fractal **sets** and many natural geometric patterns. Thus, the term "natural fractal" refers to natural phenomena that are similar to fractal sets, such as the path followed by a dust particle as it bounces about in the air.

Another good example of a natural phenomenon that is similar to a fractal is a coastline, because it exhibits three important properties that are typical of fractals. First, a coastline is irregular, consisting of bays, harbors, and peninsulas. Second, the irregularity is basically the same at all levels of magnification. Whether viewed from **orbit** high above the earth, from a helicopter, or from land, whether viewed with the naked eye, or a magnifying glass, every coastline is similar to itself. While the patterns are not precisely the same at each level of magnification, the essential features of a coastline are observed at each level. Third, the length of a coastline depends on the magnification at which it is measured. Measuring the length of a coastline on a photograph taken from **space** will only give an estimate of the length, because many small bays and peninsulas will not appear, and the lengths of their perimeters will be excluded from the estimate. A better estimate can be obtained using a photograph taken from a helicopter. Some detail will still be missing, but many of the features missing in the space photo will be included, so the estimate will be longer and closer to what might be termed the "actual" length of the coastline. This estimate can be improved further by walking the coastline wearing a pedometer. Again, a longer measurement will result, perhaps more nearly equal to the "actual" length, but still an estimate, because many parts of a coastline are made up of rocks and pebbles that are smaller than the length of an average stride. Successively better estimates can be made by increasing the level of magnification, and each successive measurement will find the coastline longer. Eventually, the level of magnification must achieve atomic or even nuclear resolution to allow measurement of the irregularities in each grain of sand, each clump of dirt, and each tiny pebble, until the length appears to become infinite. This problematic result suggests the length of every coastline is the same.

The resolution of the problem lies in the fact that fractals are properly characterized in terms of their **dimension**, rather than their length, **area**, or **volume**, with typical fractals described as having a dimension that is not an integer. To explain how this can happen, it is necessary to consider the meaning of dimension. The notion of dimension dates from the ancient Greeks, perhaps as early as **Pythagoras** (582-507 B.C.) but at least from **Euclid** (c. 300 B.C.) and his books on **geometry**. Intuitively, we think of dimension as being equal to the number of coordinates required to describe an object. For instance, a line has dimension 1, a **square** has dimension 2, and a cube has dimension 3. This is called the topological dimension. However, between the years 1875 and 1925, mathematicians realized that a more rigorous **definition** of dimension was needed in order to understand extremely irregular and fragmented sets. They found that no single definition of dimension was complete and useful under all circumstances. Thus, several definitions of dimension remain today. Among them, the **Hausdorf dimension**, proposed by **Felix Hausdorf**,

results in fractional dimensions when an object is a fractal, but is the same as the topological value of dimension for regular geometric shapes. It is based on the increase in length, area, or volume that is measured when a fractal object is magnified by a fixed scale **factor**. For example, the Hausdorf dimension of a coastline is defined as $D = \log(\text{Length Increase})/\log(\text{scale factor})$. If the length of a coastline increases by a factor of four whenever it is magnified by a factor of three, then its Hausdorf dimension is given by $\log(\text{Length Increase})/\log(\text{scale factor}) = \log(4)/\log(3) = 1.26$ Thus, it is not the length that properly characterizes a coastline but its Hausdorf dimension. Finally, then, a fractal set is defined as a set of points on a line, in a plane, or in space, having a fragmented or irregular appearance at all levels of magnification, with a Hausdorf dimension that is strictly greater than its topological dimension.

Great interest in fractal sets stems from the fact that most natural objects look more like fractals than they do like regular geometric figures. For example, clouds, trees, and mountains look more like fractal figures than they do like circles, triangles, or pyramids. Thus, fractal sets are used by geologists to **model** the meandering paths of rivers and the rock formations of mountains, by botanists to model the branching patterns of trees and shrubs, by astronomers to model the distribution of mass in the universe, by physiologists to model the human circulatory system, by physicists and engineers to model turbulence in fluids, and by economists to model the stock market and world economics. Often times, fractal sets can be generated by rather simple rules. For instance, a fractal dust is obtained by starting with a line segment and removing the center one-third, then removing the center one-third of the remaining two segments, then the center one-third of those remaining segments and so on. Rules of generation such as this are easily implemented and displayed graphically on computers. Because some fractal sets resemble mountains, islands, or coastlines, while others appear to be clouds or snowflakes, fractals have become important in graphic art and the production of special effects. For example, "fake" worlds, generated by computer, are used in science fiction movies and television series, on CD-ROMs, and in video games, because they are easily generated from a set of instructions that occupy relatively little computer memory.

FRACTIONS

Fraction is the name for part of something as distinct from the whole of it. The word itself means a small amount as, for example, when we ask someone to "move over a fraction." We mean them to move over part of the way, not all the way.

Fractional parts such as half, quarter, eighth, and so on form a part of daily language usage. When, for example, we refer to "half an hour," "a quarter pound of coffee," or "an eighth of a pie." In **arithmetic**, the word fraction has a more precise meaning since a fraction is a numeral. Most fractions are called common fractions to distinguish them from special kinds of fractions like **decimal fractions**.

A fraction is written as two stacked numerals with a line between them, e.g., $\frac{3}{4}$, which refers to three-fourths (also called three quarters). All fractions are read this way.

$\frac{5}{9}$ is called five-ninths and 5, the top figure, is known as the numerator, while the bottom figure, 9, is called the denominator.

A fraction expresses a relationship between the fractional parts to the whole. For example, the fraction $\frac{3}{4}$ shows that something has been divided into four equal parts and that we are dealing with three of them. The denominator denotes into how many equal parts the whole has been divided. A numerator names how many of the parts we are taking. If we divide something into four parts and only take one of them, we show it as $\frac{1}{4}$. This is known as a unit fraction.

Whole numbers can also be shown by fractions. The fraction $\frac{5}{1}$ means five wholes, which is also shown by 5.

Another way of thinking about the fraction $\frac{3}{4}$ is to see it as expressing the relationship between a number of items set apart from the larger group. For example, if there are 16 books in the classroom and 12 are collected, then the relationship between the part taken (12) and the larger group (16) is 12/16. The fraction 12/16 names the same number as $\frac{3}{4}$. Two fractions that stand for the same number are known as equivalent fractions.

A third way of thinking about the fraction $\frac{3}{4}$ is to think of it as measurement or as a **division** problem. In essence the symbol $\frac{3}{4}$ says: take three units and divide them into four equal parts. The answer may be shown graphically. The size of each part may be seen to be $\frac{3}{4}$.

To think about a fraction as a measurement problem is a useful way to help understand the operation of division with fractions which will be explained later.

A fourth way of thinking about $\frac{3}{4}$ is as expressing a ratio. A ratio is a comparison between two numbers. For example, 3 is to 4, as 6 is to 8, as 12 is to 16, and 24 is to 32. One number can be shown by many different fractions provided the relationship between the two parts of the fraction does not change. This is most important in order to add or subtract, processes which will be considered next.

Fractions represent numbers and, as numbers, they can be combined by **addition**, **subtraction**, **multiplication**, and division. Addition and subtraction of fractions present no problems when the fractions have the same denominator. For example, $\frac{1}{8} + \frac{5}{8} = \frac{6}{8}$. We are adding like fractional parts, so we ignore the denominators and add the numerators. The same holds for subtraction. When the fractions have the same denominator we can subtract the numerators and ignore the denominators. For example, $\frac{5}{6} - \frac{4}{6} = \frac{1}{6}$.

To add and subtract fractions with unlike denominators, the numbers have to be renamed. For example, the problem $\frac{1}{2} + \frac{2}{3}$ requires us to change the fractions so that they have the same denominator. We try to find the lowest common denominator since this makes the calculation easier. If we write $\frac{1}{2}$ as $\frac{3}{6}$ and $\frac{2}{3}$ as $\frac{4}{6}$, the problem becomes $\frac{3}{6} + \frac{4}{6} = \frac{7}{6}$.

Similarly, with subtraction of fractions that do not have the same denominator, they have to be renamed. $\frac{3}{4}$ - 1/12 needs to become 9/12 - 1/12, which leaves 8/12.

The answer to the third addition problem was $\frac{7}{6}$, which is known as an improper fraction. It is said to be improper because the numerator is bigger than the denominator. Often an improper fraction is renamed as a mixed number which is the sum of a whole number and a fraction. Take six of the parts to make a whole (1) and show the part left over as ($\frac{1}{6}$), giving the answer $1\frac{1}{6}$.

Similarly the answer to the second subtraction problem may be reduced to the lowest terms. A fraction is not changed if you can do the same operation to the numerator as to the denominator. Both the numerator and denominator of 8/12 can be divided by four to rename the fraction as $\frac{2}{3}$. Both terms can also be multiplied by the same number and the number represented by the fraction does not change. This idea is helpful in understanding how to do division of fractions which will be considered next. When multiplying fractions the terms above the line (numerators) are multiplied, and then the terms below the line (denominators) are multiplied, e.g., $\frac{3}{4} \times \frac{1}{2} = \frac{3}{8}$.

We can also show this graphically. What we are asking is if I have half of something, (e.g., half a yard) what is $\frac{3}{4}$ of that? The answer is $\frac{3}{8}$ of a yard

It was mentioned earlier that a fraction can be thought of as a division problem. Division of fractions such as $\frac{3}{4} \div \frac{1}{2}$ may be shown as one large division problem: $\frac{3}{4}$ (N) / $\frac{1}{2}$ (D).

The easiest problem in the division of fractions is dividing by one because in any fraction that has one as the denominator, e.g., $\frac{7}{1}$, we can ignore the denominator because we have 7 wholes. So in our division problem, the question becomes what can we do to get 1 in the denominator? The answer is to multiply $\frac{1}{2}$ by its reciprocal, $\frac{2}{1}$, and it will cancel out to one. What we do to the denominator we must do to the numerator. The new equation becomes

$$(\tfrac{3}{4} \times \tfrac{2}{1} / \tfrac{4}{1} - \tfrac{1}{2} / \tfrac{1}{2} \times \tfrac{2}{1}) = (\tfrac{6}{4} / \tfrac{1}{1}) = \tfrac{6}{4} / 1 = 1\tfrac{1}{2}.$$

We can also show this graphically. What we want to know is how many times will a piece of cord $\frac{1}{2}$ inch long fit into a piece that is $\frac{3}{4}$ inch long. The answer is $1\frac{1}{2}$ times.

Fractions are of immense use in **mathematics and physics** and the application of these to modern technology. They are also of use in daily life. If you understand fractions you know that 1/125 is bigger than 1/250, so that shutter speed in photography becomes understandable. A screw of 3/16 is smaller than one of $\frac{3}{8}$, so tire sizes that are shown in fractions become meaningful rather than incomprehensible. It is more important to understand the concepts than to memorize operations of fractions.

FRÉCHET, MAURICE (1878-1973)

French topologist

Maurice Fréchet was one of the creators of the 20th-century discipline of **topology**, which deals with the properties of geometric configurations; his research added a new degree of abstractness to the mathematical advances of the previous generation. He profited from a rich mathematical environment during his studies and in turn passed on a wealth of ideas to his students over a long career. Some of the mathematicians who had learned their skills before Fréchet's work appeared could not help wondering whether there were advantages to the new degree of generality in his work. The answer lay in the fruitfulness of the methods of Fréchet for addressing problems whose solution included concrete problems of long standing.

René Maurice Fréchet was born on the 10th of September in 1878 in Maligny, a small town in provincial France, where his father Jacques directed an orphanage. Soon after Frechet's birth the family moved to Paris, much to his advantage in terms of mathematical environment. His mother Zoé was responsible for a boardinghouse for foreigners, which early put Fréchet in contact with a cosmopolitan community. This may be reflected in his subsequent hospitality toward students and collaborators from all over the world. At his lycée (high school), he was singularly fortunate in learning mathematics from **Jacques Hadamard**, already a mathematician of distinction who would shortly thereafter provide a **proof** of a central result of **number theory** called the **prime number theorem**.

Fréchet's talents blossomed under Hadamard's encouragement and he was well prepared to enter the École Normale Supérieure, the great French scientific university, in 1900. Hadamard was not the only mathematical influence on the young Fréchet. After graduating from the École Normale, he began to work with mathematician **Émile Borel**, who was only seven years older than Fréchet but who had started his career so early that he may have seemed to belong to an earlier generation. Fréchet collaborated with Borel on the publication of a series of the latter's lectures and continued to be involved with the publishing of the so-called Borel collection for Gauthier-Villars. Even though Borel's role in the collection was primarily an editor's, he also wrote all the volumes to begin with, the first exception being one written by Fréchet. In turn, Fréchet undertook the editing of a series on general **analysis** published by Hermann (the other great mathematical publisher in Paris) and undertook the writing of several of the volumes as well.

Fréchet wrote his thesis under Hadamard, who had returned to Paris, and then followed Hadamard in teaching at the level of the lycée for a few years. His marriage in 1908 to Suzanne Carrive produced four children, whom he supported with professorships outside of Paris until 1928. He was officially connected with the University at Poitiers from 1910 to 1918, but World War I took him out of mathematics and into the less familiar surroundings of working as an interpreter with the British army, where his early exposure to different languages was of help. After his return from military service, he was head of the Institute of Mathematics at the University of Strasbourg, still a provincial appointment. It was not until 1928 that he was called to the University of Paris.

One of the reasons for the delay in the recognition of Fréchet's work by the French academic establishment was its revolutionary character. The notions of **set theory** as introduced by the German mathematician **Georg Cantor** in the previous century were slowly winning converts, although there were differences of opinion about which axioms ought to be accepted. What Fréchet did in his thesis and in the most influential of his subsequent work was to bring the ideas of general set theory to bear on questions of the new discipline of topology, the generalization of **geometry** that had been given a good deal of prominence in French mathematics by the work of **Jules Henri Poincaré**. The questions that Poincaré had raised were new, but they were in the context of classical mathematics, centered on **space** with standard **Cartesian coordinates** (those points commonly expressed as located along x, y, and z axes), although perhaps in more than three dimensions.

This much of a revolution the mathematical community had come to accept, but Fréchet's thesis pushed the level of abstractness to new heights. Rather than looking just at **sets** of points in Cartesian space, he was prepared to handle sets of points in arbitrary spaces—so-called abstract spaces. The important tool that he used to handle such sets was a **distance** function. The ordinary distance function for sets of points with Cartesian coordinates (x, y) comes from the **Pythagorean theorem** and involves taking the **square root** of the sums of the squares of the differences in each coordinate. Since in abstract spaces there weren't necessarily any coordinates to assign to points, the distance function had to be more general and governed by some of the principles that applied to the Cartesian version.

The advantage of the new approach of Fréchet was that complicated algebraic expressions could be replaced by general considerations about distance. Spaces with a distance function were called metric spaces and proved to be the setting for expressing many of the results hitherto considered limited to spaces with **real numbers** as coordinates. Having once introduced these ideas into topology, Fréchet proceeded to look at **calculus** in metric spaces, an area that became known as functional analysis. Again, the basis for progress on long-standing problems was the avoidance of the complicated calculations that had bedeviled earlier work and the application of general notions from topology instead. Fréchet extended the notions of derivatives and integrals from standard calculus so that they could be used in the setting of a metric space; in addition, he introduced new types of **functions** called functionals, which took real numbers as values but could operate on the points of abstract spaces. Much of his work from his thesis onward was summarized in *Les espaces abstraits,* published in 1926.

Fréchet taught at the University of Paris until 1949, and a good deal of his time there was spent on questions of probability. Just as general questions about calculus could be asked in the setting of abstract metric spaces, so the techniques of probability could be moved there as well. The application of probability to continuous quantities, as opposed to discrete quantities that took only a finite number of values, had always been dependent on calculus, and Fréchet's results showed that the extension to the abstract setting could be fruitful as well. As with functional analysis in general, the more one could move away from messy computations, the more one could hope that the idea behind a proof could be visible.

Another possible reason for Fréchet's move into probability was the hope that a more concrete area would make the techniques of abstract spaces more palatable to the part of the mathematical community uneasy about getting too far from applications. If so, the efforts proved largely unavailing, at least in France, although the level of abstractness introduced by Fréchet was one of the inspirations for the Polish mathematical school between World Wars I and II. It is perhaps indicative of the relative opinions of his work that Fréchet was elected to the Polish Academy of Sciences in 1929 but not to the French Académie Royale des Sciences until 1956. He was recognized as a member of the Legion of Honor, and some accumulation of praise could hardly be avoided as he lived into his nineties. He died in Paris on the 4th of June in 1973, having earned belated recognition of his role in bringing mathematics into the twentieth century on the wings of abstractness.

FREDHOLM, ERIK IVAR (1866-1927)
Swedish number theorist

Erik Ivar Fredholm developed the modern theory of integral **equations**. His work served as the foundation for later critical research performed by **David Hilbert**, and several concepts and theorems are attributed to him. Fredholm was born on April 7, 1866, in Stockholm, Sweden. His family was upper-middle-class; his father was a well-to-do merchant, and his mother came from a cultured background. He was privy to the highest quality education available in his country and proved gifted. In 1885 he began studies at the Polytechnic Institute in Stockholm, where he developed what turned out to be a life-long interest in problems of practical **mechanics**. He remained at the Institute for only one year and enrolled in the University of Uppsala in 1886, receiving his bachelor's degree in 1888.

Fredholm received his doctorate from Uppsala in 1898, although he conducted the bulk of his studies under Mittag-Leffler at the University of Stockholm. At that time, Uppsala was the only university in Sweden that offered a doctoral degree. Fredholm conducted his doctoral thesis on partial **differential equations**, and his work became significant to the study of deformation of anisotropic media, such as crystals.

After receiving his doctoral degree Fredholm accepted a position as lecturer in mathematical physics at the University of Stockholm and in 1906 became a professor of rational mechanics and mathematical physics. The research he conducted during this time yielded a fundamental integral equation that now bears his name. The equation, which is highly relevant in physics, was contained in a seminal research paper for which Fredholm was honored with the Wallmark Prize of the Swedish Academy of Sciences and the Poncelet Prize of the Academie des Sciences.

Much of Fredholm's research on integral equations was based on the work of American astronomer George William Hill. Fredholm laid the foundation for this renowned research in a 1900 paper, *Sur une nouvelle methode pour la resolution du problèm de Dirichlet*. It was in this paper that Fredholm developed the essential component of the theory that led to what is now called Fredholm's Integral Equation. Fredholm

then went on to develop what came to be known as the Fredholm Equation of the Second Type, which involved a **definite integral**. He also discovered the algebraic analog of his theory of integral equations. While Fredholm's contributions to **mathematics and physics** were significant, his research resume is sparse. Biographers attribute his small output to the mathematician's strict attention to detail, a characteristic that earned Fredholm an excellent reputation throughout Europe.

Fredholm's work was carried on by David Hilbert, who learned of Fredholm's work through Erik Holmgren, a colleague of Fredholm's whom Hilbert met in Göttingen. Hilbert incorporated Fredholm's ideas into his own theories, including the theory of eigen-values and the theory of spaces involving an infinite number of dimensions. These theories, in turn, laid the foundation for the study of quantum theory and the discovery of what are now termed Hilbert spaces.

Fredholm remained at the University of Stockholm until his death on August 17, 1927.

FREEDMAN, MICHAEL H. (1951-)
American topologist

Michael H. Freedman has been recognized by the American Mathematical Society, the International Congress of Mathematicians, the United States Government, and the MacArthur Foundation for his research breakthroughs in **topology**, a branch of mathematics that deals with the **invariant** properties of geometric objects rather than their sizes and shapes. Freedman's work has been fundamental in making progress with some of the most difficult problems in four–dimensional **geometry** and topology. He is perhaps best known for his **proof** of the four–dimensional **Poincaré conjecture**, a problem dating from 1904.

Michael Hartley Freedman was born in Los Angeles on April 21, 1951, to Benedict Freedman and Nancy Mars Freedman. Freedman began his post–secondary education with a year at the Berkeley campus of the University of California in 1968. He then transferred to Princeton University, where he received his Ph.D. under William Browder four years later. While in college, he pursued his hobby of rock climbing, scaling the northeast ridge of Mount Williamson alone in 1970. A decade later, he won the Great Western boulder climbing championship.

Freedman joined the Department of Mathematics at the University of California, Berkeley as a lecturer in 1972. In 1974, he spent a year at the Institute for Advanced Study (IAS) in Princeton. Then he returned to California, this time to the University of California, San Diego (UCSD) campus, where he quickly progressed through the ranks of assistant professor, associate professor, and full professor. In 1985, Freedman was appointed by UCSD as the first professor to hold the newly endowed Charles Lee Powell chair of mathematics.

Throughout the 20th century, mathematicians have made progress in understanding geometric objects in terms of associated algebraic operations. In particular, topologists have tried to use **algebra** to classify manifolds (multidimensional

Michael H. Freedman

surfaces). Visualizing surfaces in more than three-dimensional **space** is difficult. Four dimensions are somewhat intuitive if one considers the fourth **dimension** to be time. In his *Fortune* magazine description of Freedman's work, Gene Bylinsky suggested thinking of an eight-dimensional **sphere** as a ball with attached information about its age, color, temperature, weight, and bounciness.

In 1904, French mathematician **Jules Henri Poincaré** designed a system to classify manifolds. He imagined a loop of string wrapped around a surface and determined how far the loop could be shrunk. On a sphere, the loop could be shrunken to a single point. On a **torus** (doughnut-like shape), a loop encircling the hole cannot shrink smaller than the **circumference** of the hole. Thus, a sphere and a torus belong to different classifications.

Three-dimensional manifolds are especially difficult to classify because they can be stretched and folded in many different ways. Poincaré devised a series of tests that he believed could be used to identify any three-dimensional **manifold**, no matter how distorted, that was topologically equivalent to a sphere. The statement of this problem was refined over the years, but not until 1960 did Stephen Smale give the first proof of the Poincaré conjecture for all dimensions greater than four. Smale and other topologists followed algebraic guidelines in cutting the manifold apart and sewing it back together as a sphere, a technique known as surgery. However,

manifolds of dimension three or four do not have as much "room" for maneuvering. Thus, the necessary surgery is much more difficult, and the four-dimensional Poincaré conjecture remained unsolved for another two decades.

Finally, after seven years of work, Freedman solved the surgery problem for simply connected four-dimensional manifolds in 1982. His paper "The Topology of Four-Dimensional Manifolds" gives a complete classification of all simply connected, four-dimensional manifolds in terms of two quantities. In the course of proving this **theorem**, Freedman exhibited several new four-dimensional manifolds, including the first examples of such manifolds that do not support a coordinate system for **calculus**. These results, along with nearly 50 papers on the structure and classification of three– and four–dimensional manifolds, resolved many fundamental issues in these physically significant dimensions.

John Milnor, a mathematician of considerable stature, wrote in *Notices of the American Mathematical Society* that Freedman's classification theorems for important classes of four–dimensional topological manifolds "are simple to state and use, and are in marked contrast to the extreme complications that are now known to occur in the study of differentiable and piecewise linear 4-manifolds."

The four years following his proof of the famous conjecture were eventful for Freedman. In 1983, he married Leslie Blair Howland, with whom he would raise three children. In 1984, he received a five-year MacArthur Foundation Fellowship to provide financial support while he continued his research. That same year he was elected to the National Academy of Sciences, and the following year to the American Academy of Arts and Sciences. In 1986 Freedman received the Fields Medal, the highest honor in mathematics.

The American Mathematical Society awarded the 1986 **Oswald Veblen** Prize in Geometry to Freedman for his work in four–dimensional topology. In his response to this award, Freedman discussed the importance of interchange among the various branches and applications of mathematics. That statement, which was printed in the *Notices of the American Mathematical Society,* included his assertion that "Mathematics is not so much a collection of different subjects as a way of thinking. As such, it may be applied to any branch of knowledge." In particular, he praised the movement among mathematicians to voice their opinions on such topics as education, energy, economics, defense, and world peace. He noted that "Experience inside mathematics shows that it isn't necessary to be an old hand in an area to make a contribution. Outside mathematics the situation is less clear, but I can't help feeling that there, too, it is a mistake to leave important issues entirely to the experts."

Freedman pushed the limits not only in terms of his expectations of mathematicians in society, but also in his technical research. Speaking to a joint meeting of the American Mathematical Society and the Mathematical Association of America in early 1997, he described the direction of his current work on the subject of computability. Computer scientists are interested in being able to identify which problems are "hard" in the sense of not being solvable with an efficient computer **algorithm**. The classic question, known as the Traveling Salesman Problem, asks for the most efficient itinerary for a person to visit each of a certain number of sites. Analogous problems have practical applications in areas such as cryptography and computer data security. According to Barry Cipra's article in *Science* magazine, Freedman thinks key to deciding whether a problem is "hard" may be to look at the limiting case of such a problem as the number of choices approaches **infinity**. Cipra quotes Freedman as saying, "This is always an attractive situation for the pure mathematician, when there's a very clear, well-defined problem blocking understanding. It's kind of like waving a red flag at a bull!"

Although Freedman has announced the broad direction of his assault on the two-decade-old Traveling Salesman Problem, the mathematics community has yet to see a detailed description of his tactic. The considerable respect with which Freedman is regarded by fellow mathematicians generates optimism that his approach may generate at least some progress on this elusive topic.

FREGE, GOTTLOB (1848-1925)

German logician

Gottlob Frege made seminal contributions to the philosophy of language and of mathematics, yet his major intellectual project ran aground as it was in the last stages of being launched. He worked in a fairly narrow area of mathematics throughout his career, although his contributions to philosophy were uncommonly acute for a mathematician. Frege's understanding of language continues to be the focus for philosophical discussion and his project in mathematical logic, though unsuccessful, provided the foundation for much further work in the subject.

Friedrich Ludwig Gottlob Frege was born on November 8, 1848, in Wismar, Mecklenburg–Scherin (present–day Germany). His father, Alexander, was principal of a girls' high school. In 1869 Frege enrolled at the University of Jena, where he stayed for two years at Jena before transferring to the University of Göttingen. There, his studies were not limited to mathematics, but included physics, chemistry, and philosophy. In 1873 Frege earned his doctorate from Göttingen; his thesis examined the question of the geometric representation of imaginary structures in the plane. The choice of subject was in keeping with the geometrical aspect of German mathematical research at the time, one year after the inauguration of the Erlangen program by **Felix Klein**. A year later, Frege earned the right to teach at the University of Jena with a dissertation dealing with the subject of groups of a certain kind of function. Frege's work on **group theory** continued, but he never made it the focus of his courses as did some of his contemporaries.

The goal to which Frege devoted himself was the task of reducing **arithmetic** to logic. There were two main currents in mathematics which led him to find the project attractive and feasible. The first was the work of **Karl Weierstrass** and his school of **analysis**. Weierstrass had taken the work of two centuries in the area of **calculus** and determined a way to rewrite

the foundations arithmetically. The other current was the mathematization of logic in the work of **George Boole** and other algebraists. After centuries of little progress, logic had begun to take on some of the qualities of modern mathematics.

As a first step, Frege felt it necessary to devise a new notation in order to prevent hidden assumptions from creeping into arguments, which was published as *Begriffsschrift* ("Concept Notation"), in 1879. This notation was a tool for analyzing and representing proofs within mathematics, but it was cumbersome and typographically awkward. As a result, Frege's notation did not find many supporters, although the project was taken up by others, including **Bertrand Russell**, with more success.

The next step for Frege was to prove that the current philosophy of mathematics was flawed and in need of revamping, an argument he detailed in his 1884 book *Grundlagen der Arithmetik* ("The Foundations of Arithmetic"). Frege took on psychologism, the doctrine that numbers were merely objects in the mind, and was able to refute the claims of its current advocates. Similarly, he assailed the empiricism of John Stuart Mill, which sought to find numbers as objects in the world of sense and sound. Both psychologism and empiricism survived Frege's onslaught, but he had clearly demonstrated some of their shortcomings. The *Grundlagen* also included Frege's definition of number. It is tied to the notion of concept and deals with the question of the existence of a 1–1 **correspondence** between objects falling under two concepts.

Frege found it necessary to address a few philosophical issues connected with concepts, objects, and how they are represented linguistically. His essays dealing with these subjects have a continuing philosophical appeal independent of the success of his logical project. In particular, the 1892 essay, "über Sinn und Bedeutung" ("On Sense and Reference") may be the most influential paper in the history of the philosophy of language.

The first volume of the *Grundgesetze der Arithmetik* ("The Basic Laws of Arithmetic") was published in 1893 and it appeared that Frege had finally constructed a system that reduced arithmetic to logic. In 1902, however, he received a letter from Bertrand Russell, who had detected an error in Frege's system. What Russell had found was a contradiction, a derivation of both a statement and its negation. According to the rules of Frege's logic, the presence of such a contradiction rendered the system useless.

Frege's response was intellectually honest. In the second volume of the *Grundgesetze*, Frege acknowledged Russell's contribution and noted that his own system was unable to handle it. For the remainder of his career, Frege worked on issues that were connected with philosophy and logic without seeking to repair his system. He retired from the University of Jena in 1917 and died at the German resort of Bad Kleinen on July 26, 1925. The subsequent interest in Frege's work, especially among philosophical logicians, bears witness to the value of his reformulation of logic and the farsightedness of his ambition.

FROBENIUS, FERDINAND GEORG (1849-1917)

German mathematician

German mathematician Ferdinand Georg Frobenius was a number theorist who made critical contributions to the study of **group theory**.

Frobenius lived most of his life in and around Berlin. He was born in Berlin-Charlottenburg, Prussia in 1849, and after seven years in the Joachimsthal Gymnasium, or secondary school, he embarked on an academic career at the University of Göttingen at age 18. Frobenius only spent a semester at Göttingen before moving to the University of Berlin. He studied under the respected mathematician Karl Weierstrass in Berlin, and was awarded his doctorate in 1870. Four years later, he was awarded a professorship in mathematics at the University of Berlin. Frobenius only remained in the position for a year when he decided to relocate to Zürich, Switzerland to teach mathematics at the Eidgenössische Polytechnikum (Federal Polytechnic) there. Frobenius remained in Zürich until 1892, when he returned home to Berlin to resume his position at the University as mathematics professor.

While in Zürich, Frobenius began to publish his pivotal papers on group theory. In 1879, Frobenius and colleague Ludwig Stickelberger published *"Über Gruppen von vertauschbaren Elementen"* ("Concerning Groups of Permutable Elements"), which explained the mathematicians' work on abstract group theory. Several other important papers followed, including important work on group characters (1896) and group representations (1897-1899) that laid the foundation for future work on representation theory.

Frobenius also served as mentor and teacher, overseeing the doctoral studies of such promising mathematicians as **Edmund Landau** and Issai Schur. He died at the age of 67 in Berlin.

See also Group theory

FUBINI, GUIDO (1879-1943)

Italian-American mathematician

Fubini had a distinguished career in mathematics, making contributions to many different aspects of the field, particularly in **geometry** and harmonics. Many of his important theories relating to engineering were published in a text book some 11 years after his death, this was co-authored with G. Albenge.

Guido Fubini, whose father was a mathematics teacher in Venice, showed an aptitude for mathematics at an early age. He studied the subject at university, at Pisa, and by 1900 he had completed his doctoral thesis on elliptic geometry, specifically on Clifford's parallelism. This thesis coincided with the publication by Bianchi (Fubini's doctoral supervisor) on differential geometry. In this work Bianchi discussed the work carried out by Fubini, thrusting him rapidly into the mathe-

matical world spotlight. Fubini initially stayed at Pisa University where he took up a position as a lecturer. This allowed him to extend his work into the field of **harmonic analysis**, with particular reference to spaces of constant curvature. Fubini continued in this area of research at several different universities, including Sicily, Genoa, and Turin. These last three positions were as a professor, a position which he obtained at the very young age of 21. Fubini spread his work even further and included **differential equations** and various aspects of **analysis** and **analytic geometry**. With the advent of the First World War Fubini applied his skill to the practical subject of ballistics and accuracy of weapons fire. With the rise of fascism and religious persecution prior to the Second World War Fubini left his native Italy to continue work in the United States, where for five years before his death in 1943 he taught at New York University.

See also Elliptic geometry; Harmonic analysis

FUNCTION

A function represents a mathematical relationship between two **sets** of **real numbers**. These sets of numbers are related to each other by a rule which assigns each value from one set to exactly one value in the other set. The standard notation for a function $y = f(x)$, developed in the 18th century, is read "y equals f of x." Other representations of functions include graphs and tables. Functions are classified by the types of rules which govern their relationships including; algebraic, trigonometric, and logarithmic and exponential. It has been found by mathematicians and scientists alike that these **elementary functions** can represent many real-world phenomena.

History of functions

The idea of a function was developed in the 17th century. During this time, **René Descartes** (1596-1650), in his book *Geometry* (1637), used the concept to describe many mathematical relationships. The term "function" was introduced by **Gottfried Wilhelm Leibniz** (1646-1716) almost fifty years after the publication of *Geometry*. The idea of a function was further formalized by **Leonhard Euler** (pronounced "oiler" 1707-1783) who introduced the notation for a function, $y = f(x)$.

Characteristics of functions

The idea of a function is very important in mathematics because it describes any situation in which one quantity depends on another. For example, the height of a person depends on his age. The **distance** an object travels in four hours depends on its speed. When such relationships exist, one **variable** is said to be a function of the other. Therefore, height is a function of age and distance is a function of speed.

The relationship between the two sets of numbers of a function can be represented by a mathematical equation. Consider the relationship of the **area** of a **square** to its sides. This relationship is expressed by the equation $A = x^2$. Here, A, the value for the area, depends on x, the length of a side. Consequently, A is called the **dependent variable** and x is the **independent variable**. In fact, for a relationship between two variables to be called a function, every value of the independent variable must correspond to exactly one value of the dependent variable.

The previous equation mathematically describes the relationship between a side of the square and its area. In functional notation, the relationship between any square and its area could be represented by $f(x) = x^2$, where $A = f(x)$. To use this notation, we substitute the value found between the parenthesis into the equation. For a square with a side 4 units long, the function of the area is $f(4) = 4^2$ or 16. Using $f(x)$ to describe the function is a matter of tradition. However, we could use almost any combination of letters to represent a function such as $g(s)$, $p(q)$, or even $LMN(z)$.

The set of numbers made up of all the possible values for x is called the **domain** of the function. The set of numbers created by substituting every value for x into the equation is known as the **range** of the function. For the function of the area of a square, the domain and the range are both the set of all positive real numbers. This type of function is called a one-to-one function because for every value of x, there is one and only one value of A. Other functions are not one-to-one because there are instances when two or more independent variables correspond to the same dependent variable. An example of this type of function is $f(x) = x^2$. Here, $f(2) = 4$ and $f(-2) = 4$.

Just as we add, subtract, multiply or divide real numbers to get new numbers, functions can be manipulated as such to form new functions. Consider the functions $f(x) = x^2$ and $g(x) = 4x + 2$. The sum of these functions $f(x) + g(x) = x^2 + 4x + 2$. The difference of $f(x) - g(x) = x^2 - 4x - 2$. The product and quotient can be obtained in a similar way. A composite function is the result of another manipulation of two functions. The composite function created by our previous example is noted by $f(g(x))$ and equal to $f(4x + 2) = (4x + 2)^2$. It is important to note that this composite function is not equal to the function $g(f(x))$.

Functions which are one-to-one have an **inverse function** which will "undo" the operation of the original function. The function $f(x) = x + 6$ has an inverse function denoted as $f^{-1}(x) = x - 6$. In the original function, the value for $f(5) = 5 + 6 = 11$. The inverse function reverses the operation of the first so, $f^{-1}(11) = 11 - 6 = 5$.

In addition to a mathematical equation, graphs and tables are another way to represent a function. Since a function is made up of two sets of numbers each of which is paired with only one other number, a graph of a function can be made by plotting each pair on an X,Y coordinate system known as the **Cartesian coordinate system**. Graphs are helpful because they allow you to visualize the relationship between the domain and the range of the function.

Classification of functions

Functions are classified by the type of mathematical equation which represents their relationship. Some functions

are algebraic. Other functions like $f(x) = \sin x$, deal with angles and are known as trigonometric. Still other functions have logarithmic and exponential relationships and are classified as such.

Algebraic functions are the most common type of function. These are functions that can be defined using **addition**, **subtraction**, **multiplication**, **division**, **powers**, and **roots**. For example $f(x) = x + 4$ is an algebraic function, as is $f(x) = x/2$ or $f(x) = x^3$. Algebraic function are called polynomial functions if the equation involves powers of x and constants. The most famous of these is the quadratic function (quadratic equation), $f(x) = ax^2 + bx + c$ where a, b, and c are constant numbers.

A type of function which is especially important in **geometry** is the trigonometric function. Common **trigonometric functions** are **sine**, **cosine**, tangent, secant, cosecant, and cotangent. One interesting characteristic of trigonometric functions is they are periodic. That means there are an infinite number of values of x which correspond to the same value of the function. For the function $f(x) = \cos x$, the x values 90° and 270° both give a value of 0, as do $90° + 360° = 450°$ and $270° + 360° = 630°$. The value 360° is the period of the function. If p is the period, then $f(x + p) = f(x)$ for all x.

Exponential functions can be defined by the equation $f(x) = b^x$, where b is any positive number except 1. The variable b is constant and known as the base. The most widely used base is an irrational number denoted by the letter **e**, which is approximately equal to 2.71828183. Logarithmic functions are the inverse of exponential functions. For the exponential function $y = 4^x$, the logarithmic function is its inverse, $x = 4^y$ and would be denoted by $y = f(x) = \log_4 x$. Logarithmic functions having a base of e are known as natural **logarithms** and use the notation $f(x) = \ln x$.

We use functions in a wide variety of areas to describe and predict natural events. Algebraic functions are used extensively by chemists and physicists. Trigonometric functions are particularly important in architecture, astronomy, and navigation. Financial institutions use exponential and logarithmic functions. In each case, the power of the function allows us to take mathematical ideas and apply them to real world situations.

FUNDAMENTAL GROUP

Henri Poincaré (1854-1912) discovered the fundamental group (also known as the Poincare group or the first homotopy group) of a **manifold** (or more generally, any topological **space**) and used it to classify manifolds. A manifold is a topological space every point of which has a neighborhood which looks like (i.e. is homeomorphic to) n-dimensional real space R^n for some number n. For example, n-dimensional real space, circles, spheres, the surface of a donut, the universe and the complement of a knot (a knot is the mathematical analogue of a loop of string with its ends glued together) in three dimensional space are all manifolds.

If p is a point in a manifold M and c and c' are paths (i.e. continuous maps of [0,1] into M) which begin and end at p then Poincare considered them to be equivalent (i.e. in the same homotopy class) if one could be continuously deformed into the other. Precisely, they are equivalent if there exists a continuous function from the unit **square** (i.e. [0,1] x [0,1]) in $\mathbf{R}^2$ to M such that c is that path traced out by F(0,t) as t goes from 0 to 1, c' is that path traced out by F(1,t) as t goes from 0 to 1 and F(t,0) = F(t,1) = p for all t in [0,1]. If d is another path that begins and ends at p then the path which first traces out d and then traces out c is denoted by cd and is called the concatenation of c and d. It can be proved that if c and c' are equivalent then cd and c'd are equivalent. Also the path which traces out c backwards is denoted by c⁻¹. It has the property that cc⁻¹ is equivalent to the path that stays at p. The set of **equivalence** classes of paths (which begin and ends at p) of M with the operation of concatenation is called the fundamental group of M at p and is usually denoted by Pi_1(M,p).

The fundamental group of the real line or **sphere** (based at any point p) consists of a single equivalence class since any path on either space can be continuously deformed to a point. The fundamental group of the **circle**, however, is isomorphic to **Z** the group of **integers** by the function which assigns to the class of any path c in the circle the number of times it winds areound the circle clockwise minus the number of times it winds around the circle counterclockwise.

If p and p' are points in M and there is a path d from p to p' in M then Pi_1(M,p) is isomorphic to Pi_1(M,p') by the map which assigns to the class of a path c which begins and ends at p to the class of the path which first traces out d backwards then traces out c and the traces out d forwards. Hence if M is path-connected (i.e. there exists a path between any two points of M) then the fundamental group of M is unambiguously defined up to **isomorphism**. Any continuous map f from a manifold M to a manifold N induces a function denoted by f_* from Pi_1(M,p) to Pi_1(N,f(p)) that assigns the class of the path c to the class of the path f composed with c. If the map f is a homeomorphism (i.e. it is a one-to-one **correspondence** which preserves the **topology** of the spaces) then f_* is an isomorphism. Thus fundamental groups can be used to study the topological properties of manifolds.

Poincare conjectured that if two manifolds have isomorphic fundamental groups and some other conditions are satisfied (the manifolds must be closed and have the same homology) then they must be homeomorphic. This was first disproved by James Alexander (1888-1971) of Princeton who found specific three dimensional manifolds satisfying the hypothesis of the conjecture but not the conclusion. However if one assumes in addition to the above hypothesis that the fundamental groups of the manifolds contain only one equivalence class then the problem is still open (for the three dimensional case). It is known as Poincaré's conjecture. In 1989, Cameron Gordon and John Luecke proved that if K and K' are knots in $\mathbf{R}^3$ and the fundamental group of the complement of K is isomorphic to the fundamental group of the complement of K' then the two complements are homeomorphic and in fact K can be continuously deformed into K' without intersecting

itself in the process. This result resolved one of the longest standing conjectures in knot theory.

See also Poincaré's conjecture

FUNDAMENTAL THEOREM OF ALGEBRA

The fundamental **theorem** of **algebra** is the statement that every polynomial with **complex numbers** as coefficients has a complex number as a root, or equivalently, that such a polynomial has n **roots**, where n is its degree, in the complex numbers, counted with multiplicitly (that is, double roots count double and so on).

While this can be checked for **polynomials** of degree up to four, using the formulas for the roots, this approach will not work for higher degrees. Some early authors seem to have taken the fundamental theorem of algebra for granted, without realizing it required a justification. Albert Girard (1629) might have been the first to call attention to the statement of the fundamental theorem of algebra but without trying to justify it. The famous mathematician and philosopher **Gottfried Leibniz** (1702) even doubted its validity. The first serious attempt at a **proof** was made by the French mathematician **Jean Le Rond D'Alembert** in 1746 but his proof was incomplete. The first correct proof was given by the great German mathematician **Carl Frederich Gauss** in 1799. Gauss then subsequently gave three other proofs and since then there has been many more different proofs. Stricly speaking, the fundamental theorem of algebra is really a theorem in **Analysis**, since its **truth** rests on the **continuity** properties of real and complex numbers. The algebraic content of the theorem has been made explicit by the theory of real closed fields developed by Emil Artin and Otto Schreier (1926).

See also Galois theory

FUNDAMENTAL THEOREM OF CALCULUS

The Fundamental **Theorem** of **Calculus** states that the **area** under the graph of a function over an **interval** can be calculated by evaluating any antiderivative of the function at the endpoints of the interval. That is, if f is a function defined on an interval [a,b] and F is any antiderivative of f (that is, if F' = f), then the **definite integral** of f from a to b (i.e. the area under the curve y=f(x) between x=a and x=b) equals F(b) - F(a). Another formulation of this theorem states that the area function of a function f is an antiderivative of f: that is, for any integrable function f, fix a value a in its **domain** and let $A_f(x)$ be the area under the graph of f between a and x. Then $A_f'(x) = f(x)$.

What these two equivalent formulations are saying is that, in essence, computing areas and taking derivatives are inverse operations, just as are **multiplication** and **division** or **addition** and **subtraction**. In the first formulation, if we denote f as F', the theorem states that if we first differentiate

F and then calculate the area underneath that derivative, we get back the original function F (up to a constant). The second formulation gives us the reverse statement: if we start by computing the area under the graph of f and then take the derivative of the resulting function, we again get back the original function.

This remarkable theorem is called the Fundamental Theorem of Calculus because it provides the link between the two branches of calculus, **differential calculus** and **integral calculus**. Differential calculus is the study of derivatives, which can be thought of as slopes of tangent lines to curves, or as instantaneous rates of change of **functions**. Integral calculus is the study of areas under curves, which are defined in terms of **limits** of Riemann sums—the area under a curve is approximated by a collection of rectangles, and the widths of these rectangles are taken to be smaller and smaller to obtain better and better estimates of the actual area. These two branches of mathematics were developed independently of each other. The insight that they were very closely linked—for that matter, that they were linked at all—was due to Sir **Isaac Newton** and to **Gottfried Leibniz**, working independently in the late seventeenth century.

From a practical point of view, the theorem gives a simple method for calculating the area under the graph of any function whose antiderivative can be found. For example, suppose we wish to compute the area between 0 and 1 under the curve $y=a^x$. Using Riemann sums, we have to evaluate the limit as n approaches **infinity** of the sum of $a^{k/n} (1/n)$, where k ranges from 1 to n. However, the Fundamental Theorem tells us that we can simply evaluate any antiderivative of a^x at the two endpoints, 0 and 1, to find the area. Since $a^x/\ln(a)$ is an antiderivative of a^x, we can easily find the area to be $a^1/\ln(a) - a^0/\ln(a) = (a-1)/\ln(a)$.

Conversely, the theorem also gives a method of approximating the antiderivative of a function whose antiderivative cannot be expressed in closed form. For example, the antiderivative of the function $f(x) = \cos^2 x$ cannot be expressed as an algebraic combination of "common" functions (i.e., **polynomials**, exponentials, **logarithms**, and **trigonometric functions**). However, we know from the Fundamental Theorem that the function A(x) that measures the area under the graph of $y=\cos^2 t$ from, say, 0 to the **variable** endpoint x is an antiderivative of $\cos^2 x$. Thus for any specific value of x, the value of A(x) can be approximated to any desired level of accuracy by means of Riemann sums.

See also calculus; differential calculus; integral calculus

FUNDAMENTAL TRIGONOMETRIC FUNCTIONS

There are two fundamental trigonometric **functions**: **sine** and **cosine**, and four co-fundamental functions: tangent, secant, cosecant, and cotangent. These four co-fundamental functions are defined in terms of sine and cosine and their values can be quickly computed from the corresponding values of the sine

and cosine. The two fundamental **trigonometric functions** are the most basic functions in mathematics and were initially developed in the study of the measurement of triangles and their angles. Although they are currently used in such practical fields as architecture, surveying, and navigation, their importance to pure and **applied mathematics** extends far beyond these uses. These functions can be used to describe any natural phenomenon that is periodic, and, as incorporated into higher mathematics, they are useful in understanding abstract spaces.

The fundamental trigonometric functions were formulated after an early connection was made between mathematics and astronomy. The early work incorporating these functions demonstrated that the studies of spherical triangles were as important as the studies of plane triangles. The earliest work on trigonometric functions related chords of a **circle** to a given **angle**. The first known table of chords in a circle was published by **Hipparchus**, the Greek mathematician attributed with the founding of **trigonometry** in about 140 BC. Although these tables have not survived, Hipparchus wrote twelve books of tables relating chords in a circle to given angles. About 40 years later Menelaus produced six books of tables of chords and worked extensively on spherical chords. He proved the property known as *regula sex quantitatum* that involves plane triangles and corresponding spherical triangles.

The first appearance of the sine of an angle appears in the Hindu's work. In about 500 the Indian mathematician **Aryabhatta** called the sine of an angle *ardha-jya,* which meant half-cord. The word was soon shortened to *jya* or chord. Later this word was translated by the Arabs into their vowelless language as *jb.* Eventually the Arabs began adding vowels to their language again and the word was transformed into *jaib,* which was already the word for bosom or fold. Around the 12th century, when mathematics was translated into Latin, the Arabic word was translated into *sinus,* meaning bosom or fold in Latin. **Fibonacci**'s wide use of *sinus rectus arcus* soon lead to the universal use of sine. **Bhaskara** detailed a method for constructing a table of sines for any angle in 1150. In 1542 **Rheticus** published all of the trigonometry relevant to astronomy in **Copernicus**'s book. Rheticus also published substantial tables of sines and cosines after his death. All authors at the outset of its usage did not accept the abbreviation sin. E. Genter was the first to use the abbreviation sin in 1624 in a drawing; and Herigone, a French mathematician, first used the abbreviation in a book in 1634. Cosine followed a similar development in notation. Originally, Viete used the term *sinus residuae* for cosine. In 1620 Gunter suggested this be replaced by co-sinus, and eventually the abbreviated cos was employed.

The tangent function arrived via a different route from the chord approach of sine and cosine. The tangent and cotangent functions were developed together and were not at first associated with angles. These functions arose out of the study of the height of an object as it relates to the length of the shadow that the object casts. This principle was important in the functioning of the sundial, and **Thales** used this relationship to calculate the heights of the pyramids. Although the Arabs in about 860 produced the first known tables of shadows, the name tangent was first used by **Fincke** in 1583, followed by cotangens by E. Gunter in 1620. The development of the common abbreviations for these two functions followed those for sin and cos. A. Girard first used the abbreviation tan in 1626 although he used the abbreviation written over the angle. The abbreviation cot was first used by J. Morre in 1674.

The co-fundamental trigonometric functions, secant and cosecant, were not used by early astronomers or surveyors. These functions came into use in about the 15th century when navigators started to prepare tables for nautical navigation.

See also Trigonometric functions; Trigonometry

G

GALILEI, GALILEO (1564-1642)

Italian astronomer, physicist and mathematician

Galileo Galilei, known best as simply "Galileo," was a scientist at a most difficult time in history: the time of the Inquisition, when the Roman Catholic Church was still furiously resisting evidence from new discoveries. Galileo is one of the Inquisition's most famous victims; he supported **Copernicus**'s discovery that the Earth revolves around the Sun. Although he was not killed for his beliefs, Galileo was silenced and placed under house arrest from the time of his trial to the end of his life.

Galileo was born February 15, 1564, in Pisa, Italy. His father, Vincenzio Galilei, was a scientist, investigating acoustics and musical theory. His mother was Giulia Ammannati. Galileo, the oldest of seven children, began his studies with Jacopo Borghini, but had to leave his tutor when his father moved the family back to his native Florence about 1575. Galileo then studied at the monastery of Santa Maria, and entered the order as a novice. His father removed the young man from the order and tried to get him a scholarship to the University of Pisa, but failed. Galileo returned to the monastery to study until entering the University of Pisa as a medical student in 1581, following his father's wishes.

It was that year Galileo made the first of a lifetime of great discoveries. While sitting in church, he noticed a swinging lamp. No matter how great the swing's arc, they all seemed to take the same amount of time, he realized. After the service, Galileo began experimenting with different weights and lengths of string, and devised a simple device that would measure a patient's pulse. Pisa faculty members improved on the device and the pusilogia was used for many years afterward. Despite this invention, Galileo was not interested in medicine. He was a difficult student, often challenging his professors, who gave him the nickname "The Wrangler." Mathematics did interest him, however; quite possibly sparked by his father's work with acoustics, particularly the study of the effects of the length of musical strings on consonance. Although still officially a medical student, Galileo began his mathematical studies outside of the University with Ostillio Ricci, mathematician to the grand duke of Tuscany. Despite his father's objections and his suffering medical studies, Galileo persisted. He finally left the university in 1585 to study on his own, partly because his misbehavior made him unsuitable for a scholarship that would have allowed him to continue his medical studies.

Galileo went home to Florence and began tutoring students in mathematics. It was during this time that he became interested in **Archimedes**' experiment that disclosed that a crown was indeed not solid gold, but gold and a base metal (Archimedes had placed the crown in water and measured the amount of water it displaced, which turned out to be less than a solid–gold object of the same side would have). Galileo created a hydrostatic balance, a small scale that could perform the same measurement more accurately, and this invention led to his first published piece of scientific writing, a booklet describing it, which appeared in 1586.

Galileo apparently enjoyed debunking famous myths and regularly set about deflating the ideas of the ancient Greeks, particularly **Aristotle**. His work with falling bodies damaged Aristotelian physics beyond repair. Going contrary to the widely held belief of the time, Galileo demonstrated that Aristotle was incorrect in his claim that light objects fell more slowly than heavy ones. From this work arose the legend about Galileo dropping two cannon balls, one heavier than the other, off the Tower of Pisa.

In another experiment, he challenged Aristotle's idea that **force** had to be continually applied in order to keep a body in **motion**. Some had taken this idea and explained that the source of this force was hardworking angels. However, Galileo's experiment with a body rolling down an inclined plane showed that if force were continuously applied, the body would continue to accelerate.

Among Galileo's other discoveries in the field of **mechanics** were the fact that two forces can act on a body

Galileo

simultaneously, causing the body to move in a parabolic curve (the base of the mathematical science kinematics), and studies concerning the strength of materials. He discussed both of these ideas in his book *Discourses and Mathematical Demonstrations Concerning Two New Sciences*. Galileo's text forms the basis of modern physics because it shows the use of mathematics in understanding motion and the value of physical experiment and mathematical **analysis** in solving problems. Although the calculations he made regarding projectiles used low speeds, Galileo's tables and ideas would be further refined by others who came later, and made gunnery a science.

Throughout his investigations, Galileo's proofs used only the same geometric methods that had been available to the Greeks. **Algebra** and other more complicated methods of calculation would not be available until the time of **René Descartes** and **Isaac Newton**. However, Galileo believed mathematics was superior to logic. In 1592 he was named Chair of Mathematics at the University of Padua.

Galileo was not a man of all work. He had a mistress, Marina Gamba, for 10 years and had three children by her. The daughters, Virginia and Livia, entered the Franciscan convent in their mid–teens, taking the names Sister Maria Celeste and Sister Arcangela. Galileo did eventually recognize his son, Vincenzo.

Galileo's attention was diverted from his studies of mechanics by the invention of the telescope. Over the years he built hundreds of telescopes, many of which he gave as gifts to persons of influence. Some of his telescopes that have survived to this day have nearly perfect optics.

With his telescopes, Galileo observed the face of the moon and showed that it was not smooth as had been thought, but rather rough–surfaced. He found that constellations such as Orion and formations like the Pleides comprised far more stars than had ever been guessed; in Orion alone he counted 500 stars. He also observed the planets, including Jupiter and Saturn.

Galileo's discovery of Jupiter's moons was of ground-breaking importance. On January 7, 1610, he saw three small but bright stars near the planet, two east and one west. Observations on successive nights found that the stars moved—eventually with all three west of the planet. "My confusion was transformed to amazement," he wrote. "I had now decided beyond all question that three stars were wandering around Jupiter, as do Venus and Mercury around the sun." Three nights later, he found another "star" circling Jupiter. Over the next six weeks he continued to watch, and the evidence was irrefutable: here was a perfect example of smaller heavenly bodies circling a larger one, just as Copernicus had said that the Earth circled the sun.

Quickly Galileo wrote and published his observations in a book called *Siderius nuncius* (The Starry Messenger). It became the most important book of the 17th century, and set the stage for Galileo's downfall.

Galileo's discoveries made him reluctant to continue teaching his students the old Ptolemaic system (Earth–centered) and he resigned his chair at Padua and turned to Florence, where he became the Grand duke of Tuscany's mathematician and philosopher and the University of Pisa's chief mathematician, a non–teaching position. By 1614, his stance on the question of an Earth–centered solar system had earned him attacks from Church leaders. In 1616, Pope Paul V summoned Galileo to Rome and ordered him to stop disseminating this theory called Coperinicism. Angry, Galileo nevertheless obeyed; there was no other choice, for if he had defied the Church he would have been imprisoned or tortured (in 1600, philosopher Giordano Bruno had been burned at the stake for refusing to recant his scientific views).

On August 6, 1623, Urban VIII was named pope. Somehow, Galileo was persuaded that Urban would be open to his ideas, and so he wrote *Dialogue on the Two Chief World Systems*. In the *Dialogue*, Galileo pitted an Aristotelian, Simplico, against a quick–minded Salviati, who got the best of the argument. Galileo wrote the book in Italian, not Latin, the language of scholars, and so it was accessible to anyone who could read, and was quickly translated into other languages.

Unfortunately for Galileo, one of his most bitter enemies, Father Christopher Scheiner, convinced Urban that the Simplico character was a buffoonish caricature of himself. Galileo was in trouble again. The Pope called him to Rome in 1633 and forced him to renounce any views that were at odds with the Church's belief.

Galileo was nearly 70 years old at this time. With the example of Bruno in his mind, he agreed. The church placed him under house arrest in his villa at Arcetri for the remainder of his life. By Christmas 1637, Galileo became completely blind. He died January 8, 1642.

On October 31, 1992, over 350 years after Galileo's trial, the Roman Catholic Church acknowledged that it had been in error and acknowledged the validity of his work.

GALOIS, ÉVARISTE (1811-1832)
French algebraist and group theorist

Évariste Galois discovered mathematics as an adolescent and published his first original work at age 17. In his short life, Galois originated algebraic applications of finite groups, now known as Galois groups, and developed the foundations for the solvability of algebraic **equations** using rational operations and extraction of **roots**. His mathematical works have been credited as having transformed the theory of algebraic equations.

Galois was born October 25, 1811, at Bourg–la–Reine near Paris, the second of three children of Nicholas–Gabriel and Adelaide–Marie Demante Galois. His parents were well educated, although no one in the family is known to have excelled in mathematics. His father, an ardent republican and a composer of light verse, was director of a boarding school and was elected mayor of Bourg–la–Reine in 1815. His mother, who came from a family of jurists, had been trained in religion and the classics and was the only teacher Galois knew until he was age 12.

In October 1823, Galois' parents sent him to the Lycée of Louis–le–Grand in Paris. At first, Galois did well, receiving a prize in the General Concourse and three mentions. His character was generally described as "good, but singular." In 1827, after he was demoted because of deficiencies in rhetoric, he began his first mathematics class under M. Vernier. Galois read **Adrien–Marie Legendre**'s **geometry** text and **Joseph Lagrange**'s original memoirs with ease, and apparently was able to work out complicated problems without pencil or paper. As he became absorbed in mathematics, he neglected his other courses. Although Vernier remarked of Galois' "zeal and progress," he described him as "closed and original" and termed his work "inconstant." Vernier urged Galois to work methodically, but he did not take his instructor's advice. Galois was eager to enter the École Polytechnique, which trained mathematicians and engineers, but without the preparation in basics, he took the entrance exam a year early and failed.

Later in 1827 Galois enrolled in the mathematics course of Louis–Paul–Emile Richard. Richard, a distinguished teacher, found him markedly superior to other students and believed he should be admitted to the Polytechnique without examination. In April 1829, Galois' first mathematics paper, a minor work on continued **fractions**, was published in the *Annales de Gergonne*. In May and June, while still only 17, he submitted two articles on the algebraic solution of equations to the Paris Académie Royale des Sciences, with **Augustin–Louis Cauchy** as referee. It has been widely written that Cauchy

Évariste Galois

ignored and lost the manuscripts, but letters indicate that he had read and was impressed with them. Tony Rothman suggests that Cauchy encouraged Galois and recommended that he combine the two papers and submit them for the Académie's Grand Prize in Mathematics. In February 1830, Galois submitted such an entry to the Académie's secretary, **Jean Baptiste Joseph Fourier**. Unfortunately, Fourier died in May and Galois' entry was not found among his papers.

In April of 1830, Galois, now a student at the École Normale, published "An Analysis of a Memoir on the Algebraic Resolution of Equations" in the *Bulletin de Ferussac*. In June, he published "Notes on the Resolution of Numerical Equations" and "On the Theory of Numbers." These and a later memoir make up what is now called **Galois theory**. In July, the French revolution of 1830 came to a head, and the director of the École Normale, M. Guigniault, locked in his students so that they would not participate in the rioting. Galois tried to scale the walls, but failed, and wrote a letter critical of Guigniault to the *Gazette de Écoles,* which resulted in his expulsion. Galois then joined the Republican Artillery of the National Guard, but it was shortly abolished by royal decree because of its perceived threat to the King. Galois was probably also a member of the Society of the Friends of the People, a secret republican society. In January 1831, he organized a private class in **algebra** which attracted 40 students, and

at the invitation of Siméon–Denis Poisson, submitted a third version of his paper to the Académie.

On May 9, 1831, at a noisy republican banquet celebrating the acquittal of 19 guardsmen of conspiracy, Galois held a glass and an open dagger and toasted to King Louis–Phillipe. The next day he was arrested for threatening the King's life, and was imprisoned until June 15 when he was tried and acquitted. But on July 14, Bastille Day, Galois, armed with a rifle, dagger, and pistols, was again arrested, this time for wearing the illegal Artillery Guard uniform. He was sentenced to six months in prison.

In October, still incarcerated at Saint–Pélagie, he received word of Poisson's rejection of his paper. Poisson noted that Galois' argument was "neither sufficiently clear nor sufficiently developed to allow us to judge its rigor; it is not even possible for us to give an idea of this paper," and suggested a more complete account. Resolving to publish his papers privately with the aid of his friend, Auguste Chevalier, Galois gathered them and wrote a vitriolic five–page preface.

Because of a threatened cholera outbreak at the prison, Galois was transferred to the pension Sieur Faultrier in March 1832 and was set free on April 29. Just a month later, on May 30, he faced Pescheux d'Herbinville (a republican and one of the guards at the pension) in a duel. The circumstances leading to the duel are not known; however, it seems clear that it was not political, but rather a personal quarrel that involved Stéphanie du Motel, the daughter of a physician at the pension. Galois was shot in the abdomen and was unattended for hours until a peasant took him to the Hospital Cochin. Galois refused the services of a priest and died the following day, with his brother, Alfred, beside him. Galois was buried in a common burial ground in South Cemetery. No trace remains of his grave.

The night before the duel, Galois wrote letters and made notes and corrections on some of his papers, entrusting them to his friend Chevalier. "All I have written down here has been clear in my head for over a year," he wrote. "Make a public request of Jacobi or Gauss to give their opinions not as to the **truth** but as to the importance of these theorems. After that, I hope some men will find it profitable to sort out this mess."

Alfred Galois and Chevalier copied the papers and submitted them to **Karl Gauss, Karl Jacobi**, and others, and in 1846 **Joseph Liouville** edited some of Galois' manuscripts for publication in the *Journal de Mathematiques*. Liouville noted, "I saw the complete correctness of the method by which Galois proves, in particular, this beautiful **theorem**: In order that an irreducible equation of prime degree be solvable by radicals it is necessary and sufficient that all its roots be rational **functions** of any two of them."

GALOIS THEORY

Galois theory is related to **group theory**, which is a powerful method employed in the analysis of abstract and physical systems that contain **symmetry**. Group theory plays a critical role in many scientific areas: it is the fundamental basis of the space of **quantum mechanics**; classical mechanical geometric symmetries are understood using group theory; the development of **non-Euclidean geometry** by **Lagrange** in the 19th century is hinged on group theory; the development of algebraic structures of linear and **vector spaces** is based on group theory; and the analysis and understanding of molecular systems in detail is accomplished only by employing group theory. Galois theory is a generalized **field** theory that is the mathematical interpretation of group theory. Mathematically, group theory is the basis of real analysis and has had powerful implications in the development of many other areas of mathematics. **Évariste Galois**, a French mathematician, is attributed with formulating Galois theory and recorded his thoughts on the subject the night before he died in 1832. Galois theory stemmed from an undertaking by Galois for a deeper understanding of the essential conditions an equation must satisfy in order for it to be solvable by radicals.

As the name implies, group theory is concerned with the study of groups. Mathematically, a group is a basic structure of modern **algebra** and has specific properties that arise in connection with symmetry. A group consists of a set of elements, sometimes referred to as operands, and an operation or binary operator that takes any elements of the set and forms another element of the set under certain conditions. An example of a group would be the set of all numbers, including **negative numbers** and **zero**, and the operation of **addition**. The addition operation would take the sum of any two numbers of the set and form another number of the set. All groups have three properties in common: associativity, **identity element**, and inverses.

Associativity is a property that assures that the outcome of an operation is the same regardless of the order of operation. So *(-3 + 4) +1 = -3 + (4 + 1)* no matter which order the operation of addition is carried out.

All groups must also contain a special element called the identity element. It's special because when an operation is performed on the identity element and another element of the set, the operation leaves the other element unchanged. So, for example, in the group previously described, when zero is added to a number on the left of the equality sign or on the right of the equality sign, it leaves the number unchanged: *6 + 0 = 0 + 6*. In this group zero is defined as the identity element.

The last property common to all groups is the inverse property. An inverse is an element of the set that, when it is combined with the operation of the group on another element of the set, gives the identity element. In the previous example *-6* is called the inverse of *6* because when summed with *6* it yields the identity element *0*. This property must exist for all elements of the set in a group. An Abelian group is one that satisfies all these properties and satisfies one additional property. The added property is that for every pair of elements in the group, the operation can be performed in either order and yield the same result. For the group described in the previous example: *x + y = y + x* must be true for all pairs of elements in that group for that group to be Abelian. If a group contains a finite number of elements then the group is called a finite

group and the number of elements is called the order of the group.

Évariste Galois, although only 21 at the time of his death, was probably one of the most important contributors to the development of mathematics. Galois expanded the **congruence calculus** of **Gauss** to formulate the notion of a field in group theory. He noted that all that is required for a well-defined field is an Abelian group with a second binary operation that contains a unity. Basically what this means is that the Abelian group minus the identity element must be an Abelian group with respect to the second binary operation. A well-defined field is a necessity when building up the necessary theorems for calculus. The field is the fundamental element to the study of the entire base of mathematics. He recorded all of these thoughts the night before he was killed in a duel in 1832. It wasn't until 1843 that **Joseph Liouville** announced to the French Academy that he had obtained Galois' papers detailing his ideas of group theory. Eventually, in 1846, Liouville published Galois' papers in his journal, *Journal de Mathématiques Pures et Appliquées*. In 1870 **Camille Jordan**, another French mathematician, published the full Galois theory in *Traité des Substitutions*.

Galois theory depends on the concept of a group since it is the mathematical interpretation of group theory. Galois theory has been employed most often in mathematics in the study of the algebraic solvability of polynomial **equations**. In fact, the most well-known application of Galois theory is in the **proof** that radicals cannot solve the general quintic equation with rational coefficients. It is a theory that enables elegant handling of **polynomials** with minimal algebraic manipulations. Galois, using Galois theory, formulated a method of determining when radicals could solve a general polynomial equation and when they could not. This theory enabled the unification of geometry and algebra. Galois' work contributed to the transition from classical algebra to modern algebra.

GALTON, FRANCIS (1822-1911)
English anthropologist

Galton was a pioneer in the use of statistical methods as applied to biological problems. He was interested in the heredity of intelligence, and he used many statistical methods to attempt to correlate his views with the results he obtained from surveys of many different groups of people. Galton proved that a normal mixture of normal distributions is itself normal, and he also linked regression to normal distributions.

Francis Galton, who was cousin to the scientist Charles Darwin, was born to wealthy parents just outside Birmingham in England; this gave him the freedom to pursue his own areas of research. After schooling in Birmingham, Galton embarked on medical training, which was carried out by touring around Europe; this followed by further study at Birmingham and London. His studies were cut short by the death of his father in 1844. Galton became a member of the Royal Geographical Society and spent a number of years **mapping** various parts of Africa. By surveying the local pop-

Francis Galton

ulations and comparing the results with the same tests back in England, Galton felt he had proven the differences between the two groups, the white English and the black Africans. In reality many of the tests Galton used were subjective and served merely to reinforce his own preconceived notions. In 1874 Galton produced a massive statistical survey of the members of the Royal Society, *English Men of Science: Their Nature and Nurture*. This again served the purpose of proving Galton's hypothesis that people cannot improve on what they are born with. One of the mathematical concepts that came from this book was the **correlation** between two **sets** of data. For example Galton gave a numerical value to the relationship between the head size and the height of a human; this value varied between +1 and -1. By looking at these and other variables and plotting them on many charts and graphs Galton found that many human characteristics were represented by a normal distribution.

Galton devoted the latter part of his life to the study of anthropology, eugenics, and meteorology. Galton was elected a Fellow of the Royal Society in 1860 and he received their prestigious Copley medal in 1910. He was knighted some two years before his death in 1911. Galton was so well respected that his ideas on race, intelligence, and heredity were accepted until the middle of the twentieth century. Many of his ideas would now be regarded as extremely racist. Due to his pio-

neering work in statistical analysis and testing of humans, however, Galton is regarded by some as the father of psychological testing. An excellent biography of Galton was written in four volumes by his one time co-worker, **Karl Pearson**; published between 1914 and 1930, it is entitled *The Life, Letters and Labours of Francis Galton*.

See also Karl Pearson; Statistics

GAME THEORY

Game theory is a branch of mathematics concerned with the analysis of conflict situations. It involves determining a strategy for a given situation and the costs or benefits realized by using the strategy. First developed in the early twentieth century, it was originally applied to parlor games such as bridge, chess, and poker. Now, game theory is applied to a wide range of subjects such as economics, behavioral sciences, sociology, military science, and political science.

The notion of game theory was first suggested by mathematician **John von Neumann** in 1928. The theory received little attention until 1944 when Neumann and economist **Oskar Morgenstern** wrote the classic treatise *Theory of Games and Economic Behavior*. Since then, many economists and operational research scientists have expanded and applied the theory.

Characteristics of games

An essential feature of any game is conflict between two or more players resulting in a win for some and a loss for others. Additionally, games have other characteristics which make them playable. There is a way to start the game. There are defined choices players can make for any situation that can arise in the game. During each move, single players are forced to make choices or the choices are assigned by random devices (such as dice). Finally, the game ends after a set number of moves and a winner is declared. Obviously, games such as chess or checkers have these characteristics, but other situations such as military battles or animal behavior also exhibit similar traits.

During any game, players make choices based on the information available. Games are therefore classified by the type of information that players have available when making choices. A game such as checkers or chess is called a "game of perfect information." In these games, each player makes choices with the full knowledge of every move made previously during the game, whether by herself or her opponent. Also, for these games there theoretically exists one optimal pure strategy for each player which guarantees the best outcome regardless of the strategy employed by the opponent. A game like poker is a "game of imperfect knowledge" because players make their decisions without knowing which cards are left in the deck. The best play in these types of games relies upon a probabilistic strategy and as such, the outcome can not be guaranteed.

Analysis of zero-sum, two-player games

In some games there are only two players and in the end, one wins while the other loses. This also means that the amount gained by the winner will be equal to the amount lost by the loser. The strategies suggested by game theory are particularly applicable to games such as these, known as zero-sum, two-player games.

Consider the game of matching pennies. Two players put down a penny each, either head or tail up, covered with their hands so the orientation remains unknown to their opponent. Then they simultaneously reveal their pennies and pay off accordingly; player A wins both pennies if the coins show the same side up, otherwise player B wins. This is a zero-sum, two-player game because each time A wins a penny, B loses a penny and visa versa.

To determine the best strategy for both players, it is convenient to construct a game payoff **matrix**, which shows all of the possible payments player A receives for any outcome of a play. Where outcomes match, player A gains a penny and where they do not, player A loses a penny. In this game it is impossible for either player to choose a move which guarantees a win, unless they know their opponent's move. For example, if B always played heads, then A could guarantee a win by also always playing heads. If this kept up, B might change her play to tails and begin winning. Player A could counter by playing tails and the game could cycle like this endlessly with neither player gaining an advantage. To improve their chances of winning, each player can devise a probabilistic (mixed) strategy. That is, to initially decide on the percentage of times they will put a head or tail, and then do so randomly.

According to the Minimax **Theorem** of game theory, in any zero-sum, two-player game there is an optimal probabilistic strategy for both players. By following the optimal strategy, each player can guarantee their maximum payoff regardless of the strategy employed by their opponent. The average payoff is known as the minimax value and the optimal strategy is known as the solution. In the matching pennies game, the optimal strategy for both players is to randomly select heads or tails 50% of the time. The expected payoff for both players would be 0.

Nonzero-sum games

Most conflict situations are not zero-sum games or limited to two players. A nonzero-sum game is one in which the amount won by the victor is not equal to the amount lost by the loser. The Minimax Theorem does not apply to either of these types of games, but various weaker forms of a solution have been proposed including noncooperative and cooperative solutions.

When more than two people are involved in a conflict, oftentimes players agree to form a coalition. These players act together, behaving as a single player in the game. There are two extremes of coalition formation; no formation and complete formation. When no coalitions are formed, games are said to be non-cooperative. In these games, each player is solely interested in her own payoff. A proposed solution to these types of conflicts is known as a non-cooperative equilibrium. This solution suggests that there is a point at which no player can gain an advantage by changing strategy. In a game when complete coalitions are formed, games are described as cooperative. Here, players join together to maximize the total payoff for the group. Various solutions have also been suggested for these cooperative games.

Application of game theory

Game theory is a powerful tool that can suggest the best strategy or outcome in many different situations. Economists, political scientists, the military, and sociologists have all used it to describe situations in their various fields. A recent application of game theory has been in the study of the behavior of animals in nature. Here, researchers are applying the notions of game theory to describe the effectiveness of many aspects of animal behavior including aggression, cooperation, hunting and many more. Data collected from these studies may someday result in a better understanding of our own human behaviors.

GAMMA FUNCTION

A gamma function is the solution to a specific integral. It is useful for physical applications and has very little theoretical interest value for mathematicians. It is occasionally also related to the "error functions." Its simplest expression is at positive integer values, where it is the same as the **factorial** function. The factorial function is the product of the integer in question with all positive **integers** smaller than that integer. Many of its other forms are recursive as well.

The infinite or Euler limit defines the gamma function of z in terms of a limit as n approaches **infinity** of the quantity n^z $(1*2*3*...*n)/[z(z+1)(z+2)...(z+n)]$. This can be consolidated into a recursion relation, where the gamma function of z+1 is simply z times the gamma function of z. Since we know that the value of this function at **zero** and at one is one, this **definition** reduces to the factorial definition for integers. The integral definition is that the gamma function is the integral from zero to infinity of $e^{-t}t^{z-1}dt$. There are alternate forms of this definition, one of which is over the finite **interval** zero to one and integrates over $[\ln(1/t)]^{z-1}$ instead of over the quantity above. There is also a Weierstrass infinite product definition of the gamma function, which utilizes the Euler-Mascheroni constant, but it is hardly ever used in practical applications. These definitions allow for convenient calculations in widely varying situations.

See also Factorial; Function

GAUSS, JOHANN KARL FRIEDRICH (1777-1855)
German geometer and astronomer

Gauss invites comparisons only to **Isaac Newton** or perhaps Johann Wolfgang von Goethe, being the sort of endlessly

Johann Friedrich Carl Gauss

inventive mind that achieved results when put to any task. His contributions to the fields of pure and **applied mathematics** were all equally sensational in his day, and equally influential into the 20th century. Gauss' major discoveries reached back to Greek practices, either updating or employing them to novel use. His penchant for publishing only the most rigorous and polished proofs had set a standard for arguments in **symbolic logic** not seen before. Gauss' formulation of the complex number system advanced number theoryso that all possible operations could be performed on all possible numbers without needing to create new ones. His investigations into **algebra** and **geometry** paved the way for the modern disciplines of **probability theory**, **topology**, and vector **analysis**. Among Gauss' inventions and collaborations include the heliotrope(a trigonometric measuring device), a prototype of the electric telegraph, and the bifilar magnetometer. Gauss' interests also ran to crystallography, optics, **mechanics**, and capillarity.

Johann Friedrich Carl Gauss was the only son born to Gebhard Dietrich, a laborer and merchant, and Dorothea Benze Gauss, a servant. They made their home in Brunswick, capital of the Duchy of Braunschweig, Germany. As Gauss himself later calculated he was born on April 30, 1777, eight days after Ascension of that year as his mother always told him. His mother was a functional illiterate, a fact which lent a special poignancy to Gauss' later fame, which she could only

ascertain by asking others, rather than experiencing direct **proof** of it herself.

An **arithmetic** prodigy, Gauss enjoyed telling the story of catching his father's **addition** mistake at the age of three. The most famous vignette related to his youth involved an obnoxious teacher who instructed his class to add all the **integers** from one to 100. By adding them in pairs the eight–year–old boiled them down to a smaller set of 101s, fifty to be exact, and calculated the sum from there to be 5050. "Ligget se!" was all he had to say to his teacher, and showed him his slate. The formula Gauss had arrived at is given by $S = n(n+1)/2$ and was actually in use during the days of **Pythagoras.** Such adroitness was initially disparaged by his father, but Gauss was eventually rewarded by a tutor's aid and admission to secondary school in 1788. He began his higher education at Caroline College in his hometown, which offered mathematical training and lessons in Latin and High German. From there, Gauss proceeded to the University of Göttingen in 1795.

Gauss never published a proof until it was airtight, but his interests ranged so far so early in his life that he preceded Bode's Law, **Janos Bolyai** and **N.I. Lobachevsky** 's non–Euclidean geometry, **Karl Jacobi**'s double–period elliptic **functions**, **Augustin–Louis Cauchy**'s functions of a complex **variable** and **William Hamilton**'s **quaternions**. While still a teenager, Gauss constructed a with a ruler and compass a 17–sided polygon inscribed in a **circle**. This was the first true innovation in Euclidean geometry since the time of the ancient Greek mathematicians.

Gauss also discovered the law of quadratic reciprocity and the method of least squares. In 1799, he proved the **fundamental theorem of algebra**: that every polynomial equation has a root in the form of a complex number a+bi. His thesis, "Disquisitiones Arithmeticae," was completed in 1798 but not published until 1801.

The University of Helmstedt awarded Gauss a doctorate in 1799. A return to the University of Göttingen allowed for his early research and later career with the help of his benefactor, the Duke of Brunswick. He held a dual post of Professor of Mathematics and Director of Göttingen Observatory by 1807, though not before enduring a period of unemployment. His most famous work in applied mathematics, *Theoria motus corporum celestium*, followed just eight years after his first major publication. In 1801, Gauss had rediscovered the "lost" orbiting asteroid or minor planet, Ceres. He successfully calculated the object's **orbit** according to certain observations and predicted where it would next reappear—a triumph that secured his fame. This method was refined during his subsequent tenure at Göttingen Observatory into the book–length work. He was also retained by various governments to travel, making geodetic surveys at different locations. For this, Gauss invented a new measuring device, the heliotrope.

Such applied work inspired more pure mathematics, this time differential geometry of curved **space** and surfaces. Gauss' third major publication, *Disquisitiones Generales Circa Superficies Curvas*, was not published until 1827 because of the subject's far–reaching implications. What he called "intrinsic" geometry would pave the way for current differential geometry. Gauss entertained the idea of the curvature of all space, an idea that would be of central importance to **Albert Einstein**'s formulation of space time as a geometric whole.

Gauss married twice, first to Johanna Osthoff on October 9, 1805. The union produced three children, the youngest of whom died soon after birth, followed by the mother. Though a recent widower, Gauss proposed to Friederica Wilhelmina Waldeck, the daughter of a fellow professor. They married August 4, 1810, and had three children before the second Mrs. Gauss died of tuberculosis. Eugene, the eldest boy from Gauss' second marriage, grew up with the same abilities as his illustrious father. However, for reasons that can only be speculated, Gauss prevented his son from following him into the mathematical field.

This reticence also held in Gauss' relationships with his students. Despite his encouragement of a protégéé named Eisenstein who died tragically young, and correspondence with the self–taught pioneer **Sophie Germain**, Gauss never really took anyone under his wing. To Gauss, lecturing would not improve a bad student nor impress a good one. He tended to consider fellow mathematicians as rivals or distractions. He was fond of newspapers and magazines, novelties of the early 19th century. Students in the university library called him the "newspaper tiger" for his habit of staring down anyone who tried to take any newspaper he wanted first.

Nearly seventy–five official honors came to Gauss throughout his life from various countries, though Gauss made light of their accompanied ceremonies, preferring instead to make curmudgeonly jokes at the speechmakers' expense. These recognitions included being installed as a Foreign Member of the Royal Society of England, but Gauss was content to be considered without question the greatest mathematician in the world and get on with his work. He was not intellectually isolated, however. At the age of 62, Gauss taught himself Russian so he could more easily read the works of Lobachevsky. Eventually his health failed, and the loss of friends and family members through death and estrangement took a toll. Gauss died February 23, 1855, of a heart attack after suffering from an enlarged heart for some time. He was buried in Göttingen next to the simple grave of his mother.

Throughout his career, Gauss repositioned pure inquiry as the ultimate test of logic, unbuttressed by geometric or theoretical assumptions and circular arguments. He considered mathematics as a science, with arithmetic as its most important subdiscipline. Gauss avoided trivialities by realizing that one cannot study a magnitude in isolation, for true mathematics lies in the study of relationships. Moreover, he envisioned new relationships to consider among **infinite series**, hypercycles, and pseudospheres, the sort of fanciful mathematical entities that populate the imaginations of contemporary theorists.

GAUSSIAN CURVATURE

Gaussian curvature is a numerical quantity associated with an **area** of a surface that describes the intrinsic geometric prop-

erty of that area. It is different from the curvature of a curve, for that is an extrinsic geometric property defining how it is bent in a plane or **space**. The Gaussian curvature remains the same no matter how a surface is bent as long as it is not distorted. It is defined as the product of the principal curvatures, the maximum and minimum values of normal curvature at a point on a surface. Since it is the product of two curvatures, Gaussian curvature has the units of curvature squared. If the Gaussian curvature is a positive value then the surface is locally either a peak (both the maximum and minimum values are positive, hence the product is positive) or a valley (both the maximum and minimum values are negative, hence the product is positive). A negative value indicates that the surface has saddle points locally, that is either the maximum or minimum value is negative and the other is positive. A value of **zero** is indicative of a surface that is flat in at least one direction. Surfaces such as a plane and a **cylinder** have zero Gaussian curvature.

Gaussian curvature can be thought of in another way. The Gaussian curvature of a surface determines when one surface can or cannot be bent into another. The total Gaussian curvature of a region on a surface can be determined by opening up the surface and measuring the **angle** by which it is opened up. For instance, if one cuts a small portion of a **sphere** out and cuts open the portion so that the material is free to flatten out one can measure this angle. The angle between the tangents to the curve at the two sides of the cut is the total Gaussian curvature. This numerical quantity associated to an area of a surface is very closely related to angle defect. In fact, the total Gaussian curvature of a region of a **polyhedron** containing only one vertex is the angle defect at that vertex. To think of Gaussian curvature in this way it is useful to think of trying to coerce a flat sheet of paper into a sphere. It cannot be done unless some of the paper is removed or distorted in some way. The Gaussian curvature of a sphere is not equal to that of a flat sheet of paper and therefore one cannot shape a piece of paper into a sphere. On the other hand it is possible to fold a flat sheet of paper into a cylinder. Both the cylinder and flat sheet of paper have Gaussian curvatures of zero and so can be interconverted from one to the other.

The Gaussian curvature of a function describing a two-dimensional object vanishes wherever the function is locally one-dimensional. So for a plane the Gaussian curvature is zero as it is for a cylinder. The formal form for the Gaussian curvature of a regular surface in three-dimensional space at any point p is $G(p) = \det(S(p))$, Where S is the shape operator and det signifies the determinant. The shape operator, or Weingertan map or second fundamental tensor as it is sometimes called, is an extrinsic curvature. Gaussian curvature is also given by $G(p) = 1/R_1R_2 = K_1K_2$, Where R_1R_2 are the principal radii of curvature and K_1K_2 are the principal curvatures. While a surface on which the Gaussian curvature is always positive is called synclastic a surface on which the Gaussian curvature is everywhere negative is called anticlastic. Surfaces with constant Gaussian curvature include the sphere, cone, cylinder, and plane. In areas of the surface which bend highly the curvature is high, whereas areas where bending is slight the curvature is low.

Gaussian curvature is used to explore and describe curved surfaces and spaces. While doing geodetic survey work for the governments of Hanover and Denmark in 1821 a German scientist, **Carl Friedrich Gauss**, developed the concept of Gaussian curvature. He used a Gauss map to define curvature. A Gauss map is a function describing the **distance** to a sphere from an orientable surface in Euclidean space. It associates its oriented normal vector to every point on the surface. As well as playing a fundamental role in **Einstein**'s theory of gravitation, smooth surface points can be catalogued by their Gaussian and toric curvature. Since Gaussian curvature has units of curvature squared many corneal topography advocates are tempted to redefine Gaussian curvature as its **square** root. The **square root** of Gaussian curvature is known as geometric **mean** curvature or root Gaussian curvature in these situations although it is of no real use in corneal topography. Gaussian curvature is especially useful in designing pressing process for items such as car bodies. Shapes with nonzero Gaussian curvature will require that some material is stretched or deformed in the pressing process. To do this effectively there has to be a balance in the stretching and amount of material and such stress a material can tolerate.

GELFOND, ALEKSANDR OSIPOVICH
(1906-1968)
Russian mathematician

Aleksandr Gelfond made significant contributions to the theory of **transcendental numbers** and the theory of **interpolation** and approximation of the **functions** of a complex **variable**. He established the transcendental character of any number of the form $a{:}ssb{:}ks$, where a is an algebraic number different from 0 or 1 and b is any irrational algebraic number, which is now known as Gelfond's **theorem**.

Gelfond was born in St. Petersburg (later Leningrad); his father was a physician who also dabbled in philosophy. Gelfond entered Moscow University in 1924 and completed his undergraduate degree in mathematics in 1927. He pursued postgraduate studies from 1927 to 1930 under the direction of A.J. Khintchine and V.V. Stepanov.

Gelfond's first teaching assignment was at the Moscow Technological College. He quickly won a more prestigious appointment at Moscow University, where he began teaching mathematics in 1931. He became a professor of mathematics in 1931, a position he held until his death. For several years, Gelfond served as the chairman of the mathematics department specializing in the theory of numbers. His enthusiasm for the **history of mathematics** was reflected not only by his own works on **Leonhard Euler**, but by his incorporating a "history of mathematics" division into the theory department he chaired.

In 1933, Gelfond was also appointed to a post in the Soviet Academy of Sciences Mathematical Institute. He completed a doctorate in **mathematics and physics** in 1935 and

was elected a corresponding member of the Academy of Sciences of the U.S.S.R. in 1939.

Gelfond found his greatest inspiration in the past. In 1748, Euler proposed that **logarithms** of rational number with rational bases are either rational or transcendental; in 1900, **David Hilbert** developed a 23-problem series on the rationality or transcendence of the logarithms of such numbers. For three decades, mathematicians were unable to trace a solution to the puzzle posed by Hilbert's seventh problem—the assumption that $a{:}ssb{:}ks$ is transcendental if a is any algebraic number other than 0 or 1 and b is any irrational algebraic number.

In 1929, Gelfond established connections between the properties of an analytic function and the **arithmetic** nature of its values, publishing his first paper on the topic, "Sur les nombres transcendant," in 1929. He built on this discovery to unravel Hilbert's seventh riddle by using linear forms of **exponential functions**. Gelfond published the results of his work, "Sur le septieme probleme de Hilbert," in 1934. He continued his explorations, using his knowledge of functions to develop theorems related to rational **integers**, transcendental numbers (he was able to construct new classes in this area), mutual algebraic independence, and analytic theory.

Gelfond's interest in function theory was probably shaped by the Luzitania—an informal academic and social organization clustered around **Nikolai Nikolaevich Luzin**, a noted mathematician in the 1920s. Gelfond was a contemporary and colleague of **Nina Karlovna Bari**, a Luzin protegee; although Gelfond's name does not appear on the list of those who declared themselves Luzitanians. Luzin's prominence and the intellectual vigor of the students he attracted influenced the philosophy and direction of mathematics at the university. By 1930, the Luzitania movement sputtered and died, and Luzin left Moscow State for the Academy of Science's Steklov Institute.

In 1936, during the dictatorship of Josef Stalin, Luzin was charged with ideological sabotage. Luzin's trial was abruptly and surprisingly canceled, but he was officially reprimanded and withdrew from academia. Luzin's fall demonstrated—in a way that could not be ignored—the inextricable interweaving of politics and academic achievement. Gelfond was permitted to pursue his studies in peace in part because of his political connections.

"He was a member of the Communist Party," wrote Ilya Piatetski-Shapiro, for whom Gelfond was an instructor, mentor and advisor. "His father was personally acquainted with Lenin... he said that his father and Lenin had disagreements in public life, but in private life they were friends. Being a member of the Communist Party, Gelfond felt that he had some influence...." Such influence could not overcome the deep wave of anti-Semitism that swept over Russia after World War II. Despite Gelfond's recommendation, Piatetski-Shapiro, who was Jewish, was denied admission to Moscow University's graduate school by the party committee of the mathematics department.

But Gelfond "was a very warm person, very humane and sensitive to me and to the other students," Piatetski-Shapiro wrote, and Gelfond was reluctant to let a promising student—winner of the Moscow Mathematical Society award for young mathematicians—languish. Although his sponsorship could have had dire implications for his own career, Gelfond persisted, and finally secured admission to the graduate program for Piatetski-Shapiro at the Moscow Pedagogical Institute.

Gelfond's most comprehensive publications were released in 1952. *Transtsendentnye i albegraicheskie chisla* provided an overview of his work in transcendental numbers, and his work on the theory of the functions of a complex variable is compiled in *Ischislenie knoechnyko raznostey*.

In 1968, Gelfond was named a corresponding member of the International Academy of the History of Science. He also served as chair of the scientific council of the Soviet Academy of Sciences Institute of the History of Science and technology, which refereed works on the history of physics and mathematics.

Gelfond's drive to expand the understanding of mathematics theory persisted to the day of his death. "When he died... I was present in the hospital," wrote Piatetski-Shapiro. "I remember he was trying to write some formula and tell me something which was clearly related to the **zeta function**. He could not because he was already paralyzed."

Gelfond died in Moscow; most sources list the year of his death as 1968, but Piatetski-Shapiro records it as 1966.

GELFOND-SCHNEIDER THEOREM

The Gelfond-Schneider **theorem** states that if a and b are **algebraic numbers**, a is not **zero** or one, and b is irrational then a^b is a transcendental number. For example, the theorem guarantees that $2^{\sqrt{2}}$ is transcendental by applying the result with a=2 and b = $\sqrt{2}$, which is irrational. It also guarantees that, for example, the base 10 logarithm of 2 is transcendental.

The statement of the Gelfond-Schneider theorem was posed as the seventh problem of D. Hilbert's famous list of 23 problems. Hilbert is reputed to have said that he did not expect this problem to be solved in his lifetime. Contrary to his expectations, the theorem was proved independently by the Russian mathematician **Aleksandr Osipovich Gelfond** and the German mathematician Theodor Schneider in 1934. Their work is a vast extension of the work of **C. Hermite** and **F. Lindemann** who, in the 19th century, proved the transcendence of the numbers e and π, respectively. The work of Gelfond and Schneider was further developed into modern transcendence theory by many authors, notably the British mathematician A. Baker who proved a conjecture of Gelfond on linear combinations with algebraic coefficients of **logarithms** of algebraic numbers, for which he received the Fields medal in 1966.

See also Transcendental numbers

GEODESIC CURVATURE

Geodesic curvature measures how much a curve bends or turns. For example, a straight line has geodesic curvature **zero** at all of its points. The absolute value of the geodesic curvature of a **circle** is the same for all its points too and depends only on the circle's radius. A large radius means that the circle is almost straight and so the curvature is small. But a small radius means that the bending is fast and so the curvature is large. The sign of geodesic curvature depends on the direction of the curve. If we are traveling in a circle counterclockwise, the curvature is positive. But if we move in the opposite direction, the curvature is negative.

Suppose we want to find the geodesic curvature of a curve called c at a point called p. First draw the tangent line to the curve at c. One way to draw this line is first, to consider a point p(t) on the curve that is some **distance** t away from p. Draw the line through p and p(t). Next pick another point (also called p(t)) that is on the curve but closer to p and draw the line through p and the new p(t). If we keep doing this, the lines that we draw should get closer and closer to the tangent line at p. But, it can happen that these lines do not get close to any line at all. In this case the tangent line at p is not defined and neither is the curvature. The curve is said to be non-differentiable if this happens. Suppose now that we have drawn the tangent line. Now draw a circle that is tangent to the tangent line and intersects p. Choose the circle so that the part of the circle close to p is as close as possible to the part of the curve near p. In other words, choose the circle so that it approximates c near p. Then, the absolute value of the geodesic curvature is equal to one divided by the radius of this circle. If we stand at the point p and face in the direction that the curve is going, then the geodesic curvature is positive if the circle that we drew is on our left side. Otherwise, it is negative. The circle that we drew is called the osculating circle of c at p.

Geodesic curvature is defined for curves on surfaces, too. A surface is an object that locally, looks like two-dimensional **space**. Here are some examples: two-dimensional space, the **sphere**, the surface of a donut, the surface of a pretzel, a funnel. A geodesic segment in a surface is a curve between two points that is a shortest path between those points. For example, on the sphere a geodesic segment from the north pole to the south pole is an arc of a longitude line. If x and y are two points on a surface S, then the S-distance between them is the length of the shortest path in S between them. For example, the unit sphere-distance between the north and south pole is **Pi** even though the distance between those points in three-dimensional space is two. A geodesic is a path that is infinite in both directions and has the property that if two points are contained in the geodesic then the part of the geodesic between them is a geodesic segment. On the sphere, geodesics are also called great circles. Geodesics play the same role in the **geometry** of surfaces as lines do in the geometry of the plane. So, a curve on a surface is differentiable at a point if it has a tangent geodesic at that point. If p is a point on a surface S and r is a positive number, then the set of all points in S that are distance r away from p in S is called the S-circle of radius r centered at p. S-circles play the same role in the geometry of surfaces as circles do in the geometry of the plane. So, the absolute value of the geodesic curvature of a curve on a surface can be defined in the same way as for curves in the plane with the words 'geodesic' and 'S-circle' substituted for 'line' and 'circle'. In particular, the geodesic curvature of a geodesic is always zero. The sign of geodesic curvature on a surface depends on the orientation of the surface. If the surface is not oriented, then the sign is undefined. Otherwise the sign is positive when the ordered pair of vectors (v,w) is positive with respective to the orientation where v points in the direction of the velocity of the curve and w points in the direction of the acceleration of the curve.

Another way to define geodesic curvature uses the derivative (from differential **calculus**) of a curve and the some vector **analysis**. If c is a map from [0,1] to the plane, then its derivative at time t is denoted by $c'(t)$. The inner product of two vectors v and w is denoted by $<v,w>$ and equals the **cosine** of the **angle** between them multiplied by the product of their lengths. The length of v is denoted by $\|v\|$. The curvature of c at time t is given by the formula $k[c](t) = <c''(t), Jc'(t)>/\|c'(t)\|^3$. Here, J is the linear map that sends any point (x,y) to the point $(-y, x)$. If S is a surface inside three-dimensional space, then every point on the surface has a tangent plane. A vector v whose tail is on the surface can be projected on to the tangent plane of the surface to a vector $P(v)$. Then the geodesic curvature of a curve c on S at time t is given by the formula $P(c''(t)) = <k[c](t), Jc'(t)>$. It can be shown that a curve on a surface is determined by three objects: its starting point, the tangent vector at its starting point, and its geodesic curvature.

See also Differential calculus; Gaussian curvature; Geometry; Surface; Vector analysis; Velocity vector and acceleration vector

GEOMETRIC SERIES

A sequence of numbers is said to be geometric if any term after the first can be obtained by multiplying the previous number by the same constant. This constant is called the common ratio. So the sequence 1,2,4,8,16 is geometric since each number in the sequence after the first can be obtained by multiplying the previous term by 2. Here the common ratio is 2. A geometric series is the sum of the terms of a geometric sequence. Thus, 1+2+4+8+16 is a geometric series. Geometric series may be finite, such as the series in the preceding sentence, or infinite, such as 1+2+4+..., where the three dots indicate that the series follows this pattern forever. A finite series obviously has a sum, such as 1+2+4+8+16=31. If we write the general finite geometric series as $a+ar+ar^2+ar^3+...+ar^n=S_n$, then it can be shown that $S_n=a-ar^{n+1}/(1-r)$. Trying this formula for 1+2+4+8+16, we get $(1-2^5)/(1-2)=(-31)/(-1)=31$, agreeing with our answer computed in the traditional manner above. What may be surprising is that some infinite geometric series also have finite sums. This is true whenever the absolute value of r is less than 1, or equivalently, when -1 < r < 1. Here's why. When -1 < r < 1, the expression r^{n+1} approaches 0 when n is very large, so for an infinite geometric series the expression for the sum $a-ar^{n+1}/(1-r)$ approaches $a/(1-r)$.

The ancient Greek philosopher **Zeno** (5th century BC) was famous for creating paradoxes to vex the intellectuals of his time. In one of those paradoxes, he says that if you are 1 meter away from a wall, you can never reach the wall by walking toward it. This is because first you have to traverse half the **distance**, or ½ meter, then half the remaining distance, or ¼ meter, then half again, or 1/8 meter, and so on. You can never reach the wall because there is always some small finite distance left. The theory of infinite geometric series can be used to answer this paradox. Zeno is actually saying that we cannot get to the wall because the total distance we must travel is ½ + ¼ + 1/8 + 1/16 +..., an infinite sum. But according to our discussion in the preceding paragraph, this is just an infinite geometric series with first term ½ and common ratio ½, and its sum is (1/2)/(1-1/2)=1. So the infinite sum is one meter and we can indeed get to the wall.

Infinite geometric series are of major importance in **calculus** in connection with Taylor Series. A Taylor Series is an **infinite series** representation of some mathematical function. Many of the most important mathematical **functions**, such as sin x, cos x, e^x, ln x and others may be expressed as infinite Taylor series, which is useful in creating algorithms for calculators and computers to give very accurate approximations of these functions for specific values of x. Referring again to the opening paragraph of this article, we can see that $1/(1-x)=1+x+x^2+x^3+...$, whenever -1 < x < 1, and this is just the Taylor series for the function 1/(1-x). This series can then be used to derive series for other important mathematical functions.

See also Infinite series; Taylor's and Maclaurin's series

GEOMETRY

Geometry is the branch of mathematics that deals with measurements and properties of points, lines, and angles. Geometry is one of the oldest branches of mathematics, used by the Egyptians and Babylonians as early as 2000 B.C. The ancient Egyptians and Babylonians had a practical knowledge of many geometric ideas, which they applied to surveying and construction projects; the first meaning of the word "geometry," which is of Greek origin, was "measurement of the earth." Early tablets have shown that the Egyptians and Babylonians formed reasonable estimates of the value of **pi** (π), and that the Babylonians were aware of the **Pythagorean theorem**, which gives a relationship between the lengths of the sides of a right **triangle**. The Egyptians and Babylonians discovered geometric ideas largely through experimental means—from examining the measurements of physical objects.

Geometry came into its own in the time of the Greeks, who developed the notion that geometric relationships could be proven. With that idea, geometry turned from an experimental science into an intellectual pursuit, one that used the laws of logic to deduce complex geometric relationships from simpler ones. The crowning achievement of Greek geometry was **Euclid**'s *Elements,* a textbook written around 300 B.C. that developed a rigorous presentation of elementary geometry. Euclid followed an axiomatic approach, which has been enormously influential to the development of modern mathematics: he started with five basic assumptions (axioms), and used deductive reasoning to create a huge edifice of theorems that are logical consequences of those axioms. Among his theorems are, for example, the laws of congruent and similar triangles, a **proof** of the Pythagorean **theorem**, and the relationships between lengths and areas in circles.

For almost 2000 years after Euclid wrote his *Elements,* the study of geometry consisted chiefly of understanding the *Elements,* and enlarging on the ideas and techniques found within it. In the early 17th century, the mathematician and philosopher **René Descartes** made a great step forward by connecting **algebra** and geometry, through the coordinate system. Descartes realized that the location of a point in the plane could be described by an ordered pair of numbers: one number that described its position on a horizontal axis, one on a vertical axis (according to myth, Descartes conceived this idea while watching a fly walk on the ceiling). With the advent of Cartesian geometry, geometers could use the techniques of algebra, which streamlined geometric proofs enormously. Around the same time, two problems from outside of mathematics began to influence the development of geometry: the new science of mapmaking, through which geometers began to understand the very different nature of the curved earth and the flat plane, and the use of perspective in art, which stimulated the development of projective geometry.

An important question that plagued geometers was the problem of how fundamental Euclid's five axioms really were. The first four of **Euclid's axioms** dealt with such natural concepts that most geometers were willing to accept that they were truly basic assumptions for any geometric **space**. The fifth **axiom**, Euclid's **parallel postulate**, was much more controversial: it stated that, given a line and a point not contained in the line, there was one and only one line through the point that never met the original line. That statement certainly appeared to be true, but it was nowhere near as simple as the other four postulates. Its complexity led mathematicians to ask whether the fifth axiom was truly a basic assumption, or whether it was really a theorem, something that could be proven from the other four axioms. For many years this question was open, but in the 19th century, **Johann Carl Friedrich Gauss**, **János Bolyai**, and **Nikolay Lobachevsky** shocked the mathematical community by showing that there exist perfectly consistent geometries in which Euclid's fifth axiom is not true.

The discovery of **non-Euclidean geometries** spurred the development of differential geometry, the study of curved spaces. The Euclidean geometry of the *Elements* is the geometry of the flat plane, and its theorems and measurements only give information about objects in a flat space. Differential geometry, developed by Gauss and **Bernhard Riemann**, focused on measurements of lines, circles, triangles, and other objects on surfaces or spaces that are curved, like a **sphere** or a **cylinder**. In these curved spaces, the idea of a straight line is replaced by the idea of a geodesic, the shortest path between two points. On a surface, the shortest path between two points can be found by attaching a rubber band to the two points and

pulling it taut, always making sure it lies on the surface. So for example, on a sphere, a longitudinal line is a geodesic between the north and south poles. This example exposes some key differences between lines in Euclidean space and geodesics in curved spaces: there does not have to be only one geodesic between two points, and when a geodesic is extended, it may run into itself, instead of extending infinitely far. With this new notion of a "line," differential geometers can ask the same questions about triangles, circles, and angles that Euclidean geometers have asked (although the answers can be wildly different!).

One of the key notions of differential geometry is the idea of the curvature of a surface or space. Curvature is a number associated to a space, which measures the degree and the type of curving on the space. A precise **definition** requires advanced mathematical tools, but among surfaces, the following are the basic distinctions: a surface shaped like a sphere is said to be positively curved; a surface that is saddle-shaped is negatively curved; a surface that is flat, like the Euclidean plane, has **zero** curvature. A more complicated surface can have different curvature at different points on the surface.

The notions of curvature and geodesics can be extended to three- and higher-dimensional spaces. One of the most important ideas of **Einstein**'s **relativity** theory is that space has higher curvature near massive objects; it is this curvature that creates gravitational attraction. Recently, the Hubble telescope has located points in space that are connected by more than one geodesic path, raising the question of the precise geometry of the universe. Differential geometry continues to be fundamental in the attempts of modern physics to advance the quest begun by the Egyptians and Babylonians: to measure the world.

See also Euclid's axioms; Hyperbolic geometry; Non-Euclidean geometries

GERARD OF CREMONA (CA. 1114-1187)

Italian translator

Gerard of Cremona is sometimes referred to as Gerhard, Gherardo, or Gherard of Cremnona, and these variant spellings should be noted. Born in Cremona in Italy, Gerard eventually died in 1187 in Toledo, Spain. He was a translator who realized that much of the knowledge of the Arab world was still largely unknown in the West. To rectify this Gerard set himself the task of translating Arabic scientific texts into Latin, which was then the language of the educated in Europe. Gerard travelled to Toledo to learn Arabic. His first translation was of **Ptolemy**'s *Almagest.* By the end of his life Gerard had translated over 90 works from Arabic into Latin. These works covered such topics as mathematics, astronomy, and medicine. In his translations Gerard produced Latin words to cover mathematical concepts which were rendered in Arabic. From this we have the names that we now use for many mathematical terms, particularly those of **geometry**. One example of this is the derivation of the word **sine**. **Aryabhata the Elder** called it ardhaiya (half chord), which he abbreviated to iya (chord). When this was translated into Arabic the word was taken as jiba, which was written as jb. This was eventually replaced by jaib (cove or bay) and Gerard took the nearest Latin equivalent of sinus (a bent curve), and from this we get sine.

The translations of Gerard were of such high quality that they were sought out by many mathematicians for centuries to come. It is to Gerard that we owe the preservation of many original texts, or later commentaries on them (some of the Arabic texts translated were commentaries on original Greek texts that were long lost even by the twelfth century). One of the most famous works that Gerard translated was *Elements* by **Euclid of Alexandria**. There are no known examples of this work from the time of Euclid, all copies are translations of Arabic copies made by the Byzantines in the eighth century, the first known such work is by **Adelard of Bath**, with Gerard's version appearing some 30 years later. Gerard was also responsible for translating the **algebra** of **Al Khwarizmi**.

See also Adelard of Bath; Euclid of Alexandria; Ptolemy

GERGONNE, JOSEPH-DIAZ (1771-1859)
French mathematician

Joseph-Diaz (sometimes rendered as Diez) Gergonne introduced the word polar into mathematics along with the principal of duality in projective **geometry**. Gergonne also solved the problem of **Apollonius of Perga**. In 1810 Gergonne started editing and producing the first purely mathematical journal, *Annales de Mathematiques Pures et Appliquees*, which is sometimes also known as *Annales de Gergonne*.

Born in Nancy, France, Gergonne became an artillery officer and a professor of mathematics. He quickly provided an elegant solution to the problem of Apollonius, which deals with finding a **circle** that touches three given circles. In 1810, while a professor of mathematics at Nimes University, Gergonne launched his *Annales*. It was initially intended as an educational journal, but rapidly it became a home for unusual mathematics with many of the papers being written by Gergonne himself, sometimes anonymously. The journal ceased production in 1832 when Gergonne became Rector of the Academie at the University of Montpellier. He discovered the principle of duality in about 1825, which provoked ill feeling between him and **Jean Victor Poncelet**, who claimed priority in the discovery. Gergonne died at Montpellier in 1859; he was 88.

See also Appolonius of Perga; Jean Victor Poncelet

GERMAIN, SOPHIE (1776-1831)
French number theorist

Sophie Germain's foundational work on **Fermat's Last Theorem**, a problem unsolved in mathematics into the late 20th century, stood unmatched for over one hundred years.

Sophie Germain

Though published by a mentor of hers, **Adrien–Marie Legendre**, it is still referred to in textbooks as Germain's **Theorem**.

Germain worked alone, which was to her credit, yet contributed in a fundamental way to her limited development as a theorist. Her famed attempt to provide the mystery of Chladni figures with a pure mathematical **model** was made with no competition or collaboration. The three contests held by the Paris Académie Royale des Sciences from 1811 to 1816, regarding acoustics and elasticity of vibrating plates, never had more than one entry—hers. Each time she offered a new breakthrough: a fundamental hypothesis, an experimentally disprovable claim, and a treatment of curved and planar surfaces. However, even her final prizewinning paper was not published until after her death.

Marie–Sophie Germain was born April 1, 1776, in Paris to Ambroise–François Germain and Marie–Madeleine Gruguelu. Her father served in the States–General and later the Constituent Assembly during the tumultuous Revolutionary period. He was so middle class that nothing is known of his wife but her name. Their eldest and youngest daughters, Marie–Madeleine and Angelique–Ambroise, were destined for marriage with professional men. However, when the fall of the Bastille in 1789 drove the Germains' sensitive middle daughter into hiding in the family library, Marie–Sophie's life path diverged from them all.

From the ages of 13 to 18 Sophie, as she was called to minimize confusion with the other Maries in her immediate family, absorbed herself in the study of pure mathematics. Inspired by reading the legend of **Archimedes,** purportedly slain while in the depths of geometric meditation by a Roman soldier, Germain sought the ultimate retreat from ugly political realities. In order to read **Leonhard Euler** and **Isaac Newton** in their professional languages, she taught herself Latin and Greek as well as **geometry**, **algebra**, and **calculus**. Despite her parents' most desperate measures, she always managed to sneak out at night and read by candlelight. Germain never formally attended any school or gained a degree during her entire life, but she was allowed to read lecture notes circulated in the École Polytechnique. She passed in her papers under the pseudonym "Le Blanc."

Another tactic Germain used was to strike up correspondences with such successful mathematicians as **Karl Gauss** and Legendre. She was welcomed as a marvel and used as a muse by the likes of **Jean B. Fourier** and **Augustin–Louis Cauchy**, but her contacts did not develop into the sort of long–term apprenticeship that would have compensated for her lack of access to formal education and university–class libraries. Germain did become a celebrity once she dropped her pseudonym, however. She was the first woman not related to a member by marriage to attend Académie des Sciences meetings, and was also invited to sessions at the Institut de France—another first.

Some interpret Gauss' lack of intervention in Germain's education and eventual silence as a personal rejection of her. Yet this conclusion is not borne out by certain facts indicating Gauss took special notice. In 1810, Gauss was awarded one of his many accolades, a medal from the Institut de France. He refused the monetary component of this award, accepting instead an astronomical clock Germain and the institute's secretary bought for him with part of the prize. Gauss' biographer, G. Waldo Dunnington, reported that this pendulum clock was used by the great man for the rest of his life.

Gauss survived her, expressing at an 1837 celebration that he regretted Germain was not alive to receive an honorary doctorate with the others being feted that day. He alone had lobbied to make her the first such honored female in history. A hint of why Gauss valued her above the men who joined him in the Académie is expressed in a letter he sent to her in 1807, to thank her for intervening on his behalf with the invading French military. A taste for such subjects as mathematics and science is rare enough, he announced, but true intellectual rewards can only be reaped by those who delve into obscurities with a courage that matches their talents.

Germain was such a rarity. She outshone even **Joseph–Louis Lagrange** by not only showing an interest in **prime numbers** and considering a few theorems, about which Lagrange had corresponded with Gauss, but already attempting a few proofs. It was this almost reckless attack of the most novel unsolved problems, so typical of her it is considered

Germain's weak point by 20th century historians, that endeared her to Gauss.

Germain's one formal prize, the Institut de France's Gold Medal Prix Extraordinaire of 1816, was awarded to her on her third attempt, despite persistent weaknesses in her arguments. For this unremedied incompleteness, and the fact that she did not attend their public awards ceremony for fear of a scandal, this honor is still not considered fully legitimate. However, the labor and innovation Germain had brought to the subjects she tackled proved of invaluable aid and inspiration to colleagues and other mathematical professionals as late as 1908. In that year, L. E. Dickson, an algebraist, generalized Germain's Theorem to all prime numbers below 1,700, just another small step towards a complete **proof** of Fermat's Last Theorem.

Germain died childless and unmarried, of untreatable breast cancer on June 27, 1831 in Paris. The responsibility of preparing her writings for posterity was left to a nephew, Armand–Jacques Lherbette, the son of Germain's older sister. Her prescient ideas on the unity of all intellectual disciplines and equal importance of the arts and sciences, as well as her stature as a pioneer in women's history, are amply memorialized in the École Sophie Germain and the rue Germain in Paris. The house on the rue de Savoie in which she spent her last days was also designated a historical landmark.

GIBBS, JOSIAH WILLARD (1839-1903)
American mathematical chemical–physicist

J. Willard Gibbs is not as famous as the Europeans who discovered and lionized him. **James Clerk Maxwell** was the first and for a time nearly the only major scientist among his contemporaries to fully understand Gibbs' publications and what they implied. **Albert Einstein** called him "the greatest mind in American history." Gibbs' studies of thermodynamics and electromagnetics and discoveries in statistical **mechanics** made Einstein's later theories conceivable. He is also largely responsible for the field of physical chemistry, which impacted the steel and ammonia industries. Gibbs is known as the "father of **vector analysis**" for replacing **William Rowan Hamilton**'s **quaternions** in the field of mathematical physics. Thanks to him there are such ideas as the Gibbs phase rule, the Gibbs adsorption isotherm regarding surface tension, Gibbs free energy, and Gibbsian ensembles. Even two short letters to *Nature* in the late 1890s defined what is now known as the "**Gibbs phenomenon**" in the **convergence** of a **Fourier series**. Gibbs' deployment of probability set the stage for **quantum mechanics** to come about some decades after his death. For all of these achievements, he was elected to the Hall of Fame for Great Americans in 1950.

The Gibbs family was originally from Warwickshire, England, having emigrated to Boston in the 17th century. Josiah, born on February 11, 1839, bears the same name as his father but was not known as Josiah Willard Gibbs, Jr. The two men eventually differentiated themselves according to their use of initials. Gibbs' father, a professor of biblical or "sacred"

Josiah Willard Gibbs

literature at Yale University, went by J.W. Gibbs. His wife Mary Anna's maiden name was Van Cleve.

Josiah was the only boy in the family. Of his four sisters he would remain closest to Anna, who also never married, and Julia, who married Addison Van Name, a member of the Connecticut Academy that first published Josiah's articles. The Van Name home in New Haven, Connecticut, where Josiah would stay later in life until his death, was within walking distance of the house where he was born.

Josiah's childhood was marred by scarlet fever, which left him sensitive to illness in adulthood. However, his home life was otherwise supportive. It has been assumed that the young Gibbs' latent scientific talents were actually inherited from his mother, who was an amateur ornithologist. Her wit and charm were widely acknowledged, and her ingenuity extended to building dollhouses for her girls that included realistic plumbing and kitchen equipment. His father, who was considered a prodigy in his day, was an exemplary scholar and teacher in the humanities who received an honorary degree from Harvard after his retirement. Tragedy came later, as two sisters and both parents eventually died by the time Gibbs was a graduate student.

Gibbs began school at the age of nine, in a private boy's school known informally as "Mr. Farren's School." From there he transferred to Hopkins Grammar School, no more

than half a block from the family home. At Yale, it seemed likely that Gibbs' would follow his father into philology. He was a highly decorated Latin student often chosen to give orations at university functions. Like his father, he earned a bachelor's degree at 19.

The American academy at that time valued and rewarded only applied science and mathematics, so when Gibbs' continued his studies at Yale he wrote a fairly pedestrian doctoral thesis on spur gear design in 1863. This made him one of three Ph.D. recipients that year, at the first American institution to offer the degree. His doctorate in particular was the first in engineering and the second in science ever conferred in the United States. After a short stint as a Latin tutor, he returned to his chosen field. Gibbs was awarded a patent for his redesigned railway car brakes in 1866. That same year he took his one major trip abroad. For three years he attended physics lectures at Paris, Berlin, and Heidelberg's universities given by the field's foremost practitioners. Gustav Kirchhoff and Hermann von Helmholtz were of particular influence. Luckily, Gibbs could subsist on monies inherited from his parents, because upon his return to Yale he was appointed a professor of mathematical physics at no pay. This has been explained as resulting from his lack of published works. Gibbs would keep that unpaid post until an offer from Johns Hopkins University in 1880 of three thousand dollars a year forced Yale authorities to counteroffer two thousand.

In 1873 Gibbs devised a geometrical representation of the surface activity of thermodynamically active substances, and wrote "Graphical Methods in the **Thermodynamics** of Fluids," his first publication. Although the publishers of *Transactions of the Connecticut Academy* did not fully understand this or his other papers, they raised money specifically to print his material, which was sometimes lengthy. Much has been made of the fact that he was by then 34 years old. Most mathematicians or scientists peak at a much younger age. However, Gibbs had apparently been developing his theories for quite a long time, and was only just beginning to articulate his discoveries. He built on principles previously set down by Helmholtz, James Joule, and Lord Kelvin, but whereas his predecessors had been specifically concerned with an immediate example like the heat engine, Gibbs preferred to keep it mathematically general. His diagrams treated entropy, temperature, and pressure, in their relations to **volume**, as coordinates. When he considered a three–dimensional surface, the coordinates he chose were entropy, volume, and energy.

By mathematically formulating the second law of thermodynamics regarding entropy and mechanical energy, Gibbs made thermodynamics scientifically viable. His phase rule is a simple looking formula: f = n + 2 – r. In that sequence *f* represents the total degrees of freedom in temperature and pressure, *n* the total of chemical elements in the object's makeup, and *r* the number of phases the object may take over time—solid, liquid, or gas in any combination. This rule was central to his reconception of the thermodynamics of a complex of systems into a single system over time—the probability of the existence of any possible system in "phase–space" overall.

By doing so, and by applying the first and second laws of thermodynamics to complex substances, Gibbs set the theoretical basis for physical chemistry. This relatively new specialty deals with phenomena like **hydrodynamics**, and novel types of mathematical **modeling** like electrochemical simulations and genetic algorithms. Gibbs' "single system" method is now a fundamental part of statistical mechanics in the form of Gibbsian ensembles. These are defined as large numbers of thermodynamically equivalent macroscopic systems. By using these ensembles, laboratory and factory researchers save themselves the tedium and risk of trial and error experimentation when synthesizing new compounds or alloys.

"Gibbs free energy" refers to the likelihood of any one chemical reaction taking place, which takes into account both entropy—the disorder in a system—and enthalpy, its heat content. At least one biographer considers Gibbs' clarification of Rudolf Clausius' original **definition** of entropy, as it became ranked with other thermodynamic properties such as energy, temperature, and pressure. His ideas of chemical potential and free energy now take precedence in the conception of how chemical reactions take place.

Although he drew mixed reviews as a teacher, Gibbs was clearly concerned with his relationship with his students and with involving them in mathematics. In 1877 he founded the Math Club at Yale, the second of such informal groups there. He served as executive officer for ten years, and more than likely gave his first impressions of vector analysis and perhaps even the multiple **algebra** or **matrix** studies that gave birth to vectors. There are other indications that he used his classes as seminars.

For his vector analysis Gibbs drew from **Hermann Grassman** as well as William Hamilton. Even in its early stages he could use it to calculate the orbits of planets and comets such as Swift's, improving upon **Karl Gauss**' method, and also applied vectors to problems in crystallography. Gibbs accounted for most of the properties of **light** as an electromagnetic phenomenon according to Maxwell's theory, in purely theoretical terms, during the 1880s. In doing so, he succeeded in treating the relation between **force** and displacement waves in electricity in the same way as others had for mechanical and acoustic waves. All this was an outgrowth of Gibbs' courses from 1877 to 1880, including the first college–level course in vector analysis with concentration on electricity and magnetism and the first public usage of vector methods. His analyses have not had to be corrected since.

Gibbs' ideas could not be disseminated in Europe until they were translated, because no one on the Continent at that time followed American scientific or mathematical journals consistently. Consequently, both Helmholtz and Karl Planck unknowingly duplicated some of Gibbs' findings, and Jacobus Van't Hoff independently conceived of chemical thermodynamics. This situation changed when Friedrich Ostwald translated Gibbs into German in 1891. Ostwald was followed near the end of that decade by Henri Le Chatelier, who translated Gibbs into French. Edwin Wilson, who also wrote posthumous biographical commentaries on his professor, wrote a textbook published in 1901 entitled *Gibbs' Vector Analysis* that suc-

ceeded in reaching a larger audience. The book was edited from class notes used between 1881 and 1884.

Such efforts led to great fame for Gibbs near the last years of his life. After reading Gibbs in translation, one French scientist called him America's answer to Antoine Lavoisier. Gibbs was awarded honorary Doctor of Science degrees from Erlangen, Williams College, and Princeton University. Scientific and mathematical societies in Haarlem, Göttingen, Amsterdam, Manchester, and Berlin made him an honorary or foreign member. He was also given the Copley Medal by the Royal Society of London two years before he died. Before the advent of the Nobel Prize in 1901, the Copley Medal was the highest honor conferrable in the scientific world.

Gibbs became a member of the American Mathematical Society in March of 1903, one month before his death. He had also recently signed a contract to reprint his "Equilibrium" series with approximately 50 pages of additions. Because of his childhood illness, he had taken on a lifetime's regimen of mild outdoor activities such as long walks, horseback riding, and camping with close friends like fellow professor Andrew Phillips. His teaching schedule sometimes required "enforced rest" leaves of absence, but he never went against his doctor's advice. In fact, according to Wheeler's biography, Gibbs' health was excellent for more than 30 years before his final illness. However, an apparently mild illness could not be shaken off, and he died on April 28, 1903, the night before he was to resume his duties. Gibbs' last resting place, two blocks from his brother–in–law's house, is marked with a headstone identifying him only as a Yale professor.

GIBBS PHENOMENON

The non-uniform **convergence** of the **Fourier series** for discontinuous **functions** is known as the Gibbs phenomenon. In 1899 American mathematician **Josiah Willard Gibbs** noticed that near a point where a function has a jump discontinuity, the partial sums of a Fourier series show a substantial overshoot near these endpoints. Carrying out the sums of the Fourier series to a higher number of terms will not diminish the amplitude of the overshoot although the overshoot occurs over a smaller and smaller **interval**. This overshoot exhibits itself in an oscillatory behavior near the discontinuous point(s) of the function. Although Wilbraham first analyzed this phenomenon in 1848, it was Gibbs that studied it in detail and for whom the behavior is named. Later, in 1906 Bôchner generalized this phenomenon to arbitrary functions. The Gibbs phenomenon is not only observed in Fourier series but also occurring at simple discontinuities in other eigenfunction series.

The magnitude of the overshoot is dependent upon the type of discontinuity and not on the values of the function studied. Normally the overshoot is about 9% of the magnitude of the discontinuity jump but it is sometimes as high as 17.9%. Gibbs showed that is a function of x ($f(x)$) is piecewise smooth on $[-\pi, \pi]$, and x_0 is a discontinuity point, then the Fourier partial sums will exhibit a overshoot with height almost equal to $0.09(f(x_0 max)-f(x_0 min))$, where $f(x_0 max)$ is the maximum value of the function at the point of discontinuity and f($x_0$min) is the minimum value of the function at that point. The Wilbraham-Gibbs constant, usually denoted G, actually quantifies the degree to which the Fourier series of a function overshoots the function value at a discontinuity. This constant is useful but the reader needs to be aware that there are differences in its usage. This constant is not really a constant. Sometimes the limiting crest of the highest wave deviating from the actual function is denoted by 2G where G = $\int_0^\pi$ sin(θ)/θ dθ = 1.851937, whereas other times it may be denoted as 1/2+G/π = 1.089489, and still other times it is denoted as 2G/π = 1.178979.

There are complex methods to smooth the Gibbs phenomenon. One method is called the Σ-approximation or sometimes it is referred to as the Lanczos sigma **factor**. In this approximation a Σ function is multiplied by the coefficients in the Fourier partial sums. This Σ function is a complex sin function involving the period of the original function. Another form of smoothing is called Hanning smoothing. Physical scientists who often observe a complex phenomenon using a large bandwidth detection system to collect as much information as possible in the shortest amount of time usually employ this type of smoothing. The technique involves reducing the number of data points around a discontinuity in the raw data before the Fourier transform is taken. Although this smoothes the Gibbs phenomenon it also degrades the frequency resolution.

GÖDEL, KURT FRIEDRICH (1906-1978)

Austrian-born American logician

Kurt Friedrich Gödel was a mathematical logician who proved perhaps the most influential **theorem** of 20th-century mathematics—the incompleteness theorem. Although he was not prolific in his published research and did not cultivate a group of students to carry on his work, his results have shaped the development of logic and affected **mathematics and philosophy**, as well as other disciplines. The philosophy of mathematics has been forced to grapple with the significance of Gödel's results ever since they were announced. His work was as epoch-making as that of **Albert Einstein**, even if the ramifications have not been as visible to the general public. Gregory H. Moore, in *Dictionary of Scientific Biography*, related that in May of 1972 mathematician **Oskar Morgenstern** wrote that Einstein himself said that "Gödel's papers were the most important ones on relativity theory since his own [Einstein's] original paper appeared."

Gödel was born in Brünn, Moravia (now Brno, Czech Republic), on April 28, 1906, the younger son of Rudolf Gödel, who worked for a textile factory in Brunn, and Marianne Handschuh. Gödel had an older brother, Rudolf, who would study medicine and become a radiologist. The Gödels were part of the German-speaking minority in Brünn, which subsequently became one of the larger cities in the Czech Republic. The family had no allegiance to the nationalist sentiments around them, and all of Gödel's educational

experience was in German-speaking surroundings. He was baptized a Lutheran and took religion more to heart than the rest of his family.

Gödel began his education in September, 1912, when he enrolled in a Lutheran school in Brünn. In the fall of 1916 he became a student in a gymnasium, where he remained until 1924. At that point he entered the University of Vienna, planning to major in physics. In 1926, influenced by one of his teachers in **number theory**, he changed to mathematics; he did, however, retain an interest in physics, which he expressed in a number of unpublished papers later in life. He also continued his studies in philosophy and was associated with the Vienna Circle, a gathering of philosophers of science that had great influence on the English-speaking philosophical community. Gödel never was one, however, to follow a party line, and he went his own way philosophically. He felt that his independence of thought contributed to his ability to find new directions in mathematical logic.

Gödel's father died in February of 1929, and shortly thereafter his mother and brother moved to Vienna. Gödel completed the work for his dissertation in the summer of that year. He received his doctorate in February of 1930 for his **proof** of what became known as the completeness theorem. The problem that Gödel had considered was the following: Euclidean **geometry** served as an example of a kind of branch of mathematics where all the results were derived from a few initial assumptions, called axioms. However, it was hard to tell whether any particular list of axioms would be enough to prove all the true statements about the objects of geometry. Gödel showed in his dissertation that for a certain part of logic, a set of axioms could be found such that the consequences of the axioms would include all true statements of that part of logic. In other words, the collection of provable statements and the collection of true statements amounted to the same collection. This was a reassuring result for those who hoped to find a list of axioms that would work for all of mathematics.

In September of 1930, however, mathematical logic changed forever when Gödel announced his first incompleteness theorem. One of the great accomplishments of mathematical logic earlier in the century had been the work of two British mathematicians, **Alfred North Whitehead** and **Bertrand Russell**. Their three-volume work *Principia Mathematica* (Latin for "mathematical principles" and based on the title of a work by **Isaac Newton**), tried to derive all of mathematics from a collection of axioms. They examined some areas very thoroughly, and though few mathematicians bothered to read all the details, most were prepared to believe that Whitehead and Russell would be able to continue their project through the rest of mathematics.

Gödel's work was written up under the title "On Formally Undecidable **Propositions** of *Principia Mathematica* and Related Systems."In this paper, which was published in a German mathematical journal in 1931, Gödel introduced a new technique which enabled him to discuss logic using **arithmetic**. He translated statements in logic into statements involving only numbers, and he did this by assigning numerical values to symbols of logic. It had long been known that there were problems involved in self-reference; any statement that discussed itself, such as the statement "This statement is false," presented logical difficulties in determining whether it was true or false. The assumption of those who hoped to produce an axiomatization of all of mathematics was that it would be possible to avoid such self-referring statements.

Gödel's method of proof enabled him to introduce the technique of self-reference into the very foundations of mathematics; he showed that there were statements which were indisputably true but could not be proved by axiomatization. In other words, the collection of provable statements would not include all the true statements. Although the importance of Gödel's work in this area was not immediately recognized, it did not take long before those seeking to axiomatize mathematics realized that his theorem put an immovable roadblock in their path. The proof was not obvious to those who were not used to thinking in the terms that he introduced, but the technique of Gödel numbering rapidly became an indispensable part of the logician's tool kit.

Of the schools of mathematical philosophy most active at the time Gödel introduced his incompleteness theorem, at least two have not since enjoyed the same reputation. Logicism was the belief that all mathematics could be reduced to logic and thereby put on a firm foundation. **Formalism** claimed that certainty could be achieved for mathematics by establishing theorems about completeness. In the aftermath of Gödel's work, it was even suggested that his theorem showed that man was more than a machine, since a machine could only establish what was provable, whereas man could understand what was true, which went beyond what was provable. Many logicians would dispute this, but no philosophy of mathematics is imaginable which does not take account of Gödel's work on incompleteness.

Gödel was never a popular or successful teacher. His reserved personality led him to lecture more to the blackboard than to his audience. Fortunately, he was invited to join the Institute for Advanced Study at Princeton, which had opened in the fall of 1933, where he could work without teaching responsibilities. Despite the attractions of the working environment in Princeton, Gödel continued to return to Austria, and it was there that he lectured on his first major results in the new field to which he had turned attention, the theory of **sets**.

Set theory had been established as a branch of mathematics in the last half of the nineteenth century, although its development had been hindered by the discovery of a few paradoxes. As a result, many who studied the field felt it was important to produce an axiomatization that would prevent paradoxes from arising. The axiomatization which most mathematicians wanted was one which would capture the intuitions they had about the way sets behaved without necessarily committing them to points about which there was disagreement. Two of the statements about which there were disagreement were the **axiom** of choice and the continuum hypothesis. The axiom of choice said that for any family of sets there is always a function that picks one element out of each set; this was indisputable for finite collections of sets but was problematic when infinite collections of sets were introduced. The contin-

Kurt Friedrich Gödel

uum hypothesis stated that, although it was known that there were more real (rational and irrational) numbers than **whole numbers** (**integers**), there were no infinite sets in size between the **real numbers** and the whole numbers.

Gödel's major contribution in set theory was the introduction of what are known as constructible sets. These objects formed a **model** for the standard axiomatization of set theory. As a result, if it could be shown that the axiom of choice and the continuum hypothesis applied to the constructible sets, then those disputed principles had to be at least consistent with the standard axiomatization. Gödel successfully demonstrated both results, but this still left open the question of whether the two statements could be proved from the standard axiomatization. One of the major accomplishments of set theory in the second half of the century was the demonstration by **Paul Cohen** that neither the axiom of choice nor the continuum hypothesis could be proved from the standard axiomatization.

Gödel had suffered a nervous breakdown in 1934 which aggravated an early tendency to avoid society. He married Adele Porkert Nimbursky, a nightclub dancer, on September 20, 1938. He had met his wife when he was 21 years old, but his father had objected to the match, based on the difference in their social standing and the fact she had been married before. After his marriage, his domestic situation was something of a comfort in the face of the deteriorating political situation in

Austria, especially after the union of Austria and Germany in 1938, when Adolf Hitler was in power. When he returned to Vienna from the United States in June of 1939, he received a letter informing him that he was known to move in "Jewish-liberal" circles, not an attractive feature to the Nazi regime. When he was assaulted by fascist students that year, he rapidly applied for a visa to the United States. It was a sign of his stature in the profession that at a time when so many were seeking to escape from Europe, Gödel's request was promptly granted. He never returned to Europe after his hasty departure.

Gödel was appointed an ordinary member of the Institute for Advanced Study in Princeton, where he would remain for the rest of his life. His closest friends were Einstein and Oskar Morgenstern, and he took frequent walks in Einstein's company. Einstein and Gödel were of opposing temperaments, but they could talk about physics and each respected the other's work. Morgenstern was a mathematical economist and one of the founders of the branch of mathematics known as **game theory**. Gödel and his wife were content with this small social circle, remaining outside the glare of publicity which often fell on Einstein.

After his arrival in Princeton, Gödel started to turn his attention more to philosophy. His mathematical accomplishments guaranteed his philosophical speculations a hearing, even if they ran counter to the dominant currents of thought at the time. Perhaps the most popular philosophical school then was naturalism—the attempt to ground mathematics and its language in terms of observable objects and events of the everyday world. Gödel, however, was a Platonist and he believed that mathematics was not grounded in the observable world. In two influential published articles, one dealing with **Bertrand Russell** and the other with the continuum hypothesis, Gödel argued that mathematical intuition was a special faculty which needed to be explored in its own right. Although the bulk of mathematical philosophers have not followed him, they have been obliged to take his arguments into account.

Although Gödel moved away from mathematics in his later years, he contributed occasionally to the field. One of his last mathematical articles, published twenty years before his death, dealt with the attempt to formalize the approach to mathematical philosophy known as **intuitionism**. Gödel himself was not partial to that approach, but his work had wide influence among the intuitionists. American mathematician Paul Cohen was also careful to bring his work on the axiom of choice and the continuum hypothesis to him for his approval.

In his years at the Institute for Advanced Study, awards and distinctions began to accumulate. In 1950, Gödel addressed the International Congress of Mathematicians and the next year received an honorary degree from Yale; in 1951, he also received the Einstein award and delivered the Gibbs lecture to the American Mathematical Society. Harvard gave him an honorary degree in 1952 and in 1975 he received the National Medal of Science. That same year he was scheduled to receive an honorary degree from Princeton, but ill health kept him from the ceremony. By contrast, Gödel refused honors from Austria, at least as long as he lived;

however, the University of Vienna gave him an honorary doctorate posthumously.

Gödel had a distrust of medicine that amounted in his later years to paranoia. In late December of 1977 he was hospitalized and he died on January 14, 1978 of malnutrition, brought on by his refusal to eat because of his fear of poisoning. His wife survived him by three years; they had no children. Gödel's heirs were the mathematical community to which he left his work and the challenge of understanding the effects of his results. The year after his death Douglas Hofstadter's book *Gödel, Escher, Bach* became a best seller, illustrating Gödel's ideas in terms of art and music.

GOLDBACH, CHRISTIAN (1690-1764)
Russian historian and mathematician

Christian Goldbach was a Russian mathematician born in Königsberg, Prussia in 1690. Not much is known about his early life and education in Russia, but he was appointed to the position of mathematics professor and historian at the St. Petersburg Academy in 1725. Several years later, in 1728, Goldbach left the Academy to tutor Tsar Peter II. His position enabled him to tour Europe and to meet some of the greatest mathematical minds of the day. He began to correspond regularly with some of these men, including the great Swiss mathematician who had also held a professorship at St. Petersburg before moving to the Berlin Academy in 1741.

In a 1742 letter to Euler, Goldbach introduced his most famous mathematical theory, now known as **Goldbach's conjecture**, which states that every even number n that is greater than two is the sum of two **prime numbers**. Goldbach also made a second conjecture, known as the Ternary Goldbach Conjecture, which states that every odd number is the sum of three prime numbers. In 1764, Goldbach died at age 74 in Moscow, Russia, his conjectures still unproven. Soviet mathematicians Lev Shnirelman and Ivan Vinogradov made significant progress towards proving the second, ternary conjecture in the 1930s, and the Chinese mathematician Chen Jing Run made further advances in the early 1970s. However, Goldbach's original conjecture still lacks an absolute **proof** (although as of 1998 it had been verified up to 4×10^{14}).

See also Leonhard Euler; Goldbach's conjecture

GOLDBACH'S CONJECTURE

Goldbach's conjecture was formulated by **Christian Goldbach** in 1742 in a letter to the great Swiss mathematician **Leonhard Euler**. It is the assertion that every even number bigger than two is a sum of two **prime numbers**. For example, 4=2+2, 6=3+3, 10=3+7, 54=7+47=11+43 and so on.

While this has been verified for all even numbers up to 10^{14}, by J.-M. Deshouillers, H. J. J. te Riele and Y. Saouter, in 1998, it has never been proved that it holds for all even numbers. The Russian mathematician A. I. Vinogradov proved in

1937 that every large odd number was a sum of three primes and the chinese mathematician J. R. Chen proved in 1966 that every large even number is a sum of a prime and a number with at most two prime factors. As close as these results seem to the conjecture, a new breakthrough would be needed to prove it.

The Goldbach conjecture is mentioned in the text of **David Hilbert**'s eighth problem as a possible consequence of a thorough understanding of the distribution of prime numbers. Indeed, if a strong form of the **prime number theorem** could be proved that predicted the distribution of primes in short intervals, one would be able to prove that every large even number is a sum of two primes in many different ways. This is consistent with the available numerical evidence.

See also Prime number theorem

GOLDEN MEAN

The golden **mean** is a number that appears in a dazzling array of mathematical and natural structures. The golden mean has fascinated both mathematicians and amateurs since the time of **Pythagoras**, and perhaps even earlier, with the Egyptians. The great astronomer **Johannes Kepler** called the golden mean one of the "two great treasures" of **geometry** (the other was the **Pythagorean theorem**), likening it to a "precious jewel."

There are many ways to define the golden mean, which is usually written phi (ϕ). The earliest known appearance of phi among mathematicians is in the Brotherhood of Pythagoras, a group of scholars from the 6th century B.C. In order to recognize its members, the Brotherhood adopted a symbol called the pentagram, which is a pentagon with a five-pointed star inscribed in it. At the center of the star is a smaller pentagon. The golden mean is the ratio of the length of one of the rays of the star to the length of a side of the small pentagon; it is also the ratio of the length of a side of the large pentagon to the length of one of the rays. Historians speculate that the Brotherhood considered the pentagram to be the most beautiful of shapes, and that they were aware that the golden mean also appears in two of the **Platonic solids**: the dodecahedron and the icosahedron.

A second definition of the golden mean is the one given by **Euclid** in his *Elements,* which concerns dividing a line into two parts. If we ask ourselves what is the "most pleasing" way to divide a line into two parts, or a piece of music, or a painting, one way is to divide it into two equal parts; but another interesting way is to divide it so that the ratio of the larger piece to the smaller piece equals the ratio of the whole to the larger piece, a **division** that art theorists called the principle of "dynamic symmetry." With this division, the ratio of the larger piece to the smaller piece is the golden mean—it is not obvious, but it's true, that this definition gives the same number as the pentagram definition. With this definition, it is not difficult to show that $(phi)^2=1+phi$, and then the quadratic formula tells us that the value of the golden mean is $(1+\sqrt{5})/2$ (approxi-

mately 1.618034). This expression for phi shows that phi is an irrational number—it is not the ratio of two **whole numbers**.

The formula $(phi)^2=1+phi$ gives rise to some interesting representations for phi. We can rewrite the formula as phi=$\sqrt{(1+phi)}$, and that tells us that wherever we see phi, we can replace it with $\sqrt{(1+phi)}$. That means that phi=$\sqrt{(1+phi)}$=$\sqrt{(1+\sqrt{(1+phi)})}$=$\sqrt{(1+\sqrt{(1+\sqrt{(1+phi)})})}$=$\sqrt{(1+\sqrt{(1+\sqrt{(1+...)})})}$. What's more, we can rearrange the formula $(phi)^2=1+phi$ to give the new formula phi=1+1/phi, which means that wherever we see phi we can replace it with 1+1/phi; thus, phi=1+1/phi = 1+1/(1+1/phi) = 1+1/(1+1/(1+1/phi)) = 1+1/(1+1/(1+1/(1+...))). This last expression is called the continued fraction representation of phi.

The golden section also appears in connection with the **Fibonacci sequence**. That is the sequence of numbers 1, 1, 2, 3, 5, 8, 13, 21,... in which each term is the sum of the two preceding terms. It can be shown that the **ratios** of terms, 1/1, 2/1, 3/2, 5/3, 8/5, 13/8,... get closer and closer to phi, which is their limit.

The above expressions involving phi have been known for centuries; but phi continues to appear in new circumstances, as new mathematics are discovered. Their most recent appearance has been in connection with **Penrose tilings**. These are tilings of the plane that have the unusual property of being aperiodic—that is, they are not made of copies of a single patterned block. Penrose tilings can be made out of two rhombic tiles, one long and thin, the other more fat. In a Penrose tiling, the ratio of thin tiles to fat tiles is the golden mean.

The golden mean has also been important in the world of art. The Greeks and Egyptians used it in their architecture, and artists have used it through the centuries to create **proportions** in paintings and sculptures. Recently, music theorists have discovered that the golden ratio also appears as the division between sections in some of Mozart's music. This raises the question, Is the golden ratio the most aesthetically pleasing division? Through the centuries, many people have invested phi with an almost mystical significance, and it is sometimes referred to as the "divine proportion."

The golden mean occurs not just in the works of man, but also in those of nature. In a seed pod, the seeds **spiral** out from the center, with more and more seeds in each revolution. The ratio of the number of seeds in one revolution to the number of seeds in the previous revolution is the golden mean. The golden mean is also closely related to the logarithmic spiral, which is the shape that appears in seashells.

The golden mean has also been observed, with less precision, as the ratio cut off by the navel in the human body. Some have taken this, together with the other appearances of phi in nature, as evidence of a divine Creator who built an aesthetically pleasing world for humans to inhabit. The appearances of phi in seed pods and seashells may be explained more scientifically, however, in terms of the most efficient arrangements of seeds or shells. And it can also be argued that we find phi aesthetically pleasing because it appears in nature and human anatomy, instead of the other way around.

See also Fibonacci sequence; Mathematics and art; Penrose tilings

GOODMAN, NELSON (1906-1998)

American philosopher

Mathematically Nelson Goodman is known for a number of publications on **calculus** and **symbolic logic** and most particularly Goodman's paradox. It should be borne in mind that the mathematical side of Goodman, although important, is only a tiny facet of a bright and varied career.

A prolific author of both articles and books Nelson Goodman was born in 1906 in Somerville, Massachusetts. Goodman's interests spanned many fields other than mathematics and this is reflected perfectly in his varying career. In 1928 he received his bachelor's degree in Philosophy from Harvard, he received his doctorate, also from Harvard, in 1941. From 1929 to 1941 Goodman was the Director of the Walker-Goodman Art Gallery in Boston, following this he was in the miltary until the end of the Second World War. Immediately upon release from the United States Army Goodman was Instructor in Philosophy at Tufts College, Massachusetts and then Associate Professor of Philosophy at the University of Pennsylvania from 1946 to 1951. Goodman was then promoted to full professor at Pennsylvania, a position he retained until 1964 when he became Professor of Philosophy at Brandeis University Massachusetts. From 1968 to 1977 Goodman was Professor of Philosophy at Harvard University. Goodman received many awards in his life ranging from a Fellowship of the American Academy of Arts and Sciences to a Guggenheim Award, he was also involved in a large range of societies covering Philosophy and Art. In 1967 Goodman founded project **Zero** at Harvard University, this still running project investigates aesthetic education in an interdisciplinary manner. Goodman also wrote and produced a number of films and stage shows. Goodman was a true workaholic who took great pleasure in introducing others to the things that were of importance in his life. Goodman died in 1998 at Needham, Massachusetts at the age of 92. He continued working and publishing right to the end and indeed he had been booked to deliver a conference paper a few days after his death. Nelson Goodman leaves behind a legacy of over 120 articles, as well as over a dozen books. These figures do not include the articles and books that have been written about him and his work.

Books about Nelson Goodman and his impact on his chosen fields started to appear as early as 1967 and by the year 2000 there were over 45 separate books available. The readers of one set of books would not recognise the same Nelson Goodman, so wide and varied were his interests. For example in 1968 Goodman published *Languages of Art: An Approach to a Theory of Symbols* and 20 years later *Variations on Variation—or Picasso Back to Bach*, but we also have such titles as *The Calculus of Individuals and Its Uses* and *A Study of Methods of Evaluating Information Processing Systems of Weapons Systems*. Goodman is a man who accomplished whatever he set out to do irrespective of the field.

GOOGOL

Googol is the name for the number 10 to the power 100 (10^{100}). In other terminology the googol is stated as "ten billion, billion, billion, billion, billion, billion, billion, billion, billion, billion, billion". In 1938 mathematician Dr. Edward Kasner (1878-1955) supposedly asked his nine-year-old nephew, Milton Sirotta, to think of a name for a very, very big number; namely, one followed by one hundred zeros. "Googol" was the name young Milton supplied to his uncle. Googol is an incredibly large number, but at the time of its labeling Kasner decided to name an even larger, but still finite, number. In fact, the largest number so far named was called by Kasner the "googolplex," the name for 10 to the power of googol (or, $10^{(10^{100})}$). Googolplex is denoted as one followed by 10^{100} zeros. Initially, Kasner wanted the value of the googolplex to be the numeral one followed by as many zeroes that an individual could write before becoming tired of writing. But, Kasner quickly realized that people become tired at different rates.

As a comparison for the extreme size of the googol, it is estimated that there are some one hundred billion (10^{11}, or one followed by 11 zeroes) galaxies in the universe, with each averaging around one hundred billion stars. There are perhaps, then, about ten billion trillion (10^{22}, or one followed by 22 zeroes) stars in the entire universe. This number is much smaller than the googol. As another example, United States astronomer Carl Sagan (1934-1996) once said that if we randomly placed ourselves on another planet in the universe, the probability that we could locate the planet earth would be less than one in a billion trillion trillion (10^{33}, or one followed by 33 zeroes). This number still does not come close to the value of a googol. After naming the word googol, Dr. Kasner regretfully admitted that mathematicians seldom use the colorful words of googol and googolplex.

GRAPHING CALCULATOR

During the 1970s, the hand-held **calculator** became quite commonplace for use in mathematical calculations. Originally, these could perform only the basic functions of **addition, subtraction, multiplication,** and **division**, and perhaps derive the **square** of a number. Later scientific calculators, the more advanced of which were programmable, added many new functions to the abilities of these useful devices. Yet the most they could display was numerical or symbolic results that often had to be interpreted by the user.

In the late 1980s, however, graphing calculators began to combine the power of its best predecessors with the ability to visually display results of certain calculations. The 1990s saw the graphing calculator virtually become a hand-held computer with advanced liquid crystal display capabilities. Some now even allow for symbolic **algebra** and **calculus** to be performed as well as graphing of two variables in three dimensions, a feat that would otherwise require a mathematician to be something of a technical illustrator as well. There are definite advantages to using graphing calculators for learning mathematics. These include the ability to visualize examples that are more complex than those that might normally be attempted by students or instructors. By using the graphing calculator students can discover properties for themselves, allowing them to be more engaged in the learning process. Graphing calculators connect a visual representation to the functions being studied.

See also Graphs, domains, and ranges

GRAPHS, DOMAINS, AND RANGES

Graphs provide a means to visualize a mathematical set, function, or even set of **functions**. Objects depicted in a graph may be one-dimensional, such as a number line; two-dimensional, such as a **parabola**, or three-dimensional, such as a **sphere**. Graphs are used extensively throughout the fields of mathematics, **statistics**, science, engineering, business, and research in order to depict information—whether a mathematical curve of statistical data—in a form that humans find easy to process. Graphs provide a means to quickly assimilate information—and even trends present in the information—in a more intuitive way than reviewing long lists of numerical data.

Domain

The **domain** of a function, often symbolized by an italicized capital D, is the set of values over which the function exists. The set of values is generally connected (that is, continuous, such as all values in the closed **interval** [-1, 1]), but may include points or intervals of discontinuity, especially with piecewise functions. For example, the domain of the function $f(x) = x^2$ is negative **infinity** to positive infinity (that is, all **real numbers**), with no discontinuities. The domain of the function $f(x) = 1 / x$, however, is the set of all real numbers, excepting $x = 0$ since it is impossible to divide by **zero**. Similarly, the domain of the function $f(x) = 1/(x - 1)^2$ is the set of all real numbers, excepting $x = 1$, which would also result in a **division** by zero error. The domain of $f(x) = 1 / x$ is often written $(-\infty, 0) \cup (0, \infty)$ which reads "the set of all real numbers in the open interval negative infinity to zero union the set of all numbers in the open interval zero to infinity." The domain of $f(x) = 1/(x - 1)^2$ is similarly written $(-\infty, 1) \cup (1, \infty)$. It is important to note that domains may also be composed of **complex numbers** if the function is defined in the complex number plane.

Range

The **range** of a function, often symbolized by an italicized capital R, is the set of values over which the domain of a function is mapped by the function. The **mapping** may be one-to-one, which means that the function maps each unique value in the domain to a unique value in the range, or the mapping may be many-to-one, which means that the function maps two or more values in the domain to the same value in the range. Written symbolically, a one-to-one function implies that

if $f(x) = f(y)$ then $x = y$. Many-to-one functions do not have this restriction. A function's codomain, which is sometimes confused with a function's range, is any set of values that *contains* the range; thus, the codomain may or may not be equal to the range. For example, if the range of a function is [-1, 1] (such as the function $\sin x$ or $\cos x$), a possible codomain of the function is [-2, 2], which is clearly not the same as [-1, 1]. It is important to note that the domain and range of a function *may* be the same, but this condition is unlikely. For example, the domain of $f(x) = \sin x$ is $(-\infty, \infty)$, but its range is [-1, 1].

The functions $f(x) = 2x$ and $f(x) = 3x^3 - 2$ are examples of one-to-one functions with a range of $(-\infty, \infty)$. For both functions, each value x in the domain has a unique value $f(x)$ in the range. The functions $f(x) = |x|$ and $f(x) = 4x^2$, however, are many-to-one functions with a range of $[0, \infty)$. When $f(x) = |x|$, for example, $f(-1) = f(1) = 1$ and when $f(x) = 4x^2$, $f(-2) = f(2) = 16$. For these functions, each unique value x in the domain does not map to a unique value in the range. Like the domain of a function, the range may also be a disconnected set. The range of the function $f(x) = 1 / (x^2 - 1)$, for example, is $(-\infty, -1]$ $\cup (0, \infty)$, which does include the interval (-1, 0].

Types of Graphs

Many different kinds of graphs exist in order to present information more effectively. Common types include bar, **circle**, line, scatter, contour, trajectory, ternary, box and mesh plots. Each type of graph presents the function or data in a slightly different way and some graphs are more suited to a particular kind of function or data than others. Graphs can be further delineated by the type of coordinate system used to present the function or data. Use of the **Cartesian coordinate system** and **polar coordinate systems** is common.

See also Cartesian coordinate system; Complex numbers; Continuity; Dimension; Functions; Infinity; Interval; Mapping; Mathematical symbols; Negative numbers; Orthogonal coordinate system; Polar coordinate systems; Real numbers; Set; Zero

GRASSMANN, HERMANN GÜNTHER (1809-1877)

German geometer

Hermann Günther Grassmann was a gifted German thinker whose work spanned the fields of mathematics and linguistics, theology, and botany. His decision to focus on mathematics came when he was 31, but he abandoned the field 20 years later when his formulation of a geometric **calculus** did not receive the recognition it deserved. Grassmann's conception of n–dimensional vector space and multi–linear **algebra**, laid out in his monumental work *Die Lineale Ausdehnungslehre*, were ahead of his time but had great impact once they were grasped by late 19th– and early 20th–century mathematicians.

Grassmann was born on April 15, 1809, in Stettin, Prussia (present–day Szczecin, Poland), the third of 12 children. His father, Justus Günther Grassmann, taught **mathemat-**

ics and physics at the local gymnasium and wrote several basic–level mathematics textbooks. The Grassmann family was a religious one: Justus had briefly served as a Protestant minister before becoming a teacher, and Grassmann's mother, Johanne Medenwald, was the daughter of a minister.

Grassmann was schooled first by his mother and at a private academy before enrolling at the Stettin Gymnasium. He was a fine student, earning the second highest score on the final secondary school examination. During his three years at the University of Berlin, Grassmann focused his studies on theology and classical languages and literature. His work on mathematics and physics was done on his own once he returned to Stettin in 1830.

Over the next decade Grassmann took a series of examinations in order to secure a job in the scholastic community. He passed an examination in December of 1831 that allowed him to teach only at the elementary school level. The following spring, Grassmann took a position teaching at the Stettin Gymnasium. In 1834, he passed the first level of theological examinations administered by the local Lutheran church, but instead of pursuing a religious career, he took a job as senior master at the Gewerbeschule in Berlin. Grassmann changed jobs a year later to take a teaching post at the Otto Schule in Stettin. By 1840, he had completed his round of tests, passing the second–level theology examination and the mathematics examination that allowed him to teach at the secondary school level.

It was during this last mathematics examination that Grassmann first applied his geometric calculus to solve a problem on the theory of the tides. After this 1840 examination, Grassmann decided to devote his energy to mathematics, particularly to the development of the geometric calculus he had been working on since 1832. In 1844, *Die Lineale Ausdehnungslehre* was published, presenting Grassmann's geometric calculus as a combination of synthetic geometry's treatment of points (and not numbers) with analytic geometry's use of calculations. Grassmann introduced *n*–dimensional vector **space** and multi–linear algebra, concepts that paved the way for the creation of exterior algebra.

Despite containing profoundly revolutionary ideas, *Ausdehnungslehre* was largely ignored by Grassmann's contemporaries because of the work's abstract nature and unreadable style. With mathematicians such as **Julius Plücker** and **August Ferdinand Möbius** refusing to write a review about the book, the professional community generally disregarded the work. Grassmann used the concept of connectivity he established in *Ausdehnungslehre* in an 1845 paper in which he revised Ampere's fundamental law for the reciprocal effect of two infinitely small currents. Again, Grassmann's poor writing style obscured a scientifically important paper. When Grassmann applied for a position as a university professor in 1847, E. E. Kummer's critique of Grassmann's 1845 paper, which stated that it contained "commendably good material expressed in a deficient form," prevented Grassmann from landing the job.

Grassmann married Marie Therese Knappe on April 12, 1849, and the couple had 11 children, two of whom died as young children. Convinced of the importance of his geometric calculus, Grassmann revised his *Ausdehnungslehre* for republication in 1862. Although Grassmann tried a new approach in explaining his methodology, the second version met with the same reception as the first. By the mid–1860s, Grassmann was frustrated by the lack of recognition for his mathematical contributions, so he turned his full attention to his studies in linguistics and other sciences.

As early as 1849, Grassmann began studying Sanskrit, followed by Lithuanian, Russian, and older forms of Prussian and Persian. In 1854 he developed a theory about the tonal components of vowels, and his 1863 theory about aspirates and the sound shift in Germanic languages became the linguistic law that bears his name. By 1860, Grassmann had taken interest in the Hindu literary masterpiece, the *Rig–Veda*. Grassmann compiled a glossary and composed a translation of the *Rig–Veda* during the 1870s, finding the instant acclaim in linguistics that he was never awarded in mathematics.

Grassmann also tried his hand at a variety of other projects during his lifetime, including a study of colors in his 1853 *Zur Theorie der Farbenmischung* and the renaming, using German etymological roots, of plant species native to German–speaking areas in his 1870 *Deutsche Pflanzennamen*. Writing for a political newspaper in 1848, Grassmann penned a series of articles supporting a Germany united under constitutional monarchy. Grassmann even wrote folk songs in which he harmonized up to three voices.

After his 1871 election to the Göttingen Academy of Sciences, Grassmann returned to mathematics. He published several papers before he died on September 26, 1877, of heart failure. The following year a third version of *Ausdehnungslehre*, prepared before his death, was published. An appreciation of Grassmann, appearing a year after his death in the *Schulprogramm* of the Stettin Marienstifts gymnasium, said of Grassmann, "... only a quite independent spirit could dare to break his own paths in mathematics, on which others followed him only after decades...." Indeed, it was only after his death that mathematicians began drawing off Grassmann's *Ausdehnungslehre* and crediting his discoveries with the later development of linear **matrix** algebra.

GREATEST COMMON FACTOR

The greatest common **factor** (or *greatest common divisor*) of a set of natural numbers is the largest natural number that divides each member of the set evenly (with no remainder). For example, 6 is the greatest common factor of the set $\{12, 18, 30\}$ because $12 \div 6 = 2$, $18 \div 6 = 3$, and $30 \div 6 = 5$.

Similarly, the greatest common factor of a set of **polynomials** is the polynomial of highest degree that divides each member of the set with no remainder. For example, $3(x+2)^3(x-4)^2$, $12(x+2)^4(x-4)^3(x^2+x+5)$, and $6(x+2)^2(x-4)$ have $3(x+2)^2(x-4)$ for the highest common factor.

GREGORY, JAMES (1638-1675)

Scottish-born mathematician and astronomer

James Gregory's work laid the foundation for the development of **calculus**, and his work in astronomy and optics for astronomical observations influenced the works of **Isaac Newton**. He was born in Drumoak, Scotland, the son of John Gregory, a minister, and Janet Anderson Gregory. Gregory was a sickly child, and his mother guided his early education at home; she must have been an unusual woman for the 17th century, because she included **geometry** among the subjects she taught her son.

In 1651, Gregory left home for grammar school in Aberdeen, Scotland. He completed his preparatory work, then graduated from Aberdeen's Marischal College, where he focused his studies in mathematical optics and astronomy. Frustrated by the lack of scholarly opportunities in Aberdeen, Gregory traveled to London in 1662, where he met Robert Moray, an influential member of the Royal Society. In 1663, Gregory published *Optica promota*, a work that anticipated Newton's efforts in optics by suggesting the use concave mirrors in telescopes. He searched unsuccessfully to find a technician skillful enough to construct the prototype of such an instrument. Moray attempted to introduce Gregory to the Dutch mathematician **Christiaan Huygens** to help further the young man's studies, but was unsuccessful. Gregory then decided to pursue scientific studies in Italy with Stefano degli Angeli, and left London for Padua in 1664.

At the University of Padua, Gregory studied geometry, **mechanics**, and astronomy. In 1667 he published *Vera circuli et hyperbola quadratura*, in which he explored the nature of the **area** of circles and hyperbolas. *Geometriae pars universalis, inserviens quantitatum curvarum transmutationi & mensurae* followed in 1668, where Gregory introduced the concepts of convergent and divergent series. He also discussed the differences between algebraic and transcendental **functions**, and offered a series of expressions for **trigonometric functions** and a **proof** for the **fundamental theorem of calculus**.

Gregory returned to London in 1668. His work in Italy and his contacts with the Italian scientific community initially earned him considerable notice, and he was quickly elected to the Royal Society. Gregory was named as the new chairman of mathematics at St. Andrew's College, Scotland, in 1668. Gregory's time was consumed by his duties for the next several years. "I am now much taken up and hath been... this winter bypast, both with my publik lectures, which I have twice a week, and resolving doubts... gentlemen and scholars proposeth to me," he wrote in 1671.

Despite the time devoted to teaching and academic administration, Gregory managed to maintain a voluminous correspondence with John Collins, who forwarded to Gregory copies and transcripts of material from such noted scholars as **Isaac Barrow**, René–Francois de Sluse, and Newton.

Gregory waged war on the antiquated curriculum at St. Andrew's, but his efforts to incorporate contemporary science into the college's course of studies was resisted by the faculty and the college's governing board of regents. Gregory hoped to establish the first public observatory in Great Britain at St.

James Gregory

Andrew's; in 1673, he journey to London to seek advice and obtain instruments for such a facility. Unsuccessful in his efforts to secure financial backing for the project, Gregory returned to St. Andrew's to find himself an academic outcast. A student rebellion against the established curriculum pushed the board of regents to action. Gregory and his radical ideas were the obvious scapegoat; servants were forbidden to wait on him and his salary was withheld. Colleagues and students were instructed to treat him as a pariah. "Scholars of most eminent rank were violently kept from me... the masters persuading them that... they were not able to endure mathematics," he wrote.

When Edinburgh University offered Gregory its newly endowed chairmanship of mathematics, he fled St. Andrew's. Sadly, within a year of the appointment, Gregory suffered a debilitating, blinding stroke while observing Jupiter through a telescope. He died a few days later, in October 1675.

Before his death, Gregory ceased publishing papers in pure mathematics. His private papers, however, are rich in theories, proofs, and questions that might have earned him wide acclaim during his lifetime, had the materials been issued. Gregory delved into the theory of **equations** and the location of their **roots**, and attempted to solve the general quintic. His letters to Collins include his work on quadrature and rectification of the logarithmic **spiral**, his independent discovery of the general binomial expansion, several trigonometrical series (including those for the natural and logarithmic tangent and secant), and a series solution to **Johannes Kepler**'s problem in which he outlined how the series could be applied to the roots of equations.

Much of the work Gregory pursued during the last years of his life is lost. In a short paper published in 1672, he proved that atmospheric height is logarithmically related to barometric pressure. Other surviving papers demonstrate that he deduced the elliptical integral expressing the time of **vibration** in a circular pendulum and pursued his work in theoretical astronomy.

Gregory's reluctance to publish limited his success during his lifetime. His work served as the springboard for the published works of others, including Newton and Gregory's nephew, David, who did not acknowledge their debt to Gregory. The scope of Gregory's work and the extent of its influence on the 17th century scientific community were little recognized until 1939, when some of his notes—scribbled in 1671 on the back of a letter from a bookseller—were published and scholarly curiosity in Gregory's work was awakened.

GRELLING, KURT (1886-CA. 1942)
German logician, philosopher, and mathematician

German secondary school teacher Kurt Grelling was part of the Vienna Circle, a group of philosophers devoted to logical positivism, or the logical **analysis** of scientific knowledge. Logical positivists believed that knowledge can only come from logical reasoning and empirical experience, and that scientific theories are useful only if they can be proved true or false through observed experience.

In 1908, Grelling proposed Grelling's paradox. This semantic paradox was based on the idea of autological and heterological words. Autological words are self-descriptive, meaning that they possess the property they represent, such as the words one, visible, and multi-syllabic. Heterological words are non-self-descriptive, meaning that they do not contain the property they represent, such as the words edible, long, and incomplete. Grelling's paradox was whether or not heterological is in fact a heterological world. If heterological was autological, or self-descriptive, it would have to have a heterological meaning by definition, but if it had a heterological meaning, it would have to be non-self-descriptive. Thus, heterological can only be a heterological word if it is not the word heterological.

Grelling also worked with mathematician and logician **Kurt Gödel**, a key figure in the development of mathematical logic. In 1936, Grelling published an article defending Gödel's incompleteness **theorem**, which states that **arithmetic** can never be completely axiomatized.

This work on paradoxes and logical mathematical theory was significant in that it established that there are certain contridictions, or exceptions to rules, that challenge the nature of observed experience and scientific theory itself. In simplest terms, it established that consistency in mathematical theory was something that could never be taken for granted.

The exact date and circumstances of Grelling's death is unknown. Many believe he was taken to the Auschwitz concentration camp and killed, along with his wife, in 1942. One of Grelling's philosophy colleagues, Carl Hempel, tells a similar story, recounting that Grelling was attempting to cross the border from France to Spain in 1941 to escape the Nazi regime, when he was caught and sent to a Polish concentration camp.

GROTHENDIECK, ALEXANDER (1928-   )
French algebraic geometer and analyst

Alexander Grothendieck has had an influence on the mathematics of the second half of the 20th century well beyond the scale of his publications. Grothendieck started off as an especially prolific contributor to each of the areas to which he turned, including functional **analysis**, **algebraic geometry**, and category theory, only to move away from mathematics later in his career. As a result, he has had fewer students to carry on his research tradition than if he had followed a more orthodox path. Nevertheless, one of the chief activities of the mathematicians in several areas has been to recast their field in the terms introduced by Grothendieck.

Grothendieck's early years have been difficult to reconstruct, due to his reluctance to deal in ordinary reminiscences and his distrust of biographers. The generally accepted date and place of his birth are March 28, 1928, and Berlin, but the identity of his parents is less clear. At least one version that Grothendieck has given, as noted by Colin McLarty, indicates that his father was named Morris Shapiro and that he was sentenced to death for attempting to assassinate the Russian Czar in 1905. After Shapiro served a number of years in prison in Siberia, he was released by the Bolsheviks and went to Germany in 1922, about which time he met his future wife. Grothendieck was their first child and took his name from that of the governess who cared for him from 1929 to 1939. In the latter year, his mother took him to France, where he learned for the first time that he was Jewish by ancestry. His father died in the Auschwitz concentration camp and Grothendieck was saved thanks to the cooperation of Protestant and Catholic clergy in Le Chambon sur Lignon in southern France. His mother died in the 1960s.

The story becomes much clearer once Grothendieck entered the French higher education system after the war. He studied at the University of Montpellier and spent a year at the École Normale Supérieure, one of the leading traditional scientific universities in France. At this time France was undergoing a mathematical renaissance, thanks to the pedagogic efforts of the group known under the collective pseudonym of **Nicolas Bourbaki**. Among those who took part in the grand program of rewriting all of mathematics in the Bourbaki mode were Jean Dieudonné and **Laurent Schwartz**, both at the University of Nancy. Grothendieck went to work in the area of functional analysis with the two Bourbakists and rapidly produced material sufficient and appropriate for a thesis.

Grothendieck's first conspicuous success was in the area known as functional analysis. This mixed the traditional area of **calculus** with the more recent developments in **topology**, the field dealing in properties of geometric configurations, to be able to handle broad ranges of questions. The idea was to replace detailed and lengthy calculations with shorter and more insightful proofs. It is not surprising that such an area attracted Grothendieck, who did not feel that his greatest strength was in long, technical arguments. His contributions came in the area of reconsidering disciplines from new perspectives.

Grothendieck's most lasting influence came from his work in the area to which he now moved, algebraic **geometry**. This field had been in existence for many years and could be traced back to French mathematician and philosopher **René Descartes** in the 17th century. The idea of merging **algebra** and geometry to enhance the study of both received a new impetus with the accelerated development of abstract algebra in the late nineteenth century. There was a flourishing Italian school of algebraic geometry in the first half of the twentieth century, but it was effectively wiped out by the World War II. American mathematician Oscar Zariski carried on the Italian tradition in the United States, although he felt that he had added a good deal of algebraic sophistication.

Grothendieck was supported during his early investigations into algebraic geometry by the French national center for scientific research. This allowed him plenty of opportunity for travel and he spent part of the 1950s in Brazil and part in Kansas. Perhaps the most fruitful environment he found was at Harvard, where Zariski had settled. As his Harvard colleagues noted, Grothendieck was obsessed by mathematics and worked for many hours at a stretch in an unheated study, emerging with 3000-page manuscripts. On the strength of his energy and imagination, Grothendieck was able to revolutionize mathematics with his research.

One of the chief elements in Grothendieck's approach to mathematics involved the relatively recent field of category theory. **Set theory** had become an accepted part of the foundations of mathematics, but category theory sought to add a new idea to the basic notions of set and membership—the idea of function. **Functions** had long been used in mathematics, but category theory built them into the basis of the mathematical universe. One way of looking at the change was that mathematicians began to realize that what was important about the objects of mathematics was how they were connected by functions, not their composition out of basic elements.

Before the work of Grothendieck, category theory had been an active area of research but with limited applications. Grothendieck combined the ideas of category theory with the traditional studies of algebraic geometry to raise the latter to a new level of abstraction. The innovations introduced by Zariski in the previous generation shrank by comparison. As Zariski was quoted in *The Unreal Life of Oscar Zariski,* "After Grothendieck's great generalization of the field... what I myself had called abstract turned out to be a very, very concrete brand of mathematics."

In 1959 Grothendieck took a position with the Institut des Hautes Études Scientifiques (IHES), recently established in Paris upon the **model** of the Institute for Advanced Studies in Princeton, New Jersey. There Grothendieck had the chance to lecture on a regular basis on his work in algebraic geometry and to attract mathematicians from all over the world. Not surprisingly, in 1966 he received the Fields Medal from the International Mathematics Congress, the highest award that

the mathematical community can convey. Among the attractions of his work was its applicability to extending a variety of theorems that had originally been established in narrow contexts. Questions about number fields that had required immense amounts of computation to answer could be replaced by conceptually simpler questions about algebraic varieties, and the answers would have wide domains of applicability.

This golden age for algebraic geometry came to an end in 1970. Grothendieck had never been comfortable with playing the role of the "great man" and felt that the adulation of students was not good for him as a human being or as a mathematician. He also moved in a radical direction politically and hoped to be able to galvanize the mathematical community into political action. As a result, he left the IHES and taught at other French universities, particularly Montpellier, from which he retired in 1988. In the meantime, his ideas about category theory continued to supply the fuel for other areas of mathematics, including the foundations. The idea of a topos, a particular kind of category especially useful for analyzing logic, was introduced by Grothendieck for purposes of algebraic geometry. The continued fertility of topos theory adds to the fields indebted to Grothendieck's work during his contributions to algebraic geometric issues.

Grothendieck's memoir, *Récoltes et Semailles,* discusses at length his views on a number of subjects, most of which are unrelated to mathematics. More representative of his career in mathematics is the three-volume set of papers gathered for his 60th-birthday *festschrift* and published in 1990. The range of contributors includes many of the names of leaders of the mathematical community. His vision of mathematics has led not just to individual results but to a new sense of the powers of the subject.

GROUP THEORY

A group is a simple mathematical system, so basic that groups appear wherever one looks in mathematics. Despite the primitive nature of a group, mathematicians have developed a rich theory about them. Specifically, a group is a mathematical system consisting of a set G and a binary operation * which has the following properties:

[1] $x*y$ is in G whenever x and y are in G (closure).

[2] $(x*y)*z = x*(y*z)$ for all x, y, and z in G (**associative property**).

[3] There exists and element, e, in G such that $e*x=x*e=x$ for all x in G (existence of an **identity element**).

[4] For any element x in G, there exists an element y such that $x*y=y*x=e$ (existence of inverses).

Note that commutativity is not required. That is, it need not be true that $x*y=y*x$ for all x and y in G.

One example of a group is the set of **integers**, $\{...-4,-3,-2,-1,0,1,2,3,4,...\}$ under the binary operation of **addition**. Here the sum of any two integers is certainly an integer, 0 is the identity, $-a$ is the inverse of a, and addition is certainly an associative operation. Another example is the set of positive **fractions**, m/n, under **multiplication**. The product of any two positive fractions is again a positive fraction, the identity element is 1 (which is equal to 1/1), the inverse of m/n is n/m, and, again, multiplication is an associative operation.

The two examples we have just given are examples of commutative groups. (Also known as Abelian groups in honor of **Niels Henrik Abel**, a Norwegian mathematician who was one of the early users of group theory.) For an example of a non-commutative group consider the permutations on the three letters a, b, and c. All six of them can be described by

$$\begin{pmatrix} a\,b\,c \\ a\,b\,c \end{pmatrix} \begin{pmatrix} a\,b\,c \\ a\,c\,b \end{pmatrix} \begin{pmatrix} a\,b\,c \\ b\,a\,c \end{pmatrix} \begin{pmatrix} a\,b\,c \\ b\,c\,a \end{pmatrix} \begin{pmatrix} a\,b\,c \\ c\,a\,b \end{pmatrix} \begin{pmatrix} a\,b\,c \\ c\,b\,a \end{pmatrix}$$

$$\text{I} \qquad \text{P} \qquad \text{Q} \qquad \text{R} \qquad \text{S} \qquad \text{T}$$

I is the identity; it sends a into a, b into b, and c into c. P then sends a into a, b into c, and c into b. Q sends a into b, b into a, and c into c and so on. Then P*Q=R since P sends a into a and Q then sends that a into b. Likewise P sends b into c and Q then sends that c into c. Finally, P sends c into b and Q then sends that b into a. That is the effect of first applying P and then Q is the same as R.

Following the same procedure, we find that Q*P=S which demonstrates that this group is not commutative. A complete "multiplication" table is as follows:

	I	P	Q	R	S	T
I	I	P	Q	R	S	T
P	P	I	R	Q	T	S
Q	Q	S	I	T	P	R
R	R	T	P	S	I	Q
S	S	Q	T	I	R	P
T	T	R	S	P	Q	I

From the fact that I appears just once in each row and column we see that each element has an inverse. Associativity is less obvious but can be checked. (Actually, the very nature of permutations allows us to check associativity more easily.) Among each group there are *subgroups*-subsets of the group which themselves form a group. Thus, for example, the set consisting of I and P is a subgroup since P*P=I. Similarly, I and T form a subgroup.

Another important concept of group theory is that of *isomorphism*. For example, the set of permutations on three letters is isomorphic to the set of *symmetries* of an equilateral **triangle**. The concept of **isomorphism** occurs in many places in mathematics and is extremely useful in that it enables us to show that some seemingly different systems are basically the same.

The term "group" was first introduced by the French mathematician **Evariste Galois** in 1830. His work was inspired by a **proof** by Abel that the general equation of the fifth degree is not solvable by radicals.

GULDIN, PAUL (1577-1643)
Swiss mathematician

The work of Guldin is covered in four separate volumes which he published during his life, they are entitled *De Centro Gravitas*. Volume one considers centres of gravity with particular reference to the centre of gravity of the Earth. The second volume contains what is now known as Guldin's second rule or Guldin's **theorem**. Volume three considers cones and cylinders.

Paul Guldin was born Habakuk Guldin in St Gall (now Sankt Gallen), Switzerland. Initially a goldsmith Guldin became a Catholic (his parents were Protestants but of Jewish descent) aged 20 when he joined the Jesuits. On joining the order Guldin changed his name to Paul. In 1609 he was sent to college in Rome where he subsequently taught mathematics. He also taught at Graz in Austria, a position he gave up in 1623 to become professor of mathematics in Vienna. Eventually in 1637 Guldin returned to Graz to teach mathematics. From 1627 Guldin kept up a lengthy correspondence with **Johannes Kepler**, this was initially mostly about religion and only the later letters deal with mathematics and astronomy.

Guldin's theorem is also known as Pappus's theorem and it was originally written down by **Pappus of Alexandria** sometime in the fourth century. It is believed that this is merely coincidence. Although Guldin had access to translations of the work of Pappus, and he regularly quoted from them, various historians have shown that they were incomplete and lacked Pappus's theorem. Also during his lifetime the theorem was recognised as being that of Guldin, by amongst others Kepler. Guldin ensured that Kepler had a telescope when Kepler was persecuted for religious reasons and Kepler includes a grateful acknowledgement to Guldin in *The Dream* which was published posthumously by Kepler's son. The letters between Guldin and Kepler are published in *Johannes Kepler Gesammelte Werke*. Gulidn eventually died in 1643 in Graz, Austria aged 66.

GUNTER, EDMUND (1581-1626)
English mathematician

Edmund Gunter was responsible for introducing the words **cosine** and cotangent to the world as well as producing a seven figure table of **logarithms** of sines and tangents. Gunter also manufactured and used a precursor of the **slide rule**, called the Gunter scale. As if this were not enough Gunter also invented a device called Gunter's chain which was used for surveying, he published a work on navigation, studied magnetic declination and was the first to observe the secular variation. Whilst not one of the most original mathematicians Gunter had a very applied mind, and he contributed a number of practical uses for mathematical techniques and devices.

Gunter was born in Hertfordshire in Southern England in 1581. He attended Westminster School in London as a child and then carried on his education at Christ Church in Oxford where he remained until 1615. Gunter graduated initially in

1603 with a bachelors degree and then he gained a masters in 1605 and finally he obtained a divinity degree in 1615. After his ordination Gunter became the rector of St George's Church, in Southwark, London. He retained this position until his death aged 45 in 1626. From 1619 until 1626 Gunter was also the Professor of Astronomy at Gresham College, London. In 1620 Gunter published *Canon Triangulorum* or the *Table of Sines and Tangents*. This book was a list of logarithms of sines and tangents to seven decimal places. To aid in calculations using these figures Gunter made a precursor to the slide rule. This was called Gunter's scale and it gave the user the ability to multiply the logs together using a single scale on a copper rod in conjunction with a pair of dividers. Logarithms were plotted on this rod on a straight scale and **multiplication** could be carried out rapidly by the **addition** or **subtraction** of different lengths using the dividers. This device was extensively used by seamen as a navigation aid and its description was published in his 1624 book *Description and Use of the Sector, the Crosse Staffe and Other Instruments*. The slide rule as we now know it is essentially two Gunter scales which are slid one against another. The advantage of Gunter's scale was that it could be used to provide a rapid answer of sufficient accuracy for navigation at sea. For more accurate results lengthy calculations would have to be entered into and these were often unnecessary for the job in hand.

In 1623 Gunter published *New Projection of the Sphere* which looked at navigation and the magnetic effects of the Earth thereon. In 1624 Gunter published a book on a number of sundials he had installed in Whitehall, this was at the direct request of the future Charles I. In 1626 Gunter died at the age of 45 in London.

See also Logarithm; Slide rule

H

HADAMARD, JACQUES (1865-1963)
French analyst

Widely considered the preeminent French mathematician of the 20th century, Jacques Hadamard has made an impact on many fields of mathematics. Although an analyst and a student of theoretical **calculus** by training, he has influenced **topology**, **number theory**, and even psychology. His work on defining **functions** won him the Grand Prix of the Académie des Sciences early in his career, and his **proof** of the **prime number theorem** solidified his importance in the mathematical world. He wrote several textbooks on a variety of mathematical subjects, including one which explained a mathematician's thought processes. Hadamard was first and foremost a teacher, however, and he used his position to help both students and colleagues alike see the connections between seemingly unrelated fields.

Born in Versailles on December 8, 1865, Jacques-Salomon Hadamard was the son of two teachers. His mother, Claude-Marie Picard, taught piano, while his father, Amédée, taught Latin at a prominent Paris high school. In 1884, at the age of eighteen, Hadamard began studying at the École Normale Supérieure. His first teaching job was at a high school in Paris, the Lycée Buffon, in 1890. When he was not teaching, he worked on his doctoral dissertation, and the research he did during this period led to his first breakthrough in mathematics.

Hadamard's dissertation concerned determining the shape of a function and finding certain points on that function where **division** by **zero** was involved in the original equation. Such functions had previously been considered undefined and unsolvable, but Hadamard found a way to solve them using a set of **fractions** known as the Taylor series. Published in 1892, his work was so revolutionary that the French Académie des Sciences immediately awarded him its highest honor, the Grand Prix. This was also the year Hadamard married Louise-Anna Trenel, with whom he would have five children. In 1892

Hadamard also accepted a position as lecturer at the Faculté des Sciences of Bordeaux, where he continued his work. Although his accomplishments in defining functions had been important to the mathematical community at large, for Hadamard it was just another step toward a larger goal. He wanted to find a proof of the prime number **theorem**. For years, some of the world's best mathematicians had attempted to prove that the total number of primes could be defined and that individual primes could be determined by something other than the endless testing of possible factors. Many had discovered estimates and close guesses, but no one had achieved accurate results.

Hadamard used his work on the Taylor series as a guide, and he established that the number of primes below any given number could be determined by using **complex numbers**, also known as **imaginary numbers**. While his theory only works when the numbers used are sufficiently large, mathematicians generally only concern themselves with primes when such large numbers are involved. Later attempts to improve upon or generalize Hadamard's 1896 prime number theorem by such noted mathematicians as **S. I. Ramanujan** have failed.

Following publication of the proof of the prime number theorem, Hadamard left Bordeaux for a lectureship at the Sorbonne in Paris. A return to the intellectual center of Paris also meant greater involvement with the mathematical community, in which Hadamard had earned a high place. While many mathematicians were content to specialize in a small area of mathematics, Hadamard saw the importance of finding connections between the various fields. He was openly critical of mathematicians who limited their work to their immediate subject. In 1902 he argued, for example, that the definitions **Vito Volterra** had offered for the calculus terms *continuity, derivative,* and *differential* were inadequate because they could not be generalized to other fields, especially the relatively new area of topology. Instead of merely criticizing Volterra, however, Hadamard applied himself to generalizing **analysis** so it would be more applicable to other fields. His creation and **definition** of the term *functional,* first put forth in 1903, is one

Jacques Hadamard

result of this generalization. Though Hadamard had used standard analysis to come up with functionals, the application of the idea to topology was important to establishing the validity of that field.

Hadamard's work forming connections between topology and analysis was interrupted in 1904 by a debate over mathematical logic which raged through the mathematical community. **Ernst Zermelo**, a German mathematician, had proposed that given an infinite number of **sets**, it would be possible to select exactly one, definable item from each set. This proposal was called the **axiom** of choice. Zermelo argued that it was obvious and thus needed no proof, but many of the most prominent mathematicians of the time, including **Émile Borel**, **Jules Henri Poincaré**, and **Henri Lebesgue**, disputed it. As Morris Kline describes the controversy in *Mathematics: The Loss of Certainty:* "The nub of the criticism was that, unless a definite law specified which element was chosen from each set, no real choice had been made, so the new set was not really formed." Yet the axiom of choice was necessary to establish sections of abstract **algebra**, topology, and standard analysis. Hadamard supported Zermelo. He rejected the

idea that the item taken from the set could necessarily be defined, yet he felt that any theory which allowed mathematics to progress should be accepted, with or without formal proof.

In 1908, Hadamard spoke at the Fourth International Congress of Mathematicians in Rome, where he met the famous German topologist **L. E. J. Brouwer**. They began a correspondence relating to the mathematical ideas of their time, and the exchange of letters was crucial to Brouwer. The German mathematician used Hadamard's ideas as a springboard to some of his most important topological discoveries. In 1909, Hadamard left the Sorbonne for a more prestigious appointment as professor at the École Centrale des Arts et Manufactures. He would remain there, teaching concurrently at the Collège de France after 1920, until his retirement at the age of 71.

In 1912, Hadamard's friend and colleague Jules Henri Poincaré died. Poincaré, like Hadamard, had been involved in several different fields of mathematics, and his work had greatly influenced Hadamard's interest in generalization. Saddened by the loss of this great mathematician, Hadamard devoted a great deal of his research time after Poincaré's death to writing biographical works of his friend. Hadamard did his last piece of major research in the field of calculus in 1932, when he addressed a problem posed by the French mathematician **Augustin-Louis Cauchy**. But even after his retirement in 1937, Hadamard continued to ponder some of the questions that had concerned him throughout his career. The old controversy over the axiom of choice became the basis of a new book on the importance of accepting intuition for the sake of mathematical progress. He published *The Psychology of Invention in the Mathematical Field* in 1945, at the age of 80, and it was widely considered an innovative attempt at understanding how mathematicians come up with their ideas. Some of the work on this book was done in the United States, where he was a visiting professor at Columbia University in New York in 1941. Unlike many European mathematicians, however, Hadamard did not stay in America. He returned home to France, living out the rest of his life quietly. He died in Paris on October 17, 1963, at the age of 97.

HAKEN, WOLFGANG (1928-   )
German mathematician

Wolfgang Haken is most known for his work on three dimensional manifolds. Such is his prominence in this field that three dimensional manifolds are often referred to as Haken manifolds. Haken has also published extensively on **four color map theorem**, a branch of graph theory. In 1976 he produced a complete solution for the four color map theorem, this was the first major theorem to be proven using a computer.

Wolfgang Haken was born in Berlin in 1928. He attended university in Kiel where he studied mathematics, philosophy and physics, eventually receiving his doctorate in 1953. From 1954 to 1962 Haken worked for Siemens in Munich, this was in the research and development section, specialising in the application of microwave technology. At this time Haken was still carrying out research in mathematics and his discovery of a mathematical technique for discovering if a knot is knotted or not resulted in an invitation to become a visiting professor at the University of Illinois. In 1965 he was given a full chair at the University of Illinois. In 1990 he was made a member of the Center for Advanced Study in the United States and in 1993 he was given an honrary doctorate from Frankfurt University.

Haken is still active in his research interests in has maintained links with the University of Illinois where he is emeritus professor.

See also Four color map theorem; Manifolds

HAMILTON, WILLIAM ROWAN (1805-1865)

Irish algebraist

William Rowan Hamilton was an Irish mathematician and astronomer of the 19th century, considered by some to be near in intellect to **Isaac Newton**. He created a novel system of **algebra** for operating on **complex numbers**, coined the term "vector," founded **vector analysis**, developed icosian **calculus**, and made important contributions to the understanding of **light** and optics.

Born in Dublin at the stroke of midnight, between the 3rd and 4th of August 1805, Hamilton was the only son of Archibald Hamilton and Sarah Hutton. In 1808, possibly because of his family's difficult financial condition, Hamilton was sent to live with his uncle, the Reverend James Hamilton, head of a diocesan school in the village of Trim, 40 miles northwest of Dublin. Hamilton had little contact with his parents during childhood, but his four sisters lived with him intermittently, continuing into adulthood. His mother died in 1817 and his father in 1819. Hamilton began his education as soon as he arrived in Trim, quickly revealing himself as a child prodigy. His uncle was eager to pour as much knowledge into him as was possible and at the age of 10, Hamilton was supposedly schooled in 13 foreign languages. His mathematical education began around the same time, with a study of **Euclid**. Hamilton went on to read Newton's *Universal Arithmetic* and *Principia*, **analytic geometry**, calculus, and **Pierre Laplace**'s *Mecanique Celeste*, all independently. In 1824, Hamilton entered Trinity College in Dublin; that same year he published a paper correcting a mistake in Laplace's work. Hamilton was a top scholar, and the college awarded him two separate *optimes*, or "off–scale grades," for his performance on examinations. So rare was such an honor that no *optimes* had been awarded in the previous 20 years.

While still a student, Hamilton wrote the first part of a pivotal treatise on optics and delivered a paper on the subject to the Royal Irish Academy. His work dealt with the patterns of light produced by reflection and refraction. Hamilton demonstrated that light travels by the path of least action and he used algebraic **functions** to express its path mathematically.

William Rowan Hamilton

He also predicted the phenomenon of conical refraction, an idea which was later proved experimentally by the astronomer Humphrey Lloyd. Hamilton's reputation in mathematics and astronomy became widespread. In 1827, Dr. Brinkley, a professor of astronomy, resigned from Trinity to become a bishop. Hamilton was unanimously voted in as Brinkley's replacement and agreed to take the professorship, despite the fact that he had never applied for the post. His new job included an appointment as Royal Astronomer of Ireland and the directorship of the Dunsink Observatory. As head of the observatory, Hamilton lived at Dunsink. In 1828, he completed his manuscript on light patterns, *The Theory of Systems of Rays*, which was then published by the Royal Irish Academy. Hamilton retained his position as Royal Astronomer and the overseer of Dunsink throughout his life. He performed his duties diligently, but his first love was always pure mathematics.

In the early 1830s, Hamilton began searching for a way in which to interpret complex numbers geometrically. He started by developing algebraic rules for working with complex numbers of the class $a + ib$, where a and b are **real numbers** and i is imaginary. He characterized this class of complex numbers in relatively simple terms, as ordered pairs, or "couples" of real numbers a and b. Hamilton represented these complex numbers geometrically, as line segments having both length and direction. He coined the term vector to describe the segments, and applied algebraic rules to their analysis. Hamilton showed that the sum of two complex numbers could be represented by a parallelogram. His initial work demonstrated that complex numbers could be useful tools for dealing with the concept of **rotation** in plane, or two–dimensional **geometry**.

Hamilton's *Theory of Algebraic Couples*, published in 1835 by the Royal Irish Academy, expounded on his application of algebra to complex numbers. In the following years, realizing its potential importance for three–dimensional geometry, Hamilton tried to develop an algebra for dealing with complex numbers as triples, that is, of the type $a + ib + jc$, where a, b, and c are real numbers and i and j are imaginary. On an October day in 1843, while walking along the Royal Canal in Dublin, Hamilton had the sudden insight that algebraic operations on such complex numbers required the imposition of a fourth **dimension** in geometric **space**. At that moment, he conceived of the idea of operating on quadruples, number groupings he later termed **quaternions**. With a knife, he carved the equation $i^2 = j^2 = k^2 = ijk = -1$ in a stone of the Brougham Bridge. This proved to be the formula for operating on complex numbers of the type $a + ib + jc + kd$, where a, b, c, and d are real numbers and i, j, and k are imaginary. Among the important realizations Hamilton made about the algebra of his quaternions was that **multiplication** is not commutative. That is, for example, ij does not equal ji. This realization was especially important because ultimately it forced mathematicians to abandon the belief that the commutative law of multiplication was axiomatic. A plaque commemorating Hamilton's inspiration about quaternion algebra is installed today at the site on the bridge where he carved out his discovery.

During the decade after Hamilton's initial discovery, he lectured widely on his theory of quaternions. He believed quaternion algebra would transform the field of mathematical physics and become as important historically as the invention of the calculus. Although his expectations were never realized, Hamilton's ideas did play a historical role in the development of **matrix** algebra and certainly, in the development of vector **analysis**. In 1853, Hamilton's *Lectures on Quaternions* was published. He devoted most of the remainder of his life to refining his quaternion theory and investigating its application. His *Elements of Quaternions* was published posthumously in 1866.

Outside of his study of quaternions, the only mathematical work Hamilton pursued in his latter years was that of the icosian calculus. In 1856 he developed this new calculus, based originally on the geometric properties of the icosahedron, a 20–sided solid. Hamilton's discoveries about the vertices and edges of icosahedrons and other solids provided groundwork for what became known in the 20th century as graph theory.

Hamilton was recognized throughout his life for his contributions to mathematics and astronomy. When he was 30 years old, he became an officer of the British Association for the Advancement of Science. In that same year he was knighted by the lord lieutenant of Ireland and received the Royal Medal of the Royal Society. In 1837, he was elected president of the Royal Irish Academy, and six years later the British Government awarded him a Civil List life pension of 200 pounds a year. Shortly before his death, Hamilton was voted the first foreign member of the National Academy of Sciences in the United States. This last honor, bestowed on him for his work on quaternions, is the one that he reportedly cherished the most.

Hamilton married Helen Marie Bayly, the daughter of a country parson, in 1833. An illness left her a semi invalid for most of her life. Together they bore two sons, William Edwin, born in 1834, who squandered his father's money on various business schemes, and Archibald Henry Hamilton, born in 1835, who became a clergyman. A daughter, Helen Eliza Amelia Hamilton, was born in 1840. Hamilton's wife was known as a pious, but shy and timid woman, whose intellect did not match that of her husband. Hamilton's friends were critical of his marriage, though for his part, he behaved dutifully as a husband and never complained. Nonetheless, in the latter part of his life he indulged in romantic fantasies and a prolonged and secretive correspondence with a woman who had rejected him as a young man. In the summer of 1865, having suffered from occasional bouts of gout over the years, Hamilton took seriously ill and died on September 2, 1865, in Dublin. His wife survived him by four years.

Hamilton was good friends with **Augustus De Morgan** and physicist John Herschel, as well as poets Samuel Taylor Coleridge and William Wordsworth. He was known as a convivial and jovial man, though as he grew older he developed a severe drinking problem. In his private life, his habits were apparently slovenly, and after his death his library was found to be strewn not only with unpublished manuscripts and disheveled papers, but desiccated food and dirty dishes.

HARDY, GODFREY HAROLD (1877-1947)

English pure mathematician

Godfrey Harold Hardy was one of the foremost mathematicians in England during the early part of the 20th century. He was primarily a pure mathematician, specializing in branches of mathematics that study the behavior of numbers (such as **number theory** and **analysis**). He also made important contributions to areas of **applied mathematics**, and is known for formulating the Hardy-Weinberg law of population genetics. He taught at both Cambridge and Oxford and published over 350 research papers, either alone or in collaboration with other mathematicians—most notably John Edensor Littlewood and **S. I. Ramanujan.**

Born on February 7, 1877, in Cranleigh, England, Hardy was the elder of two children of Isaac and Sophia Hall Hardy. Both his parents came from poor families and were unable to afford university education for themselves, but they were people with a taste for intellectual and cultural pursuits and had made a place for themselves as schoolteachers. Hardy's father was the geography and drawing master at Cranleigh School, where he also gave singing lessons, edited the school magazine, and played soccer. His mother taught piano lessons there and helped run a boarding house for the younger students. They took great pains to educate their children well, and both Hardy and his sister Gertrude inherited their parent's love for education and the intellect. A gifted student, Hardy displayed a special talent and interest for mathematics from a very young age. When he was just two, he was writing down numbers into the millions, a common sign of future numerical ability. Rather than attend regular classes in mathematics, he was coached by a private tutor, and he completed sixth form at Cranleigh when he was only thirteen—about five years younger than the usual age—ranking second in class. He then won a prestigious scholarship to attend Winchester College, a private secondary school where he spent six years before graduating in 1896, winning another scholarship to attend Trinity College at Cambridge University.

Hardy initially chose to attend Cambridge rather than Oxford because of its standing in mathematics, and Trinity College was the premier institution for the subject in England. During his first years at Cambridge, however, he very nearly gave up mathematics altogether, in disgust over the examination system then in existence. Mathematics students had to take the Tripos examination, which consisted of eight days of solving problems. Hardy disliked the system because, rather than gauging the ability and insight of the student, he believed it tested endurance and the ability to memorize formulae and **equations**. Special private coaches trained students for Tripos, while lecturers at the universities pursued their own mathematical research. Hardy considered Tripos an utter waste of time, and he tried to change his course of study to history. What kept him in the field was his professor, A. E. H. Love, who recognized Hardy's affinity for pure mathematics and recommended that he read a book by the French mathematician **Camille Jordan**.

Entitled *Cours d'analyse de l'Ecole Polytechnique,* Jordan's book kept Hardy in mathematics, and he persevered through Tripos, putting real mathematics aside for two years. In his autobiographical book, *A Mathematician's Apology*, Hardy wrote of his career after reading Jordan's book: "From that time onwards I was in my way a real mathematician, with sound mathematical ambitions and a genuine passion for mathematics." Despite his acute distaste for Tripos, Hardy ranked fourth in the first examination in 1898, and he scored the highest points in the second part of the examinations two years later. Upon his graduation in 1899, he was named a fellow of Trinity College at Cambridge.

As a fellow, Hardy was finally free to devote his time to pure mathematics, and he did so with great enthusiasm and fervor. Over the next ten years he produced several papers on number series that established his reputation as an analyst, and in 1908 he published a book, *A Course of Pure Mathematics.* This was the first mathematical textbook in the English language to explain rigorously the fundamental concepts of the subject. Until then, books and teachers had merely provided these formulae and moved on to using them in various practical applications. Continuing his interest in mathematical education, Hardy joined a panel and tried to reform Tripos as the

first step—he hoped—to abolishing it altogether. Although this latter goal proved futile (Tripos is still in existence nearly a century later), the panel did succeed in eliminating the worst features of the system.

Also in 1908, Hardy made his only contribution to applied mathematics in the form of a letter to the American journal *Science*. Mendelian genetics being the subject of much debate at that time, an article that recently appeared in *The Proceedings of the Royal Society of Medicine* had disputed some of Mendel's theories of inheritance of various traits. In his letter, Hardy used simple algebraic principles to prove the error in the article, and he set down an equation that predicted the patterns of inheritance. In the same year, a German physician named Wilhelm Weinberg devised a similar mathematical method for prediction, and the principle was named the Hardy-Weinberg law in honor of them both. Widely used in the study of the genetic transmission of blood groups and rare diseases, this law appears today as a fundamental principle of population genetics.

Despite what many saw as his productivity during the years between 1900 and 1910, Hardy himself felt that he did not do too much of value, and he said so in *A Mathematician's Apology*: "I wrote a great deal during the next ten years but very little of any importance; there are not more than four or five papers I can still remember with some satisfaction." He believed that his best work came later, out of his associations with John Edensor Littlewood and S. I Ramanujan.

Hardy began his collaboration with the mathematician J. E. Littlewood in 1911. The partnership, which lasted for over 35 years and resulted in the publication of over one hundred papers, was described by C. P. Snow in his foreword to Hardy's *A Mathematician's Apology* as "the most famous collaboration in the **history of mathematics**. There has been nothing like it in any science or in any other field of creative activity." Some eight years younger than Hardy, Littlewood was a brilliant mathematician who had already made a name for himself in the mathematical community.

Not much is known about exactly how the two men worked together. At the height of their combined productivity, from 1920 to 1931, they were not even at the same university. Hardy had moved to Oxford by then, while Littlewood remained in Cambridge. According to Snow, Hardy always maintained that Littlewood was the better, more powerful mathematician, although at meetings and conferences it was always Hardy who presented the papers. Indeed, Littlewood seldom attended these meetings, and other mathematicians were known to have joked that they doubted whether he even existed.

Hardy's second collaboration began in 1913 with a letter from India. The writer of the letter, Srinivasa Iyengar Ramanujan, was then an unknown clerk in Madras, who had received no formal education or training in mathematics but who claimed to have made some original discoveries while working on his own. After a single paragraph of introduction, Ramanujan plunged into his mathematics, providing page after page (the letter was over ten pages long) of theorems and results written out neatly by hand. Hardy's first instinct was to disregard the letter as a crank, filled as it was with wild claims and bizarre theorems without offering any proofs to support them. Indeed, two other Cambridge mathematicians had done just that, having received similar letters from Ramanujan earlier.

But something about the letter, perhaps its very strangeness, also intrigued Hardy, and he decided it was worth a closer look. He invited Littlewood to join him. After three hours of perusing the papers, the two men decided that the work was that of a genius. Hardy then wrote back to Ramanujan asking for proofs for some of his results, but untrained as he was Ramanujan did not or could not furnish these. In fact, when he wrote again it was to present Hardy with even more results and theorems. To Hardy, who had reintroduced the concept of rigor in **proof** to England, Ramanujan's intuitive reasoning and unorthodox methods were very frustrating. He invited Ramanujan to England, where from 1914 to 1919 the two men worked together and published many important papers. Hardy personally trained Ramanujan in modern mathematics and analysis; as he did in his collaboration with Littlewood, he also wrote most of the papers and presented the talks. Ramanujan himself returned to India and died in 1920, having contracted tuberculosis in England, but Hardy continued to promote his work long after his death.

Meanwhile, Hardy was growing disenchanted with life at Cambridge, and controversies surrounding World War I had much to do with this. In his foreword to *A Mathematician's Apology*, Snow describes the years from 1914 to 1918 as the "dark years" for Hardy. Most of his friends were away at the war. His work with Littlewood was also suffering, as the latter had gone away to serve as a second lieutenant in the army. In 1916, **Bertrand Russell**, the noted philosopher, mathematician, and pacifist, was dismissed from his lectureship at Trinity for his antiwar activities. Hardy was a close personal friend of Russell's; outraged at this dismissal, he fought bitterly with many of his mathematical colleagues. In 1918, the university dismissed yet another person for their antiwar views, this time a librarian, upsetting Hardy even more, and he actively opposed the firing. Snow writes in *A Mathematician's Apology* that "it was the work of Ramanujan which was Hardy's solace during the bitter college quarrels."

Adding to his discontent was the fact that his duties at Cambridge were becoming increasingly administrative, leaving him little time for research. In 1919, he moved to Oxford University as Savilian Professor of **Geometry** at New College. Here, he reached the pinnacle of his career, setting up a flourishing research school and enjoying the best years of his collaboration with Littlewood. His flamboyance, radical antiwar views, and outspokenness were appreciated at Oxford. Hardy had an exceptional gift for working well with other people, and besides Ramanujan and Littlewood he collaborated with many other leading mathematicians of the day. He also spent one year as an exchange professor at Princeton University. In 1931, he returned to Cambridge as Sadleirian Professor of Pure Mathematics. He retired in 1942, after which he continued to live in his rooms at Trinity. Shortly before his death in 1947, the Royal Society awarded him their highest honor, the Copley Medal.

There was nothing in life Hardy cared about more than mathematics, but he did have other interests. During his early years at Cambridge, he was part of several social groups, including a secret intellectual society known as the Apostles. This society met weekly to discuss and debate philosophical issues, and over the years it boasted some brilliant minds among its membership. During Hardy's time the philosophers Bertrand Russell and G. E. Moore were members, and it was through this association that many of his closest friendships were fostered.

Hardy was intensely fond of sports, particularly cricket. He followed cricket matches and scores with great attention. Hardy was not above bringing his passion for cricket into the classroom, describing the quality of mathematical work he considered exceptionally good to be in the "Bradman" or "Hobbs" class. As both men were cricketeers, not mathematicians, such references were apt to confuse unsuspecting students. To the end of his days he remained passionately interested in cricket, and when he died he was listening to his sister read to him from a book on the history of Cambridge University cricket.

Hardy was also a talented writer. He was often called upon to write obituaries of famous mathematicians. In addition to numerous mathematical texts, he also wrote *Bertrand Russell and Trinity*, a recounting of the wartime controversy, and *A Mathematician's Apology*, a treatise describing his love for the subject. In this book he offered a justification for his choice of career: "I have never done anything 'useful'.... Judged by all practical standards the value of my mathematical life is nil." But he adds, "The case for my life is this: that I have added something more to knowledge, and helped others to add more."

HARMONIC ANALYSIS

Harmonic **analysis** is the branch of mathematics which developed from the study of **Fourier series**, Fourier transforms, and other related operators. One of the most basic problems in harmonic analysis is to determine how periodic **functions** can be written as a sum of **trigonometric functions**. Suppose, for example, that $f(x)$ is a real or complex valued function defined for all **real numbers** x, and f is also *periodic* with period 1. By periodic with period 1 we mean that f satisfies the identity $f(x+1) = f(x)$ for all real numbers x. It follows then that $f(x+n) = f(x)$ for all **integers** n. By considering a graph of $y = f(x)$ it is apparent that f is completely determined by its behavior on any **interval** of length 1, for example, the interval [0,1]. Now some of the simplest examples of periodic functions with period 1 are the trigonometric functions

$$\cos 2\pi x, \cos 4\pi x, \cos 6\pi x,..., \cos 2\pi l x,...,$$

where l is a positive integer, and

$$\sin 2\pi x, \sin 4\pi x, \sin 6\pi x,..., \sin 2\pi m x,...,$$

where m is a positive integer. The constant function 1 is also periodic with period 1 and it is convenient to include this as $\cos 2\pi l x$ with $l = 0$. If we form a linear combination of these functions, such as

$$\sum_{l=0}^{L} a_l \cos 2\pi l x \;+\; \sum_{m=1}^{M} b_m \sin 2\pi m x$$

where a_0, a_1, a_2,..., a_L and b_1, b_2, b_3,..., b_M are real or complex coefficients, then we continue to get a function which is periodic with period 1. Now it is natural to ask if more general functions $f(x)$, which are periodic with period 1, can be written as sums of this sort. Since we want to reach as many functions $f(x)$ as possible, we should also allow **infinite series** expansions rather than just finite sums.

In order to describe this problem and its solution more easily, it is a good idea to replace the functions $\cos 2\pi l x$ and $\sin 2\pi m x$ with the complex **exponential functions**
$e^{2\pi i n x} = \cos 2\pi n x + i\sin 2\pi n x$ where $n =$... -3, -2, -1, 0, 1, 2, 3,....

These functions are also periodic with period 1 and allow us to work with cos and sin simultaneously. Now the basic problem can be restated as follows: if $f(x)$ is periodic with period 1, does there exist a sequence of real or **complex numbers**... c_{-2}, c_{-1}, c_0, c_1, c_2,... such that

$$\lim_{N \to \infty} \sum_{n=-N}^{N} c_n e^{2\pi i n x} = f(x) \;?$$

And if these numbers exist, then how can we compute them from our knowledge of f? Here is one solution: assume that f is a continuous function and then define the complex number c_n by the formula

$$c_n = \int_0^1 f(x) e^{-2\pi i n x} \; dx.$$

Also, assume that

$$\sum_{n=-\infty}^{\infty} |c_n| < \infty,$$

that is, the infinite series formed with the numbers c_n is absolutely convergent. Under these assumptions we have

$$\lim_{N \to \infty} \sum_{n=-N}^{N} c_n e^{2\pi i n x} = f(x)$$

at each real number x. The coefficients c_n are called the *Fourier coefficients* of f. It is obvious that they are determined by the function $f(x)$ and to indicate this they are often written

$$\hat{f}(n).$$

Notice that we may regard:

$$\hat{f} \; \mathsf{Z} \to \mathsf{C}$$

as a function from the integers to the complex numbers that has been determined by the original periodic function $f(x)$.

Here is a second solution: assume that f is a continuous function and let the complex numbers

$$c_n = \hat{f}(n)$$

be defined as before. In this case we make no assumption about the absolute **convergence** of the infinite series formed with the numbers

$$\hat{f}(n).$$

Then we have

$$\lim_{N \to \infty} \sum_{n=-N}^{N} \left(1 - \frac{|n|}{N+1}\right) \hat{f}(n) e^{2\pi i n x} = f(x)$$

at each real number x. Notice that in this second solution we have introduced the additional factors

$$\left(1 - \frac{|n|}{N+1}\right)$$

into the sum. Some modification of this sort is necessary because of the following surprising fact. There exists a continuous function $f(x)$ and a real number ξ such that

$$\lim_{N \to \infty} \sum_{n=-N}^{N} \hat{f}(n) e^{2\pi i n \xi} \quad \text{does not exit.}$$

Here is an example: let $g(x) = |\sin \pi x|$. Then it is easy to check that g is periodic with period 1 and that g is continuous. The Fourier coefficients of g are given for each integer n as the value of the integral

$$\int_0^1 |\sin \pi x| e^{-2\pi i n x} \; dx = \hat{g}(n) = \frac{2}{\pi(1 - 4n^2)}.$$

Since the infinite series

$$\sum_{n=-\infty}^{\infty} |\hat{g}(n)| < \infty,$$

is convergent, we have

$$\lim_{N \to \infty} \sum_{n=-N}^{N} \frac{2}{\pi(1 - 4n^2)} e^{2\pi inx} = |\sin \pi x|$$

at each real number x.

The process of going from the periodic function $f(x)$ to the function:

$$\hat{f} \; \mathbf{Z} \to \mathbf{C}$$

is an example of the Fourier transform. Then a basic problem of harmonic analysis is to learn how to recover the function $f(x)$ from knowledge of

$$\hat{f}.$$

This process can also be formulated in other settings. Suppose that $f: \Re \to \mathbf{C}$ is a Lebesgue integrable function, (see the article on the **Lebesgue integral**). In this case we define the Fourier transform of f to be the function:

$$\hat{f}$$

$\Re \to \mathbf{C}$ defined by the Lebesgue integral

$$\hat{f}(t) = \int_{-\infty}^{\infty} f(x) e^{-2\pi ixt} \; dx.$$

Although we have continued to use the same notation as for periodic functions and their Fourier coefficients, in fact the situation now is quite different. The function $f(x)$ is defined for real numbers x and is not assumed to be periodic, and the function

$$\hat{f}(t)$$

is also defined for real numbers t. Because f is an integrable function, it can be shown that

$$\hat{f}(t)$$

is a continuous function and also satisfies

$$\lim_{t \to \pm\infty} \hat{f}(t) = 0.$$

In this setting we may also ask how the function $f(x)$ can be recovered from its Fourier transform

$$\hat{f}(t).$$

We can give a reasonably simple answer by imposing two additional conditions which in practice are often satisfied. Assume that $f(x)$ is continuous and that

$$\hat{f}(t)$$

in integrable on $\Re$. Then with these assumptions we have

$$f(x) = \int_{-\infty}^{\infty} \hat{f}(t) e^{2\pi int} \; dx.$$

This is called the *Fourier inversion formula*. Notice that the inversion formula is nearly the same as the formula defining the Fourier transform.

Here is an example: let $h(x) = e^{-2\pi|x|}$. Then the Fourier transform of h is the function $\hat{h}: \Re \to \mathbf{C}$ defined for each real number t by

$$\int_{-\infty}^{\infty} e^{-2\pi|x|} e^{-2\pi ixt} \; dx = \hat{h}(t) = \frac{1}{\pi(1 + t^2)}.$$

Because h is continuous and $\hat{h}$ is integrable, the Fourier inversion formula provides the identity

$$\int_{-\infty}^{\infty} \frac{1}{\pi(1 + t^2)} e^{2\pi ixt} \; dt = h(x) = e^{-2\pi|x|}$$

for all real numbers x.

HAUSDORFF DIMENSION

Around the turn of the century, the intuitive notion of **dimension** was made mathematically rigorous for the first time. The new definitions did not simplify matters however, because there are now more then ten different definitions of dimension: topological dimension, Hausdorff dimension, capacity dimension, self-similarity dimension, information dimension, box-counting dimension, and more. Many of these concepts were inspired by **Felix Hausdorff**'s pioneering work. In 1919, he introduced the Hausdorff dimension. Although important for theoretical purposes, the Hausdorff dimension of most objects is too difficult to compute to be practical.

To define the Hausdorff dimension, we need a few concepts from Euclidean **geometry**. The **distance** between points $(x_1,...,x_n)$ and $(y_1,...,y_n)$ (in Euclidean n-dimensional **space** E^n) is the **square** root of $(x_1-y_1)^2 +... + (x_n - y_n)^2$. The open ball centered at x with radius r is the set of all points in E^n that are a distance less than r from x. An open subset is a union of open balls. The **diameter** of an open subset is a number k such that all pairs of points in the set are less than a distance k apart and k is the smallest real number with this property. An open cover of a subset A in E^n is a collection of open **sets** whose union contains A. Such a cover has diameter k if every open set in the cover has diameter less than or equal to k. The d-dimensional Hausdorff measure of a subset A is a real number, denoted by $H^d(A)$, with two properties. First, if y is a number such that for every k, there is a k-diameter open cover of A such that the sum of the dth **powers** of the diameters of the open cover is less than y, then y is greater than or equal to $H^d(A)$. Second, $H^d(A)$ is greater than or equal to any number that satisfies the first property. Hausdorff proved that if A is a smooth curve, then $H^1(A)$ is the length of A. If A is a smooth surface, then $H^2(A)$ is the **area** of A. Also, there exists a number called $D(A)$, such that if x is less than $D(A)$ then $H^x(A)$ is **infinity** and if x is greater than $D(A)$ then $H^x(A)$ is **zero**. This number, $D(A)$, is called the Hausdorff dimension of A.

For many **fractals**, such as the **snowflake curve**, the Menger curve, Sierpinski's **triangle** and carpet, the Hausdorff dimension is equal to the self-similarity dimension (which is also equal to the capacity dimension).

See also Dimension; Fractal; Sierpinski triangle, carpet, and sponge; Snowflake curve

HAUSDORFF, FELIX (1868-1942)

German topologist

Felix Hausdorff laid the foundations of set theoretic **topology**, which has evolved into an elaborate discipline that interacts with nearly every other field of mathematics. He precisely developed such basic notions as **limits**, continuous maps, connectedness, and compactness, which have become fundamental in building many kinds of mathematical structures. One of Hausdorff's revolutionary ideas, spaces of non–integer dimension,plays an important role in various topics, including geometric measure theory, the theory of dynamical systems, and in the description of the popularized notion of **fractals**. He was also a philosopher and author.

Hausdorff was born on November 8, 1868, in Breslau, Germany, which is now Wrocaw, Poland. His mother was Johanna Tietz Hausdorff; his father, Louis Hausdorff, was a dry goods merchant. The family moved to Leipzig, Germany, in 1871. The young Hausdorff eventually attended Leipzig University, where he studied astronomy and mathematics, earning his Ph.D. in 1891. His early research concentrated in the areas of optics and astronomy. After graduation, Hausdorff volunteered to serve in the German infantry. He achieved the rank of vice-sergeant before removing himself from consideration for further promotion in 1894. Hausdorff was Jewish, and no acknowledged Jews had been commissioned as officers in the German military for nearly 15 years. In 1896, following his father's death, Hausdorff succeeded him as a partner in the publishing firm Hausdorff and Company, which produced the leading trade magazine for spinning, weaving, and dyeing. That same year, he was accepted as a lecturer at Leipzig University.

Hausdorff had a lively interest in the fine arts and in philosophy. An accomplished pianist, he occasionally composed songs. Like many others of his generation, Hausdorff was deeply influenced by the philosophy of Friedrich Nietzsche, though he maintained a critical distance to certain parts of Nietzsche's work.

The first of Hausdorff's four full-length literary works was published in 1897. He wrote under the pseudonym Dr.

Paul Mongréso that he could express himself freely without jeopardizing his university position. The first book, *Sant' Ilario: Thoughts from the Landscape of Zarathustra,* was primarily a collection of aphorisms relating to Nietzsche's influential volume *Thus Spake Zarathustra*. It was published by the same company that had published Nietzsche's works and was even produced with a similar book cover.

Chaos in Cosmic Choice, the second book written under the name Mongré, dealt with relationships between **space** and time and was intended as a radical continuation of Immanuel Kant's criticism of traditional metaphysics. Hausdorff presented the same concepts in "The Space Problem," his 1903 inaugural lecture after being appointed as an associate professor at Leipzig University. Mongré's third major literary work was *Ecstasies*, a volume of sonnets and poems published in 1900. He also wrote *The Doctor's Honor*, a satirical play that was successfully produced in Hamburg and Berlin.

In 1897, Hausdorff began publishing papers on topics in mathematics, including **non-Euclidean geometry**, **complex numbers**, and probability. He became interested in **Georg Cantor**'s work on **set theory**, and during the summer semester of 1901 he taught what may have been the first course on set theory to be presented in Germany. Also, about this time, **David Hilbert** was publishing work applying set theory to geometry; this work may have been the inspiration for Hausdorff's greatest mathematical accomplishment.

In 1910, Hausdorff accepted a position as associate professor at the University of Bonn. Although he had written one or two articles on set theory each year for two decades, he published nothing from 1910 until 1914; apparently this was a period of intense work on the creation of point set topology. Hausdorff moved to Greifswald in 1913 to become a professor at the university there, and the following year, he published his monumental *Grundzüge der Mengenlehre* ("Basic Features of Set Theory")

The *Grundzüge* was a comprehensive text dealing with set theory,point set topology (now more commonly called set theoretic topology), and real **analysis**. Although the book was written for students at the advanced undergraduate level, Hausdorff noted in the preface that the volume also offered new ideas and methods to his professional colleagues. By organizing point set theory with just the right choice of axioms, he so thoroughly revised the related existing work that his book became the foundation on which modern topology has been developed.

Topology generalizes concepts such as **continuity** and limits to **sets** other than real and complex numbers. A topological space is free of all imposed structure not relevant to the continuity of **functions** defined on it. While Hausdorff's definitions and axioms were so general that an unlimited variety of geometric interpretations was possible, he developed the Euclidean plane as a special case by adding appropriate postulates. As Carl B. Boyer wrote in *A History of Mathematics,* "Topology has emerged in the twentieth century as a subject that unifies almost the whole of mathematics, somewhat as philosophy seeks to coordinate all knowledge."

Because of its generality, topology gives rise to apparent paradoxes that violate intuition, two of which Hausdorff addressed in the *Grundzüge*. One involves the **transfinite numbers** developed by Cantor, in which there are different magnitudes of **infinity**, and an infinite, proper subset may have the same number of elements as its superset. The other, now called the Hausdorff Paradox, shows that the surface of a **sphere** can be decomposed into three equal, nonintersecting sets so that the original sphere may be represented as the union of any two of them.

In 1919, Hausdorff introduced another revolutionary concept. He generalized the notion of **dimension** (e.g., a two–dimensional **triangle** or a three–dimensional cube) to include the possibility of objects with fractional dimensions. This has proven highly fruitful in various areas of mathematics, besides being popularized in the form of computer-generated fractal images.

Hausdorff returned to the University of Bonn in 1921, where he worked as a professor for the rest of his career. He was respected as the most capable mathematician in Bonn and as a professor whose lectures were well reasoned and clearly delivered. He taught until 1935, when he reached the mandatory retirement age of 67. Hausdorff continued to publish mathematical papers until 1938.

In 1899, Hausdorff married Charlotte Goldschmidt; his only child, a daughter named Lenore (usually called Nora), was born the following year. Although Charlotte came from a Jewish family, she had been baptized a Protestant Christian in 1896, and Lenore was similarly baptized. The family moved to Bonn in 1921, and the street on which they lived would be renamed Hausdorffstrasse in 1949.

The anti-Semitism that had blocked Hausdorff's promotion in the infantry and threatened to prevent his promotion at Leipzig University continued to plague him throughout his lifetime. For instance, a young professor whose appointment Hausdorff had supported in 1926 became openly anti-Semitic in 1933, repudiating any former contact with Jews and refusing to join the rest of the faculty in attending seminars given by Jewish mathematicians. Some of Hausdorff's Jewish friends left Germany to escape the persecution; others whose emigration was thwarted committed suicide.

Suicide was a topic often addressed by Nietzsche; consequently, it had been a subject for reflection by Hausdorff. *Zarathustra* advocated "voluntary death" as a consummation of life for the noble man. "Death and Return," an 1899 essay by Mongré, broached the subject in the form of a letter to a fictitious, depressed friend. In it, the author advised that "this final remedy really helps, that it does not [merely] plunge one into a futile expense for morphine or revolver cartridges." Apparently, Hausdorff viewed suicide as an effective impediment to the Nazis' strategy of destroying their victims' human dignity: in death, the individual had no future, but his past was indestructible.

The infamous pogrom of November 1938 called Kristallnacht, in which 20,000 Jews were arrested and gov-

ernment-sanctioned attacks resulted in 25 million marks' worth of damage to hundreds of Jewish homes, shops, and synagogues, occurred the day after Hausdorff's seventieth birthday. Charlotte Hausdorff and her sister, Edith Pappenheim (who had come to live with them a few months earlier), tried to bolster Hausdorff's spirits. He continued to work on his mathematics, but he put his writings into storage rather than publishing them.

In mid–January of 1942, the Hausdorffs were ordered to report to an internment camp located at a former monastery; this would probably be followed by deportation to a concentration camp. After organizing their affairs and leaving property disposal and cremation instructions with trusted friends, Hausdorff, his wife, and her sister committed suicide on January 26, 1942, by taking an overdose of sedatives.

On January 25, 1980, a memorial plaque honoring Hausdorff was placed at the entrance of the Mathematical Institute at the University of Bonn. In 1992, an exhibition of photographs and personal, literary, and mathematical documents was held at the University of Bonn, commemorating the fiftieth anniversary of Hausdorff's death.

HEAWOOD, PERCY JOHN (1861-1955)
English mathematician

Percy John Heawood is most well known for his work on the **four color map theorem** on which he published extensively during his life. His other mathematical interests were chiefly in **geometry**, approximation theory, continuous **fractions**, and **quadratic equations**.

P. J. Heawood was born in Ipswich in Southern England in 1861. After initial schooling at Ipswich he was awarded a scholarship to study at Oxford University in 1880. Whilst at Oxford Heawood received several prizes including both Junior and Senior Mathematical Scholarships (1882 and 1886 respectively) and the Lady Herschell prize in 1886. The following year Heawood took up his life long post as lecturer in mathematics at Durham University, being given a chair in 1911. 1890 saw Haewood publish the first of many papers on four color map theorem. Heawood was regarded as an eccentric in the circles in which he mixed. His only passion outside of mathematics was Durham Castle (a Norman castle in the center of Durham City in the North east of England) which in 1928 was found to be in a dangerous state of repair. Haewood raised the vast majority of the money required to correct the problems with the foundations and for his efforts he was awarded an OBE in 1939 (Order of the British Empire). 1939 also saw the retirement of Heawood although he remained active in research and continued to publish many papers, his last being in 1949, some six years before his death at the age of 93.

See also Four color map theorem; Quadratic equations

HEINE, HEINRICH EDUARD (1821-1881)

German number theorist

Heinrich Eduard Heine was one of the most productive mathematical writers in 19th-century Germany, having published approximately 50 papers on a wide range of topics. His most notable achievements in mathematics were the formulation of the concept of uniform continuity and the Heine–Borel **theorem**.

Born in Berlin, Heine was the eighth child in a family of nine. His father, Karl, was a banker. Henriette Märtens, his mother, stayed home and tended to the children. Heine was privately schooled at home before enrolling in the Friedrichswerdersche Gymnasium, and in 1838 graduated from the Köllnische Gymnasuim in Berlin. He attended Göttingen University and often sat in on the lectures of **Karl Gauss** and Moritz Stern. In 1842, Heine received his Ph.D. in mathematics from Berlin University, where he was a student of **Peter Dirichlet**. He taught at Bonn University as a *privat-dozent,* or unpaid lecturer, and as professor before moving to Halle University, where he remained throughout his career. Heine held the office of rector for the university in 1864–1865.

Heine is most noted for his work regarding the Heine–Borel Theorem, which is defined as the following by Carl Boyer in his *A History of Mathematics*: "If a closed set of points on a line can be covered by set intervals so that every point of the set is an interior point of at least one of the intervals, then there exists a finite number of intervals with this covering property." Heine formulated the notion of uniform **continuity** and proved its existence in continuous **functions**. What has been noted often is that Heine's was the essential discovery and **Émile Borel**'s reduction of uniform continuity to the covering property was secondary. Heine also studied Bessel functions, Lamé functions, and spherical functions, also known as Legendre **polynomials**, and in 1861 he published his most influential work, *Handbuch der Kugelfunctionen*, which was considered the authoritative text on spherical functions well into the turn of the 20th century.

In 1850, Heine married Sophie Wolff and had four children. He was an active member of the Prussian Academy of Sciences as well as a member of the Göttingen Gesellschaft der Wissenschaften. In 1875, Heine declined the offer to chair the mathematics department at Göttingen, but received the Gauss Medal in 1877. Heine died in 1881 in Halle.

HEISENBERG, WERNER KARL (1901-1976)

German physicist

Nobel prize winning physicist Werner Heisenberg was born in 1901, the second child of Annie and August Heisenberg. Heisenberg's father, August, was a professor of Greek languages at the University of Munich and a teacher at a secondary school, and as such, encouraged both his sons to excel in their academic and musical studies. Heisenberg completed his

Werner Heisenberg

early schooling at the Maximillan Gymnasium in Munich, and entered the University of Munich as a physics students in 1920 at the age of 19. After earning his doctorate, he moved to the University of Göttingen, where he served as an assistant to physicist Max Born. In 1924, Heisenberg worked with Niels Bohr at the University of Copenhagen before returning to Göttingen in 1925.

In 1925 in a paper entitled *"Uber quantentheoretische Umdeutung kinematischer und mechanischer Beziehungen"* ("About the Quantum-Theoretical Reinterpretation of Kinetic and Mechanical Relationships"), Heisenberg redefined Bohr's **model** of the planetary atom and paved the way for the development of **quantum mechanics** itself. Heisenberg departed from Bohr's idea of invisible particles moving in unseen predefined paths, instead focusing on a theory that expressed the physical variables in atomic systems as abstract arrays of numbers, or matrices. These matrices, or **matrix equations**, observed the same rules as matrix **algebra**. Heisenberg, Born, and colleague Pascual Jordan went on to publish *The Physical Principles of Quantum Theory*, a paper detailing these new matrix mechanics, in 1928. It was for this groundbreaking work in quantum physics and the later applications of his theory in discovering certain forms of hydrogen that Heisenberg was awarded the Nobel Prize for Physics in 1932.

In 1926, Heisenberg returned to the University of Copenhagen to once again study under **Neils Bohr**. The following year, he was appointed the Professor of Theoretical Physics at the University of Leipzig. It was here that Heisenberg published *"On the perceptual content of quantum theoretical kinematics and mechanics,"* which outlined his famed uncertainty principle, or indeterminacy principle of matrix mechanics.

Heisenberg moved to Berlin in 1939 shortly after the outbreak of World War II, after being called on by the Army Ordnance Office to develop a nuclear fission program with a number of his physicist colleagues (called "the uranium club"). While working on the nuclear program, Heisenberg was appointed Professor of Physics at the University of Berlin and Director of the Kaiser Wilhelm Institute for Physics in 1942. Shortly after Germany's surrender in 1945, he was interned at an allied prisoner-of-war camp in Versailles for his role in the German war effort, and then moved to England for security reasons. However, the British viewed Heisenberg and his colleagues, who had never joined the Nazi party, as a crucial part of their long-term plan for an economic and scientific revival of what was eventually to be an autonomous West German state, and in 1946, the scientists were returned to Germany, although they remained English detainees. Heisenberg was reunited with his family later that year and with allied approval, was named as the head of the Max Planck Institute for Physics and Astrophysics in Göttingen. In 1958, both Heisenberg and the Institute relocated to Munich, where he would spend the remainder of his life. He succumbed to kidney and gall bladder cancer in 1976 at the age of 75.

See also Matrix; Quantum mechanics

HERMITE, CHARLES (1822-1901)

French mathematician

Charles Hermite was one of the founders of analytic **number theory**. This discipline uses the techniques of **analysis** (the **calculus**) to handle questions about positive **whole numbers**. Hermite is also remembered for having shown that one of the central constants of mathematics, **e**, the base of natural **logarithms**, belongs in the class of **transcendental numbers**.

The son of Ferdinand Hermite and Madeleine Lallemand, Hermite was born on Christmas Eve, 1822. His ancestry was both French and German and the town Dieuze, Hermite's birthplace, was at one time claimed by both France and Germany. Nevertheless, Hermite considered himself French all his life and became one of the mainstays of the French academic establishment.

Hermite attended the Collège Henri IV and proceeded from there to the Collège Louis-le-Grand, where he was taught mathematics by the same instructor who had supervised the work of the ill-fated French genius **Èvariste Galois**. When Hermite decided to continue his studies at the Ècole Polytechnique, he was admitted 68th in his class, thanks to his

having neglected **geometry**. Throughout his life Hermite had a dislike for examinations and preferred to pick up material spontaneously rather than under the pressure of a deadline. Hermite enjoyed corresponding with the best mathematicians of Europe, including **Karl Jacobi**, and some of the material in Hermite's letters is remarkably sophisticated. In particular, Hermite generalized a result of **Niels Abel** that applied elliptic **functions** to the class of hyperelliptic functions as well.

Hermite's family life reflected his increasingly central position in the French mathematical establishment. His wife was the sister of the mathematician Joseph Bertrand, and one of his daughters married the eminent analyst Emile Picard (who was to edit Hermite's works after his father-in-law's death). From the time he became admissions examiner at the Ècole Polytechnique in 1848, Hermite devoted much of his effort to working with students at every level. In addition to Picard, his other distinguished students included **Henri Poincare**, **Camille Jordan**, and Paul Painleve. This record attests to his eagerness in welcoming students as colleagues. **Emile Borel** is said to have remarked that no one made people love mathematics so deeply as Hermite did.

On a professional level, Hermite accomplished some formidable tasks with the analytic apparatus which he had mastered. The solution of the general quadratic (second-degree) equation had been known since ancient times. Solutions to the cubic and **quartic equations** had been developed during the Italian Renaissance. When Galois showed that ordinary algebraic methods could not solve the general quintic (fifth-degree) equation, the subject appeared to have reached a dead end. Using once again the techniques of elliptic functions, Hermite showed that fifth-degree equations could be solved after all.

The single result for which Hermite is best known was the transcendence of e. A number is said to be algebraic if it is the solution of a polynomial equation with integer coefficients. For example, the **square root** of 2 is algebraic, since it is a solution of the equation x+b2-2=0. A real number that is not algebraic is called transcendental. The French mathematician **Joseph Liouville** had shown that there were transcendental numbers but no familiar examples were known. Hermite was able to show that e could not be written as the solution of a polynomial equation and therefore had to be transcendental. His technique was used shortly thereafter to show that π was also transcendental, although Hermite does not seem to have recognized just how useful his technique was.

After his appointment as professor analysis at both the Ècole Polytechnique and the Sorbonne, Hermite took to writing textbooks that were widely used and appreciated. Although he resigned his chair at the Ècole Polytechnique after only seven years in 1876, he remained at the Sorbonne for another 21 years. Hermite attached a great value to insight and did not include much rigor in his teaching of elementary material. If his papers suffer from a fault, it is the occasional tendency to allow the details to get in the way of the overall picture. A large number of his ideas were developed by others and the complex generalization of quadratic forms named for him proved to be central in the formulation of **quantum mechanics**.

Charles Hermite

At the time of his 70th birthday, the adulation Hermite received from across Europe attests to his reputation as an elder mathematical statesman. His interests were never narrow, and he was awarded an impressive collection of decorations both at home and abroad. Hermite died on January 14, 1901, leaving a solid basis of mathematics and an unmatched collection of students to carry on his work.

HERO OF ALEXANDRIA (CA. 65-125)

Greek geometer and engineer

Hero (or Heron) of Alexandria was a Greek mathematician and engineer whose major contributions to mathematics were Hero's formulaand the first approximation in Greece of a number's **square root**. He also wrote a number of books on mathematics, including *Metrica*, is a treatise on **geometry**. While much of his writings in mathematics and **mechanics** stem from earlier authors, Hero is credited as being one of the earliest and most comprehensive and detailed recorders of ancient technology, especially through such works as *Penumatica*and *Automata*. Hero also designed many mechanical instruments, which earned him the name "the mechanic" or "the machine man."

Other than his writings, nothing is known about the life of Hero, who is sometimes referred to as Heron. The date of his birth is unknown, and estimates of when he lived vary by 150 years. However, Hero mentions an eclipse of the moon visible from Alexandria in his book *Dioptra*. Modern astronomers date this eclipse as having occurred in 62 A.D., thus providing the major clue as to the era when Hero lived. Scholars agree that Hero probably lived and worked in Alexandria. However, there is some question as to his nationality, which may have been Egyptian.

Despite the lack of historical records on Hero's life, the breadth of his writings on mathematics and mechanics leave little doubt that he was well educated. Hero was strongly influenced by the writings of Ctesibius of Alexandria and may even have been a student of the ancient mechanical engineer. His works draw on a wide range of sources, including, Greek, Latin, and Egyptian. Unlike most of his contemporaries, Hero makes no mention of working for a Roman patron.

Hero's writings in mathematics and especially in mechanics reveal that he was practical by nature, often using ingenious means to attain his goal, like his design for a steam engine, catapults for war, and various machines for lifting that used compound pulleys and winches. Hero was also precise in dictating the types of materials to be used to make the machine function properly. Interestingly, Hero designed several mechanical devices to simulate "temple miracles," including a device attached to the temple door which made a trumpet play when the door was opened.

Hero's background in mechanical engineering is clearly evident in his practical rather than theoretical approach to mathematics and geometry. Although credited with the first approximations of square **roots** and his famous formula for calculating the **area** of a **triangle**, Hero's primary contribution to geometry stem from a series of treatises in which he freely incorporated the writings and findings of others to compile a coherent body of work on the subject.

The most famous of these works is *Metrica*, which consists of three books focusing on the calculation of areas and volumes and their **division**. This book was lost until the last century, and scholars knew of its existence only through a 6 A.D. commentary by Eutocius. Then, in 1894, historian Paul Tannery discovered a fragment of the book in Paris. A completed copy was found by R. Schöne in Constantinople (Istanbul) in 1896. Book I of *Metrica* contains the famous **Hero's formula**, which he used to calculate the area of a triangle when the sides are given and provided surveyors of his day with a formula for determining the area of land lots. Like most of Hero's work, this formula may came from an earlier source. Although an ancient Islamic manuscript credit's **Archimedes** with developing the famous equation, no writings of Archimedes are known to contain the formula.

Hero's other works include *Definitions*, basically a catalogue of geometrical terms; *Geometrica*, an introduction to geometry, and *Sterometrica*, which focuses on **solid geometry** for spheres, cubes, pyramids, and other figures. He is also believed to have written a commentary on the famous Greek mathematician **Euclid**. Hero's emphasis on the practical use of geometry is evident in the types of problems he tackles. For example, he provides a method for calculating a theater's seating capacity and for determining the number of jars that could be stored in a ship. In the theoretical realm of mathematics, Hero is credited as the first Greek mathematician to use systematic geometrical terminology and symbols. Although the Babylonians had developed a formula for approximating square roots nearly 2,000 years before Hero, he was the first Greek to develop methods for finding approximate numerical square and cube roots.

Hero's contributions to the field of mechanics are wide ranging, and he achieved considerable fame in his own day for some of his inventions. For the most part, these inventions focused on the practical, like keeping time with a water clock and developing a compressed air catapult for war. His most famous mechanical design was for the aeolipile, which used steam to rotate a **sphere** and has been compared to the modern–day jet engine.

Much of Hero's fame today lies in the fact that most of his treatises on mechanics have survived throughout the centuries. As a result, he is considered the definitive source on ancient Greek and Roman technology. His design for the aeolipile appear in his *Penumatica*, which focuses on designs for machines powered by compressed air, siphons, and steam pressure. Although the treatise contains some theoretical components, the most important aspect of the book is Hero's ability to combine different schools of mechanical thought into a cohesive treatise. In *Mechanica*, Hero focuses on basic mechanics and their application, including the levers, pulleys, screws, and various tools. Hero was also interested in mechanical "gadgets," primarily used to produce "miracles" in a religious context. He describes these devices in both *Pneumatica* and *Automata*, which includes designs for making miniature mechanized puppet theaters. A book on machines for lifting heavy objects, *Baroulkos*, is also lost.

The wide range of Hero's interests are evident in his other works, which include *Dioptra*, at treatise on surveying. In *Catoptrica*, Hero discusses mirrors, including the theory of refraction and the various kinds of mirrors, such as flat, concave, and complex. He also wrote lost works on topics like large and hand catapults (*Baelopoeica* and *Cheiroballistra*), time keeping devices, and vault construction.

Hero's stature as a historical figure in mathematics and mechanics has grown with the rapid advance of technology in the last two centuries. Hero was more than a mere chronicler of ancient devices and how they should be built, he also exhibit scrupulous attention as to their construction and the materials to be used. Although Hero never built many of his models, they presented interesting design problems that grew in importance with the industrial revolution.

Hero is not without his critics for a number of reasons. First and foremost, his works contain some notable errors, primarily in the area of mathematics. Hero is also known to have gathered much of his knowledge from previous writers. Despite these criticisms, Hero revealed an advanced understanding of harnessing power, especially in the form of wind and steam. As a scholar, he conducted comprehensive research

and took a systematic approach to revealing the basic of many useful devices, including a coin–operated machine. As with his birth, the date of Hero's death is unknown.

HERO'S FORMULA

Hero's formula, also called Heron's formula, relates the **area** of a **triangle** to the measures of its three sides. Allowing a, b, and c to denote the lengths of the sides of a triangle, s, the semi-perimeter of the triangle, becomes $s = (a + b + c)/2$. Hero's formula states simply that the area of a triangle, A, can be expressed as $A = $ sqrt $[s(s - a)(s - b)(s - c)]$. As a result, the area of the triangle can be determined without knowing its perpendicular height, also known as the triangle's **altitude**.

Accredited to the Greek geometer **Hero**, the formula is derived in Hero's most important manuscript, *Metrica*. Hero's work was discovered in fragmentary form in 1894 and recovered fully in 1896. Recent scholarship, however, suggests that **Archimedes** (c. 212 BC) may have previously derived the formula.

Although the formula itself is deceptively simple, Hero's original **proof** is both clever and arduous, utilizing the properties of inscribed quadrilaterals (cyclic quadrilaterals) and right triangles. Today, there are numerous elegant algebraic, geometric, and trigonometric proofs for Hero's formula.

One geometric proof of Hero's formula involves inscribing a **circle** inside a triangle, and then summing the areas of the three triangles formed by connecting each vertex of the triangle to the center of the circle. Since radii of the inscribed circle are perpendicular to the points of tangency, segment extensions and properties of similar triangles can be used to complete the proof.

Trigonometric and algebraic proofs of Hero's formula generally use the law of cosines, cos $C = (a^2 + b^2 - c^2)/2ab$. One can then solve for sin C and use the identity that the area of a triangle is equal to $(1/2) * a * b * $ sin C to complete the proof.

Hero's formula is particularly useful when attempting to determine the area of a triangle for which the length of the sides—but not the perpendicular height—is known, making it ideal for use by surveyors and natural scientists collecting data in the field, such as ecologists and wildlife experts. Indeed, Hero noted the appropriateness of this formula for surveying land in his work *Dioptra* and even developed an instrument called the diopter for the same purpose. Contemporary surveyors use a device very similar-and based on the same mathematical principles—as Hero's diopter.

The side-side-side **theorem** from Euclidean **geometry** states that the triangle is uniquely determined by its side lengths—that is, if a, b, and c are the lengths of a triangle's sides, there exists exactly one triangle that can be composed from these segments, and that all other triangles composed of equal segments will be congruent to the original. This uniqueness provides that *any* information about the triangle—area, inradius (the radius of the circle inscribed by a triangle), circumradius (the radius of the circle inscribing the triangle), etc.—can be determined by knowing only the lengths of its sides. However, the "synthetic" proof of the side-side-side theorem (that is, a proof based on **deduction** from geometric axioms) does not suit the *algebraic* needs of a surveyor who wants to actually compute the area. Thus Hero's formula, though it appears more complicated than the side-side-side theorem, fills a practical gap that the other cannot.

Additionally, Hero's formula can be used to solve for the perpendicular height(s) of a triangle given only the measures of its sides through simple algebraic manipulation, which allows one to easily determine the shortest **distance** from the vertex of a triangular area to its base. Applying Hero's formula to the special case of a right triangle leads to the **Pythagorean Theorem** of $c^2 = a^2 + b^2$ (that is, the **square** of the **hypotenuse** of a right triangle is equal to the sum of the squares of its sides). The Pythagorean Theorem, then, can be considered a degenerate case of Hero's formula.

See also Archimedes; Pythagorean theorem; Perimeter; Triangle

HILBERT, DAVID (1862-1943)

German number theorist

By the end of his career, David Hilbert was the best known mathematician in the world, as well as the most influential. His contributions did not merely affect but decisively altered the directions taken in many fields. In some ways, however, his career ended in disappointment; he had inherited one of the great mathematical centers for research and teaching, but from his retirement he had to watch its glory disappear under the ideological onslaught from the Nazi government. Nevertheless, the heritage of the contributions he made to mathematics, as well as the students he trained, has outlasted the disruptions of World War II.

Hilbert was born in Wehlau, near Königsberg, on January 23, 1862. His family was staunchly Protestant, although Hilbert himself was later to leave the church in which he was baptized. Otto Hilbert, his father, was a lawyer of social standing in the society around Königsberg, and his mother's family name was Erdtmann. The name "David" ran in the family—a fact Hilbert had subsequently to verify to the Nazi regime, which suspected that anyone with the name was of Jewish ancestry. Hilbert's early education was in Königsberg, which he would always consider his spiritual home.

In 1880, Hilbert entered the University of Königsberg, where he received his Ph.D. in 1885. By the next year he had become a privatdozent, and by 1892 Hilbert had been appointed to the equivalent of an assistant professorship at Königsberg, rising in the ranks to a professorship the next year. In 1895 he took a chair at Göttingen, where he remained until his retirement. As this rapid progress attests, Hilbert knew enough about academic politics to advance through the complexities of the German system. In this he had the guidance of a mathematician with great political skills, named

David Hilbert

Felix Klein, who had devoted much of his life to building the University of Göttingen into the world's mathematical center.

Hilbert made his mathematical reputation on the strength of his research into **invariant** theory. The notion of an invariant had been created in the nineteenth century as an expression of something that remains the same under various sorts of transformations. As a simple instance, if all the coefficients in an equation are doubled, the solutions of the equation remain the same. A good deal of work had been done in classifying invariants and in trying to prove what sorts of invariants existed. The results were massive calculations, and books on invariant theory were made up of pages completely filled with symbols. Hilbert rendered most of that work obsolete by taking a path that did not require explicit calculation. Those who had been practicing invariant theory were taken aback by Hilbert's effrontery, and one of them described Hilbert's approach as "not mathematics, but theology." Invariant theory quickly disappeared from the center of mathematical interest, as Hilbert's work required some time to be absorbed. Only much later was the field reopened, as invariant theorists at last were ready to proceed from his calculations.

Perhaps the mathematician closest to Hilbert was **Hermann Minkowski**, two years younger than Hilbert but well-known at an even earlier age. At first, Hilbert's family did not approve of their friendship because Minkowski was the son of a Jewish rag merchant. Hilbert nonetheless kept in close contact with Minkowski, who had won a prize from the French Academy while still in his teens. Hilbert eventually managed to bring Minkowski to Göttingen.

In 1893 the German Mathematical Association appointed Hilbert and Minkowski to summarize the current state of the theory of numbers. **Number theory** was the oldest branch of mathematics, as it dealt with the properties of **whole numbers**. Much new work had been done by **Karl Friedrich Gauss**, and throughout the second half of the nineteenth century further progress had been made. The accessibility of the statement of problems in number theory made them attractive as an object of investigation, although the work of **Leopold Kronecker**, perhaps the biggest influence on Hilbert, was already well beyond what the nonmathematician could easily follow. Minkowski withdrew from the project, and in 1897 Hilbert submitted a report called *Der Zahlbericht* ("Number Report"). His work advanced the subject to a more technical level, one which has been maintained throughout the 20th century. Many of the results included still bear Hilbert's name, a tribute to the longevity of his influence.

The next direction in which Hilbert pursued his research was somewhat unexpected. After all his work in **algebra**, he began to look at the foundations of **geometry**. **Euclid** had already laid the foundations more than 2,000 years before, but detailed examination of some of Euclid's proofs revealed gaps in his presentation; he had made assumptions that were neither explicit nor justified by what had been proven earlier. In addition to problems posed by these gaps, another source for a new approach to geometry was the discovery during the 19th century of **non-euclidean geometries**. These shared some axioms or assumptions with Euclid's system, but differed in other respects. For example, in Euclidean geometry the sum of the angles of a **triangle** was equal to 180 degrees, while in non-euclidean geometries the sum could be greater or less. One of the reasons that it had taken so long to develop non-euclidean geometries was a general disagreement over what was true about geometrical objects. Hilbert felt that the only way to make progress was to be entirely explicit about each **proof** and not to trust to unspoken assumptions.

The safest way to avoid these assumptions was to regard the terms of the subject as defined only by the axioms in which they were used. Mathematicians might think they know what a "line" means and may be tempted to use this mental image in trying to prove a fact about lines. But that mental image could easily add something to the notion of line beyond what is given in the axioms. Taken in this way, the axioms can be considered as a kind of **definition** for the terms used in them. As Hilbert noted, the question of which theorems followed from which axioms had to be unchanged if all the technical terms of the subject (like point, line, or plane) were replaced by words from some other area. It was the form of the **axiom** that mattered, not what the objects were. This brought Hilbert into conflict with **Gottlob Frege**, one of the founders of mathematical logic. The controversy between Hilbert and Frege involved issues about the philosophy of mathematics that remained central to the field for much of the

20th century. In general, it can be claimed that Hilbert's perspective has been more helpful in enabling mathematicians to pursue the foundations of geometry.

One of the highlights of Hilbert's career came in 1900, when he was invited to address the International Congress of Mathematicians in Paris. His talk consisted of the statement of twenty-three problems, which he challenged his peers to solve in the 20th century. Although not all of the problems have proved to be of the same importance, by posing them Hilbert created an agenda that has been followed by many distinguished mathematicians. The first problem dealt with the question of how many **real numbers** there were, compared to the number of whole numbers. It was not resolved until 1963, when it was shown that the answer depends on which axioms are selected as the basis for the theory of **sets**. Hilbert's seventh problem dealt with the irrationality of certain real numbers, an area to which Hilbert himself had contributed. A number is rational if it is the ratio of two whole numbers, irrational if can not be so expressed, and transcendental if it is not the solution of a polynomial equation with whole number coefficients. Although Hilbert was not the first to prove that the numbers **e** and π (the base of natural **logarithms** and the ratio of the **circumference** of a **circle** to its **diameter**, respectively) were transcendental, he had simplified the proofs considerably. **A. O. Gelfond** solved his seventh problem, by establishing that a whole class of numbers was transcendental. Hilbert's tenth problem, on the solubility of certain **equations**, required much progress in mathematical logic before it could be solved. Entire books and conferences have been devoted to the state of the solutions to **Hilbert's problems**.

In addition to the study of the foundations of geometry, Hilbert turned to mathematical **analysis** and left a decisive imprint on this field as well. The previous generation of mathematicians had found defects in one of the standard principles from earlier in the century. Hilbert showed that the principle could be preserved, and he proceeded from there to make great progress in the study of integral equations. Hilbert has been credited with the creation of functional analysis, and although there was more foundational work to be done after him, his brief involvement in the area had once again altered it irrevocably.

In the fall of 1910, the second Bolyai Prize was awarded to Hilbert as a confirmation of his mathematical stature. The best-known images of him come from the period surrounding World War I. His distinctive appearance, from Panama hat to bearded chin, and his sharp voice set the tone for mathematics in Germany and the world. During the war he refused to sign the "Declaration to the Cultural World," which claimed that Germany was innocent of alleged war crimes. He was also willing to put mathematics before nationality, and he included an obituary for a French mathematician in the journal *Mathematische Annalen* (the showpiece of German mathematics) during the war. These acts made him unpopular with German nationalists. In the same way, he also took pleasure in fighting the academic establishment over the rights of **Emmy Noether**, who was both a woman and a Jew. Of the 69 students who wrote their theses under Hilbert (an enormous number for any time), there were several women.

After a brief dalliance with theoretical physics (an area Hilbert felt too important to leave to the physicists but to which he made few lasting contributions), Hilbert returned to questions of the philosophy of mathematics that had arisen earlier during his work on geometry. He was eager to pursue a program that could result in the establishment of secure foundations for mathematics. While he was willing to grant some importance to finite mathematics, he felt that the infinite required special treatment. In his account, called **formalism**, he set out to prove the consistency of mathematics. This enterprise put him in conflict with the other philosophies of mathematics most frequently advocated. He gave expression to his views most notably in an address "On the Infinite" in 1925; he was challenged by many, including **L. E. J. Brouwer** and **Hermann Weyl**, but it was the incompleteness **theorem** of the young Austrian mathematician **Kurt Gödel** which threatened the entire program Hilbert was pursuing. Certain narrow interpretations of formalism were put to rest by Gödel's work, and some of Hilbert's views were included among these. On the other hand, Hilbert's account of the foundations of mathematics changed during his career, and a good part of the work he did in the area has survived within post-Gödel logic under the heading of proof theory.

Hilbert married Kathe Jerosch in 1892. While he was willing to be casual with regard to his appearance, his wife helped prevent at least some of his sartorial excesses. She also proved a source of strength to Hilbert in his disappointments, one of which was their only son Franz, who never lived up to his father's expectations and probably suffered from a mental disorder. The last years of Hilbert's life were also darkened by the advent of National Socialism and its dire effects on Germany's intellectual community. Hilbert was proud of receiving honorary citizenship from Königsberg in 1930, the year of his retirement from Göttingen, but nothing could assuage his grief over the sequence of losses that the university suffered from the departure of many of its leading minds. Hilbert turned 71 in 1933, the year the Nazis came to power, and it was too late for him to look for a new home. Many of his students had found academic homes abroad, and nothing could rebuild the university in the face of racial laws and hatred of the intellect. It is a measure of the state of German mathematics and the political atmosphere in Göttingen that at Hilbert's death on February 14, 1943, no more than a dozen people attended his funeral.

Hilbert space

A Hilbert **space** is an infinite dimensional functional space. That is to say, it is a vector space composed of an infinite set of orthogonal **functions**. The orthogonality of a Hilbert space is defined by the integral over the appropriate **interval** of the product of two functions from the space, with a weighting **factor**. If this integral equals the **Kronecker delta**, the space is orthogonal; usually the weighting factor includes a normaliza-

tion factor as well. Because of the centrality of this relation, sometimes a Hilbert space is called a complete inner product space. Most proofs concerning Hilbert spaces use the orthogonality or orthonormality property extensively.

As a functional space, the Hilbert space can have linear operations performed on the space itself. The whole infinite space can thus be shifted. This result sometimes produces another Hilbert space and sometimes just produces an interesting group of functions, depending on which linear operator is used.

One of the most common uses of a Hilbert space is for **transformation** purposes. An arbitrary function can be expressed as the sum of components of a Hilbert space, with each component having a weighting factor overall. The famous Fourier transformation, essential to many parts of physics and engineering, is exactly this type of transformation. The orthogonality condition is used to determine generalized coefficients of the integrals of each Hilbert space function with the original function being transformed. While the **sine** and **cosine** functions used in the Fourier transformation are fairly common, polynomial expansions of functions and many other functional transformations useful in myriad applications also work this way.

See also David Hilbert; Orthogonal coordinate system

Hilbert's paradox

David Hilbert presented his paradox (also known as Hilbert's hotel paradox) as an explanation of **infinity** and an extension of Cantor's continuum hypothesis. In revealing his paradox, Hilbert posited a hotel with infinitely many rooms, of which a finite number n are full. If a finite number x of new guests show up at the hotel and request rooms, all the newcomers can be accommodated if each of the n guests are moved x rooms down the hall. Since the hotel has infinitely many rooms, all newcomers are welcome, and this adjustment holds true even if *all* rooms at the inn are full. Hilbert's paradox, however, broadens this idea to a hotel with infinitely many rooms, all of which are full. If a new group of guests (of which there are infinitely many) arrive, it would seem that they could not be accommodated, since all rooms are full. Suppose, however, that each of the current guests is moved to an even-numbered room; that is, the guest in room 1 is moved to room 2, the guest in room 2 is moved to room 4, the guest in room 3 is moved to room 6, and so on. Then, the infinitely many newcomers can be accommodated, with each assigned to an odd-numbered room. Since there are infinitely many odd-numbered rooms and infinitely many even-numbered rooms, there are sufficient rooms to house all guests requesting lodging, despite the fact that all rooms are full.

Hilbert's hotel paradox and similar questions pose challenges to mathematicians (and philosophers, for that matter) who focus on **set theory**. In particular, Hilbert's paradox naturally leads one to consider whether there exist different types of infinity, particularly different types of infinity that vary in size. The set of all even numbers, for example, seems intuitively "smaller" than the set of all **integers**, yet each is infinitely large and share the same **cardinal number**. Similarly, it seems intuitive that the set of **real numbers** contained within the closed **interval** 0 to 10 is "larger" than the set of real numbers contained within the closed interval 0 to 1. However, if one were to try to make a list of such numbers, one would quickly realize that each interval contains an infinitely large set of real numbers. Further, the set of all integers—i.e., 1, 2, 3 is well-ordered, whereas the set of all real numbers between 0 and 1 is not well-ordered. Each set is infinitely large and yet has fundamentally different characteristics.

Implications of Hilbert's paradox are particularly interesting for mathematical problems that involve taking **limits** or summing elements of a sequence. Although questions of whether infinities of different relative size still abound, it is clear that some sequences approach infinity "faster" than others—$2n^3$, for example, grows "faster" than $100n^2$. If such sequences represent the number of computations performed by a programming **algorithm**, comparing the rate of growth is an important factor in determining the relative efficiency of each.

Paradoxes such as Hilbert's have fascinated philosophers and mathematicians since ancient times; well-known cases of logical inconsistencies and contradictions provide opportunities to better (or more strictly) define mathematical properties, making them more useful. Hilbert's paradox illustrated the dilemma encountered in attempting to verify or negate Cantor's continuum hypothesis, the first of but 23 unsolved problems Hilbert posed to the International Congress of Mathematicians in the summer of 1900. Hilbert's paradox specifically challenged the accepted axioms of set theory. As a result of Hilbert's questioning, mathematicians revisited the fundamental axioms on which this branch of mathematics is based, redefining the underlying principles of the field.

Hilbert's discourse on Cantor's continuum hypothesis and the other 22 problems was a call to the community to better formalize the canons of both mathematics *and* logic, to make them more rigorous and thereby increase their utility. Hilbert expressly desired to formalize both reasoning and **deduction** within mathematics, constructing a formal methodology for solving mathematical problems. New paradoxes, however, were quickly encountered (and old paradoxes resurfaced) that negated the idea that human thought and logic can be reduced to a finite number of truths from which all other deduction and reasoning could be generated.

A general consensus formed that language itself—the basis for human communication and the expression of human ideas—was too ambiguous to allow for such **formalism**. Some areas of mathematics—arithmetic, for example—did indeed lend themselves nicely to formalism, but other ideas based on abstract concepts—such as **number theory** and set theory—could not be formalized exactly. More rigorous definitions, and therefore more precise axioms, were developed in many mathematics disciplines as a result of Hilbert's 23 problems, but undecided statements—(i.e., statements that cannot be

proven true or false using a formal system) remain today. Research on Hilbert's 23 problems, however, is hardly irrelevant, as more precise definitions and axioms are now applied to many areas of mathematics—including the development of algorithms that are then programmed into a computer—to make them more efficient.

See also Cardinal number; Continuum hypothesis; David Hilbert; Hilbert's problems; Hilbert's program; Number theory; Set theory; Tristam Shandy paradox; Voting paradox; Zeno's paradoxes

HILBERT'S PROBLEMS

In an address to the International Council of Mathematicians in 1900, **David Hilbert** (1863-1942), a professor of mathematics at the University of Goettingen, outlined 23 significant problems in mathematics for the community to research in the new century. The problems cross many areas of mathematics, including **set theory**, **arithmetic**, **geometry**, **group theory**, variable **calculus**, **algebra**, and others. Some problems were relatively straightforward and were thus quickly solved, but others were expansive and may never be completely resolved. Some mathematicians consider Hilbert's address as one of the most influential ever made in contemporary study of the field.

Beyond stating the problems Hilbert felt important for the mathematics community to address in the coming years, Hilbert expressed his address in the context of his particular philosophy about the field. Hilbert believed that the development of mathematics stemmed from two sources—practicality and reason—with each intertwined since ancient times. The idea of a straight line as the shortest **distance** between two points, for example, arose from practicality—taking a straight path or using a straight-edge measurement is often the most efficient course of action. Other developments in mathematics, however, are based on reason, either inductive or deductive. Such inductive or deductive reason may be based on observations of the surrounding environment or simply the logical consideration of an active mind. Each gain of new knowledge, however, grows with others, and it becomes difficult, if not impossible, to retrace absolutely the steps of discovery. Hilbert referred to this phenomena as the "ever-recurring interplay between thought and experience."

Fundamentally, Hilbert wished to provide greater precision to the reasoning and **deduction** of mathematics, developing basic axioms on which all other axioms could be based. He contended that all mathematical problems could be correctly solved by applying logical reasoning to a finite number of processes. In short, he demanded rigor in reasoning, with this rigor based on exactly formulated statements. Hilbert hoped that by making the logic of mathematics itself more exacting, fewer complications and paradoxes would be encountered in the study of mathematics. He also noted that forcing the rigor of mathematical proofs to an well-understood and developed chain of axioms, the simplest (and thus more understandable) proofs would be developed. Hilbert particularly opposed the idea that the axioms of geometry, through which greater understanding of the axioms of arithmetic have been achieved, are not disposed to such rigorous proofs, disputing that rigorous **proof** is confined to concepts of **analysis**, or perhaps arithmetic alone.

While such fundamental axioms were possible in some fields of mathematics—particularly arithmetic—more abstract fields of mathematics did not lend themselves easily to this approach. As a result, Hilbert's objectives to derive all areas of mathematics from a few fundamental principles are not likely to be realized. **Formalism**, it seems, has limits when apply to abstraction. Important strides, however, have been made on the governing axioms in many areas of mathematics, leading to the efficiency and simplicity for which Hilbert strived.

Briefly stated, Hilbert's 23 problems are as follows:

I. Problems related to the foundation of mathematical science

- The **cardinal number** of the continuum (based on Cantor's **continuum hypothesis**)
- The consistency of the axioms of arithmetic
- The equality equality of volumes of two tetrahedra with equal heights and bases
- The systematic review and development of the axioms of geometry
- The establishment of axioms for group theory independent of the assumption of differentiability of **functions**
- The application of mathematical rigor to the axioms of mechanical physics

II. Problems related to particular branches of mathematical science (arithmetic and algebra)

- The relationship of irrational **exponents** and **transcendental numbers**
- The distribution of **prime numbers** (based on the Riemann hypothesis)

III. Problems related particular branches of mathematical science (**number theory**)

- The extension of the law of reciprocity to the general case for any number field
- The development of a universal process for determining solutions to **Diophantine equations**
- The solution of **quadratic equations** with algebraic numerical coefficients

IV. Problems related particular branches of mathematical science (algebra and functions)

- The extension of Kroneker's **theorem** to any algebraic field
- The determination of whether functions of two arguments can be used to solve an arbitrary 7th-order equation
- The determination of finiteness of complete systems of form with the theory of algebraic invariants
- Application of rigorous proof to Schubert's enumerative geometry

- The **topology** of algebraic surfaces and curves
- The representation of definite forms by squares of forms

V. Problems related particular branches of mathematical science (geometry)

- The use of congruent polyhedra to construct spaces
- The relationship of analytic solutions to the problems of variable calculus

VI. Problems related particular branches of mathematical science (analytic functions)

- The generalization of solutions of boundary value problems
- The existence of a Fuchsian linear differential equation, given singularities and a monodromic group
- The use of automorphic functions to uniformize nonanalytic algebraic functions
- The extension of calculus of variations

See also Algebra; Arithmetic; Formalism; Geometry; Group theory; David Hilbert; Hilbert's paradox; Hilbert's program; Lie group; Number theory; Set theory

HILBERT'S PROGRAM

David Hilbert (1862-1943), a mathematician from the University of Göttingen, launched his program during an address to the International Congress of Mathematicians in the summer of 1900. Hilbert contended that all mathematical principles, or axioms, should be derived from first-order statements, with the logic of **deduction** and reasoning leading to final conclusions regarding the problem at hand. Using this methodology, the foundations of mathematics could be developed exactly and with certainty; these fundamental axioms would form the basis for all other research into the varied branches of mathematics. In his address, Hilbert famously outlined 23 "unsolved problems" that he characterized as both interesting and important for further study in the coming century. Hilbert hoped that his "program" of **formalism** would be applied to each.

Hilbert's program, as his initiative as come to be called, was a call to the mathematics community to formalize the basic tenets of the science. Hilbert viewed the new century with great optimism for the study of mathematics, feeling that the 23 unsolved problems demonstrated both the vitality of the field and the opportunity for growth and innovation within it. When a field of study ceases to hold an abundance of interesting, albeit difficult problems, Hilbert felt that the extinction of independent development in the science itself was near. Bernoulli's problem of the "line of quickest descent" (in its simplest form, the idea that an object, when traveling under its own power down a hill or similar surface, will follow the steepest path of travel and thus attain the maximum speed—or rate of change of **distance** over time—possible) was a difficult problem of its time, yet its usefulness cannot be questioned.

Although his "call to action" involved many areas of mathematics, Hilbert particularly emphasized the relationship of **arithmetic** axioms to geometric principles and contended that the former could be used to formalize the latter, or at the very least, the formalization of arithmetic could serve as a **model** for the formalization of **geometry**. He asserted that development of a mathematical concept would not be complete until it could be easily explained to a layperson—that is, that mathematics should be clear and concise to the average reader. Hilbert further stated that many problems of mathematics were born of the intertwined relationship between pure reason and the reality of observation and intuition. Both pure reason and the observable world pose questions for study; it is through the examination of each that gains in knowledge are made. While Hilbert felt that both reason and phenomena were crucial instruments of learning, he contended that care must be taken that axioms, and thereby proofs, are based on progressive logic. **Proof** by exhaustion should be avoided at all costs, and problems requiring such proof such be simplified by revisiting from new vantage angles.

Perhaps most importantly, Hilbert declared that the correct solution to any mathematical problem should be solved with a finite number of steps. By solving problems with a finite number of processes, Hilbert asserted that rigor in reasoning would be maintained and thus deduction would apply. To Hilbert, this methodology ensured not only completeness of a concept, but also ensured humanity's philosophical need to understand the concept was attained. New problems, Hilbert felt, must be solved with extensions of axioms developed from old problems, ensuring a continuity of reason and intellectual development.

Hilbert was most ardently concerned with the concept of **infinity**, especially as it related to the continuum hypothesis. Indeed, the very first of his 23 unsolved problems posed to the mathematics community was based on verifying the concepts of infinity as defined in Cantor's axioms for **set theory**. At the time of Hilbert's address, the concept of physics and mathematics were reaching significant points of departure; experiments in physics were demonstrating that energy was not infinite; that matter consisted of a finite number of sub-atomic particles and could not be infinitely divided; and that time itself, while extending from an unfathomably distant point in the past and an incomprehensibly distant point in the future, may not be infinite. It was Hilbert's hope that by perfecting the axioms of *finite* mathematics that the formal system necessary to define *infinite* mathematics. Thus, Hilbert contended, the axioms of finite mathematics, once robustly stated, could be extended to prove the consistency of infinite mathematics.

Sometimes called a "glorious failure," Hilbert's program, particularly as applied to infinity, did not meet Hilbert's expectations. In 1938, **Gödel** demonstrated that the **continuum hypothesis** is consistent with **Zermelo-Fraenkel set theory**; in 1963, **Paul Cohen** demonstrated that the negation of the continuum hypothesis is also consistent. Using current axioms of set theory, the question is undecided. Hilbert's program and its emphasis on rigor, however, can hardly be considered immaterial, as it still influences scholars' approach to investigations into contemporary mathematics.

See also Cantor, Georg Ferdinand Ludwig Philipp; Continuum hypothesis; Formalism; Kurt Friedrich Gödel; David Hilbert; Hilbert's paradox; Set theory; Zermelo-Fraenkel set theory

HIPPARCHUS OF RHODES (CA. 180 B.C.-CA. 125 B.C.)
Greek trigonometer and astronomer

Hipparchus of Rhodes was a renowned Greek astrologer whose mathematical computations to chart the sun, moon, and stars led to his being named the founder of **trigonometry**. Hipparchus is also considered the founder of Greek astronomy for his systematic approach to the discipline, which based it on a solid scientific foundation. In his book, the *Almagest*, the famous Greek astronomer and mathematician **Claudius Ptolemy** holds Hipparchus in the highest regard, referring to him as the "man who loved work and truth."

Details of Hipparchus' life remain lost. What little is known about him can be found in the writings of Strabo and Ptolemy. According to several ancient sources, Hipparchus was born in the city of Nicaea in a part of northwestern Asia Minor that is now Turkey. The era in which he lived has been determined primarily from the dates of his astronomical observations. In the *Almagest*, Ptolemy refers to Hipparchus making observations from Rhodes in Greece, indicating that he spent the latter part of his life there. According to some ancient references, Hipparchus also made astronomical observations from Alexandria, Egypt.

While Hipparchus the man remains a mystery, his reputation as a pioneer in astronomy and mathematics led to his image appearing on second– and third–century Nicaean coins showing a man contemplating a globe. The few surviving anecdotes concerning Hipparchus may be truth or legend. According to one anecdote, Hipparchus once caused a stir among fellow theater goers when he came in wearing a cloak to protect himself from a storm he had predicted. In another questionable anecdote handed down by Pliny, Hipparchus is credited with predicting eclipses of the sun and moon for 600 years.

Like many ancient Greeks, Hipparchus' interest in mathematics stemmed from his devotion to astronomical studies. Pliny reports that Hipparchus became the first Western astronomer to record the observation of a nova, or new star, probably around 133 or 134 B.C. Hipparchus discovered the star in the constellation Scorpio during one of his nightly observations of the skies. The sighting marked a phenomenal event in his time since the Greeks believed that all stars were fixed, with their position and number firmly established. The discovery led Hipparchus to create a catalogue of stars so future generations of astronomers could accurately determine the appearance of new stars and record changes in their positions and brightness.

Before Hipparchus, Greek astronomy was based primarily on observation and a few geometrical models. In Hipparchus' time, it was believed that the stars were located on a single **sphere** and revolved around the Earth. To create his catalogue of stars, Hipparchus developed the idea that their positions and **distance** could be determined by taking three stars at a time and forming a spherical **triangle**. Since **Euclid**'s mathematical theorems focused on plane triangles, Hipparchus developed his own special theorems for working with spherical triangles. These efforts led him to create a table of chords, or straight lines in a **circle**. Using these chords, Hipparchus developed a method for finding general solutions for trigonometrical problems and, in effect, became the founder of trigonometry.

Hipparchus is also credited with introducing a new unit of measurement, the degree, which is still used to measure arcs and angles, and proposed the idea of dividing a circle into 360 degrees. He may also have been the first to use letters to indicate the points of a triangle. Substituting mathematical models for mechanical ones, Hipparchus extended the work of **Apollonius of Perga** on epicyclesand eccentrics.

In the realm of astronomy, Hipparchus' most noted achievement is the discovery of the precession of the equinoxes. By comparing early astronomers' measurement of the positions of stars with his own measurements, Hipparchus found that the stars seemed to have shifted systematically in the same direction. He went on to establish that this phenomenon was due to a shift in the position of the equinoxes.

Hipparchus' other accomplishments include accurately establishing the distance to the moon and calculating the length of the year to within 6.5 minutes of the modern year. He also improved upon or invented several observational instruments, including one for measuring the diameters of the sun and the moon. Hipparchus may also have invented the plane astrolabe and stereographic **projection**. In the area of geography, which was his other major field of study, Hipparchus advocated the use latitudes and longitudes.

Although Hipparchus was reported to have written several works on astronomy and geography, his only surviving work is the *Commentary on the Phaenomena of Eudoxus and Aratus*. This three–book treatise criticizes earlier astronomers' estimates of the position of stars and constellations. Other reported works by Hipparchus include *Geography*, *On the Length of the Year*, *On the Displacement of the Solstitial and Equinoctial Points*, and *On Bodies Carried down by Their Weight*. He also wrote *Against the Geography of Eratosthenes*, attacking the ancient Greek geographer's works, including **Eratosthenes**' treatise *On the Measurement of the Earth*.

Fortunately, Hipparchus' ideas were saved and reported by those who followed. Ptolemy credits Hipparchus with providing much of the foundation for the *Almagest*, which was long considered the "Bible of Astronomy." Unquestionably, Hipparchus was an open–minded seeker of truth who helped establish astronomy and mathematics as evolving disciplines. Ironically, Ptolemy's *Almagest* was such a comprehensive achievement that it nearly obscured Hipparchus' achievements. If Ptolemy had not acknowledged his debt to this pioneering thinker, Hipparchus may have been relegated to footnote status in the history of Greek science and thought.

Hipparchus

HIPPOCRATES OF CHIOS (CA. 470 B.C.-CA. 410 B.C.)
Greek geometer

Hippocrates of Chios was a Greek merchant turned mathematician who wrote the first textbook on **geometry**. He is also noted for his efforts in elucidating the properties of circles and the quadrature of the lune. Despite turning to mathematics later in life, Hippocrates, who was also interested in astronomy, has been called the greatest mathematician of the fifth century B.C.

Hippocrates was born on the island of Chios but little else is known about his life. He is not to be confused with the more famous Hippocrates of Cos, who was born in the same century and is known as the father of medicine. According to information passed down by **Aristotle**, Proclus, Simplicius, and others, Hippocrates' first calling was as a merchant. However, unlike the famous Greek mathematician **Thales**, who made a fortune in commerce, Hippocrates' endeavors in trade led to his financial ruin.

It is speculated that Hippocrates was a thriving merchant until an unfortunate turn of events. According to one legend, he lost most of his wares when attacked by Athenian pirates near Byzantium. Another version points to dishonest customs men who threatened him with imprisonment and then bilked him of almost everything he owned. Hippocrates is said to have traveled to Athens to recoup his losses in a court of law. Required to stay in Athens for an extended period of time, Hippocrates began to attend lectures on **mathematics and philosophy**. He eventually became proficient enough in mathematics to open his own school. Although Aristotle characterized Hippocrates as a "competent geometer," he also—perhaps unfairly—said Hippocrates lost his fortune because he was "stupid and lacking in sense."

Hippocrates is believed to have been greatly influenced by the Pythagorean school of mathematics, named after the famed Greek mathematician and philosopher **Pythagoras**. Whether he came under this influence in his home of Chios, which is close to Samos where Pythagoras was born, or in Athens is debatable. Hippocrates' concept of proportion and his astronomical theories are both related to the Pythagorean school of thought.

Fortunately, Hippocrates' misfortunes in commerce had a silver lining. Although the Pythagoreans believed it was taboo to earn money from their knowledge, Hippocrates was reportedly allowed to establish a school in Athens because of his financial troubles. Hippocrates went on to write the first mathematical textbook, called the *Elements of Geometry*. This work precedes the better known *Elements* written by **Euclid** more than a century later.

Although Hippocrates' book is lost, it had a profound influence on the mathematicians who followed him. Through his pioneering book, Hippocrates was the first to develop geometrical theorems from axioms and postulates in a scientifically precise and logical manner. His book may also have contained the first written accounts of Pythagorean mathematics since the Pythagoreans themselves did not believe in written texts. Although Euclid's book was far superior in its approach to geometry and went on to become the most famous textbook of all time, Euclid most certainly based some of his work on that of his predecessor, including much of what appears in Books I and II of his *Elements*.

One of the most famous problems faced by ancient Greek mathematicians was doubling the cube, also called the Delian Problem. According to one legend, the Delian Problem arose when a Greek concluded that a typhoid plague was a scourge sent by the god Apollo, who was displeased with his altar. Apollo ordered a second altar built in his honor that would be double in size but have the same cubical form as the first altar. Mistakenly, the Athenians thought the problem was solved simply by doubling each of the old altar's edges. As the legend goes, the plague continued and the problem of doubling the cube became a preeminent mathematical problem in ancient Greece.

In his attempts to solve the problem of doubling the cube, Hippocrates used the method of reduction. Although **Plato** developed a method of reduction for philosophical prob-

lems, Proclus credits Hippocrates as the first to use such an approach in geometry. Basically, this method operated by altering a difficult problem into a simpler form, solving this simpler form of the problem, and then attempting to apply the solution to the more difficult problem. In the case of the Delian Problem, Hippocrates proposed that a cube could be doubled by finding the two **mean** proportionals (geometric means) between two given lines or between a number and its double. While Hippocrates never completely solved the problem of doubling the cube, others followed up on his directive and went on to develop several solutions to this ancient geometric puzzle.

Besides his book, Hippocrates' most noteworthy contribution to ancient mathematics was his quadrature of the lune, a figure bounded by two crescent-shaped arcs of unequal radii. Hippocrates' interest in the quadrature of the lune probably stemmed from his attempt to solve another popular problem of ancient Greece, namely the squaring of the **circle**. Hippocrates' based his work on the **theorem** that the areas of two circles are the same as the ratio of the squares of their **diameter**, or radii. According to some accounts, Hippocrates falsely claimed that his work on the quadratures of lunes led him to discover how to **square a circle**.

Among Hippocrates' other contributions to mathematics were geometrical solutions to quadratic **equations** and an early method of integration. Through his theorems on circles, Hippocrates may have also introduced the indirect method of **proof** to mathematics, also known as the *reductio ad absurdum*. In essence, this approach first assumes the opposite of what is wanted to be proved is true. By proving the opposite to be false, the alternative is then considered true.

Like many of his contemporaries, Hippocrates was also enamored by the heavens. Chios had long been a center of astronomical studies, and Hippocrates is believed to have formed his own theories concerning comets and the galaxy. In keeping with the Pythagorean view of the heavens, Hippocrates believed that there was only one comet, which was a planet that appeared at long intervals. He added the belief that the comet's tail was a type of mirage caused by the comet taking up moisture when it neared the sun. Some ancient commentators also say Hippocrates created a similar theory to explain the appearance of the galaxy.

Since Hippocrates turned to math and astronomy rather later in his life, it is noteworthy that he apparently attained great renown in a relatively short period of time. Much of his work is known only through commentaries by later mathematicians and historians, like Simplicius, who based his work on the *History of Geometry* by Eudemus. As a result, certain claims pertaining to Hippocrates are difficult to substantiate, including that he was the first to use letters in geometric figures and that he established the technical meaning of the word "power," which is now used in **algebra**. However, his reputation as an excellent geometer who influenced the course of mathematical thought is well-founded. Nothing is known about his death.

HISTORY OF MATHEMATICS

The earliest development of mathematics is closely tied to the development of counting and numbers. For thousands of years before there was a system for writing numbers, ancient people used a variety of methods to either estimate or guess at the amounts of things they dealt with in their daily lives. At some point, they began to invent ways of keeping a tally of their possessions, such as livestock or sacks of grain. The simple methods for doing this varied widely from one part of the world to another, from civilization to civilization, but inevitably they led to the development of numbers. The first we know about came from Sumeria (around the area now known as Iraq) about 3000 B.C. Number systems also varied from one area to another, just as the early Sumerian system was very different from what we use today, but even their early attempts to improve how they worked with quantities helped prepare the way for the development of one of the most important uses for numbers, mathematics.

The Sumerians, Babylonians and, the ancient Egyptians were the first mathematicians and, not by coincidence, the first highly skilled scientists. They took the use of numbers beyond the everyday basic tasks of counting or keeping track of flocks and grain, to trading with "bills of sale" instead of direct bartering for goods and services, and solving the practical problems of agriculture, building, and astronomy. At first, in some areas of the world, such as Greece, the use of mathematics, like many advanced skills developed by early civilizations, was only taught to the wealthy or powerful, but it eventually became clear that, in many was, civilization itself could benefit so much from the knowledge of mathematics that its use needed to be available to more than just the elite. The ability to create an accurate **calendar** in order to predict the seasonal cycles was vital to survival. As civilizations grew, whole economies based on trade and more complicated building practices also demanded the use of fairly advanced mathematics. The more complex their lives became, the more important were their abilities to use mathematics. The ability to do calculations had to grow with these needs.

About 3000 B.C., in China, the first **abacus** was invented and became the first "adding machine," allowing simple **arithmetic** calculations to be performed without the use of objects or fingers. About 1000 years later, the Babylonians used **multiplication** tables so by that time the first branch of mathematics, arithmetic, was being used. **addition** and **subtraction** had been used in many different ways for much longer, and multiplication had been carried out very early as well by combining symbolic representations of numbers called hyroglyphs. But the use of actual multiplication tables was a true advancement. The Babylonians had chosen to use the cycle of the moon's phases as the basic unit of their calendar. This means they had to use **fractions** in order to arrange the number of lunar cycles within a year. About 1675 B.C., the oldest mathematical record was written by a scribe named **Ahmes**. Among other things, this document, called the **Rhind papyrus**, tells how to use fractions to divide daily food rations.

It is clear, then, that by 1600 years B.C., arithmetic was being used for a number of practical purposes.

As these very basic aspects of mathematics came to be commonly used, other methods also began to benefit the people through the use of basic **geometry**. Literally the Greek word for "earth measuring," geometry was used in about 240 B.C. to actually measure the **circumference** of the earth with amazing accuracy. The astronomer who performed this feat was a Greek named **Eratosthenes of Cyrene**. Greece produced a number of great mathematical minds, including the philosopher and mathematician **Pythagoras** (c. 580-500 B.C.). He is famous for the **theorem** which carries his name to this day. It states that the sum of the squares of two sides of a right **triangle** is equal to the **square** of the third side. This seemingly simple concept and others developed during the centuries to come would advance mathematics into the future. But long before the time or Pythagoras, geometry was used in many ways. Perhaps the most famous and accurate use from the ancient world was the determination of the angles and heights of the sides for the pyramids of Egypt. The sides of the Great Pyramid are the same to within one inch and the angles to within a fraction of a degree. This took not only very accurate measurements, but also an accurate understanding of geometry.

Through astronomy and other sciences, mathematics began, very early to help people better understand their world and the universe. Complicated calculations had to be made in order to predict the **motion** of the sun and moon. Out of necessity, mathematics continued to develop because it was needed in order to survive, but also to increase understanding of the physical world. In about 330 B.C., the Greek mathematician, **Euclid of Alexandria** published his ideas on geometry and **number theory** in which he collected the works of others including Pythagoras, recording, for the first time the basic rules of mathematics as they were known until that time. Geometry students used Euclid's book as their mathematics text for more than 2000 years and his ideas still influence mathematicians today. Various aspects of arithmetic, **algebra**, geometry and **trigonometry** continued to be applied as to matters of daily life, commerce, architecture, and the sciences both old and new throughout this time, with ever greater complexity and accuracy. But in the 17th century, the first really dramatic addition to mathematics would come. Devised independently by two mathematicians, **Gottfried Wilhelm Leibniz** in Germany, and **Isaac Newton** in England, **calculus** provided new techniques for dealing with very small changes in **distance**, area, or times. This new form of mathematics allowed for the determination of areas and volumes of complex shapes, among other things. With calculus, scientists now strive to understand, calculate and create models for virtually everything both large and small, from the tiniest sub-atomic particle to the universe as a whole.

See also Babylonian mathematics; Egyptian mathematics

HOOKE, ROBERT (1635-1703)

English mathematician and scientist

Robert Hooke was a typical scientist of his times—his interests ranged widely over all areas of science. Some of his main contributions were in the fields of optics and **mechanics**. Hooke was instrumental in the thinking of **Isaac Newton** as he formulated his laws of planetary **motion**. Hooke is also responsible for Hooke's law of elasticity and he was the inventor of the conical pendulum.

Robert Hooke was born in Freshwater in England in 1635. His initial education was carried out at Westminster School in London and then from 1653 at Christ Church, Oxford. In 1655 Hooke was employed by **Robert Boyle** in the construction of an air pump. 1660 was the year Hooke discovered his law of elasticity. In 1663 Hooke was elected a Fellow of the Royal Society (London). In 1665 Hooke became professor of **geometry** at Gresham College, Oxford. In 1665 Hooke published *Micrographica* which was a catalogue of objects studied by Hooke under a microscope he had constructed himself. Using this machine Hooke was the first person to observe and recognise cells in living material. Although he was initially on friendly terms with Newton and other noted scientists of the day (he carried on a lengthy correspondence with Newton discussing the motion of heavenly bodies as a problem of dynamics) Hooke was jealous and vain. He accused Newton and **Christiaan Huygens** of stealing his work. In addition to the above Hooke also made numerous astronomical observations using a self made telescope, he was also assistant to Christopher Wren in the rebuilding of London after the great fire of 1666. Hooke also had several watches made to his own designs, as well as making the most accurate spring balance of its time. He also made a mathematical calculating machine and he invented a primitive type of telegraph machine. Hooke also recognised that fossils were petrified versions of once living organisms and that the fossil record was actually a record of changes that had occurred throughout the life of the planet. Hooke is perhaps best known for his contributions to biology in *Micrographica* rather than his extensive efforts in other fields. Hooke died in London at the age of 66 in 1703.

See also Christiaan Huygens; Mechanics; Isaac Newton

HORNER, WILLIAM GEORGE (1786-1837)

English mathematician

William George Horner is best known for what is now called Horner's method. Horner's method is a technique used for finding the **root** of a polynomial equation. His method was first published by in the Philosophical Transactions of The Royal Society in 1819 although it was independently arrived at by the Italian mathematician **Paolo Ruffini** in 1804. As a result of this Horner's method is also known as the Ruffini Horner method. In reality this technique was first published

some 500 years by Chinese mathematicians and was possibly known even earlier.

William George Horner was born in Bristol in south west England in 1786. After an initial education at Kingswood school, Bristol, Horner became an assistant teacher there in 1800. By 1804 Horner was headmaster of the school. At this point he was aged 18. In 1809 Horner founded his own school in Bath. In 1819 Horner submitted his method for solving algebraic **equations** to the Royal Society (London). Horner continued as headmaster of his own school until his death in 1837, aged 51.

HUBBARD, JOHN R. (1945-   )

American mathematician

John Hubbard is perhaps most well known for his early work on **Mandelbrot sets** in which he discovered the so called pictures in 1976. As part of this work Hubbard also had a great interest in **polynomials** and **exponential functions**. His more recent research work has concentrated on **analysis** and implementation of numerical **algorithm**s, **probability theory**, computing, and object orientated programming and database systems.

John R Hubbard attended Cornell University, the University of Rochester, the University of Michigan, and Penn State. From these he obtained a Bachelors degree and a doctorate in mathematics and a masters degree in computer science. Since 1983 Hubbard has been employed at the University of Richmond working on both mathematics and computer science. His outside interests include music (he performs in both the University and the faculty orchestras), genealogy, travel, history, and philately. Hubbard has a thriving graduate student program that produces many mathematical and computer science graduates.

See also Mandelbrot sets

HUSSERL, EDMUND (1859-1938)

German philosopher

Edmund Husserl was a German philosopher who founded a philosophical movement known as phenomenology, and explored the psychological basis and objective truths of mathematics.

Husserl was born in the Austrian town of Moravia to a middle-class merchant and his wife in 1859. After attending Gymnasium (or secondary school) in Vienna, Husserl entered University, attending first the University of Leipzig and then the University of Berlin to study both mathematics and science. In 1881, he transferred to the University of Vienna where he earned his doctorate, completing a dissertation on the **calculus** of variations. While in Berlin, he studied mathematics under the respected German mathematician **Karl Weierstrass**, and served as his assistant for a brief period before returning to Vienna to study philosophy.

In his first book, *Philosophie der Arithmetik* (*Philosophy of Arithematic*), Husserl outlines a psychological theory of mathematics that asserts that mathematical truths have objective validity regardless of an individual's subjective views towards them, a view that would change in later works of the philosopher. *The Philosophy of Arithematic* was published in 1891, while Husserl was a philosophy lecturer at the University of Halle, and combines his mathematical expertise with a growing passion for philosophy.

Husserl moved to the University of Göttingen in 1901, where he taught philosophy for 15 years until being appointed to a professorship at the University of Freiburg in 1916. It was at Göttingen that Husserl wrote his most influential philosophical work, *Ideas: A General Introduction to Pure Phenomenology* (1913), which is based on the idea that the meaning or essence of an event or object is not in the thing itself, but in the intentional consciousness of the person who perceives it. Husserl's concept of phenomenological reduction refers to a study of conscious reflection, of the mind itself and how it perceives real or imagined things.

Husserl remained at Freiburg for the remainder of his career, where he published several more philosophical works before retiring in 1928. He passed away in 1938 at the age of 69.

See also Mathematics and philosophy

HUYGENS, CHRISTIAAN (1629-1695)
Dutch astronomer and mathematical physicist

Christiaan Huygens is best known for his work in astronomy and physics, but he also did important work in mathematics. Although he advanced no new theories, Huygens made improvements to existing methods of calculation and applied them to solving problems in the natural sciences. Had he not been diverted from mathematics by astronomy and physics, he might have be one of history's greatest mathematicians. However, the modern world might be lacking the pendulum clock and the wave theory of **light**, Huygens' greatest accomplishments.

Born April 14, 1629, at The Hague in the Netherlands, Christiaan Huygens was the son of a family of diplomats. His grandfather, also named Christiaan, had been secretary to William the Silent and Prince Maurice, and his father, Constantjin, had been in service to Prince Frederic Henry and the House of Orange his whole life. Young Christiaan was expected to follow the family path (his brother Constantjin, like his father, served the House of Orange), but he was far too interested in the sciences.

Huygens studied law and mathematics at the University of Leiden and the Collegium Arausicum at Breda. There, he studied classical mathematics and the more modern techniques of **René Descartes** and others. At the end of these formal studies, Huygens was able to live at home from 1650 to 1666, thanks to support from his father. Those sixteen years at home were the most fruitful of his scientific career.

Christiaan Huygens

During this period, Huygens made a number of trips to Paris and London. On his first journey, to Paris in September 1655, he met the men who would become the core of the Académie Royale des Sciences, which was founded in 1666 (Huygens soon became a member). He traveled to Paris again and stayed from October 1660 to March 1661; he then spent two months in London, attending meetings at Gresham College and meeting some of England's great thinkers. Huygens eventually settled in Paris and lived there from 1666 to 1680, but political unrest and ill health made it essential for him to return to The Hague, where he died in 1695.

Huygens first studied mathematics, concentrating on determinations of quadraturesand cubatures. He published *Theoremata de quadratura hyperboles, ellipses et ciculi*; in it, he related the quadriture and the center of gravity of these various shapes. His next book, *De circuili magnitudine inventa*, appeared in 1654.

After Huygens heard about **Blaise Pascal**'s work in probability, he began to investigate it. He published *Tractatus de ratiociniis in alea ludo* (a book about the theory of probabilty) in 1657, and applied the theory to calculating life expectancy. It remained the only book on the subject until the 18th century.

Huygens was known as a Cartesian—that is, he followed the ideas of **René Descartes**. As such, he did not believe in the action of forces as **Isaac Newton** proposed, believing instead that there could be a mechanical explanation for everything. Huygens' great gift to others was demonstrating how mathematics could be applied to natural problems, and his work, *Holorogium oscillatorium*, shows the strength of this approach.

Huygen's first major achievement was the development of the pendulum clock. Years before, **Galileo** had observed the movement of a lamp in a church and realized that the time of each swing was nearly the same, no matter the extent of the swing. Despite his brilliance, Galileo never managed to develop a working pendulum, which would have led to more accurate timekeeping (of increasing importance in this era of scientific revolutions). Huygens realized that the circular arcing swings that Galileo had observed were not identical; rather, they were nearly so. For a pendulum to move in swings of equal time, its movement would have to follow a curve known as a cycloid. Huygens designed a clock with such a pendulum that had weights attached near its fulcrum that would make it move in this nearly circular, cyloidian arc; he then attached the pendulum to a clock's works and used a system of falling weights that would keep the pendulum moving despite friction and air resistance. The mechanisms of the modern grandfather clock have changed little from Huygens' first one, which he presented to the Dutch estates general. He described the clock in his work *Horologium*, published in 1658.

More important than this invention was Huygens' development of the wave theory of light. At the time, it was believed only particles that traveled in a straight line could cast shadows as light did; Newton had advanced the idea of light consisting of such straight–traveling particles. Because of the examples of water waves, it was thought that waves bent around objects. In an effort to correct Newton, Huygens proved that in some cases waves could indeed travel in straight lines. He believed that light was a series of shock waves that disturbed particles existing in the "ether." When struck by the light, these closely packed particles would move and form new wave fronts. Where these wave fronts overlapped, there was light—a concept known as Huygen's principle. Although his theory allowed him to explain refraction and reflection, and predicted that light would more slowly in a denser medium such as water, it did not explain polarization. Newton's theory remained at the forefront; Huygens' idea laid dormant until it was rediscovered and improved on by Thomas Young in the 19th century.

Huygens also worked in optics, creating fine telescopes that led to important astronomical discoveries. Thanks to his improved instrument, he was able to correct Galileo's error that stated Saturn was a triple planet. Galileo's telescope did not have sufficient resolution; Saturn's rings appeared to him as two planets snuggled up to the main body. Huygens was able to discern a single ring around the planet.

Huygens also was the first to guess at the distances of the stars. Assuming that Sirius was as bright as the sun, he calculated it to be 2.5 trillion miles from Earth. We now know that Sirius is far brighter than the sun and its true **distance** is 20 times Huygens' approximation. Huygens also was the first to notice surface markings on Mars and the largest of Saturn's moons, Titan.

HYDRAULICS

Hydraulics is the practical application of fluid **mechanics**. It deals with engineering and scientific devices that control the flow, or movement, of liquids, such as valves, pipes, pneumatic controls, and nozzles; the manmade and natural environments that store, release, or channel liquids, such as dams, pumps, tanks, turbines, rivers, and canals; and the instruments that measure the flow of liquid through particular containers, channels, or systems, such as flow meters and pressure gauges. In large part, the theory of hydraulics is based on fluid mechanics, a dynamic field of science that has enjoyed explosive growth in the mid-1800s, and under certain conditions, hydraulics can also be applied to the flow of gases through a system. This application of concepts of hydraulics to gases generally occurs when the densities of the gases are kept within parameters at which the principles of fluid mechanics apply. Hydraulics encompasses the study of the flow of any liquid, but it is often applied to oil or water.

Investigations in hydraulics are dependent primarily on the mathematical concepts of density (the mass of an object over a particular **volume**), **force**, pressure (the force exerted by an object over a particular area), and mass. In turn, these concepts are governed by the known principles of classical physics, such **Newton's laws**, the laws of **thermodynamics**, conservation of mass, conservation of energy, and laws of **motion**. The principles of hydraulics on which modern hydraulic-power systems, including hydraulic lifts and brakes, are based is derived largely from the works of French scientist **Blaise Pascal** (1623-1662) and Swiss physicist **Daniel Bernoulli** (1700-1782).

Pascal's Law

While pursuing medical studies and in particular studying the flow of blood through the body, Pascal declared in 1652 that pressure applied to any confined, incompressible liquid will be applied equally to the liquid and the container holding the liquid in all directions, with no loss. This statement is known as Pascal's law (or Pascal's principle) and can be succinctly written that $pressure_{in}$ = $pressure_{out}$. It implies that even if pressure is applied to just a portion of the liquid, the pressure will be distributed over the entire volume of the liquid and the surface containing that liquid. For example, if an incompressible liquid is contained in a cylindrical container and then pressure is applied to just one of the circular faces of the **cylinder**, the change in pressure will distribute equally within the liquid and the cylinder's surface. An equilibrium will eventually result due to the distribution of pressure. Of course, if the pressure exceeds the strength of the material containing the liquid, the cylinder will burst, distributing the pressure over a larger volume.

Hydraulic Lever

The hydraulic lever is based on the concept of Pascal's law. In a simple hydraulic lever, a finite amount of incompressible liquid is placed between two pistons of unequal surface area. Pressure applied to the smaller piston results in a force being applied to the smaller piston, and as a result, a proportional force is applied to the larger piston. The relationship in applied force is described by the equation force$_{out}$ = force$_{in}$ * (area$_{out}$ / area$_{in}$). The amount of work, which is equal to force times **distance**, exerted on both systems is equal, providing the useful result that a small force exerted over a large distance can be transformed into a larger force exerted over a small distance. A car jack operates on this principle, as one has to pump the handle many times (that is, move the handle a significant distance) to lift the car a short distance off the ground.

Bernoulli's Principle

Simply stated, Bernoulli's principle (also known as Bernoulli's equation) asserts that the speed of moving fluid in inversely proportional to the pressure of the fluid (if traveling horizontally). If the moving liquid must transverse an elevation change, Bernoulli's equation must account for gravitational force, becoming $p + 1/2\rho v^2 + \rho g v = k$, where p is pressure, ρ is density, g is the gravitational constant, v is the velocity of the liquid, and k is a constant. As a result, fluid moving from a wider pipe to a smaller pipe increases in speed, and fluid moving from a smaller pipe to a larger pipe decreases in speed.

The principles of hydraulics have widespread applicability to science, engineering, and every day life, especially when applied to systems that require the transfer of power. Aircraft controls, for example, are based on hydraulic concepts, as are many braking systems. Modern industrial processes often use hydraulics to perform activities requiring more force than a human can exert.

See also Area; Bernoulli, Daniel; Euler's laws of motion; Force; Mass and energy; Mathematics and physics; Motion; Newton's laws; Newton, Isaac; Pascal, Blaise; Proportions; Thermodynamics; Volume

HYDRODYNAMICS

Hydrodynamics is a branch of fluid **mechanics** that involves mathematical **analysis** of the forces occurring at a fluid-object interface, e.g., between a submarine and the surrounding water or between a pipe wall and the flowing water. In general, it does not matter mathematically whether the object or the fluid is moving, rather it is the relative **motion** between the two that is important. The fluid is considered to be composed of particles that flow in layers. These layers become distorted and slide over one another (like playing cards) when the fluid intercepts an object.

Historically, hydrodynamics focused on theoretical **equations** for the flow of a fictitious (ideal) fluid that was assumed to be inviscid (i.e., totally without viscosity). In other words, this fluid would experience no frictional effects between its own moving layers or between fluid layers and object surfaces. These equations, though mathematically correct, indicated that a submerged object would experience **zero** drag (i.e., no resistance from the fluid to motion), a phenome-

non that was known to be false from practical experience. This contradiction was called d'Alembert's paradox, after **Jean le Rond d'Alembert**, an eighteenth-century scientist.

In 1904, Ludwig Prandtl published the boundary layer theory, which showed that flow around a submerged solid object contained two distinct regions with very different flow characteristics. Flow in a thin boundary layer around the object was affected by the surface of the object and by effects of internal friction due to the fluid's viscosity. This is called the viscous effect. Outside of the thin boundary layer, the viscous effect was negligible, and the free-flowing fluid behaved like an ideal fluid. Prandtl's boundary layer concept solved d'Alembert's paradox and laid the groundwork for modern hydrodynamics (and aerodynamics, in which air is the "fluid"). Boundary layer formation also occurs in flow over spillways and through pipes, channels, nozzles, and orifices.

The velocity of flow at the very bottom of the boundary layer is zero, because flow is retarded by viscous forces (which are also called shear stresses) and by attractive forces that pull water particles to the object surface (i.e., surface tension). These forces lessen with increasing **distance** from the object surface, so that velocity increases within the boundary layer until it is the same as the free-flow velocity. Thus, a velocity **field** or velocity gradient exists within the boundary layer.

Fluid-object interactions are often illustrated using flow lines to represent flow behavior. Prior to the interface, the approaching flow is a pattern of smooth parallel flow lines. This is called rectilinear flow. Velocity along each line is constant (although different lines may have different velocities). At the front surface of the object, the flow lines part to move around it. For a smooth streamlined object with its axis aligned with flow (e.g., a thin plate situated parallel to flow), the disturbance in the flow lines is minor, and they quickly retake a rectilinear pattern after leaving the trailing edge of the object. This flow is considered one-dimensional, because all of the velocity vectors remain virtually parallel. Mathematical analysis becomes considerably more complex when two- and three-dimensional flow fields are considered.

Consider the flow lines for a flat plate situated perpendicular to flow. They approach the plate in rectilinear flow and then are forced widely apart upon impact. They move across the plate surface (where the boundary layer forms), but are unable to make the sharp turn following the plate edge. The boundary layer separates from the plate surface and plunges into the free-flow area. An area of extreme turbulence occurs directly behind the plate, characterized by swirling eddies and vortices (i.e., a wake). The eddies dissipate their energy with increasing distance from the plate and eventually flatten out into uniform flow lines again.

Flow within the boundary layer along the plate surface can be laminar (uniform) or turbulent. This is important because the shear forces (and thus the drag forces) are higher in areas of turbulent flow. For the parallel-flow plate, flow is usually laminar near the plate's leading edge, and then becomes turbulent at some point along the surface. Quantitatively, this is represented by the dimensionless Reynolds Number (N_{Re}). For any distance x along the plate,

$N_{Re,x} = xv_\infty\rho/\mu$, where v_∞ is the bulk fluid velocity, ρ is the fluid density, and μ is the fluid viscosity. The $N_{Re,x}$ value at which flow becomes turbulent varies between 10,000 and 3,000,000, depending on the smoothness of the plate and the turbulence of the approaching flow. If the transition from laminar to turbulent flow occurs nearer the plate's leading edge, then the total drag is higher than it would be for a transition located farther downstream.

Besides shear, drag forces can also be due to pressure, gravity, and compressibility effects. Although determination of drag forces is very complex for most flow situations, drag coefficients can be calculated for certain regular-shaped objects immersed in low-velocity flows. For example, the drag coefficient for a smooth **sphere** subjected to flow at very low Reynolds number is $C_D=24/N_{Re}$, which is a derivation of Stokes' law. The drag **force** is then calculated from $D=C_D A\gamma v^2/2g$, where A is the area of the sphere projected onto a plane perpendicular to flow, γ is the specific weight of the fluid, v is velocity, and g is gravitational acceleration. Drag coefficients for plates and cylinders can be similarly calculated or approximated from tables if the Reynolds number is known.

Other hydrodynamic forces acting at the fluid-object interface can be determined using quantitative parameters including the Froude Number(which is a ratio of inertial forces to gravitational forces), the Weber Number (which is a ratio of inertial forces to surface tension forces), the pressure coefficient (which is a ratio of pressure forces to inertial forces), and the Mach Number (which is a ratio of inertial forces to compressibility forces).

HYPATIA OF ALEXANDRIA (CA.370-415)

Greek geometer, astronomer and philosopher

Hypatia, the earliest known woman mathematician, wrote commentaries on several classic works of mathematics. The daughter of a mathematician, she was trained in **mathematics and philosophy** and became head of the Neoplatonic school at Alexandria, where she taught philosophical doctrines dating back to **Plato**'s Academy. Hypatia was a respected teacher and influential citizen of Alexandria, greatly admired for her knowledge as well as for her decorum and dignity. Although Hypatia's original work has not survived, she is known from the letters of her student Synesius of Cyrene. She is also mentioned in the fifth–century *Ecclesiastical History* of Socrates Scholasticus, and in the tenth–century *Lexicon* of Suda (or Suidas). Information about Hypatia is fragmentary and oblique, fact and fiction have mingled, and her life has become the stuff of legend, inconsistencies, and conflicting opinions.

Hypatia was born in Alexandria, Egypt; the year is generally thought to be 370, although some scholars argue for an earlier date, 355. Founded on the Nile River by Alexander the Great in 332 B.C., Alexandria had been the center of scholarly attainment in science, and during Hypatia's time was the third largest city in the Roman Empire. Hypatia's father, **Theon**, was a member of the Museum, a place of residence, study, and teaching similar to a modern university. A mathematician and astronomer, Theon had predicted eclipses of the sun and the moon which were observed in Alexandria, and his scholarship included commentaries on **Euclid** and **Claudius Ptolemy**. Hypatia was taught by Theon, collaborated with him, and did independent work. Whereas Theon also produced poetic work and a treatise on the interpretation of omens, Hypatia's works seem to have been strictly mathematical.

Hypatia was recognized as a gifted scholar and eloquent teacher, and by 390 her circle of influence was well–established. By 400, she was head of the Neoplatonic school, for which she received a salary. Socrates Scholasticus, the Byzantine church historian, wrote that Hypatia was so learned in literature and science that she exceeded all contemporary philosophers. Philostorgius, another historian, noted that she surpassed her father in mathematics, and especially in astronomy. From Synesius' letters to and about her, it is clear Hypatia had extensive knowledge of Greek literature. Her students were aristocratic young men, both pagan and Christian, who rose to occupy influential civil and ecclesiastical positions. They came from elsewhere in Egypt, and from as far away as Cyrene, Syria, and Constantinople to study privately with Hypatia in her home. They were united through intellectual pursuits and considered Hypatia their "divine guide" into the realm of philosophical and cosmic mysteries, which included mathematics. Hypatia combined the principles of free thinking with the ideal of pure living. She was known for her prudence, moderation and self–control, for her ease of manner, and for her beauty. She chose to remain a virgin and to devote her life to pursuit of knowledge and the philosophical ideal. According to an account in Suda, which may be apocryphal, when one of her students fell in love with her, she threw at him a rag that was the equivalent of a sanitary napkin, saying "You are in love with this, not with [the Platonic ideal of] the Beautiful."

By wearing a tribon, the characteristic rough white robe of the philosopher, Hypatia indicated that she did not wish to be treated as a woman. She traveled freely about the city in her chariot, instructed her students in Platonic and Aristotelian philosophy, visited and lectured at public and scientific institutions. She had exerted political influence and may have held a political position. In one of his letters to her, Synesius asks Hypatia to intervene with her powerful friends to restore the property of two young men. Throughout his life Synesius remained devoted to Hypatia, praised her erudition, and asked for advice on his own writings. He must have visited her the several times he visited Alexandria, including in 410, when he was consecrated as Bishop of Ptolemais by Theophilus, the patriarch of Alexandria.

Although several of Theon's mathematical and astronomical works have survived, Hypatia's have not. It is known that Hypatia wrote a treatise titled *Astronomical Canon*, presumably on the movements of the planets, and a commentary on the algebraic work of **Diophantus of Alexandria**, which contains the beginnings of **number theory**. Diophantus, who lived in the third century A.D., is quoted by Theon, and some scholars believe that the survival of most of Diophantus' original thirteen books of the *Arithmetica* is due to the quality of Hypatia's work. The surviving texts, including six in Greek and four translated into Arabic, contain notes, remarks, and interpolations that may come from Hypatia's commentary. Hypatia also wrote *On the Conics of Apollonius*, in which she elaborated on **Apollonius**' third–century B.C. theory of **conic sections**.

In collaboration with Theon, Hypatia also worked on Ptolemy's *Almagest*, the second–century work which brought together disparate works of early Greeks in 13 volumes and served as the standard reference on astronomy for more than 1,000 years. In the *Almagest*, Ptolemy introduced a method of classifying stars, and used Apollonius' mathematics to construct a masterful (though incorrect) theory of epicycles to explain the movement of the sun, moon and planets in a geocentric system. Hypatia may have corrected not only her father's commentary but also the text of *Almagest* itself, and may also have prepared a new edition of Ptolemy's *Handy Tables*, which appears in the work of Hesychius under the title *The Astronomical Canon.*

Synesius' letters reveal Hypatia's interest in scientific instruments. In one instance he asks her to have a hydroscope (an instrument for measuring the specific gravity of a liquid) made for him. In another, he consults her about the construction of an astrolabe, an instrument used to measure the position of the stars and planets.

In Hypatia's time, Christianity became the official religion of the Roman empire, and Greek temples were converted to Christian churches. In 411, Cyril succeeded Theophilus as bishop of Alexandria. One of his actions, following Jewish–Christian riots, was to expel Jews from the city. Orestes, the civil governor, disapproved of Cyril's actions and the growing encroachment of the Christian church on civil authority. Cyril roused negative sentiment toward Orestes, and Orestes was attacked by five hundred Nitrian monks, who lived in monasteries outside the city. The monk Ammonius threw a stone that wounded Orestes. Intervention by the populace saved Orestes, who then ordered Ammonius tortured to the extent that he died. Cyril applauded Ammonius' actions as admirable.

Hypatia fell victim to these political hostilities. She was a close associate of Orestes, and undoubtedly was defamed by Cyril. Admiration for her turned to resentment, and she was perceived as an obstacle to the conciliation of Orestes and Cyril. In March of 415, during Lent, as Hypatia rode in her chariot through the streets of Alexandria, she was attacked by a fanatical mob of antipagan Christians. The mob dragged Hypatia into the Caesareum, then a Christian church, where she was stripped naked and murdered. According to ancient accounts, Hypatia's flesh was stripped from her bones, her body mutilated and scattered throughout the streets, then burned piecemeal at a place called Cinaron.

Following Hypatia's murder many of her students migrated to Athens, where they contributed to the Athenian school, which in 420 acquired a considerable reputation in mathematics. The Neoplatonic school at Alexandria continued until the Arab invasion of 642. The books in the library at

Hypatia

Alexandria were subsequently used as fuel for the city's baths, where they lasted six months. Hypatia's works were probably among them.

HYPERBOLA

A hyperbola is a curve formed by the intersection of a right circular cone and a plane. When the plane cuts both nappes of the cone, the intersection is a hyperbola. Because the plane is cutting two nappes, the curve it forms has two U-shaped branches opening in opposite directions.

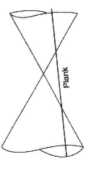

Figure 1.

A hyperbola can be defined in several other ways, all of them mathematically equivalent:

1. A hyperbola is a set of points P such that $PF_1 - PF_2 = \pm C$, where C is a constant and F_1 and F_2 are fixed points called the "foci." That is, a hyperbola is the set of points the difference of whose distances from two fixed points is constant.

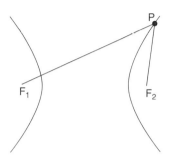

Figure 2.

The positive value of ± C gives one branch of the hyperbola; the negative value, the other branch.

2. A hyperbola is a set of points whose distances from a fixed point (the "focus") and a fixed line (the "directrix") are in a constant ratio (the "eccentricity"). That is, PF/PD = e.

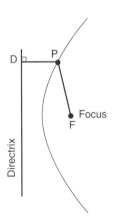

Figure 3.

For this set of points to be a hyperbola, e has to be greater than 1. This **definition** gives only one branch of the hyperbola.

3. A hyperbola is a set of points (x,y) on a Cartesian coordinate plane satisfying an equation of the form $x^2/A^2 - y^2/B^2 = \pm 1$. The equation $xy = k$ also represents a hyperbola, but of **eccentricity** $\sqrt{2}$ only. Other second-degree **equations** can represent hyperbolas, but these two forms are the simplest. When the positive value in ± 1 is used, the hyperbola opens to the left and

right. When the negative value is used, the hyperbola opens up and down.

When a hyperbola is drawn as in Figure 4, the line through the foci, F_1 and F_2, is the "transverse axis."

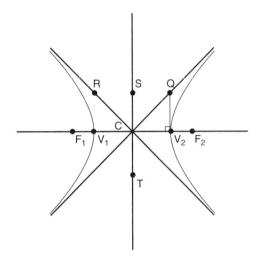

Figure 4.

V_1 and V_2 are the "vertices," and C the "center." The transverse axis also refers to the **distance**, V_1V_2, between the vertices.

The ratio CF_1/CV_1 (or CF_2/CV_2) is the "eccentricity" and is numerically equal to the eccentricity e in the focus-directrix definition.

The lines CR and CQ are asymptotes. An **asymptote** is a straight line which the hyperbola approaches more and more closely as it extends farther and farther from the center. The point Q has been located so that it is the vertex of a right **triangle**, one of whose legs is CV_2, and whose **hypotenuse** CQ equals CF_2. The point R is similarly located.

The line ST, perpendicular to the transverse axis at C, is called the "conjugate axis." The conjugate axis also refers to the distance ST, where $SC = CT = QV_2$.

A hyperbola is symmetric about both its transverse and its conjugate axes.

When a hyperbola is represented by the equation $x^2/A^2 - y^2/B^2 = 1$, the x-axis is the transverse axis and the y-axis is the conjugate axis. These axes, when thought of as distances rather than lines, have lengths 2A and 2B respectively. The foci are at $\sqrt{(A^2 + B^2)}, 0$ and $\sqrt{(A^2 + B^2)}, 0$; the eccentricity is $\sqrt{(A^2 + B^2)} \div A$.

The equations of the asymptotes are y = Bx/A and y = -Bx/A. (Notice that the constant 1 in the equation above is positive. If it were -1, the y-axis would be the transverse axis and the other points would change accordingly. The asymptotes would be the same, however. In fact, the hyperbolas $x^2/A^2 - y^2/B^2 = 1$ and $x^2/A^2 - y^2/B^2 = -1$ are called "conjugate hyperbolas.") Hyperbolas whose asymptotes are perpendicular to each other are called "rectangular" hyperbolas. The hyperbolas $xy = k$ and $x^2 - y^2 = \pm C^2$ are rectangular hyperbolas. Their eccentricity is $\sqrt{2}$. Such hyperbolas are geometrically similar, as are all hyperbolas of a given eccentricity.

If one draws the **angle** F_1PF_2 the tangent to the hyperbola at point P will bisect that angle.

Hyperbolas can be sketched quite accurately by first locating the vertices, the foci, and the asymptotes. Starting with the axes, locate the vertices and foci.

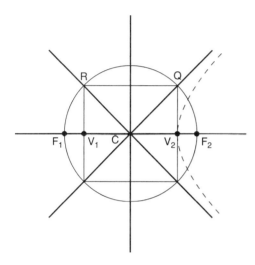

Figure 5.

Draw a **circle** with its center at C, passing through the two foci. Draw lines through the vertices perpendicular to the transverse axis. This determines four points, which are corners of a rectangle. These diagonals are the a symptotes.

Using the vertices and asymptotes as guides, sketch in the hyperbola as shown in Figure 4. The hyperbola approaches the asymptotes, but never quite reaches them. Its curvature, therefore, approaches, but never quite reaches, that of a straight line.

If the lengths of the transverse and conjugate axes are known, the rectangle in Figure 5 can be drawn without using the foci, since the rectangle 's length and width are equal to these axes.

One can also draw hyperbolas by plotting points on a coordinate plane. In doing this, it helps to draw the asymptotes, whose equations are given above.

Hyperbolas have many uses both mathematical and practical. The hyperbola $y = 1/x$ is sometimes used in the definition of the natural logarithm. In Figure 6 the logarithm of a number n is represented by the shaded area, that is, by the area bounded by the x-axis, the line $x = 1$, the line $x = n$, and the hyperbola.

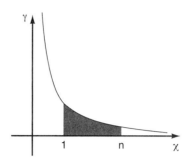

Figure 6.

Of course one needs **calculus** to compute this area, but there are techniques for doing so.

The coordinates of the point (x,y) on the hyperbola $x^2 - y^2 = 1$ represent the hyperbolic **cosine** and hyperbolic **sine functions**. These functions bear the same relationship to this particular hyperbola that the ordinary cosine and sine functions bear to a unit circle:

$$x = \cosh u = (e^U + e^{-u})2$$

$$y = \sinh u = (e^U - e^{-u})\,2$$

Unlike ordinary sines and cosines, the values of the **hyperbolic functions** can be represented with simple **exponential functions**, as shown above. That these representations work can be checked by substituting them in the equation of the hyperbola. The parameter u is also related to the hyperbolas. It is twice the shaded area in Figure 7.

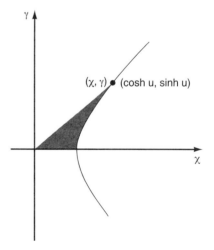

Figure 7.

The definition $PF_1 - PF_2 = \pm C$, of a hyperbola is used directly in the LORAN navigational system. A ship at P receives simultaneous pulsed radio signals from stations at A and B. It can't measure the time it takes for the signals to arrive from each of these stations, but it can measure how much longer it takes for the signal to arrive from one station than from the other. It can therefore compute the difference PA - PB in the distances. This locates the ship somewhere along a hyperbola with foci at A and B, specifically the hyperbola with that constant difference. In the same way, by timing the difference in the time it takes to receive simultaneous signals from stations B and C, it can measure the difference in the distances PB and PC. This puts it somewhere on a second hyperbola with B and C as foci and PC - PB as the constant difference. The ship's position is where these two hyperbolas cross.

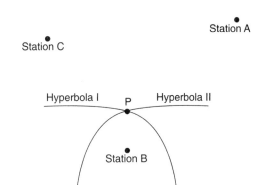

Figure 8.

Maps with grids of crossing hyperbolas are available to the ship's navigator for use in areas served by these stations.

HYPERBOLIC FUNCTIONS

The hyperbolic **functions** are similar to the **trigonometric functions**, often called circular functions, in that they play an important role in problems in mathematics and the mathematics of physics. The hyperbolic functions are involved in mathematical problems in which integrals involving $\sqrt{(1 + x^2)}$ arise whereas trigonometric functions involve functions with integrals of the type $\sqrt{(1 - x^2)}$. Although the hyperbolic functions are similar to trigonometric functions, their definitions are much more straightforward. As the trigonometric functions yield parameters describing the unit **circle**, the hyperbolic functions yield parameters describing the standard **hyperbola**: $x^2 - y^2 = 1$ for $x > 1$. For every trigonometric function there is a corresponding hyperbolic function, though they are not necessarily identical.

Like the trigonometric functions the hyperbolic functions are periodic because they are all elliptic functions. Elliptic functions can be doubly periodic, singly periodic, or non-periodic, which is often referred to as trivially periodic. The exponential function, on which the hyperbolic functions are based, is a singly periodic elliptic function that means the modulus of periodicity is 2π. This means that 2π is the smallest constant for which the identity is true. The two main hyperbolic functions are hyperbolic **sine**, sinh x, and hyperbolic **cosine**, cosh x. They are defined as: sinh $x = (e^x - e^{-x})/2$ and cosh $x = (e^x + e^{-x})/2$. Analogous to the trigonometric functions sinh x is an odd function with sinh $0 = 0$, and cosh x is an even functions with cosh $0 = 1$. Because of the properties of e^x, it follows that sinh $x < 0$ for $x < 0$, sinh $x > 0$ for $x > 0$, and

cosh $x > 0$ for all x. The four other hyperbolic functions, hyperbolic tangent, hyperbolic cotangent, hyperbolic secant, and hyperbolic cosecant, can all be defined in terms of sinh x and cosh x: tanh $x = $ sinh x / cosh x, coth $x = $ cosh x / sinh x, sech $x = 1$ / cosh x, and csch $x = 1$ / sinh x. Just as there are **trigonometric identities** there are hyperbolic identities, the most important being the analogous version of the **Pythagorean theorem**: $(\cosh x)^2 - (\sinh x)^2 = 1$.

Although **Johann Heinrich Lambert**, a French mathematician, is often credited for the introduction of the hyperbolic functions, it was actually Vincenzo Riccati, an Italian mathematician, who did so in the mid 18th century. He studied these functions and employed them to obtain solutions of **cubic equations**. Riccati found the standard **addition** formulas, similar to trigonometric identities, for hyperbolic functions as well as their derivatives. He revealed the relationship between the hyperbolic functions and the exponential function. This work, some done jointly with Saladini, was published between 1757 and 1767. In these publications Riccati used Sh. and Ch. for hyperbolic sine and cosine similar to his abbreviations for the circular functions, Sc. and Cc. Later, in 1768, Lambert published further developments concerning the theory of hyperbolic functions. He used sin h and cos h to abbreviate hyperbolic sine and hyperbolic cosine at that time. In 1771 Lambert began using sinh, the Latin *sinus hyperbolus*, and cosh to represent these functions. Although this notation is widely used today to abbreviate the hyperbolic functions, in some instances hysin and hycos are used. These notations were introduced in 1902 by George Minchin and published in the journal *Nature* that year.

The hyperbolic functions are important to many mathematical and physical problems. Hyperbolic sine arises in the formulation of the gravitational potential of a **cylinder** and the calculation of the Roche limit. Hyperbolic cosine is connected with the **catenary** form, meaning chain. This shape is formed when a flexible inelastic chain of uniform density is suspended by its two ends and is subjected to only the influence of gravity. This shape imparts great strength to structures when built in the shape of a catenary. The hyperbolic tangent is involved in the calculation of the magnetic moment as well as the rapidity of special relativity. The rapidity of special relativity is defined as: $\Theta = tanh^{-1}B$, where $B = v/c$, and v is the speed of a particle and c is the speed of **light** in a vacuum. The hyperbolic secant is involved in the profile of a laminar jet and the hyperbolic cotangent is connected with the Langevin function for magnetic polarization. As can be seen the hyperbolic functions are not only important in mathematics but in physics and engineering.

See also Trigonometric functions

HYPERBOLIC GEOMETRY

Hyperbolic geometry is an alternative **geometry** to Euclidean geometry, in which Euclid's first four axioms are true but the fifth **axiom**, the **parallel postulate**, is not. Hyperbolic geometry was discovered almost accidentally in the 19th century. Its discoverers were not looking for an alternative geometry, but instead were trying to prove that such a geometry could not exist. Ever since **Euclid** first set out his five basic assumptions (axioms) of geometry around 300 BC, a debate had raged over whether the fifth axiom is truly a basic assumption, or whether it can in fact be proven using the other four axioms; this would make it not an assumption, but a **theorem**. Many mathematicians tried to prove the fifth axiom from the other four, but sooner or later each such "proof" was found to have some hole in it.

In the middle of the 19th century the mathematicians **Lobachevsky**, **Bolyai** and Beltrami set out to prove the fifth axiom using a new technique: they tried to show that if the fifth axiom were *false*, all sorts of absurd and contradictory consequences would result. The fifth axiom states that, given a line and a point P that is not on the line, there is one and only one line through P that never meets the original line. So, Lobachevsky, Bolyai and Beltrami, each working separately, tried to see what would be the consequences if, given a line and a point P not on the line, there could be more than one line through P that never met the original line.

Lobachevsky, Bolyai and Beltrami expected that this assumption would lead to ridiculous results, but it did not. Instead, they got one reasonable result after another, discovering many interesting and consistent properties of this alternative geometry. They succeeded in describing the geometry of its triangles, its **area** measurements, and its rules of **trigonometry**. Eventually they decided that this new geometry was every bit as consistent as Euclidean geometry, and they gave it a name: hyperbolic geometry.

The mathematical community was not immediately ready to accept the existence of hyperbolic geometry. In 1868, however, Beltrami won over the critics by describing a concrete **model** for hyperbolic geometry: a new way of measuring lengths and angles in a disk (a **circle** together with its interior), that would have all the properties that he and his colleagues had discovered for hyperbolic geometry.

In the hyperbolic disk, the shortest path between two points is not an ordinary straight line, but rather is an arc of a semicircle whose center is on the boundary of the disk; diameters of the disk are also shortest paths, but other straight lines are not. Thus, these arcs of semicircles are referred to as the straight lines of hyperbolic geometry, and to an inhabitant of the hyperbolic disk, these arcs would appear straight, since they are the most direct paths between points.

The famous Dutch artist **M. C. Escher**, who is renowned for the mathematical themes of his work, has exploited the unusual properties of hyperbolic geometry in several of his etchings, including *Angels and Devils*. Using Euclidean length measurements, the angles and devils are many different sizes. Using hyperbolic measurements, however, all the angels are exactly the same size; likewise for the devils. An inhabitant of the hyperbolic disk would see a vast tiling of the disk by identical angel tiles and identical devil tiles, in much the same way that in Euclidean geometry we have tilings by identical squares. From Escher's etching it is evident that using hyper-

bolic measurements, the area of the disk is infinite, since it contains infinitely many tiles, and each tile is the same size. Furthermore, the **distance** to any point on the boundary is infinite, since to get to the boundary it is necessary to cross an infinite number of tiles.

There are many other counterintuitive aspects to hyperbolic geometry. In a **triangle**, for example, the measures of the interior angles always add up to *less than* 180 degrees, in contrast to Euclidean geometry in which the angles add up to exactly 180 degrees. The reason for this is that hyperbolic "straight lines" are arcs of semicircles, so hyperbolic triangles become very pinched at the corners. Another strange phenomenon is that as the radius of a circle grows, the area of the circle grows much faster than its Euclidean counterpart. For example, in Euclidean geometry a circle whose radius is, say, 47 units will have area roughly 22,000 **square** units. In Beltrami's hyperbolic disk, on the other hand, a circle whose radius is 47 units will have area roughly 1×10^{20}, 1 followed by 20 zeros.

The hyperbolic disk is in some ways an excellent model of hyperbolic **space**, but it suffers from the extreme distortion that makes two angels that are the same hyperbolic size appear very different to the eye. It would be desirable to construct a model in which this distortion does not occur. In 1901, however, the great German mathematician **David Hilbert** demonstrated that there is no way to build a complete hyperbolic surface in ordinary three-dimensional space without some distortion; loosely speaking, the difficulty is that the area of hyperbolic circles grows so quickly that a hyperbolic surface would get very wavy around the edges as it grew, like a floppy wide-brimmed hat, and eventually it would start bumping into itself.

Although hyperbolic surfaces cannot be built in ordinary Euclidean space, hyperbolic geometry is every bit as concrete and consistent as Euclidean geometry. In fact, physicists and mathematicians speculate whether the universe might actually be governed by hyperbolic geometry, rather than Euclidean geometry. At first glance this seems nonsensical. However, on a very small scale, measurements in Euclidean and hyperbolic geometry are almost identical; it is only on a large scale that noticeable differences emerge. At present, human beings are only able to make measurements in a small portion of the universe. This opens up the possibility that the universe is hyperbolic; a question that can only be answered when human beings develop the technology to make measurements in the far reaches of the universe.

HYPOTENUSE

In a right **triangle**, the hypotenuse is the side of the triangle which is opposite the right **angle**. It is also the longest side of the triangle. The other two sides are sometimes called the "legs" of the right triangle. The Greek word *hupoteinousa*, from which we derive "hypotenuse," means, roughly, a stretched cord, invoking the image of a rope stretched tightly between two stakes in the ground. The hypotenuse plays an essential role in what is perhaps the most famous of all mathematical theorems, the **Pythagorean Theorem**. This **theorem**, attributed to the Greek philosopher and mathematician, **Pythagoras** (c 569-c 475 BC), states that if the legs of a right triangle are denoted by a and b and the hypotenuse by c, then $c^2=a^2+b^2$, or, in English, the **square** of the hypotenuse is equal to the sum of the squares of the other two sides. The hypotenuse became a source of great consternation to Pythagoras, who was something of a mystic in his study of numbers. Pythagoras and his followers believed that numbers ruled the universe, and by "numbers" they meant the natural numbers 1,2,3, etc. For certain right triangles, the hypotenuse was indeed one of these natural numbers. For example, a right triangle with legs of lengths 3 units and 4 units has a hypotenuse of length 5 units, since $3^2+4^2=5^2$. For this reason the numbers 3, 4, and 5 are collectively called a Pythagorean triple. Other examples include 5, 12, and 13; 8, 15, and 17; and 7, 24, and 25. In fact, there are an infinite number of Pythagorean triples or, to put it another way, there are an infinite number of right triangles with the lengths of all three sides being natural numbers. These triangles were very pleasing to Pythagoras as they fit in very nicely with his theories about the world being ruled by natural numbers. Unfortunately, for Pythagoras, there are also an infinte number of right triangles which do not have this property. In particular, Pythagoras noticed that a right triangle whose legs are each 1 unit long will have a hypotenuse with length $\sqrt{2}$, which is not a natural number. Furthermore, $\sqrt{2}$ is a non-repeating infinite decimal and cannot be expressed as the ratio of two natural numbers. For this reason, Pythagoras called such numbers incommensurable, and he was evidently quite troubled by their appearance as lengths. How could such a number represent a length? Where exactly does one find such a number on the real number line? The incommensurables, which we now call **irrational numbers**, did not fit nicely with Pythagoras' doctrine that all quantities should be natural numbers or **ratios** of natural numbers. Legend says that Pythagoras instructed his followers not to discuss the incommensurability of the hypotenuse with the legs in certain right triangles. As far as we know, Pythagoras never tried to develop a theory which would bring the irrational numbers into his view that "number" was the ruling concept of the universe.

The hypotenuse is also involved in the **definition** of the so-called trigonometric ratios for right triangles. The **sine** of an angle is defined to be the ratio of the length of the side opposite the angle to the length of the hypotenuse. The **cosine** of an angle is the ratio of the length of the side adjacent to the angle to the length of the hypotenuse. The tangent ratio is then defined to be the ratio of the sine to the cosine. These ratios combined with the Pythagorean theorem may be used to determine the lengths of sides or the measure of angles in a right triangle when other sides and angles are known. For example, if we know that the hypotenuse of a right triangle is 20 meters long and that one angle of the triangle has a measure of $30°$, then we can determine the length of the adjacent side, a, to the angle using the equation $a/20=\sin(30°)$, from which it follows that $a=20\sin(30°)=10$ meters. Similarly, we could determine the length of the opposite side from the angle by using the cosine ratio. Then, as a check of our computations, we could add the square of the adjacent side's length to the square of the opposite side's length which should give us 400 or the square of the length of the hypotenuse.

We might also note that the familiar "distance formula" for finding the **distance** between two points in the plane is nothing more than a computation of the length of the hypotenuse of the right triangle whose legs are formed by drawing perpendicular segments from the two points to the intersection of the two segments. Thus, the distance between the points (2,3) and (4,7), by the distance formula is $\sqrt{((4-2)^2+(7-3)^2)}=\sqrt{20}$; but if we constuct the right triangle with vertices (2,3), (4,7), and (4,3), we see that its legs have lengths of 2 and 4 units, so that the Pythagorean Theorem gives $(\text{hypotenuse})^2=2^2+4^2=4+16=20$. Therefore, the hypotenuse has length $\sqrt{20}$, which agrees with the distance formula computaion.

I

IDENTITY ELEMENT

Any mathematical object that, when applied by an operation, such as **addition** or **multiplication**, to another mathematical object, such as a number, leaves the other object unchanged is called an identity element. The two most familiar examples are 0, which when added to a number gives the number, and 1, which is an identity element for multiplication.

More formally, an identity element is defined with respect to a given operation and a given set of elements. For example, 0 is the identity element for addition of **integers**; 1 is the identity element for multiplication of **real numbers**. From these examples, it is clear that the operation must involve two elements, as addition does, not a single element, as such operations as taking a power.

Sometimes a set does not have an identity element for some operation. For example, the set of even numbers has no identity element for multiplication, although there is an identity element for addition. Most mathematical systems require an identity element. For example, a group of transformations could not exist without an identity element that is the **transformation** that leaves an element of the group unchanged.

IDENTITY PROPERTY

When a set possesses an **identity element** for a given operation, the mathematical system of the set and operation is said to possess the identity property for that operation. For example, the set of all **functions** of a **variable** over the **real numbers** has the identity element, or identity function, $I(x) = x$. In other words, if $f(x)$ is any function over the real numbers, then $f(I(x)) = I(f(x)) = f(x)$.

IMAGINARY NUMBERS

Like **real numbers**, imaginary numbers are a subset of the set of all **complex numbers**. Imaginary numbers are typically represented as either the constant i (in mathematics) or j (in engineering and the sciences), where $i^2 = -1$ or sqrt(-1) $= i$. Like the real numbers, imaginary numbers have relative magnitude and can be plotted along a number line, with $3i > i$. Imaginary numbers may also be negative, with $3i > i > -i > -3i$. Any imaginary number ki can be written as the complex number $0 + ki$.

Properties of Imaginary Numbers

Imaginary numbers follow the same rules of **addition** and **subtraction** available to real numbers. For example, $3 + 2 = 5$ and $4 - 6 = -2$ in the real number plane. Similarly, $3i + 2i = 5i$ and $4i - 6i = -2i$ in the imaginary number plane. Note, however, that $3 + 3i$ (6, as $3 = 3 * $ sqrt(1) and $3i = 3 * $ sqrt(-1)), which are not equal.

Scalar multiples of imaginary numbers also follow the conventions of real numbers. For example, $2 \times 3 = 6$. Similarly, $2 \times 3i = 6i$. Note, however, that if the scalar **factor** is imaginary itself, such as $2i$, the product $2i \times 2i = 4i^2 = -4$. If an imaginary scalar number is multiplied against an imaginary number, then, the product will be a real number; the product of any two imaginary numbers is real.

Division within imaginary numbers also follows the rules of real numbers, but the quotient will also be a real number. For example, $4i / 2i = 2$ and $6i / 2i = 3$. If an imaginary number is divided by a real number (in effect, multiplying by the reciprocal of the division), then the number will be imaginary. For example $20i / 4 = 20i \times 1/4 = 5i$.

Just as i can be multiplied by itself to form a **square**, so can i be raised to any power. A pattern quickly develops:

- $i^1 = i$
- $i^2 = -1$
- $i^3 = i \times i^2 = -i$
- $i^4 = i^2 \times i^2 = 1$

- $i^5 = i \times i^4 = i$
- $i^6 = i^2 \times i^4 = -1...$

It is also possible to raise i to a power that is not a positive integer, such as -1 or 1/2. Interestingly, $i^{-1} = 1/i = 1/i \times i/i = i/i^2 = -i$, which provides the interesting result that the reciprocal of i is equal to its opposite, a useful relationship in engineering. Further, it is straightforward to show that sqrt(i) = (1 + i)/sqrt(2), indicating that i itself is a square.

One of the first known explorations of imaginary numbers occurred in Heron's *Stereometica*. In his work, Heron (also known as **Hero**) developed the expression sqrt(81 - 144), for which no understanding existed at the time. Later, mathematicians seeking solutions to quadratic **equations** encountered equations for which they contended no solution existed, such as $x^2 + 1 = 0$ and $x^2 + 2x^2 + 2 = 0$. The equations do, however, have solutions as valid as the solutions to the equation $x^2 - 1 = 0$ (that is, $x = 1$ and $x = -1$. In the former case, the solutions are $x = i$ and $x = -i$; in the latter case, the solutions are $x = -1 + i$ and $x = -1 - i$. This understanding of imaginary numbers had led to many mathematical and scientific advances; imaginary numbers are used widely in science and engineering applications across the world today.

See also Addition; Complex numbers; Division; Hero of Alexandria; Multiplication; Negative numbers; Numbers and numerals; Quadratic equations; Products and quotients; Real numbers; Subtraction

IMPLICATION

Implication is a binary propositional connective read most often as an "if-then" statement. An "If-then" statement is composed of an antecedent (what comes after the "if"), and a consequent (what comes after the "then"). Implication is defined as follows: the antecedent S implies the consequent P if and only if it is impossible for S to be true while P is false. For example, the proposition "If there are objects, then there is space" expresses an implication. **Space** is implied by objects. If it is true that there are objects, then it must also be true that there is space, since space is a **necessary condition** of the existence of objects. This type of implication is called *material* implication, and it can also be expressed as a necessary implication, as it is with respect to validity.

Implication is a form of **inference** closely related to the concept of validity. In an argument, a set of true **propositions** that implies or entails another true proposition with necessity is a valid argument. However, whereas validity is a feature of certain deductive arguments, implication expresses the relationship between propositions and **sets** of propositions.

Not all "if-then" propositions express material or necessary implication. For example, "If Lee Harvey Oswald had not killed Kennedy, then someone else would have killed Kennedy," expresses a counter-factual implication—that is, an "if-then" statement whose antecedent is false. The implication is only probable at best, because if it is false that Lee Harvey Oswald had killed Kennedy, then it is only probable that

someone else would have killed Kennedy. There is no good reason to infer that someone necessarily would have killed Kennedy if Oswald had not.

See also Deduction; Induction; Inference; Proof by deduction; Proof by induction

IMPLICIT DIFFERENTIATION

Implicit differentiation is a technique used in **calculus** to compute the derivative of **functions** even when an explicit formula for the function is unknown. This technique is a consequence of a general **theorem**, called the implicit function theorem, which enables mathematicians to produce a limitless supply of smooth curves and surfaces.

A **variable** y is said to be an explicit function of another variable, x, if there is an equation $y = f(x)$ relating y to x. Here f represents a function, in other words a rule that assigns to each value of x one and only one value of y. But the **equations** of many classical curves, such as the **circle** $x^2 + y^2 = 1$, are not written in the standard $y = f(x)$ format. Instead, x and y are mutually dependent, and so it is not immediately clear whether y can actually be considered to be a function of x.

In fact, in the cited example each value of x corresponds to two values of y, namely $y = \sqrt{(1-x^2)}$ and $y = -\sqrt{(1-x^2)}$. Therefore y is not a function of x over the entire curve. However, over a small piece of the circle—for example, a small arc containing the point (3/5, 4/5)—the variable y is indeed uniquely defined as a function of x. It is said to be "implicitly defined" by the equation $x^2 + y^2 = 1$.

Not only is y a function of x in this restricted sense, but its derivative can be computed by differentiating the equation for the circle and applying the **Chain Rule**, to give the new equation $2x + 2y(dy/dx) = 0$. Solving this equation for dy/dx gives $dy/dx = -x/y$. At the point (3/5, 4/5) mentioned above, it follows that $dy/dx = -3/4$. This method of computing the derivative is much easier than the alternative of solving for y as a function of x and differentiating explicitly.

In general, the implicit function theorem states that an equation $g(x, y) = C$ (where g is a continuously differentiable function and C is any constant) locally defines y as an implicit function of x in a neighborhood of any point where the partial derivative of g with respect to y is nonzero. At any such point, the derivative $dy/dx = -(\partial g/\partial x) \div (\partial g/\partial y)$. The theorem extends as well to equations of three or more variables. It can be shown that an equation $g(x, y) = C$ defines a smooth curve at all points except where both partial derivatives of g are 0, and similarly that an equation $g(x, y, z) = C$ defines a smooth surface at all points except where all three partial derivatives are 0. Such points are called singular points of g.

IMPROPER INTEGRALS

An improper integral is an integral of one of two types: either the **interval** over which the interval is taken is unbounded, or the integr and becomes unbounded in a neighborhood of one or more points in the interval (including the endpoints). The common thread in these two types of integrals is that both have the potential to be infinite.

The first type of integral is an integral over either a half-infinite interval, i.e. from a finite value a to **infinity** or from negative infinity to a, or over the interval from negative infinity to positive infinity. Let us first consider the first of these possibilities. Suppose f is a function that is defined for all x greater than or equal to a. We are trying to measure an **area** whose "horizontal dimension" is infinite. The mathematical method for doing computations involving infinity is to take a limit of finite quantities. For any number b greater than a, the integral of f from a to b (that is, the area under the graph of f between x=a and x=b) is a well-defined, finite quantity. If there is in fact a finite total area under the graph of f from a to infinity, then as b gets larger and larger the area between a and b will approach the total area from a to infinity. Therefore we define the integral of f from a to infinity to be the limit as b approaches infinity of the integral of f from a to b, if this limit exists.

To look at a concrete example, consider the integral of the function f(x) = 1/(x²) from, say, 1 to infinity. For any finite b > 1, the integral of f from 1 to b equals 1 - 1/b. As b approaches infinity, this quantity approaches 1, and hence the improper integral from 1 to infinity is defined to be equal to 1. In other words, there is one unit of area under the graph of y = 1/(x²) to the right of x=1. To understand how this can be, look at this graph. Although the graph stretches infinitely far to the right, it is shrinking rapidly as it does so. Intuitively speaking, an improper integral over an infinite interval will be finite if, on its graph, the height shrinks "faster" than its width grows. For an example of a function that shrinks as x approaches infinity but does not shrink fast enough, consider the integral of f(x) = 1/x from 1 to infinity. The integral from 1 to b equals ln(b), which approaches infinity as b approaches infinity. Thus the improper integral does not exist.

An example of an improper integral does not exist even though the area in question does not become infinite is the integral of sin x from 0 to infinity. As b approaches infinity, the area under y=sin x between 0 and b perpetually fluctuates between 0 and 1, and consequently the limit does not exist.

An improper integral from negative infinity to some finite value a is defined in the analogous way, by taking **limits** of proper integrals as their left endpoints approach negative infinity. An improper integral from negative infinity to positive infinity is defined by taking any value a and adding the values of the improper integrals from negative infinity to a and from a to positive infinity, provided *both* exist.

Now suppose f is a function, defined on a half-open interval (a,b], whose values approach infinity as x approaches a. In contrast to the integrals considered above, here we are trying to measure an area whose horizontal **dimension** is finite but whose vertical dimension is infinite. Intuitively, whether or not this area is finite will depend on the rate of growth of the function f as x approaches a. Formally, the integral of f from a to b is defined by computing the integral of f from c to

b, where c is a value between a and b, and taking the limit as c approaches a. For example, the integral of f(x) = x^(-1/3) from 0 to 1 is improper because f(x) approaches infinity as x approaches 0. For any c between 0 and 1, the integral of f from c to 1 equals (3/2)(1 - c^(2/3)). As c approaches 0, this quantity approaches 3/2, so this is the value of the improper integral.

The **definition** is similar for an integral that is improper at its right endpoint instead of its left. For an integral from a to b that is improper at both endpoints, a value c is chosen between a and b, and the improper integrals from a to c and from c to b are added together, provided they both exist. Finally, an integral from a to b can also be improper at a value c between a and b. In this case, once again the improper integrals from a to b and from c to b are added, if they both exist.

INDEPENDENT VARIABLE

An independent **variable** is the variable that is acted upon by a function. In the traditional variable set-up of an equation like y = 3x + 7 or y = x³, x is the independent variable. That is, y depends upon x. Often, **equations** can be rewritten so that the independent variable appears to be the dependent one and vice versa. However, in statistical and scientific contexts, it usually becomes clear from the context of the application of the equation which is which.

In an experimental setting, the independent variable is the one that a researcher can set himself or herself. For example, if one is examining the magnitude of gravitational attraction between bodies as a function of the mass of one of those bodies, the mass would be the independent variable, the part the researcher can change with impunity. In other words, this is the source of direct change in the experiment. It is also sometimes called the "control" variable, while the **dependent variable** is the "experimental" variable.

In many situations, there will be more than one independent variable for a dependent variable. These are usually designed so that each can be changed independently of each other, with whatever convolutions are necessary. Each function can have many different partial derivatives that are taken with respect to the separate independent variables. In a properly designed experiment, the independent variable parameters will be varied separately from each other so that the effects of each are clear. Real world situations rarely reflect a properly designed experiment.

See also Dependent variable

INDETERMINATE LIMITS

Indeterminate **limits** are mathematical expressions that cannot be assigned a unique value. Although these expressions have no meaning per se, they frequently arise in limit problems in **calculus**. For example, to compute the derivative (**slope**) of the **sine** curve at any point on the curve, one needs to evaluate the limit of sin(x)/x as x approaches 0. Simple substitution of x =

0 into this expression does not produce a satisfactory result: It gives the indeterminate limit 0/0. But a more careful approach, using the "sandwich principle" of calculus, shows that the correct value of the limit is actually 1.

Generally speaking, a limit of a quotient of two **functions**, $f(x)/g(x)$, is said to be indeterminate of the form 0/0 if both $f(x)$ and $g(x)$ approach 0. The actual limit of $f(x)/g(x)$ in such a case depends on more detailed information about these two functions - intuitively, which one approaches 0 "faster." A powerful tool for evaluating such limits is L'Hospital's Rule, which provides the needed information by comparing the derivatives of $f(x)$ and $g(x)$.

Besides 0/0, the other commonly occurring indeterminate limits are (infinity)/(infinity), (**infinity**) - (infinity), $1^{\text{(infinity)}}$, (infinity)^0, and 0^0. **L'Hôpital's Rule** applies directly to the first of these; the other four generally need to be transformed algebraically before l'Hôpital's Rule can be applied.

See also L'Hôpital's Rule

INDUCTION

In mathematics as well as logic, the word induction means "generalization." In logic, it is a process of reasoning from particular situations or conditions to general ones in order to arrive at a conclusion about other similar situations. This is similar to its use in mathematics. Mathematical induction is the use of a formula to prove and analyze. In both fields, the justification for the usefulness of induction is the assumption that if something is true in a number of observed situations, it must also be true in similar but, as yet, unobserved or unproven situations. The probability that an outcome arrived at through induction is accurate depends, in part, on the number of situations observed that show a particular outcome. In mathematics, the induction principle says that if you can prove every other prediction about a series or equation, then the assumption in question must also be true. One of the simplest examples of induction is the interpretation of opinion polls, in which the answers given by a small percentage of the total population are assumed to be the same answers that would be given by the entire country. If 60 out of 100 people use Brand X laundry detergent to wash their clothes, then by induction it can be assumed that 600,000 out of a million do also. Of course, a poll that samples many more than 100 people and people form different parts of the country and many different income letters would povide a more accurate prediction.

INEQUALITIES

Inequalities are among the most important technical tools in mathematics. The concept of an inequality is at least as old as the concept of number, and yet it is only in relatively modern times that inequalities have been studied in a systematic way. Inequalities often occur implicitly in the statement that a function is convex or that a function of two variables defines a metric.

To begin with we will consider inequalities that involve finite **sets** of **real numbers**. Let $x_1, x_2,..., x_N$ and $y_1, y_2,..., y_N$ be two sets of real numbers. Then *Cauchy's inequality* asserts that

$$\left| \sum_{n=1}^{N} x_n y_n \right| \leq \left(\sum_{n=1}^{N} x_n^2 \right)^{\frac{1}{2}} \left(\sum_{n=1}^{N} y_n^2 \right)^{\frac{1}{2}},$$

and *Minkowski's inequality* is

$$\left(\sum_{n=1}^{N} (x_n + y_n)^2 \right)^{\frac{1}{2}} \leq \left(\sum_{n=1}^{N} x_n^2 \right)^{\frac{1}{2}} + \left(\sum_{n=1}^{N} y_n^2 \right)^{\frac{1}{2}}$$

Both of these inequalities have important geometrical interpretations. Suppose that **x** is a vector in $\Re^N$ with coordinates $x_1, x_2,..., x_N$ and **y** is a vector in $\Re^N$ is coordinates $y_1, y_2,..., y_N$. Then the Euclidean norm of **x** is defined by

$$\|\mathbf{x}\|_2 = \left(\sum_{n=1}^{N} x_n^2 \right)^{\frac{1}{2}},$$

and the norm of **y** is defined in a similar manner. The inner product of **x** and **y** is written $\langle \mathbf{x}, \mathbf{y} \rangle$ and defined by

$$\langle \mathbf{x}, \mathbf{y} \rangle = \sum_{n=1}^{N} x_n y_n.$$

We recall that two vectors **x** and **y** in $\Re^N$ are *orthogonal* if $\langle \mathbf{x}, \mathbf{y} \rangle = 0$. Using these concepts Cauchy's inequality can be expressed as

$|\langle \mathbf{x}, \mathbf{y} \rangle| \leq \|\mathbf{x}\|_2 \|\mathbf{y}\|_2,$

and Minkowski's inequality becomes the **triangle** inequality

$\|\mathbf{x} + \mathbf{y}\|_2 \leq \|\mathbf{x}\|_2 + \|\mathbf{y}\|_2$

for the Euclidean norm. There are also inequalities in which the inner product $\langle \mathbf{x}, \mathbf{y} \rangle$ is replaced by some other bilinear form in **x** and **y**. One of the most important of these is *Hilbert's inequality*. One form of Hilbert's inequality states that

$$\sum_{m=1}^{N} \sum_{n=1}^{N} \frac{x_m y_n}{m + n} \leq \pi \|\mathbf{x}\|_2 \|\mathbf{y}\|_2.$$

The constant π which occurs in this inequality is the smallest possible constant, independent of N, for which the inequality holds for all vectors **x** and **y**.

Both Cauchy's inequality and Minkowski's inequality can be generalized by introducing the *p-norms*, where $1 \leq p \leq \infty$. If $1 \leq p < \infty$ and **x** is a vector in $\mathfrak{R}^N$ we define

$$\|\mathbf{x}\|_p = \left(\sum_{n=1}^{N} |x_n|^p \right)^{\frac{1}{p}},$$

and in case $p = \infty$ we set
$\|\mathbf{x}\|_\infty = \max\{|x_1|, |x_2|,..., |x_N|\}.$

Of course the Euclidean norm is the special case $p = 2$. If $1 \leq p \leq \infty$ we define q in the **range** $1 \leq q \leq \infty$ by $p^{-1} + q^{-1} = 1$. Here we use the convention that if $p = 1$ then $q = \infty$, and if $p = \infty$ then $q = 1$. Then *Hölder's inequality* states that
$|\langle \mathbf{x}, \mathbf{y} \rangle| \leq \|\mathbf{x}\|_p \|\mathbf{y}\|_q,$
and the general form of Minkowski's inequality for the p norm is
$\|\mathbf{x} + \mathbf{y}\|_p \leq \|\mathbf{x}\|_p + \|\mathbf{y}\|_p.$

There is also an important inequality comparing these norms. In order to state this inequality we assume that $0 < r < s < \infty$, but we make no further assumptions about r and s. Then if **x** is a vector in $\mathfrak{R}^N$ we have

$$\left(\sum_{n=1}^{\infty} |x_n|^s \right)^{\frac{1}{s}} \leq \left(\sum_{n=1}^{\infty} |x_n|^r \right)^{\frac{1}{r}}.$$

Usually an inequality involving a finite sum has an analogous, but more general, statement in which the finite sum is replaced by a suitable **infinite series** or by an integral. Suppose, for example, that $1 < p < \infty$, $1 < q < \infty$, $p^{-1} + q^{-1} = 1$, and both

$$\sum_{n=1}^{\infty} |x_n|^p < \infty \quad \text{and} \quad \sum_{n=1}^{\infty} |y_n|^p < \infty.$$

Then it follows from Hölder's inequality that the infinite series

$$\sum_{n=1}^{\infty} x_n y_n$$

converges absolutely and satisfies

$$\sum_{n=1}^{\infty} |x_n y_n| \leq \left(\sum_{n=1}^{\infty} |x_n|^p \right)^{\frac{1}{p}} \left(\sum_{n=1}^{\infty} |y_n|^q \right)^{\frac{1}{q}}.$$

Let $(u,v) \subseteq \mathfrak{R}$ be an open **interval**, and here we include the possibility of an infinite interval. A function $\varphi:(u,v) \rightarrow \mathfrak{R}$ is said to be *convex* on the interval (u,v) if it satisfies the inequality
$\varphi(\theta x_1 + (1 - \theta)x_2) \leq \theta\varphi(x_1) + (1 - \theta)\varphi(x_2)$
for all real numbers x_1 and x_2 in the interval (u,v), and all real numbers θ with $0 < \theta < 1$. The graph of a convex function has a simple geometrical property: if φ is convex then the

straight line segment connecting two distinct points on the graph of $y = \varphi(x)$ always lies above the graph of $y = \varphi(x)$. If $\varphi: (u,v) \rightarrow \infty$ has a second derivative at each point of the interval (u,v), then φ is convex on the interval if and only if $\varphi''(x) \geq 0$ at each point x in (u,v). Using this criterion it is easy to see that the exponential function $\exp(x)$ (also written e^x) is convex on $\mathfrak{R}$. Therefore we get the inequality
$\exp(\theta x_1 + (1 - \theta)x_2) \leq \theta \exp(x_1) + (1 - \theta) \exp(x_2)$
for all real numbers x_1 and x_2, and all θ with $0 < \theta < 1$. Alternatively, if we write $y_1 = \exp(x_1)$ and $y_2 = \exp(x_2)$ we obtain the inequality

$$y_1^\theta y_2^{(1-\theta)} \leq \theta y_1 + (1 - \theta)y_2,$$

for all positive real numbers y_1 and y_2, and all θ with $0 < \theta < 1$. This is one form of the *arithmetic mean–geometric mean* inequality. A more common form of the inequality, which also follows from the fact that the exponential function is convex, is

$$\left(y_1 y_2 \cdots y_N \right)^{1/N} \leq \frac{y_1 + y_2 + \cdots + y_N}{N},$$

which holds for all finite sets of nonnegative numbers $y_1, y_2,... ,y_N$. The left hand side of the inequality is often called the geometric **mean** of the nonnegative numbers $y_1, y_2,... ,y_N$, while the right hand side is the **arithmetic** mean of the numbers $y_1, y_2,... ,y_N$.

Let $A = (a_{mn})$ be an $N \times N$ **matrix** with real entries. As usual we suppose that $m = 1, 2,..., N$ indexes rows and $n = 1, 2,..., N$ indexes columns. *Hadamard's inequality* provides an upper bound for the determinant of the matrix A in terms of the Euclidean norms of the columns (or rows) of the matrix. The precise statement of Hadamard's inequality is

$$|\det A| \leq \prod_{n=1}^{N} \left(\sum_{m=1}^{N} |a_{mn}|^2 \right)^{\frac{1}{2}}.$$

An alternative formulation of the inequality can be given by writing $\mathbf{a}_1, \mathbf{a}_2,..., \mathbf{a}_N$ for the N columns of the matrix A. Then the Euclidean norm of the vector $\mathbf{a}_n$ in $\mathfrak{R}^N$ is given by

$$\|\mathbf{a}_n\|_2 = \left(\sum_{m=1}^{N} |a_{mn}|^2 \right)^{\frac{1}{2}}.$$

In terms of these norms Hadamard's inequality can be written as

$$|\det A| \leq \prod_{n=1}^{N} \|\mathbf{a}_n\|_2.$$

Now consider the parallelepiped in $\Re^N$ having the vectors $\mathbf{a}_1, \mathbf{a}_2,..., \mathbf{a}_N$ as its edges. The **volume** of this parallelepiped is exactly $|\det A|$. And then $\|\mathbf{a}_n\|_2$ is the length of the edge determined by the vector $\mathbf{a}_n$. Thus we see that Hadamard's inequality has an important geometrical interpretation: the volume of a parallelepiped is less than or equal to the product of the lengths of its edges. If the vectors $\mathbf{a}_1, \mathbf{a}_2,..., \mathbf{a}_N$ are orthogonal then the parallelepiped is rectangular and its volume is equal to the product of the lengths of its edges. This shows that there is equality in Hadamard's inequality in case the vectors $\mathbf{a}_1, \mathbf{a}_2,..., \mathbf{a}_N$ are orthogonal.

INFERENCE

In argumentation, inference is a reasoning process expressed by an argument. In mathematics, inference is the reasoning process expressed by computation, calculation, and measurement. In general, inference is a rational movement from one concept to another. It is most easily understood by reference to a hypothetical, or "if-then" proposition. For example, the proposition, "If it is raining, then it is wet," illustrates an inference made from the concept of rain to the concept of wetness. We infer that it is wet from the assertion that it is raining. Another way of expressing inferential reasoning is by the logical rule of inference called **implication**. Without the ability to make inferences, rational processes would be disconnected.

The process of inference is the derivation of a proposition (the conclusion) from one or more **propositions** (the premises) in an argument. Depending upon the validity of the inference, there is good reason to assert that the premise or premises do indeed support the conclusion. The proposition, "Larry is sleeping," can be inferred from the propositions, "If Larry is at home then he is sleeping" and "Larry is at home," with good reason because the relationships between the terms in the premises provide inferential support for the terms in the last proposition. This means that, if the premises are true, the conclusion must be true on pain of contradicting the premises.

Similarly in mathematics, inferences are made based on the relationships between numbers. For example, we can infer the number twelve when we add together the numbers five and seven. By themselves the latter numbers do not imply the number twelve, but the process of **addition** allows us to combine five and seven to conclude twelve.

In logic there are two basic forms of inference. One is **deduction** and the other is **induction**. Deductive inferences are necessary because, as the Larry example shows, if the premises are true then the conclusion may not be false without contradicting the premises. Inductive inferences, on the other hand, are only probable. In an inductive argument, the **truth** of the premises means that it is improbable for the conclusion to be false. If the conclusion is false, however, while the premises are true, there is no contradiction between premises and conclusion. For example, "Larry will be at home today," follows with probability from "Larry has been at home for the last few days." The conclusion is not inferred with necessity, however, because it is possible that Larry is not at home today.

See also Implication; Proof by deduction; Proof by induction

INFINITE REGRESS

An infinite regress is a series of causes and effects that go on indefinitely. If every event must have a cause, then any cause will also have a cause, and so on endlessly. Suppose someone asks what makes the grass grow. An explanation is provided by way of a cause. In turn, the cause of the cause of grass growing is questioned, and so on. Another, related way, of defining infinite regress is in terms of explanations that are not causal. Still, the explanation is dependent on a previous explanation, and so on. Many thinkers stop the infinite regress by supposing a first cause, or ultimate explanation such as God. The infinite is typically defined as that which is endless, boundless, without limit. Such definitions imply that **infinity** is a numerical and physical concept, yet by definition, it cannot be measured.

The problem of infinite regress for mathematics is best illustrated, perhaps, by one of **Zeno's paradoxes**. Although it is not unsolvable, it may help to explain the basic problem of a mathematical infinite regress. In one paradox, Achilles races a tortoise, and because the tortoise is supposed to be much slower than Achilles, the tortoise is allowed a head start. According to **Zeno**, before Achilles can reach and pass the tortoise, he must first reach the place from which the tortoise started. By the time Achilles has reached that point, the tortoise has advanced to another, farther point. Now, Achilles must get to this advanced point, and so forth. The **distance** traveled is the sum of an infinite number of distances, and the problem is that before any distance is traveled, an infinite number of distances must be traveled. This represents an infinite divisibility of **space** (and also implies the same for time and number).

INFINITE SERIES

An infinite sequence is an ordered listing of an infinite number of **real numbers**, such as ½, ¼, 1/8, 1/16,... In more formal mathematical language, an infinite sequence is a function whose **domain** is the set of natural numbers (sometimes with 0 included) and whose **range** is the set of real numbers. An infinite series is just the sum of all the terms of an infinite sequence. For example, $1/2+1/4+1/8+...+1/2^n+...$ designates the infinite series whose nth term has the form $1/2^n$. In general if the nth term of an infinite series is a_n, then the series is written $a_1+a_2+a_3+...+a_n+...$, which is also written as $\sum a_n$ where the capital Greek letter $\sum$ stands for the summation of all terms of the form a_n. Of interest to mathematicians is which infinite series have finite sums and which grow without bound as each new term is added. It may surprise someone who is unfamiliar with infinite series to hear that some of them actually do have finite sums. A naïve view might be that because they have an infinite number of terms being added together their sums must always be infinite; but consider the infinite series above: $1/2+1/4+1/8+...+1/2^n+...$. Mathematicians define a sequence of "partial sums", S_1, S_2, S_3,... where, in general, S_n is the sum of the first n terms of the infinite series. Thus, for our series, $S_1=1/2$, $S_2=3/4$, $S_3=7/8$, and, in general, $S_n=(2^n-1)/2^n=1-1/2^n$. Now as n grows without bound, S_n gets ever closer to 1. Clearly S_n can never become greater than 1, so mathematicians agree that the sequence of partial sums for this infinite series approaches 1 and this is defined to be the sum of the series $1/2+1/4+1/8+...+1/2^n+...$. Thus, in general, the sum of an infinite series is the limit of the sequence of partial sums as n grows without bound, provided that such a limit exists. In this case, the infinite series is said to converge to this limit. If the sequence of partial sums continues to grow without bound as n grows without bound, then the infinite series does not have a sum and is said to diverge. In **calculus**, a good deal of time is devoted to determining which infinite series converge and which do not. The problem is that it is not always simple or even possible to get a formula for the sequence of partial sums, so mathematicians have needed to devise a number of **convergence** tests and to prove that they work for certain classes of infinite series.

The reason that questions of convergence are so important is that infinite series are used as representations of many of the most familiar mathematical **functions** to approximate values of these functions. Such approximations are valid only if they are carried out within the **interval** of convergence of the series being used. The interval of convergence is just the set of all real numbers for which the infinite series converges. The use of infinite series as function representations is due to the work of the English mathematician **Brook Taylor** (1685-1731), who showed how to construct such series using techniques of calculus. Such series are now called Taylor series. An example is the Taylor series for the familiar **sine** function from **trigonometry**. Using Taylor's methods, one can show that the infinite series representation of $\sin(x)$ is $x-x^3/3!+x^5/5!-x^7/7!+...+(-1)^{n+1}x^{2n+1}/(2n+1)!+...$, where $n=0,1,2,3...$. It can also be shown that the interval of convergence for this infinite series is the set of all real numbers. Algorithms based upon series expansions of functions are used in calculators and computers to give highly accurate approximations of those functions. Thus when one pushes the sin button on a **calculator** followed by a number, the calculator does not search in some infinite table of values to find the sine of just this number. Rather it computes the sum of a sufficient number of terms of a series expansion for the sine function. Taylor's theory also allows the mathematician or engineer to determine how many terms of an infinite series expansion must be used to get a desired number of decimal places of accuracy.

Infinite series may be used to derive a number of remarkable mathematical results. For example, it can be shown that the Taylor series expansion of the inverse tangent function, denoted $\arctan(x)$ is $x-x^3/3+x^5/5-x^7/7+...$ Now the inverse tangent of 1 is $\pi/4$, which means that $\pi/4=1-1/3+1/5-1/7+...$or $\pi=4-4/3+4/5-4/7+...$ Early attempts to give π to a specifed number of decimal places made use of this infinite series. The only problem with this method is that this infinite series converges to π very slowly, so that modern computer programs for generating millions of decimal places of π use more sophisticated algorithms. On a different topic, a famous paradox posed by the Greek philosopher **Zeno** (5th century BC) can be explained by the theory of infinite series. Zeno said that if one is standing at point A one meter from point B, one can never reach point B by traversing the line segment connecting points A and B. Zeno's reasoning was that to get from A to B, one would first have to travel half the **distance** or ½ meter, then half the remaining distance or ¼ meter, then half again or 1/8 meter, and again and again and again..., so that one's total distance would be computed by ½+1/4+1/8+1/16+... Since this is the sum of an infinite number of terms, Zeno claimed that the distance could never be traversed. As we saw above, however, the partial sums of Zeno's infinite series approach 1 as a limit, so that the by the mathematical **definition** of the sum of an infinite series, this total distance is 1. The subject of **infinity** has vexed mathematicians throughout recorded history, prompting them to seek carefully thought out explanations for the paradoxes, such as Zeno's, that occur when the term "infinity" is not treated with precise mathematical definitions. The work of Taylor and other mathematicians in infinite series has been just one example of this quest to understand the infinite.

See also Geometric series

INFINITESIMAL

An infinitesimal is a numerical quantity small enough that its limit in some condition is **zero**. Infinitesimals are widely used in **calculus**, where the formal, limit-based **definition** of a derivative depends upon an arbitrary additive quantity going to zero. In **integral calculus**, the **definite integral** can be thought of as the sum of infinitesimal slices of the area. In fact, calculus was originally thought to be infeasible because it required the use of infinitesimals, which were then thought to be mathematically sketchy at best. Calculus proofs were worked out in epsilon-delta notation, but the intuitive explanations of calculus were most commonly used with infinitesimal rectangles, infinitely small straight lines making up a curve, and other such infinitesimal-related imagery.

In the 1960s, Abraham Robinson founded nonstandard **analysis**, a branch of mathematics that constructs a hyperreal set of numbers, or a set of numbers that is an extension to the usual set of **real numbers** and does not follow all of the same axioms. These numbers can be used to construct mathematically viable infinitesimals, and the system thus constructed can be used to write calculus proofs using the same intuitive ideas as **Leibniz**, **Newton**, and **Cauchy**.

The term infinitesimal is used somewhat more colloquially, even in math circles, to mean "exceptionally small" or "so small that it may be neglected." The **distance** between real numbers can be called infinitesimal in this usage, and a term in a series can be so comparably small with respect to the other terms that it is "practically infinitesimal." However, this non-rigorous usage of the word is not recommended for composing mathematical proofs.

INFINITY

The term infinity conveys the mathematical concept of large without bound, and is given the symbol ∞. As children, we learn to count, and are pleased when first we count to 10, then 100, and then 1,000. By the time we reach 1,000, we may realize that counting to 2,000, or certainly 100,000, is not worth the effort. This is partly because we have better things to do, and partly because we realize no matter how high we count, it is always possible to count higher. At this point we are introduced to the infinite, and begin to realize what infinity is and is not.

Infinity is not the largest number. It is the term we use to convey the notion that there is no largest number. We say there is an infinite number of numbers.

There are aspects of the infinite that are not altogether intuitive, however. For example, at first glance there would seem to be half as many odd (or even) **integers** as there are integers all together. Yet it is certainly possible to continue counting by twos forever, just as it is possible to count by ones forever. In fact, we can count by tens, hundreds, or thousands, it does not matter. Once the counting has begun, it never ends.

What of **fractions**? It seems that just between **zero** and one there must be as many fractions as there are positive integers. This is easily seen by listing them, 1/1, 1/2, 1/3, 1/4, 1/5, 1/6, 1/7, 1/8,.... But there are multiples of these fractions as well, for instance, 2/8, 3/8, 4/8, 5/8, 6/8, 7/8, and 8/8. Of course many of these multiples are duplicates, 2/8 is the same as 1/4 and so on. It turns out, after all the duplicates are removed, that there is the same number of fractions as there are integers. Not at all an obvious result.

In addition to fractions, or **rational numbers**, there are **irrational numbers**, which cannot be expressed as the ratio of **whole numbers**. Instead, they are recognized by the fact that, when expressed in decimal form, the digits to the right of the decimal point never end, and never form a repeating sequence. Terminating decimals, such as 6.125, and repeating decimals, such as $1.33\overline{3}$ or $6.53\overline{4}$ (the bar over the last digits indicates that sequence is to be repeated indefinitely), are rational. Irrational numbers are interesting because they can never be written down. The instant one stops writing down digits to the right of the decimal point, the number becomes rational, though perhaps a good approximation to an irrational number.

It can be proved that there are infinitely more irrational numbers than there are rational numbers, in spite of the fact that every irrational number can be approximated by a rational number. Taken together, the rational and irrational numbers form the set of **real numbers**.

The word infinite is also used in reference to the very small, or **infinitesimal**. Consider dividing a line segment in half, then dividing each half, and so on, infinitely many times. This procedure would results in an infinite number of infinitely short line segments. Of course it is not physically possible to carry out such a process; but it is possible to imagine reaching a point beyond which it is not worth the effort to proceed. We understand that the line segments will never have exactly zero length, but after a while no one fully understands what it means to be any shorter. In the language of mathematics, we have approached the limit.

Beginning with the ancient Greeks, and continuing to the turn of the twentieth century, mathematicians either avoided the infinite, or made use of the intuitive concepts of infinitely large or infinitely small. Not until the German mathematician, **Georg Cantor** (1845-1918), rigorously defined the **transfinite numbers** did the notion of infinity finally seem fully understood. Cantor defined the transfinite numbers in terms of the number of elements in an infinite set. The natural numbers have $\aleph_0$ elements (the first transfinite number). The real numbers have $\aleph_1$ elements (the second transfinite number). Then, any two **sets** whose elements can be placed in 1-1 **correspondence**, have the same number of elements. Following this procedure, Cantor showed that the set of integers, the set of odd (or even) integers, and the set of rational numbers all have $\aleph_0$ elements; and the set of irrational numbers has $\aleph_1$ elements. He was never able, however, to show that no set of an intermediate size between $\aleph_0$ and $\aleph_1$ exists, and this remains unproved today.

INFORMATION

Information is the fundamental concept of **information theory**, the mathematical study of symbolic communication. Information theory was founded in 1948 by one seminal paper, Claude Shannon's "The Mathematical Theory of Communication," and has since proved essential to the development of computers, telecommunications, digital music recordings, and many other technologies of the "Information Age." It has also been applied fruitfully in cryptography, genetics, linguistics, and other disciplines.

Shannon, a quirky mathematician and electrical engineer famous for juggling while riding his unicycle down the hallways of Bell Laboratories in the 1940s and 50s, was not the first theorist to ponder the subject of "information." He was, however, the first to define the term rigorously and to specify its unit of measure—the now-famous "bit" (short for "binary digit").

Shannon defined "information" in what might at first seem an odd way: not as a substance that can exist in fixed quantities, although this is how information is usually spoken of, but as a change in another, more fundamental quantity: as a reduction in uncertainty. "Uncertainty" must thus be defined before "information." Consider the following scenario:

Someone is waiting to receive a message over some communications channel, perhaps a telephone. The message will specify a number chosen somehow at the other end of the channel. Before the message arrives there is, obviously, a certain degree of uncertainty, the precise amount of which depends on how many numbers the sender is choosing from (the more choices, the more uncertainty) and on how likely each choice is (if one choice is a million times more likely than all the others put together, there is little uncertainty). Unlike psychological uncertainty or doubt, which is a state of mind, this "uncertainty" is a quantity that can be calculated precisely and expressed in units of bits.

When the message is received, this original uncertainty is reduced. By how much? There are two basic possibilities: 1) If the communications channel is noiseless (perfectly reliable), then the uncertainty has been reduced to zero, for the person receiving the message now knows what number was sent. 2) If the channel is noisy, the person receiving the message cannot be perfectly sure that the number they have received is that was sent. In this case uncertainty has been reduced by receipt of the message, but not to zero; some doubt may remain. Either way, something has been learned: uncertainty has decreased. Shannon defined this decrease as the information gained by the receiver.

Because the word "information" has so many everyday meanings, a few cautions are in order. First, numbers as such (or other symbols) are not information. A string of numbers may be perfectly random, containing no information at all. Only symbols that reduce uncertainty about a well-defined question convey information in the mathematical sense. Second, as Shannon himself warned in 1948, information is irrelevant to meaning. Meaning is a mood or quality perceived subjectively by persons, and it cannot be discussed mathematically. A single bit of information may answer a life-and-death question or an utterly trivial one. Regardless of what meaning it conveys, however, it remains 1 bit of information, and that is all that can be said about it mathematically.

Until Shannon defined information and its unit of measure, it was not possible to give an exact definition of the capacity of a communications channel to transmit information. With these tools in hand, he was able to write down at once an exact expression for the maximum amount of information conveyable by any channel (the "channel capacity"); furthermore, he was able to prove that it is possible, while operating under this limit, to transmit information with as few errors as desired by paying a toll of redundancy.

With these upper and lower limits clearly in view, the design of practical systems could at last move forward on a systematic basis. Similarly, with an intuitive knowledge of fluids one might design hoses and cylinders, but until the technical concept of "pressure" is invented hydraulic design can only proceed by trial and error. In a sense, nothing was done after Shannon's invention of information theory that was not done before (codes, computers, and electronic communications systems had all been built prior to 1948), but afterward, it was done systematically.

Shannon's very general (and thus portable) approach transformed large areas of technology and science radically in just a few years. Information theory quickly became a key tool in telecommunications, encryption, error detection and correction, radar and sonar, seismology, radio astronomy, and other fields. Efforts were also made to apply its concepts in traditionally less mathematical disciplines such as economics, biology, linguistics, and psychology. Not all such attempts were successful. The study of the psychology of speech perception and attention, for example, was dominated by information theory throughout the 1950s, but a shortage of consistent experimental results led for the most part to abandonment of information-theoretic approaches in these fields.

Perhaps the discipline most profoundly connected to information theory is **thermodynamics**. Shannon's uncertainty measure is formally equivalent to the thermodynamic expression for entropy, and the two turn out to be intimately related. Just as a change in symbolic uncertainty corresponds to symbolic information, so a change in a physical system's entropy (roughly speaking, its state of disorder) corresponds to "thermodynamic information." In 1951, Leon Brillouin proved the extraordinary theorem that the observation of one bit of information by any possible physical system requires a minimum of $k_B T \ln(2)$ ergs of energy, where k_B is **Boltzmann**'s constant and T is the absolute temperature. As Dennis Gabor, the Nobel prize-winning inventor of holography, put it: "You cannot get something for nothing, not even an observation."

Today, applications of information theory are ubiquitous. Error-correction coding schemes ultimately derived from Shannon's work enable deep-space probes to transmit data across the Solar System despite high noise, immense distances, and signal strengths of only a few watts. They also allow near-perfect reproduction of music from compact discs in the presence of dust, scratches, and fingerprints.

INFORMATION THEORY

While researching methods for how to more efficiently transmit information over noisy communications channels, Claude Shannon, an electrical engineer, published "A Mathematical Theory of Communication" in 1948, which spawned two disciplines—information theory and coding theory. Shannon's paper captured the basic mathematical principles relevant to transmitting, receiving, and processing information via unspecified communications media, as well as fundamentally redefining how communications engineers and specialists perceive information. At its essence, information theory is a combination of elements of communications theory, probability, and **statistics**. Beyond providing a means to numerically measure the quantity of information to be transmitted, information theory also encompasses how the information is to be represented (that is, coded) during transmission, as well as the capacity of the communications system to transmit, receive, process, and store the information.

At its most basic level, a communications system consists of a message source (such as a telegraph, broadcast television station, or radio antenna), a device for encoding the message (which may be hardware, software, or a combination of both and which Shannon termed the *transmitter*), a communications channel over which the message is transmitted, a device for decoding the message (which generally performs an **inverse function** to the encoder and which Shannon termed the *receiver* of a message), and destination of the message (the person or thing for whom the message is intended). The message can take nearly any form imaginable, such as a sequence of alphanumeric characters (a document or electronic mail, for example), a sequence of numbers, (a bit stream such as 1 0 1 1 1 0 0 1, for example, or credit card number), or even ultra-high frequency (UHF) radio waves. Communication channels that attenuate, or distort, the message during transmission are called *noisy*, and the amount of noise added to the transmitted message will vary according to the construction and design of the communications channel, and perhaps even atmospheric conditions. Generally speaking, less noise results in fewer errors when the message is decoded, and information theory provides a means to mathematically evaluate how much noise a particular communications channel is adding to the transmitted message, as well as how much noise can be accepted before the message becomes garbled beyond recognition. It also provides a mechanism for determining the bandwidth—or capacity—required for transmitting a particular message.

Information theory rests on the idea that of the entire message to be transmitted, only the parts which provide new, or non-redundant, data are relevant. Telegraphers, for example, often omitted the words "the," "a," and "an" from messages. Including these articles was both expensive and redundant—the message recipient could easily determine where these articles should be re-inserted into the message. Omitting redundant articles, then, is a rudimentary form of data compression. Since only the data that is not redundant is relevant and in thus transmitted, information can be considered the probability that the actual message is selected from the set of all possible messages once the redundancies are removed.

In basic digital communication systems, information is transmitted in bits, an abbreviation for *bi*nary dig*its*. Under a uniform **binary number system**, one has two choices—zero or one—both of which are equally likely and thus each is assigned a probability of one-half. This means that for any given message of length one bit, the chance that a **zero** was transmitted is 50 percent and the chance that a one was transmitted is also 50 percent. If there are N equally likely possibilities, however, the number of bits needed to transmit the information equals the base 2 logarithm of N (that is, $\log_2 N$). If the N possibilities have unequal probabilities, then the number of bits associated with the message is the sum of base 2 **logarithms** of the reciprocals of the probabilities. If the N possibilities have probabilities $p_1, p_2,..., p_N$, for example, the number of bits associated with the message will be $\log_2 (1 / p_1) + \log_2 (1 / p_2) +... + \log_2 (1 / p_N)$. The expected value of the number of bits of information (that is, the number of bits of information, on average, that each message will hold) is known as the entropy.

One of the most useful results of information theory is Shannon's relationship between information capacity I (the amount of information—or number of bits—that can be trans-

mitted over a given communications channel in a specific amount of time), bandwidth B (the size of the communications channel), and the signal-to-noise ratio *(S / N)* (the strength of the message relative to the amount of noise introduced by the communications channel, with both signal strength and noise strength often measured in watts). Shannon concluded that $I = B * \log_2 [1 + (S / N)]$. This relationship provides an upper bound for the information capacity of communications channels that is still used today.

See also Binary number system; Binary, octal, and hexadecimal numeration; Logarithms; Probability theory; Set theory; Sets; Statistics; Zero

INSTANTANEOUS EVENTS

In the realm of mathematics, "instantaneous events" refer not to events that occur in an incredibly short amount of time, but to events that occur at an exact point within a defined vector **space**, including the **space-time** continuum. The duration of these events—that is, the time that elapses between the start of the event and the end of the event is not only infinitesimally small, it is **zero**. Instantaneous events capture a single, exact point in time. A good, but not perfect, analogy is a still picture produced by a camera. The picture captures a particular scene or object, or particular expression, at a single moment in time for later viewing. The conditions under which the photograph was taken can likely be reproduced with similar results—that is, results so similar to the original that differences are entirely negligible—but the event itself cannot be *exactly* reproduced.

Instantaneous events are closely related to the mathematical concepts of derivative and **limits**. Instantaneous values can provide a means to express the rate of change of one parameter of an object relative to another parameter of the same object. Instantaneous velocity, for example, provides an expression of how an object's position in space is changing relative to the existence of the object in time. This expression may encompass a number of sub-parameters, which can be considered dimensions in themselves.

For example, at an given moment in time, the position of a runner along a straight track is generally measured in just one direction—along the straight line, the axis of travel (call it x). The instantaneous event is the runner's speed and direction *at that exact moment in time*. A child playing hopscotch, however, moves horizontally—along both the straight line of the sidewalk (that is, along the x-axis)—*and* vertically (that is, in the air along a y-axis) as a result of the "hopping" **motion**. The instantaneous event captures the rates of change in two directions with respect to time. Similarly, the instantaneous velocity of an airplane 20 seconds after take-off will capture the rates of change in three directions (along the x-, y-, and z-axes, if the **Cartesian coordinate system** is used) with respect to time.

Since instantaneous values can describe the change of one parameter of an object *relative* to another, it is common to use a direction in space or time as the reference parameter. An emergency crew, for example, may be interested in the height of a flooding river over its normal water level. This instantaneous value may be critical to determining whether an area must be evacuated for safety.

Interestingly enough, it is not often possible to measure instantaneous values, even with the most sensitive measuring equipment. If a thermometer is placed in a refrigerator for a number of hours and then placed at room temperature, the mercury level (or digital display) will take a finite amount of time to register the change—perhaps seconds or even minutes. Like all scientific instruments, the thermometer is simply not sensitive enough to measure the room temperature *immediately* or instantaneously. It must measure the value over a finite amount of time and then report the average value of the measurement. Even the most sophisticated scientific instruments have limits on their accuracy. Speeds of an object, for example, may be measured to the nearest picosecond (1 x 10^{-12} second) but even then, the value is not instantaneous. Instead, an average value over an extremely small time **interval** is given.

For two events to occur instantaneously, they must somehow overlap over a finite period of time, which may be so small as to appear negligible to the viewer. In digital computer architectures, the concept of instantaneous events is most often referred to as a *collison* and is defined as the arrival of two pieces of information (such as a bit) at the same location during the same time *interval*. This collision occurs because even the fastest computers operate according to an internal clock, processing a bit of information in a finite period of time. If two bits of information are traveling along independent paths that then merge, the bits have the potential to collide since the processor in *after* the merge point handles bits as discrete objects, not continuous objects. If so much data is being processed that the bit stream must be queued, it is possible that two items of data will try to "merge" into the same place in the queue during the same time intervals. These happenstances are collisions.

See also Derivatives and differentials; Infinitesimal; Instantaneous velocity; Limits; Space-time; Vector spaces; Zero

INSTANTANEOUS VELOCITY

Instantaneous velocity is the vector expression of the speed an object is moving at any given second. That is, it takes into account both how fast the object is moving and in what direction. Instantaneous velocity can be calculated by taking the derivative of the position with respect to time and evaluating the answer at a known time.

Most of the time, average velocity is used rather than instantaneous velocity in everyday applications. For example, a speedometer in an automobile does not have an abstract position function, of which it would take the derivative. Instead, it measures the **distance** traveled in fractional amounts of time and reports the speed of the automobile as a time average on a very short time. For the velocity reported to

truly be the instantaneous velocity, the time frame would have to become **infinitesimal**.

However, the instantaneous velocity can be found in situations where the position function is well known. For example, an object in free-fall can easily have its instantaneous velocity determined, since the equation for its position as a function of time is well known. In quantum mechanical applications, the instantaneous velocity is involved in the **Heisenberg** uncertainty relation of position and momentum, and thus it and the position cannot be simultaneously determined if the mass is known. However, in most aspects of physics, the instantaneous velocity is a clear demonstration of the relation of physics to **calculus**.

See also Differentiating distance and velocity; Mathematics and physics

INTEGERS

Integers are **whole numbers**, including positive numbers, **negative numbers**, and **zero**.

...-3, -2, -1, 0, 1, 2, 3...

Integers have unique properties that have always interested mathematicians. For example, anyone who adds, subtracts, or multiplies two integers will always have an integer as the result. However, dividing two integers can result in a noninteger.

Throughout history, integers have been the building blocks for many advances in mathematics. The group—or set—of positive numbers, negative numbers and zero can also be called rational integers. Many think of positive integers as the first system humans developed for counting possessions. Ancient people probably used their fingers or small stones to count whole numbers. Positive integers are also called the counting numbers.

Records 4,000 years old show that Babylonians looked at the properties of positive integers (now thought of as 'pure' **arithmetic**). **Diophantus of Alexandria**, Greece, revised the system of **algebra** first developed by the Babylonians. Under the Diophantine method of **analysis** that evolved around A.D. 250, **equations** were solved with integers. The problems could involve single equations or **systems of equations** containing two or more unknowns, but the solutions were always whole numbers.

Many advances in the European study of mathematics began with the study of integers. **Carl Friedrich Gauss** (1777-1855), one of the most influential scholars in mathematics, enrolled at Göttingen University in Germany in 1795. While Gauss was an undergraduate, he spent three years writing a classic book on the mathematics of integers, *Arithmetical Disquisitions*. **Albert Einstein** later praised the importance of Gauss's many contributions to the field of mathematics and showed how Gauss's work helped Einstein develop the theory of **relativity**.

Other European mathematicians were intrigued with the challenge of defining integers. This was a problem that had to be resolved as part of the process of developing definitions of rational arithmetic. The mathematician **Leopold Kronecker** (1823-1891) tried to devise a system of math that depended on positive whole numbers and was so convinced of the key role of integers that he declared, only partly in jest, "God made the integers, all the rest is the work of man."

The importance of mathematics to other fields of study was aptly illustrated by the research of scientists John Dalton (1766-1844) and Dmitry Ivanovitch Mendeleev (1848-1907). Dalton built his work on the study of the atom in ancient Greece and noted that the atomic weights for all of the compounds he measured were **ratios** of simple integers, such as 4:1. The ratios were always whole numbers, never **fractions**. This observation led Dalton to conclude that elements could only combine in certain ways to form compounds. Building on Dalton's work, Mendeleev observed the relationships of valences and chemical elements and helped to develop the periodic law for the properties of chemical elements. The curiosity about whole numbers and measurements of elements and compounds led directly to the periodic table used by contemporary chemists. While the discoveries of Dalton and Mendeleev were not mathematically complex, they were of great scientific significance. Mathematics became more than a tool for analyzing numbers and was used in new ways to advance physics and chemistry.

INTEGRAL CALCULUS

The study of **calculus** can be separated into two components: derivative or **differential calculus** and integral calculus. Integral calculus can be thought of as dealing with the **area** under curves, but it also touches on two- and three-dimensional situations with path and surface integrals. On the most basic level, with finding the area under curves, integrals can be the reverse of derivatives—hence their other name, anti-derivatives. There are two major types of integrals on the lowest level: the definite and the indefinite. The **definite integral** provides a numerical **range** over which the integral is to be calculated; the indefinite integral is looking for a closed-form solution to which any range can be applied.

Integral calculus is often used in conjunction with **differential equations**. A differential equation will, as its name indicates, provide information about the rate of change of a function in its differential form. The process of solving the equation then usually results in an equation with an integral on each side. These one-dimensional integrals are solved with a variety of direct, numerical, and series form methods.

A change to two or three dimensions in an integral will work as if they are nested unrelated problems as long as the **variable** over which one integrates is not a vector. However, there are some situations in which the direction is important. Path integrals can be solved along two- or three-dimensional surfaces, but the direction of integration is important, in that the rate of change (the differential inside the integral) may vary according to how one moves from point A to point B. Therefore, a path of integration must be specified. Surface

integrals generalize integration to three dimensions and deal with all three at once, so the direction of integration and the perpendicular direction are both important when calculating these integrals.

All of these major studies within integral calculus have uses in the physical world as well as in mathematical theory. The study of complex variables requires a firm basis of knowledge of integral calculus, and the physical sciences use both its anti-differential closed form and in its calculation of cumulative results or total areas in its definite form. Integral calculus provides both a test of experimental results and a way to formulate information about new theories, when it is applied to the sciences. In computer science, the numerical side of integral calculus is more important for understanding of many internal calculations, but the whole discipline is still vital for success in computer theory. In mathematics, its tenets and associated skills are assumed even when they are not directly used—and they are directly used frequently.

INTEGRATING ACCELERATION AND VELOCITY

The ideas of acceleration and velocity are quite familiar to anyone who drives a car, and most people use their integration indirectly, without knowing they are doing so. Acceleration is the rate of change of the speed of an object, and velocity is the rate of change of position of an object. Both rates of change are with respect to some time. Since they are both time rates of change, the integral over each will give the value of the quantity that is changing. That is, the integral of acceleration over time will yield velocity, and the integral of velocity over time results in position.

Recognizing the integral relationship of these three kinematic quantities can give insight into the physical system they represent. If the boundary conditions—the values of the integrated quantity at the beginning and end of the integration period—are known, then numerical averages can be calculated. This is more complicated and more precise than figuring out how far a car has gone by multiplying its speed times the amount of time traveled, but the underlying concept is the same. If the boundary conditions are not known, then a general equation for either velocity or position (depending) will result. This general equation can be analyzed for patterns of behavior: free fall, simple harmonic **motion**, and other common situations will be easily recognizable.

See also Differentiating distance and velocity

INTERPOLATION

Interpolation involves defining a curve that passes through a set of defined points in a plane. The curve that passes through those points is said to interpolate those points, and the curve is called an interpolating curve for those particular points. In order to define a curve passing through a defined set of points an estimation of a value of a function or series between two known values must be made. These estimated points must conform to the law of the series of known points. The word interpolate is derived from the Latin *interpolat* meaning to touch up or refurbish and is extracted from *interpolis*.

Several different methods are available in order to accomplish estimating values of a function at positions between given values. Some of the most common methods include linear, spline, and cubic spline interpolation methods. Linear interpolation is the simplest form of interpolation where a function is estimated by drawing a straight line between the nearest neighboring given points on either side of a required position. It assumes a constant rate of change between two points. Spline interpolation uses a polynomial that incorporates information from neighboring points to obtain a degree of overall smoothness. It is a bit more complex than simple linear interpolation but yields a smooth fit to the given data points. The last mentioned method of interpolation, the cubic spline interpolation, employs piecewise third-order **polynomials** which pass through the set of given values to obtain the unknown values in between. The second derivative of each polynomial is usually set equal to **zero** at the endpoints. This provides a boundary condition that completes the system of polynomial **equations** so that they can then be solved to give the coefficients of the polynomials and yield an interpolating curve for the given data points. This method is more complex than both the linear and spline interpolation methods but it yields a superior curve fitting the given set of values. There are many other interpolation methods but they are generally more complex than the ones mentioned here but in some cases may yield better fits to the data.

INTERVAL

An interval is a set containing all the **real numbers** located between any two specific real numbers on the number line. It is a property of the set of real numbers that between any two real numbers, there are infinitely many more. Thus, an interval is an infinite set. An interval may contain its endpoints, in which case it is called a closed interval. If it does not contain its endpoints, it is an open interval. Intervals that include one or the other of, but not both, endpoints are referred to as half-open or half-closed.

Notation

An interval can be shown using set notation. For instance, the interval that includes all the numbers between 0 and 1, including both endpoints, is written $\{x \mid 0 \leq x \leq 1\}$, and read "the set of all x such that 0 is less than or equal to x and x is less than or equal to 1." The same interval with the endpoints excluded is written $\{x \mid 0 < x < 1\}$, where the less than symbol ($<$) has replaced the less than or equal to symbol ($\leq$). Replacing only one or the other of the greater than or equal to signs designates a half-open interval, such as $\{x \mid 0 \leq x < 1\}$, which includes the endpoint 0 but not 1. A shorthand notation, specifying only the endpoints, is also used to designate inter-

vals. In this notation, a **square** bracket is used to denote an included endpoint and a parenthesis is used to denote an excluded endpoint. For example, the closed interval $\{x \mid 0 \leq x \leq 1\}$ is written [0,1], while the open interval $\{x \mid 0 < x < 1\}$ is written (0,1). Appropriate combinations indicate half-open intervals such as [0,1) corresponding to $\{x \mid 0 \leq x < 1\}$.

An interval may be extremely large, in that one of its endpoints may be designated as being infinitely large. For instance, the set of numbers greater than 1 may be referred to as the interval $\{x \mid -1 < x < \infty\}$, or simply (-1,∞). Notice that when an endpoint is infinite, the interval is assumed to be open on that end. For example the half-open interval corresponding to the nonnegative real numbers is [0,∞), and the half-open interval corresponding to the nonpositive real numbers is (-∞,0].

Applications

There are a number of places where the concept of interval is useful. The solution to an inequality in one **variable** is usually one or more intervals. For example, the solution to $3x + 4 \leq 10$ is the interval (-∞,2].

The interval concept is also useful in **calculus**. For instance, when a function is said to be continuous on an interval [a,b], it means that the graph of the function is unbroken, no points are missing, and no sudden jumps occur anywhere between x = a and x = b. The concept of interval is also useful in understanding and evaluating integrals. An integral is the **area** under a curve or graph of a function. An area must be bounded on all sides to be finite, so the area under a curve is taken to be bounded by the function on one side, the x-axis on one side and vertical lines corresponding to the endpoints of an interval on the other two sides.

INTUITIONISM

Intuitionism is a philosophy of mathematics which regards the objects of mathematical discourse to be mental constructions based upon intuitively self-evident ideas. Thus, for the intuitionist, mathematical "objects" are purely mental and have no independent physical existence. Founded by the Dutch mathematician **L. E. J. Brouwer** (1881-1966), the intuitionst philosophy emerged as a reaction to **Georg Cantor**'s theory of infinite **sets** and the application of standard logic to such sets. In particular, Brouwer required that all mathematical **proof** be "constructive." By this, he meant that in order to prove the existence of any mathematical object, one must provide the instructions for mentally constructing the object in a finite number of steps. Traditionally, mathematicians have been satisfied to show that a mathematical entity exists by denying its existence and showing that such a denial leads to a contradiction of some known mathematical **truth**. This method of proof, called indirect proof, is unacceptable to intuitionists because it provides no finite step-by-step method for constructing the entity in question. Indirect proof is based upon the "law of the excluded middle" from standard logic, which intuitionists reject when it is applied to infinite sets. The law of the excluded middle says that a statement S is either true or false.

There is no middle ground, hence the term "excluded middle." Intuitionists, on the other hand, regard any unproved statement about infinite sets to be neither true nor false. The "intuition" that Brouwer and his followers claim to rely on is the intuition we have about the natural numbers. Here is an infinite set that can be constructed by our intuitive understanding of succession. The set of natural numbers can be built up by starting with the number 1, constructing its successor, 2, and then 3, and so on. Now, Brouwer claims, all the rest of what we regard as legitimate mathematics must come from this intuitive set of natural numbers. All the familiar concepts of mathematics must be built constructively from the natural numbers. Paraphrasing the mathematician **Leopold Kronecker** (1823-1891), intuitionists seem to say that God created the natural numbers and all the rest of mathematics is the work of humans. While Brouwer regarded the set of natural numbers as a completed infinite set, there are intuitionists who do not believe in the concept of completed infinite sets. They believe only in the concept of "potential infinity," an idea that can be traced back to **Aristotle**. Other factions of intuitionism deny the existence of infinite sets altogether. To them, **infinity** is not an intuitive concept.

The appeal of intuitionism is that if we base our mathematics only upon those ideas which are intuitively obvious or self-evident, then we may hold to the conclusions of our mathematics with more certainty than if we use ideas about which we have some uneasiness. If all of our proofs are purely constructive, then the evidence for their truth is spelled out in a finite step-by-step sequence for all to see. It is not unreasonable to suggest that a completely constructive proof is more convincing than an indirect proof. Nevertheless, intuitionists make up a very small subset of working mathematicians. The reason for this is that to be a thoroughgoing intuitionist one must give up much of classical mathematics or reconstruct it from scratch. Constructivist proofs are longer and more difficult than proofs which allow the law of excluded middle and Cantor's theory of infinite sets. The more radical forms of intuitionism that deny the existence of any infinite sets, are even more crippling to the actual practice of mathematics. As a practical matter, most working mathematicians choose to ignore intuitionism even if they think that it has some merit philosophically. It is simply too much trouble to sacrifice the toolkit of classical mathematics for the sake of a little more certainty. The great German mathematician, **David Hilbert** (1862-1943) said that taking away the law of the excluded middle from the mathematician would be like forbidding the astronomer to use a telescope. This is probably the view of most modern mathematicians, even though, in the latter half of the 20th century, intuitionists have managed to revise much of classical mathematics using constructivist methods. One such example is the 1967 work *Foundations of Constructive Analysis* by Errett Bishop in which the author develops a large part of 20th century real **analysis** using intuitionist principles. It remains to be seen whether intuitionist mathematics will gain an equal footing or surpass classical mathematics in the 21st century. A case can be made that intuitionism is more compatible with the algorithmic approach of computer science than is traditional mathematics. If so, then perhaps intuitionism has not yet seen its heyday.

INVARIANT

In mathematics a quantity is said to be invariant if its value does not change following a given operation. For instance, **multiplication** of any real number by the **identity element** (1) leaves it unchanged. Thus, all **real numbers** are invariant under the operation of "multiplication by the identity element (1)." In some cases, mathematical operations leave certain properties unchanged. When this occurs, those properties that are unchanged are referred to as invariants under the operation. **Translation** of coordinate axes (shifting of the origin from the point (0,0) to any other point in the plane) and **rotation** of coordinate axes are also operations. Vectors, which are quantities possessing both magnitude (size) and direction, are unchanged in magnitude and direction under a **translation of axes**, but only unchanged in magnitude under rotation of the axes. Thus, magnitude is an invariant property of vectors under the operation of rotation, while both magnitude and direction are invariant properties of a vector under a translation of axes.

An important objective in any branch of mathematics is to identify the invariants of a given operation, as they often lead to a deeper understanding of the mathematics involved, or to simplified analytical procedures.

Geometric invariance

In **geometry**, the invariant properties of points, lines, angles, and various planar and solid objects are all understood in terms of the invariant properties of these objects under such operations as translation, rotation, reflection, and magnification. For example, the **area** of a **triangle** is invariant under translation, rotation and reflection, but not under magnification. On the other hand, the interior angles of a triangle are invariant under magnification, and so are the proportionalities of the lengths of its sides.

The **Pythagorean Theorem** states that the **square** of the **hypotenuse** of any right triangle is equal to the sum of the squares of its legs. In other words, the relationship expressing the length of the hypotenuse in terms of the lengths of the other two sides is an invariant property of right triangles, under magnification, or any other operation that results in another right triangle.

Very recently, geometric figures called **fractals** have gained popularity in the scientific community. Fractals are geometric figures that are invariant under magnification. That is, their fragmented shape appears the same at all magnifications.

Algebraic invariance

Algebraic invariance refers to combinations of coefficients from certain **functions** that remain constant when the coordinate system in which they are expressed is translated, or rotated. An example of this kind of invariance is seen in the behavior of the **conic sections** (cross sections of a right circular cone resulting from its intersection with a plane). The general equation of a conic section is $ax^2 + bxy + cy^2 + dx + ey + f = 0$. That is, each of the **equations** of a **circle**, or an **ellipse**, a **parabola**, or **hyperbola** represents a special case of this equation. One combination of coefficients, (b^2-4ac), from this equation is called the **discriminant**. For a parabola, the value of the discriminant is **zero**, for an ellipse it is less than zero, and for a hyperbola is greater than zero. However, regardless of its value, when the axes of the coordinate system in which the figure is being graphed are rotated through an arbitrary **angle**, the value of the discriminant (b^2-4ac) is unchanged. Thus, the discriminant is said to be invariant under a rotation of axes. In other words, knowing the value of the discriminant reveals the identity of a particular conic section regardless of its orientation in the coordinate system. Still another invariant of the general equation of the conic sections, under a rotation of axes, is the sum of the coefficients of the squared terms $(a+c)$.

INVERSE FUNCTION

An inverse function is one which reverses the action of a previous function. That is, it takes the value from the previous function and returns it to its previous value for all values of that function. Many inverse **functions** are familiar —for example, if the function in question was $y = 2x$, it would be clear that $z = y/2$ would return $z = x$. Some functions are their own inverse—for example, $y = 1/x$ results in $z = 1/y$ as an inverse. Some functions lack inverses.

Other inverse functions present special problems. For example, cyclic functions like the **sine** and **cosine** function require a specified **interval**, since the sine and cosine will **range** between **zero** and one in exactly the same way from zero to **infinity** in both directions, positive and negative. In this case, a two **pi** interval is required to secure the desired inverse value. For other functions, there will only be two or three ranges that can be selected out of all numbers. For example, if one is taking the **square** root of a number, one must specify whether one wants the positive or negative value of the **square root**. This specification is what is necessary to make the inverse function a function, **mapping** on a one-to-one basis, rather than a mere number relation, which might have two or more valid results. In practical situations, it is usually clear which values apply.

INVERSE HYPERBOLIC FUNCTIONS

The inverse hyperbolic **functions** are simply the inverse of the different **hyperbolic functions** cosh x, coth x, csch x, sech x, sinh x, and tanh x. The inverse of a function is that function reflected about the line $y = x$ and usually denoted by $f^{-1}(x)$. The inverse hyperbolic functions are denoted as sinh^{-1} x (or arcsinh x), cosh^{-1} x (or arccosh x), tanh^{-1} x (or arctanh x), csch^{-1} x (or arccsch x), sech^{-1} x (or arcsech x), and coth^{-1} x (or arccoth x). Since the hyperbolic functions are based on the exponential

function it is not surprising that the inverse hyperbolic functions are based on the natural logarithmic function, ln.

Like the hyperbolic functions the inverse hyperbolic functions are periodic since they are elliptic functions. Corresponding to the two main hyperbolic functions are the inverses of these functions. They are inverse hyperbolic **sine**, sinh^{-1} x, and inverse hyperbolic **cosine**, cosh^{-1} x and are defined as: sinh^{-1} $x = \ln(x + \sqrt{(x^2 + 1)})$ and cosh^{-1} $x = \ln(x +/- \sqrt{(x^2 - 1)})$. The inverse hyperbolic cosine has the $+/-$ sign because it is not a one to one function, that is the **domain** either has to be restricted or the $+/-$ sign has to be used. Inverse hyperbolic tangent is defined as: tanh^{-1} $.x = 1/2 \ln((1 + x)/(1 - x))$. The three other inverse hyperbolic functions, inverse hyperbolic cotangent, inverse hyperbolic secant and inverse hyperbolic cosecant, can be described in terms of inverse hyperbolic tangent, inverse hyperbolic sine and inverse hyperbolic cosine: coth^{-1} $x = $ tanh^{-1} $1/x$, sech^{-1} $x = $ cosh^{-1} $1/x$, and csch^{-1} $x = $ sinh^{-1} $1/x$. Just as the hyperbolic functions have identities the inverse hyperbolic functions have similar identities and negative argument formulas. The negative argument formulas are some of the more useful of these relations.

See also Hyperbolic functions

INVERSE TRIGONOMETRIC FUNCTIONS

Inverse trigonometric **functions** are the inverse of the different **trigonometric functions sine**, **cosine**, tangent, cotangent, secant, and cosecant. The inverse of a function is simply the reflection of that function about the line $x = y$ and for inverse trigonometric functions is expressed as $f^{-1}(x)$. The inverse trigonometric functions are expressed as sin^{-1}x (or arcsin x), cos^{-1}x (or arccos x), tan^{-1}x (or arctan x), cot^{-1}x (or arccot x), sec^{-1}x (or arcsec x), and csc^{-1}x (or arccsc x). The trigonometric functions are based on complex exponentials and so the inverse of such functions can be expressed in terms of the natural logarithmic function, ln. Although this is true it is much more practically useful to define the inverse functions in terms of side measurements of triangles as they relate to the angles in the **triangle** when solving geometric problems.

Inverse trigonometric functions are periodic since they are derived from trigonometric functions that are circular in nature. Since they are not one to one functions the **domain** must be restricted for each inverse trigonometric function just as in the case of the trigonometric functions. The restricted domains for each function are as follows:

- sin$^{-1}x = y$ for $-1 \leq x \leq 1$ and for $-\pi/2 \leq y \leq \pi/2$
- cos$^{-1}x = y$, for $-1 \leq x \leq 1$ and for $1 < y$
- tan$^{-1}x = y$, for any x and for $-\pi/2 < y < \pi/2$
- cot$^{-1}x = y$, for any x and for $0 < y < \pi$
- sec$^{-1}x = y$, for ($x(\geq 1$ and for $0 \leq y < \pi/2$ or $\pi \leq y < 3\pi/2$
- csc$^{-1}x = y$, for ($x(\geq 1$ and for $0 < y \leq \pi/2$ or $\pi < y \leq 3\pi/2$.

The inverse trigonometric functions are examples of transcendental functions. Although these inverse functions are usually employed in problems that require finding angles from side measurements in triangles, when they are written in terms of natural logarithmic functions they are also useful as anti-derivatives for a wide variety of functions. In this sense they are often utilized in solving **differential equations** that arise in mathematics, engineering and physics.

IRRATIONAL NUMBERS

Irrational numbers are numbers that are neither **whole numbers** (like 2, 0, or -3) nor **ratios** of whole numbers. Irrational numbers are **real numbers** in the sense that they appear in measurements of geometric objects—for example, the number **pi** (π), which is the ratio of the **circumference** of a **circle** to the length of its **diameter**, is an irrational number. However, irrational numbers cannot be represented as decimals, unlike **rational numbers**, which can be expressed either as finite decimals or as infinite decimals that eventually follow a repeating pattern. For instance, the decimal 6.412121212... is equal to 6348/990. By contrast, irrational numbers have infinitely long decimal expansions that never form a repeating pattern. Thus, the number pi can never be written down exactly in decimal form, it can only be approximated, by decimals such as 3.14159.

Irrational numbers were discovered in the school of **Pythagoras**, a great Greek mathematician who founded a Brotherhood of mathematicians and philosophers in the Italian port town of Cortona in the 6th century B.C. Pythagoras and his followers were not looking for the irrational numbers; on the contrary, they did not expect such numbers to exist. Preceding the discovery of the irrational numbers, Pythagoras had found many natural phenomena that were described by rational numbers. He had realized, for example, that in musical instruments, strings whose lengths are related by rational ratios have harmonious pitches. This observation led him to believe that the harmony of the world was closely related to that of the rational numbers, and that every natural phenomenon could be expressed in terms of these numbers—hence the name "rational." Great was his dismay, therefore, when he realized that one of the simplest geometric quantities of all, the sidelength of a **square**, could not be compared to the length of the square's diagonal by a rational ratio.

The Brotherhood had previously discovered the famous **Pythagorean theorem**, which states that in a right **triangle**, the sum of the square **powers** of the lengths of the legs is equal to the square power of the length of the **hypotenuse**. The diagonal of a square divides it into two right triangles. Suppose that the sides of the square are one unit long, and let's call the length of the diagonal d. The Pythagorean **theorem** says that $1^2 + 1^2 = d^2$, or $d^2 = 2$. Thus, d is a number whose square is equal to 2. According to Pythagoras' credo, this number should be rational. But if it is a rational number, which one is it?

The **square root** of 2 is in fact not a rational number, as was discovered by the Pythagorean Brotherhood. According to one legend, this idea was so shocking to Pythagoras that he put the discoverer to death. Another version says that Pythagoras declared the irrationality of the square root of 2 to be a secret, not to be revealed to anyone outside the Brotherhood. When one of the members dared to defy Pythagoras' decree and inform the outside world of the discovery, he was killed.

The **proof** that the Pythagoreans discovered has not survived, and the earliest written proof that has been found is in **Euclid**'s *Elements,* written more than 2 centuries after the time of Pythagoras. The proof that Euclid gives is one of the most famous in mathematics, renowned for its simplicity and elegance. Euclid follows a mathematical strategy called reductio ad absurdum: to show that something is true, assume that the opposite is true, and follow the logical consequences of that assumption until you reach an absurdity.

To use this technique to show that the square root of 2 is irrational, Euclid starts by supposing that the opposite is true: that the square root of 2 is a rational number. He writes the square root of 2 as a ratio a/b of whole numbers a and b, in lowest terms (a fraction a/b is in lowest terms if a and b have no factors in common; so 2/3 is in lowest terms, but 6/9 is not, because 6 and 9 are both divisible by 3). Since a/b is supposed to be the square root of 2, we have $(a/b)^2 = 2$, or $a^2/b^2 = 2$. Multiplying both sides of this equation by b^2 gives the new equation $a^2 = 2b^2$. This means that a^2 is an even number, since it is 2 times a number. Now, if a^2 is even then a itself must be even, since only even numbers have even squares. Thus a is equal to 2 times another number c: a=2c.

If we substitute 2c for a in the equation relating a and b, we get $(2c)^2 = 2b^2$, or $4c^2 = 2b^2$. Dividing both sides of the equation by 2 yields $2c^2 = b^2$. This tells us that b^2 is also an even number, since it is 2 times a number. And since b^2 is even, b is even as well.

We have come to the conclusion that both a and b are even numbers. But a and b were supposed to be in lowest terms, and now we have discovered that they are both divisible by 2. This is a contradiction. So starting with the assumption that the square root of 2 is rational, we have arrived at an absurdity. Hence the square root of 2 must be irrational.

Since the discovery of the irrational numbers, mathematicians have gone on to prove that pi is irrational, and likewise that e, the base of the natural logarithm, is irrational. In fact, it has been established that there are infinitely many irrational numbers. After the mathematician **Georg Cantor** developed a way to compare the size of infinitely large **sets**, mathematicians proved that there are more irrational numbers than rational numbers, even though both sets are infinite: it is possible to "count" the rational numbers, but not the irrational numbers. What's more, almost all numbers are irrational—in other words, if you pick a real number completely randomly, you are virtually guaranteed that the number will be irrational, not rational.

ISOMORPHISM

An isomorphism is a one to one **mapping** of the elements of one set onto another such that the result of an operation on the elements of one set are identical to the result of the same operation on the elements of the other set. An isomorphism is also known as bijective morphism, a mapping of one set onto another set without the loss of information. Isomorphism is derived from the two Greek words *iso* and *morphism*. The Greek word *iso* means equal or identical. The Greek word *morphism* is actually composed of two other Greek words, *morph* meaning form or shape and *ism* indicating an action. Combining these two words together into *morphism* yields a word meaning to form or shape. The symbol indicating isomorphic behavior is usually denoted as $\cong$ and is written A $\cong$ B, meaning A is isomorphic to B. This symbol is also sometimes used to mean geometric **congruence** in mathematics.

Isomorphism was first discovered by German scientist Eilhard Mitscherlich in 1819. Mitscherlich observed that chemical compounds having the same number of atoms per molecule are predisposed to form crystals that have identical angles. He called this property isomorphism and continued to advance the theory of isomorphism, which describes a relationship between crystalline structure and chemical composition. After his work was verified it became one of the basic levers of the atomic theory. The concept of isomorphism was extended to mathematics to mean two different systems that are the same but does not mean that they are equal. For example, for two numerical systems to have the same structure, each system must contain a number that has a counterpart in the other system. For the two systems to be isomorphic: 1) there is a mapping of one system to the other that put them into a one to one **correspondence** and 2) in this mapping, the results of mathematical operations, such as **addition** and **multiplication**, are preserved. An example of two number systems that are isomorphic to one another are Arabic and Roman.These systems use different symbols to represent numbers but the results of addition and multiplication are equal in the two systems.

The identities and inverses of a group are preserved in an isomorphism. Isomorphism can be applied to spaces, shapes and groups. A **vector space** in which addition and scalar multiplication are preserved is a space isomorphism. There are some specific isomorphisms that have unique names describing them. An automorphism is the result of an isomorphism of a group onto itself. Mapping a geometric figure or topological space onto another figure or space is called a homeomorphism. This **equivalence** relation is continuous in both directions and as such is also known as a continuous **transformation**. Isometry is a type of homeomorphism in which distances are also preserved. Although all of these types of transformation are specific to the details of each situation they are all still isomorphisms.

ITERATION

Iteration consists of repeating an operation of a value obtained by the same operation. It is often used in making successive approximations, each one more accurate than the one that preceded it. One begins with an approximate solution and substitutes it into an appropriate formula to obtain a better approximation. This approximation is subsequently substituted into the same formula to arrive at a still better approximation, and so on, until an exact solution or one that is arbitrarily close to an exact solution is obtained.

An example of using iteration for approximation is finding the **square** root. If s is the exact **square root** of A, then A $|6\text{-}8|$ s = s. For example, since 8 is the square root of 64, it is true that 64 $|6\text{-}8|$ 8 = 8. If you did not know the value of $|5\ 14|$ 64, you might guess 7 as the value. By dividing 64 by 7, you get 9.1. The average of 7 and 9.1 would be closer. It is 8.05.

Now you make a second iteration by repeating all the steps but beginning with 8.05. Carry out the **division** to the hundredths place; 64 $|6\text{-}8|$ 8.05 = 7.95. The average of 8.05 and 7.95 is 8. A third iteration shows that 8 is the exact square root of 64.

Finding the roots of an equation

Various methods and formulas exist for finding the **roots** of **equations** by iteration. One of the most general methods is called the method of successive bisection. This method can be used to find solutions to many equations. It involves finding solutions by beginning with two approximate solutions, one that is known to be too large and one that is known to be too small, then using their average as a third approximate solution. To arrive at a fourth approximation, it is first determined whether the third approximation is too large or too small. If the third approximation is too large it is averaged with the most recent previous approximation that was too small, or the other way around; if approximation number three is too small it is averaged with the most recent previous approximation that was too large. In this way, each successive approximation gets closer to the correct solution. Testing each successive approximation is done by substituting it into the original equation and comparing the result to **zero**. If the result is greater than zero then the approximation is too large, and if the result is less than zero, then the approximation is too small.

Iteration has many other applications. In **proof**, for example, mathematical **induction** is a form of iteration. Many computer programs use iteration for looping.

J

Jacobi, Karl Gustav Jacob (1804-1851)

German mathematical physicist

As the impact of the American and French Revolutions was felt across Europe, a social atmosphere arose that encouraged ground breaking work in mathematics. Karl Gustav Jacob Jacobi, who attracted early attention from luminaries such as **Adrien–Marie Legendre** and **Karl Gauss**, appeared alongside and sometimes worked with a handful of innovative contemporaries like **William Hamilton, Augustin–Louis Cauchy, Peter Dirichle**t, and **Niels Abel**. Jacobi's own contributions range across older subjects in math such as number theoryand newer fields like **analysis**. Two mathematical terms he devised now bear Jacobi's name. He electrified the teaching profession with an unprecedented practice of opening up his theoretical notes to his students, thereby inventing the research seminar now common in universities. While the political instabilities of post–revolutionary times sometimes threatened Jacobi's livelihood, his sheer genius was always enough to attract a new protector to sponsor his continued work. Jacobi's mathematics had an immediate effect on the classical **mechanics** of **Isaac Newton**, **Pierre Laplace** and **Joseph Lagrange**, and later on the **quantum mechanics** and relativity theories of the 20th century.

Jacobi, whose first name is sometimes spelled "Carl," was born in Potsdam, Prussia (Germany) into a wealthy and well–educated Jewish family on December 10, 1804. He was one of two boys who would both gain a measure of fame during their lifetimes. Karl's older brother, Moritz, would later be celebrated—and even further on dismissed—as the founder of an experimental concept in electricity known as "galvanoplastics." The other children in the family were Eduard and Therese. All that is generally known of their mother is her own family name, Lehmann, and that her brother tutored Karl until he was 12. His father, Simon Jacobi, was a banker in Berlin.

Young Jacobi was a prodigy. After being home schooled in the classics and in mathematics, he entered Potsdam Gymnasium in 1816. He promptly rose to the top of his class within a few months, proving he was ready for university training. He was still only 12 years old. Early entry to higher education was forbidden by the authorities, so the Gymnasium kept him until the legal age of sixteen. The rector of the school described him as "a universal mind," expressing high hopes for his future.

The University of Berlin had to make way for Jacobi, who rebelled against his teacher, Heinrich Bauer, and preferred to read **Leonhard Euler** and Lagrange on his own. Jacobi graduated within a year with top marks for classical languages, history, and mathematics. As a graduate student Jacobi majored in philology, since he had no peer at Berlin in mathematics. Instead, he continued to read on his own and correspond with Gauss at Göttingen. Jacobi qualified for a teaching position at age 19, which he took without pay the next year, after completing his doctorate on partial **fractions**. Jacobi achieved all this as a Jew in a highly anti-semitic environment; but for perhaps other reasons he converted to Christianity at age 20.

During his student days Jacobi continued to write to his uncle and former teacher, and corresponded with other mathematicians, a habit he continued throughout his career. To one who complained of the physical costs of intellectual labor, Jacobi replied, "Only cabbages have no nerves." How do they benefit, he asked rhetorically, from such well–being? This riposte shows Jacobi's lack of concern for his own health, which may well have contributed to his early death from a combination of illnesses. Some of his vacations were forced, whenever his schedule threatened to wear him down.

Jacobi came into his own as a lecturer. After six months at Berlin he transferred to the University of Königsberg on the recommendation of Legendre, who was impressed with Jacobi's additions to his own pioneering work in elliptic integrals. Jacobi combined new mathematical ventures and classwork to show his new inventions as they took shape. More importantly, he presented himself as one who knew little and

Karl Gustav Jacob Jacobi

desired to know more, inspiring his students to forge ahead according to his example. You do not have to "meet" all subjects in math, he argued by analogy, before "marrying" one.

Jacobi exclusively studied (at least for a while) the work of Legendre on elliptic integrals. Independently of Abel in Norway, Jacobi fully developed the new area of elliptic **functions** and introduced the notation used today for these functions. He investigated hyperelliptic integrals, making important discoveries about the generalizations of elliptic functions that were later to become known as Abelian functions. Because of his work in mathmatical physics, the ellipsoids of equilibrium for rotating liquid masses are known as Jacobi ellipsoids. Jacobi's work in functions inspired Cauchy and also **Joseph Liouville**. Other interests included **determinants**, especially those used in relation to partial **differential equations** and called the Jacobian determinants. Jacobi published three papers on determinants in 1841. For the first time he gave an algorithmic **definition** of the determinant that applied to cases when the entries were either numbers or functions. These papers helped to make the idea of a determinant more widely known. Partial differential **equations** came into play in dynamics, a subject which interested Jacobi. Here, he parlayed the findings of Hamilton into results later applied to quantum mechanics. To classical mechanics he contributed work on the three–body problem

and other dynamical problems. In **number theory**, Jacobi proved an assertion of **Pierre de Fermat**'s regarding the expression of **integers** as sums of squares.

Simon Jacobi died in 1832, and the family was able to live on his bequest for another eight years. In 1840, however, financial troubles forced them into bankruptcy. These difficulties likely contributed to Jacobi's subsequent collapse from overwork. Jacobi relinquished his chair of Ordinary Professor of Mathematics at Königsberg in 1842. He thereafter subsisted on a pension granted by Frederick William IV from the Prussian government, staying in Italy for a time due to ill health. For personal reasons, Jacobi ran for local office in Berlin as a liberal in 1848, leading to a temporary suspension of his Prussian funding, a great sorrow to Jacobi's wife and seven children. However, he was soon back in favor. He taught intermittently at Berlin meanwhile, but died February 18, 1851. Suffering from diabetes, he developed a fatal case of smallpox, contracted after a bout of influenza. Jacobi was memorialized by his friend and colleague **Lejeune Dirichlet** in 1852 with a special lecture given at the Berlin Academy of Sciences. In it, Dirichlet called Jacobi the greatest Academy member since Lagrange.

JORDAN CURVE

A Jordan curve is a curve that falls in a plane and is topologically equivalent to the unit **circle** in that it is simple, closed, and does not cross itself. A Jordan curve is a homeomorphic image of the unit circle such that the unit circle can be deformed into a Jordan curve by a continuous, invertible **mapping**. The curve divides the plane into exactly two parts, one part that is inside of the curve and another part that is outside of it. It is impossible to pass continuously from one part to the other without crossing the curve. A figure-8 curve is not a Jordan curve.

There are some interesting features of Jordan curves. For sufficiently smooth Jordan curves it is known that all four vertices of some **square** are contained inside of the curve. Also, for every **triangle** T and Jordan curve C, C has an inscribed triangle similar to T. These are important properties of simple, closed curves.

The term Jordan curve arose in the book *Cours d'analyse de l'École Polytechnique*, 3 vol. written by **Camille Jordan**, a French mathematician, in 1882. In this book Jordan examined the theory of **functions** from a modern viewpoint, dealing with the bounded variation function. He applied this function to the curve known as Jordan's curve. The Jordan curve **theorem** states that any simple closed curve divides the points of the plane not on the curve into two distinct domains, with no points in common, of which the curve is the common boundary. Jordan's **proof** for this theorem was very complicated and, as it turns out, invalid. It has been a difficult theorem to prove because of the generality of the concept of a simple closed curve. A simple closed curve is not restricted to the class of **polygons** or smooth curves but also includes all curves that are topological images of a circle. The first correct proof of this theorem appeared in 1905 in an article written by **Oswald Veblen**. Homology theory has been used to simplify the proof. One of the earliest uses of the term Jordan curve was in W. F. Osgood's article "On the Existence of the Green's Function for the Most General Simply Connected Plane Region" that appeared in July 1900.

JORDAN, MARIE ENNEMOND CAMILLE (1838-1922)

French mathematician

Camille Jordan published papers in all branches of mathematics. In **analysis** he discovered the bounded function. In **topology** he investigated the relationship between a plane and a closed curve. However it is for **algebra** that Jordan is best known. Jordan was particularly eminent in the field of **group theory**. He is known as the originator of Jordan curves, Jordan algebra and the Jordan Holder Theory. Jordan also worked on solvable groups and movements in three dimensional **space**.

Marie Ennemond Camille Jordan, or Camille Jordan as he is more commonly known, was born in Lyon, France in 1838. Jordan carried out his initial mathematical studies (specifically in engineering) at the Ecole Polytechnique, Paris.

In 1870 Jordan published *Traite des substitutions et des equations algebriques* (Treatise on Substitutions and Algebraic Equations) which was an excellent overview of **Galois theory**. As well as being an overview of Galois theory it also established several important results in group theory. This book was awarded the prestigious Poncelet Prize of the Academy of Sciences. From 1873 Jordan was a lecturer in mathematics at the Ecole Polytechnique, Paris, and in 1876 he was made a professor. He remained as a professor until 1912. During much of his time as professor of mathematics Jordan continued working as an engineer, a position he did not relinquish until his retirement in 1885. The work of **Georg Cantor** was pushed to the fore in the second edition of Jordan's *Cours d'analyse de l'Ecole Polytechnique* (A Course of Analysis of the Polytechnic School) published in 1893 (the first edition was published in 1882). It is the third edition (1909-1915) of this book which introduces the **Jordan curve** theorem. In the first edition the point set topology of Cantor was included in the appendix to the third volume, in the second edition it was given prominence by its inclusion in the first chapters.

Jordan died in Milan, Italy in 1922, afer having published over 120 papers, but his work was carried on by many of his students, most famously and **Marius Sophus Lie**.

See also Jordan curve

K

KANTOROVICH, LEONID VITALYEVICH (1912-1986)

Russian economist and mathematician

Leonid Vitalyevich Kantorovich was a Soviet mathematician who is considered notable not only for his contributions to mathematics, but to economics, as well. Kantorovich developed and applied mathematical problems to the area of economic planning. His work, which was sometimes considered to be controversial, was so influential that he was awarded the Nobel Prize for economics in 1975.

Kantorovich was born in St. Petersburg, Russia, on January 19, 1912. It became evident early in his life that he was mathematically gifted. He was widely recognized as a brilliant mathematician by the time he received a doctorate in mathematics from Leningrad State University in 1930, when he was 18 years old. He was named a professor there in 1934, and remained in that position until 1960.

Although he was trained only in mathematics, Kantorovich had a keen interest in, and profound understanding of economics. His work in both fields has been widely recognized, and, in 1961 he was named the head of the department of mathematics and economics in the Siberian branch of the United Socialist Soviet Republic Academy of Sciences. He was honored in 1964 when he was named to the highly regarded Academy of Sciences of the Soviet Union, and he received the prestigious Lenin Prize in 1965. In 1971, Kantorovich became the director of research at the Institute of National Economic Planning in Moscow, a position he held for five years.

While serving as director at the Institute of National Economic Planning, Kantorovich received the 1975 Nobel Prize for economics for his work on the optimal allocation of scarce resources. He shared the prize with Tjalling C. Koopmans, a Dutch-born economist who was teaching at Yale University. Kantorovich and Koopmans used the mathematical technique of **linear programming** as a tool in economic planning.

Linear programming, for which Kantorovich pioneered a **model** in 1939, is a specific class of mathematical problems, and has various applications. It was developed as a discipline in the 1940s as a means of solving difficult planning problems in wartime operations. Kantorovich and Koopmans applied the technique to economic planning, ultimately showing that prices of goods should be based on the relative scarcity of resources.

Although Kantorovich was highly regarded as a mathematician and economist, his views, which were sometimes critical of the Soviet economic policy, often were in conflict with those of his colleagues. He was considered to be a reform economist, with ideas that frequently ran contrary to traditional Marxist thinking.

Kantorovich's best-known work is *The Best Use of Economic Resources*, which he wrote in 1959. He also wrote *Functional Analysis* in 1977. He died in the Soviet Union on April 7, 1986, at the age of 74.

KATO, TOSIO (1917-1999)

Japanese-born American mathematical physicist

Tosio Kato, whose career in mathematics ranged over more than 40 years, made major contributions to the field of mathematical physics. A prolific writer, he produced hundreds of published articles during his career. His most important research, on perturbation theory, won him acclaim and awards in both his native country of Japan and in the United States, where he spent most of his career.

Kato was born on August 25, 1917, in Tochigiken, Japan, the son of Shoji and Shin (Sakamoto) Kato. He attended the University of Tokyo, where he received a bachelor's degree in 1941; he would receive a doctor of science degree from the university ten years later. In 1943, he began teaching at the University of Tokyo, and due to the stability this position provided he was able to marry Mizue Suzuki the following year.

Even before his appointment to full professorship at the university, which he achieved upon receiving his doctorate in 1951, Kato had begun to publish the beginnings of his research in perturbation theory. Perturbation theory is the study of a system which deviates slightly from a less complex, ideal system. This is an important field of research because most systems that mathematicians and physicists study are not ideal. Kato examined only the perturbation theory which relates to linear operators (**functions**). The groundbreaking work in the field had been accomplished by John Rayleigh and Erwin Schrödinger in the 1920s.

Kato's contributions to perturbation theory were three-fold. First, he laid the mathematical foundation for the theory, applying ideas in modern **analysis** and function theory. Second, he established the selfadjointness of Schrödinger operators—in other words, he showed that Schrödinger operators are symmetric. This was significant because these operators are a fundamental tool of quantum physics and knowledge of their **symmetry** makes their manipulation much simpler. Finally, he began the study of the spectral properties of the operators, which describes the variety of simple effects which the operators can have when applied to specific elements of a given set. This is important as it allows the operators to be described in terms of many simple effects which can be combined into larger ones. The term *spectral* is related to the way in which a complicated operator splits into distinct effects, as **light** can be split into distinct colors.

The culmination of his work was his definitive book on the subject, *Perturbation Theory for Linear Operators,* which he began writing in Japan and completed in the United States, after accepting a professorship at the University of California at Berkeley in 1962. Although the book did not appear until 1966, his home country had already recognized him as a leading researcher in his field, presenting him with the Asahi Award in 1960.

In the United States, Kato continued his research into perturbation theory and used functional analysis to solve problems in **hydrodynamics** and evolution **equations**. Even though Kato was well known in the American mathematical community by his hundreds of published articles, acknowledgement of the importance of his work came late in the United States. It was not until 1980 that the American Mathematical Society and the Society for Industrial and Applied Mathematics jointly awarded him the Norbert Wiener Prize for applied mathematics. In Japan, Kato continued to receive recognition. On the occasion of his retirement from the University of California in 1989, the University of Tokyo held a conference in his honor. The conference, entitled "The International Conference on Functional Analysis in Honor of Professor Tosio Kato," paid homage to Kato's many contributions in the field of mathematical physics. Kato died on October 2, 1999, of a heart attack in Oakland, California, at the age of 82.

KEEN, LINDA (1940-)

American geometric analyst

Linda Keen devotes her time and energies to some of the hottest topics of the day. Her work in **complex analysis** and dynamical systems deals with the mathematics responsible for the vibrant graphics seen in science shows, **fractal** art, and life-like computer animations. For more than 30 years, her research has been funded by grants and fellowships from the National Science Foundation (NSF). Keen has helped evaluate other postdoctoral fellowships for the NSF as well as for NATO. She is also active in the mathematical community, through her professional participation on various editorial boards and steering committees. Her influence has been felt at local and national levels regarding such pressing issues as women and minority involvement, librarianship, project funding, educational and test standards, research goals, and professional ethics. Among her most visible posts have been the presidency of the Association for Women in Mathematics (1985–1986), and the vice–presidency of the American Mathematical Society (1992–1995).

Linda Keen was born Linda Goldway in New York City on August 9, 1940. Her father was an English teacher who did not take his daughter's interest in the comparatively obscure language of mathematics personally. In fact, he encouraged her to study at a local magnet school, the Bronx High School of Science. She would stay in New York throughout her early academic career, earning a B.S. in 1960 from City College, an M.S. from New York University (NYU) two years later, and a Ph.D. from the Courant Institute of Mathematical Sciences in 1964.

Keen was fortunate to have as her Ph.D. advisor the celebrated Lipman Bers. "Lipa," as he was called, was a political refugee throughout the 1930s, coming to the U.S. from Latvia via Prague. At NYU, he was a colleague and friend as much as an authority on complex **analysis**. In 1964, it was still quite rare for a male mathematician to be even tolerant of women. Yet, as Bers said himself in an interview with Donald Albers and Constance Reid, it never occurred to him that women could be intellectually inferior to men. While studying with Bers, Keen focused on the analytic aspects of Riemann surfaces.

Keen's obituary for Bers highlighted his lack of pretense and approachability. Classes were held Friday afternoons, so Bers extended the time with his students to include lunch—the "children's lunch" as he called it. Keen, who was almost always the youngest at the table, would be given the check to divide in her head. Once, when the "children" all insisted that the oldest person take the chore, Bers was unprepared and his calculations were incorrect.

Bers' commitment to political activism and human rights, resulting from his years of living beneath the shadow of dictatorships, seems to have influenced Keen as well. From 1992 to 1996, she chaired the special advisory committee to write Ethical Guidelines and Procedures for the American Mathematical Society (AMS) Committee on Professional Ethics (COPE). The document developed by the special advi-

Linda Keen

sory committee provides professional mathematicians with guidance about ethical issues including giving credit for new findings, refereeing papers responsibly, protecting "whistle blowers," and social responsibility.

Among the issues with which the committee's report grappled are those arising from work in industry and with the government, as well as standards of conduct within professional organizations such as the AMS. Protecting confidentiality, anonymity, and privileged information is given top priority. As the guidelines state, "Freedom to publish must sometimes yield to security concerns, but mathematicians should resist excessive secrecy demands whether by government or private concerns." Those AMS members who advise graduate students are now expected to paint a realistic picture of employment prospects, and not to exploit their students by giving them heavy workloads at low pay. The guidelines also include a standard nondiscrimination policy. The special advisory committee's proposed guidelines were ratified in 1995 by a 25 to 3 vote.

Keen has served her profession in similar capacities a number of times. She began her involvement with COPE in 1986, and became a member of various policy boards for the AMS throughout the 1990s. Keen has also worked with the International Mathematics Union. She was of a member of the panels charged with evaluating the mathematics departments

of the State University of New York–Potsdam and Rutgers University–Newark, and the minority program at the University of Minnesota. Keen was also a charter member of the Mayor's Commission for Science and Technology of the City of New York, serving on this commission from 1984–1985.

Keen's professional career has taken her to various institutions in her home state, including Hunter College and the City University of New York. When Lehman College, formerly the Bronx campus of Hunter, became independent in 1968, Keen remained on the faculty. She was promoted to full professor at Lehman in 1974 and presently holds a dual appointment in the Graduate Center Doctoral Faculties in Computer Science and in Mathematics. Keen has also held visiting professorships at the University of California at Berkeley, Columbia University, Boston University, Princeton, and MIT, as well as at mathematical institutions in several foreign countries, including Germany, Brazil, Denmark, Great Britain, and China. Keen's editorial services are equally international. Currently she serves on editorial boards for the *Journal of Geometric Analysis* and the *Annales* of the Finnish Academy of Sciences.

Throughout her career, Keen has preferred working collaboratively with other mathematicians to working alone. During the 1980s she worked with Caroline Series on the geometric aspects of Riemann surfaces, and, more recently, she contributed to the field of dynamical systems in cooperation with Paul Blanchard, Robert Devaney, and Lisa Goldberg. As Keen puts it, "I am basically a social person and enjoy people." She currently counts her husband and two children as her chief supporters, worthy successors in this regard to her father and Lipman Bers.

KEPLER, JOHANNES (1571-1630)

German astronomer and mathematician

Johannes Kepler is best known for his discovery of the three laws of planetary **motion**, demonstrating that the universe operates according to fixed, natural laws. A disciple of the great astronomer **Copernicus**, Kepler refined and advanced Copernican theory. After the death of **Tycho Brahe**, a leading astronomer of the 16th century, Kepler was appointed Imperial Mathematician in Prague by Emperor Rudolph II. His many published works describe his search for the mathematical accordances on which the universe is structured, and his study laid the foundation for the understanding of the cosmos, out of which grew the science of modern astronomy. **Isaac Newton**'s work in universal gravitation was influenced by Kepler's discoveries and the standards he set for scientific inquiry. Kepler's research in **light** refraction resulted in two important books on the subject of opticsand improvements in the astronomic telescope. As a mathematician, his efforts to complete Brahe's planetary tables were furthered by his discovery of the usefulness of **logarithms** in making computations. Although logarithm formulas and tables were published by **John Napier**, a Scottish mathematician, in 1614, Kepler used a modified

Johannes Kepler

system of his own creation, and these early computations prefigure the invention of **calculus**. Kepler's published works were numerous, and of great influence to those astronomers who followed him. Some of his most prominent writings were: *Mysterium cosmographicum*, *Astronomiaepars optica*, *Astronomia nova*, *Dioptrice*, *Harmonice mundi*, *Epitome astronomiae Copernicanae*, *Tabulae Rudolphinae*, and *Ephemerides novae*.

Kepler was born December 27, 1571, in Weil der Stadt, Germany. He was the eldest son of Heinrich and Katharina (Guldenmann) Kepler. His grandfather was the local mayor. Although the family was of the Lutheran faith, Kepler's father became a mercenary soldier in the service of the Catholic Duke of Alva, fighting a Protestant reform movement in the Netherlands when Johannes was a small boy. In 1588, he abandoned his family. All accounts describe his mother as a difficult, quarrelsome woman, and in later life, Kepler was obliged to defend her against charges of witchcraft. Katharina influenced her son in a more positive way, when as a small boy of six, he was taken by her to observe the comet of 1577. Kepler married twice; first to Barbara Müller (April 27, 1597), with whom he had five children, two of whom died young. His first wife died in 1611 of typhus. On October 30, 1613, he married Susanna Reuttinger. They had seven children, five of whom died in infancy.

Kepler's elementary schooling began at Leonberg, when he entered a church school. His religious nature was apparent at a young age, and he began to prepare for a life in the church. After two years at a monastery school in Adelberg, he enrolled at Maulbronn, preparing for the University of Tübingen. He entered the University in 1589. There, Michael Maestlin, the astronomy professor at Tübingen, was to open the Copernican door that led to Kepler's future.

A brilliant student, Kepler was seen as destined for great accomplishments. In August 1591, he received a master's degree from the University and began his theological studies. However, his plans to enter the ministry were forever altered by the death of Georgius Stadius, teacher of mathematics at the Lutheran high school in Graz. Searching for a replacement, the school turned to the University for a recommendation. Kepler was nominated, and accepted the post. His duties were to teach mathematics and oversee the annual publication of astrology almanacs—weather forecasts and other predictions based on astrological rules. It may seem ironic that a learned scientist would consent to dabble in the astrological arts. But, although he regarded astrology as popular superstition and at best a lesser form of astronomy, Kepler's observation of and profound feeling for cosmic harmonies made him tolerant of astrology's attraction as a guiding force for the common individual.

Kepler's teaching at Graz was not very fulfilling to him. But inspiration struck one day as he pondered such questions as: Why are there only six planets? How can their **distance** from the sun be determined? Why do planets farthest from the sun move the slowest? He theorized that these questions could be answered by relating their placement and movement to the five regular solids of **Euclidean geometry**. This insight resulted in his first major publication, *Mysterium cosmographicum*, in 1596. In it, Kepler describes a mathematical relationship between a planet's distance from the sun and its **orbit** periodicity. From this intriguing beginning was to evolve **Kepler's laws** of planetary motion.

Seeking to disseminate his findings, Kepler sent copies of *Mysterium cosmographicum* to other European scientists. Tycho Brahe, a leading observational astronomer, was impressed by Kepler's mathematical theories (if not their Copernican basis) and invited him to come to Prague to work with him. In 1598, Kepler and all other Protestant teachers were forced by the Catholic authorities to leave Graz on short notice. Newly married, and in need of work, he accepted Brahe's invitation. Although his wife was from a wealthy family, her inheritance was tied to property in Graz. The Keplers arrived in Prague in 1600 without assets and were to be plagued with financial problems throughout their marriage.

Kepler's first assignment was to study the orbit of Mars. This work was to occupy him for eight years, and resulted in the discovery of his first two laws of planetary motion. The first law shows that every planet's orbit is an **ellipse** with the sun at one focus. This was revolutionary, because until Kepler proved otherwise, it was believed that a planet's orbit was circular. The second law shows that the line from the sun to a planet sweeps across equal areas in equal time. Kepler's great work, *Astronomia nova*, published in 1610, presents these findings. During the years spent on the Mars orbital study, Kepler also began his work in optic research. This interest came about when he built a pinhole camera to observe a solar eclipse in 1600. His optic discoveries formed the basis of modern geometrical optics. In 1604, Kepler published his theories in *Astronomiae pars optica*. In 1611, his *Dioptrice* applied those optical laws to the telescope.

Brahe had died in 1601 and Kepler assumed the position of Imperial Mathematician and the obligation to complete and publish his mentor's tables of planetary motion. Although their working relationship had many times been strained, Kepler took on this task. Called the *Tabulae Rudolphinae* after his patron, the Emperor Rudolph, the work was eventually published in 1627.

After the death of his first wife in 1611, Kepler took a position as a mathematics teacher in Linz, Austria. He remarried, and was to remain in Linz for fourteen years, during which he produced *Harmonice mundi*, his theories on the harmonies of the universe. This book contained Kepler's third law: that the **square** of a planet's period measured in years equals the cube of its distance from the sun measured in astronomical units. His textbook series of seven books, *Epitome astronomiae Copernicanae*, described all that was then known of heliocentric astronomy, and included Kepler's three laws. During this time, he also produced a work called *Stereometria doliorum vinariorum*, stemming from his efforts to measure the **volume** in a wine cask. This work is considered to be a preliminary basis of calculus.

Kepler left Linz during the Counter Reformation and settled his family in Regensburg. After the publication of the *Tabulae Rudolphinae*, he began once again to look for employment. In 1628, Kepler was promised a post by the king of Bohemia, Ferdinand III, in the newly acquired duchy of Sagan. It was there that he published the last of his works, *Emphemerides pars II* and *Emphemerides pars III* (astronomical charts and weather observations); and a curious book of science fiction, *Somnium seu astronomia lunari*, about a voyage to the moon.

Financial hardship (he did not receive the promised salary from King Ferdinand) caused Kepler to return to Regensburg, bringing with him all his books and research papers of a lifetime of work. There, he became ill with a fever and died on November 15, 1630. He was buried in the Regensburg Protestant cemetery.

Kepler has been called a mathematical mysticist. His deep religious beliefs led him to search for harmony between humanity and the mathematical principles by which the universe was ordered.

KEPLER'S LAWS

Johannes Kepler made it his life's work to create a heliocentric (sun-centered) **model** of the solar system which would accurately represent the observed **motion** in the sky of the Moon and planets over many centuries. Models using many geometric curves and surfaces to define planetary orbits, including one with the orbits of the six known planets fitted inside the five perfect solids of **Pythagoras**, failed.

Kepler was able to construct a successful model with the Earth the third planet out from the Sun after more than a decade of this trial and error. His model is defined by the three laws named for him. He published the first two laws in 1609 and the last in 1619. They are:

1. The orbits of the planets are ellipses with the Sun at one focus (F_1) of the **ellipse**.

2. The line joining the Sun and a planet sweeps out equal areas in the planet's **orbit** in equal intervals of time.

3. The squares of the periods of revolution "P" (the periods of time needed to move 360°) around the Sun for the planets are proportional to the cubes of their **mean** distances from the Sun. This law is sometimes called Kepler's Harmonic Law. For two planets, planet A and planet B, this law can be written in the form: $(P^2/_A \div P^2/_B) = (a^3/_A \div a^3/_B)$.

A planet's mean **distance** from the Sun (a) equals the length of the semi-major axis of its orbit around the Sun.

Kepler's three laws of planetary motion enabled him and other astronomers to successfully match centuries-old observations of planetary positions to his heliocentric solar system model and to accurately predict future planetary positions. Heliocentric and geocentric (Earth-centered) solar system models which used combinations of off-center circles and epicycles to model planetary orbits could not do this for time intervals longer than a few years; discrepancies always arose between predicted and observed planetary positions.

The fact remained, however, that, in spite of Kepler's successful **modeling** of the solar system with his three laws of planetary motion, he had discovered them by trial and error without any basis in physical law. More than 60 years after Kepler published his third law, **Isaac Newton** published his *Principia*, in which he developed his three laws of motion and his Theory and Law of Universal Gravitation. By using these laws, Newton was able to derive each of Kepler's laws in a more general form than Kepler had stated them, and, moreover, they were now based on physical theory. Kepler's laws were derived by Newton from the basis of the two-body problem of celestial **mechanics**. They are:

1. The orbits of two bodies around their **center of mass** (barycenter) are **conic sections** (circles, ellipses, parabolas, or hyperbolae) with the center of mass at one focus of each conic section orbit. Parabolas and hyperbolas are open-ended orbits, and the period of revolution (P) is undefined for them. One may consider a circular orbit to be a special case of the ellipse where the two foci of the ellipse, F_1 and F_2, coincide with the ellipse's center (C), and the ellipse becomes a **circle** of radius (a).

2. The line joining the bodies sweeps out equal areas in their orbits in equal intervals of time. Newton showed that this generalized law is a consequence of the conservation of angular momentum (from Newton's Third Law of Motion) of an isolated system of two bodies unperturbed by other forces.

3. From his Law of Universal Gravitation, which states that two bodies of masses, M_1 and M_2, whose centers are sep-

arated by the distance "r" experience equal and opposite attractive gravitational forces (F_g) with the magnitude $F_g = GM_1M_2 \div r^2$, where G is the Newtonian gravitational **factor**. And from his Second Law of Motion, Newton derived the following generalized form of Kepler's Third Law for two bodies moving in elliptical orbits around their center of mass where π is the ratio of the **circumference** of a circle to its **diameter**, "a" is the semi-major axis of the *relative* orbit of the body of smaller mass, M_2, around the center of the more massive body of M_1: $P^2 = [4\pi^2 \div G(M_1 + M_2)]a^3$.

Some of the applications of these generalized Kepler's laws are briefly discussed below.

Let us first consider applications of Kepler's Third Law in the solar system. If we let M_1 represent the Sun's mass and M_2 represent the mass of a planet or another object orbited the Sun, then if we adopt the Sun's mass ($M_1 = 1.985 \times 10^{30}$ kg) as our unit of mass, the astronomical unit (a.u.; 1 a.u. = 149,597,871 km) as our unit of length, and the sidereal year (365.25636 mean solar days) as our unit of time, then $(4\pi2/G) = 1$, $(M_1 + M_2) = 1$ (we can neglect planet masses M_2 except those of the Jovian planets in the most precise calculations), and the formidable equation above is reduced to the simple algebraic equation $P^2 = a^3$ where "P" is in sidereal years and "a" is in astronomical units for a planet, asteroid, or comet orbiting the Sun. Approximately the same equation can be found from the first equation if we let the Earth be Planet B, since F_B=1 sidereal year and a_B is always close to 1 a.u. for the Earth.

Let us return to the generalized form of Kepler's Third Law and apply it to planetary satellites; except for the Earth-Moon and Pluto-Charon systems (these are considered "double planets"), one may neglect the satellite's mass (M_2=0). Then, solving the equation for M_1: $M_1 = (4\pi^2 \div G)(a^3 \div P^2)$.

Measurements of a satellite's period of revolution (P) around a planet and of its mean distance "a" from the planet's center enable one to determine the planet's mass (M_1). This allowed accurate masses and mean densities to be found for Mars, Jupiter, Saturn, Uranus, and Neptune. The recent achievement of artificial satellites of Venus have enabled the mass and mean density of Venus to be accurately found. Also the total mass of the Pluto-Charon system has been determined.

Now we consider the use of Kepler's laws in stellar and galactic astronomy. The equation for Kepler's Third Law has allowed masses to be determined for double stars for which "P" and "a" have been determined. These are two of the orbital elements of a visual doublestar; they are determined from the doublestar's true orbit. Kepler's Second Law is used to select the true orbit from among the possible orbits that result from solutions for the true orbit using the doublestar's apparent orbit in the sky. The line joining the two stars must sweep out equal areas in the true and apparent orbits in equal time intervals (the time rate of the line's sweeping out area in the orbits must be constant). If the orbits of each star around their center of mass can be determined, then the masses of the individual stars can be determined from the sizes of these orbits. Such doublestars give us our only accurate informa-

tion about the masses of stars other than the Sun, which is very important for our understanding of star structure and evolution.

In combination with data on the motions of the Sun, other stars, and interstellar gases, the equation for Kepler's Third Law gives estimates of the total mass in our Milky Way galaxy situated closer to its center than the stars and gas studied. If total mass ($M_1 + M_s$) is constant, the equation predicts that the orbital speeds of bodies decrease with increasing distance from the central mass; this is observed for planets in the solar system and planetary satellites. The recently discovered fact that the orbital speeds of stars and gas further from the center of the Milky Way than the Sun are about the same as the Sun's orbital speed and do not decrease with distance from the center indicates much of the Milky Way's mass is situated further from the center than the Sun and has led to a large upward revision of the Milky Way's total estimated mass. Similar estimates of the mass distributions and total masses of other galaxies can be made. The results allow estimates of the masses of clusters of galaxies; from this, estimates are made of the total mass and mean density of detectable matter in the observable part of our universe, which is important for cosmological studies.

When two bodies approach on a parabolic or hyperbolic orbit, if they do not collide at their closest distance (pericenter), they will then recede from each other indefinitely. For parabolic orbit, the relative velocity of the two bodies at an infinite distance apart (**infinity**) will be **zero**, and for a hyperbolic orbit their relative velocity will be positive at infinity (they will recede from each other forever).

The parabolic orbit is important in that a body of mass M_2 that is insignificant compared to the primary mass, M_1 (M_2=0) that moves along a parabolic orbit has just enough velocity to reach infinity; there it would have zero velocity relative to M_1. This velocity of a body on a parabolic orbit is sometimes called the parabolic velocity; more often it is called the "escape velocity." A body with less than escape velocity will move in an elliptical orbit around M_1; in the solar system a spacecraft has to reach velocity to orbit the Sun in interplanetary **space**. Some escape velocities from the surfaces of solar system bodies (ignoring atmospheric drag) are 2.4 km/sec for the Moon, 5 km/sec for Mars, 11.2 km/sec for Earth, 60 km/sec for the cloud layer of Jupiter. The escape velocity from the Earth's orbit into interstellar space is 42 km/sec. The escape velocity from the Sun's photosphere is 617 km/sec, and the escape velocity from the photosphere of a white dwarf star with the same mass as the Sun and a photospheric radius equal to the Earth's radius is 6,450 km/sec.

The last escape velocity is 0.0215 the vacuum velocity of **light**, 299,792.5 km/second, which is one of the most important physical constants and, according to the Theory of **Relativity**, is an upper limit to velocities in our part of the universe. This leads to the concept of a black hole, which may be defined as a **volume** of space where the escape velocity exceeds the vacuum velocity of light. A black hole is bounded by its Schwartzchild radius, inside which the extremely strong **force** of gravity prevents everything, including light, from escaping to the universe outside. Light and material bodies can fall into a black hole, but nothing can escape from it, and theory indicates that all we can learn about a black hole inside its Schwartzchild radius is its mass, net electrical charge, and its angular momentum. The Schwartzchild radii for the masses of the Sun and the Earth are 2.95 km and 0.89, respectively. Black holes and observational searches for them have recently become very important in astrophysics and cosmology.

Hyperbolic orbits have become more important since 1959, when space technology had developed enough so that spacecraft could be flown past the Moon. Spacecraft follow hyperbolic orbits during flybys of the Moon, the planets, and of their satellites.

KHAYYAM, OMAR (CA. 1048-CA. 1131)
Persian mathematician, poet, astronomer, and philosopher

Known in the West as Omar Khayyam, Ghiyāth al–Dī n Abu' l–Fath 'Umar ibn Ibrāhī m al–Nī sāburī al–Khayyāmī was born and died in Nī shāpur, Khorāsān, Persia (now Iran). Khayyam's given name was 'Umar, while "al–Khayyāmī " means tent–maker, which may have been the family trade. He was one of the most brilliant figures of Islamic civilization. His passionate and thought–provoking *Ruba'iyat* ("Quatrains"), in the West far better known than his extraordinary work as a mathematician, is a much–anthologized verse collection that has been praised as one of the treasures of world literature. As a mathematician, Khayyam is noted for his work in **cubic equations**.

Around 1070, Khayyam traveled to Samarkand, subsequently proceeding to Isfahan, upon the invitation of the Seljuk sultan Jalal–al–Din Malik–shah. Employed by the sultan as the court astronomer, Khayyam supervised a team of royal astronomers whose task it was to compile astronomical tables. Among Khayyam's accomplishments in Isfahan was a projected **calendar** reform. Khayyam's calendar is more accurate than the Gregorian calendar, accumulating an extra day in 3,770 years (as opposed to 3,333 years in the Gregorian calendar currently in use). His calculation of the length of the year—365.24219858156 days—is more precise than the Gregorian 365.2425 days. During this period, Khayyam also wrote commentaries on **Euclid**, as well as philosophical treatises on such fundamental questions as being, existence, and universal science.

In 1092, Khayyam found himself in a difficult situation following the sultan's assassination. He not only lost a patron but acquired powerful enemies. The observatory lost its funding, the calendar reform was halted, and Khayyam, who may have not been religious in a traditional sense, had to defend himself against charges of atheism. Nevertheless, Khayyam stayed in Isfahan until 1118, working hard to convince the late sultan's successors to provide support for science and education. In Merv, the new Seljuk capital, Khayyam continued his scientific and literary work. He eventually returned to Nishapur toward the end of his life.

Omar Khayyam

Khayyam's great contribution to **algebra**, described in his book *Algebra*, is his method of solving cubic **equations**. In his approach to **quadratic equations**, described in his *Elements*, Euclid relied on **geometry**. In fact, Euclid used line segments to indicate magnitudes which modern algebra represents by letters. In other words, Greek algebra was geometric, i.e., mathematicians used geometry to map algebraic problems. For example, the quadratic equation $x^2 - ax + b = 0$, transposed as $x^2 + b = ax$, would be translated into the geometrical problem of creating a **square** (x^2) so that the sum of x^2 and the given **area** b equals the area of the rectangle xa, a being a given side. Continuing in the tradition of Greek algebra, Khayyam successfully applied the geometric method for the solution of quadratic equations to cubic equations. In doing so, he had to imagine cubes and parallelepipeds instead of squares and other rectangles. However, cubic equations required geometry that cannot be done by straightedge and compass only, and Khayyam, drawing on Greek sources, arrived at a solution by using **conic sections**.

Although relying on geometry to solve equations, Khayyam used numbers, instead of line segments, exemplifying, as Carl B. Boyer has written, the seemingly clairvoyant efforts of Arabic mathematics "to close the gap between numerical and geometric algebra," a problem which will occupy mathematicians for several centuries. According to historians, Khayyam's work with cubic equations is one the great accomplishments of Arabic mathematics.

In his book *Commentaries on the Difficulties in the Postulates of Euclid's Elements*, Khayyam attempted to prove the "parallel postulate," which states that, given a line *a*, only one line parallel to *a* can be drawn through a point which is not on *a*. Euclid's fifth **postulate**, as it is known, is not, despite its apparent simplicity and transparence, easy to prove, and has therefore posed a great challenge to mathematicians. The fifth postulate is, as mathematician and musicologist Edward Rothstein calls it, "a troublesome axiom," which indeed explains why many generation of mathematicians tried to prove it. As Rothstein points out, this **axiom** is troublesome because it cannot be derived from the other axioms of the Euclidean systems. And it can be contradicted, by imagining an **infinity** of lines drawn through a point (not on line *a*) which are parallel to *a*. This new situation is possible in a new geometric system. According to Rothstein, "Properly reinterpreted, the new axiomatic system actually creates a different geometric universe, one that is called 'non–Euclidean'."

Unsatisfied by previous efforts to prove the fifth postulate, Khayyam attacked the problem by constructing a quadrilateral with two equal sides that are perpendicular to their base. Accepting that the quadrilateral's upper angles are by definition equal, Khayyam considered three hypotheses, namely, that these angles are right, acute, or obtuse. The second and third hypotheses had to be rejected, because, in accordance with the fifth postulate, converging lines will intersect. Although Khayyam dismissed the problematic hypotheses, he entered a theoretical universe in which, centuries later, **non–Euclidean geometry** was accepted as possible.

In his commentary on Euclid, Khayyam also discusses the theory of **proportions**. According to Euclid, the validity of a ratio, for example, *a* : *b* = *c* : *d*, can only be established if the alphabetical symbols stand for commensurable numbers, i.e., numbers that can be divided into units. Thus, the Euclidean definition accepts only **rational numbers**, (numbers that are **integers** or a quotient of two integers). Khayyam found Euclid's definition too narrow, suggesting instead that the idea of proportion be expressed purely numerically, without reference to the notion of unit. Khayyam's conception of proportion implies that numbers, but not only rational numbers, can represent proportion. Thus, by examining and expanding the Euclidean concept of viable number, Khayyam anticipated a line of theoretical work which culminated in the definition of the real number concept by **Julius Dedekind** in the 19th century.

Possessing a truly encyclopedic mind, Khayyam also applied his mathematical insights to musical theory. In his "Discussion on Genera Contained in a Fourth," he tackled the problem, already studies by Greek mathematicians, of dividing the **interval** of a fourth, which **Pythagoras** defined as the ratio of 3 : 4 (the integers represent string lengths) into three intervals which would basically fit into tonal systems whose

building–blocks are whole tones and half–tones (diatonic) or only half–tones (chromatic). What made the problem interesting was not just a mathematical challenge, but also the fact that the Greek scale was *enharmonic*, unlike the modern scale of equally–spaced 12 half–tones constituting an octave. This means that, depending on the tuning of a string instrument, (c-sharp and d-flat), which on a piano would be represented by the same key, would not necessarily have the same pitch. Khayyam divided the fourth in 22 ways, three being new.

To readers of poetry Khayyam is known as the author of the *Ruba'iyat*, widely praised as one of the great treasures of world poetry. A *ruba'i* (plural: *ruba'iyat*) is a two–line stanza. As each line consists of two hemistiches, the stanza has four hemistiches, which explains the term *ruba'i*—the Arabic word meaning *foursome*. The *Ruba'iyat*, which consist of over 200 stanzas attributed to Khayyam, address such ultimate questions as death, eternity, reality, time, as well as the paradox of human existence. Rich in imagery and profound wisdom. Khayyam's poetry has, through numerous translation, found enthusiastic readers all over the world, and inspired almost cultic devotion in England.

Among the numerous philosophical, scientific, and literary discussions generated by Khayyam's poetry throughout the centuries possibly the most fascinating, and inconclusive, one is the question concerning his possible adherence to Sufism, the venerable mystical tradition which developed within Islam. While some scholars have found no traces no Sufism in the works of the mathematician–poet, writers such as Idries Shah have identified the timeless quality of Khayyam's poetry as typical of the Sufi world view. However, a scientist's or poet's writings, as commentators often realize, may not reveal his (or her) deepest beliefs, and this probably, as B. W. Robinson has remarked, explains why Khayyam eludes attempts at classification, scholars offering no "reliable gauge of Omar Khayyam's own position in relation to the life of the spirit." Robinson concludes: "Life is lived very publicly in the East. As a result, though the heart is much talked of, men have learnt how to conceal what is really in it and, while verses might beguile a moment of leisure among friends and win applause for the poet's skill, they very rarely, as any student of Persian lyrical poetry knows, reveal precisely what the poet believed or did not believe."

KHINCHIN, ALEXSANDR YAKOVLEVICH (1894-1959)

Russian mathematician

Aleksandr Yakovlevich Khinchin (or Khintchine) is best known as a mathematician in the fields of **number theory** and **probability theory**. He is responsible for Khinchin's constant and the Khinchin Levy constant (sometimes referred to as the Levy Khinchin constant). These are both constants used in the calculation of fraction or decimal expansions. Several constants have been subsequently calculated from the Khinchin constant including those of Robinson in 1971 looking at non-standard **analysis**.

Aleksandr Yakovlevich Khinchin was born in Kondrovo in Russia in 1894. As a student at Moscow University he worked with Nikolai Lusin where they both studied number theory and probability, they also extended the large number work of **Emile Borel**. In 1927 Khinchin became Professor of Mathematics at Moscow University. The **proof** behind Khinchin's constant was published in 1934 in *Continued fractions*. Much of his work is summarised in his book of 1957 (English translation) *Mathematical Foundations of Information Theory*. This was published 2 years before his death in Moscow in 1959. Khinchin spent all of his academic life at Moscow University.

See also Number theory; Probability theory

KLEIN BOTTLE

The Klein bottle is a one-sided surface named after the great 19th-century mathematician **Felix Klein**, who was one of the first mathematicians to explore its unusual properties. The Klein bottle is formed from a **square** piece of paper by gluing opposite edges according to the following rules. The top and bottom edges are glued to each other by a reverse gluing; that is, the top edge is twisted 180 degrees before it is glued to the bottom edge. Performing only this gluing will turn the square into a surface called the **Möbius strip**. The left and right edges are glued in a very simple way: each point on the left edge gets glued to the opposite point on the right edge. Performing only this gluing will turn the sheet of paper into a **cylinder**.

Each of these gluings separately is easy to carry out. However, it is impossible to perform both gluings unless you allow the piece of paper to pass through itself, as in the depicted figure. Mathematicians have proven that there is no way to "embed" the Klein bottle in 3-dimensional **space**: no matter how much it is pulled or twisted, it will always intersect itself.

Although the Klein bottle cannot be constructed in space without self-intersection, mathematicians are still able to study the shape of the Klein bottle by looking at it intrinsically, considering what life would be like for a two-dimensional creature living on the Klein bottle. From this point of view, it is possible to ignore the specific way in which the Klein bottle is placed in space (the extrinsic point of view), and simply consider what a creature would experience living on a sheet of paper that has the appropriate gluings. Looking at the Klein bottle's intrinsic structure, it becomes clear that the Klein bottle is an example of a non-orientable surface, that is, a surface on which it is impossible to distinguish between left-handedness and right-handedness. Consider what happens if you draw a right shoe somewhere on the sheet of paper. If you move it to the right, when it gets to the edge of the paper it will come out on the other end, from the left edge, since these edges are glued together. It will eventually return to the spot where it started, just the way a person walking along the equator of the Earth will eventually return to his starting place. This process will not change the appearance of the shoe. If instead you move the shoe upward until it reaches the top edge, it will come out again from the bottom edge. However, since these edges are glued in reverse, the shoe will come out as the mirror image of a right shoe—that is, as a left shoe. Thus, there is no consistent way to define the difference between a left shoe and a right shoe on the Klein bottle.

The Klein bottle has another surprising property: although it is a closed surface, with no holes or boundary, it does not divide 3-dimensional space into an inside and an outside. If you stand on one side of the piece of paper and walk towards the top edge, then since the top edge is glued to the bottom edge with a twist, when you walk across the glued edge you will find yourself on the other side of the sheet of paper. Surfaces that have this property are known as one-sided. In its one-sidedness the Klein bottle resembles the Möbius strip, the simplest example of a one-sided surface. In fact, the Klein bottle can be formed from two Möbius strips, by gluing them together along their boundaries. The Klein bottle is also closely related to another non-orientable surface, the projective plane, which is formed from a round sheet of paper by gluing together opposite points on the boundary **circle** (like the Klein bottle, this surface cannot actually be constructed without the sheet of paper passing through itself). A Klein bottle is the "connected sum" of two projective planes: it can be formed by cutting a hole out of each projective plane and then running a tube from one hole to the other, connecting the two projective planes.

The Klein bottle cannot sit in 3-dimensional space without passing through itself, but in 4-dimensional space it is possible to build a Klein bottle that never intersects itself. In 3-dimensional space, the Klein bottle intersects itself along a circle. In 4-dimensional space, it is possible to "lift" one of the two circles of intersection off the other, into the fourth **dimension**. This is difficult to visualize, but it is similar to a lower-dimensional phenomenon: if you take a shoelace and lay it on the ground (a 2-dimensional surface) in a figure-8 pattern, the shoelace will cross itself at the center point. If however you are allowed to move part of your shoelace off the ground (into 3-dimensional space) then at the crossing point you can lift the top piece of shoelace up a bit, resulting in an "embedded" shoelace, that it, one that never touches itself. Mathematicians have used these techniques to show that every surface can be embedded in a higher-dimensional space.

KLEIN, CHRISTIAN FELIX (1849-1925)

German mathematician

Felix Klein is arguably one of the most influential mathematicians of the 19th century. He is best known for building the mathematical community at the University of Göttingen which became a **model** for research facilities in mathematics worldwide.

Christian Felix Klein was born on November 25, 1849 in Dusseldorf, the son of an official in the local finance department. Klein graduated from Gymnasium (the German equivalent of an academic high school) in Dusseldorf and began studying at the University of Bonn in 1865. At Bonn, he fell under the influence of **Julius Plücker**, one of the best-known geometers of the century. Plücker had moved the center of his interest to physics, and it had been in physics that Klein originally wanted to work, but Plücker returned to his original interest in **geometry** and took Klein with him. After Plücker's death in 1867, Klein became responsible for finishing a manuscript of Plücker's, which gave him an early introduction to the scholarly community and, in particular, to Alfred Clebsch, another prominent geometer of the time.

After receiving his doctorate in 1868 Klein spent a year traveling between Göttingen, Berlin, and Paris. Of the three, he enjoyed Göttingen immensely, did not like Berlin, and had to leave Paris ahead of schedule because of the outbreak of the Franco-Prussian War. Some of his travels were spent with the young Norwegian mathematician **Marius Sophus Lie**, whose ideas on geometry and **analysis** were much in common with Klein's. Klein's patriotism led him to enlist as a medical orderly during the war, but before the year was over he had returned to Dusseldorf, suffering from typhoid fever. The next year Klein qualified as a lecturer at Göttingen, but the following year he accepted a chair at the University of Erlangen. The complexities of academic promotion within the German university system at the time frequently required moving about from one university to another, merely for the sake of promotion within the original university.

It was the custom for a new professor to deliver an inaugural address at a German university, and in 1872 Klein followed suit at Erlangen. At the time, it was difficult to speak of one geometry, as recent developments had led to a collection of geometries whose relation to one another was unclear. There was the familiar **Euclidean geometry**, based on the ordinary axioms including the **parallel postulate** (which stipulated that there was exactly one parallel to a line through a point not on that line). There were at least two **non-Euclidean geometries**, one denying the existence of any parallels through a point not on a line, the other allowing the existence of an infi-

nite number of parallels. Finally, projective geometry, which had been known since the 17th century, had been given a more quantitative turn in the work of **Arthur Cayley**, among others.

As outlined by Klein, geometry is the study of the properties of figures preserved under the transformations in a certain group. Which group of transformations one started with determined the geometry in which one was working. For example, if the transformations were limited to rigid motions, then one had Euclidean geometry. If projections were allowed, then one had projective geometry. If an even wider class were included, then one could end up with **topology**. This view (called the Erlangen or **Erlanger program**) has infused the spirit, not just of geometry, but of mathematics as a whole ever since.

Also, in 1872 Klein took over editing *Mathematische Annalen* after the death of Clebsch. Under his editorship this was the leading mathematical journal in the world and it was to remain so until World War II. By 1875, Klein had left Erlangen for the Technische Hochschule in Munich and then in 1880 he went to the University of Leipzig. In 1884 he was invited to take the place of **James Joseph Sylvester** at Johns Hopkins University in Baltimore, but he declined. He did make several visits to the United States subsequently, where both his personal influence and those of his students were strong. Finally, in 1886, Klein achieved the goal of a chair at Göttingen.

Two factors in particular led Klein to successfully create a mathematical center at Göttingen. One was personal, as he was married to Anne Hegel, a descendant of the German philosopher Georg Wilhelm Friedrich Hegel. Her striking beauty may have been a draw even for those who were not yet convinced of the mathematical attractions of her husband. In the course of their married life the Kleins had one son and three daughters.

The other factor was not so pleasant. One of the subjects on which Klein had been working while at Leipzig were automorphic **functions**, transformations of the complex plane into itself that satisfied certain conditions. Unfortunately, for Klein the year 1884 turned into a competition with the younger French mathematician **Henri Poincare** seeking fundamental results. Although Klein's work during this period was of a high quality, he felt that he had not lived up to expectations and suffered a nervous breakdown.

Thereafter, Klein immersed himself in creating a major mathematical center at Göttingen. The mathematical discussions did not stop with the classroom walls, but continued at the Kleins' home or on walks into the woods around Göttingen. One feature of the institute was a room filled with geometrical models to help with visualization. The presence of such a room was a reminder of Klein's antipathy to the abstract style of analysis favored by **Karl Weierstrass** at Berlin. Klein wanted his mathematics to have intuitive content, which explains why he was anathema to Weierstrass. Klein attracted many of the leading German mathematicians to Göttingen, the most outstanding being **David Hilbert**. Göttingen's creative atmosphere encouraged the presence of women in the lecture hall and foreign visitors.

At the time of Klein's retirement shortly before the outbreak of World War I, he could take pride in having brought together a mathematical research community the like of which the world had never seen. In 1912, he received the Copley medal of the Royal Society, one of just many honors. His last years were saddened by the death of his son on the battlefield during the war, and he died on January 22, 1925.

Within ten years of his death the Nazi government had undertaken the dismantling of the research community in Göttingen. When the Institute for Advanced Studies was founded at Princeton in the 1930s, it modeled itself after Göttingen. The dream which Klein had brought into reality lived on.

KOLMOGOROV, ANDREI NIKOLAEVICH
(1903-1987)
Russian probabilist and educator

Andrei Nikolaevich Kolmogorov made major contributions to almost all areas of mathematics and many fields of science and is considered one of the 20th century's most eminent mathematicians. He was the founder of modern **probability theory**, having formulated its axiomatic foundations and developed many of its mathematical tools. Kolmogorov also helped make advances in many applied sciences, from physics to linguistics. A great teacher, he did much to keep the Soviet Union in the forefront of research in theoretical and **applied mathematics** and was responsible for reforms in mathematics education at the elementary and high–school levels.

Kolmogorov was born in the town of Tambov in central Russia on April 25, 1903. His father, Nikolai Kataev, became a professional agriculturalist and was killed during World War I. His mother, Mariya Yakovlevna Kolmogorova, was not formally married to his father and died during his birth. Her sister, Vera Yakovlevna Kolmogorova, adopted and raised the boy in the family's home village of Tunoshna. As a child, young Kolmogorov and his friends attended a school run by his two aunts. At the age of five, he made his first mathematical discovery by noticing the pattern that $1=1^2$, $1+3=2^2$, $1+3+5=3^2$, etc.

In 1920, at the age of 17, Kolmogorov enrolled in Moscow University. To help support himself while he attended the university, he worked as a secondary school teacher. He took an active role in the school, and he is said to have been more proud of that work than of the honors he garnered for his own academic progress. Within two years, Kolmogorov had completed a study in the theory of operations on **sets**, which was eventually published in 1928. A second project he also completed in 1922 brought immediate recognition: He formulated the first known example of an integrable function with a **Fourier series** that diverged almost everywhere (he soon extended that result to *everywhere*). The international mathematics community took notice of the bright 19–year–old. During his years as a university student, he published 18 mathematical papers including the strong law of large numbers, generalizations of **calculus** operations, and discourses in intu-

Andrei Nikolaevich Kolmogorov

itionistic logic. In 1925, Kolmogorov received a doctoral degree from the department of physics and mathematics and became a research associate at Moscow University. At the age of 28, he was made a full professor of mathematics; two years later, in 1933, he was appointed director of the university's Institute of Mathematics. In 1942, Kolmogorov married Anna Dmitrievna Egorova.

While he was still a research associate, Kolmogorov published a paper, "General Theory of Measure and Probability Theory," in which he gave an axiomatic representation of some aspects of probability theoryon the basis of measure theory. His work in this area, which a younger colleague once called the "New Testament" of mathematics, was fully described in a monograph that was published in 1933. The paper was translated into English and published in 1950 as *Foundations of the Theory of Probability*. Kolmogorov's contribution to probability theory has been compared to **Euclid**'s role in establishing the basis of **geometry**. He also made major contributions to the understanding of **stochastic** processes (involving random variables), and he advanced the knowledge of chains of linked probabilities.

Kolmogorov developed many applications of probability theory, creating a powerful technique for using probability to make observa- tion–based predictions in the face of randomness and researching statistical inspection methods for mass production. One of the applications that Kolmogorov developed is known as reaction–diffusion theory, which deals with the manner in which an event spreads through a given population. It is now used to study the **dispersion** of epidemics, cultural changes, effects of advertising, and a variety of other situations in such fields as biology and chemistry. He was largely responsible for demonstrating the applicability of the emerging fields of probability and **statistics** on substantial problems in science and engineering. In fact, Kolmogorov contributed to the war effort in the 1940s by solving statistical problems relating to artillery fire and by applying stochastic theory to suggest the most effective placement of barrage balloons for protecting Moscow from Nazi bombing assaults.

An appointment as Chair of Theory of Probability at Moscow University in 1937 served as official recognition of Kolmogorov's achievements in probability theory. Communication with the West was sporadic, however, and it was not until the late 1950s that Western mathematicians discovered that Kolmogorov had already determined the nature of many issues in probability theory that they were still working to discover. In 1939, at the age of 36, Kolmogorov became one of the youngest full members elected to the Soviet Academy of Sciences. He was later appointed academician–secretary of the Academy's department of physical and mathematical sciences. These honors were in recognition not only of his work in probability theory but of his contributions to other areas of theoretical and applied mathematics.

Kolmogorov made significant contributions to **set theory**, measure theory, integration theory, **topology**, functional **analysis**, constructive logic, **differential equations**, and the theory of approximation of **functions**. Among his many accomplishments in applied fields, Kolmogorov developed important results in such areas as biological statistics, econometrics, mathematical linguistics, and the theory of fluid turbulence. His work in fluid turbulence was so profound that in 1946 he was chosen to head the Turbulence Laboratory at the Academy Institute of Theoretical Geophysics. He continued to work in this field for many years, sailing around the world in 1970–1972 to study ocean turbulence. He helped construct the Kolmogorov–Arnold–Moser (KAM) **theorem**, which is used to analyze stability in dynamic systems. Kolmogorov introduced the concept of entropy (a theoretical measure of unavailable energy in a thermodynamic system) as a measure of disorder. In the first detailed solution of the three-body problem, which had been proposed by **Isaac Newton,** Kolmogorov analyzed the interactions of two celestial bodies orbiting in the same plane, one with an elliptical **orbit** and the other with a circular path.

In an obituary published in *Physics Today,* V. I. Arnold of the Steklov Mathematical Institute wrote that "Kolmogorov considered his most difficult achievements to be his work from 1955 to 1957 on the 13th **[David] Hilbert** problem." The problem involves finding a way to represent a function of many variables in terms of a combination of functions having fewer variables.

In a remembrance of Kolmogorov published in *Statistical Science,* A. N. Shiryaev wrote that "One sensed that

he had continuously intensive brain activity." His active intellect led him to investigate questions in a wide range of mathematical fields as well as a number of applied subjects including meteorology, **hydrodynamics**, celestial **mechanics**, genetics, history, and linguistics. Shiryaev wrote, "According to his own words, Kolmogorov had a lively interest in a problem only until it became clear what the answer should be. As soon as the picture became clear he tried to avoid writing down the results and proofs; he would look for someone else to take over." Indeed, by the time Kolmogorov solved a problem, he would have identified other topics to investigate. Shiryaev quoted the mathematician **A. Ya. Khinchin** as saying, "The most important and most fascinating feature of [Kolmogorov] as a mathematician is the wealth of his ideas. Each sentence of his about any work could become the basis for a Ph.D. dissertation." Also speaking at a 50th birthday tribute to Kolmogorov, **Aleksandr Gelfond** said, "The fact that mathematics is viewed as a unified discipline is due to a large extent to Kolmogorov."

Kolmogorov was also actively interested in mathematical education in the U.S.S.R., working as the chairman of the Academy of Sciences Commission of Mathematical Education. He played a pivotal role in overhauling the teaching of mathematics during the 1960s, and his leadership in mathematics education for secondary schools and universities helped move the U.S.S.R. to the forefront of mathematics internationally during the following decades. In fact, being of the opinion that no mathematician could possibly do meaningful research after the age of 60, Kolmogorov retired in 1963 and spent the following 20 years teaching high school. During his final years, he compiled his collected works; an English version was published in the United States in 1991. He died in Moscow on October 20, 1987, at the age of 84.

KOVALEVSKAYA, SONYA VASILIEVNA [SOFIA VASILEVNA KOVALEVSKAIA] (1850-1891)
Russian mathematician and educator

Sonya Vasilievna Kovalevskaya has been applauded by some as the most astounding mathematical genius to surface among women in the last two centuries, and one of the first women to make contributions of high quality to the field.

The middle of three children, Kovalevskaya was born on January 15, 1850, in Moscow. Her father, Vasilli Korvin–Krukovski, was an Artillery General in the Russian army. He was a educated and disciplined man who was fluent in English and French, and was a stern but benevolent parent. Her mother, Elizaveta Fyodorovna Schubert, came from a family of German scholars who had emigrated to Russia in the mid–1700s. Kovalevskaya's grandfather, Fyodor Fyodorovich Schubert, and great–grandfather, Fyodor Ivanovich Schubert, were noted mathematicians.

A singular incident in Kovalevskaya's childhood seems to have been a portent of her devotion to the study of mathematics. While living at the family estate, she came upon a room where the wallpaper consisted of sheets of Mikhail Ostrogradsky's lithographed lectures on the differential and the **integral calculus**. The child spent hours trying to decipher the formulae. Years later, at the age of 15, she astonished her tutor with how quickly she grasped and assimilated the conceptions of **differential calculus**. Kovalevskaya wrote in her memoirs that she "vividly remembered the pages of Ostrogradsky... and the conception of **space** seemed to have been familiar to me for a long time."

It was also rather exceptional that her father allowed Kovalevskaya to study with a tutor at all. She described her father as one who "harbored a strong prejudice against learned women." It has been suggested that the best explanation is that Kovalevskaya's own fierce determination was the catalyst for changing her father's mind. Once the tutelage was over, however, she faced an uncertain future for obtaining advanced education. She knew her father would never agree to sending her to a university. During that time, women were not allowed to attend the universities in Russia and most fathers, including Kovalevskaya's, were unwilling to give consent to daughters to study abroad. Again, Kovalevskaya's determination was stronger than her father's will.

The device she used to get her way was a popular one at the time; she began searching for a husband. The type of husband she was looking for had to agree to sign papers allowing Kovalevskaya to travel, live apart from him, and pursue an education. The agreement also came with the understanding that the marriage was a platonic one, without the marital rights usually afforded a husband. She found such a man in Vladimir Kovalensky, who made his living translating and publishing books while pursuing a degree in paleontology. Along with his high intellect, Kovalensky also distinguished himself by supporting liberal causes.

The resistance by Kovalevskaya's family to the marriage was anticipated and overcome by using the same sort of guile that had created the situation. She sent notes to a number of distinguished family friends happily announcing her impending marriage, thereby forcing her father to either give public approval or publicly admit that his daughter was rebellious. To make certain her father would not renege on the announcement, Kovalevskaya ensconced herself in Kovalensky's apartment, refusing to leave, until she felt secure that the marriage would indeed take place. The couple was married in September 1868.

In early 1869 the newlyweds left Russia and settled in Heidelberg, Germany. This was where Kovalevskaya was to fulfill her dream of a higher education. Because she was a woman, the officials at the University of Heidelberg demanded that she secure the written permission of each of her professors before full admittance was granted. She undertook a class schedule of 22 hours per week, 16 of which were spent studying mathematics with Paul Du Bois–Reymond and Leo Köenigsberger, both of whom were students of the renowned mathematician **Karl Weierstrass**. After three successful semesters of study at the university, Kovalevskaya left for Berlin,

Sonya Vasilievna Kovalevskaya

seeking out Weierstrass. Their initial meeting marked a personal as well as professional relationship that lasted a lifetime.

Kovelevskaya did not arrive in Berlin unannounced, however. The praises of her professors at Heidelberg preceded her, and this did much to persuade Weisertrass' decision to become Kovelevskaya's mentor. In addition, Weierstrass had written a paper where he gave credit to Kovalevskaya's grandfather, Fyodor Schubert, for a mathematical maxim eight years before meeting Kovelevskaya. Unfortunately, winning Weierstrass' acceptance was not the only obstacle to the continuation of her studies—the university forbade women from attending Weierstrass' formal lectures. The obstacle was removed when Weierstrass agreed to teach Kovalevskaya privately twice a week, giving her the same courses as his regular university students.

In the beginning, Weierstrass never imagined that Kovalevskaya would want a formal degree in mathematics, believing that a married woman would have no use for one. In the fall of 1872 however, Kovalevskaya confided the **truth** about her marriage and he began to steer her toward work on a dissertation. By the spring of 1874, Kovalevskaya had written three doctoral dissertations, each of them in Weierstrass' opinion worthy of a degree, and one so outstanding that both were confident of forthcoming recognition. They were not disappointed. Weierstrass submitted Kovalevskaya's work to the

University of Göttingen and she was awarded her doctoral *summa cum laude* in the fall of 1874, becoming the first woman to earn a doctorate in mathematics.

Elated but exhausted by her labors, Kovalevskaya and her husband returned to Russia to relax with friends and family. Both were also hoping to secure positions in the academic world, but for a combination of reasons neither was welcomed to university posts. Kovalevskaya found herself discriminated against because of her gender, and Kovalensky's liberal activities spawned suspicion among Russian academics. Kovalevskaya and her husband decided to consummate their relationship, and Kovalevskaya's only child, Sofia, was born in 1878. For the next five years, Kovalevskaya and her husband put aside their respective fields of study and concentrated on trying to make a living at various commercial endeavors.

During this time it became apparent that Kovalevskaya was as gifted at writing as she was at mathematics and for a time her heart was torn between the two pursuits. The fiction she produced, including the novella *Vera Barantzova*, were met with acclaim and translated into several foreign languages. Meanwhile, Kovalensky was involved in questionable financial dealings. Faced with prosecution from charges of mishandling stock, Kovalensky committed suicide in April of 1883.

Kovalevskaya returned to her study of mathematics and through the efforts of a friend and fellow student of Weierstrass, Gosta Mittag–Leffler, Kovalevskaya was offered a position at Stockholm University as a *privatdozent* (a licensed lecturer who could receive payment from students but not from the university) in 1884. Five years later, Kovalevskaya became the first female mathematician to hold a chair at a European university. This appointment was accompanied by the editorship of the journal *Acta Mathematica*, where she came in contact with the leading European mathematicians of the day. In 1888, Kovalevskaya's paper on the study of the **motion** of a rigid body received the Prix Bordin, given by the French Académie Royale des Sciences.

Kovalevskaya died in 1891 of influenza when she was only 41 years old, at the height of her mathematical career. Although she published only ten papers during her lifetime, Kovalevskaya's work has withstood the test of time. The research that won her the Prix Bordin is now known as the Kovelevskaya top, and her doctoral dissertation on partial **differential equations** lives on as the Cauchy–Kovelevskaya **Theorem**.

KRONECKER DELTA

A Kronecker delta is a compact notation that indicates the equivalent of a dot product in vector notations. In this usage, it takes the form of a Greek letter delta with two subscripts which indicate that the components are to be multiplied if the subscripts are equal and thrown out if the subscripts are different. That is, the Kronecker delta notation indicates that the x-components should be multiplied together, the y-compo-

nents should be multiplied together, and the product of x- and y-components should be ignored.

The Kronecker delta is not only used in vector **equations**, however. It can be used with a set of subscripted **functions** to indicate the orthogonality of that set. If an arbitrary set K_n of functions is orthogonal, then the integral of the product of those functions K_m and K_l with their appropriate weighting function will equal the Kronecker delta with respect to l and m: 1 if l = m and 0 if l and m are not equal. This flexible notation can also be used to mean that only those functions which "match" or have the same subscripts should be considered in product, or that only repeated subscripts should be considered. It can indicate that, even if the products would otherwise not be **zero**, there are considerations that leave them out of the problem at hand. The Kronecker delta is not something that can be proven or disproven—it merely indicates, without a lot of excess verbiage, which of several potential components the mathematician is using.

See also Leopold Kronecker

KRONECKER, LEOPOLD (1823-1891)
German mathematician

Leopold Kronecker's approach to mathematics can best be described in perhaps his most famous words, spoken in 1886 in Berlin: "The integer numbers were made by God, everything else is the work of man." Kronecker believed that all mathematics could be reduced to and explained by finite **whole numbers** only, and that irrational and **transcendental numbers** and any related proofs and theories involving them did not exist.

Kronecker was born into a well-to-do family, and through inheritance and business interests, he remained independently wealthy throughout his life. He developed a keen interest in mathematics when in school at the Liegnitz Gymnasium (secondary school). Here, he met teacher Ernst Kummer, whom he would work with for the greater part of his mathematical career.

In 1841, Kronecker enrolled at Berlin University to study mathematics, philosophy, and astronomy. After a brief stint at the University of Breslau, where he once again studied under Kummer, Kronecker returned to Berlin to earn his doctorate in mathematics in 1845, studying under Peter Dirichlet and writing his final dissertation on algebraic **number theory**.

Instead of pursuing his academic career after earning his doctorate, Kronecker left Berlin for the family estate, assisting in business affairs and managing the family property for the next decade. He continued to work on mathematical theory, but as a hobby rather than a serious academic pursuit. However, in 1855, Kronecker returned to Berlin to once again work with his mentor, Kummer, who had taken a position at the University of Berlin. Kronecker did not seek a teaching position in Berlin, and financially, did not need one. He instead spent his time on mathematical studies and research. It was here that he also met and worked with respected mathematician **Karl Weierstrass**.

During this period, Kronecker published a large body of research on number theory, elliptic **functions**, and algebraic **equations**. After Kummer nominated him to the Berlin Academy in 1860, he began to lecture at the University. Although he would by choice remain a lecturer and researcher with no official teaching position for over twenty years, Kronecker's reputation as a serious and respected mathematical researcher grew amongst his colleagues, as did his influence in the mathematical community. In addition to the Berlin Academy, he was elected to a number of prominent mathematical societies, including the Paris Academy and the Royal Society of London. He was also appointed to the editorial staff of *Crelle's Journal*, a high profile German mathematical journal. In 1880, Kronecker took over as editor of the journal, expanding his sphere of influence even further.

Kronecker was extremely vocal in his views on number theory and what he perceived to be the invalidity of the use of anything other than finite numbers in number theory. He used his considerable influence to speak out against the views of mathematicians such as **Heine, Cantor, Dedekind**, and even colleague Weierstrass, and may have been responsible for blocking the publication of some of their work in *Crelle's Journal*. It is thought that Kronecker also used his influence at the University to block the appointment of Cantor to a teaching position in Berlin because of the mathematician's ideas about **set theory** and **irrational numbers**.

Upon Kummer's retirement in 1883, Kronecker was appointed to the chair of mathematics in Berlin, his first official teaching position, and to the position of codirector of Berlin's mathematical seminar. He died in 1891 at the age of 68, several months after his wife Fanny was killed in a climbing accident.

See also Algebraic numbers; Number theory

KUPERBERG, KRYSTYNA (1944-   )

Polish mathematician

Krystyna Kuperberg is a researcher and educator best known for disproving the famous Seifert conjecture in **topology**. Her counterexample, first announced in the mid-1990s, was termed a "small miracle" of **geometry** by Ian Stewart. It was quickly generalized and should prove central to the continued development of dynamic systems theory, by way of the vector fields used to study physical and statistical phenomena.

Kuperberg was born Krystyna M. Trybulec in Tarnow, a city in southern Poland, on July 17, 1944. Her parents, Jan W. and Barbara H. (Kurlus) Trybulec, were both trained pharmacists. Her brother, Ardrzej, also became a mathematician. After receiving a master's degree from Warsaw University in 1966, Kuperberg had to wait until settling in the United States to earn her Ph.D. This was awarded by Rice University in 1974. Upon graduating she accepted her first post at Auburn University. Kuperberg remains a member of the faculty at Auburn, and has been a full professor there since 1984. She has also held visiting positions at Oklahoma State University,

the Courant Institute of Mathematical Sciences in New York, the Mathematical Sciences Research Institute at Berkeley, and l'Universite de Paris-Sud, Centre d'Orsay.

In 1974, the first counterexample to the famous Seifert conjecture was found by P. A. Schweitzer. His "plug" was devised to cancel out any circular **orbit**, but it broke down to two minimal **sets**. Kuperberg, who had begun publishing papers in 1971, was already interested in dynamical systems. She resolved a conjecture about fixed points in 1981 with Coke Reed, and built upon the methods used in this work to find a new kind of counterexample with only one minimal set. What she eventually found served to disprove the Siefert conjecture for all three-dimensional manifolds.

The Seifert conjecture is a higher-dimensional extension of the "hairy billiard ball" **theorem** for the two-dimensional surface of a **sphere**. The idea that you cannot comb down all the hairs on a fully hairy ball without getting a cowlick is really a geometric statement about a dynamical system. The one-dimensional version, the **circle** or "1-sphere," is "combable," allowing for a smooth vector **field**. The 2-sphere is combable because it contains at least one "bald spot" consisting of a fixed point around which the trajectories can flow. The fact that there is always some place on Earth where the wind is not blowing is a real-world example of this "bald spot."

More complex surfaces proved more difficult to analyze. In the case of a torus-shaped vector field or "hairy donut," for instance, whether a trajectory is fixed or not depends upon how it advances along the **circumference** of the **torus** as it flows. This explains why it was impossible to prove a simple conjecture about one of the three-dimensional shapes for more than 40 years. In 1950, Herbert Seifert had proposed that in the three-dimensional case any smooth vector field will have at least one "closed" or periodic orbit. It was already known that 3-spheres did not have any "bald spots," but it seemed reasonable enough to think that they would have at least one closed orbit.

Kuperberg disproved this conjecture in 1993 by constructing a smooth vector field with no closed orbits, and her construction applies not only to 3-spheres, but to all three-dimensional manifolds. To do this she used a Wilson plug, a kind of topological tool, to break up any closed orbits that might be present. This plug is a three-dimensional shape that traps the trajectories of one or more formerly closed orbits inside itself. The trick is to apply the plug without creating any new closed orbits. To accomplish this, Kuperberg modified the plug so that it "eats its own tail" like a snake. Thus, the trajectories that enter get trapped in an infinite **spiral** and no new closed orbits can be formed. In addition to disproving the Siefert conjecture, Kuperberg's construction produces a "minimal set" that may be of an entirely new kind, according to John Mather, a dynamical systems theorist at Princeton. Since minimal sets are basic components of dynamical systems, Kuperberg's plug may help mathematicians better understand the range of things that can happen in these systems.

Since her success in disproving the Siefert conjecture, Kuperberg has been especially in demand as a speaker at events devoted to topology or dynamical systems, and at honorary symposia worldwide. She delivered the MSRI-Evans lecture at Berkeley in 1994, and addressed the American Mathematical Society and Mathematical Association of America meetings in 1995 and 1996.

Kuperberg's husband and frequent collaborator, Wlodzimierz, received his Ph.D. in mathematics from Warsaw University. He is also a professor at Auburn University. Krystyna and Wlodzimierz were married in Poland and lived there until 1969. Their son, Greg, born in Poland in 1967, is also a mathematician; he received a Ph.D. in mathematics from the University of California at Berkeley. Their daughter,

Anna, born two years later in Sweden, holds a M.F.A. from the San Francisco Art Institute.

In addition to awards from Auburn University for her research and professorship, and National Science Foundation grant support, Kuperberg also won the Alfred Jurzykowski Foundation Award in 1995 from the Kosciuszko Foundation in New York. In 1996, Kuperberg was elected to the American Mathematical Society council as a member at large for a three-year term. She also currently edits the *Electronic Research Announcement* of the American Mathematical Society.

L

LAGRANGE, JOSEPH–LOUIS (1736-1813)
Italian-born French algebraist and number theorist

Comte Joseph–Louis Lagrange is considered by many historians to be the foremost mathematician of 18th century Europe. He invented the **calculus** of variations, laid the foundation for modern **mechanics**, and made major contributions to the fields of **algebra** and **number theory**. He is also credited with establishing the standard of the **metric system**, now in widespread use throughout the world. Lagrange is highly regarded both for the originality of his work and the rigor and generality of his mathematical proofs.

Lagrange was born in Turin, Italy, on January 25, 1736. His father, a Frenchman, served as Treasurer of War in Sardinia, and his mother, Marie–Therese Gros, was the daughter of a wealthy Italian physician. Lagrange was the youngest of 11 children and the only one to survive past infancy. His early schooling focused on the classics, and although he read the works of **Euclid** and **Archimedes**, Lagrange displayed little initial interest in mathematics. It was not until he came across an article about calculus and its superiority over ancient Greek **geometry** that his interest was kindled. Its author, English astronomer and mathematician Edmund Halley, was cited years later by Lagrange as the individual who had most influenced his decision to pursue mathematics. By the time Lagrange was 16 years old, he had mastered so much of the subject that he was appointed as a professor of mathematics at the Artillery School in Turin.

At the age of 19, Lagrange began working on a solution to several isoperimetric problems that were then under discussion among the leading European mathematicians of the era. In the course of deriving his proofs, Lagrange developed what was to become his most important intellectual achievement, the creation of the calculus of variations. In 1756, he sent a letter describing his work to **Leonhard Euler**, then the director of the mathematics division at the Berlin Academy of Sciences. Euler praised and encouraged Lagrange, realizing at once the

significance of his results, and the two mathematicians began a long correspondence. By 1759, through Euler's influence, Lagrange was elected to the Berlin Academy.

Lagrange continued teaching in Turin. Pulling together his most talented students, he organized a research society that evolved into the Turin Academy of Sciences. In 1759, the Academy published its first volume of memoirs. Lagrange contributed three papers, one introducing his calculus of variations, another on the application of **differential calculus** to the field of probability, and a third regarding the theory of sound. In this last paper, he provided a mathematical description of string **vibration**, using a partial differential equation. His result settled an ongoing controversy about the subject between Euler and French mathematician **Jean le Rond d'Alembert**, favoring Euler's position.

Throughout 18th century Europe, the learned academies encouraged research in celestial mechanics, frequently offering the incentive of a prize, since knowledge in this area was valuable to navigation. In 1764, Lagrange entered a competition sponsored by the French Académie Royale des Sciences to determine the gravitational forces that caused the moon to present a relatively consistent face to earth. For his calculations he received the Grand Prize. Two years later Lagrange again won the Grand Prize from the Académie, this time for deriving a partial solution to a more complicated gravitational problem involving the planet Jupiter, its four then–known satellites, and the sun. That same year, Lagrange received an invitation from King Frederick of Prussia to become director of mathematics at the Berlin Academy, replacing Euler, who had departed for St. Petersburg. Accepting the appointment in November of 1766, Lagrange entered a prolific period, composing memoirs nearly every month on subjects ranging from probability to the theory of **equations**.

In 1767, Lagrange published *On the Solution of Numerical Equations*, a treatise in which he explored universal methods for reducing equations from higher to lower degree. This work would help set the stage for the development of modern algebra. Lagrange also made early contributions to

Joseph–Louis Lagrange

It was not until after the French Revolution that Lagrange finally began to emerge from his apathy and end his ambivalence toward mathematics. The monarchy had fallen and the Académie lost its royal patronage. Many of Lagrange's colleagues were beheaded, but Lagrange himself had managed to remain a politically neutral figure. He was granted a pension by the revolutionists and eventually appointed to a government committee that was charged with establishing standards for weights and measures. It was in this position that he persuaded the French to adopt the metric system. In 1795, Lagrange became a professor of mathematics at the newly formed École Normale and when it closed two years later, he assumed a professorship at the École Polytechnique. In an attempt to clarify the topic of calculus for his students he wrote *Theory of Analytic Functions* in 1797 and *Lessons on the Calculus of Functions* in 1801. In these texts Lagrange attempted to reduce calculus to an algebraic system and was unsuccessful. Still, the works proved valuable as a catalyst to 19th century mathematicians who would refine his ideas and develop a more coherent calculus.

When Lagrange reached his seventies, he began revising and extending his *Mecanique Analytique*, completing a second edition. The long hours of work diminished his strength and energy and he suffered increasingly from fainting spells. On April 10, 1813, Lagrange died. His body was brought to rest in the Pantheon, as a tribute to the contributions he made to France.

Lagrange was known for his gentle demeanor and his diplomatic skills. At the Berlin Academy, he fell into favor with Frederick the Great, who had been highly critical of Lagrange's predecessor, Leonhard Euler. When Lagrange arrived at the French Académie he was doted upon by Queen Marie–Antoinette, yet he also managed to remain on good terms with leaders of the French Revolution, Napoleon Bonaparte, who made Lagrange a Senator, a Count of the Empire, and a Grand Officer of the Legion of Honor, consulted him frequently on philosophical and technical matters. Shortly after accepting his appointment at the Berlin Academy in 1766, Lagrange married a young woman to whom he was related. In his correspondence with his friend and colleague d'Alembert, he referred to the marriage as "inconsequential" and one of convenience. Nonetheless, when his wife took ill and died a few years later, Lagrange was reportedly heartbroken. He did not marry again until he was living in Paris and suffering from his depression. Then, at the age of 56 he took as his second wife the teenage daughter of his friend, astronomer Pierre–Charles Lemonnier. His new bride was devoted to him, and it is said she helped Lagrange regain his interest in mathematics. Neither marriage resulted in children.

number theory, solving several of Fermat's theorems. He continued his investigation of gravitational interactions among planetary bodies, winning, in 1772, his third Grand Prize from the French Académie for a memoir on attractions among the sun, moon, and Earth. In 1774 and in 1778 he again won Grand Prizes, first for work related to lunar movement, then for a study of the perturbations of comets.

During his tenure at the Berlin Academy, Lagrange worked steadily on the topic of mechanical **analysis**, employing his calculus of variations. Through his efforts, the study of fluid and solid mechanics was to be unified. His *Mecanique Analytique*, finally published in 1788, applied calculus to the mechanics of rigid bodies and contained the general equations for describing **motion** in mechanical systems. This work is universally considered his most important masterpiece. Ironically, Lagrange lost interest in the subject, and in mathematics in general, in the years before its publication. He began suffering from depression around 1780. When Frederick the Great died in 1786 an indifference toward science and a resentment toward foreigners arose in Berlin. Lagrange sought and obtained a position with the French Académie, where he was well–received. In Paris, his depression worsened. He was known to stare out the window for long periods of time and he hardly spoke. In letters to his friends and colleagues he wrote that mathematics was no longer important.

LAGRANGE'S THEOREM

Lagrange's **theorem** is one of the fundamental theorems of finite **group theory**. Formally stated, the order of a subgroup of a group divides the order of the original group. This means that the number of elements in any subgroup of a finite group must divide evenly into the number of elements in the group. This would mean that a group with 50 elements could not have a subgroup of 8 elements since 8 does not divide 45 evenly. The subgroup could have 2, 5, 10, or 25 elements since these numbers are all divisors of 50. This theorem is also referred to as Lagrange's group theorem or sometimes Lagrange's lemma. The converse of Lagrange's theorem is not generally true.

In the 1770s **Joseph Louis Lagrange**, one of the greatest mathematicians on the 18th century, formulated the theorem named in his honor. He was one of the first mathematicians to study group structure. Lagrange, who was prolific in his career studying **mathematics and physics**, had a special talent for **number theory** and developed several theories dealing with numbers and their relations. Lagrange's theorem is just one of his many theories concentrated on number theory and more specifically group theory.

The converse of Lagrange's theorem is an interesting question in group theory. In general the converse of Lagrange's theorem is not true; that is that if a number n is a divisor of the order of a group that the group must have a subgroup of order n. There are however special cases where this is true. These cases are the focus of Sylow theorems.

LAMBERT, JOHANN HEINRICH (1728-1777)

German statistician, geometer, and analyst

Johann Heinrich Lambert stood largely outside the academic environment of his time. Although his ideas did not become popular until the next century and some of his other work were lost for many years, his scientific investigations gave rise to questions about how to handle data, from which he developed some basic ideas of **statistics**.

Lambert was born on August 25, 1728 in the town of Mulhouse, at that time a free city allied with Switzerland. Lambert's father, Lukas, was a poor tailor and his mother was named Elisabeth Schmerber. At the age of 12 Lambert left school in order to help his father. Within the next few years Lambert went through a variety of professions, including clerk at an ironworks and (at the age of 17) secretary to the editor of a Basel newspaper. Lambert's father died in 1747 and the next year Lambert was hired by the von Sails family as a tutor for their children. From 1752 until his death, Lambert kept a diary which furnishes a guide to the direction of his thoughts and the subjects which he was investigating.

From 1756 through 1758 Lambert traveled throughout Europe with his pupils. It may have been intended as an educational experience for the children, but Lambert used it as an opportunity to connect with various scientific societies in the towns through which they traveled. He was elected a corresponding member of the Learned Society at Göttingen but by the end of the trip he was looking for a permanent scientific position. After an opportunity to organize a Bavarian Academy of Sciences failed, Lambert was offered a position at St. Petersburg. He was reluctant to follow some of his countrymen to the Russian court, however, and preferred to wait until something arose at the Prussian Academy instead. He was proposed for membership in 1761 and began receiving a salary in 1765.

During his time as an employee of the Prussian Academy, Lambert's wrote more than 150 papers. He based his investigations from some of the same mathematical perspectives as **Gottfried Leibniz** and followed some of his philosophical roads as well. One of Leibniz's goals had been to create an ideal language for carrying out reasoning and Lambert imitated his predecessor with no greater success. It was not until the work of **Gottlob Frege** in the 19th century that technical developments allowed for the creation of a language useful for imitating reasoning.

Lambert's best-known work in mathematics was connected with an ancient geometrical problem, **squaring the circle**. In an effort to construct by geometrical means a **square** the same size as a circle, it was crucial to know as much about the number π, the ratio of the **circumference** of a circle to its **diameter**, as possible. One fundamental question was whether π was a rational number (that is, could be expressed as the ratio of two **whole numbers**) or not. Certain **irrational numbers** like the **square root** of two were well known, but π was not similar to those irrationals. Finally, Lambert was able to prove that π was not a rational number by looking at a continued fraction expanion. A continued fraction is one with a denominator within a denominator and so forth, leading to an expression that is cumbersome to express typographically but which can be used to stand for an infinite process. The same technique could also be employed to establish the irrationality of **e**, the base of natural **logarithms**.

Another area in which Lambert investigated is what was to become known as **non-Euclidean geometry**. In **Euclidean geometry**, there is only one line parallel to a given line through a point not on the line. By altering that condition, one can end up with geometries with different properties. Lambert was attempting to prove the **parallel postulate** of **Euclid** and stumbled on these non-Euclidean consequences in the process. Unfortunately, his work was not discovered until after non-Euclidean geometry had been more thoroughly explored.

Perhaps the work with the longest-term consequences which Lambert published was a general theory of errors. In a work on the measurement of **light** he discussed the problem of determining the probability distribution for errors and formulated a method analogous to that of maximum likelihood as developed by statisticians more than a century later. He viewed weather as though it were produced by an infinite number of unknown causes, like the outcomes of a game of chance, and argued that if positive and negative errors are equally possible, then they will occur with equal frequency. As the importance of errors in measurement became more widely recognized, Lambert's work on the theory of errors was applied by **Karl Gauss** and others.

Lambert's work on map projections was among the first to give the theory of the subject, although the details of the formulation were improved by others. He allowed for the possibility of probabilities that did not simply add up, thereby imitating the structure of what has become known as "belief functions." In short, Lambert made contributions to many sciences as well as to mathematics and statistics. What is discouraging is to realize how much of Lambert's work disappeared, only to be rediscovered much later. At least his work on the irrationality of π and e became part of a living tradition leading to the **proof** of their transcendence as executed in the 19th century. Lambert died in Berlin on September 25, 1777.

LANDAU, EDMUND (1877-1938)
German number theorist

Edmund Landau profoundly influenced the development of **number theory**. His primary research focused on analytic number theory, especially the distribution of **prime numbers** and prime ideals. An extremely productive author of at least 250 publications, Landau's writings had a distinct style. His prose was carefully crafted, highlighted by lucid, comprehensive argumentation and a thorough explanation of the background knowledge required to understand it. Landau's writing style became more succinct over the course of his career. He was forced to retire from teaching at the behest of Nazi anti-Semitic policies.

Born in Berlin on February 14, 1877, Landau was the son of Leopold, a gynecologist, and Johanna (Jacoby) Landau. Johanna Landau came from a wealthy family with whom the Landaus lived in an affluent section of Berlin. Although Leopold Landau was an assimilated Jew and a German patriot, in 1872 he helped found an Judaism academy in Berlin. Landau himself studied in Berlin at the *Französische Gymnasium* (French Lycee), graduating two years early at age 16. He promptly began studying at Berlin University. Landau had published twice before receiving his Ph.D; both pieces explored chess related mathematical problems.

Under the tutelage of Georg Frobenius, Landau was awarded his doctorate at Berlin University in 1899 at the age of 22 years old. His dissertation dealt with what became his life's work: number theory. Landau began teaching at Berlin in 1901, when he earned the advanced degree which allowed him to teach mathematics. He proved to be a popular lecturer at the university because of his personal excitement of the carefully prepared material he presented to his students.

Landau's first major accomplishment as a mathematician came in 1903, when he simplified and improved upon the **proof** for the **prime number theorem** conjectured by **Karl Gauss** in 1796, and demonstrated independently by **Jacques Hadamard** and **C.J. de la Vallee-Poussin** in 1896. In Landau's proof, the theorem's application extended to algebraic number fields, specifically to the distribution of ideal primes within them.

Landau married Marianne Ehrlich (daughter of Paul Ehrlich, a friend of Landau's father, who won the 1908 Nobel prize in medicine or physiology) in 1905 at Frankfurt-am-Main, and fathered two daughters and two sons (one of whom died before age five). He served as a professor of mathematics at Berlin until 1909.

Landau published his first major work in 1909, the two-volume *Handbuch der Lehre von der Vertiolung der Prizahalen*. The volumes were the first orderly discussion of analytic number theory, and were used for many years in universities as a research and teaching tool. Landau's texts are still considered important documents in the **history of mathematics**.

In the same year, Landau became a full professor at the University of Göttingen. Although the faculty at Berlin tried twice to keep Landau on staff, the government wanted to make Göttingen a center of German mathematical learning. They succeeded in their objective, and Landau stayed there until 1934. In 1913, Landau even declined an offer from a university in Heidelberg for a chair position. Although he was still a charismatic, inspiring teacher by the 1920s he was criticized for his rigid, almost perfectionistic lecture style. A demanding lecturer, he insisted that one of his assistants sit through his presentations so any errors could be immediately corrected.

Landau continued his father's support of Jewish institutions. In 1925, he gave a lecture on mathematics in Hebrew at the Hebrew University in Jerusalem, an institution Landau heartily embraced. His activities there continued when he took a sabbatical from Göttingen and taught a few mathematics classes in 1927-28. Landau even contemplated staying in Jerusalem at one point.

Landau also published another important treatise in 1927, the three volumes of *Vorlesungen über Zahlentheorie*. In these texts, Landau brought together the various branches of number theory in one comprehensive text. He throughly explored each branch from its origins to the then-current state of research. Two years later, the widely respected Landau received a honorary doctorate of philosophy from the University of Oslo in Norway. The next year, Landau published another landmark book, entitled *Grundlagen der Analysis*. Beginning with **Giuseppe Peano**'s axioms for natural numbers, this volume presented **arithmetic** in four forms of numbers: whole, rational, irrational, and complex.

The Nazi Party and their policies of discrimination against Jews led to a premature end to Landau's academic career. In late 1933, he was forced to cease teaching at Göttingen, although he was one of the last Jewish professors to be purged from that institution. While technically not subject to the 1933 non-Aryan clause attached to Nazi civil servant laws, all Jewish mathematical professors were forced to leave Göttingen. Landau stayed on through the summer and fall terms of 1933, but he could only teach classes through assistants. Landau would sit in the back of every class, ready to teach at any moment if his ban was raised.

On November 2, 1933, Landau attempted to resume teaching his class. The students, alerted to this impropriety in advance, boycotted his lecture. SS Guards were stationed at the entrance in case a student did not want to boycott; only one got in. When it was clear he would not be allowed to lecture, Landau returned to his office. The boycotting students explained by letter that they no longer wanted to be taught by a Jew and be indoctrinated in his mode of thought.

In 1934, Landau was given his retirement leave, and he and his family moved back to Berlin. Although he never taught in Germany again, he did lecture out of the country at universities such as Cambridge in 1935 and Brussels in 1937. Landau died in Berlin of natural causes on February 19, 1938, and was buried in the Berlin-Weissensee Jewish cemetery.

LAPLACE, PIERRE SIMON (1749-1827)

French mathematical physicist, statistician, and astronomer

Pierre Simon Laplace's work in both celestial **mechanics** and probability represent pinnacles of intellectual achievement. Laplace extended the work of **Isaac Newton,** explaining variations in the planetary orbits and establishing the stability of the solar system. In addition, Laplace's nebular hypothesis used natural law to explain the origins of the solar system, providing a new cosmogony. His work in celestial mechanics led him into the area of statistical inferenceand probability, to which he made substantial contributions. In physics, Laplace's theory of the potential enabled advances in the understanding of electromagnetismand other phenomena. He also predicted the existence of black holes.

Laplace was born on March 23, 1749, in Beaumont–en–Auge, Normandy, France, to Pierre and Marie–Anne Sochon Laplace. His father was in the cider business and an official of the local parish; his mother's family were well–to–do farmers from Tourgéville. Laplace had one sister, Marie–Anne, born in 1745. From age 7 to 17 he was a day student at the Benedictine school at Beaumont–en–Auge, where his paternal uncle, Louis Laplace, an unordained abbé, was a teacher. Laplace was recognized for his intelligence, prodigious memory, and skill at argument.

In 1766, Laplace entered the University of Caen for theological training. His mathematical interests and talents were readily apparent, however, and he was encouraged in mathematics by his teachers, Christophe Gadbled and Pierre Le Canu. In 1768, at the age of 19, Laplace went to Paris to meet the mathematician **Jean d'Alembert,** carrying a letter of recommendation from Le Canu. Legend has it that d'Alembert twice gave Laplace mathematical problems and Laplace solved them overnight. In any case, d'Alembert, who was permanent secretary of the Paris Académie Royale des Sciences, was sufficiently impressed with Laplace's abilities that he used his influence to secure a job for him at the Military School, where Laplace taught elementary mathematics to young cadets. Several years later, in 1785, 16–year–old Napoleon Bonaparte attended the school and Laplace examined and passed him.

Settled in Paris, Laplace began to submit mathematics papers to the Académie. Nicolas Condorcet, the secretary, wrote that never had the Académie received in so short a time so many important papers on such varied and difficult topics. Laplace was proposed for membership in 1771, and was elected to the Académie on March 31, 1773, at age 24.

Pierre Simon Laplace

Laplace's first contributions involved adapting **integral calculus** to the solution of difference **equations.** He then addressed problems in mathematical astronomy. He explained the observed shrinking **orbit** of Jupiter and the expanding orbit of Saturn, demonstrating that the orbital eccentricies are self–correcting and that **mean** motions of the planets are invariable. Laplace also addressed and explained the acceleration of the moon around the Earth, cometary orbits, and the perturbations produced in the **motion** of the planets by their satellites.

From 1799 to 1825, Laplace produced his monumental five–volume *Traité de mécanique céleste (Treatise on Celestial Mechanics).* In it, he completed Newton's work and extended **Joseph Louis Lagrange**'s planetary work. Although Newton had believed that divine intervention would be necessary periodically to "reset" the solar system, Laplace showed that Newton's law of universal gravitation implied its long–term stability. When Napoleon observed that Laplace's voluminous treatise did not mention God as creator of the universe, Laplace replied, "Sir, I do not need that hypothesis."

In 1796, Laplace published *Exposition du systéme du monde,*in which he stated in an extended footnote the idea that the solar system condensed from a rotating cloud of gas. This was the nebular hypothesis originally proposed by the philosopher Immanuel Kant without mathematical elaboration. In his analysis of the gravitational field surrounding a **sphere**, Laplace's equations predicted the concept of the black hole, whose characteristics were deduced much later in **Albert Einstein**'s general theory of **relativity**.

With Antoine Laviosier, his colleague at the Académie, Laplace worked on several problems in physics, including thermal conductivity and capillary action. He is best known to physicists for his theory of the potential, which is useful in studying gravity as well as electromagnetic interactions, acoustics, and **hydrodynamics**.

Laplace contributed greatly to the field of probability and statistical **inference**. In 1774, using principles similar to those developed by **Thomas Bayes**, he derived essentially the same result involving integrals for determining probability, given empirical evidence. His *Théorie analytique des probabilités* ("Analytic Theory of Probability"), published in 1812 and expanded in 1814 with his *Philosophical Essay*, summarized all the materials known at that time in the area of probability, including the theory of games, geometrical probabilities, the theory of least squares, and solutions of **differential equations**. In his probability work Laplace introduced the idea of the Laplace transform, a simple and elegant mathematical technique for solving integral equations. Some of Laplace's contributions to probability were derived from questions in astronomy, for example, the **central limit theorem** which applied to the inclination of the orbits of comets. Laplace believed that through probability mathematics could be brought to bear on the social sciences, and suggested applications in insurance, demographics, decision theory, and the credibility of witnesses. In 1786, he published a study of the vital **statistics** of Paris, using probability techniques to estimate the population of France.

In May 1790 the revolutionary government passed a law requiring standardization of weights and measures in France. The Académie Royale de Sciences was charged with making recommendations, and Laplace, along with Lagrange and **Gaspard Monge**, served on the committee appointed to consider the issue. They made recommendations for units of length, area, **volume**, and mass, with decimal subdivisions and multiples. They also devised decimal systems for money, angles, and the **calendar**. The basic unit of length as it was defined was named the "meter" at Laplace's suggestion. The decimalization of angles and the calendar lasted only a few years, but the other metric units were gradually accepted around the world.

In the late 18th century and into the 19th century, Laplace dominated the Académie Royale des Sciences, imposing his scientific preferences and deterministic ideology on younger colleagues. He presided over the Bureau des Longitudes, which addressed the needs of astronomy and navigation, taught at the École Normale when it was opened briefly in 1795, and in 1800 was instrumental in creating the governing body for the École Polytechnique (founded in 1794), where he served as a graduation examiner. Laplace was known for the "rapidity" of his teaching, and in his writing he became known for his use of the phrase "it is easy to see," by

which he skipped steps in his explanations, confounding some of his later readers and translators.

Laplace's prominence in science brought him conspicuously into the political arena, where he served on numerous blue–ribbon commissions, including a commission that investigated hospital care. In 1799, Napoleon appointed him minister of the interior, but dismissed him after six weeks to place his brother in the position and appointed Laplace instead to the senate, where he eventually became chancellor. Later, Napoleon made Laplace a count of the empire and conferred upon him France's highest honors, the Grand Cross of the Legion of Honor and the Order of the Reunion. At one time, however, Napoleon noted that Laplace was a "mathematician of the first rank," but a mediocre administrator. As a member of the senate, Laplace voted against Napoleon in 1814, supporting Louis XVIII instead. After the restoration of the Bourbon monarchy in 1815, Laplace was rewarded with the title of marquis and was appointed president of the committee to oversee the reorganization of the École Polytechnique. His political opportunism allowed him to prosper and to continue his scientific work.

Laplace retired to his country estate in Arcueil, near Paris, where he and Claude Berthollet, a chemist and physician who also boasted a distinguished career in the service of science and France, formed an informal school. The discussions of the Société d'Arcueil were published between 1807 and 1817.

Although Laplace has been criticized for his arrogance, unreliability in giving credit to others, and wavering allegiances, he is considered to be France's most illustrious scientist in its golden age, and one of the most influential scientists of all time. Laplace had an acute mind and was generally healthy and vigorous until the end. After a short illness, he died on March 5, 1827, at Arcueil, just short of his 78th birthday. At his funeral, he was eulogized by many, including his student Siméon–Denis Poisson, who called him "the Newton of France." Laplace was buried at Pére Lachaise. In 1878, the monument erected to him was moved to Beaumont–en–Auge, and his remains were transferred to the small village of St. Julien de Mailloc. Laplace's papers were destroyed by a fire at the château of Mailloc in 1925, which was then owned by his great–great grandson, the comte de Colbert–Laplace.

LEAST COMMON DENOMINATOR

A common denominator for a set of **fractions** is simply the same (common) lower symbol (denominator). In practice the common denominator is chosen to be a number that is divisible by all of the denominators in an **addition** or **subtraction** problem. Thus for the fractions 2/3, 1/10, and 7/15, a common denominator is 30. Other common denominators are 60, 90, etc. The smallest of the common denominators is 30 and so it is called the least common denominator.

Similarly, the algebraic fractions $x/2(x+2)(x-3)$ and $3x/(x+2)(x-1)$ have the common denominator of $2(x+2)(x-3)(x-1)$ as well as $4(x+2)(x-3)(x-1)(x^2+4)$, etc. The polynomial of the least degree and with the smallest numerical coefficient is the least common denominator. Thus 2(x+2)(x-3)(x-1) is the least common denominator.

The most common use of the least common denominator (or L.C.D.) is in the addition of fractions. Thus, for example, to add 2/3, 1/10, and 7/15, we use the L.C.D. of 30 to write

2/3 + 1/10 + 7/15 as 2x10/3x10 + 1x3/10x3 + 7x2/15x2
which gives us 20/30 + 3/30 + 14/30 or 37/30

Similarly, we have

x/2(x+1)(x-3) + 3x/(x+2)(x-1) = x(x-1)/2(x+1)(x-3)(x-1) + 6x(x-3)/2(x+2)(x-1)(x-3)= [x(x-1)+6x(x-3)]/2(x+1)(x-3)(x-1).

LEBESGUE, HENRI (1875-1941)

French mathematician

In the first decade of the 20th century, Henri Lebesgue developed a new approach to **integral calculus** in order to overcome the restrictions of previous theories. At that time, integration was used to calculate the **area** under a curve, but if the curve was discontinuous the theory was difficult to apply and left some questions unanswered. Lebesgue's theory of integration circumvented the problems caused by these discontinuities and was compatible with other basic mathematical operations.

Henri Léon Lebesgue was born in Beauvais, France, on June 28, 1875. His father was a typographical worker, and his mother was an elementary school teacher. He entered the École Normale Supérieure in 1894 and quickly demonstrated mathematical talent along with an irreverent attitude that gave him the tendency to ignore subjects that did not interest him. For instance, he passed his chemistry course only by mumbling his answers to the examiner who was hard of hearing. Even in mathematics he graduated third in his class in 1879. Nevertheless, his questioning of the traditional methods of mathematics was the basis for his reexamination of the concepts of length, area, and **volume**. He stayed on to work in the library for two years after his graduation in 1879.

Lebesgue inherited the solid foundation for the theory of **calculus** that was laid by the mathematical giants of the 19th century. **Karl Friedrich Gauss**, **Augustin-Louis Cauchy**, **Niels Henrik Abel**, and others of this period had introduced rigorous definitions of **convergence**, limit, and **continuity**. They had also formulated a precise **definition** of the integral, one of the two central concepts of calculus. Just as **addition** gives the total of a finite set of numbers, the integral pertains to the limiting case of the sum of a quantity that varies at every one of an infinite set of points. Such a quantity is described mathematically by a function, and integration of a function can represent the area bounded by a curve, the total work done by a **variable force**, the **distance** a planet travels in its elliptical **orbit** around the Sun, among other possibilities. Cauchy's definition of the integral applied to **functions** that were continuous, that is, curves without any jumps. It could also handle a finite number of discontinuities, points where jumps occurred.

In 1854 **Georg Riemann** introduced an extension of the concept of integration which found its way into most calculus books. Unfortunately, the Riemann integral was unsatisfactory for dealing with some sequences of functions. Even if the sequence of functions approached a limit and each function in the sequence was continuous, the limit might not be a function that could be handled by Riemann integration. The problem was to find a definition for integration that would be compatible with taking the limit of a sequence of functions.

Lebesgue was influenced by the work of René Baire (another recent graduate of the École Normale Supérieure) and **Émile Borel**. By 1898, when Lebesgue published his first results, Baire had formulated an insightful theory of discontinuous functions. In that same year Borel published a theory of measure that generalized the concepts of area to new types of regions obtained as **limits**.

Lebesgue taught at the Lycée Central in Nancy from 1899 to 1902. During that time he developed the ideas for his doctoral thesis at the Sorbonne. The work, "Intégrale, longueur, aire," extended Borel's theory of measure, defined the integral geometrically and analytically, and established nearly all the basic properties of integration. J. C. Burkill notes in the obituary of Lebesgue for the *Journal of the London Mathematical Society:* "It cannot be doubted that this dissertation is one of the finest which any mathematician has ever written."

Lebesgue's consideration of discontinuous curves and nonsmooth surfaces was shocking to some of his contemporaries. **Camille Jordan** cautioned Lebesgue that he should not expect other scholars to appreciate his work. Fortunately, the usefulness of his new ideas quickly overcame any resistance, and Lebesgue received a university appointment as maître des conférences at Rennes in 1902. During his first year he gave lectures for the Cours Peccot at the Collège de France on his new integral and the next year on its application to trigonometric series. Lebesgue published these lectures in the series of tracts edited by Borel and was the first author in this series of monographs other than Borel himself. He gave a thorough exposition of the historical background of the problems leading up to the properties an integral should satisfy, including the compatibility with the limit of a sequence of functions.

In 1906 Lebesgue left Rennes to become chargé de cours for the faculty of sciences, and later a professor, at Poitiers. In 1910 he was maître des conférences at the Sorbonne, and in 1921 he became professor at the Collège de France. Among his many prizes and honors was his election to the Académie des Sciences in 1922. By that time Lebesgue had nearly 90 publications on measure theory, integration, **geometry**, and related topics. Although his ideas were ignored at the great centers of mathematics such as Göttingen, his integral was presented to undergraduates at Rice Institute as early as 1914, and served as an inspiration to the founders of the Polish schools of mathematics at Lvov and Warsaw in 1919.

During the last 20 years of his life, Lebesgue's work became widely known, and his approach to integration evolved as a standard tool of **analysis**. Lebesgue himself began to concentrate more on the historical and pedagogical issues associated with his work. He believed that mathematicians should work from the problems that motivate theory and resist being bound to tradition. He felt that mathematical education should follow this same principle. He freely used the words "deception" and "hypocrisy" to describe the lack of connection between students' natural intuition of numbers and geometry and the manner in which these subjects were taught. In "Sur le mesure des grandeurs," Lebesgue complained about teaching of mathematics: "An infinite amount of talent has been expended on little perfections of detail. We must now attempt an overhaul of the whole structure."

Lebesgue died in Paris on July 26, 1941. Even during the last months of his terminal illness, he continued his course on geometrical constructions at the Collège de France and dictated a book on **conic sections**. He was survived by his wife, mother, a son, and a daughter. His view of mathematics can be summed up in the concluding words of "Sur le développement de la notion d'intégrale": "A generalization made not for the vain pleasure of generalizing but in order to solve previously existing problems is always a fruitful generalization. This is proved abundantly by the variety of applications of the ideas that we have just examined."

LEBESGUE INTEGRAL

The Lebesgue integral is one of the most important and powerful tools in mathematical **analysis**. The Lebesgue integral can be defined in very general settings and the **vector space** of real or complex valued Lebesgue integrable **functions** can be organized into a **Banach space**. In order to define the Lebesgue integral of a function it is necessary to develop the most basic concepts of measure theory.

Let Ω be a nonempty set. A collection A of subsets of Ω is called a σ-algebra if it satisfies the following conditions:

- (1) both the empty set ϕ and the set Ω are elements of A,
- (2) if $A \subseteq \Omega$ belongs to A then its compliment $\Omega \setminus A$ also belongs to A,
- (3) if $A_1, A_2,...$ is a countable collection of **sets** in A then both

$$\bigcup_{n=1}^{\infty} A_n \quad \text{and} \quad \bigcap_{n=1}^{\infty} A_n \quad \text{belong to } \mathcal{A}.$$

The pair (Ω, A) is called a *measurable space*. Here is an example of a measurable space. Let $\mathfrak{R}$ be the set of **real numbers**. Then let $\mathfrak{B}$ be the intersection of all σ-algebras of subsets of $\mathfrak{R}$ that contain the open intervals. There is at least one σ-algebra in $\mathfrak{R}$ that contains the open **interval**, namely, the σ-algebra of all subsets of $\mathfrak{R}$. And it can be shown that the intersection of an arbitrary family of σ-algebras is again a σ-algebra. Thus the collection $\mathfrak{B}$ exists and is a σ-algebra of subsets of $\mathfrak{R}$ containing the open intervals. Because each open subset of $\mathfrak{R}$ is a countable union of open intervals, it follows that $\mathfrak{B}$ contains all

the open subsets of $\mathfrak{R}$. Since each closed subset of $\mathfrak{R}$ is the compliment of an open set, we also find that $\mathfrak{B}$ contains all the closed subsets of $\mathfrak{R}$. The σ-algebra $\mathfrak{B}$ is called the σ-algebra of *Borel sets* in $\mathfrak{R}$. Thus the pair $(\mathfrak{R}, \mathfrak{B})$ forms a measurable space.

A function $\mu:A \to [0,\infty]$ is called a *measure* if it satisfies the conditions:

- (4) $\mu(\phi) = 0$,
- (5) if $A_1, A_2,...$ is a countable collection of *disjoint* sets in A then

$$\mu\left(\bigcup_{n=1}^{\infty} A_n\right) = \sum_{n=1}^{\infty} \mu(A_n).$$

Notice that the values taken on by a measure may include the symbol ∞ and this must be treated appropriately. For example, in condition (5) it may happen that $\mu(A_n) < \infty$ for each $n = 1, 2,...$, that $\mu(\cup A_n) = \infty$ and then the sum on the right hand side of (5) must diverge to ∞. The triple (Ω, A, μ) is called a *measure space*. It can be shown that there exists a unique measure λ defined on the σ-algebra $\mathfrak{B}$ of Borel sets in $\mathfrak{R}$ such that for each interval $I \subset \mathfrak{R}$ the value of $\lambda(I)$ is the length of the interval. In particular, $\lambda(I) = \infty$ if I is in an infinite interval. Of course λ is defined for all sets in $\mathfrak{B}$, not just the intervals, in such a way that it satisfies the requirements of a measure. This particular measure on $\mathfrak{B}$ is called *Lebesgue measure*. Thus the triple $(\mathfrak{R}, \mathfrak{B}, \lambda)$ forms a measure space. There is also a larger σ-algebra $\mathcal{L}$ in $\mathfrak{R}$ called the σ-algebra of *Lebesgue measurable sets* that contains $\mathfrak{B}$ as a sub-σ-algebra. And the measure λ can be extended to a measure on $\mathcal{L}$ so that $(\mathfrak{R}, \mathcal{L}, \lambda)$ is a measure space.

A function $f: \Omega \to \mathfrak{R}$ is said to be *measurable* with respect to the σ-algebra A if $f^{-1}(B)$ is contained in A for each Borel set B in $\mathfrak{B}$. When there is only one σ-algebra being considered in the **domain** Ω of such a function, we say more simply that the function is *measurable*. For example, let A be a set in A, that is, A is a measurable subset of Ω. Then define a function χ_A: $\Omega \to \mathfrak{R}$ by $\chi_A(x) = 1$ if x is a point in A, and $\chi_A(x) = 0$ if x not a point in A. The function χ_A is called the *characteristic function* of A, and it is easy to see that χ_A is a measurable function. More generally, if $A_1, A_2,... A_n$ is a sequence of measurable subsets of Ω and $c_1, c_2,... c_N$ is a corresponding set of nonnegative real numbers, then the function $\varphi: \Omega \to \mathfrak{R}$ defined by

$$\varphi(x) = \sum_{n=1}^{N} c_n \chi_{A_n}(x)$$

is a measurable function. Here φ is an example of a *simple* function, that is, a measurable function that takes on finitely many nonnegative values. It can be shown that every simple function can be written as a finite linear combination of characteristic functions of measurable sets in the same manner that we have defined φ.

We are now in position to define the Lebesgue integral. First of all, let A be a measurable subset of Ω. Then the

Lebesgue integral over the space Ω of the characteristic function χ_A of A is defined to be the measure of A. That is, we define

$$\int_{\Omega} \chi_A(x)\, d\mu(x) = \mu(A).$$

More generally, if φ is a simple function written as before, then the Lebesgue integral of φ over the space Ω is defined by

$$\int_{\Omega} \varphi(x)\, d\mu(x) = \sum_{n=1}^{N} c_n \mu(A_n).$$

Here it is important to recognize the convention that if $c_n = 0$ and $\mu(A_n) = \infty$ for some integer n, then $c_n \mu(A_n) = 0$. Next we suppose that $f: \Omega \to [0, \infty)$ is a measurable function. In this case we define

$$\int_{\Omega} f(x)\, d\mu(x) = \sup\left\{ \int_{\Omega} \varphi(x)\, d\mu(x) \right\}$$

where the supremum is taken over the set of all simple functions φ such that $0 \le \varphi(x) \le f(x)$ at each point x in Ω. Notice that the value of the integral of a nonnegative measurable function f could be the symbol ∞. This is convenient for establishing the basic properties of the integral.

Now assume that $f: \Omega \to \Re$ is measurable. Define

$f^+(x) = \frac{1}{2}|f(x)| + \frac{1}{2} f(x)$ and $f^-(x) = \frac{1}{2}|f(x)| - \frac{1}{2} f(x)$

so that f^+ and f^- are nonnegative valued measurable functions,

$|f(x)| = f^+(x) + f^-(x)$ and $f(x) = f^+(x) - f^-(x)$.

We say that f is *integrable* over the measure space (Ω, A, μ) if

$$\int_{\Omega} |f(x)|\, d\mu(x) < \infty.$$

If f is integrable it follows that

$$\int_{\Omega} f^+(x)\, d\mu(x) < \infty. \quad \text{and} \quad \int_{\Omega} f^-(x)\, d\mu(x) < \infty.$$

Therefore, if f is integrable we define the value of the Lebesgue integral of f over the measure space (Ω, A, μ) by

$$\int_{\Omega} f(x)\, d\mu(x) = \int_{\Omega} f^+(x)\, d\mu(x) - \int_{\Omega} f^-(x)\, d\mu(x).$$

Clearly the value of the integral of an integrable function is always a real number and never the symbol ∞. If f and g are both integrable, if α and β are real numbers, then $\alpha f + \beta g$ is integrable and

$$\int_{\Omega} \alpha f(x) + \beta g(x)\, d\mu(x) = \alpha \int_{\Omega} f(x)\, d\mu(x) + \beta \int_{\Omega} g(x)\, d\mu(x).$$

Thus the Lebesgue integral is a linear **transformation** on the vector space of integrable functions. Here is an example of one of the important **convergence** theorems for the Lebesgue integral: suppose that $f_1(x)$, $f_2(x)$,... is a sequence of measurable functions such that

$$\lim_{n \to \infty} f_n(x) = F(x) \quad \text{exists}$$

for each point x in Ω. Also assume that there exists a nonnegative valued integrable function $g: \Omega \to [0, \infty)$ such that $|f_n(x)| \le g(x)$ for all x in Ω and all positive **integers** n. Then each function f_n is integrable, the function $F: \Omega \to \Re$ is measurable, F is integrable, and

$$\lim_{n \to \infty} \int_{\Omega} f_n(x)\, d\mu(x) = \int_{\Omega} F(x)\, d\mu(x).$$

We have already noted that $(\Re, \mathcal{L}, \lambda)$ is an example of a measure space and therefore the concepts described here for the Lebesgue integral apply in this particular measure space. For example, let $[u, v] \subset \Re$ be a closed interval and let $f: [u, v]: \to \Re$ be a Riemann integrable function. It will be convenient to set $f(x) = 0$ if x is a real number not in the interval $[u, v]$. Then it can be shown that f is measurable with respect to the σ-algebra $\mathcal{L}$, f is integrable over the measure space $(\Re, \mathcal{L}, \lambda)$, and the value of the Lebesgue integral of f on $\Re$ is equal to the value of the Riemann integral of f on the interval $[u, v]$. On the other hand, there exist measurable functions $g: \Re \to \Re$, with $g(x) = 0$ for real numbers x not in $[u, v]$, and such that g is Lebesgue integrable over $(\Re, \mathcal{L}, \lambda)$ but the restriction of g to the interval $[u, v]$ is *not* Riemann integrable. Thus the Lebesgue method of integration extends the Riemann method of integration to a wider class of functions. However, the Riemann method of integration is certainly not obsolete. Suppose that $h: [u, v] \to \Re$ is a continuous function, and therefore both Riemann and Lebesgue integrable. If we wish to find an approximate numerical value for the integral of h over $[u, v]$ the principles used to define the Riemann integral of h may well be more useful.

LEGENDRE, ADRIEN–MARIE (1752-1833)

French geometer, number theorist, and elliptic function theorist

Adrien–Marie Legendre is best known as the author of *Éléments de géométrie*, a simplification of **Euclid**'s *Elements*. Published in 1794, it went through numerous editions and translations, and served as a standard **geometry** text for the next hundred years. Legendre made significant contributions to several other fields, however, including number theory and celestial **mechanics**. In addition, he is generally regarded as the founder of the theory of elliptic **functions**.

Legendre was born in Paris on September 18, 1752. Although little is known of his early life, Legendre apparently came from a well–to–do family. He studied at the Collége Mazarin in Paris, where he received an unusually progressive education in mathematics. One of his professors there was Abbé Joseph–François Marie, a highly regarded mathematician. At age 18, Legendre successfully defended his theses in **mathematics and physics**, and at age 22 he issued his first publication, a treatise on mechanics.

Legendre's modest family fortune was sufficient to support his research. Nevertheless, in 1775 he accepted a position at the École Militaire in Paris. He taught there until 1780, but he had yet to make an impact on the scientific world. Then in 1782, he won a prize awarded by the Berlin Academy for an essay on the path of projectiles taking the resistance of air into account. Legendre's winning essay, which was published in Berlin, attracted the attention of **Joseph–Louis LaGrange,** one of the finest mathematicians of the day. LaGrange, in turn, sought more information on the young author from another great French mathematician, **Pierre Simon Laplace.**

In January 1783, Legendre presented his first major paper before the French Académie Royale des Sciences. This paper dealt with the attraction of planetary spheroids (objects that have a sphere–like shape, but are not perfectly round). He also submitted to Laplace essays on various topics, including the properties of continued **fractions** and the **rotation** of bodies subject to no accelerating **force**. As a result of these efforts, Legendre was elected to the Académie on March 30, 1783.

As Legendre strove to make his work better known, his career continued to flourish. In 1784 he presented another paper on celestial mechanics. This one dealt with the form of equilibrium of a sphere–like mass of rotating liquid. The paper introduced the famous "Legendre polynomials," which are solutions to a particular kind of differential equation that still plays an important role in **applied mathematics**. Legendre's interest in celestial mechanics eventually led to two further papers, one on the attraction of certain ellipsoids, and the other on the form and density of fluid planets.

A second line of productive research by Legendre involved **number theory**, and a third comprised elliptic functions. Legendre began both investigations in the mid–1780s, although it was not until later that he made his most significant contributions. Legendre's first paper on number theory dates from 1785. Among other things, it set forth the law of quadratic reciprocity, by which any two odd **prime numbers** can be related. His first publication on elliptic functions came the following year, which addressed the integration of elliptic curves.

In 1787 Legendre was assigned to work on a project in geodesy (a branch of applied mathematics dealing with the measurement of the Earth). This project was a joint undertaking of observatories in Paris and Greenwich, England. The most important result was a **theorem** that stated how a spherical **triangle** may be treated as a plane, provided certain corrections are made to the angles.

By the close of the 1780s, however, Legendre's scientific progress was being impeded by the French Revolution. He was particularly affected by the suppression of the French Académie Royale des Sciences, which forced him to publish some of his research himself. Legendre's small fortune dwindled, and he had to seek work to support himself and his new bride, Marguerite Couhin. Beginning in 1791, Legendre served on several public commissions, including one that converted the measurement of angles to the decimal system. In 1794, he became a professor of mathematics at the short–lived Institut de Marat in Paris. Legendre succeeded Laplace as the examiner in mathematics of students assigned to the artillery in 1799, a position he held until 1815.

During this same period, Legendre published the popular text *Éléments de géométrie*, which dominated geometry instruction in many parts of the world for the next century. Later editions of the book also contained the elements of **trigonometry**, as well as proofs of the irrationality of π and π^2. In addition, an appendix on the theory of parallel lines was issued in 1803. Although this book was among Legendre's more mundane achievements, it is one for which he is still remembered.

Another obstacle to Legendre's success was the jealousy of Laplace. In *A Short Account of the History of Mathematics*, W.W. Rouse Ball describes the situation this way: "The influence of Laplace was steadily exerted against [Legendre's] obtaining office or public recognition, and Legendre, who was a timid student, accepted the obscurity to which the hostility of his colleague condemned him."

These distractions may have slowed Legendre's progress somewhat, but they could not completely stifle his creativity. In 1798, he published the first edition of *Essai sur la théorie des nombres*, in which he further explored number theory. Among other topics, Legendre returned to the law of quadratic reciprocity, which he had stated 13 years before. (**Karl Gauss** offered the first rigorous demonstration of this law in 1801.) Appendices were added to the book in 1816 and 1825. The book's third edition, issued as two volumes in 1830, is considered a standard text in the field.

In 1811, Legendre published the first volume of another major work, *Exercices de calcul intégral*, much of which dealt with elliptic functions. Second and third volumes followed by 1817. However, Legendre saved his most influential work for last. From 1825 through 1828, he published three volumes of *Traité des fonctions elliptiques*, in which he expanded upon his elliptic function research. Unfortunately, Legendre treated this subject merely as a problem in **integral calculus**, failing to realize that it might be considered as a higher trigonometry and, thus, a distinct branch of **analysis**. The discovery of superior methods for handling elliptic functions were left to a younger generation of mathematicians, including **Karl Gustav Jacob Jacobi** and **Niels Henrik Abel.**

Legendre recognized the superiority of Jacobi's and Abel's methods at once. In the final years before his death, Legendre issued three supplements to the third volume of *Fonctions elliptiques*. In these supplements, he discussed the younger mathematicians' ideas. As Ball notes in *A Short*

Account of the History of Mathematics: "Almost the last act of [Legendre's] life was to recommend those discoveries which he knew would consign his own labours to comparative oblivion."

Legendre's contributions did not go entirely unheralded. He was made a member of the French Legion of Honor and was granted the title of Chevalier de l'Empire. This was a minor honor compared to the title of count, however, which was bestowed upon Laplace and LaGrange. When LaGrange died in 1813, his post at the Bureau des Longitudes was given to Legendre, who held it for the rest of his life.

Legendre died in Paris on January 9, 1833, after a long, painful illness. He and his wife had never had children. Legendre's widow made a cult of his memory, carefully preserving his belongings. Upon her death in 1856, she left to the village of Auteuil (now part of Paris) the last country house where the couple had lived.

LEGENDRE SYMBOL

The Legendre symbol is a notation used for stating a central **theorem** of elementary **number theory**, the *quadratic reciprocity law*. This theorem was first proved by **Carl Friedrich Gauss** in 1801, after the French mathematician **Adrien-Marie Legendre** had published two incorrect proofs. As a sort of consolation prize, tradition has named the notation after Legendre.

The quadratic reciprocity law gives an effective procedure for determining whether a number is a perfect **square** in **modular arithmetic**. For example, 2 is a square modulo 7, because it is congruent to the number $9 = 3^2$. By contrast, 3 is not a square modulo 7. It is congruent to the numbers 10, 17, 24, 31,..., none of which is the square of a whole number.

In the real number system, it is easy to tell squares and non-squares apart. Squares are positive or **zero**, and non-squares are negative. But it is much less obvious how to tell whether a number x is a square (modulo n). As a first step, it turns out to be sufficient to determine whether x is a square modulo each of the prime factors of n. If so, then it is also a square modulo n.

Hence the key problem is to determine when x is a square modulo a prime number p. This is where the Legendre symbol enters the problem. The Legendre symbol, written (x/p), is defined as follows: $(x/p) = 1$ if x is a square modulo p, and $(x/p) = -1$ if x is not a square modulo p. Note that the parentheses are an essential part of the symbol, and that the Legendre symbol (x/p) has nothing to do, in general, with the fraction x/p. For example, $(2/7) = 1$ and $(3/7) = -1$.

The Legendre symbol possesses some algebraic properties that make it a useful device for solving the "key problem" mentioned above. First, it is multiplicative: $(xy/p) = (x/p)(y/p)$. Stated in words, this means: a square times a non-square is a square; a square times a non-square is a non-square; and a non-square times a non-square is a square. (Compare this with the arithmetic of positive and negative numbers.) Second, the Legendre symbol satisfies the law of quadratic reciprocity. *Assuming that both x and p are odd primes*, their roles can be reversed. More specifically, $(x/p) = (p/x)$ if either x or p is 1

greater than a multiple of 4; but if x and p are both 3 greater than a multiple of 4, then $(x/p) = -(p/x)$.

As an example, to compute $(3/7)$, we switch the 3 and 7. Noting that 3 and 7 are both 3 greater than a multiple of 4, we conclude that $(3/7) = -(7/3)$. But since 7 is congruent to 1 modulo 3, and 1 is a square $(1 = 1^2)$, 7 is also a square, and hence $(7/3) = 1$. This means $(3/7) = -1$, so 3 is not a square modulo 7.

There is no truly simple explanation for why quadratic reciprocity works; Gauss himself proved it six times, looking for a more satisfying answer. But quadratic reciprocity and its generalizations have led to many deep results in number theory, including **Dirichlet**'s theorem on primes in arithmetic progressions; factoring techniques such as the "quadratic sieve"; and the entire field of algebraic number theory. Legendre symbols themselves provide the most basic example of a multiplicative function from a group (here, the numbers modulo p) into the **real numbers**. Such **functions** are now called *characters*, and play a fundamental role in both number theory and **group theory**.

See also Number theory

LEIBNIZ, GOTTFRIED WILHELM VON (1646-1716)
German logician and philosopher

Gottfried Wilhelm Leibniz' restless intellect ranged from mathematics, physics and engineering to politics, linguistics, economics, law and religion. He is best known for his contributions in metaphysics and logic, and for his invention, independent of **Isaac Newton**, of **differential** and **integral calculus**, which is an indispensable mathematical tool. Leibniz received his doctoral degree when he was 19 years old, declined an early offer of an academic career, and became a lifelong employee of noblemen. Although he had met and corresponded with many influential and learned men, his death was marked by indifference. A nationalistic debate over priority in developing the **calculus** during Leibniz's lifetime resulted in a century–long rift between English and continental mathematics.

Leibniz was born in Leipzig, Saxony (now Germany), on July 1, 1646, four years after the birth of Newton. His father, Friedrich Leibnütz, was a lawyer and professor of moral philosophy at the University of Leipzig; Gottfried's mother, Catherina Schmuck, was Friedrich's third wife. Both sides of the family enjoyed social standing and scholarly reputations. Leibniz (who changed the spelling of his name) had a half–brother, Johann Friedrich; a half–sister, Anna Rosina; and a sister, Anna Catherina, whose son, Friedrich Simon Löffler, became his sole heir. His father died when he was six, but young Gottfried had already begun to demonstrate a passion for knowledge and omnivorous reading. He studied his father's library of classic, philosophical and religious works, and his school syllabus included German literature and history, Latin, Greek, theology and logic. By age 12, Leibniz read Latin and was adept at writing Latin verse. He began to for-

Gottfried Wilhelm Leibniz

mulate his own ideas in logic, among them the ideas of an alphabet of human thought and a universal encyclopedia.

In 1661, at age 15, Leibniz entered the University of Leipzig, where he studied with philosophy professor Jakob Thomasius. He received his bachelor's degree in 1663, then spent the summer in Jena studying with the mathematician Erhard Weigel, who introduced him to elementary **algebra** and Euclidean **geometry**. After receiving his master's degree in 1664, Leibniz wrote a dissertation for the Doctor of Law degree, but because of his youth, the university refused to award it to him. Leibniz subsequently entered the University of Altdorf in Nuremburg, where his dissertation was accepted and his degree was awarded in 1666. The university then offered him a professorial position, but he declined.

Leibniz's first job was in Nuremburg, as secretary of a society of intellectuals interested in alchemy. By chance he met Baron Johann Christian von Boineburg, a former Chief Minister of the Elector of Mainz. Boinburg became Leibniz' patron, and helped him obtain a position as assistant to the Elector's legal adviser; later Leibniz was promoted to Assessor in the Court of Appeals and to a diplomatic position. Leibniz spent the next five years in Mainz and Frankfurt. In addition to writing legal and position papers for the Elector, he wrote papers on religious subjects relating to the reunion of the Protestant and Catholic churches, and continued to

develop his philosophy, to which logic was central. Leibniz began a voluminous correspondence with hundreds of people on every conceivable topic. More than 15,000 of his letters survive.

France had been encroaching on the Rhineland, and in the winter of 1671–1672, Leibniz devised a plan to distract the French by encouraging them to conquer Egypt and build a canal across the isthmus of Suez. The plan eventually came to nothing, but it allowed Leibniz to accompany a diplomatic mission to Paris, meet prominent philosophers and scholars, and immerse himself in Parisian salons. This was the time of Leibniz' greatest advances in mathematics, and he formed a lifelong friendship with **Christiaan Huygens**, who was in the employ of the Académie Royale des Sciences. He was also able to obtain and copy unpublished manuscripts of **Blaise Pascal** and **René Descartes**, and he devised a calculating machine that could add, subtract, divide, multiply and extract **roots**.

In 1673, Leibniz visited London on a diplomatic mission, and at a meeting of the Royal Society of London displayed a **model** of his calculating machine. He met Society secretary Heinrich Oldenburg, with whom he had corresponded, and was elected to membership in the Society. He also talked with the chemist **Robert Boyle**, the microscopist **Robert Hooke**, and the mathematician **John Pell**. The latter pointed out the gaps in Leibniz' mathematical knowledge. Leibniz left London in the grip of mathematical ideas, and returned to Paris to study higher geometry under Huygens, beginning the work that led him to the discovery of differential and **integral calculus**. In 1673 he developed his general method of tangents.

In the meantime, both Leibniz' patron and the Elector had died and Leibniz' future with his successor was uncertain. He wanted to remain in Paris, and hoped that his reputation might win him a paid position at the Académie Royale des Sciences or the professorship vacated by the death of Roberval at the Collége Royal but was disappointed. Leibniz visited London again briefly, and in 1676 left Paris for Hanover to act as advisor to the Duke of Brunswick and take charge of the ducal library. En route to Hanover, where he would spend most of the next forty years, Leibniz stopped in Holland for discussions with mathematician Jan Hudde and microscopists Jan Swammerdam and Antoni van Leeuwenhoek, and at The Hague to talk with the philosopher Spinoza.

In 1684, Leibniz' brief paper, "A New Method for Maxima and Minima, as well as Tangents, which is impeded neither by Fractional nor Irrational Quantities, and A remarkable Type of Calculus for This" appeared in *Acta Eruditorum*. Leibniz used the term "calculus" to mean rules. This paper addressed differential calculus; he introduced integral calculus in 1686 in the same publication. Newton's similar work was published in 1689; however, because of its approach, it was not immediately obvious that it was the same as Leibniz'. Newton had used a geometrical approach; Leibniz' was algebraic. Early in the 1700s, supporters of Newton in Great Britain and of Leibniz on the Continent began to argue about the merits of the two systems, and about the priority of their discovery. Newton and Leibniz were eventually drawn into the animosities, and the consequences lasted for more than a century. For a hundred years, Continental mathematicians using Leibniz' methods and superior notation moved ahead in the theory of the calculus, while the English were held back by Newton's more cumbersome method. It is clear that Newton and Leibniz developed calculus independently; however, Leibniz has provided its notation and its name.

Leibniz was a versatile and prolific contributor to mathematics. In *On the Secrets of Geometry and Analysis of Indivisible and Infinite Quantities* (1686) Leibniz first used the integral sign. In a 1693 letter, Leibniz used multiple indices to state the result of three **linear equations**. He also worked on the problem of elimination in the general theory of **equations** and laid the foundations of the theory of **determinants**. In another letter, he suggested expanding cube roots into **infinite series**. He introduced terminology, including the term "function," borrowed by Bernoulli.

In 1702, Leibniz published *A Justification of the Calculus of the Infinitely Small*, in which he attempted to justify his algorithms for differentation and integration. Between the years 1702–1703 he integrated rational **fractions** in trigonometric and logarithmic **functions**, produced a theory of special curves, and stated equations important in navigation. Leibniz was one of the first to work out the properties of the **binary number system**; he anticipated the central concerns of modern computer science, bringing human reasoning under mathematical law.

In 1686, Leibniz began a study of the genealogy of the House of Brunswick that would occupy the remainder of his life. He began his research with a three–year trip to Bavaria, Austria, and Italy, along the way meeting with scholars and scientists and writing treatises on other topics. In 1696 he was promoted to Privy Councillor at Brunswick–Wolfenbüttel. He had promoted the establishment of scientific academies in several countries, and in 1700, through the influence of the Electress Sophie Charlotte, the Berlin Academy was founded with Leibniz as president. He was also elected an external member of the Paris Académie Royale des Sciences in 1700.

As a young man, Leibniz had a reputation as a savant, a wit, and an elegant courtier. In his last years he was increasingly disregarded. On November 14, 1716, after a week in bed suffering from arthritis, gout and colic, Leibniz died in Hanover in the presence of his secretary, Johann Georg Eckhart. His funeral and burial took place on December 14 in Neustüdter Church. No one from the Court attended his funeral, and his grave went unmarked for 50 years. An elegy was read at the Académie Royale des Sciences a year after Leibniz' death, but neither the Royal Society of London nor the Berlin Academy published an obituary.

Lemniscate of Bernoulli

First examined by **Jacob Bernoulli** (1654-1705) in 1694, the lemniscate of Bernoulli has the general form of a figure eight, or the mathematical symbol for **infinity**. Bernoulli's conceptu-

alization of the lemniscate followed his earlier research on parabolas, logarithmic spirals, and epicycloids. Bernoulli originally termed the curve *lemniscus*, which is Latin for "a pendant ribbon," to describe the unique shape of the curve. Roughly defined, the lemniscate of Bernoulli is the infinite set of all points (called a **locus**) satisfying the property that the product of the distances between the point and two foci separated by $2a$ is exactly a^2. In other words, if $(a,0)$ and $(-a,0)$ are the two foci of the curve, each point (x,y) along the curve satisfies the condition that sqrt$[(x - a)^2 + (y - 0)^2]$ * sqrt$[(x - (-a))^2 + (y - 0)^2] = a^2$. This stipulation is expressed here in **Cartesian coordinates**, but can be translated to other coordinate systems (such as polar or cylindrical) as well.

Interestingly enough, Bernoulli's lemniscate is actually a special case of a **family of curves** known as Cassini's ovals (and alternatively known as Cassini's curve). Named for the astronomer Gian Domenico Cassini (1625-1712), Cassini first described his ovals in 1680 while researching the relative **motion** of the earth and the sun. Cassini's curves are more general than Bernoulli's lemniscate; instead of describing just a single curve, Cassini described a family of curves in which the locus of each curve satisfies the condition that the product of the distances to each foci is exactly a constant, generally denoted $c2$. In other words, if $(a,0)$ and $(-a,0)$ are the two foci of the curve, each point (x,y) along the curve satisfies the condition that sqrt$[(x - a)^2 + (y - 0)^2]$ * sqrt$[(x - (-a))^2 + (y - 0)^2] = c^2$. Bernoulli's lemniscate is the special case in which $c = a$. If $c > a$, the curve has the form of two distinct loops; if $c < a$, the curve appears almost peanut-shaped, and when $a = 0$, the curve forms an oval. Cassini's ovals were apparently unknown to Bernoulli at the time of Bernoulli's investigation.

In Cartesian coordinates, the lemniscate is commonly expressed by the equation $(x^2 + y^2)^2 = 2a^2(x^2 - y^2)$, where $2a$ is the **distance** between the two foci. The expression, however, is much simpler in polar coordinates. In that system, two common forms of Bernoulli's lemniscate are $r^2 = c^2\sin($ and $r^2 = c^2\cos($. In this case, c represents the radius of each petal of the lemniscate, and c is related to a by $c =$ sqrt(2) * a. Of course, Bernoulli's lemniscate can also be written parametrically. One common form is $x = c\cos t / (1 + \sin^2 t)$ and $y = c\sin t\cos t / (1 + \sin^2 t)$.

Both Gauss and Euler studied the arc length of the lemniscate curve. Much like the relationship between **pi** and the **diameter** of a **circle**, Bernoulli's lemniscate has a unique constant that relates its arc length to the distance between the foci. Specifically, **Gauss** determined for a leminscate in which the distance between the two foci is 2 (that is, $a = 1$), the arc length can be represented by $L = 2c$ times the integral over the **interval** [0,1] of the quantity $(1 - t^4)^{-(1/2)}dt = 1 /$ sqrt(2 * pi) * [Gamma(1 / 4)] 2 "5.24411.... Then, the lemniscate constant is $L/2$, which is related to Gauss' constant M by $L = 2 *$ pi $/ M$.

See also Archimedes' spiral; Daniel Bernoulli; Jacob Bernoulli; Johann Bernoulli; Cartesian coordinate system; Cosine; Cycloids; Leonhard Euler; Fermat's spiral; Johann Carl Friedrich Gauss; Locus; Parabola; Polar coordinate system; Sine; Spiral

LESNIEWSKI, STANISLAW (1886-1939)

Russian logician

Stanislaw Lesniewski cofounded the Warsaw school of logic and served as one of its top representatives. Along with a student, **Alfred Tarski**, and colleague, **Jan Lukasiewicz**, he formed a triangle of expertise that made the University of Warsaw the world center for research in formal logic in the 1920s and 1930s. Lesniewski's main contributions to mathematical logic comprise three distinct yet interrelated systems that he dubbed prototethic, ontology, and mereology. Together, these systems provide a logical foundation for all branches of classical mathematics.

Born in Serpukhov, Russia on March 30, 1886, Lesniewski was the son of a communications engineer who served as one of the chief builders of the trans-Siberian railroad. Beginning in 1896, the boy attended school in Trojskosawski, near the Mongolian border, which was near where his father was working. In 1899, Lesniewski began studying at a school in Irkutsk, leaving in 1904 to take classes here and there in continental Europe, but mainly in Germany. He achieved his doctoral degree, a philosophical treatise that focused on the analysis of existential **propositions**, in 1912 at the University of Lviv (then in Austria, now in Ukraine).

Lesniewski's first important work, on the theory of collective **sets**, appeared in 1916. At that time he also began teaching his first course in mathematical logic at the University of Lviv. Until this time, Lesniewski's career had been tending toward an interest in philosophy, but this changed when he read Lukasiewicz's book *On the Principle of Contradiction in Aristotle*. Lesniewski became an impassioned student of logic when he perceived that the antimonies (paradoxes) in mathematics and logic discussed in the book were a dire threat to the foundations of deductive science. His desire to eliminate this threat eventually yielded the contributions for which Lesniewski is still remembered.

From 1916 to 1927, Lesniewski engaged in intensive research, although he did not publish much during this creative period. One work, *Foundations of the General Theory of Sets,* appeared in 1916. In 1917, Lesniewski took a teaching post at the Warsaw Gymnasium, but in 1919 he followed Lukasiewicz to the University of Warsaw and began working as professor of mathematical philosophy. Their dynamism attracted many gifted students from all over the world.

Beginning in 1927, Lesniewski finally started publishing some of his work, although reluctantly because, as a perfectionist, he did not feel the manuscripts were ready for the world to see. Part of the problem, he felt, was that everyday language was inadequate for the discussion of ontological issues. This was the primary motivation for Lesniewski's focused search for a new kind of language that would be suitable for science. In the early part of the decade, he had started experimenting with using symbolic language instead of natural language for this purpose. Eventually, in order to explain his prototethic, ontology, and mereology systems, Lesniewski came up with a general theory of semantic groups that was similar to the "meaning categories" of his contemporary, **Edmund Husserl**.

Lesniewski's systems of logic came under close scrutiny later in his life as he began to publish more of his research. The basis of his most famous logistic theory, prototethic (from the Greek *protos,* or "first"), is still considered the most complete theory on the relationships among propositions, statements that affirm or deny something so they can then be characterized as true or false. The second part, ontology (from the Greek *on,* or "being), concerns the logic of names. When combined with prototethic, ontology yields all the theorems of logical and syllogistic ("if-then") **algebra**, in addition to the logic of relations and sets. The last component, mereology (from the Greek *meros,* or "part"), is a general theory of the relationship between a whole and its parts. Lesniewski developed these separate yet complementary systems of logic using a precision and clarity that forever raised the bar for standards of mathematical rigor, as well as establishing the Warsaw school of logic as the world's most advanced.

Lesniewski was given a full professorship at the University of Warsaw in 1936, after which he spent several months of that year touring the scientific centers of Europe. He died suddenly of thyroid cancer on May 13, 1939 in Warsaw just as the Warsaw school was reaching the height of its influence. Virtually all of Lesniewski's manuscripts, including those that he never published, were destroyed during World War II. Fortunately, Tarski—through his own research—did much to publicize Lesniewski's work after the elder man's death.

See also Jan Lukasiewicz; Alfred Tarski

LEVI-CIVITA, TULLIO (1873-1941)

Italian geometer and theoretical physicist

Tullio Levi–Civita's most crucial contribution to mathematics lay in his development of tensor **calculus** (originally called absolute **differential calculus**). This new calculus had a widespread effect. Although **Albert Einstein**'s theory of **relativity** depended on its development, it had many mathematical applications. Related to his tensor calculus, Levi–Civita developed theorems about the curvature of spaces and Riemannian geometry's covariant differentiation. Levi–Civita also propagated work in pure **geometry**, **hydrodynamics**, engineering, celestial and analytical **mechanics**. In addition, Levi–Civita was a prolific author of more than 200 publications as well as a gifted and well–liked teacher.

Levi–Civita was born on March 29, 1873, in Padua, Italy. He was the son of Giacomo Levi–Civita, a lawyer and senator. Educated in his hometown, Levi–Civita was an outstanding student in secondary school there. He also studied at the University of Padua from 1891 to 1895, under the tuteledge of Gregario Ricci Curbastro. Levi–Civita received his BA in 1894, and published while an undergraduate. He began his teaching career at a Pavia–based teacher–training college in 1895. A year later, he published his first piece dealing with what would be his most important contribution to mathematics: tensor calculus. Entitled "Sulle trasformazioni delle equazioni dinamiche," Levi–Civita stretched the work of his mentor,

Ricci Curbasto, by using Curbastro differential geometry methods in what became known as absolute differential calculus.

Levi–Civita began teaching at the University of Padua in 1898 and he served as the chair of Mechanics. Around 1900, Levi–Civita and Ricci Curbasto jointly published the progress they had made in the development of absolute differential calculus in the *Méthodes de calcul différéntiel absolu et leurs applications* ("Methods of the Absolute Differential Calculus and Their Applications"). They had been working on this calculus together for several years. As developed by them, absolute differential calculus was an unparalled breakthrough in part because it was applicable to many fields. It could be applied to three key mathematical spaces: both Euclidean and non–Euclidean, as well as Riemannian curved spaces. Albert Einstein used absolute differential calculus in formulating his theory of general relativity 15 years later. For several years after Einstein's discovery, however, Levi–Civita was uncertain about parts of the theory of relativity, but he eventually embraced it.

In 1902, Levi–Civita became a professor of rational mechanics at Padua, a position he held until 1918. Levi–Civita's contributions to science were not limited to absolute differential calculus. Two subjects in which he showed extreme interest in and did the most work on while at Padua were celestial mechanics and hydrodynamics. From 1903–1916, Levi–Civita studied various aspects of celestial mechanics. He was especially concerned with the three body problem (how three bodies move when considered as mass centers and under **Newton**'s mutual attraction theories). He achieved results on the problem from 1914–1916, but graciously admitted another scholar, Karl F. Sundmann, had reached the same conclusions in a roundabout manner. Levi–Civita also published relevant work on hydrodynamics, a subject of which he was particularly fond, from 1906 on. He focused on concepts such as how an immersed solid's translational movements relate to the resistance of a liquid.

Levi–Civita married one of his students, Libera Trevisani, in 1914. Before leaving Padua with his wife in 1917, Levi–Civita published his most important solo work in absolute differential calculus. In it, he promulgated the idea of parallel displacement and its place in curved **space**. It was this breakthrough that helped Einstein with the mathematical models of relativity. Parallel displacement had relevancy in other areas as well; in pure mathematics, it affected the growth of the theory of modern differential geometry as applied to generalized spaces in **topology** and the geometry of paths. The scholarly discourse that rose from Levi–Civita's absolute differential calculus influenced its evolution into tensor calculus. Tensor calculus is used by mathematicians in the derivation of gravitation and **electromagnetism** unified theories.

Levi–Civita left Padua in 1918 to become a professor of higher analysisat University of Rome. In 1920, he became a Rational Mechanics Professor, a post he served at until he was forced to retire in 1938. Levi–Civita's work did not go unnoticed by his international peers. In 1922, Levi–Civita won the Sylvestor Medal from the Royal Society of London, which elected him a foreign member in 1930.

In Rome, Levi–Civita published a series of important works. He wrote *Lezioni di meccanica razionale* ("Lessons in Rational Mechanics") in 3 volumes, from 1923–27. In 1924, he produced *Questioni di meccanica classica e relativistica* ("Questions of Classical and Relativistic Mechanics"). A year later, his *Lezioni di calcolo differenziale assoluto* ("The Absolute Differential Calculus") appeared. He also published more on hydrodynamics and his theory of canal waves in the same year. Levi–Civita returned to questions raised by Einstein in 1929, when he published *A Simplified Presentation of Einstein's Unified Field Equations*. During the 1920s, Levi–Civita also was inspired by the rise of atomic physics to explore related mathematical problems.

An outspoken opponent of fascism, Levi–Civita was forced out of his professorship and into retirement in 1938 because of anti–Semitic laws. His membership in scientific societies of Italy was also revoked. Soon after his forced removal, Levi–Civita's health began to decline and he suffered from heart problems. A stroke ultimately caused his death in Rome on December 29, 1941.

L'HÔPITAL'S RULE

L'Hôpital's rule provides an efficient technique for using derivatives to evaluate **limits** of quotients of **functions** that can otherwise not be determined. It uses derivatives to find limits of functions and different forms can be applied to the two main types of indeterminate forms: ∞/∞ and $0/0$.

L'Hôpital's rule is usually written as follows: Let f and g be continuous functions defined on an **interval** $[a, b]$ and differentiable on (a, b). Suppose that $g'(x) \neq 0$ for $a < x < b$ and $\lim_{x \to a}+f(x) = \lim_{x \to a}+g(x) = 0$. Then if $\lim_{x \to a}+f'(x)/g'(x)$ exists, and $\lim_{x \to a}+f(x)/g(x)$ also exists and the two are equal: $\lim_{x \to a}+f'(x)/g'(x) = \lim_{x \to a}+f(x)/g(x)$. This is also true if $\lim_{x \to a}+$ is replaced by $\lim_{x \to b}-$, or by $\lim_{x \to c}$ where c is any number in $[a, b]$. In the case where $\lim_{x \to c} f$ and g need not be differentiable at c.

L'Hôpital's rule can be applied to many different situations. It can be applied to the two types of indeterminate forms, $0/0$ and ∞/∞. These forms arise when $\lim_{x \to a}+f(x) = 0 = \lim_{x \to a}+g(x)$. Then it is said that $\lim_{x \to a}+f(x)/g(x)$ has the indeterminate form $0/0$ because the limit may or may not exist. The same idea applies if $\lim_{x \to a}+$ is replaced by $\lim_{x \to b}-$, $\lim_{x \to c}$, $\lim_{x \to \infty}$, or $\lim_{x \to -\infty}$. L'Hôpital's rule can also be applied to the following situations: $\lim f(x)*g(x)$, where $f(x)$ approaches **infinity** and $g(x)$ approaches **zero**; $\lim f(x)-g(x)$, where both $f(x)$ and $g(x)$ approach infinity; $\lim f(x)^{g(x)}$, where both $f(x)$ and $g(x)$ approach infinity; $\lim f(x)^{g(x)}$, where $f(x)$ approaches infinity and $g(x)$ approaches zero; and $\lim f(x)^{g(x)}$, where $f(x)$ approaches one and $g(x)$ approaches infinity.

French mathematician Marquis Guillaume F. A. de l'Hôptal (see **L'Hospital** next) is attributed with formulating l'Hôptal's rule. The rule first formally appeared in l'Hôptal's book *analyse des infiniment petits pour l'intelligence des lignes courbes* published in 1696 and is the first text book concerned with differential **calculus**. Although the rule first appeared here it should really be called Bernoulli's rule because it first appears in a letter from **Johann Bernoulli** to l'Hôptal in 1694. L'Hôpital and Bernoulli had made an agreement under which l'Hôpital paid Bernoulli a monthly fee for solutions to particular problems. It was in one of these correspondences that the rule was first formulated. L'Hôpital's first use was in connection to solving: $\lim_{x \to a} [\sqrt{(2a^3 x - x^4)} - a^3\sqrt{(a^3 x)}]/[a - \sqrt[4]{(ax^3)}]$. The generalized **mean value theorem** plays an important role in proving l'Hôpital's rule.

L'HOSPITAL, GUILLAUME FRANCOIS ANTOINE (1661-1704)
French mathematician

Guillaume L'Hospital is perhaps most famous for his 1696 book *Analyse des infiniment petits pour l'intelligence des lignes courbes (Analysis of the Infinitely Small to Learn about Curved Lines)*, the first textbook ever on **calculus**. Today he is remembered mainly through **L'Hôpital's rule**, the rights to which he is rumored to have bought from **Johann Bernoulli**.

L'Hospital, whose name some sources also spell "L'Hôpital," was the marquis of St. Mesme. He was born in Paris in 1661 to Anne-Alexandre de L'Hospital and Elizabeth Gobelin. As a boy, L'Hospital's mathematical talents were readily apparent. By the time he was 15, he had stunned esteemed academics by solving a problem concerning **cycloids** that **Blaise Pascal** had posed.

That success set the stage for a long career of distinguished accomplishments in mathematics, except for a brief interruption during which L'Hospital served as a cavalry officer in the French Army. Due to his serious near-sightedness, however, he was forced to resign his commission. Afterward, L'Hospital devoted all of his attention to mathematics. He even hired Bernoulli in 1691 to live at his family chateau to teach him the new discipline of **differential calculus**.

L'Hospital caused something of a controversy in the mathematics world when he published his now famous book five years later. Bernoulli claimed it was based on his (Bernoulli's) original research, although L'Hospital cited his legitimate purchase of the main concept, that "a quantity, which is increased or decreased only by an infinitely smaller quantity, may be considered as remaining the same," from Bernoulli. *Analyses des infiniment* contained L'Hospital's rule, which stated a method for finding the limit of a rational function whose denominator and numerator **zero** at a point. L'Hospital's reputation suffered somewhat by the discovery in 1921 of an early Bernoulli manuscript on which L'Hospital seemed to have based his book.

L'Hospital is also known for his solution of the brachystochrone, the curve of fastest descent in a gravitational **field** of a weighted particle moving between two points. Although the problem was solved independently by such other contemporaneous mathematicians as **Isaac Newton**, **Jacob Bernoulli**, and **Gottfried Wilhelm von Leibniz**, this only served to establish L'Hospital's place among the eminent mathematicians of the day.

L'Hospital's book was widely responsible for bringing differential notation into widespread use in Europe. He also wrote a treatise on analytical conics, published posthumously in 1707, that for almost 100 years was considered the standard reference on the topic.

L'Hospital died in Paris on February 2, 1704. Of his personality, we know that, unlike many of his mathematical peers, he was charismatic, generous, and modest. He had three children with his wife, Marie-Charlotte de Romilley de La Chesnelaye.

LHUILIER, SIMON ANTOINE JEAN (1750-1840)
Swiss teacher and author

Simon Jean Antoine Lhuilier (also written L'Huilier) is best remembered for making mathematical advances that were key in the subsequent development of **topology**, a branch of **geometry**. His textbooks on geometry and **algebra** were staples in the curricula of many European schools in the late 1700s and early 1800s.

Lhuilier was born on April 24, 1750 in Geneva, Switzerland, the fourth child of a family of goldsmiths and jewelers. Even as a young man, Lhuilier knew that mathematics would play a large role in his future; he turned down a wealthy relative's offer to leave him a fortune if he would pursue a religious career. Lhuilier was an outstanding student in secondary school, after which he began concentrating on mathematics with courses at the Calvin Academy.

Through a distinguished family connection, Lhuilier found a job in 1771 as tutor for the children of the wealthy Rilliet-Plantamour family. However, in 1773 he published an article in the *Journal encyclopedique* called "A Letter in Response to Objections against Newtonian Gravitation." This drew the attention of many of the top physicists of the day, and some of them tried to persuade Lhuilier to enter a contest sponsored in 1775 by the Polish monarch to create textbooks on physics for Polish schools. There was also a royal contest for mathematics textbooks, though, which Lhuilier preferred to try his hand at. Not only did he win, but Prince Adam Czartoryski asked Lhuilier to tutor his son at the royal Polish family's castle.

The mathematician spent from 1777 to 1788 as tutor to the royal family, but he also found time to write more on mathematics despite his work and the many social obligations that went along with his position. His most famous textbook, *The Elements of Arithmetic and Geometry*, was published in 1778. In 1786 Lhuilier won a competition put on by the Berlin Academy on the theory of mathematical **infinity**. His work in this area would become one of Lhuilier's most important legacies to his field. Titled *Exposition elementaire des principes des calculs superieurs*, Lhuilier's memoir contained important work on **Leonhard Euler**'s polyhedra formula. This contribution was particularly important in the later evolution of **topology**, with its fresh insights into the concept, interpretation, and use of limit.

When he finally returned to Switzerland in 1789, it was only to find that the country was in a state of intense political unrest. Thus Lhuilier decided to stay with a friend in Germany, where he remained and continued his mathematical research and teaching until 1794. Tempted by an offer of a professorship at the University of Leiden the following year, Lhuilier instead threw his hat into the ring in a competition for a professorship at the Geneva Academy. He won the post and began working there later in 1795, also serving as chair of the mathematics department until 1823. He married Marie Cartier in 1795; the couple would eventually have two children together.

Lhuilier was not solely concerned with mathematics he was also closely involved in Swiss politics, and served as president of the Legislative Council in 1796. In the meantime, however, he remained as active in the mathematics world as ever. Also in 1796, Lhuilier published an algebraic solution to a general problem posed by **Pappus** and four influential articles on probabilities. In 1804, Lhuilier wrote *The Elements of Algebraic Reasoning,* a textbook for his students at the Geneva Academy. The mathematician's last major work, which appeared in 1809, concerned geometric loci in **space** and in a plane. From 1810 to 1813 Lhuilier worked as editor of *The Annals of Pure and Applied Mathematics.*

Despite the serious nature of his research, Lhuilier acquired over his lifetime a reputation for whimsicality, at least in relation to other mathematicians. For instance, he enjoyed writing songs about **square roots** and the number three in particular, as well as putting geometric theorems into verse. He died in Geneva on March 28, 1840.

LIE GROUPS

Lie groups, named after the Norwegian mathematician **Sophus Lie** (1842-1899), stand at the intersection of several different branches of mathematics: **algebra**, **analysis**, **geometry** and **topology**. They are defined as groups that have the additional geometric structure of a smooth **manifold**. To tie the algebraic structure together with the geometric structure, the operations of **multiplication** and inversion are required to be differentiable.

At first glance, Lie groups might seem like the mathematical analogue of a centaur—an unnatural combination of two different creatures. But unlike centaurs, Lie groups really do exist. Some examples include:

- The *Euclidean group,* consisting of all rigid motions of the plane (including rotations, translations and reflections).

- The *Lorentz group,* consisting of all **Lorentz transformations** of special relativity that map one inertial reference frame to another.

- The *general linear group GL(n),* consisting of all *n*-by-*n* invertible matrices. The group operation is **matrix** multiplication, and the formulas for **matrix multiplication** and inversion from **linear algebra** prove that these operations are differentiable.

- The *special orthogonal group SO(n),* which consists of all rotations of *n*-dimensional **space**. These can be represented as *n*-by-*n* matrices.

- The *unitary* and *special unitary groups U(n)* and *SU(n),* which are analogous to *SO(n)* but contain *n*-by-*n* matrices of **complex numbers** instead.

- The *symplectic groups Sp(n),* whose elements can be thought of either as 4*n*-by-4*n* real matrices, or 2*n*-by-2*n* complex matrices, or *n*-by-*n* quaternionic matrices.

The crucial role of Lie groups stems from their interpretation as symmetries. **Felix Klein**, in 1872, argued that classical **Euclidean geometry** is simply the study of the Euclidean group. Similarly, the predictions of special relativity, such as time dilation and the impossibility of faster-than-light travel, all derive from the Lorentz group. In each case, the Lie group consists of all the transformations that preserve the essential features of the geometry being studied.

Lie groups are at the opposite extreme of abstract algebra from finite groups. Because Lie groups always contain infinitely many elements, the parts of **group theory** that deal with counting—for example, the order of an element, or the order and index of a subgroup—are nearly useless. But the geometric structure provides useful alternatives. The "size" of a group is now measured by its **dimension** as a manifold, in other words the number of degrees of freedom involved in specifying one element of the group. For example, the Lorentz group is 10-dimensional, because any inertial reference frame is defined by a location in spacetime (4 coordinates), a choice of coordinate axes (3 coordinates), and a velocity (3 coordinates). For compact Lie groups, a more significant measure of size is the rank of the group. For the classical matrix groups, which all represent rigid motions of Euclidean space, the rank can be interpreted as the largest number of independent motions that avoid interfering with one another. That is, they can be done in either order and produce the same result; in the language of abstract algebra, they *commute*. The rank of *SO(3)* is 1 because in 3-space (as anybody who has played with a Rubik's cube can tell) only rotations about a single axis will commute. In *SO(4)*, on the other hand, the rank is 2: It is possible for 2 rotations in 4-space to commute, provided that they are rotations in completely orthogonal planes.

Finite-dimensional Lie groups can always be viewed as matrix groups. By the **definition** of a manifold, a Lie group *G* has a tangent plane *V* at the **identity element** 1. The elements of this tangent plane, called the *Lie algebra* of the group, can be thought of as "**infinitesimal**" rotations. The Lie group acts on itself by an operation called adjoint map, $\mathrm{ad}_x(y) = xyx^{-1}$, and hence it also acts on the infinitesimal rotations in the same way. But *V* is a vector space, and linear maps on finite-dimensional **vector spaces** can always be represented by matrices. Thus the group element *x* corresponds to the matrix that represents ad_x.

In fact, a Lie group typically has many different "representations" as matrix groups, which can be constructed from a small set of building blocks, called *irreducible* representations. The adjoint representation of a group, described above, is usually far from the simplest. Henri Cartan, in the 1910s, and Hermann Weyl, in the 1920s and 1930s, developed a beautiful theory to explain how to decompose any representation of a Lie group into irreducible ones.

As in finite group theory, a fundamental problem in Lie group theory is to classify the simple compact Lie groups—in other words, the ones that cannot be decomposed into smaller groups. The answer is again more elegant than one would have a right to hope for: There are only four families of simple compact Lie groups—namely the families $SU(n)$, $SO(n)$ for odd n, $Sp(n)$, and $SO(n)$ for even n. In addition, there is a rogue's gallery of exactly five exceptional Lie groups, called G_2, F_4, E_6, E_7, and E_8. (Here the subscripts refer to the rank of the groups.) The exceptional groups have dimensions 14, 52, 78, 133, and 248 respectively.

A surprising number of recent developments in **mathematics and physics** are tied to Lie groups. The classification of finite simple groups, completed in 1980, was certainly inspired by the much simpler classification of simple compact Lie groups. Moreover, many of the groups that appear in this classification are "finite groups of Lie type"—in other words, finite groups that are modeled after the classic matrix groups. The "fake **R**⁴"s discovered by Simon Donaldson in 1983 (in other words, four dimensional universes that are topologically the same as Euclidean four-space but cannot be mapped to it smoothly, like a shirt that is too wrinkled to be ironed flat again) were based on properties of the exceptional group E_8.

Finally, quantum physicists have used representations of Lie groups repeatedly to predict new subatomic particles. Perhaps the most spectacular example was Murray Gell-Mann's theory of the "eightfold way," which predicted the existence of quarks. He named the theory both for the eight paths to enlightenment in Buddhist philosophy, and for the eight dimensions of the group $SU(3)$, which he used to describe the symmetries of the weak nuclear **force**. Each irreducible representation of this group corresponds to a different combination of quarks, or equivalently to a different subatomic particle. Gell-Mann received the Nobel Prize in physics for his work in 1969.

See also Algebra; Group; Marius Sophus Lie; Manifold; Matrix; Matrix multiplication; Hermann Weyl

LIE, MARIUS SOPHUS (1842-1899)
Norwegian geometer

Marius Sophus Lie was one of the first prominent Norwegian scientists and among the last of the great 19th–century mathematicians. His main contribution was his theory of groups. Lie groups and Lie algebras are fundamental tools in many parts of 20th century mathematics, from the theory of **differential equations** to the understanding of elementary particle physics. Although he was an isolated academic who generally lacked regular contact with colleagues or interested students, Lie produced his finest work in collaboration with **Felix Klein** and later Friedrich Engel.

Lie was born on December 17, 1842, in Nordfjordeide, Norway, the youngest of six children. His father, Johann

Herman Lie, was a Lutheran pastor. Lie's education was standard: he first studied in Moss, moving onto Kristiania (present–day Oslo) to study at Nissen's Private Latin School from 1857 to 1859. For the next six years, Lie studied mathematics and science at Kristiania University, graduating without distinction in 1865.

Lie tutored other students for the next few years and pursued his own interests in astronomy and **mechanics**, but he did not find a professional field that appealed to him until 1868, when he was introduced to writings by **Jean–Victor Poncelet** and **Julius Plücker**. Lie was particularly intrigued by Plücker's proposal of a new kind of **geometry** that would use lines and curves instead of points as the elements of a given **space**.

After publishing his first paper in 1869, Lie was awarded a scholarship to study in Berlin, where he met Felix Klein, another mathematician indebted to Plücker's ideas. Lie and Klein spent the following summer in Paris, publishing several papers together, and it was here that Lie developed his idea of contact transformations.

Midway through the summer of 1870, the Franco–Prussian war broke out. En route to Italy, Lie was arrested by the French on charges that he was a Prussian spy. The arresting officers alleged that Lie's mathematical notes were in fact coded messages. Through the intervention of J.G. Darboux, a French mathematician he had worked with earlier that summer, Lie was released after spending a month in jail.

Lie returned to Kristiania the following year to teach at two of his alma maters: the university and Nissen's Private Latin School. At this time, Lie formulated his integration theory of partial differential **equations**, sharing his results with Adolph Mayer, who was concurrently working on the same method. Lie was not recognized for this development at the time, perhaps because he did not write about his discovery using the accepted contemporary analytical language.

Lie earned his Ph.D. in 1872 and married Anna Sophie Birch two years later, having two sons and a daughter with her. During the 1870s, Lie worked on **transformation** groups, which he called finite continuous groups. These groups, later called **Lie groups**, possessed a fixed number of parameters but could be differentiated in any desired order. Lie applied his theory of transformation groups to show that a majority of the known methods of integration could be introduced all together by means of **group theory**. He also used transformation groups to help classify ordinary differential equations and to give a unified method of solution using group–theoretic considerations. Lie later used his finite continuous groups in 1890 to identify the defects in Hermann von Helmholtz's application of group theory to the foundations of geometry.

While Lie was fond of taking hikes and admiring the Norwegian landscape, he found that Kristiania offered him little contact with other mathematicians or interested students. In 1886, his friend Klein, with whom he had maintained correspondence over the years, suggested that Lie take his place as professor of geometry at Leipzig University. Lie accepted, and his 12 years at Leipzig proved to be his most prolific.

Beginning in 1884, Lie teamed up with Friedrich Engel, a student recommended by Klein and Mayer, and together they produced a three–volume work on transformation groups. *Theorie der Transformationsgruppen*, published between 1888 and 1893, owes much of its completion to Engel, as Lie failed to finish most of his other works on his own. It was during this time that Lie began suffering an acute nervous breakdown, diagnosed by doctors at the time as "neurasthenia." Although he was treated in a mental hospital in 1890, Lie remained severely depressed. In the third part of the *Theorie der Transformationsgruppen*, Lie wrote that "I am no pupil of Klein, nor is the opposite the case, although this might be closer to the truth." Klein was deeply hurt by the remark but still welcomed Lie into his home after the incident.

In 1898, Lie was awarded the first International Lobachevsky prize, and in September of that year he returned to teach at Kristiania University. The university had created a special mathematics chair to lure back the Norwegian native, but Lie occupied it for only a short while. He died on February 18, 1899, at the age of 56, of pernicious anemia.

Lie's influence continued well after his death, with mathematicians all over Europe continuing to work on Lie groups. Wilhelm Killing began classifying Lie groups during Lie's lifetime, only to garner Lie's criticism when his work produced errors. Killing's final work, which **Elie Joseph Cartan** later revised and based much of his work on, was still an important contribution. **Hermann Weyl** breathed new life into Lie's groups in his papers from 1922 and 1923, and subsequent generalizations of Lie's groups gave them a greater role in quantum physics and **quantum mechanics**.

LIGHT

Light can be narrowly defined as the visible portion of the electromagnetic spectrum. A broader **definition** would include infrared, ultraviolet, and x-ray wavelengths, which are not visible to the eye. The nature of light has been the subject of controversy for thousands of years. Even today, while scientists know how light behaves, they do not always know why light behaves as it does.

The Greeks were the first to theorize about the nature of light. Led by the scientists **Euclid** and **Hero** (first century A.D.), they came to recognize that light traveled in a straight line. However, they believed that vision worked by intromission—that is, that light rays originated at the eye and traveled to the object being seen. Despite this erroneous hypothesis, the Greeks were able to successfully study the phenomena of reflection and refraction and derive the laws governing them. In reflection, they learned that the angles of incidence and reflection were approximately equal; in refraction, they saw that a beam of light would bend as it entered a denser medium (such as water or glass) and bend back the same amount as it exited.

The next contributor to the embryonic science of optics was the Arab mathematician and physicist Alhazen (965-1039), who is sometimes called the greatest scientist of the Middle Ages. Experimenting around the year 1000, he showed that light comes from a source (the Sun) and reflects from an object to the eyes, thus allowing the object to be seen. He also studied mirrors and lenses and further refined the laws of reflection and refraction.

By the 12th century, scientists felt they had solved the riddles of light and color. The English philosopher Francis Bacon (1561-1626) contended that light was a disturbance in an invisible medium which could be detected by the eye; subsequently, color was caused by objects "staining" the light as it passed. More productive research into the behavior of light was sparked by the new class of realistic painters, who strove to better understand perspective and shading by studying light and its properties.

In the early 1600s, the refracting telescope was perfected by **Galileo** and **Johannes Kepler**, providing a reliable example of the laws of refraction. These laws were further refined by **Willebrord Snell**, whose name is most often associated with the **equations** for determining the refraction of light. By the mid-1600s, enough was known about the behavior of light to allow for the formulation of a wide range of theories.

The renowned English physicist and mathematician **Isaac Newton** was intrigued by the so-called "phenomenon of colors"—the ability of a prism to produce colors from white light. It had been generally accepted that white was a single color, and that a prism could somehow combine white light with others to form a multicolored mixture. Newton, however, doubted this assumption. He used a second prism to recombine the rainbow spectrum back into a beam of white light; this showed that white light must be a combination of colors, not the other way around.

Newton performed his experiments in 1666 and announced them shortly thereafter, subscribing to the corpuscular (or particulate) theory of light. According to this theory, light travels as a stream of particles that originate from a bright source and are absorbed by the eye. Aided by Newton's reputation, the corpuscular theory soon became accepted throughout Great Britain and in parts of Europe.

In the European scientific community, many scientists believed that light, like sound, traveled in waves. This group of scientists was most successfully represented by the Dutch physicist **Christiaan Huygens**, who challenged Newton's corpuscular theory. He argued that a wave theory could best explain the appearance of a spectrum as well as the phenomena of reflection and refraction.

Newton immediately attacked the wave theory. Using some complex calculations, he showed that particles, too, would obey the laws of reflection and refraction. He also pointed out that, if truly a wave form, light should be able to bend around corners, just as sound does; instead it cast a sharp shadow, further supporting the corpuscular theory.

In 1660, however, Francesco Grimaldi examined a beam of light passing through a narrow slit. As it exited and was projected upon a screen, faint fringes could be seen near the edge.

This seemed to indicate that light did bend slightly around corners; the effect, called diffraction, was adopted by Huygens and other theorists as further **proof** of the wave nature of light.

One piece of the wave theory remained unexplained. At that time, all known waves moved through some kind of medium—for example, sound waves moved through air and kinetic waves moved through water. Huygens and his allies had not been able to show just what medium light waves moved through; instead, they contended that an invisible substance called ether filled the universe and allowed the passage of light. This unproven explanation did not earn further support for the wave theory, and the Newtonian view of light prevailed for more than a century.

The first real challenge to Newton's corpuscular theory came in 1801, when English physicist Thomas Young discovered interference in light. He passed a beam of light through two closely spaced pinholes and onto a screen. If light were truly particulate, Young argued, the holes would emit two distinct streams that would appear on the screen as two bright points. What was projected on the screen instead was a series of bright and dark lines—an interference pattern typical of how waves would behave under similar conditions.

If light is a wave, then every point on that wave is potentially a new wave source. As the light passes through the pinholes it exits as two new wave fronts, which spread out as they travel. Because the holes are placed close together, the two waves interact. In some places the two waves combine (constructive interference), whereas in others they cancel each other out (destructive interference), thus producing the pattern of bright and dark lines. Such interference had previously been observed in both water waves and sound waves and seemed to indicate that light, too, moved in waves.

The corpuscular view did not die easily. Many scientists had allied themselves with the Newtonian theory and were unwilling to risk their reputations to support an antiquated wave theory. Also, English scientists were not pleased to see one of their countrymen challenge the theories of Newton; Young, therefore, earned little favor in his homeland.

Throughout Europe, however, support for the wave nature of light continued to grow. In France, Etienne-Louis Malus (1775-1826) and Augustin Jean Fresnel (1788-1827) experimented with polarized light, an effect that could only occur if light acted as a transverse wave (a wave which oscillated at right angles to its path of travel). In Germany, Joseph von Fraunhofer (1787-1826) was constructing instruments to better examine the phenomenon of diffraction and succeeded in identifying within the Sun's spectrum 574 dark lines corresponding to different wavelengths.

In 1850 two French scientists, Jéan Foucault and Armand Fizeau, independently conducted an experiment that would strike a serious blow to the corpuscular theory of light. An instructor of theirs, Dominique-Françios Arago, had suggested that they attempt to measure the speed of light as it traveled through both air and water. If light were particulate it should move faster in water; if, on the other hand, it were a wave it should move faster in air. The two scientists performed

their experiments, and each came to the same conclusion: light traveled more quickly through air and was slowed by water.

Even as more and more scientists subscribed to the wave theory, one question remained unanswered: through what medium did light travel? The existence of ether had never been proven—in fact, the very idea of it seemed ridiculous to most scientists. In 1872, **James Clerk Maxwell** suggested that waves composed of electric and magnetic fields could propagate in a vacuum, independent of any medium. This hypothesis was later proven by Heinrich Rudolph Hertz, who showed that such waves would also obey all the laws of reflection, refraction, and diffraction. It became generally accepted that light acted as an electromagnetic wave.

Hertz, however, had also discovered the photoelectric effect, by which certain metals would produce an electrical potential when exposed to light. As scientists studied the photoelectric effect, it became clear that a wave theory could not account for this behavior; in fact, the effect seemed to indicate the presence of particles. For the first time in more than a century there was new support for Newton's corpuscular theory of light.

The photoelectric effect was explained by **Albert Einstein** in 1905 using the principles of quantum physics developed by Max Planck. Einstein claimed that light was quantized—that is, it appeared in "bundles" of energy. While these bundles traveled in waves, certain reactions (like the photoelectric effect) revealed their particulate nature. This theory was further supported in 1923 by Arthur Holly Compton, who showed that the bundles of light—which he called photons—would sometimes strike electrons during scattering, causing their wavelengths to change.

By employing the quantum theories of Planck and Einstein, Compton was able to describe light as both a particle and a wave, depending upon the way it was tested. While this may seem paradoxical, it remains an acceptable **model** for explaining the phenomena associated with light and is the dominant theory of our time.

LIMITS

The limit of a function or a sequence in mathematics is the number the function or sequence approaches as the **independent variable** (x) approaches a particular number. The **definition** is as follows: If f(x) becomes arbitrarily close to a unique number L as x approaches c from either side, the limit of f(x) as x approaches c is L. This is often expressed lim x → c f(x) = L.

There are three major areas where the limit of a function or sequence does not exist. In order for a function or sequence to have a limit, the function or sequence must be approaching the said value of the function or sequence from the left and right sides of x. Left and right hand behavior must agree. If it does not, the function does not have a limit. The behavior of the function or sequence cannot approach positive or negative **infinity** from the left or right of the x value in question. If this is the case, the limit does not exist as x approaches c. This is considered unbounded behavior. The third way a limit will not exist for a function or sequence is when the value of the function or sequence oscillates between two fixed values as x approaches c.

The limit value may not be a value of the function, it is simply the value the function approaches. It is often necessary to simplify a function before ruling out the lack of a limit at a particular point. It is possible to evaluate a limit numerically (e.g., looking at a table), graphically (analyzing a graph at a particular value of x), or analytically (solving the limit algebraically). This might mean simply plugging the x value into the function and simplifying, or by simplifying the expression and then evaluating at the given value of x.

If the limit of a function does not exist because the left and right hand behavior do not agree, it is possible to talk about one-sided limits. This is when the behavior of a function is examined as x approaches c from only one side. If the function is a step function, it is possible to obtain one limit value as x is approached from the left and a completely different limit value when it is approached from the right. The limit at the x as it approaches c does not exist, but the one-sided limit does exist.

Limits play a key role in the development of the derivative in **calculus**. A derivative represents the **slope** of the line tangent to a curve at any given point. Using the limit process, it is possible to find the exact slope of the tangent line at a given point. The concept of the limit minimizes the **distance** between the point on the graph in question and a second point used to calculate the slope. As the distance between the two points collapses, the slope of the function at a given point is identified.

See also Derivatives and differentials

LINDELÖF, ERNST LEONARD (1870-1946)
Finnish mathematician

Renowned for the clarity and comprehensibility of his textbooks, Ernst Leonard Lindelöf distinguished himself as a mathematician with his research on **differential equations** and function theory. His work became the basis for the study of mathematics history in his native land.

Lindelöf, the son of a professor of mathematics, was born in Helsingfors, Sweden (now Helsinki, Finland) on March 7, 1870. Interested in mathematics and science from a young age, he studied at his father's university from 1887 to 1900, although he took breaks during that period to study at prestigious schools in Stockholm, Paris, and Göttingen, Germany. In 1890, while only twenty, Lindelöf published his first professional paper, which outlined solutions for **differential equations**. After receiving his degree at Helsingfors in 1895, Lindelöf stayed on to continue his studies and teach mathematics classes as a private tutor.

Soon recognizing the talent of the young mathematician, the university gave Lindelöf a post as assistant professor in 1902, promoting him only a year later, impressively, to full

professor of mathematics. By then, Lindelöf's reputation as a mathematician had been firmly established in the academic world. In the meantime, he had decided to abandon any ambitions for creative scientific research in favor of his responsibilities as a professor.

In 1905 Lindelöf published *Calculation of Residues and Their Application to Function Theory,* which analyzed the role of **Augustin-Louis Cauchy**'s residue theory as applied to function theory. The book was well received, both because of its brilliant content and its accessibility and lucidity.

Beginning in 1907, Lindelöf served as a member of the editorial board for *Acta Mathematica.* In the latter part of his career, he devoted himself to writing textbooks. Among these were the four-volume *Differential and Integral Calculus and Their Application* (1920-1946) and *An Introduction to Function Theory* (1936). Lindelöf died in Helsinki on June 4, 1946.

LINDEMANN, CARL LOUIS FERDINAND VON (1852-1939)
German analyst and geometer

The classic problem of squaring the **circle** had intrigued mathematicians since the time of **Euclid**. Only in 1882, however, when Ferdinand Lindemann proved that π is a transcendental number, was this problem finally resolved. While Lindemann is best known for this one result, he also played an important role in the development of mathematics in Germany during the turn of the 20th century.

Lindemann was born in Hanover, Germany, on April 12, 1852. His father was a teacher of modern languages and later a manager of a gas works while his mother was the daughter of a famous teacher of classical languages, so it is not surprising that their son finished first in his class upon graduating from his *gymnasium* in 1870. France and Germany had recently gone to war, but Lindemann's poor health prevented him from being called into the army. Instead, he enrolled at the University of Göttingen to study mathematics.

Göttingen attracted many of Europe's leading mathematicians. During his time there Lindemann attended lectures by Alfred Clebsch on analytic spatial **geometry**, algebraic curves, elliptic **functions**, and the theory of algebraic forms. He also met **Felix Klein**, who was then a lecturer at the university. In 1872 Klein became a full professor at the University of Erlangen; Lindemann joined him as Klein's second Ph.D. student, receiving his degree in 1873 with a thesis on **non–Euclidean geometry** and its connection with **mechanics**. In addition, after Clebsch's sudden death in 1872 and with Klein's encouragement, Lindemann edited and revised Clebsch's geometry lectures which he published as a textbook in 1876. The Clebsch–Lindemann text won wide acclaim and was used for several decades.

Lindemann spent part of the 1876–77 academic year in Paris, where he began a long friendship with **Charles Hermite**. Because of the success of the Clebsch–Lindemann text, he was introduced to many of the leading French mathematicians. He returned from Paris to become associate professor at the University of Freiburg after a promised position at the University of Würzburg never materialized. During his six years at Freiburg, Lindemann published several minor papers on special functions and **Fourier series**, and also wrote a paper on the **vibration** of strings, inspired by the recent invention of the microphone. But his main success came with his work on the number π. During Lindemann's visit to Paris in 1876, Hermite had shown him his **proof** that the number e is transcendental, that is, that e is not the root of any polynomial with integer coefficients. Building upon his friend's earlier work, Lindemann finally succeeded in 1882 in proving that π is also transcendental. He sent his paper "Über die Zahl π" (Concerning the number π) to Klein for publication in the *Mathematische Annalen*. Klein sent the paper to **Georg Cantor**, who could find no errors, and who passed the paper on to **Karl Weierstrass** in Berlin for final verification of the proof. With Lindemann's permission, on June 22, 1882, Weierstrass presented the result to the Berlin Academy of Sciences to great acclaim.

The problem of **squaring the circle**, that is, constructing a **square** with the same **area** as that of a given circle, fascinated mathematicians for more than two thousand years. A solution had been found by Dinostratus around 350 B.C., but no one had ever been able to find a solution using just the classical Euclidean tools of the straightedge and compass. Mathematicians knew that if a number was transcendental, and hence not algebraic, then no line of that length could be constructed using these tools. Lindemann's proof that π was transcendental, and hence unconstructible, finally established unequivocally that the squaring of the circle was impossible by means of straightedge and compass alone.

With the fame of his work on π freshly behind him, Lindemann accepted an appointment as full professor at the University of Königsberg in 1883. After ten years, he moved one final time to take a chair in mathematics at the University of Munich. He never again published a paper to rival the importance of his work on π. Nevertheless, Lindemann had a successful career as a teacher, an advisor of students, and an administrator. He supervised more than 60 German and foreign Ph.D. students, including **Hermann Minkowski** and **David Hilbert**. During his years in Munich, Lindemann served as dean of the arts and sciences, as rector of the university (an elected position comparable to that of president), and for 25 years as the director of the university's administrative committee. For several years Lindemann was also a confidential advisor to the king's court. In 1918, he received the Knight's Cross of the Order of the Bavarian Crown, an honor that granted nobility and the right to be known as Ferdinand Ritter von Lindemann.

In 1887, Lindemann married Lisbeth Küssner, a successful actress from Königsberg. They had two children, both born in Königsberg, a son in 1889 and a daughter in 1891. Their son died tragically at the age of 22 during a mountain climbing accident in the Alps. Lisbeth apparently had mathematical as well as acting talents as she collaborated with her husband in translating and revising some of the works of the French mathematician **Henri Poincaré**. Lindemann died on

March 6, 1939, three years after his wife. In his article on the man who discovered the transcendence of π, Fritsch writes that "he still published mathematical papers and thought about problems up to the day before his death."

LINEAR ALGEBRA

The study of linear **algebra** includes the topics of **vector algebra**, **matrix** algebra, and the theory of **vector spaces**. Linear algebra originated as the study of linear **equations**, including the solution of simultaneous **linear equations**. An equation is linear if no **variable** in it is multiplied by itself or any other variable. Thus, the equation $3x + 2y + z = 0$ is a linear equation in three variables. The equation $x^3 + 6y + z + 5 = 0$ is not linear, because the variable x is raised to the power 3 (multiplied together three times); it is a cubic equation. The equation $5x - xy + 6z = 7$ is not a linear equation either, because the product of two variables (xy) appears in it. Thus linear equations are always degree 1.

Two important concepts emerge in linear algebra to help facilitate the expression and solution of systems of simultaneous linear equations. They are the vector and the matrix. Vectors correspond to directed line segments. They have both magnitude (length) and direction. Matrices are rectangular arrays of numbers. They are used in dealing with the coefficients of simultaneous equations. Using vector and matrix notation, a system of linear equations can be written, in the form of a single equation, as a matrix times a vector.

Linear algebra has a wide variety of applications. It is useful in solving network problems, such as calculating current flow in various branches of complicated electronic circuits, or analyzing traffic flow patterns on city streets and interstate highways. Linear algebra is also the basis of a process called **linear programming**, widely used in business to solve a variety of problems that often contain a very large number of variables.

The collection of theorems and ideas that comprise linear algebra have come together over some four centuries, beginning in the mid-1600s. The name linear algebra, however, is relatively recent. It derives from the fact that the graph of a linear equation is a straight line. In fact the beginnings of linear algebra are rooted in the early attempts of 16th- and 17th-century mathematicians to develop generalized methods for solving **systems of linear equations**. As early as 1693, **Gottfried Leibniz** put forth the notion of matrices and their **determinants**, and in 1750, **Gabriel Cramer** published his rule (it bears his name today) for solving n equations in n unknowns.

The concept of a vector, however, was originally introduced in physics applications to describe quantities having both magnitude and direction, such as **force** and velocity. Later, the concept was blended with many of the other notions of linear algebra when mathematicians realized that vectors and one column (or one row) matrices are mathematically identical.

Finally, the theory of vector spaces grew out of work on the algebra of vectors.

An equation is only true for certain values of the variables called solutions, or **roots**, of the equation. When it is desired that certain values of the variables make two or more equations true simultaneously (at the same time), the equations are called simultaneous equations and the values that make them true are called solutions to the system of simultaneous equations.

The graph of a linear equation, in a rectangular coordinate system, is a straight line, hence the term linear. The graph of simultaneous linear equations is a set of lines, one corresponding to each equation. The solution to a simultaneous system of equations, if it exists, is the set of numbers that correspond to the location in **space** where all the lines intersect in a single point.

Since the solution to a system of simultaneous equations, as pointed out earlier, corresponds to the point in space where their graphs intersect in a single point, and since vectors represent points in space, the solution to a set of simultaneous equations is a vector. Thus, all the variables in a system of equations can be represented by a single variable, namely a vector.

A matrix is a rectangular array of numbers, and is often used to represent the coefficients of a set of simultaneous equations. Two or more equations are simultaneous if each time a variable appears in any of the equations, it represents the same quantity. For example, suppose the following relationship exists between the ages of a brother and two sisters: Jack is three years older than his sister Mary, and eleven years older than his sister Nancy, who is half as old as Mary. There are three separate statements here, each of which can be translated into mathematical notation, as follows:

Let: j = Jack's age, m = Mary's age, n = Nancy's age.

Then: j = m + 3 (1)

j = n + 11 (2)

2n = m (3)

This is a system of three simultaneous equations in three unknowns. Each unknown age is represented by a variable. Each time a particular variable appears in an equation, it stands for the same quantity. In order to see how the concept of a matrix enters, rewrite the above equations, using the standard rules of algebra, as:

1j - 1m - 0n = 3 (1')

1j + 0m - 1n = 11 (2')

0j - 1m + 2n = 0. (3')

Since a matrix is a rectangular array of numbers, the coefficients of equations (1'), (2'), and (3') can be written in the form of a matrix, A, called the matrix of coefficients, by letting each column contain the coefficients of a given variable (j, m, and n from left to right) and each row contain the coef-

ficients of a single equation (equations (1'), (2'), and (3') from top to bottom. That is,

$$1 \ -1 \ 0$$

$$A = 1 \ 0 \ -1$$

$$0 \ -1 \ 2$$

Matrix multiplication is carried out by multiplying each row in the left matrix times each column in the right matrix. Thinking of the left matrix as containing a number of "row vectors" and the right matrix as containing a number of "column vectors," matrix **multiplication** consists of a series of vector dot products. Row 1 times column 1 produces a term in row 1 column 1 of the product matrix, row 2 times column 1 produces a term in row 2 column 1 of the product matrix, and so on, until each row has been multiplied by each column. The product matrix has the same number of rows as the left matrix and the same number of columns as the right matrix. In order that two matrices be compatible for multiplication, the right must have the same number of rows as the left has columns. The matrix with 1s on the diagonal (The diagonal of a matrix begins in the upper left corner and ends in the lower right corner) and all other elements **zero**, is the **identity element** for multiplication of matrices, usually denoted by I. Thus the inverse of a matrix A is the matrix A^{-1} such that $AA^{-1} = I$. Not every matrix has an inverse, however, if a **square** matrix has an inverse, then $A^{-1}A = AA^{-1} = I$. That is, multiplication of a square matrix by its inverse is commutative.

Just as a matrix can be thought of as a collection of vectors, a vector can be thought of as a one-column, or one-row, matrix. Thus, multiplication of a vector by a matrix is accomplished using the rules of matrix multiplication. For example, let the variables in the previous example be represented by the vector j = (j,m,n). Then the product of the coefficient matrix, A, times the vector, j, results in a three-row, one-column matrix, containing terms that correspond to the left-hand side of each of equations (1'), (2'), and (3').

[1 -1 0 | j] [1j - 1m + 0n]

[1 0 -1 | m] = [1j + 0m - 1n]

[0 -1 2 | n] [0j - 1m + 2n]

Finally, by expressing the constants on the right-hand side of those equations as a constant column vector, c, the three equations can be written as the single matrix equation: Aj = c. This equation can be solved using the inverse of the matrix A. That is, multiplying both sides of the equation by the inverse of A provides the solution: $j = A^{-1}c$. The general method for finding the inverse of a matrix and hence the solution to a system of equations is given by **Cramer's rule**.

Applications of linear algebra have grown rapidly since the introduction of the computer. Finding the inverse of a matrix, especially one that has hundreds or thousands of rows and columns, is a task easily performed by computer in a relatively short time. Virtually any problem that can be translated into the language of linear mathematics can be solved, provided a solution exists. Linear algebra is applied to problems in transportation and communication to route traffic and information; it is used in the fields of biology, sociology, and ecology to analyze and understand huge amounts of data; it is used daily by the business and economics community to maximize profits and optimize purchasing and manufacturing procedures; and it is vital to the understanding of physics, chemistry, and all types of engineering.

LINEAR EQUATIONS

A linear equation is one which when graphed yields a straight line. This is true if the linear equation has two unknowns or variables. The solution to a linear equation is an ordered pair. The standard form of a linear equation in two variables is $a_1x + a_2y = b$, where a_1, a_2 and b are constants and x and y are the variables. The ordered pair (x,y) is a solution to the linear equation. A linear equation in n variables x_1, x_2, x_3,...x_n is expressed as $a_1x_1 + a_2x_2 +...a_nx_n = b$, where a_1, a_2,...a_n and b are real constants. The solution to such an equation is a sequence of n numbers s_1, s_2,...s_n such that the equation is satisfied when $x_1 = s_1$, $x_2 = s_2$,...$x_n = s_n$. A solution set is a set of all solutions of the equation. A linear equation does not involve any **roots** or products of variables. Variables occur only to the first power and do not appear in trigonometric, logarithmic, or **exponential functions**.

A finite set of linear **equations** is called a system of linear equations or a linear system. A system of linear equations has the variables x_1, x_2, x_3,...x_n and a sequence of numbers s_1, s_2,...s_n is called a solution of the system if $x_1 = s_1$, $x_2 = s_2$,...$x_n = s_n$ is a solution of every equation in the system. So if a set of numbers satisfies only a limited number of equations in the system then that set is not a solution to the system of equations. Not all **systems of linear equations** have solutions. These **systems of equations** are said to be inconsistent. If there is at least one solution the system of equations is called consistent. Graphically, inconsistent systems of equations contain lines that do not intersect whereas consistent systems of equations contain lines that do intersect, therefore having exactly one solution, or coincide, therefore having infinitely many solutions. No matter how many linear equations make up a system of linear equations the system has either no solutions, exactly one solution, or infinitely many solutions. A generic system of m linear equations in n variables is usually represented by $a_{1\,1}x_1 + a_{1\,2}x_2 +...a_{1\,n}x_n = b_1$ $a_{2\,1}x_1 + a_{2\,2}x_2 +...a_1 {}_nx_n = b_2$... $a_{m\,1}x_1 + a_{m\,2}x_2 +...a_{m\,n}x_n = b_m$ Many times augmented matrices are employed to keep track of the constants and are often used to solve systems of linear equations. The basic method for solving a system of linear equations is to replace a system by a new system that has the same solution set but that is easier to solve.

Systems of linear equations can be classified as homogeneous or nonhomogeneous. Homogeneous systems of linear equations are ones in which all of the constant terms are **zero**, that is all of the bs are zero. Every system of homogeneous lin-

ear equations is consistent since all of the variables can be set to zero. This solution is called the trivial solution and if there are other solutions they are called nontrivial solutions. In these systems there is either one solution or infinitely many solutions since homogeneous systems of linear equations must be consistent. Also, if a homogeneous system of linear equations involves more unknowns than equations then the system is assured of having nontrivial solutions. Those systems of linear equations that do not have all of the constant terms equal to zero are referred to as nonhomogeneous.

LINEAR FUNCTIONS

A linear function is a function of the form $f(x) = mx + b$ whose graph is a straight line. When written in this form, the **slope** of the graph is m and the y-intercept is b. Except for constant **functions**, linear functions are the simplest of all mathematical functions. This is because the largest exponent of the **variable** x is 1 in a linear function. So a linear function is sometimes said to be a polynomial function of degree 1. In general, polynomial functions that have degree higher than 1 have more complicated graphs and require a higher level of mathematical **analysis** than do linear functions. The simplest linear functions are those which pass through the origin. They have y-intercept equal to 0, so their **equations** take the form $f(x) = mx$, or, equivalently, $y = mx$. The slope m in this equation is calculated as follows: Pick two points on the line and calculate the difference in values of the y-coordinates of those points and divide it by the difference in the x-coordinates of the points. This calculation is sometimes called "rise over run" or "rise divided by run." In any case, the slope measures the rate of change of the y-coordinates as one moves from left to right along the graph. Essentially, the absolute value of the slope measures the "steepness" of the line. The bigger the absolute value of the slope, the steeper the line. Note also that if the slope is positive, then the line "rises" from left to right; if the slope is negative, then the line "falls" from left to right; and if the slope is 0, then the line is horizontal.

Because of their relative simplicity, linear functions are often the first choice of a mathematician or scientist for **modeling** data collected from real-world situations. If a linear **model** is a good fit for the data, then the analysis of the real-world situation is likely to be much simpler than if the model has higher degree. Thus statisticians have developed a carefully worked out theory for deciding when a linear function is a "good enough" fit for a data set. The theory is called linear regression and allows the statistician to determine the slope and the y-intercept of the so-called "line of best fit." It also provides for the calculation of a number called the "correlation coefficient" that tells how close this best fit line actually comes to "capturing" the trend of the data. Therefore, this line of best fit is also sometimes called a "trend line." In some cases, even when a scatter plot of a data set does not look linear, it is possible to use mathematical transformations to "linearize" the data. Then the techniques of linear regression can be used with the linearized model to determine the form of the best fit function for the original data. This, for example, is the theory behind semilogarithmic graph paper or the "exponential regression" feature found on modern graphing calculators. If the raw data appears to be more exponential than linear, a plot of first coordinates against the **logarithms** of the second coordinates will appear more linear. We say that a logarithmic **transformation** has been done on the second coordinates of the original data set. Then linear regression is carried out on the transformed data set, which allows the statistician to arrive at an appropriate exponential model for the original data. A graphing **calculator** with an exponential regression feature does all of this automatically, returning the best fit exponential function model within a few seconds. Similar techniques may be carried out on non-linear data which appear to be logarithmic, power, higher degree polynomial, or trigonometric. The point is that the basis of all these different types of regression analysis is linear regression.

There are many other examples of the use of linear functions to approximate non-linear functions in mathematics. For instance, to approximate the value of a non-linear function at a given point on its curve, one can use the linear function which is tangent to the curve at a nearby known point. The techniques of **calculus** can be used to determine this tangent line function and so long as the point of tangency is "close" to the point whose second coordinate is being approximated, this approximation should be quite good. In fact, it can be shown that the tangent line to a curve at a point is the best linear approximator to the curve at that point. The repeated use of this technique is the basis for Euler's method for numerically solving **differential equations** whose solution curves are non-linear.

In summary, although linear functions are themselves relatively simple to analyze, that simplicity has made them the basis for analysis of much more complicated non-linear phenomena. Thus linear functions are arguably among the most important functions in mathematics.

LINEAR PROGRAMMING

Linear programming was developed by applied mathematicians and operations research specialists as a means to solve real-world problems using linear methods. Based on the fundamentals of **matrix algebra**, linear programming seeks to find an optimal solution, using quantitative methods, to a particular problem given a finite number of constraints. It is used extensively in managerial science and has widespread utility to business, government, and industry. It is applied especially to problems in which decision-makers wish to minimize costs or maximize profits under a given operating construct. In many cases, linear programming will affect decisions regarding materials used in manufacturing and construction or even the hiring of personnel or particular skill **sets**. It is an excellent tool for decisions regarding *resource allocation.*

Linear programming is based on linear equations—equations of variables raised only to the first power. Many variables—such as x, y, and z) or more generically, x_1, x_2,..., x_k—may be used in a single equation, known as a *linear com-*

bination. Generally, each **variable** represents a quantifiable item, such as the number of carpenters' hours available to a business during a week or number of sleds available for sale during a given month. It is important to note that in basic linear programming, the coefficients of the variables neither must equal each other nor equal any particular value—for example, $5x_1 - 2.7x_2 + 0x_3 - 9x_4 \leq 7$ is a perfectly valid linear combination. A special subset of linear programming, however, does specify that the solution should contain only zeros and ones for the values assigned to the variables.

Linear programming uses a finite number of linear combinations, combining them into a system of linear **equations**. Each of the equations is generally referred to as a *constraint*. Once the constraints are determined, linear programming seeks to find an optimal solution for either maximizing or minimizing the *objective function*. The objective function is a linear combination of all the variables present in the system of **linear equations** with pre-determined coefficients. It dominates the linear programming problem. A business manufacturing two types of computer chips, for example, may wish to determine the optimal mix of production that meets customer demand and maximizes profits. Similarly, a fast-food restaurant may wish to determine the optimal number of cashiers and grill cooks that meets customer demand while minimizing labor costs. In the first case, the computer chip manufacturer desires to *maximize* the objective function, which represents profit. In the second case, the restaurant manager wishes to *minimize* the objective function, which represents labor cost.

Like all **systems of linear equations**, linear programming problems will have either one solution, no solution, or many solutions. A problem with no solution is called unfeasible; when this situation occurs, there are no values that can be assigned to the variables in the objective function that meet the criteria established by the constraints. This case is very similar to an over-determined matrix with more variables than equations. In the case of many solutions, there exist at least two sets of values (perhaps infinitely many sets) that either maximize or minimize the objective function while abiding by the constraints. This case is similar to an under-determined matrix that has fewer variables than equations.

Solutions to linear programming problems may have either integer or real-numbered values. Solutions with real-numbered values are the norm. However, these solutions are often impractical to implement, as they may suggest buying 2.7 fishing boats or 10.5 sea kayaks to a marina. In these cases, an "integer-only" constraint is also attached to the linear programming problem. The solution will not be exact, but it will fall within a specified standard of the optimal solution, generally within 90, 95, or 99 percent. The new, "integer-only" solution may closely mimic the real-valued solution, or provide unexpected results—such as a recommendation to buy no fishing boats and 21 sea kayaks.

Oddly enough, nutritionists for large-scale medical facilities—such as retirement centers, nursing homes, extended care facilities, and hospitals—were among the first to extensively use linear programming as a professional aid. Nutritionists working in such centers are charged with ensuring their patients receive a well-balanced diet. This diet contains a minimum and maximum number of calories, as well as a minimum and maximum allowable amount of vitamins (such as B12 or C) and minerals (such as iron and zinc) that must be included daily. The number of calories consumed from proteins, carbohydrates, and fats must also be strictly balanced and may even vary by patient based on unique medical conditions and treatments. Within these constraints, nutritionists attempt to develop meal plans that both vary foods and minimize cost for the facility. Using caloric, vitamin, and mineral intake as constraints and relying on large databases of nutrition and cost data, nutritionists determine an optimal mix of foods and beverages to be served at meals at minimal cost.

See also Applied mathematics; Equations; Linear algebra; Linear equations; Linear functions; Matrix; Powers; Systems of linear equations; Variable

LINEAR SPACE

A linear **space**, also called a vector space, is a set with well-defined properties. First, a linear space must be defined over a **field**. This field is often the field of **real numbers** or the field of **complex numbers**, but may be other fields as well. The field elements are known as scalars, and the elements of the linear space are known as vectors. (These vectors should not be confused with physical vectors of two- or three-dimensions—although those vectors are elements of linear spaces.) This space needs to have defined at least two operations, **addition** and scalar **multiplication**, that fit the following criteria:

- Addition must be commutative. This means that for all pairs x,y in a linear space L, x + y = y + x.
- Addition must also be associative. For all x,y,z in L, (x + y) + z = x + (y + z).
- Zero exists. There must be some element within the vector space so that x + 0 = x. Under these conditions, most randomly chosen groups of numbers are not linear spaces, as many do not contain zero.
- There is an opposite element to each element. That is, for every x, there must be a y such that x + y = 0.
- There must be an element for unitary multiplication. When using a vector space, one wants to have 1x = x.
- Multiplication must hold in three ways. First, for any scalar a,b and any vector x, (ab)x = a(bx). Then, it must also be true that (a + b)x = ax + bx. Finally, for a scalar a and any vectors x and y, it must hold that a(x + y) = ax + ay.

Most of the rules that define a linear space seem self-evident and somewhat redundant to people who are familiar with the workings of real numbers, complex numbers, or even two- or three-dimensional physical vectors. Matrices and **polynomials** also follow the rules of linear spaces. However, many spaces may be defined that do not fulfill these properties. Further, there are many familiar operations that do not have to be defined in a linear space. For one of the most obvious, ele-

ments of a linear space do not have to have multiplication between them defined.

There are many other features that may apply to linear spaces. A subspace may be defined by taking some elements from the original linear space and making sure that the above requirements are still met. The basis of a linear space is the set of linearly independent elements that can be linearly combined to form all other elements of the linear space. The **dimension** of a linear space is the number of elements in its basis.

Linear spaces can be used within linear **algebra** to determine properties of a system. An understanding of linear spaces will help with **matrix** manipulation and solving systems of **equations**, and is necessary for most higher and abstract mathematics. Further, comprehending linear spaces conceptually can be used in various quantum mechanical systems. Most quantum systems can be expressed as linear spaces, often giving insight into deep or interesting properties.

See also Vector spaces

LIOUVILLE, JOSEPH (1809-1882)

French number theorist

Although Joseph Liouville's primary contribution to mathematics was the first **proof** of the existence of **transcendental numbers** (**real numbers** that are not **roots** of **polynomials** with integer coefficients), he had a wide range of mathematical interests. Liouville's publishing and teaching activities were vital to French mathematics during the 19th century. He made critical contributions to **number theory**, differential **geometry**, celestial **mechanics**, and rational mechanics. An enthusiastic lecturer, Liouville held numerous teaching positions, usually two or more at a time. Simultaneously, he was a prolific author, publishing 400 papers in his lifetime, including over 200 on number theory alone.

Liouville was born in St. Omer, Pas-de-Calais, France, on March 24, 1809. He was the second son of an army captain, Claude–Joseph Liouville, and his wife, Thérèse (nee Balland). Liouville received his early education in Commery and Toul, before being accepted at the l'École Polytechnique. He began studying there in 1825, when he was 16 years of age, and left that institution two years later to enter l'École des Ponts et Chaussées. Liouville switched schools because of changing interests; he wanted to be an engineer. When he graduated in 1830, he was offered an engineering position, but by then he had decided he wanted to study mathematics full–time in the French center for mathematics, Paris.

Even before Liouville graduated from l'École des Ponts et Chaussées, he began to publish articles as early as 1828. He produced articles and notes on electricity and heat in scholarly journals. This early activity marked the beginning of a lifetime of fruitful scholarship. Liouville's primary publishing phase lasted until 1857. In that time, he had 100 or more treatises printed on mathematical **analysis**, as well as such topics as geometry, physics, **algebra**, and number theory.

Several years after graduation, in 1830, Liouville married his maternal cousin, Marie–Louise Balland. They had a family of three daughters and one son. To support his family, Liouville took on as many teaching positions as possible in secondary schools, sometimes teaching 34 hours or more a week. His publishing activities combined with his teaching experience led to his post as a *répétiteur* at the l'École Polytechnique in 1831. Liouville became a lecturer at l'École Centrale des Arts et Manufactures (an engineering school) in 1833. His enthusiasm for teaching led him to earn his doctorate in 1836 so he could hold professorships at the university level.

In this same year, 1836, Liouville filled a gaping hole in French mathematics academia when he founded the *Journal de mathématiques pures et appliquées* (also known as the *Journal de Liouville* or *Liouville's Journal*). There had been no forum for publishing French mathematical papers for the five years previous to his founding of the *Journal*. Although Liouville had no editorial experience, he edited this important publication for almost 40 years. He used his editorial power judiciously, publishing the best of the contemporary greats and helping young mathematicians get their first works in print. Liouville relinquished control of the *Journal* in 1874 when Henry Résal took over the editorship.

At the same time his teaching and publishing activities bloomed, Liouville experienced the most fruitful research period of his career. From 1832 to 1833, he concentrated on algebraic **functions**, specifically looking at integrals and their analytic behavior. This work led to his 1844 discovery of the proof of the existence of transcendental numbers, one of his most influential contributions to mathematics. In 1836–1837, he published influential papers with fellow mathematician Charles–François Stürm in the *Journal de mathématiques*. They delineated what became known as the Stürm–Liouville theory, based on Liouville's methodology for boundary–value problems. This theory became important in physics and integral equation theory. In analysis, Liouville is well known for proving that a bounded entire analytic function on the complex plane must be a constant function. The **fundamental theorem of algebra** follows as a simple corollary of Liouville's **theorem**.

Liouville continued to hold multiple teaching positions throughout his career. He resigned from l'École Centrale des Arts et Manufactures in 1838 when he became the professor and chair of Analysis and Mechanics at l'École Polytechnique. The previous year, 1837, Liouville began teaching at the Collège de France as an assistant for a professor there. He resigned in 1843 in protest of the Collège's choice for mathematics chair, Count Libri–Carrucci.

Liouville's contributions to French mathematics did not go unnoticed by his peers. In 1839, he became a member of the French Académie Royale des Sciences. A year later he became a member of Bureau des Longitudes, and served as its director at one point.

In 1848, Liouville took an unexpected turn into politics after the French Revolution of 1848. He won a seat on the constituent assembly as a moderate republican, but lost in an election for a position on the Legislative Assembly. This loss marked the end of his brief foray into elective politics.

Liouville resigned from l'École Polytechnique in 1851 and returned to the Collège de France. Libri–Carrucci had left France and Liouville was appointed to the mathematics chair in his place. Liouville taught at the Collège nearly continuously until 1879 or until his death in 1882 (depending on the source). He still maintained multiple teaching posts late in his career. He served as a professor of rational mechanics at the Sorbonne from 1857 to 1874, although he often used substitute teachers as his health began to decline. After his appointment at the Sorbonne, Liouville's research focused almost exclusively on two specialized topics in number theory.

Liouville's final years were marked by pain and suffering. From 1876 until his death, Liouville suffered from intense insomnia and gout. His wife and only son both died in 1880. Liouville himself died in Paris on September 8, 1882.

LIOUVILLE NUMBERS

We define a number z to be *algebraic* if it satisfies a polynomial with integer coefficients. And we say that z is *algebraic of degree n* if n is the degree of the smallest polynomial satisfied by z. For example, the degree of $\sqrt{2}$ is 2, since it satisfies the polynomial $x^2 - 2 = 0$ and no polynomial of smaller degree. Similarly, $\sqrt{2} + \sqrt{3}$ is algebraic of degree 4, as we invite the reader to show by inventing a 4th-degree polynomial which this number satisfies, and also convincing himself that there is no smaller.

This raises the question: Are there numbers which are not algebraic, that is, which do not satisfy any polynomial with coefficients in Z?

This question was answered in the affirmative in 1844 by **Joseph Liouville** (1809-1882). He first proved that if a real number z is algebraic of degree n we can say about it that there exists a constant M such that for all **rational numbers** p/q, with p,q contained in Z, the **distance** from z to p/q is greater than M/q^n, where n is the degree of the minimal polynomial satisfied by z. He then defined a class of numbers, which we now call *Liouville numbers*, to be a set of numbers that, it turns out, fail to have this property. Specifically, a number z is a Liouville number if it is real and irrational and if for any positive integer n, there is a rational number p/q such that the distance from z to p/q is less than $1/q^n$.

By formally proving what is almost apparent, that a Liouville number cannot satisfy the property above characterizing an algebraic number, we prove that a Liouville number cannot be algebraic.

To demonstrate that the class of Liouville numbers is not vacuous, Liouville easily showed that $\Sigma \, 1/10^{n!}$ is a Liouville number and thus is not algebraic. Here we take the sum from $n = 1$ to $n = $ **infinity**. The number 10 may just as well be replaced by any other, and $n!$ may be replaced by any sequence of numbers that diverges rapidly to infinity. A number that is not algebraic is said to be transcendental.

This was the first number shown to be transcendental and preceded **Georg Cantor**'s non-constructive **proof** based on a **cardinality** argument by thirty years.

Other properties of Liouville numbers are as follows:

- 1) The Liouville numbers are dense in R.
- 2) The Liouville numbers are of second category.
- 3) The Liouville numbers comprise a subset of the real line having Lebesgue measure 0.
- 4) The Liouville numbers have s-dimensional (Hausdorff measure) equal to 0, for all positive s.
- 5) The real line can be partitioned into a set of Lebesgue measure 0 and a set of first category by taking the Liouville numbers and their complement.

These concepts will be explained but not proved.

We say that a subset E of R is dense in R if every **interval** in R contains a member of E. We say that a set F is *nowhere dense* if it is dense in no interval, that is, if every interval has a subinterval contained in the complement of F. We say that a set is of *first category* if it is a countable union of nowhere dense **sets**. (A set is said to be countable if there is a one to one **mapping** from the set onto the integers.) And a set is of *second category* if it is not of first category. For example, Q, the rational numbers, is of first category since it may be represented as a countable union of singletons — {p/q}. Note however that Q is not itself nowhere dense. Indeed, Q is dense in R.

We define a set E to be of *Lebesgue measure 0* if for any preassigned quantity, however small, we can cover E with a countable collection of intervals whose total length is less than this preassigned quantity. If there exists a positive real number s so that if we can raise the length of each individual interval to the power s, and if again their sum is arbitrarily small, then we say E has *s-dimensional Hausdorff measure 0*. Symbolically, $\Sigma |I| < \varepsilon$, where ε is our arbitrarily small quantity, and $E \subset \cup I$. It can be shown that this property characterizes the Liouville numbers for all positive **real numbers** s, and thus the Liouville numbers have s-dimensional Hausdorff measure 0 for all positive real numbers s.

Calling the Liouville numbers E, then E, together with its complement, is of course all of R, so the real line then can be partitioned in to a set of Lebesgue measure 0, namely E and a set of first category, namely the compliment of E.

This conclusion is quite remarkable, as a set of Lebesgue measure 0 is thought of, with some justice, as small or thin, and so is a set of first category, yet together they make up the entire real line.

Both the **algebraic numbers** and the Liouville numbers have Lebesgue measure 0. Cantor demonstrated that the set of algebraic numbers is a countable set, and we know that such a set must have Lebesgue measure 0. What remains—the set of **transcendental numbers** which are not Liouville numbers— must have positive measure, measure 1, say in [0,1]. Thus the Liouville numbers amount only to the shoreline of the sea of transcendental numbers. It is however the only large general class of transcendental numbers with which we were initially acquainted and afforded the first techniques of creating specific transcendental numbers.

LISSAJOUS FIGURES

When two simple **sine** waves are combined, the resultant wave depends on the frequencies and phases of the original waves. Consider the sine waves $y_1 = \sin(2\pi f_1 t)$ and $y_2 = \sin(2\pi f_2 t + \varphi)$, where f_1 and f_2 are frequencies (in units of cycles per second) and the independent variable is the time t. The angle φ (phi) is called the phase difference between the two waves. If the frequencies of the two waves are the same, the sum of the two waves will still be a sine function. And because the sine function has a period of 2π radians, the waves will coincide if φ is an integral multiple of 2π. If the frequencies of f_1 and f_2 are different, the resultant wave will be what is called a complex wave, instead of a simple sine function. Sine waves like these can be used to describe the oscillations of the electromagnetic spectrum, which ranges from the low frequencies of radio through infrared radiation and visible light to the high frequencies of x rays and gamma rays.

We have been describing two sine waves that travel parallel to one another and combine to form a resultant wave traveling in the same direction. We can also consider each of these waves to represent the path of a particle, or a point, that is moving on the curve represented by the sine function. Now suppose that the particle is moving in a plane, with the two perpendicular components of its motion represented by sine **functions** in the x and y coordinates of a **Cartesian coordinate system**. The particle now traces out a curve called a Lissajous figure, which is represented by the parametric equations x = $\sin(2\pi f_1 t)$ and y = $\sin(2\pi f_2 t + \varphi)$.

The easiest way to create Lissajous figures is to introduce separate alternating current voltages to the horizontal and vertical inputs of a cathode-ray tube oscilloscope. The interaction of the two signals can then be viewed on the screen. If the perpendicular waves have the same frequency and the phase difference is **zero**, the figure resulting from the combination of the waves is a straight line. For all other values of phase difference, the result is an **ellipse**. When the phase difference is exactly 90 degrees (or $\pi/2$ radians), a special case of the ellipse, a **circle**, is formed. If the frequencies of the two waves are not the same, the Lissajous figure becomes more complex. For example, if one wave has a frequency that is twice that of the other, a sort of "figure eight" pattern results. It could be said that there is one twist, or cross-over point in the pattern at this point. Each time the frequency of one wave becomes a multiple of the other a definite pattern results with more and more twists in it as the frequency difference increases. Such figures are the result of two simple oscillating motions at right angles to each other reaching a harmonic state. When harmonics are reached, the screen of the oscilloscope shows Lissajous figures.

A very common practical use of all of this is found in phase and frequency measurements and adjustments of electronic circuits. By comparing the visual information provided by Lissajous figures, with known or expected patterns, or performance parameters, technicians can repair electronic equipment and scientists can study important phenomena. In addition, now that it is so common to create music electronically, such sine wave comparisons and indicators can also aid in the arts by providing information about the relationships between sounds that musicians use in compositions or performances.

See also Oscillating motion

LOBACHEVSKIAN GEOMETRY

Since **Euclid** many people have tried in vein to prove that the **parallel postulate** is a logical consequence of Euclid's other axioms. The parallel **postulate** states that given any line in the plane and a point not on that line then there exists a unique line through that point that does not intersect the given line. In the 1820s and 30s several people (including Janos Bolyai, **Johann Carl Friedrich Gauss**, and **Nikolai Lobachevsky**) independently realized that the parallel postulate is not implied by the other axioms. In fact the negation of the postulate leads to what is now called Lobachevskian **geometry** (or more commonly **hyperbolic geometry**). Up to this time, physicists had assumed that **Euclid's axioms** were true characterizations of physical **space**. They had thus based their theories upon those axioms. The existence of the new geometry implied that those assumptions may be false. This disturbed physicists and mathematicians so much that they believed that some mistake had been made and so they ignored Lobachevskian geometry for about thirty years. In 1868, Beltrami proved that the non-Euclidean geometry was as logically consistent as Euclidean geometry. Then Riemmann's work on curvature yielded the fact that the Lobachevsky plane is the surface of constant negative curvature. When **Albert Einstein** used Riemmann's theories in his papers on relativity, Lobachevskian geometry could no longer be ignored. In fact, its study became prerequisite to learning modern physics. With the work of Thurston, Gromov, and others in the 1970s up to the present, profound applications of hyperbolic geometry to the theory of complex variables, surfaces, three dimensional manifolds (these are mathematical analogues of alternate universes), and finitely presented groups were found. These are still very active fields of research.

Here is an approximate **model** of the hyperbolic plane which you can make. Cut out forty or more equilateral triangles (all of the same size) from felt or paper. Sew or glue them edge to edge so that seven fit around a point. The result is a very wavy piece of fabric. A "circle" in this plane can be constructed as follows: stick a pin with a string tied to it into the fabric or paper. Hold the string flat against the model and pull taut as you rotate it 360 degrees. The figure traced out by the tip of the string is a **circle** in the hyperbolic plane. Notice that it has more **area** than **pi** squared times the string's length and a longer **circumference** than twice pi times the string's length. In fact, the area and the circumference are **exponential functions** of the string's length. Since the angles around a point must add up to 360 degrees, the **angle** interior to each **triangle** is 360/7. So the sum of the angles of the triangle add up to about 169 degrees. In general, the angles of a hyperbolic triangle add up to less than 180 and the area of the triangle is given by the **pi**—the sum of the angles (in radians) of the triangle. If you think this is weird, consider the **sphere** of radius

one. Draw a triangle with one corner on the north pole and two on the equator. This triangle has at least two right angles and its area is equal to the sum of its angles (in radians) minus pi. A geodesic segment on Lobachevsky's plane is the analogue of a line segment in Euclidean geometry and is defined to be the shortest path between two points. A geodesic then is the analogue of a line and is defined as a union of geodesic segments such that each segment in the union contains the first one and so that this union is infinite in both directions. Using the model, try to get an idea of why Euclid's parallel postulate does not hold here. For any point in the hyperbolic plane and any geodesic that does not contain that point, there are an infinite number of geodesics that contain the point but never intersect the given geodesic.

Another model of the hyperbolic plane, called the Poincare disk model, was used by Escher to create his drawing "Angels and Devils." In it, all angels are the same (hyperbolic) size although they have different sizes from the viewer's perspective. This model has as boundary, the circle of radius one with center at the origin in the Euclidean plane. Like a map of the world, it is distorted. The hyperbolic **distance** d(P1,P2) between two points P1 and P2 is obtained from their Euclidean coordinates by the formula: $\cosh(d(P1,P2)) = 1 + 2*|P1 - P2|/(1 - |P1|^2)*(1 - |P2|^2)$ where |P1| means the Euclidean distance from the origin to P1 and cosh is the hyperbolic **cosine** function. One effect of these definitions is that geodesics in the Poincare model appear as arcs of circles that intersect the boundary circle at right angles or as line segments that pass through the origin. (Two circles intersect at right angles if the tangent lines to the circles at either point of intersection meet at right angles). Draw a small circle at right angles to the boundary circle. How many lines can you draw through the origin that miss the small circle? No matter which circle you drew, the answer is as many as there are **real numbers** between 0 and 1. This shows that Euclid's parallel postulate fails in hyperbolic space.

See also Bolyai, Janos; Euclid of Alexandria; Euclidean geometries; Gauss, Johann Carl Friedrich; Gaussian curvature; Lobachevsky, Nikolai

LOBACHEVSKY, NIKOLAI IVANOVICH (1793-1856)
Russian geometer and educator

Nikolai Ivanovich Lobachevsky is the first mathematician to publicly publish a system of **non–Euclidean geometry**. Although **Karl Friedrich Gauss** preceded him in the late 18th century and **János Bolyai** had devised a similar (though less analytical) conclusions around the same time, Lobachevsky showed that Euclid's Fifth postulate (also known as the **Parallel postulate**) could not be proved on the basis of the other postulates, and in turn created a new way of looking at geometry and geometric problems. Most of Lobachevsky's contemporaries scoffed at his conclusions, and he only became credited with his discoveries after his death. In fact,

Lobachevsky sought credibility by publishing in different languages, but only a few of his colleagues supported his findings, including Gauss. Lobachevsky also did relevant research in other areas, including **infinite series** theory, integral **calculus**, probability, and the approximation of **roots** of algebraic **equations**.

Lobachevsky was born on December 1, 1792, in Nizhny Novgorod (known as Gorky from 1932 to 1990), Russia. His father, Ivan Maksimovich Lobachevsky, was a peasant of Polish descent who worked as a clerk in a provincial land–surveying office. His mother was named Praskovia Aleksandrovna Lobachevskaya. Lobachevsky's father died when he was about six or seven, depending on the source, and his mother took him and his two brothers, Alexander and Alexei, to Kazan where he spent the rest of his life. He attended the local *gymnasium* on scholarship, then entered the University of Kazan when he was 14. Lobachevsky began his higher education in medicine, but when he began to study mathematics with Johann Bartels (a friend of Gauss), Lobachevsky switched majors. He earned his Master's degree in both **mathematics and physics** in 1811 or 1812.

Soon after graduation, in 1814, Lobachevsky joined Kazan's faculty as a lecturer. Later that year, he was promoted to associate professor of mathematics, then became an extraordinary professor in 1816. In these early years, Lobachevsky attempted to deduce the Fifth **postulate** as a **theorem**, but he found that he could not make it work. He then started critically analyzing various versions of the postulate.

As Lobachevsky came to his earthshattering non–Euclidean conclusions, he also played a huge role in the development of his university. In 1820, he began administrative duties at Kazan. He served as dean of the mathematics and physics' faculty twice, from 1820 to 1821, then again in 1823 to 1825. In 1822, he became a member of the committee that supervised new building construction on campus and studied architecture to better understand the duties involved. He became the chair of this committee in 1825. Lobachevsky then became the university librarian from 1825 to 1835, and was elected University Rector (equivalent to president) in 1827, a post he held until 1846. His skills in administration did much to improve Kazan, taking it from a chaotic state, and in addition to restoring innovation, raised both academic standards and the faculty's equanimity. Lobachevsky also founded an academic journal *Uchenye zapiski* ("Scientific Memoirs").

In addition to these many administrative duties, Lobachevsky held a full professorship at Kazan from 1823 to 1846 and still found time to accomplish important mathematical progresses. He wrote *Geometriya* in 1823, in which he outlined his initial ideas that developed into his non–Euclidean geometry. This book remained unpublished as written until 1909, another example of his contemporaries' disregard of his ideas. Still, Lobachevsky described his ideas in a lecture for the Kazan faculty in 1826.

In 1829, the first complete published account of non–Euclidean geometry appeared in a Kazan–based publication called the *Kazan Messenger* after the leading scientific journals in Russia declined to print it. In 1837 he published his article "Géométrie imaginaire" ("Imaginary Geometry"), and the most important and full version of his new geometry was published in 1840 as *Geometrische Untersuchungen zur Theorie der Parellellinien* ("Geometric Investigations of the Theory of Parallel Links"). In these articles, Lobachevsky demonstrated that logical possibility of non–Euclidean geometry, which he called "imaginary geometry," as an analogy to **imaginary numbers**. In particular, Lobachevsky proved that Euclid's Fifth postulate was not a deducible result of the rest of Euclid's postulates.

Lobachevsky did not marry until he was 40 years old in 1832. He married the daughter of an aristocrat, Lady Varvara Aleksivna Moisieva. It was a generally unhappy marriage, and, although she came into the marriage with wealth, their economic situation deteriorated over the rest of Lobachevsky's life.

Lobachevsky also published in other areas of mathematics. In 1834, Lobachevsky devised a way to approximate the roots of algebraic equations. He also published a paper called "Algebra ili ischislenie konechnykh" ("Algebra, or Calculus of Finites"), which concerned the theory of infinite series, and a paper on the **convergence** of trigonometric series in which he proposed a general **definition** of a function similar to that later suggested by **Peter Gustav Lejeune Dirichlet**.

Though much of his work was dismissed in his lifetime, Lobachevsky was given a heredity nobility in 1837. Perhaps this honor was bestowed on him in part for such actions as personally saving lives during the cholera epidemic of 1830.

Despite his valiant actions as administrator, teacher, and citizen of the university, Lobachevsky was relieved of his professorship in 1846. Some sources suggest this event occurred for political reasons. In that year, Lobachevsky became a government official, working in the Kazan educational district as an assistant trustee (or assistant guardian), where he remained until 1855 when he left because of his deteriorating health. Lobachevsky was not limited to his many university activities. He was a longtime member of the Kazan Economic Society, and was interested in agriculture as a hobby, which tied in with the society.

Lobachevsky's last publication of note, *Pangéométrie* (1855–56) was dictated by him in both French and Russian. (He spent the last years of his life blind or nearly so, because of cataracts.) The title means Pangeometry because he rightly thought non–Euclidean geometry had universal characteristics and applications. The text sums up his work in this field.

Lobachevsky died in Kazan, on February 24, 1856. After his death, his work continued to be reprinted and translated elsewhere, which helped to spread his work and his reputation. Finally, Lobachevsky was given his due and put in a canon with other scientific greats. In the beginning of a book on his life and work, the author quotes W. K. Clifford: "What Vesalius was to Galen, what Copernicus was to Ptolemy, that was Lobachevsky to Euclid. There is, indeed, a somewhat instructive parallel between the last two cases. Copernicus and Lobachevsky were both of Slavic origin. Each of them has brought about a revolution in scientific ideas so great that it can only be compared with that wrought by the other. And the reason of the transcendent importance of these two changes is that they are changes in the conception of the Cosmos."

LOCUS

A locus is a set of points that contains all the points, and only the points, that satisfy the condition, or conditions, required to describe a geometric figure. The word *locus* is Latin for place or location. A locus may also be defined as the path traced out by a point in **motion**, as it moves according to a stated set of conditions, since all the points on the path satisfy the stated conditions. Thus, the phrases "locus of a point" and "locus of points" are often interchangeable. A locus may be rather simple and appear to be obvious from the stated condition. Examples of loci (plural for locus) include points, lines, and surfaces. The locus of points in a plane that are equidistant from two given points is the straight line that is perpendicular to and passes through the center of the line segment connecting the two points (figure 1a).

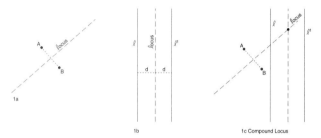

Figure 1.

The locus of points in a plane that are equidistant from each of two parallel lines is a third line parallel to and centered between the two parallel lines (see figure 1b). The locus of points in a plane that are all the same **distance** r from a single point a **circle** with radius r. Given the same condition, not confined to a plane but to three-dimensional **space**, the locus is the surface of a **sphere** with radius r. However, not every set of conditions leads to an immediately recognizable geometric object.

To find a locus, given a stated set of conditions, first find a number of points that satisfy the conditions. Then, "guess" at the locus by fitting a smooth line, or lines, through the points. Give an accurate description of the guess, then prove that it is correct. To prove that a guess is correct, it is necessary to prove that the points of the locus and the points of the guess coincide. That is, the figure guessed must contain all the points of the locus and no points that are not in the locus. Thus, it is necessary to show that (1) every point of the figure is in the locus and (2) every point in the locus is a point of the figure, or every point not on the figure is not in the locus.

In some cases, a locus may be defined by more than one distinct set of conditions. In this case the locus is called a compound locus, and corresponds to the intersection of two or more loci. For example, the locus of points that are equidistant

from two given points and also equidistant from two given parallel lines (see figure 1c), is a single point. That point lies at the intersection of two lines, one line containing those points equidistant from the two points, and one line containing all those points equidistant from the parallel lines.

There are many other interesting loci, for example the **cycloid**.

Figure 2.

The cycloid is the locus of a point on a circle as the circle rolls in a straight line along a flat surface. The cycloid is the path that a falling body takes on a windy day in order to reach the ground in the shortest possible time. Some interesting loci can be described by using the moving point **definition** of locus. For example, consider this simple mechanism.

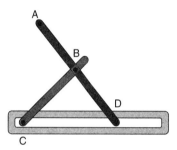

Figure 3.

It has a pencil at point A, pivots at points B and C and point D is able to slide toward and away from point C. When point D slides back and forth, the pencil moves up and down drawing a line perpendicular to the base (a line through C and D). More complicated devices are capable of tracing figures while simultaneously enlarging or reducing them.

LOGARITHMS

A logarithm is an exponent. The logarithm (to the base 10) of 100 is 2 because $10^2 = 100$. This can be abbreviated $\log_{10} 100 = 2$.

Because logarithms are **exponents**, they have an intimate connection with **exponential functions** and with the laws of exponents.

The basic relationship is $b^x = y$ if and only if $x = \log_b y$. Since $2^3 = 8$, $\log_2 8 = 3$. Since, according to a table of logarithms, $\log_{10} 2 = .301$, $10^{.301} = 2$.

The major laws of logarithms and the exponential laws from which they are derived are as follows:

I. $\log_b (xy) = \log_b x + \log_b y \mid b^n \cdot b^m = b^{n+m}$

II. $\log_b (x/y) = \log_b x - \log_b y \mid b^n/b^m = b^{n-m}$

III. $\log_b x^y = y \cdot \log_b x \mid (b^n)^m = b^{(nm)}$

IV. $\log_b x = (\log_b a)(\log_a x)$ | If $x = b^r$; $b = a^p$, then $x = a^{pr}$

V. $\log_b b^n = n$ | If $b^n = b^m$, then $n = m$

VI. Log $1 = 0$ (any base) | $b^0 = 1$

In all these rules, the bases a and b and the arguments x and y are limited to positive numbers. The exponents m, n, p, and r and the logarithms can be positive, negative, or **zero**.

Because logarithms depend on the base that is being used, the base must be clearly identified. It is usually shown as a subscript. There are two exceptions. When the base is 10, the logarithm can be written without a subscript. Thus log 1000 means $\log_{10} 1000$. Logarithms with 10 as a base are called "common" or "Briggsian." The other exception is when the base is the number **e** (which equals 2.718282...). Such logarithms are written ln x and are called "natural" or "Napierian" logarithms.

In order to use logarithms one must be able to evaluate them. The simplest way to do this is to use a "scientific" **calculator**. Such a calculator will ordinarily have two keys, one marked "LOG," which will give the common logarithm of the entered number, and the other "LN," which will give the natural logarithm.

Lacking such a calculator, one can turn to the tables of common logarithms which are to be found in various handbooks or as appendices to various statistical and mathematical texts. In using such tables one must know that they contain logarithms in the **range** 0 to 1 only. These are the logarithms of numbers in the range 1 to 10. If one is seeking the logarithm of a number, say 112 or .0035, outside that range, some accommodation must be made.

The easiest way to do this is to write the number in **scientific notation**:

$$112 = 1.12 \times 10^2$$

$$.0035 = 3.5 \times 10^{-3}$$

Then, using law I

$$\log 112 = \log 1.12 + \log 10^2$$

$$\log .0035 = \log 3.5 + \log 10^{-3}$$

Log 1.12 and log 3.5 can be found in the table. They are .0492 and .5441 respectively. Log 10^2 and log 10^{-3} are simply 2 and -3 according to law V: therefore

$$\log 112 = .0492 + 2 = 2.0492$$

$$\log .0035 = .5442 - 3 = -2.4559$$

The two parts of the resulting logarithms are called the "mantissa" and the "characteristic." The mantissa is the decimal part, and the characteristic, the integral part. Since tables of logarithms show positive mantissas only, a logarithm such as -5.8111 must be converted to .1889 - 6 before a table can be used to find the "antilogarithm," which is the name given to the number whose logarithm it is. A calculator will show the antilogarithm without such a conversion.

Tables for natural logarithms also exist. Since for natural logarithms, there is no easy way of determining the characteristic, the table will show both characteristic and mantissa. It will also cover a greater range of numbers, perhaps 0 to 1000 or more. An alternative is a table of common logarithms, converting them to natural logarithms with the formula (from law IV) ln x = 2.30285 × log x. Logarithms are used for a variety of purposes. One significant use—the use for which they were first invented—is to simplify calculations. Laws I and II enable one to multiply or divide numbers by adding or subtracting their logarithms. When numbers have a large number of digits, adding or subtracting is usually easier. Law III enables one to raise a number to a power by multiplying its logarithm. This is a much simpler operation than doing the exponentiation, especially if the exponent is not 0, 1, or 2.

At one time logarithms were widely used for computation. Astronomers relied on them for the extensive computations their work requires. Engineers did a majority of their computations with slide rules, which are mechanical devices for adding and subtracting logarithms or, using log-log scales, for multiplying them. Modern electronic calculators have displaced slide rules and tables for computational purposes—they are quicker and far more precise—but an understanding of the properties of logarithms remains a valuable tool for anyone who uses numbers extensively.

If one draws a scale on which logarithms go up by uniform steps, the antilogarithms will crowd closer and closer together as their size increases. They do this in a very systematic way. On a logarithmic scale, as this is called, equal intervals correspond to equal **ratios**. The **interval** between 1 and 2, for example, is the same length as the interval between 4 and 8.

Logarithmic scales are used for many purposes. The pH scale used to measure acidity and the decibel scale used to measure loudness are both logarithmic scales (that is, they are the logarithms of the acidity and loudness). As such, they stretch out the scale where the acidity or loudness is weak (and small variations noticeable) and compress it where it is strong (where big variations are needed for a noticeable effect). Another example of the advantage of a logarithmic scale can be seen in a scale which a sociologist might construct. If he were to draw an ordinary graph of family incomes, an increase of a dollar an hour in the minimum wage would seem to be of the same importance as a dollar-an-hour increase in the income of a corporation executive earning a half million dollars a year. Yet such an increase would be of far greater importance to the family whose earner or earners were working at the minimum-wage level. A logarithmic scale, where equal intervals reflect equal ratios rather than equal differences, would show this.

Logarithmic **functions** also show up as the inverses of exponential functions. If $P = ke^t$, where k is a constant, represents population as a function of time, then $t = K + \ln P$, where $K = -\ln k$, also a constant, represents time as a function of population. A demographer wanting to know how long it would take for the population to grow to a certain size would find the logarithmic form of the relationship the more useful one.

Because of this relationship logarithms are also used to solve exponential **equations**, such as 3 - = 2x as or 4e k = 15.

The invention of logarithms is attributed to **John Napier**, a Scottish mathematician who lived from 1550 to 1617. The logarithms he invented, however, were not the simple logarithms we use today (his logarithms were not what are now called "Napierian"). Shortly after Napier published his work, Briggs, an English mathematician, met with him and together they worked out logarithms that much more closely resemble the common logarithms that we use today. Neither Napier nor Briggs related logarithms to exponents, however. They were invented before exponents were in use.

LOGIC CIRCUITS

Electronics, which relies on switching between two states (on or off, voltage or no voltage, high voltage or low voltage, for example), is made up of what are called logic circuits. This name comes from the fact that the circuits deal with two possibilities. In the social science of logic two such opposing states are often verbalized as "yes or no," "true or false," and in slightly more sophisticated reasoning, "if yes (or no), then some action (or other) will be taken."

Computers operate electronically using such circuitry or "logic gates." These are built into integrated circuits that are part of the central processing unit and other computer "chips" on the mother board and most of the cards mounted to it. The most basic logic gate has one input and one output. Logic gate inputs or outputs represent one bit of information in one of two states, usually thought of as a 1 for "on," "yes," or "true," and a 0 for "off," "no," or "false." Individual gates are connected in large numbers to produce the sophisticated computing power available to computers owners today.

The most common logic gates currently in use make decisions or, more correctly, turn on or off, given one or two conditions. This type of action follows four basic logical operations for which the gates are named, NOT, AND, OR, and EXCLUSIVE OR. These form the basis for computer operations and are built upon and expanded to create sophisticated programs and computer controlled actions. The NOT gate is the simplest of the four. It is composed of one input and one output. The "logic table" for this gate is simple as well. In words, it can be written as follows: If there is no input, there is an output. If there is an input, there is no output. The name NOT makes sense if we write this in a way that is not quite as correct grammatically, but perhaps more illustrative: If NOT input, then output. If input, then NOT output.

The AND gate is a bit more complex, having two inputs and one output. It requires that both inputs be present, or on, in order for there to be an output. As its name implies, one must have input a AND input b in order for an output to exist. With the OR gate, only one input or the other need be present in order for an output to exist (if a OR b, then output). But the OR gate also allows output if BOTH inputs are present, just as with an AND gate. The EXCLUSIVE OR gate does away with this last situation because it only allows output if ONLY input

a OR b are present, but not both. More complex gates can be made by adding more inputs, combining gates with others of the same type of different types.

While this may all seem confusing to read about, mathematically, and physically as well, it is really quite simple. Logic circuits that use the most basic **mode** of operation (when voltage is applied the input or output is on, when there is no voltage, they are off) may be though of, almost literally, as being composed of tiny switches. What is truly difficult to comprehend for most people is how a combination of millions, or perhaps billions, of tiny switches going on and off in some precise, planned patterns millions of times per second can produce the amazing things made possible by modern computers.

See also Binary number system; Computers and mathematics

LOGICAL SYMBOLS

Logical symbols are part of a modern logical system for expressing rational thought and common patterns of reasoning. These symbols are used to clearly represent oftentimes highly complex logical relationships between statements. There are five special symbols that are employed as statement connectives or operators; these are the logical symbols. The five logical symbols are all truth-functional connectives and will be discussed in more detail below. The symbols are ~ (also symbolized as ¬), & (also symbolized as • or ^), v, → (also symbolized as ⊃), and <-> (also symbolized as ≡). Logical symbols are used in a language that has several parts. **Propositions** are the statements that can be either true or false. These individual statements are usually represented by capital letters of the alphabet and are called statement constants. They are normal declarative sentences but are represented by statement constants for convenience. The statement connectives are the logical symbols whose function is to form new compound statements. These connectives can be reapplied to the resulting compound statements to form new compounds of compounds.

Each symbol represents a common logical English expression. The ~ is the symbol for logical negation meaning simply that it reverses the **truth** value of any statement in front of which it appears. The statement can be simple or compound. So if the original statement is true then placing a ~ in front of the statement means that the statement is now false. The common English expression "it is not the case that..." can be substituted for ~. Sometimes just the word not can be used.

The & symbolizes logical **conjunction**. A compound statement formed using this connective is true only if both individual statement between which it occurs are true. If one or both of the individual statements is false then the whole compound is false. The English conjunctions "and" and "but" can be substituted for the symbol &.

The v symbolizes inclusive disjunction. That is the compound statement is true whenever either or both of its individual statements are true. Also, if both individual statements are false then the compound statement is false. On most occasions the common English expression either....or.... can be substituted for the v symbol but note that this is not the case if both individual statements are true. There is no real common English word for substitution if both statements are true.

The → symbol is used to denote a relationship called material **implication**. This type of compound statement is always true except for the situation where the individual statement on the left, the antecedent, is true and the individual statement on the right of the symbol, the consequent, is false. In all other situations the compound statement is true. A common English expression for conditional statements, such as if....then..., can be substituted here.

The last connective symbol, the <-> symbol, defines material **equivalence**. The compound statement is true only when the individual statements composing the compound have the same truth value, that is either both are true or both are false. The English connective phrase "if and only if" can be substituted for this symbol.

The five logical symbols are all truth functional in that the truth or falsity of each compound statement formed by using them is wholly determined by the truth-value of the individual statements and meaning of the connective. Truth tables can be developed as a convenience tool to easily define the meaning of each statement connective. An important logical feature is present when compound statements are formed using the five logical symbols. Independent of how long a compound statement is, the truth or falsity of the entire statement depends solely upon the truth-value of its individual statements and the meaning of the connectives it uses.

LORENTZ, HENDRIK ANTOON (1853-1928)

Dutch physicist

The work of Hendrik Antoon Lorentz went far in explaining and elaborating on the electromagnetic theory first proposed by **James Clerk Maxwell**. Not only did Lorentz explain the Zeeman effect and name the electron, but he also advanced a theory about electromagnetic phenomena that would reach its full flower in Albert Einstein's theory of relativity.

Hendrik Antoon Lorentz was born in Arnhem, The Netherlands, on July 18, 1853. He was the son of Gerrit Frederik Lorentz, a nursery owner, and Geertruida van Ginkel. His mother died when he was four, and his father remarried when he was nine.

Young Hendrik was a bright child, interested in physical science and mathematics from an early age. He mastered the logarithm tables when he was just nine. He also had a tremendous aptitude for languages, and later read and wrote widely in French, English and German as well as his native Dutch. In addition to attending regular school in Arnhem, where he excelled in science, he also attended a special evening school, where he determined his own path of study.

In 1870 he went to the University of Leiden, and by 1871 he was in the doctoral program. While he pursued his own studies, he taught high school in Arnhem. In 1875 he earned his doctoral degree, and his work concerned a new the-

Hendrik Antoon Lorentz

ory that few other scientists at the time grasped as well as the young Dutchman did: the electromagnetic theory of James Clerk Maxwell.

In 1877, the 24-year-old Lorentz became the first chair of theoretical physics at the University of Leiden. Among the young physicists he would influence during his career there was **Albert Einstein**.

He married Alletta Vaiser, the niece of one of his former professors, P. Vaiser, in 1881. Together they had two daughters, Geertruida (1885) and Johanna (1889) and a son Rudolf, born in 1895. A fourth child, a boy born between Johanna and Rudolf, died in infancy.

Lorentz enjoyed a long and productive career. He was the man who gave us the term "electron," being the first to give a name to the electrically charged particles that are part of all matter. Beginning in 1892, he began publishing three or four papers on his electron theory, a pace he would maintain through 1904. He produced **calculus** and physics textbooks that were reprinted numerous times.

He also developed new ideas concerning electromagnetic theory. He had studied the work of Maxwell while preparing for his doctoral degree, and was one of the few scientists who could make sense of Maxwell's rather cagey phrasing, which warned readers regarding their perception of the nature of electricity by saying: "it is, or is not, a substance, or that it is, or is not, a form of energy, or that it belongs to any known category of physical quantities."

Lorentz based his theory on the activity of electrons that reacted with each other through a stationary "ether." He successfully applied his theory to a previously unexplained phenomenon that Pieter Zeeman had observed: the change of spectral lines in a strong magnetic **field**. In 1902, Lorentz and Zeeman shared the Nobel Prize for their work.

However, Lorentz did not stop there. He turned his attention to the action of velocity of the system on electromagnetic phenomena, and determined that the phenomena are independent; that is, whether the system is stationary or moving at the speed of **light**, electromagnetic activity remains unchanged. This is known as Lorentz's principle of **correlation**. Lorentz's work served as the basis for much of Einstein's work on relativity.

From 1909 to 1921, Lorentz served as president of the physics section of the Royal Netherlands Academy of Arts and Sciences. He also chaired the Solvay conferences in physics from 1911 to 1927. Lorentz died after a brief illness on February 4, 1928, in Haarlem, the Netherlands. He was immensely popular and respected, and as a show of that respect, all telephone and telegraph service was suspended for three minutes on the day of his funeral. Einstein delivered the eulogy for his former teacher, whom he called, "the greatest and noblest man of our times."

LORENTZ TRANSFORMATIONS

The Lorentz transformations are a group of **equations** that yield the **geometry** of special **relativity**. They are the basis of **Einstein**'s special theory of relativity formulated in 1905. The main objective in formulating the equations that comprise the Lorentz transformations was to preserve the constancy of the velocity of light and hence the invariance of the **space-time** interval between all reference frames. Using these transformations, it is possible to correlate the space-time coordinates of a moving system with the space-time coordinates of any other system. These coordinates are four-dimensional, representing three dimensions of **space** and one **dimension** of time. The position of a point in space-time is written as a four-vector, a four-element vector.

The Lorentz transformations replace the non-relativistic Galilean transformations used in classical **mechanics**. The Galilean transformations are a set of equations that involve a transformation of coordinates from one frame of reference to another frame of reference that is moving with constant velocity with respect to the first. In fact, the Galilean transformations are just the special case of the Lorentz transformations as the velocity of light is made to approach **infinity**, instead of being constant.

If we assume that the velocity of light is the constant c, we can derive the Lorentz transformations, which are formulated as: $x' = (x - vt)/(\sqrt{(1-(v^2/c^2))})$ $y' = y$, $z' = z$, $t' = (t-(vx)/c^2)/(\sqrt{(1-(v^2/c^2))})$, where v is the velocity, c is the speed of light, x', y', z', t' refer to one frame of reference and x, y, z, t refer to the other frame of reference. It is clear from these

equations that it is impossible to travel faster than the speed of light since if v is greater than c then there would be **imaginary numbers** in the denominators of the x' and t' transformations. These transformations make the speed of light independent of the motion of either of the frames of reference. They combine time dilation and length contraction, two ideas principle to the special theory of relativity, into a single transformation.

The Lorentz transformations, considered the mathematical tool of relativity, gives a method that adjusts to the differences in the classical predictions of motion relating to time and the Michelson-Morley null experiment, conducted for the first time in 1881. The Michelson-Morley null experiment was one whose results pointed to the conclusion that the velocity of light must be constant, contrary to the universal belief at the time. It was this experiment that led to the formulation of Einstein's special theory of relativity several years later. After Michelson's first attempt at the null experiment in 1881 he repeated the experiment in 1887, obtaining the same result that the velocity of light was independent of the velocity of the observer. In 1887 Voigt, while studying the Doppler shift, wrote down a set of transformations and showed that certain equations were **invariant** under them. These transformations were again written down by Larmor in 1898 in an article *Ether and matter*. Finally in 1899 Lorentz wrote the same transformations, which with a different scale **factor** (the relativity factor), were named in his honor by Poincaré in 1905. To formulate the transformations Lorentz made two assumptions: that the speed of light was a constant for any observers regardless of their respective motions, and that the transformation between observers in different inertial frames of reference is linear. Working from these two assumptions, Lorentz formulated his transformations that are capable of correlating the space and time coordinates of one moving system with the known space and time coordinates of any other system. In 1905 Einstein first published his special theory of relativity in which he adopted the Lorentz transformation equations. Einstein gave the transformations an entirely new interpretation. The transformations describe the increase of mass, the shortening of length, and the time dilation of a body moving at speeds approaching that of the velocity of light. In 1912 Lorentz and Einstein were jointly proposed for a Nobel Prize for their work concerning special relativity. It is Lorentz who is considered to have been the first to find the mathematical content of the relativity principle.

LUCAS, FRANCOIS-EDOUARD-ANATOLE (1842-1891)
French mathematician

The discoverer of what was then the first new **Mersenne number** in more than a century, Francois-Edouard-Anatole Lucas made many contributions to **number theory** and wrote a book on recreational mathematics that is still considered a classic today.

Lucas was born in 1842 in Amiens, France, where he received an education at the Ecole Normale. A gifted student and researcher, Lucas quickly found a job as an assistant at the prestigious Paris Observatory after graduating. However, like many other young men his age at the time, he had to interrupt his promising career to serve in the army during the Franco-Prussian War. Lucas was commissioned as an artillery officer.

After the war, he accepted a position as mathematics professor in Paris at the Lycée Saint-Louis, eventually moving from there to take the same post at the city's Lycée Charlemagne. At both schools, Lucas was known for being an accessible and entertaining teacher.

In 1876, Lucas gained a lot of academic attention for developing a new method of testing primality (whether a number is prime). He used this technique to show that the Mersenne number 2^{127}-1 is prime. He accomplished this completely without benefit of an electronic computer, and it remains today the biggest prime number ever found by hand calculation alone. Lucas was extremely fond of calculation, but in his spare time often worked on plans to create a large-capacity binary-scale computer.

A major figure in the esoteric world of recreational mathematics, Lucas is perhaps best remembered for his creation of the **Towers of Hanoi** puzzle in 1883. In the meantime, he had begun what would become a four-volume work entitled *Recreational Mathematics* that he would finish in 1894. Lucas was also recognized for his original work on **Fibonacci sequences** (e.g., 1,1,2,3,5,8,13, etc.), in which every number in the sequence besides the first two is the sum of the two just before it. His work resulted in what is now known as Lucas numbers.

Lucas died a bizarre and unfortunate death on October 3, 1891, several days after a shard of broken china cut his cheek at a dinner party. Erysipelas, an acute, fast-moving streptococcal infection, was blamed for his demise.

LUKASIEWICZ, JAN (1878-1956)

Polish logician

One of the three leading members of the Warsaw school of logic during the 1920s and 1930s, Jan Lukasiewicz also developed the first nonclassical logical **calculus** (polivalent logic). He had a profound influence on the succeeding generation of mathematical logicians by introducing Polish notation, which became the basis of much of **Alfred Tarski**'s work.

Lukasiewicz, the son of a Roman Catholic Austrian Army captain, was born on December 21, 1878 in Lvov, Austrian Galicia (now Ukraine). He attended the University of Lvov, where he studied **mathematics and philosophy** and earned his doctorate with highest honors in 1902. Lukasiewicz remained there as a private tutor in logic and philosophy, giving the first Polish lectures on mathematical logic in 1907-1908. He remained there until 1915, when he accepted a position as lecturer at the University of Warsaw, which then was in German-occupied territory.

Beginning in 1910, when he published *On Aristotle's Principle of Contradiction,* which discussed noncontradiction and excluded middle (aspects of calculus), Lukasiewicz

worked toward his development in 1917 of a three-valued propositional calculus. This became the basis of his subsequent breakthroughs on many-valued, or what he called "non-Aristotelian," logic. At about this time, Lukasiewicz also created his new system of notation, which is distinguished by the absence of parentheses.

In 1919, Lukasiewicz served as education minister for the government of independent Poland. The following year he became a professor at the University of Warsaw, where, along with Tarski and **Stanislaw Lesniewski**, he formed a triumvirate of expertise in mathematical logic that attracted attention worldwide. As one of the founders of this Warsaw school of logic, Lukasiewicz used his influence to make mathematical logic a required class in all Polish universities. He viewed the subject as a vital tool of inquiry into the methodology of empirical science and the foundations of mathematics.

Lukasiewicz remained at the University of Warsaw until 1939, and during that time used modern formal techniques to look anew at ancient and medieval forms of logic, especially regarding interpretation of **Aristotle**'s **syllogisms** and the Stoics' propositional calculus. His work in this area is credited with transforming modern academics' view of the history of logic.

According to autobiographical writings, Lukasiewicz and his wife endured a great deal of suffering during World War II. When the war finally ended in 1946, Lukasiewicz was living as an exile in Belgium. However, when he considered returning to his homeland, he found that he could not accept Poland's new Soviet-enforced political system, so he took a professorship at the Royal Irish Academy in Dublin. Lukasiewicz remained there until his death on February 13, 1956.

See also Stanislaw Lesniewski; Alfred Tarski

LUSIN, NIKOLAI NIKOLAIEVICH (1883-1950)

Russian mathematician

Although he came to mathematics relatively late in life, Nikolai Nikolaievich Lusin made significant contributions in the area of **functions**, measure theory, and set **topology**. In fact, he is considered one of the founders of the famous Moscow school of function theory, which had its heyday in the 1920s. Lusin was also known in academic circles for his brilliant lectures and unconventional approach to solving problems.

Lusin was the only son of a trade official and his wife, born in Irkutsk on December 9, 1883. In about 1894 the family moved to Tomsk so the boy could attend the city's excellent *gymnasium* (a classical college-preparatory school). However, the move was something of a disaster, since Lusin began to do so poorly in mathematics that his father had to hire a tutor. It soon became apparent that Lusin was gifted in mathematics, but that he had difficulty solving problems using the prescribed methods. Instead, he would use his own unconven-

tional techniques to come up with the answers, which puzzled and infuriated his teachers.

Lusin left the *gymnasium* in 1901 when his father sold his business and moved the family to Moscow. With Moscow University now close at hand, Lusin enrolled in the School of Physics and Mathematics with the intention of becoming an engineer. The family fell on hard times soon thereafter when Lusin's father lost his savings, forcing the young man to move into a boarding house. This misfortune did not seem to affect Lusin unduly, since he was able to continue his studies at the university.

As at the *gymnasium*, though, Lusin did only mediocre work. One professor, however, noticed his unusual way of thinking and began inviting Lusin to his home for private lessons away from the inhibiting atmosphere of the formal classroom. The professor, D. F. Egorov, posed many difficult problems to Lusin and was astounded to see how quickly the student could solve them, albeit in unconventional ways. Thanks in part to this special help and guidance, Lusin graduated from Moscow University in 1905.

At this point, Lusin was unsure whether to dedicate himself to mathematics. He went through a period in which he wanted to become a doctor, precipitated partially by his observation of the many prostitutes and other impoverished people in Moscow and partly because of the effects of Russia's 1905 Revolution. He felt that studying mathematics was perhaps too trivial a way for him to spend his time. Nevertheless, Lusin's true nature soon reasserted itself, and he began studying **number theory** at the university in 1908. He also married his wife that year.

By about 1909 Lusin had completely devoted his energies to the study of mathematics. He began working on his master's thesis, and in 1910 became an assistant lecturer in pure mathematics at the university. With Egorov, Lusin wrote joint papers on function theory that launched the Moscow school. In 1915 Moscow University awarded Lusin a doctoral degree after he defended his thesis, "The Integral and Trigonometric Series," although the work was intended to fulfill only his master's degree. While some members of the doctoral committee were not impressed with the dissertation, those who were prevailed in granting Lusin the more advanced degree.

In 1917, Lusin accepted an appointment as full professor of pure mathematics at the university. By 1927, he and Egorov had established an elite research group there, which their students fondly dubbed "Luzitania." Many of Russia's best mathematicians of the period emerged from this collection of researchers. From this time on, Lusin dedicated himself to the study of descriptive **set theory**—particularly concentrating on the question, "Can we regard a line atomistically as a set of points?" In 1919, he proved an important result concerning the invariance of **sets** of boundary points under conformal mappings.

From 1922 to 1926 the Moscow school was at its height. Beginning in 1926, however, Lusin spent much of his time writing a monograph on function theory. Deprived of his energizing influence, his students began drifting off into different research areas. The prestigious Soviet Academy of Sciences elected Lusin as a member in 1927, and he spent the rest of his life working there.

In 1931, Lusin entered a new field of study: **differential equations** and their application to control theory and **geometry**. This eventually led him to investigate the bending of surfaces (the foundation of topology). From 1935, Lusin also served as chairperson of the Steklov Institute's Department of the Theory of Functions of Real Variables.

Lusin died in Moscow on February 28, 1950.

LUSIN'S THEOREM

Lusin's theorem is an important technical result in measure theory. Let $[\alpha,\beta]$ be a closed, nonempty **interval** of **real numbers**. Then let $f: [\alpha,\beta] \rightarrow \Re$ be a Lebesgue measurable, real valued **function** defined at each point of the interval (see the article on the **Lebesgue integral**). In general a function that is a Lebesgue **measurable** can be very complicated and, for example, it can be discontinuous at each point of the interval. However, it is often useful to know that f is in some sense not too far from being continuous. Lusin's theorem provides information of this sort. The precise statement of Lusin's theorem is as follows. Given the Lebesgue measurable function f and a positive number ε, there exists a continuous function $g: [\alpha,\beta] \rightarrow \Re$ such that the Lebesgue measure of the set $\{x \in [\alpha,\beta]: f(x) \neq g(x)\}$ is less than ε.

There is also a more general version of Lusin's theorem that applies to *regular* measures μ defined on the **Borel** subsets of a locally compact **Hausdorff** space X (see the article on **topology**). In this setting a measure on the σ-algebra of Borel sets in X is said to be regular if it satisfies the following conditions:

- (1) $\mu(C) < \infty$ for all compact subsets C in X,
- (2) for all Borel sets B in X, $\mu(B) = \inf\{\mu(O): B \subseteq O$ and O is open$\}$,
- (3) for all open sets O in X, $\mu(O) = \sup\{\mu(C): C \subseteq O$ and C is compact$\}$.

Now suppose that X is a locally compact Hausdorff space and μ is a regular measure defined on the Borel subsets of X. Then Lusin's theorem shows that in a certain sense a measurable function can be approximated by a continuous function. The statement of Lusin's theorem in this setting is as follows. Let $f:X \rightarrow \mathbf{C}$ be a Borel measurable, complex valued function, let B be a Borel set in X such that $\mu(B) < \infty$ and $f(x) = 0$ for all x not in B. Then for every $\varepsilon > 0$ there exists a continuous function $g: X > \rightarrow \mathbf{C}$ having compact support such that $\mu(\{x \in X: f(x) \neq g(x)\}$ is less than ε. Moreover, the function g can be selected so that $\sup\{|g(x)|:x \in X\} \leq \sup\{|f(x)|:x \in X\}$.

LYAPUNOV, ALEXSANDR MIKHAILOVICH
(1857-1918)
Russian mathematician

Known for his advances in mathematical **mechanics**, Alexsandr Mikhailovich Lyapunov is credited in particular with introducing ways to determine the stability of **sets** of regular **differential equations**. He is remembered today through his highly regarded academic books and the Lyapunov function for **differential equations**.

Born into a family of distinguished academics and artists in Yaroslavl, Russia on June 6, 1857, Lyapunov (some sources say "Liapunov") was the son of an astronomer. He was educated at home for the first part of his boyhood and then took lessons from an uncle. In 1870, Lyapunov moved with his mother and other siblings to Nizhny, Novgorod (now Gorky), where he began attending the *gymnasium*. He graduated from that school in 1876, then enrolling at St. Petersburg University as a physics and math student under **Pafnuty Lvovich Chebyshev**.

After earning his degree at the university in 1880, Lyapunov decided to heed the encouragement of his professors and stayed on as a member of the school's mechanics department in preparation for a professorial career. He published his first two scientific papers a year later, both of them dealing with hydrostatics. Having achieved a master's degree in 1885, Lyapunov began teaching mechanical physics at Kharkov University as a private tutor.

In 1888 he began publishing more of his original works, and in 1892 his classic "The General Problem of the Stability of Motion" appeared. He used this as the basis for his doctoral dissertation at Moscow University later that year. In 1893, Lyapunov accepted a professorship at Kharkov University, where he taught mechanics and mathematics. There, from 1896 to 1902 he conducted research in mathematical physics, introducing methods to determine the stability of differential equation sets in 1899. He also studied **probability theory** from 1900 to 1901.

Lyapunov left Kharkov in 1901 when he received an appointment as an associate at the St. Petersburg Academy of Sciences. At this point he devoted himself exclusively to scientific work, concentrating mainly on the theory of figures of equilibrium of heavy rotating liquids and their stability. Prior to his death, Chebyshev had challenged his student to investigate this topic, and Lyapunov spent most of the rest of his professional life on it.

In 1917 Lyapunov took his wife to Odessa for treatment for her severe tuberculosis. He took a job as a lecturer at the university there, but in 1902 his wife died of her illness, and the scientist shot himself the same day. He died three days later, on October 31, 1918.

See also Pafnuty Lvovich Chebyshev